D0907165

ANALYSIS AND DESIGN OF ANALOG INTEGRATED CIRCUITS

ANALYSIS AND DESIGN OF ANALOG INTEGRATED CIRCUITS

PAUL R. GRAY

ROBERT G. MEYER

University of California, Berkeley

JOHN WILEY & SONS

New York • Santa Barbara • London • Sydney • Toronto

To Liz and Judy

Library of Congress Cataloging in Publication Data:

Gray, Paul R 1942–
Analysis and design of analog integrated circuits.

Includes bibliographical references and index.
1. Linear integrated circuits. I. Meyer, Robert G., 1942– joint author. II. Title.
TK7874. G688 621.381'73 77-7211
ISBN 0-471-01367-6

Printed in the United States of America

10 9 8 7 6 5 4 3 2 1

SYMBOL CONVENTION

Unless otherwise stated, the following symbol convention is used in this book. *Bias* or *dc* quantities, such as transistor collector current I_C and collector-emitter voltage V_{CE}, are represented by upper-case symbols with upper-case subscripts. *Small-signal* quantities, such as the incremental change in transistor collector current i_c, are represented by lower-case symbols with lower-case subscripts. Elements such as transconductance g_m in small-signal equivalent circuits are represented in the same way. Finally, quantities such as *total* collector current I_c, which represent the sum of the bias quantity *and* the signal quantity, are represented by an upper-case symbol with a lower-case subscript.

PREFACE

This book deals with the analysis and design of analog integrated circuits. Analog circuits are circuits whose inputs and/or outputs are continuously varying analog signals in which the information is conveyed by the instantaneous value of the waveform. We want this book to be useful both as a text for class use and also as a reference book for practicing engineers. For class use, there are numerous worked problems in each chapter; the problem sets at the end of each chapter illustrate the practical applications of the material in the text. Both authors have had extensive industrial experience in integrated circuit (IC) design as well as in the teaching of courses on this subject, and this experience is reflected in the choice of text material and in the problem sets.

Although this book is concerned largely with the design of ICs, a considerable amount of material is also included on applications. In practice these two subjects are very closely linked and a knowledge of both is essential for both designers and users. The latter compose the larger group by far, and it has been our experience that a working knowledge of IC design is a great advantage to an IC user. This is particularly apparent when the user must choose from among a number of competing designs to satisfy a particular need. An understanding of the IC structure is then extremely useful in evaluating the relative desirability of the different designs under extremes of environment or in the presence of variations in supply voltage. In addition, the IC user is in a much better position to interpret a manufacturer's data if he has a working knowledge of the internal operation of the integrated circuit.

The contents of this book originated largely in two courses on analog integrated circuits given at the University of California at Berkeley. The first of these is a senior level elective and the second a graduate course. The book is structured so that such a two-course sequence can be taught using it as the basic text. The more advanced material is found at the end of each chapter or in an appendix, so that a first course in analog integrated circuits can omit this material without loss of continuity. An outline of each chapter is given below together with suggestions for material to be covered in such a first course. It is assumed that the course consists of three hours of lecture per week over a 10-week quarter and that the students have a working knowledge of Laplace transforms and frequency-domain circuit analysis. It is also assumed that the students have had an introductory course in electronics so that they are familiar with the principles of transistor operation and with the functioning of simple analog circuits. Unless otherwise stated, each chapter requires about three lecture hours to cover.

Chapter 1 contains a summary of bipolar transistor and junction field-effect transistor (JFET) device physics. We suggest spending one week on selected topics from this chapter, the choice of topics depending on the background of the students. The material on JFETs could be omitted from a first course. The material of Chapters 1 and 2 is quite important in IC design because there is significant interaction between circuit and device design, as will be seen in later chapters. A thorough understanding of the influence of device fabrication on device characteristics is essential.

Chapter 2 is concerned with the technology of IC fabrication and is largely descriptive. Two lectures on this material should suffice if the students are assigned to read the chapter.

Chapter 3 deals with the characteristics of elementary transistor connections. The material on one-transistor amplifiers should be review for students at the senior and graduate level, and can be assigned as reading. The section on two-transistor amplifiers can be touched on lightly and assigned as reading. The section on emitter-coupled pairs is most important and should be covered in full in class, requiring two to three hours. If time allows, the material on JFET source-coupled pairs or on offset voltage and current can be covered.

In Chapter 4 the important topics of current sources and active loads are considered. These configurations are basic building blocks in modern analog IC design and this material should be covered in full with the exception of the appendices.

Chapter 5 is concerned with output stages and methods of delivering output power to a load. Integrated-circuit realizations of Class A, Class B, and Class AB output stages are described, as well as methods of output-stage protection. A selection of topics from this chapter should be covered.

Chapter 6 deals with operational-amplifier design. As an illustrative example the dc and ac analysis of the 741 operational amplifier (op amp) is performed in detail and the limitations of this basic op amp are described. The design of op amps with improved characteristics is then considered, leading to a discussion of JFET and super-β input stages. The material in this chapter up to Section 6.4 should be covered.

In Chapter 7 the frequency response of ICs is considered. The zero-value time constant technique is introduced for the calculation of the -3-dB frequency of complex circuits. The material of this chapter should be covered in full.

Chapter 8 describes the analysis of feedback circuits and should be covered in full with the section on voltage regulators assigned as reading.

Chapter 9 deals with the frequency response and stability of feedback circuits and should be covered up to the section on root locus. Time does not permit a detailed discussion of root locus but some introduction to this topic can be given.

In a 10-week quarter, coverage of the above material leaves only one week for Chapters 10 and 11. A selection of topics from these chapters can be chosen as follows. Chapter 10 deals with nonlinear analog circuits and portions of this chapter up to Section 10.3 could be covered in a first course. Chapter 11 is a comprehensive treatment of noise in integrated circuits and material up to and including Section 11.4 is suitable.

The material in this book has been greatly influenced by our association with Professor D.O. Pederson of the University of California, Berkeley, and we acknowledge his contributions. Ms. Bettye Fuller and Mrs. Marie Carey typed the manuscript and we appreciate their outstanding work.

Berkeley, California, 1977

Paul R. Gray
Robert G. Meyer

CONTENTS

ANALYSIS AND DESIGN OF ANALOG INTEGRATED CIRCUITS

CHAPTER 1

MODELS FOR INTEGRATED-CIRCUIT ACTIVE DEVICES

1.1 INTRODUCTION

The analysis and design of integrated circuits depend heavily on the utilization of suitable models for integrated-circuit components. This is true in hand analysis, where fairly simple models are generally used, and in computer analysis, where more complex models are encountered. Since any analysis is only as accurate as the model used, it is essential that the circuit designer have a thorough understanding of the origin of the models commonly utilized and the degree of approximation involved in each.

This chapter deals with the derivation of large-signal and small-signal models for integrated-circuit devices. The treatment begins with a consideration of the properties of *pn* junctions, which are basic parts of most integrated-circuit elements. Since this book is primarily concerned with circuit analysis and design, no attempt has been made to produce a comprehensive treatment of semiconductor physics. The emphasis is on summarizing the basic aspects of semiconductor-device behavior and indicating how these can be modeled by equivalent circuits.

1.2 DEPLETION REGION OF A *pn* JUNCTION

The properties of reverse-biased *pn* junctions have an important influence on the characteristics of many integrated-circuit components. For example, in conventional integrated-circuit technology all elements are isolated by reverse-biased *pn* junctions, and these junctions contribute a voltage-dependent parasitic capacitance to each element. In addition a number of important characteristics of bipolar transistors, such as breakdown voltage and output resistance, depend directly on the properties of the depletion region of a reverse-biased *pn* junction. Finally, the basic operation of the junction field-effect transistor is controlled by the width of the depletion region of a *pn* junction. Because of its importance and application to many different problems, an analysis of the depletion region of a reverse-biased *pn* junction is considered below. The properties of forward-biased *pn* junctions are treated in Section 1.3 when bipolar-transistor operation is described.

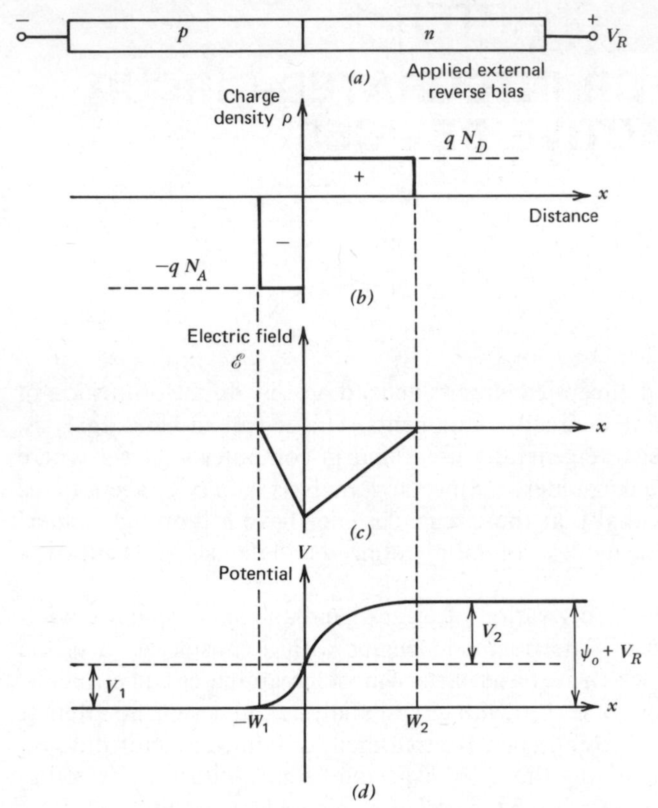

Fig. 1.1 The abrupt junction under reverse bias V_R. (*a*) Schematic. (*b*) Charge density. (*c*) Electric field. (*d*) Electrostatic potential.

Consider a *pn* junction under reverse bias as shown in Fig. 1.1. Assume *constant doping densities* of N_D atoms/cm^3 in the *n*-type material and N_A atoms/cm^3 in the *p*-type material. (The characteristics of junctions with nonconstant doping densities will be described later.) Due to the difference in carrier concentrations in the *p*-type and *n*-type regions, there exists a region at the junction where the mobile holes and electrons have been removed, leaving the fixed acceptor and donor ions. Each acceptor atom carries a negative charge and each donor atom carries a positive charge so that the region near the junction is one of significant space charge and resulting high electric field. This is called the *depletion* region or *space-charge* region. It is assumed that the edges of the depletion region are sharply defined as shown in Fig. 1.1, and this is a good approximation in most cases.

For zero applied bias, there exists a voltage ψ_0 across the junction called the *built-in potential.* This potential opposes the diffusion of mobile holes and electrons across the junction in equilibrium and has a value[1]

$$\psi_0 = V_T \ln \frac{N_A N_D}{n_i^2} \tag{1.1}$$

where $$V_T = \frac{kT}{q} \simeq 26 \text{ mV at } 300^\circ\text{K}$$

The quantity n_i is the intrinsic carrier concentration in a pure sample of the semiconductor and $n_i \simeq 1.5 \times 10^{10}$ cm^{-3} at 300°K for silicon.

In Fig. 1.1 the built-in potential is augmented by the applied reverse bias, V_R, and the total voltage across the junction is $(\psi_0 + V_R)$. If the depletion region penetrates a distance W_1 into the *p*-type region and W_2 into the *n*-type region then we require

$$W_1 N_A = W_2 N_D \tag{1.2}$$

because the total charge per unit area on either side of the junction must be equal in magnitude but opposite in sign.

Poisson's equation in one dimension requires that

$$\frac{d^2V}{dx^2} = -\frac{\rho}{\epsilon} = \frac{qN_A}{\epsilon} \quad \text{for} \quad -W_1 < x < 0 \tag{1.3}$$

where ρ is the charge density, q is the electron charge (1.6×10^{-19} coulomb), and ϵ is the permittivity of the silicon (1.04×10^{-12} farad/cm). The permittivity is often expressed as

$$\epsilon = K_S \epsilon_0 \tag{1.4}$$

where K_S is the dielectric constant of silicon and ϵ_0 is the permittivity of free space (8.86×10^{-14} farad/cm). Integration of (1.3) gives

$$\frac{dV}{dx} = \frac{qN_A}{\epsilon} x + C_1 \tag{1.5}$$

where C_1 is a constant. However the electric field $\mathscr{E}$ is given by

$$\mathscr{E} = -\frac{dV}{dx} = -\left(\frac{qN_A}{\epsilon} x + C_1\right) \tag{1.6}$$

Since there is zero electric field outside the depletion region, a boundary condition is

$$\mathscr{E} = 0 \quad \text{for} \quad x = -W_1$$

and use of this condition in (1.6) gives

$$\mathscr{E} = -\frac{qN_A}{\epsilon}(x + W_1) = -\frac{dV}{dx} \quad \text{for} \quad -W_1 < x < 0 \tag{1.7}$$

Thus the dipole of charge existing at the junction gives rise to an electric field that varies linearly with distance.

Integration of (1.7) gives

$$V = \frac{qN_A}{\epsilon}\left(\frac{x^2}{2} + W_1 x\right) + C_2 \tag{1.8}$$

If the zero for potential is arbitrarily taken to be the potential of the neutral p-type region, then a second boundary condition is

$$V = 0 \quad \text{for} \quad x = -W_1$$

and use of this in (1.8) gives

$$V = \frac{qN_A}{\epsilon}\left(\frac{x^2}{2} + W_1 x + \frac{W_1{}^2}{2}\right) \quad \text{for} \quad -W_1 < x < 0 \tag{1.9}$$

At $x = 0$, we define $V = V_1$ and then (1.9) gives

$$V_1 = \frac{qN_A}{\epsilon}\frac{W_1{}^2}{2} \tag{1.10}$$

If the potential difference from $x = 0$ to $x = W_2$ is V_2, then it follows that

$$V_2 = \frac{qN_D}{\epsilon}\frac{W_2{}^2}{2} \tag{1.11}$$

and thus the total voltage across the junction is

$$\psi_0 + V_R = V_1 + V_2 = \frac{q}{2\epsilon}(N_A W_1{}^2 + N_D W_2{}^2) \tag{1.12}$$

Substitution of (1.2) in (1.12) gives

$$\psi_0 + V_R = \frac{qW_1{}^2 N_A}{2\epsilon}\left(1 + \frac{N_A}{N_D}\right) \tag{1.13}$$

From (1.13), the penetration of the depletion layer into the p-type region is

$$W_1 = \left[\frac{2\epsilon(\psi_0 + V_R)}{qN_A\left(1 + \frac{N_A}{N_D}\right)}\right]^{1/2} \tag{1.14}$$

Similarly

$$W_2 = \left[\frac{2\epsilon(\psi_0 + V_R)}{qN_D\left(1 + \frac{N_D}{N_A}\right)}\right]^{1/2} \tag{1.15}$$

Equations 1.14 and 1.15 show that the depletion regions extend into the *p*-type and *n*-type regions in *inverse* relation to the impurity concentrations and in proportion to $\sqrt{\psi_0 + V_R}$. If either N_D or N_A is much larger than the other, the depletion region exists almost entirely in the *lightly doped* region.

EXAMPLE

An abrupt *pn* junction in silicon has doping densities $N_A = 10^{15}$ atoms/cm^3 and $N_D = 10^{16}$ atoms/cm^3. Calculate the junction built-in potential, the depletion-layer depths and the maximum field with 10 V reverse bias.

From (1.1)

$$\psi_0 = 26 \ln \frac{10^{15} \times 10^{16}}{2.25 \times 10^{20}} \text{ mV} = 638 \text{ mV at } 300^\circ\text{K}$$

From (1.14) the depletion-layer depth in the *p*-type region is

$$W_1 = \left(\frac{2 \times 1.04 \times 10^{-12} \times 10.64}{1.6 \times 10^{-19} \times 10^{15} \times 1.1}\right)^{1/2} = 3.5 \times 10^{-4} \text{ cm}$$
$$= 3.5\ \mu \text{ (where } 1\ \mu = 1 \text{ micron} = 10^{-6} \text{ m)}$$

The depletion-layer depth in the more heavily doped *n*-type region is

$$W_2 = \left(\frac{2 \times 1.04 \times 10^{-12} \times 10.64}{1.6 \times 10^{-19} \times 10^{16} \times 11}\right)^{1/2} = 0.35 \times 10^{-4} \text{ cm} = 0.35\ \mu$$

Finally, from (1.7) the maximum field that occurs for $x = 0$ is

$$\mathscr{E}_{\max} = -\frac{qN_A}{\epsilon} W_1 = -1.6 \times 10^{-19} \times \frac{10^{15} \times 3.5 \times 10^{-4}}{1.04 \times 10^{-12}}$$
$$= -5.4 \times 10^4 \text{ V/cm}$$

Note the large magnitude of this electric field.

1.2.1 Depletion-Region Capacitance

Since there is a *voltage-dependent charge* associated with the depletion region we can calculate a small-signal capacitance, C_j, as follows

$$C_j = \frac{dQ}{dV_R} = \frac{dQ}{dW_1}\frac{dW_1}{dV_R} \tag{1.16}$$

Now

$$dQ = AqN_A\, dW_1 \tag{1.17}$$

where A is the cross-sectional area of the junction.

Differentiation of (1.14) gives

$$\frac{dW_1}{dV_R} = \left[\frac{\epsilon}{2qN_A\left(1 + \frac{N_A}{N_D}\right)(\psi_0 + V_R)}\right]^{1/2} \tag{1.18}$$

Use of (1.17) and (1.18) in (1.16) gives

$$C_j = A\left[\frac{q\epsilon N_A N_D}{2(N_A + N_D)}\right]^{1/2} \frac{1}{\sqrt{\psi_0 + V_R}} \tag{1.19}$$

The above equation was derived for the case of reverse bias V_R applied to the diode. However it is valid for positive bias voltages as long as the forward current flow is small. Thus, if V_D represents the bias on the junction (positive for forward bias, negative for reverse bias), then (1.19) can be written as

$$C_j = A\left[\frac{q\epsilon N_A N_D}{2(N_A + N_D)}\right]^{1/2} \frac{1}{\sqrt{\psi_0 - V_D}} \tag{1.20}$$

$$= \frac{C_{j0}}{\sqrt{1 - \frac{V_D}{\psi_0}}} \tag{1.21}$$

where C_{j0} is the value of C_j for $V_D = 0$.

Equations 1.20 and 1.21 were derived using the assumption of constant doping in the *p*-type and *n*-type regions. However many practical diffused junctions more closely approach a *graded* doping profile as shown in Fig. 1.2. In this case a similar calculation yields

$$C_j = \frac{C_{j0}}{\sqrt[3]{1 - \frac{V_D}{\psi_0}}} \tag{1.22}$$

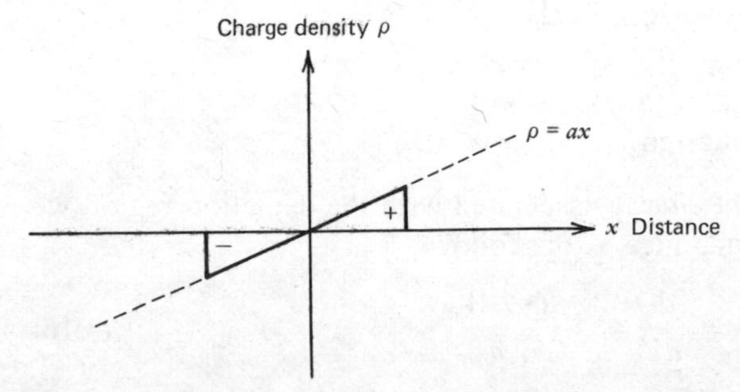

Fig. 1.2 Charge density versus distance in a graded junction.

Note that both (1.21) and (1.22) predict values of C_j approaching infinity as V_D approaches ψ_0. However the current flow in the diode is then appreciable and the equations no longer valid. A more exact analysis[2,3] of the behavior of C_j as a function of V_D gives the result shown in Fig. 1.3. For forward bias voltages up to about $\psi_0/2$ the values of C_j predicted by (1.21) are very close to the more accurate value. As an approximation, some computer programs approximate C_j for $V_D > \psi_0/2$ by a linear extrapolation of (1.21) or (1.22).

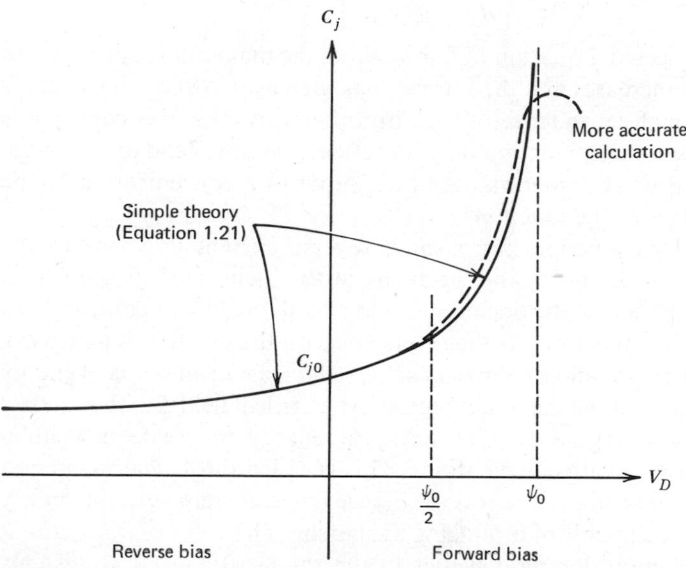

Fig. 1.3 Behavior of *pn* junction depletion-layer capacitance C_j as a function of bias voltage V_D.

EXAMPLE

If the zero-bias capacitance of a diffused junction is 3 pF and $\psi_0 = 0.5$ V, calculate the capacitance with 10 V reverse bias. Assume the doping profile can be approximated by an abrupt junction.

From (1.21)

$$C_j = \frac{3}{\sqrt{1 + \dfrac{10}{0.5}}} \text{ pF} = 0.65 \text{ pF}$$

1.2.2 Junction Breakdown

From Fig. 1.1*c* it can be seen that the maximum electric field in the depletion region occurs at the junction, and for an abrupt junction (1.7) yields a value

$$\mathscr{E}_{max} = -\frac{qN_A}{\epsilon} W_1 \tag{1.23}$$

Substitution of (1.14) in (1.23) gives

$$|\mathscr{E}_{max}| = \left[\frac{2qN_AN_DV_R}{\epsilon(N_A + N_D)}\right]^{1/2} \tag{1.24}$$

where ψ_0 has been neglected. Equation 1.24 shows that the maximum field increases as the doping density increases and the reverse bias increases. Although useful for indicating the functional dependence of $\mathscr{E}_{max}$ on other variables, this equation is strictly valid for an ideal plane junction only. Practical junctions tend to have edge effects that cause somewhat higher values of $\mathscr{E}_{max}$ due to a concentration of the field at the curved edges of the junction.

Any reverse-biased *pn* junction has a small reverse current flow due to the presence of minority carrier holes and electrons in the vicinity of the depletion region. These are swept across the depletion region by the field and contribute to the leakage current of the junction. As the reverse bias on the junction is increased, the maximum field increases and the carriers acquire increasing amounts of energy between lattice collisions in the depletion region. At a critical field $\mathscr{E}_{crit}$ the carriers traversing the depletion region acquire sufficient energy to create new hole-electron pairs in collisions with silicon atoms. This is called the *avalanche process* and leads to a sudden increase in the reverse-bias leakage current since the newly created carriers are also capable of producing avalanche. The value of $\mathscr{E}_{crit}$ is about 3×10^5 V/cm for junction doping densities in the range 10^{15}–10^{16} atoms/cm^3 but it increases slowly as the doping density increases and reaches about 10^6 V/cm for doping densities of 10^{18} atoms/cm^3.

A typical I-V characteristic for a junction diode is shown in Fig. 1.4, and the effect of avalanche breakdown is seen by the large increase in reverse current, which occurs as the reverse bias approaches the breakdown voltage BV. This corresponds to the maximum field $\mathscr{E}_{max}$ approaching $\mathscr{E}_{crit}$. It has been found empirically[4] that if the normal reverse bias current of the diode is I_R with no avalanche effect, then the actual reverse current near the breakdown voltage is

$$I_{RA} = MI_R \tag{1.25}$$

where M is the *multiplication factor* defined by

$$M = \frac{1}{1 - \left(\frac{V_R}{BV}\right)^n} \tag{1.26}$$

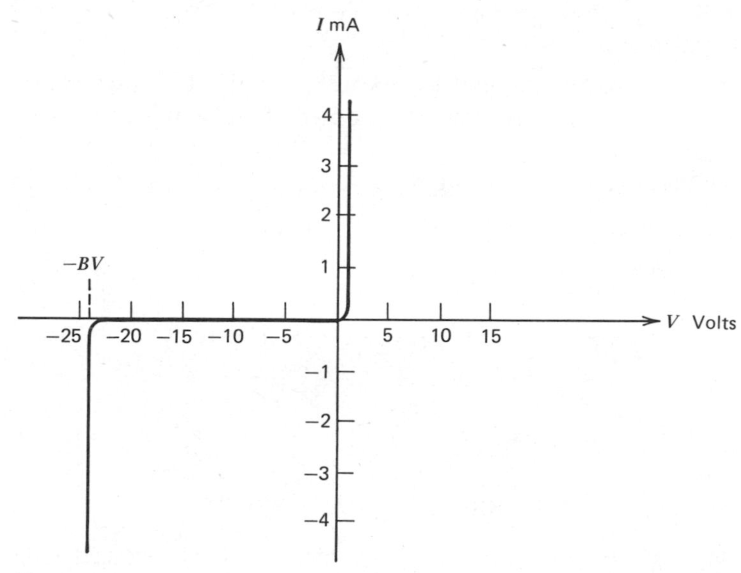

Fig. 1.4 Typical I-V characteristic of a junction diode showing avalanche breakdown.

In this equation, V_R is the reverse bias on the diode and n has a value between 3 and 6.

The operation of a *pn* junction in the breakdown region is not inherently destructive. However the avalanche current flow must be limited by external resistors in order to prevent excessive power dissipation from occurring at the junction and causing damage to the device. Diodes operated in the avalanche region are widely used as voltage references and are called *Zener diodes.* There is another related process called *Zener breakdown*[5], which is different from the avalanche breakdown described above. Zener breakdown occurs only in very heavily doped junctions where the electric field becomes large enough (even with small reverse-bias voltages) to strip electrons away from the valence bonds. This process is called *tunneling,* and there is no multiplication effect as in avalanche breakdown. Although the Zener breakdown mechanism is only important for breakdown voltages below about 6 V, all breakdown diodes are commonly referred to as Zener diodes.

The calculations above have been concerned with the breakdown characteristic of plane abrupt junctions. Practical diffused junctions differ in some respects from these results and the characteristics of these junctions have been calculated and tabulated for use by designers[5]. In particular, edge effects in practical diffused junctions can result in breakdown voltages as much as 50 percent below the value calculated for a plane junction.

EXAMPLE

An abrupt plane *pn* junction has doping densities $N_A = 5 \times 10^{15}$ atoms/cm^3 and $N_D = 10^{16}$ atoms/cm^3. Calculate the breakdown voltage if $\mathscr{E}_{crit} = 3 \times 10^5$ V/cm.

The breakdown voltage is calculated using $\mathscr{E}_{max} = \mathscr{E}_{crit}$ in (1.24) to give

$$BV = \frac{\epsilon(N_A + N_D)}{2qN_AN_D}\mathscr{E}_{crit}^2$$

$$= \frac{1.04 \times 10^{-12} \times 15 \times 10^{15}}{2 \times 1.6 \times 10^{-19} \times 5 \times 10^{15} \times 10^{16}} \times 9 \times 10^{10} \text{ V}$$

$$= 88 \text{ V}$$

1.3 LARGE-SIGNAL BEHAVIOR OF BIPOLAR TRANSISTORS

In this section, the large-signal or dc behavior of bipolar transistors is considered. Large-signal models are developed for the calculation of total currents and voltages in transistor circuits, and such effects as breakdown voltage limitations, which are usually not included in models, are also considered. Second-order effects, such as current-gain variation with collector current and Early voltage, can be important in many circuits and are treated in detail.

The sign conventions used for bipolar transistor currents and voltages are shown in Fig. 1.5. All bias currents for both *npn* and *pnp* transistors are assumed positive going into the device.

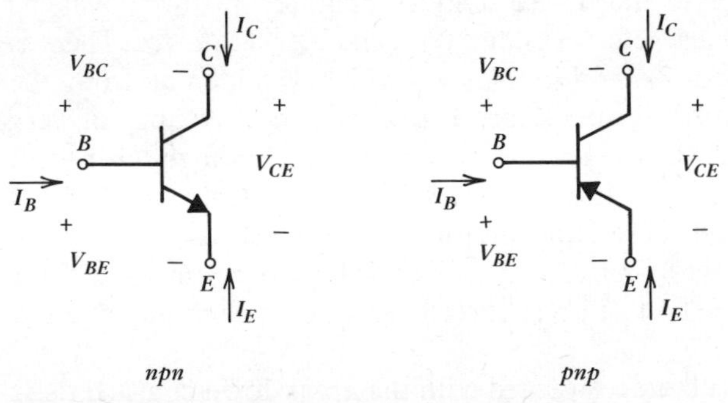

Fig. 1.5 Bipolar transistor sign convention.

1.3.1 Large-Signal Models in the Forward-Active Region

A typical *npn* planar bipolar transistor structure is shown in Fig. 1.6*a* where collector, base, and emitter are labeled *C*, *B*, and *E*, respectively. The method of

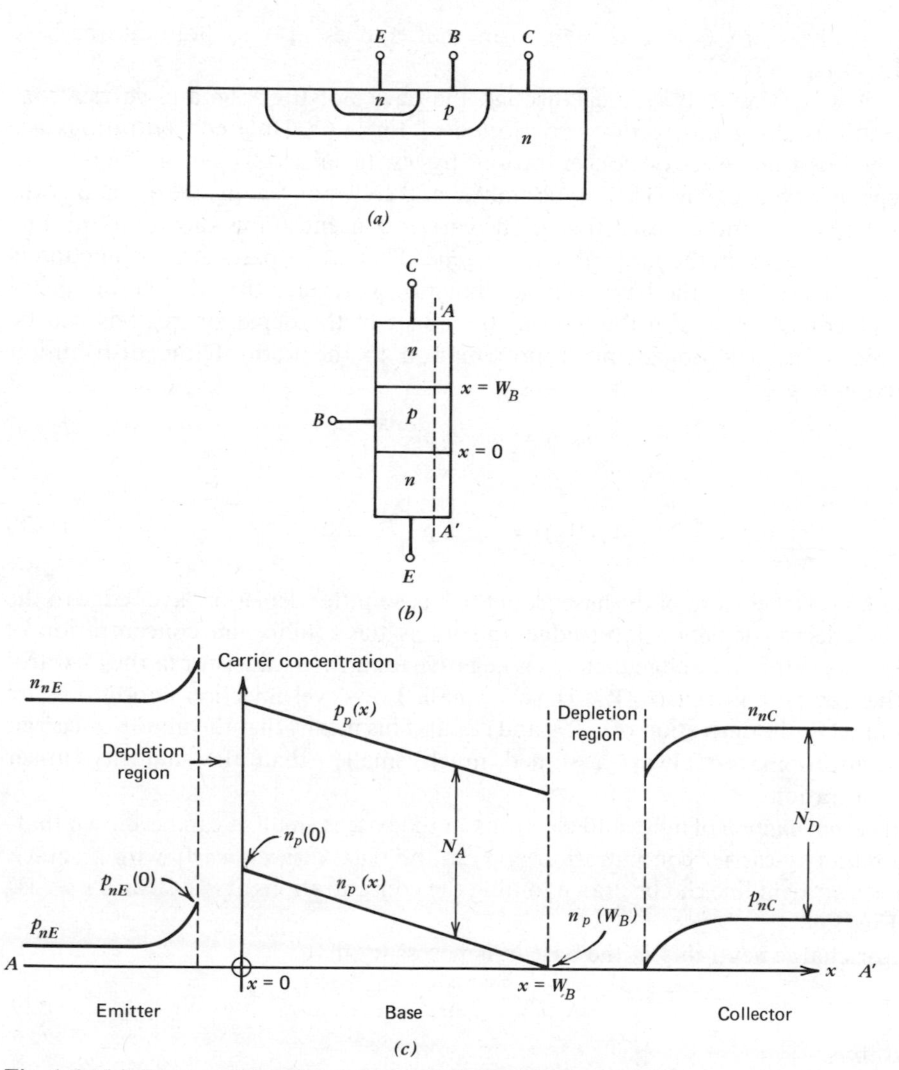

Fig. 1.6 (*a*) Cross section of a typical *npn* planar bipolar transistor structure. (*b*) Idealized transistor structure. (*c*) Carrier concentrations along the cross-section AA' of the transistor in (*b*). Uniform doping densities are assumed. (Not to scale.)

fabricating such transistor structures is described in Chapter 2. It is shown there that the impurity doping density in the base and the emitter of such a transistor is not constant but varies with distance from the top surface. However, many of the characteristics of such a device can be predicted by analyzing the idealized transistor structure shown in Fig. 1.6*b*. In this structure the base and emitter doping densities are assumed constant, and this is sometimes called a "uniform-base" transistor. Where possible in the following analyses, the equations for the uniform-base analysis are expressed in a form that applies also to nonuniform base transistors.

A cross-section AA' is taken through the device of Fig. 1.6*b* and carrier concentrations along this section are plotted in Fig. 1.6*c*. Hole concentrations are denoted by p and electron concentrations by n with subscripts p or n representing p-type or n-type regions. The n-type emitter and collector regions are distinguished by subscripts E and C, respectively. The carrier concentrations shown in Fig. 1.6*c* apply to a device in the *forward-active region.* That is, the base-emitter junction is forward biased and the base-collector junction is reverse biased. The minority-carrier concentrations in the base at the edges of the depletion regions can be calculated from a Boltzmann approximation to the Fermi-Dirac distribution function to give[6]

$$n_p(0) = n_{po} \exp \frac{V_{BE}}{V_T} \tag{1.27}$$

$$n_p(W_B) = n_{po} \exp \frac{V_{BC}}{V_T} \simeq 0 \tag{1.28}$$

where W_B is the width of the base from the base-emitter depletion layer edge to the base-collector depletion layer edge and n_{po} is the equilibrium concentration of electrons in the base. Note that V_{BC} is negative for an *npn* transistor in the forward-active region and thus $n_p(W_B)$ is very small. Low-level injection conditions are assumed in the derivation of (1.27) and (1.28). This means that the minority-carrier concentrations are always assumed much smaller than the majority-carrier concentration.

If *recombination* of holes and electrons in the base is small, it can be shown that[7] the minority-carrier concentration $n_p(x)$ in the base varies *linearly* with distance. Thus a straight line can be drawn joining the concentrations at $x = 0$ and $x = W_B$ in Fig. 1.6*c*

For charge neutrality in the base, it is necessary that

$$N_A + n_p(x) = p_p(x) \tag{1.29}$$

and thus

$$p_p(x) - n_p(x) = N_A \tag{1.30}$$

where $p_p(x)$ is the hole concentration in the base and N_A is the base doping density that is assumed constant. Equation 1.30 indicates that the hole and electron concentrations are separated by a constant amount and thus $p_p(x)$ also varies linearly with distance.

Collector current is produced by minority-carrier electrons in the base diffusing in the direction of the concentration gradient and being swept across the collector-base depletion region by the field existing there. The diffusion current density due

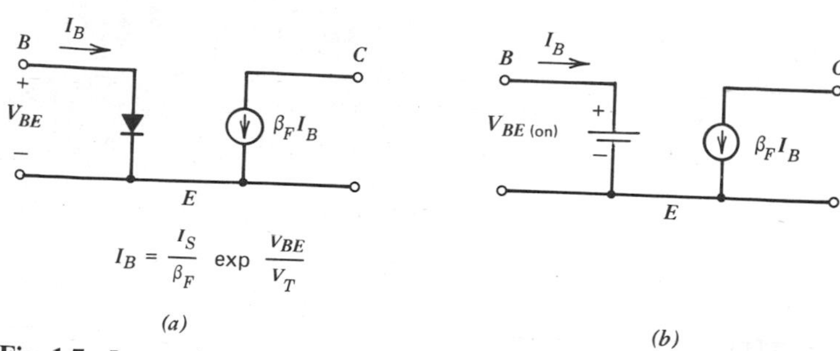

Fig. 1.7 Large-signal models of *npn* transistors for use in bias calculations. (*a*) Circuit incorporating an input diode. (*b*) Simplified circuit with an input voltage source.

transistor) injected into the base from the emitter to the total hole and electron current crossing the base-emitter junction. Ideally $\gamma \rightarrow 1$ and this is achieved by making N_D/N_A large and W_B small. In that case very little reverse injection occurs from base to emitter.

The term α_T in (1.51) is called the *base transport factor* and represents the fraction of carriers injected into the base (from the emitter) that reach the collector. Ideally $\alpha_T \rightarrow 1$ and this is achieved by making W_B small. It is evident from the above development that fabrication changes that cause α_T and γ to approach unity also maximize the value of β_F of the transistor.

The results derived above allow formulation of a large-signal model of the transistor suitable for bias-circuit calculations with devices in the forward-active region. One such circuit is shown in Fig. 1.7 and consists of a base-emitter diode to model (1.46) and a controlled collector-current generator to model (1.47). Note that the collector voltage ideally has no influence on the collector current and the collector node acts as a high-impedance current source. A simpler version of this equivalent circuit which is often useful is shown in Fig. 1.7*b* where the input diode has been replaced by a battery with a value $V_{BE(\text{on})}$, which is usually 0.6 to 0.7 V. This represents the fact that in the forward-active region the base-emitter voltage varies very little because of the steep slope of the exponential characteristic. In some circuits the temperature coefficient of $V_{BE(\text{on})}$ is important and a typical value for this is -2 mV/°C. The equivalent circuits of Fig. 1.7 apply for *npn* transistors. For *pnp* devices the corresponding equivalent circuits are shown in Fig. 1.8.

1.3.2 Effects of Collector Voltage on Large-Signal Characteristics in the Forward-Active Region

In the analysis of the previous section, the collector-base junction was assumed reverse biased and ideally had no effect on the collector currents. This is a useful approximation for first-order calculations, but is not strictly true in practice.

to electrons in the base is

$$J_n = qD_n \frac{dn_p(x)}{dx} \tag{1.31}$$

where D_n is the diffusion constant for electrons. From Fig. 1.6*c*

$$J_n = -qD_n \frac{n_p(0)}{W_B} \tag{1.32}$$

If I_C is the collector current and is taken as positive flowing *into* the collector, it follows from (1.32) that

$$I_C = qAD_n \frac{n_p(0)}{W_B} \tag{1.33}$$

where A is the cross-sectional area of the emitter. Substitution of (1.27) into (1.33) gives

$$I_C = \frac{qAD_n n_{po}}{W_B} \exp \frac{V_{BE}}{V_T} \tag{1.34}$$

$$= I_S \exp \frac{V_{BE}}{V_T} \tag{1.35}$$

where

$$I_S = \frac{qAD_n n_{po}}{W_B} \tag{1.36}$$

and I_S is a constant used to describe the transfer characteristic of the transistor in the forward-active region. Equation 1.36 can be expressed in terms of the base doping density by noting that[8] (see Chapter 2)

$$n_{po} = \frac{n_i^2}{N_A} \tag{1.37}$$

and substitution of (1.37) in (1.36) gives

$$I_S = \frac{qAD_n n_i^2}{W_B N_A} = \frac{qA\bar{D}_n n_i^2}{Q_B} \tag{1.38}$$

where $Q_B = W_B N_A$ is the number of doping atoms in the base per unit area of the emitter and n_i is the intrinsic carrier concentration in silicon. In this form (1.38) applies to both uniform and nonuniform base transistors and D_n has been replaced by $\bar{D}_n$, which is an average effective value of the electron diffusion constant in the base. This is necessary for nonuniform base devices because the diffusion constant is a function of impurity concentration. Typical values of I_S as given by (1.38) are 10^{-14}–10^{-15} A.

Equation 1.35 gives the collector current as a function of base-emitter voltage. The base current, I_B, is also an important parameter and, at moderate current levels, consists of two major components. One of these (I_{B1}) represents recombination of holes and electrons in the base and is proportional to the minority-carrier charge Q_e in the base. From Fig. 1.6c, the minority-carrier charge in the base is

$$Q_e = \tfrac{1}{2} n_p(0) W_B q A \tag{1.39}$$

and we have

$$I_{B1} = \frac{Q_e}{\tau_b} = \frac{1}{2}\frac{n_p(0) W_B q A}{\tau_b} \tag{1.40}$$

where τ_b is the minority-carrier lifetime in the base. I_{B1} represents a flow of majority holes from the base lead into the base region. Substitution of (1.27) in (1.40) gives

$$I_{B1} = \frac{1}{2}\frac{n_{po} W_B q A}{\tau_b} \exp\frac{V_{BE}}{V_T} \tag{1.41}$$

The second major component of base current (usually the dominant one in integrated-circuit *npn* devices) is due to injection of holes from the base into the emitter. This current component depends on the gradient of minority carrier holes in the emitter and is[9]

$$I_{B2} = \frac{qAD_p}{L_p} p_{nE}(0) \tag{1.42}$$

where D_p is the diffusion constant for holes and L_p is the diffusion length (assumed small) for holes in the emitter. $p_{nE}(0)$ is the concentration of holes in the emitter at the edge of the depletion region and is

$$p_{nE}(0) = p_{nEo} \exp\frac{V_{BE}}{V_T} \tag{1.43}$$

If N_D is the donor atom concentration in the emitter (assumed constant) then

$$p_{nEo} \simeq \frac{n_i^2}{N_D} \tag{1.44}$$

The emitter is deliberately doped much more heavily than the base, making N_D large and p_{nEo} small, so that the base-current component, I_{B2}, is minimized. Substitution of (1.43) and (1.44) in (1.42) gives

$$I_{B2} = \frac{qAD_p}{L_p}\frac{n_i^2}{N_D} \exp\frac{V_{BE}}{V_T} \tag{1.45}$$

The total base current, I_B, is the sum of I_{B1} and I_{B2}

$$I_B = I_{B1} + I_{B2} = \left(\frac{1}{2}\frac{n_{po} W_B q A}{\tau_b} + \frac{qAD_p}{L_p}\frac{n_i^2}{N_D}\right) \exp\frac{V_{BE}}{V_T} \tag{1.46}$$

Although this equation was derived assuming uniform base and emitter doping, it gives the correct functional dependence of I_B on device parameters for practical

double-diffused nonuniform-base devices. Second-order components of I_B, which are important at low current levels, are considered later.

Since I_C in (1.35) and I_B in (1.46) are both proportional to exp (V_{BE}/V_T) in this analysis, the base current can be expressed in terms of collector current as

$$I_B = \frac{I_C}{\beta_F} \tag{1.47}$$

where β_F is the forward current gain. An expression for β_F can be calculated by substituting (1.34) and (1.46) in (1.47) to give

$$\beta_F = \frac{\dfrac{qAD_n n_{po}}{W_B}}{\dfrac{1}{2}\dfrac{n_{po} W_B q A}{\tau_b} + \dfrac{qAD_p n_i^2}{L_p N_D}} = \frac{1}{\dfrac{W_B^2}{2\tau_b D_n} + \dfrac{D_p}{D_n}\dfrac{W_B}{L_p}\dfrac{N_A}{N_D}} \tag{1.48}$$

where (1.37) has been substituted for n_{po}. Equation 1.48 shows that β_F is maximized by minimizing the base width W_B and maximizing the ratio of emitter to base doping densities N_D/N_A. Typical values of β_F for *npn* transistors in integrated circuits are 50 to 500, whereas lateral *pnp* transistors (to be described in Chapter 2) have values 10 to 100. Finally the emitter current is

$$I_E = -(I_C + I_B) = -\left(I_C + \frac{I_C}{\beta_F}\right) = -\frac{I_C}{\alpha_F} \tag{1.49}$$

where

$$\alpha_F = \frac{\beta_F}{1 + \beta_F} \tag{1.50}$$

The value of α_F can be expressed in terms of device parameters by substituting (1.48) in (1.50) to obtain

$$\alpha_F = \frac{1}{1 + \dfrac{1}{\beta_F}} = \frac{1}{1 + \dfrac{W_B^2}{2\tau_b D_n} + \dfrac{D_p}{D_n}\dfrac{W_B}{L_p}\dfrac{N_A}{N_D}} \simeq \alpha_T \gamma \tag{1.51}$$

where

$$\alpha_T = \frac{1}{1 + \dfrac{W_B^2}{2\tau_b D_n}} \tag{1.51a}$$

$$\gamma = \frac{1}{1 + \dfrac{D_p}{D_n}\dfrac{W_B}{L_p}\dfrac{N_A}{N_D}} \tag{1.51b}$$

The validity of (1.51) depends on $W_B^2/2\tau_b D_n \ll 1$ and $(D_p/D_n)(W_B/L_p)(N_A/N_D) \ll 1$, and this is always true if β_F is large [see (1.48)]. The term γ in (1.51) is called the *emitter injection efficiency* and is equal to the ratio of the electron current (*npn*

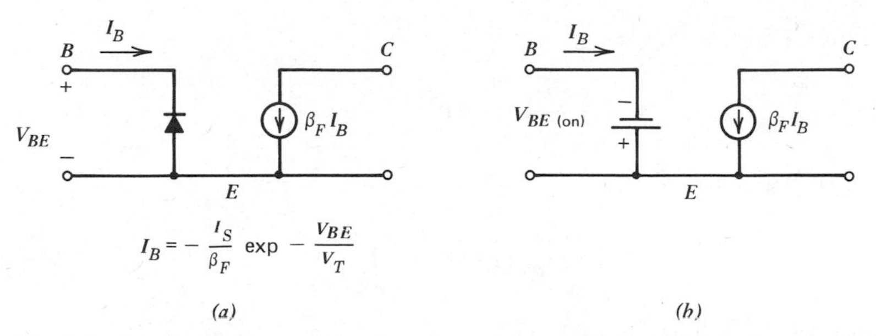

Fig. 1.8 Large-signals models of *pnp* transistors corresponding to the circuits of Fig. 1.7

There are occasions where the influence of collector voltage on collector current is important and this will now be investigated.

The collector voltage has a dramatic effect on the collector current in two regions of device operation. These are the saturation (V_{CE} approaches zero) and breakdown (V_{CE} very large) regions that will be considered later. For values of collector-emitter voltage V_{CE} between these extremes, the collector current increases slowly as V_{CE} increases. The reason for this can be seen from Fig. 1.9, which is a sketch of the minority-carrier concentration in the base of the transistor. Consider the effect of changes in V_{CE} on the carrier concentration for constant V_{BE}. Since V_{BE} is constant, the change in V_{CB} equals the change in V_{CE} and this causes an increase in the collector-base depletion-layer width as shown. The change in the base width of the transistor, ΔW_B, equals the change in the depletion-layer width and causes an increase, ΔI_C, in the collector current.

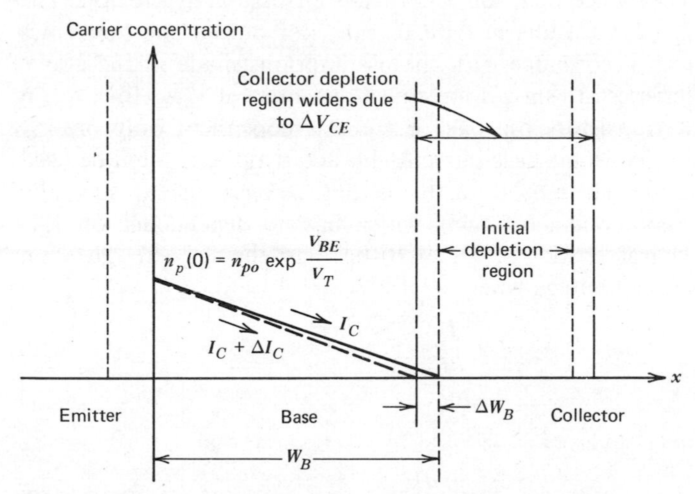

Fig. 1.9 Effect of increases in V_{CE} on the collector depletion region and base width of a bipolar transistor.

From (1.35) and (1.38) we have

$$I_C = \frac{qA\bar{D}_n n_i^2}{Q_B} \exp \frac{V_{BE}}{V_T} \tag{1.52}$$

Differentiation of (1.52) yields

$$\frac{\partial I_C}{\partial V_{CE}} = -\frac{qA\bar{D}_n n_i^2}{Q_B^2} \left(\exp \frac{V_{BE}}{V_T} \right) \frac{dQ_B}{dV_{CE}} \tag{1.53}$$

and substitution of (1.52) in (1.53) gives

$$\frac{\partial I_C}{\partial V_{CE}} = -\frac{I_C}{Q_B} \frac{dQ_B}{dV_{CE}} \tag{1.54}$$

For a uniform-base transistor $Q_B = W_B N_A$ and (1.54) becomes

$$\frac{\partial I_C}{\partial V_{CE}} = -\frac{I_C}{W_B} \frac{dW_B}{dV_{CE}} \tag{1.55}$$

Note that since the base width *decreases* as V_{CE} increases, dW_B/dV_{CE} in (1.55) is negative and thus $\partial I_C/\partial V_{CE}$ is positive. The magnitude of dW_B/dV_{CE} can be calculated from (1.18) for a uniform-base transistor. This equation predicts that dW_B/dV_{CE} is a function of the bias value of V_{CE}, but the variation is typically small for a reverse-biased junction and dW_B/dV_{CE} is often assumed constant. The resulting predictions agree adequately with experimental results.

Equation 1.55 shows that $\partial I_C/\partial V_{CE}$ is proportional to the collector-bias current and inversely proportional to the transistor base width. Thus narrow-base transistors show a greater dependence of I_C on V_{CE} in the forward-active region. The dependence of $\partial I_C/\partial V_{CE}$ on I_C results in typical transistor output characteristics as shown in Fig. 1.10. In accordance with the assumptions made in the above analysis, these characteristics are shown for constant values of V_{BE}. However in most integrated-circuit transistors the base current is dependent only on V_{BE} and not on V_{CE}, and thus constant-base-current characteristics can often be used in the following calculation. The reason for this is that the base current is usually dominated by the I_{B2} component of (1.45), which has no dependence on V_{CE}. Extrapolation of the characteristics of Fig. 1.10 back to the V_{CE} axis gives an intercept V_A called the Early voltage where

$$V_A = \frac{I_C}{\dfrac{\partial I_C}{\partial V_{CE}}} \tag{1.56}$$

Substitution of (1.55) in (1.56) gives

$$V_A = -W_B \frac{dV_{CE}}{dW_B} \tag{1.57}$$

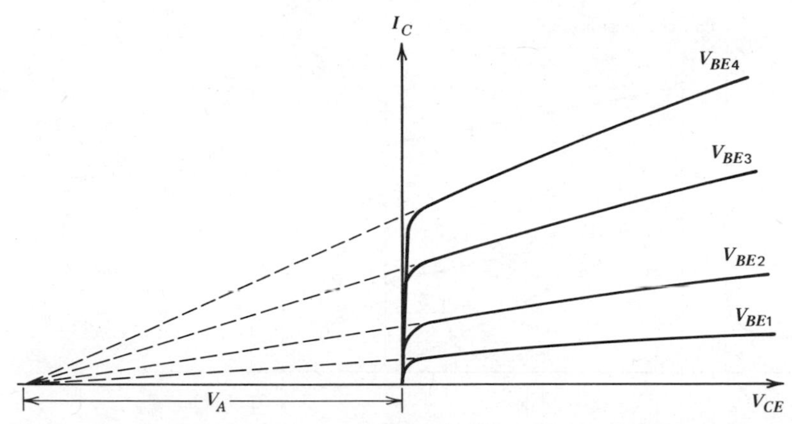

Fig. 1.10 Bipolar transistor output characteristics showing the Early voltage, V_A.

which is a constant, independent of I_C. Thus all the characteristics extrapolate to the same point on the V_{CE} axis. The variation of I_C with V_{CE} is called the Early effect and V_A is a common model parameter for circuit-analysis computer programs. Typical values of V_A for integrated-circuit transistors are 50 to 100 V. The inclusion of Early effect in dc bias calculations is usually limited to computer analysis because of the complexity introduced into the calculation. However the influence of the Early effect is often dominant in small-signal calculations for high-gain circuits and this point will be considered later.

Finally, the influence of Early effect on the transistor large-signal characteristics in the forward-active region can be represented approximately by modifying (1.35) to

$$I_C = I_S\left(1 + \frac{V_{CE}}{V_A}\right)\exp\frac{V_{BE}}{V_T} \tag{1.58}$$

This is a common means of representing the device output characteristics for computer simulation.

1.3.3 Saturation and Inverse Active Regions

Saturation is a region of device operation that is usually avoided in analog circuits because the transistor gain is very low in this region. Saturation is much more commonly encountered in digital circuits where it provides a well-specified output voltage that represents a logic state.

In saturation, both emitter-base and collector-base junctions are forward biased. Consequently, the collector-emitter voltage V_{CE} is quite small and is usually in the range 0.05 to 0.3 V. The carrier concentrations in a saturated *npn* transistor

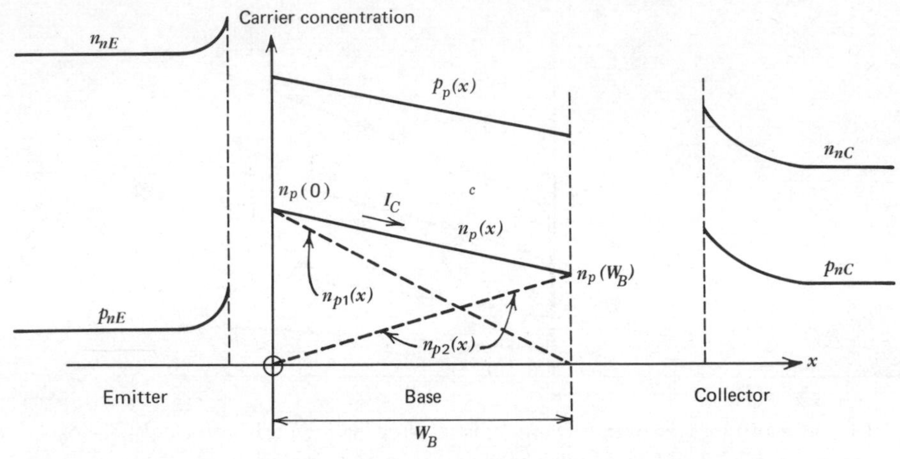

Fig. 1.11 Carrier concentrations in a saturated *npn* transistor. (Not to scale.)

with uniform base doping are shown in Fig. 1.11. The minority-carrier concentration in the base at the edge of the depletion region is again given by (1.28) as

$$n_p(W_B) = n_{po} \exp \frac{V_{BC}}{V_T} \tag{1.59}$$

but since V_{BC} is now positive, the value of $n_p(W_B)$ is no longer negligible. Consequently, changes in V_{CE} with V_{BE} held constant (which cause equal changes in V_{BC}) directly affect $n_p(W_B)$. Since the collector current is proportional to the slope of the minority-carrier concentration in the base [see (1.31)] it is also proportional to $[n_p(0) - n_p(W_B)]$ from Fig. 1.11. Thus changes in $n_p(W_B)$ directly affect the collector current and the collector node of the transistor appears to have a *low impedance*. As V_{CE} is decreased in saturation with V_{BE} held constant, V_{BC} increases as does $n_p(W_B)$ from (1.59). Thus from Fig. 1.11 the collector current decreases because the slope of the carrier concentration decreases. This gives rise to the saturation region of the $I_C - V_{CE}$ characteristic shown in Fig. 1.12. The slope of the $I_C - V_{CE}$ characteristic in this region is largely determined by the resistance in series with the collector lead due to the finite resistivity of the *n*-type collector material. A useful model for the transistor in this region is shown in Fig. 1.13 and consists of a fixed voltage source to represent $V_{BE(\text{on})}$, and a fixed voltage source to represent the collector-emitter voltage $V_{CE(\text{sat})}$. A more accurate but more complex model includes a resistor in series with the collector. This resistor can have a value ranging from 20 to 500 Ω, depending on the device structure.

An additional aspect of transistor behavior in the saturation region is apparent from Fig. 1.11. For a given collector current, there is now a much larger amount of stored charge in the base than there is in the forward-active region. Thus the base-current contribution represented by (1.41) will be larger in saturation. In

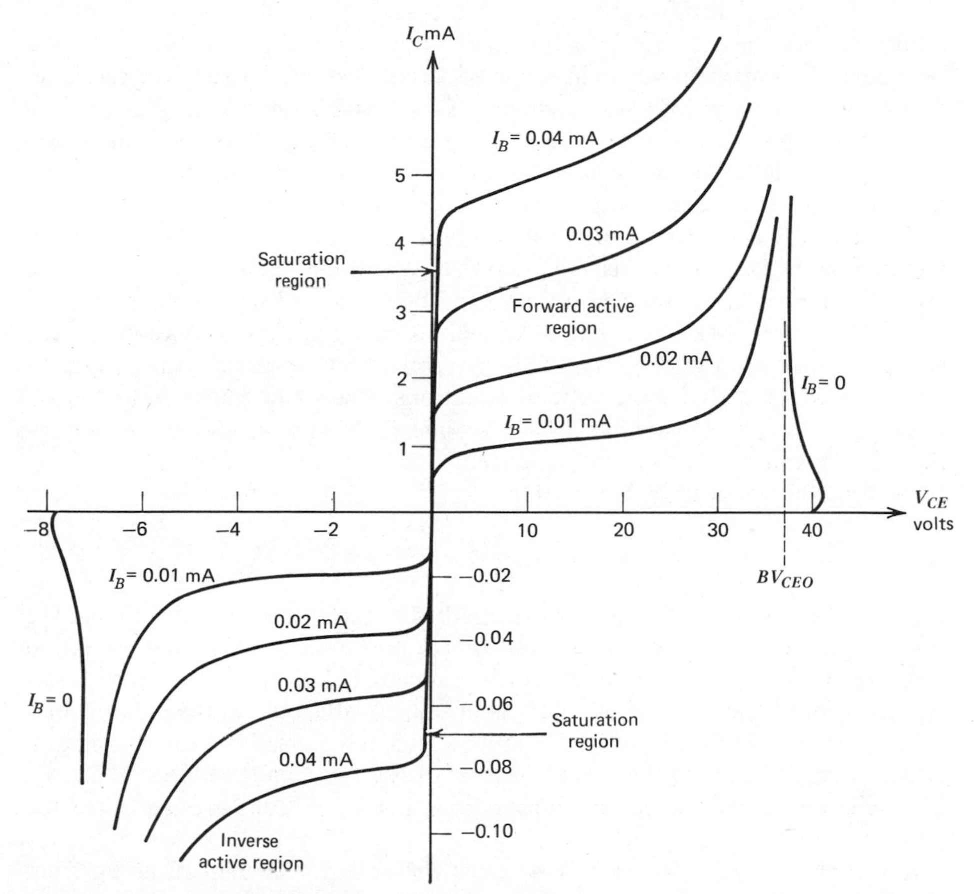

Fig. 1.12 Typical I_C–V_{CE} characteristics for an *npn* bipolar transistor. Note the different scales for positive and negative currents and voltages.

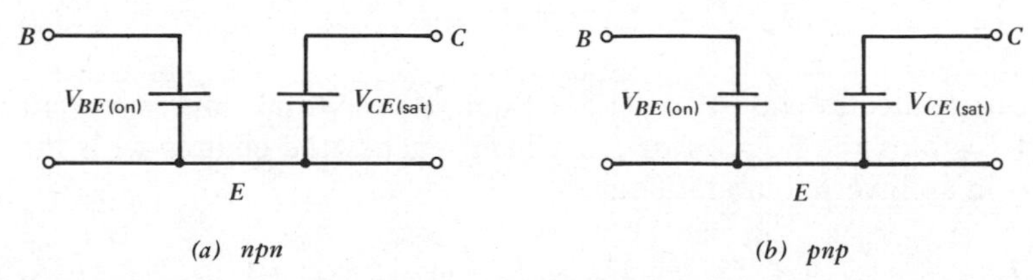

Fig. 1.13 Large-signal models for bipolar transistors in the saturation region.

addition, since the collector-base junction is now forward biased, there is a new base current component due to injection of carriers from the base to the collector. These two effects result in a base current, I_B, in saturation, which is larger than in the forward-active region for a given collector current, I_C. Ratio I_C/I_B in saturation is often referred to as the *forced* β and is always less than β_F. As the forced β is made lower with respect to β_F, the device is said to be more *heavily saturated.*

The minority-carrier concentration in saturation shown in Fig. 1.11 is a straight line joining the two end points assuming that recombination is small. This can be represented as a linear superposition of the two dotted distributions as shown. The justification for this is that the terminal currents depend *linearly* on the concentrations $n_p(0)$ and $n_p(W_B)$. This picture of device carrier concentrations can be used to derive some general equations describing transistor behavior. Each of the distributions in Fig. 1.11 is considered separately and the two contributions are combined. The *emitter* current that would result from $n_{p1}(x)$ above is given by the classical diode equation

$$I_{EF} = -I_{ES}\left(\exp\frac{V_{BE}}{V_T} - 1\right) \tag{1.60}$$

where I_{ES} is a constant that is often referred to as the "saturation current" of the junction (no connection with the transistor saturation described above). Equation 1.60 predicts that the junction current is given by $I_{EF} \simeq I_{ES}$ with a reverse-bias voltage applied. However in practice (1.60) is applicable only in the forward-bias region, since second-order effects dominate under reverse-bias conditions and typically result in a junction current several orders of magnitude larger than I_{ES}. The junction current that flows under reverse-bias conditions is often called the "leakage current" of the junction.

Returning to Fig. 1.11, we can describe the *collector* current resulting from $n_{p2}(x)$ alone as

$$I_{CR} = -I_{CS}\left(\exp\frac{V_{BC}}{V_T} - 1\right) \tag{1.61}$$

where I_{CS} is a constant. The total collector current I_C is given by I_{CR} plus the fraction of I_{EF} which reaches the collector (allowing for recombination and reverse emitter injection). Thus

$$I_C = \alpha_F I_{ES}\left(\exp\frac{V_{BE}}{V_T} - 1\right) - I_{CS}\left(\exp\frac{V_{BC}}{V_T} - 1\right) \tag{1.62}$$

where α_F has been defined previously by (1.51). Similarly the total emitter current is composed of I_{EF} plus the fraction of I_{CR} which reaches the emitter with the transistor acting in an inverted mode. Thus

$$I_E = -I_{ES}\left(\exp\frac{V_{BE}}{V_T} - 1\right) + \alpha_R I_{CS}\left(\exp\frac{V_{BC}}{V_T} - 1\right) \tag{1.63}$$

where α_R is the ratio of emitter to collector current with the transistor operating *inverted* (i.e., with the collector-base junction forward biased and emitting carriers into the base and the emitter-base junction reverse biased and collecting carriers). Typical values of α_R are 0.5 to 0.8. An inverse current gain β_R is also defined

$$\beta_R = \frac{\alpha_R}{1 - \alpha_R} \tag{1.64}$$

and has typical values 1 to 5. This is the current gain of the transistor when operated inverted and is much lower than β_F because the device geometry and doping densities are designed to maximize β_F. The inverse-active region of device operation occurs for V_{CE} negative in an *npn* transistor and is shown in Fig. 1.12. In order to display these characteristics adequately in the same figure as the forward-active region, the negative voltage and current scales have been expanded. The inverse-active mode of operation is rarely encountered in analog circuits.

Equations 1.62 and 1.63 describe *npn* transistor operation in the saturation region when V_{BE} and V_{BC} are both positive, and also in the forward-active and inverse-active regions. These equations are the *Ebers–Moll* equations. In the forward-active region, they degenerate into a form similar to that of (1.35), (1.48), and (1.50) derived earlier. This can be shown by putting V_{BE} positive and V_{BC} negative in (1.62) and (1.63) to obtain

$$I_C = \alpha_F I_{ES}\left(\exp\frac{V_{BE}}{V_T} - 1\right) + I_{CS} \tag{1.65}$$

$$I_E = -I_{ES}\left(\exp\frac{V_{BE}}{V_T} - 1\right) - \alpha_R I_{CS} \tag{1.66}$$

Equation 1.65 is similar in form to (1.35) except that leakage currents that were previously neglected have now been included. This minor difference is significant only at high temperatures or very low operating currents. Comparison of (1.65) with (1.35) allows us to identify $I_S = \alpha_F I_{ES}$ and it can be shown[10] in general that

$$\alpha_F I_{ES} = \alpha_R I_{CS} = I_S \tag{1.67}$$

where this expression represents a reciprocity condition. Use of (1.67) in (1.62) and (1.63) allows the Ebers-Moll equations to be expressed in the general form

$$I_C = I_S\left(\exp\frac{V_{BE}}{V_T} - 1\right) - \frac{I_S}{\alpha_R}\left(\exp\frac{V_{BC}}{V_T} - 1\right) \tag{1.62a}$$

$$I_E = -\frac{I_S}{\alpha_F}\left(\exp\frac{V_{BE}}{V_T} - 1\right) + I_S\left(\exp\frac{V_{BC}}{V_T} - 1\right) \tag{1.63a}$$

This form is often used for computer representation of transistor large-signal behavior.

The effect of leakage currents mentioned above can be further illustrated as follows. In the forward-active region, we have, from (1.66),

$$I_{ES}\left(\exp\frac{V_{BE}}{V_T}-1\right)=-I_E-\alpha_R I_{CS} \tag{1.68}$$

Substitution of (1.68) in (1.65) gives

$$I_C=-\alpha_F I_E+I_{CO} \tag{1.69}$$

where

$$I_{CO}=I_{CS}(1-\alpha_R\alpha_F) \tag{1.69a}$$

and I_{CO} is the collector-base leakage current with the emitter open. Although I_{CO} is given theoretically by (1.69a), in practice, surface leakage effects dominate when the collector-base junction is reverse biased and I_{CO} is typically several orders of magnitude larger than the value given by (1.69a). However, (1.69) is still valid if the appropriate measured value for I_{CO} is used. Typical values of I_{CO} are 10^{-10}–10^{-12} A at 25°C, and the magnitude doubles about every 8°C. As a consequence, these leakage terms can become very significant at high temperatures. For example, consider the base current I_B. From Fig. 1.5 this is

$$I_B=-(I_C+I_E) \tag{1.70}$$

If I_E is calculated from (1.69) and substituted in (1.70) the result is

$$I_B=\frac{1-\alpha_F}{\alpha_F}I_C-\frac{I_{CO}}{\alpha_F} \tag{1.71}$$

But from (1.50)

$$\beta_F=\frac{\alpha_F}{1-\alpha_F} \tag{1.72}$$

and use of (1.72) in (1.71) gives

$$I_B=\frac{I_C}{\beta_F}-\frac{I_{CO}}{\alpha_F} \tag{1.73}$$

Since the two terms in (1.73) have opposite signs, the effect of I_{CO} is to *decrease* the magnitude of the external base current at a given value of collector current.

EXAMPLE

If I_{CO} is 10^{-10} A at 24°C, estimate its value at 120°C.

Assuming that I_{CO} doubles every 8°C we have

$$I_{CO}\,(120°\text{C})=10^{-10}\times 2^{12}$$
$$=0.4\ \mu\text{A}$$

1.3.4 Transistor Breakdown Voltages

In Section 1.2.2 the mechanism of avalanche breakdown in a *pn* junction was described. Similar effects occur at the base-emitter and base-collector junctions of a transistor and these effects limit the maximum voltages that can be applied to the device.

First consider a transistor in the common-base configuration shown in Fig. 1.14*a* and supplied with a constant emitter current. Typical $I_C - V_{CB}$ characteristics for an *npn* transistor in such a connection are shown in Fig. 1.14*b*. For $I_E = 0$ the collector-base junction breaks down at a voltage BV_{CBO}, which represents collector-base breakdown with the emitter open. For finite values of I_E, the effects of avalanche multiplication are apparent for values of V_{CB} below BV_{CBO}. In the example shown, the effective common-base current gain $\alpha_F = I_C/I_E$ becomes larger than unity for values of V_{CB} above about 60 V. Operation in this region (but below BV_{CBO}) can however be safely undertaken if the device power dissipation is not excessive. The considerations of Section 1.2.2 apply to this situation, and, neglecting leakage currents, we can calculate the collector current in Fig. 1.14*a* as

$$I_C = -\alpha_F I_E M \tag{1.74}$$

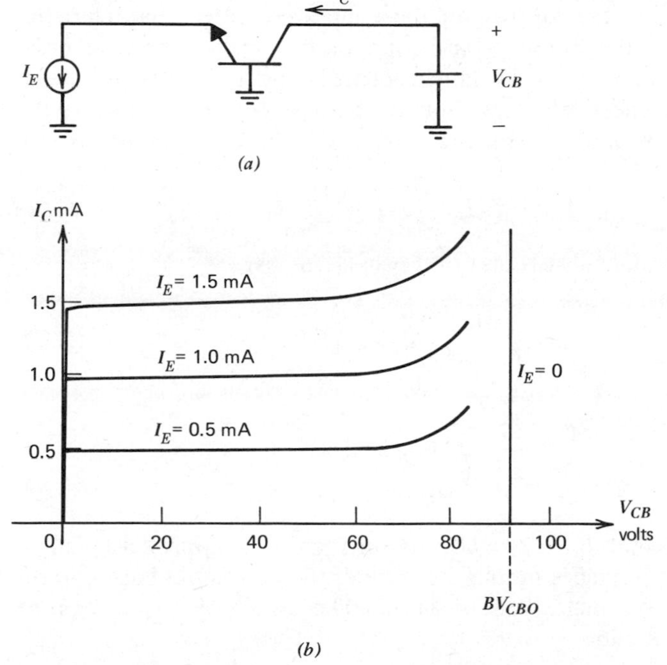

Fig. 1.14 Common-base transistor connection. (*a*) Test circuit. (*b*) I_C–V_{CB} characteristics.

where M is defined by (1.26) and thus

$$I_C = -\alpha_F I_E \frac{1}{1 - \left(\dfrac{V_{CB}}{BV_{CBO}}\right)^n} \tag{1.75}$$

One further point to note about the common-base characteristics of Fig. 1.14*b* is that for low values of V_{CB} where avalanche effects are negligible, the curves show very little of the Early effect seen in the common emitter characteristics. Base widening still occurs in this configuration as V_{CB} is increased, but unlike the common-emitter connection it produces little change in I_C. This is because I_E is now fixed instead of V_{BE} or I_B, and, in Fig. 1.9, this means the slope of the minority-carrier concentration at the emitter edge of the base is fixed. Thus the collector current remains almost unchanged.

Now consider the effect of avalanche breakdown on the common-emitter characteristics of the device. Typical characteristics are shown in Fig. 1.12 and breakdown occurs at a value BV_{CEO}, which is sometimes called the sustaining voltage LV_{CEO}. As in previous cases, operation near the breakdown voltage is only destructive to the device if the current (and thus the power dissipation) becomes excessive.

The effects of avalanche breakdown on the common-emitter characteristics are more complex than in the common-base configuration. This is because hole-electron *pairs* are produced by the avalanche process and the holes are swept into the base where they effectively contribute to the base current. In a sense the avalanche current is then *amplified* by the transistor. The base current is still given by

$$I_B = -(I_C + I_E) \tag{1.76}$$

Equation 1.74 still holds and substitution of this in (1.76) gives

$$I_C = \frac{M\alpha_F}{1 - M\alpha_F} I_B \tag{1.77}$$

where

$$M = \frac{1}{1 - \left(\dfrac{V_{CB}}{BV_{CBO}}\right)^n} \tag{1.78}$$

Equation 1.77 shows that I_C approaches infinity as $M\alpha_F$ approaches unity. That is, the effective β approaches infinity because of the additional base-current contribution from the avalanche process itself. The value of BV_{CEO} can be determined by solving equation

$$M\alpha_F = 1 \tag{1.79}$$

If we assume that $V_{CB} \simeq V_{CE}$ this gives

$$\frac{\alpha_F}{1 - \left(\dfrac{BV_{CEO}}{BV_{CBO}}\right)^n} = 1 \tag{1.80}$$

and this results in

$$\frac{BV_{CEO}}{BV_{CBO}} = \sqrt[n]{1 - \alpha_F}$$

and thus

$$BV_{CEO} \simeq \frac{BV_{CBO}}{\sqrt[n]{\beta_F}} \tag{1.81}$$

Equation 1.81 shows that BV_{CEO} is less than BV_{CBO} by a substantial factor. However, the value of BV_{CBO}, which must be used in (1.81), is the *plane* junction breakdown of the collector-base junction, neglecting any edge effects. This is because it is only collector-base avalanche current actually under the emitter that is amplified as described in the above calculation. However, as explained in Section 1.2.2 the measured value of BV_{CBO} is usually determined by avalanche in the curved region of the collector, which is remote from the active base. Consequently, for typical values of $\beta_F = 100$ and $n = 4$, the value of BV_{CEO} is about one half of the measured BV_{CBO} and not 30 percent as (1.81) would indicate.

Equation 1.81 explains the shape of the breakdown characteristics of Fig. 1.12, if the dependence of β_F on collector current is included. As V_{CE} is increased from zero with $I_B = 0$, the initial collector current is approximately $\beta_F I_{CO}$ from (1.73), and, since I_{CO} is typically several picoamperes, the collector current is very small. As explained in the next section, β_F is small at low currents and thus from (1.81) the breakdown voltage is high. However as avalanche breakdown begins in the device, the value of I_C increases and thus β_F increases. From (1.81) this causes a *decrease* in the breakdown voltage and the characteristic bends back as shown in Fig. 1.12 and exhibits a negative slope. At higher collector currents, β_F approaches a constant value and the breakdown curve with $I_B = 0$ becomes perpendicular to the V_{CE} axis. The value of V_{CE} in this region of the curve is usually defined to be BV_{CEO}, since this is the maximum voltage the device can sustain. The value of β_F to be used to calculate BV_{CEO} in (1.81) is thus the *peak* value of β_F. Note from (1.81) that high-β transistors will thus have low values of BV_{CEO}.

The base-emitter junction of a transistor is also subject to avalanche breakdown. However the doping density in the emitter is made very large to ensure a high value of β_F [N_D is made large in (1.45) to reduce I_{B2}]. Thus the base is the more lightly doped side of the junction and determines the breakdown characteristic. This can be contrasted with the collector-base junction where the collector is the more lightly doped side and results in typical values of BV_{CBO} of 80 V or more. The base is typically an order of magnitude more heavily doped than the collector and thus the base-emitter breakdown voltage is much less than BV_{CBO} and is typically about 6 to 8 V. This is designated BV_{EBO}. The breakdown voltage for inverse-active operation shown in Fig. 1.12 is approximately equal to this value because the base-emitter junction is reverse biased in this mode of operation.

The base-emitter breakdown voltage of 6 to 8 V provides a convenient reference voltage in integrated-circuit design, and this is often utilized in the form of a *Zener* diode. However care must be taken to ensure that all other transistors in a circuit are protected against reverse base-emitter voltages sufficient to cause breakdown. This is due to the fact that, unlike collector-base breakdown, base-emitter breakdown *is* damaging to the device. It can cause a large degradation in β_F, depending on the duration of the breakdown-current flow and its magnitude.[11] If the device is used purely as a Zener diode this is of no consequence, but if the device is an amplifying transistor the β_F degradation may be serious.

EXAMPLE

If the collector doping density in a transistor is 2×10^{15} atoms/cm^3 and is much less than the base doping, calculate BV_{CEO} for $\beta = 100$ and $n = 4$. Assume $\mathscr{E}_{crit} = 3 \times 10^5$ V/cm.

The plane breakdown voltage in the collector can be calculated from (1.24) using $\mathscr{E}_{max} = \mathscr{E}_{crit}$

$$BV_{CBO} = \frac{\epsilon(N_A + N_D)}{2qN_AN_D}\mathscr{E}^2_{crit}$$

Since $N_D \ll N_A$, we have

$$BV_{CBO}\bigg|_{plane} = \frac{\epsilon}{2qN_D}\mathscr{E}^2_{crit} = \frac{1.04 \times 10^{-12}}{2 \times 1.6 \times 10^{-19} \times 2 \times 10^{15}} \times 9 \times 10^{10}\text{ V} = 146\text{ V}$$

From (1.81)

$$BV_{CEO} = \frac{146}{\sqrt[4]{100}}\text{ V} = 46\text{ V}$$

1.3.5 Dependence of Transistor Current Gain β_F on Operating Conditions

Although most first-order analyses of integrated circuits make the assumption that β_F is constant, this parameter does in fact depend on the operating conditions of the transistor. It was shown in Section 1.3.2, for example, that increasing the value of V_{CE} increases I_C while producing little change in I_B, and thus the effective β_F of the transistor increases. In Section 1.3.4 it was shown that as V_{CE} approaches the breakdown voltage, BV_{CEO}, the collector current increases due to avalanche multiplication in the collector. Equation 1.77 shows that the effective current gain approaches infinity as V_{CE} approaches BV_{CEO}.

In addition to the effects just described, β_F also varies with both temperature and transistor collector current. This is illustrated in Fig. 1.15, which shows

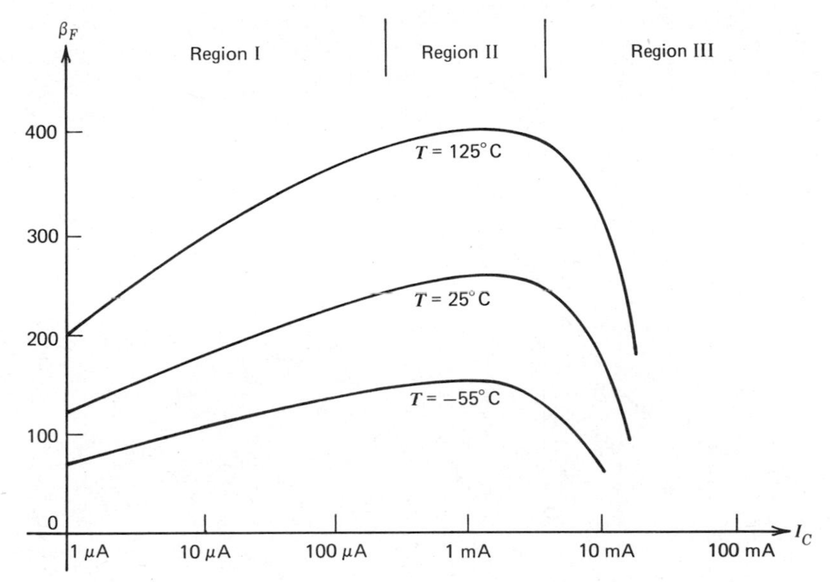

Fig. 1.15 Typical curves of β_F versus I_C for an *npn* integrated-circuit transistor.

typical curves of β_F versus I_C at three different temperatures for an *npn* integrated circuit transistor. It is evident that β_F increases as temperature increases, and a typical temperature coefficient for β_F is +7000 ppm/°C (where ppm signifies *parts per million*). This temperature dependence of β_F is due to the effect of the extremely high doping density in the emitter,[12] which causes the emitter injection efficiency γ to increase with temperature.

The variation of β_F with collector current, which is apparent in Fig. 1.15, can be divided into three regions. Region I is the low-current region where β_F decreases as I_C decreases. Region II is the midcurrent region where β_F is approximately constant. Region III is the high-current region where β_F decreases as I_C increases. The reasons for this behavior of β_F with I_C can be better appreciated by plotting base current I_B and collector current I_C on a log scale as a function of V_{BE}. This is shown in Fig. 1.16, and, because of the log scale on the vertical axis, the value of ln β_F can be obtained directly as the distance between the two curves.

At moderate current levels represented by region II in Figs. 1.15 and 1.16, both I_C and I_B follow the ideal behavior, and

$$I_C = I_S \exp \frac{V_{BE}}{V_T} \tag{1.82}$$

$$I_B \simeq \frac{I_S}{\beta_{FM}} \exp \frac{V_{BE}}{V_T} \tag{1.83}$$

where β_{FM} is the maximum value of β_F and is given by (1.48).

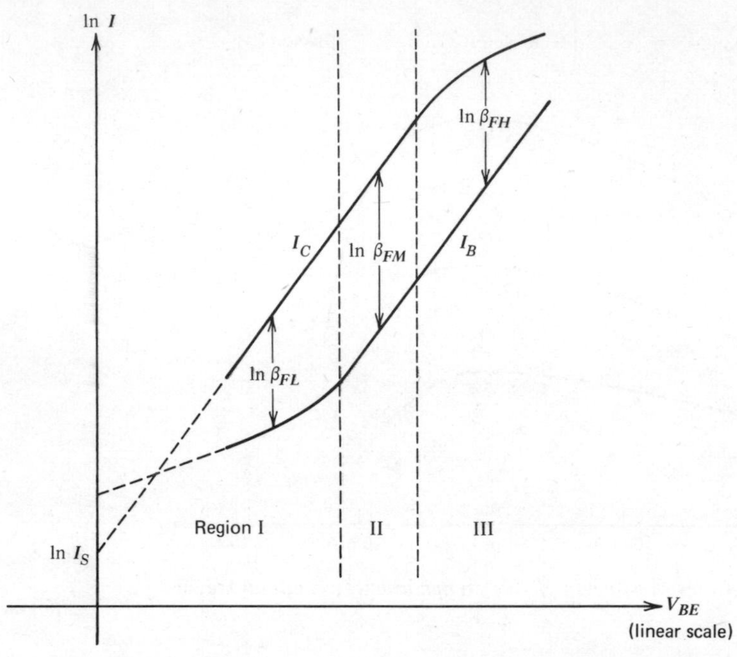

Fig. 1.16 Base and collector currents of a bipolar transistor plotted on a log scale versus V_{BE} on a linear scale. The distance between the curves is a direct measure of ln β_F.

At low current levels, I_C still follows the ideal relationship of (1.82), and the decrease in β_F is due to an additional component in I_B, which is mainly due to recombination of carriers in the base-emitter depletion region and is present at any current level. However at higher current levels the base current given by (1.83) dominates and this additional component has little effect. The base current resulting from recombination in the depletion region is[5]

$$I_{BX} \simeq I_{SX} \exp \frac{V_{BE}}{mV_T} \tag{1.84}$$

where $$m \simeq 2$$

At very low collector currents, where (1.84) dominates the base current, the current gain can be calculated from (1.82) and (1.84) as

$$\beta_{FL} \simeq \frac{I_C}{I_{BX}} = \frac{I_S}{I_{SX}} \exp \frac{V_{BE}}{V_T}\left(1 - \frac{1}{m}\right) \tag{1.85}$$

Substitution of (1.82) in (1.85) gives

$$\beta_{FL} \simeq \frac{I_S}{I_{SX}} \left(\frac{I_C}{I_S}\right)^{[1-(1/m)]} \tag{1.86}$$

If $m \simeq 2$, then (1.86) indicates that β_F is proportional to $\sqrt{I_C}$ at very low collector currents.

At high current levels, the base current, I_B, tends to follow the relationship of (1.83) and the decrease in β_F in region III is due mainly to a decrease in I_C below the value given by (1.82). (In practice the measured curve of I_B versus V_{BE} in Fig. 1.16 may also deviate from a straight line at high currents due to the influence of voltage drop across the base resistance.) The decrease in I_C is due partly to the effect of high-level injection and at high current levels the collector current approaches[7]

$$I_C \simeq I_{SH} \exp \frac{V_{BE}}{2V_T} \tag{1.87}$$

The current gain in this region can be calculated from (1.87) and (1.83) as

$$\beta_{FH} \simeq \frac{I_{SH}}{I_S} \beta_{FM} \exp\left(-\frac{V_{BE}}{2V_T}\right) \tag{1.88}$$

Substitution of (1.87) in (1.88) gives

$$\beta_{FH} \simeq \frac{{I_{SH}}^2}{I_S} \beta_{FM} \frac{1}{I_C}$$

Thus β_F decreases rapidly at high collector currents.

In addition to the effect of high-level injection, the value of β_F at high currents is also decreased by the onset of the Kirk effect,[13] which occurs when the minority-carrier concentration in the collector becomes comparable to the donor-atom doping density. The base region of the transistor then stretches out into the collector and becomes greatly enlarged.

1.4 SMALL-SIGNAL MODELS OF BIPOLAR TRANSISTORS

Analog circuits often operate with signal levels that are small compared to the bias currents and voltages in the circuit. In these circumstances, *incremental* or *small-signal* models can be derived that allow calculation of circuit gain and terminal impedances without the necessity of including the bias quantities. A hierarchy of models with increasing complexity can be derived, and the more complex ones are generally reserved for computer analysis. Part of the designers' skill is knowing which elements of the model can be omitted when performing hand calculations on a particular circuit, and this point is taken up again later.

Consider the bipolar transistor in Fig. 1.17*a* with bias voltages V_{BE} and V_{CC} applied as shown. These produce a quiescent collector current, I_C, and a quiescent base current, I_B, and the device is in the *forward-active region.* A "small-signal" input voltage, v_i, is applied in series with V_{BE} and produces a small variation in base current i_b and a small variation in collector current i_c. Total values of base and collector currents are I_b and I_c, respectively, and thus $I_b = (I_B + i_b)$ and $I_c =$

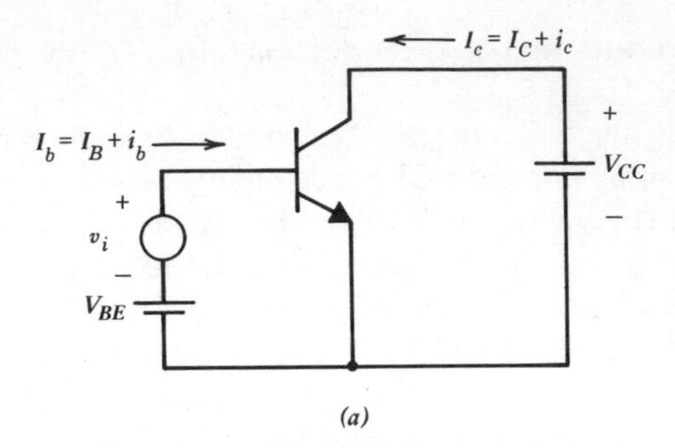

(a)

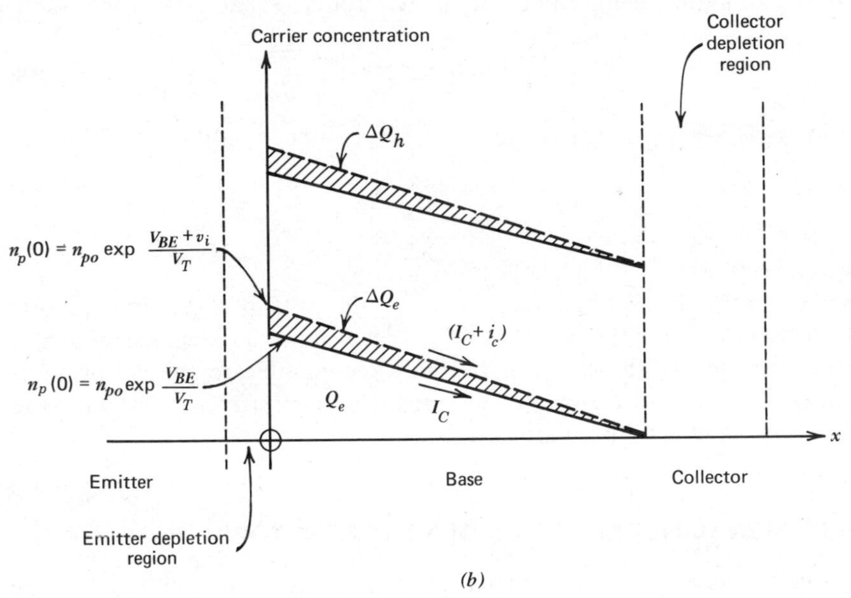

(b)

Fig. 1.17 Effect of a small-signal input voltage applied to a bipolar transistor. (*a*) Circuit schematic. (*b*) Corresponding changes in carrier concentrations in the base when the device is in the forward-active region.

$(I_C + i_c)$. The carrier concentrations in the base of the transistor corresponding to the situation in Fig. 1.17*a* are shown in Fig. 1.17*b*. With bias voltages only applied, the carrier concentrations are given by the solid lines. Application of the small-signal voltage, v_i, causes $n_p(0)$ at the emitter edge of the base to increase, and produces the concentrations shown by the dotted lines. These pictures can now be used to derive the various elements in the small-signal equivalent circuit of the bipolar transistor.

1.4.1 Transconductance

The transconductance is defined as

$$g_m = \frac{dI_C}{dV_{BE}} \tag{1.89}$$

Since

$$\Delta I_C = \frac{dI_C}{dV_{BE}} \Delta V_{BE}$$

we can write

$$\Delta I_C = g_m \, \Delta V_{BE}$$

and thus

$$i_c = g_m v_i \tag{1.90}$$

The value of g_m can be found by substituting (1.35) in (1.89) to give

$$g_m = \frac{d}{dV_{BE}} I_S \exp \frac{V_{BE}}{V_T} = \frac{I_S}{V_T} \exp \frac{V_{BE}}{V_T} = \frac{I_C}{V_T} = \frac{qI_C}{kT} \tag{1.91}$$

The transconductance thus depends linearly on the bias current, I_C, and is 38 mA/V for $I_C = 1$ mA at 25°C.

In order to illustrate the limitations on the use of small-signal analysis, the above relation will be derived in an alternative way. The total collector current in Fig. 1.17*a* can be calculated using (1.35) as

$$I_c = I_S \exp \frac{V_{BE} + v_i}{V_T} = I_S \exp \frac{V_{BE}}{V_T} \exp \frac{v_i}{V_T} \tag{1.92}$$

But the collector bias current is

$$I_C = I_S \exp \frac{V_{BE}}{V_T} \tag{1.93}$$

and use of (1.93) in (1.92) gives

$$I_c = I_C \exp \frac{v_i}{V_T} \tag{1.94}$$

If $v_i < V_T$, the exponential in (1.94) can be expanded in a power series

$$I_c = I_C \left[1 + \frac{v_i}{V_T} + \frac{1}{2}\left(\frac{v_i}{V_T}\right)^2 + \frac{1}{6}\left(\frac{v_i}{V_T}\right)^3 + \cdots \right] \tag{1.95}$$

Now the incremental collector current is

$$i_c = I_c - I_C \tag{1.96}$$

and substitution of (1.96) in (1.95) gives

$$i_c = \frac{I_C}{V_T} v_i + \frac{1}{2} \frac{I_C}{V_T^2} v_i^2 + \frac{1}{6} \frac{I_C}{V_T^3} v_i^3 + \cdots \tag{1.97}$$

If $v_i \ll V_T$, (1.97) reduces to (1.90) and the small-signal analysis is valid. The criterion for use of small-signal analysis is thus $v_i = \Delta V_{BE} \ll 26$ mV at 25°C. In practice, if ΔV_{BE} is less than 10 mV the small-signal analysis is accurate within about 10 percent.

1.4.2 Base-Charging Capacitance

Fig. 1.17*b* shows that the change in base-emitter voltage $\Delta V_{BE} = v_i$ has caused a change $\Delta Q_e = q_e$ in the minority-carrier charge in the base. By charge-neutrality requirements, there is an equal change $\Delta Q_h = q_h$ in the majority-carrier charge in the base. Since majority carriers are supplied by the base lead, the application of voltage, v_i, requires the supply of charge q_h to the base and the device has an apparent input capacitance

$$C_b = \frac{q_h}{v_i} \tag{1.98}$$

The value of C_b can be related to fundamental device parameters as follows. If (1.39) is divided by (1.33) we obtain

$$\frac{Q_e}{I_C} = \frac{W_B^2}{2D_n} = \tau_F \tag{1.99}$$

The quantity, τ_F, has the dimension of time and is called the base transit time in the forward direction. Since it is the ratio of the charge in transit (Q_e) to the current flow (I_C) it can be identified as the average time per carrier spent in crossing the base. To a first order it is independent of operating conditions and has typical values 0.1 to 1 nsec for integrated *npn* transistors and 20 to 40 nsec for lateral *pnp* transistors. Practical values of τ_F tend to be somewhat lower than predicted by (1.99) for diffused transistors that have nonuniform base doping.[14] However the functional dependence on base width W_B and diffusion constant D_n is as predicted by (1.99).

From (1.99)

$$\Delta Q_e = \tau_F \, \Delta I_C \tag{1.100}$$

But since $\Delta Q_e = \Delta Q_h$ we have

$$\Delta Q_h = \tau_F \, \Delta I_C \tag{1.101}$$

and this can be written

$$q_h = \tau_F i_c \tag{1.102}$$

Use of (1.102) in (1.98) gives

$$C_b = \tau_F \frac{i_c}{v_i} \tag{1.103}$$

and substitution of (1.90) in (1.103) gives

$$C_b = \tau_F g_m \tag{1.104}$$

$$= \tau_F \frac{qI_C}{kT} \tag{1.105}$$

Thus the small-signal, base-charging capacitance is proportional to the collector bias current.

1.4.3 Input Resistance

In the forward-active region, the base current is related to the collector current by (1.47) as

$$I_B = \frac{I_C}{\beta_F} \tag{1.47}$$

Small changes in I_B and I_C can be related using (1.47)

$$\Delta I_B = \frac{d}{dI_C}\left(\frac{I_C}{\beta_F}\right)\Delta I_C \tag{1.106}$$

and thus

$$\beta_0 = \frac{\Delta I_C}{\Delta I_B} = \frac{i_c}{i_b} = \left[\frac{d}{dI_C}\left(\frac{I_C}{\beta_F}\right)\right]^{-1} \tag{1.107}$$

where β_0 is the *small-signal* current gain of the transistor. Note that if β_F is constant, then $\beta_F = \beta_0$. Typical values of β_0 are close to those of β_F, and in subsequent chapters little differentiation is made between these quantities. A single value of β is often assumed for a transistor and then used for both ac and dc calculations.

Equation 1.107 relates the change in base current i_b to the corresponding change in collector current i_c, and the device has a small-signal input resistance given by

$$r_\pi = \frac{v_i}{i_b} \tag{1.108}$$

Substitution of (1.107) in (1.108) gives

$$r_\pi = \frac{v_i}{i_c}\beta_0 \tag{1.109}$$

and use of (1.90) in (1.109) gives

$$r_\pi = \frac{\beta_0}{g_m} \tag{1.110}$$

Thus the small-signal input shunt resistance of a bipolar transistor depends on the current gain and is inversely proportional to I_C.

1.4.4 Output Resistance

In Section 1.3.2 the effect of changes in collector-emitter voltage V_{CE} on the large-signal characteristics of the transistor was described. It follows from that treatment that small changes ΔV_{CE} in V_{CE} produce corresponding changes ΔI_C in I_C where

$$\Delta I_C = \frac{\partial I_C}{\partial V_{CE}} \Delta V_{CE} \tag{1.111}$$

Substitution of (1.55) and (1.57) in (1.111) gives

$$\frac{\Delta V_{CE}}{\Delta I_C} = \frac{V_A}{I_C} = r_o \tag{1.112}$$

where V_A is the Early voltage and r_o is the small-signal output resistance of the transistor. Since typical values of V_A are 50 to 100 V, corresponding values of r_o are 50–100 kΩ for $I_C = 1$ mA. Note that r_o is inversely proportional to I_C, and thus r_o can be related to g_m as are many of the other small-signal parameters.

$$r_o = \frac{1}{\eta g_m} \tag{1.113}$$

where

$$\eta = \frac{kT}{qV_A} \tag{1.114}$$

If $V_A = 100$ V, then $\eta = 2.6 \times 10^{-4}$ at 25°C. Note that $1/r_o$ is the slope of the output characteristics of Fig. 1.10.

1.4.5 Basic Small-Signal Model of the Bipolar Transistor

Combination of the above small-signal circuit elements yields the small-signal model of the bipolar transistor shown in Fig. 1.18. This is valid for both *npn* and *pnp* devices in the forward-active region and is called the *hybrid-π* model. Collector, base, and emitter nodes are labeled *C*, *B*, and *E*, respectively. The elements in this

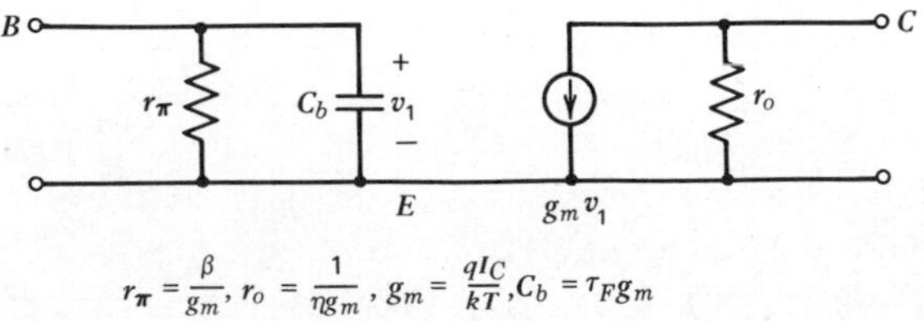

Fig. 1.18 Basic bipolar transistor small-signal equivalent circuit.

circuit are present in the equivalent circuit of *any* bipolar transistor and are specified by relatively few parameters (β, τ_F, η, I_C). Note that in the evaluation of the small-signal parameters for *pnp* transistors, the *magnitude only* of I_C is used. In the following sections, further elements are added to this model to account for parasitics and second-order effects.

1.4.6 Collector-Base Resistance

Consider the effect of variations in V_{CE} on the minority charge in the base as illustrated in Fig. 1.9. An increase in V_{CE} causes an increase in the collector depletion-layer width and consequent reduction of base width. This causes a reduction in the total minority-carrier charge stored in the base and thus a reduction in base current I_B due to a reduction in I_{B1} given by (1.40). Since an increase ΔV_{CE} in V_{CE} causes a *decrease* ΔI_B in I_B, this effect can be modeled by inclusion of a resistor r_μ from collector to base of the model of Fig. 1.18. If V_{BE} is assumed held constant, the value of this resistor can be determined as follows.

$$r_\mu = \frac{\Delta V_{CE}}{\Delta I_{B1}} = \frac{\Delta V_{CE}}{\Delta I_C}\frac{\Delta I_C}{\Delta I_{B1}} \tag{1.115}$$

Substitution of (1.112) in (1.115) gives

$$r_\mu = r_o \frac{\Delta I_C}{\Delta I_{B1}} \tag{1.116}$$

If the base current I_B is composed entirely of component I_{B1}, then (1.107) can be used in (1.116) to give

$$r_\mu = \beta_0 r_o \tag{1.117}$$

This is a lower limit for r_μ. In practice I_{B1} is typically less than 10 percent of I_B [component I_{B2} from (1.42) dominates] in integrated *npn* transistors, and since I_{B1} is very small the change ΔI_{B1} in I_{B1} for a given ΔV_{CE} and ΔI_C is also very small. Thus a typical value for r_μ is greater than 10 $\beta_0 r_o$. For lateral *pnp* transistors, recombination in the base is more significant and r_μ is in the range 2 $\beta_0 r_o$ to 5 $\beta_0 r_o$.

1.4.7 Parasitic Elements in the Small-Signal Model

The elements of the bipolar transistor small-signal equivalent circuit considered so far may be considered basic in the sense that they arise directly from essential processes in the device. However, technological limitations in the fabrication of transistors give rise to a number of parasitic elements that must be added to the equivalent circuit for most integrated-circuit transistors. A cross section of a typical *npn* transistor in an integrated circuit is shown in Fig. 1.19. The means of fabricating such devices is described in Chapter 2.

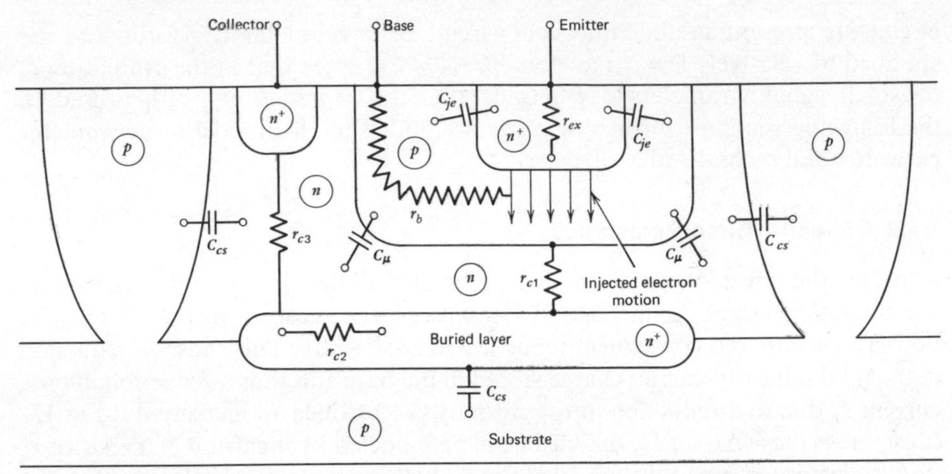

Fig. 1.19 Integrated-circuit *npn* bipolar transistor structure showing parasitic elements. (Not to scale.)

As described in Section 1.2 all *pn* junctions have a voltage-dependent capacitance associated with the depletion region. In the cross section of Fig. 1.19 three depletion-region capacitances can be identified. The base-emitter junction has a depletion-region capacitance, C_{je}, and the base-collector and collector-substrate junctions have capacitances C_μ and C_{cs}, respectively. Of these three, the base-emitter junction most closely approximates a graded junction and the variation of C_{je} with bias voltage is well approximated by (1.22). The collector-base junction behaves like a graded junction for small bias voltages since it is formed by diffusion and the doping density is a function of distance near the junction. However for larger reverse bias values (more than about a volt) the junction depletion region spreads into the collector, which is uniformly doped, and the junction tends to behave like an abrupt junction with uniform doping. The collector-base capacitance, C_μ, thus tends to follow (1.22) for very small bias voltages and (1.21) for large bias voltages. In practice, measurements show that the variation of C_μ with bias voltage can be approximated by

$$C_\mu = \frac{C_{\mu 0}}{\left(1 - \dfrac{V}{\psi_0}\right)^n}$$

where V is the forward bias on the junction and n is an exponent between 0.33 and 0.5. In this book we use $n = 0.5$ for the collector-base junction. The third parasitic capacitance in a monolithic *npn* transistor is the collector-substrate capacitance C_{cs}, and for large reverse bias voltages this varies according to the abrupt junction equation (1.21).

As described in Chapter 2, lateral *pnp* transistors have a parasitic capacitance C_{bs} from base to substrate in place of C_{cs}. Note that the substrate is always connected to the most negative voltage supply in the circuit in order to ensure that all isolation regions are separated by reverse-biased junctions. Thus the substrate is an ac ground and all parasitic capacitance to the substrate is connected to ground in an equivalent circuit. Typical zero-bias values of these parasitic capacitances for an *npn* transistor are $C_{je0} = 0.2$ to 1 pF, $C_{\mu 0} = 0.2$ to 1 pF, $C_{cs0} =$ 1 to 3 pF. Values for other devices are summarized in Chapter 2.

The final elements to be added to the small-signal model of the transistor are resistive parasitics. These are produced by the finite resistance of the silicon between the top contacts on the transistor and the active base region beneath the emitter. As shown in Fig. 1.19 there are significant resistances r_b and r_c in series with the base and collector contacts, respectively. There is also a resistance r_{ex} of several ohms in series with the emitter lead that can become important at high bias currents. (Note that the collector resistance r_c is actually composed of three parts labeled r_{c1}, r_{c2}, and r_{c3}.) Typical values of these parameters are $r_b = 50$ to 500 Ω, $r_{ex} = 1$ to 3 Ω, and $r_c = 20$ to 500 Ω. The value of r_b varies significantly with collector current because of *current crowding.*[15] This occurs at high collector currents where the dc base current produces a lateral voltage drop in the base that tends to forward bias the base-emitter junction preferentially around the edges of the emitter. Thus the transistor action tends to occur along the emitter periphery rather than under the emitter itself, and the distance from the base contact to the active base region is reduced. Consequently the value of r_b is reduced, and, in a typical *npn* transistor, r_b may decrease 50 percent as I_C increases from 0.1 mA to 10 mA.

The value of these parasitic resistances can be reduced by changes in the device structure. For example a large-area transistor with multiple base and emitter stripes will have a smaller value of r_b. The value of r_c is reduced by inclusion of the low-resistance buried $n+$ layer beneath the collector.

The addition of the resistive and capacitive parasitics to the basic small-signal circuit of Fig. 1.18 gives the complete small-signal equivalent circuit of Fig. 1.20. The internal base node is labeled B' to distinguish it from the external base contact, B. The capacitance, C_π, contains the base-charging capacitance, C_b, and the emitter-base depletion layer capacitance, C_{je}.

$$C_\pi = C_b + C_{je} \tag{1.118}$$

Note that the representation of parasitics in Fig. 1.20 is an approximation in that lumped elements have been used. In practice, as suggested by Fig. 1.19, C_μ is distributed across r_b and C_{cs} is distributed across r_c. This lumped representation is adequate for most purposes but can introduce errors at very high frequencies. It should also be noted that while the parasitic resistances of Fig. 1.20 can be very

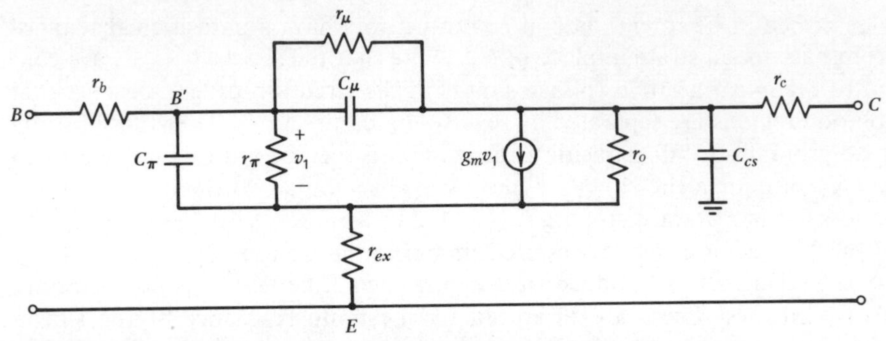

Fig. 1.20 Complete bipolar transistor small-signal equivalent circuit.

important at high bias currents or for high-frequency operation, they are usually omitted from the equivalent circuit for low-frequency calculations, particularly for collector bias currents less than 1 mA.

EXAMPLE

Derive the complete small-signal equivalent circuit for a bipolar transistor at $I_C = 1$ mA, $V_{CB} = 5$ V, and $V_{CS} = 10$ V. Device parameters are $C_{je0} = 0.5$ pF, $C_{\mu 0} = 0.5$ pF, $C_{cs0} = 2$ pF, $\beta_0 = 200$, $\tau_F = 0.38$ ns, $V_A = 100$ V, $r_b = 300\ \Omega$, $r_c = 50\ \Omega$, $r_{ex} = 2\ \Omega$, $r_\mu = 10\ \beta_0 r_o$.

Since the base-emitter junction is forward biased, the value of C_{je} is difficult to determine for reasons described in Section 1.2.1. A value can either be determined by computer or a reasonable estimation is to double C_{je0}. Using the latter approach we estimate

$$C_{je} = 1 \text{ pF}$$

Using (1.21) gives, for the collector-base capacitance,

$$C_\mu = \frac{C_{\mu 0}}{\sqrt{1 + \dfrac{V_{CB}}{\psi_0}}} = \frac{0.5}{\sqrt{1 + \dfrac{5}{0.55}}} = 0.16 \text{ pF}$$

where a typical value of $\psi_0 = 0.55$ V for the collector-base junction has been used from the data in Chapter 2. The collector-substrate capacitance can also be calculated using (1.21)

$$C_{cs} = \frac{C_{cs0}}{\sqrt{1 + \dfrac{V_{CS}}{\psi_0}}} = \frac{2}{\sqrt{1 + \dfrac{10}{0.52}}} = 0.44 \text{ pF}$$

where a typical value of $\psi_0 = 0.52$ V for the collector-substrate junction has been

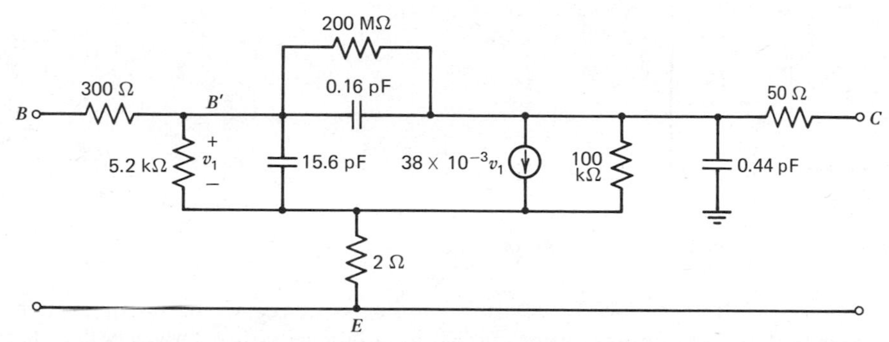

Fig. 1.21 Complete small-signal equivalent circuit for a bipolar transistor at $I_C = 1$ mA, $V_{CB} = 5$ V, and $V_{CS} = 10$ V. Device parameters are $C_{je0} = 0.5$ pF, $C_{\mu 0} = 0.5$ pF, $C_{cs0} = 2$ pF, $\beta_0 = 200$, $\tau_F = 0.38$ ns, $V_A = 100$ V, $r_b = 300\ \Omega$, $r_c = 50\ \Omega$, $r_{ex} = 2\ \Omega$, $r_\mu = 10\beta_o r_o$.

used from the data in Chapter 2. From (1.91) the transconductance is

$$g_m = \frac{qI_C}{kT} = \frac{10^{-3}}{26 \times 10^{-3}}\ \text{A/V} = 38\ \text{mA/V}$$

From (1.104) the base-charging capacitance is

$$C_b = \tau_F g_m = 0.38 \times 10^{-9} \times 38 \times 10^{-3}\ \text{F} = 14.6\ \text{pF}$$

The value of C_π from (1.118) is

$$C_\pi = (14.6 + 1)\ \text{pF} = 15.6\ \text{pF}$$

The input resistance from (1.110) is

$$r_\pi = \frac{\beta_0}{g_m} = 200 \times 26\ \Omega = 5.2\ \text{k}\Omega$$

The output resistance from (1.112) is

$$r_o = \frac{100}{10^{-3}}\ \Omega = 100\ \text{k}\Omega$$

and thus the collector-base resistance is

$$r_\mu = 10\ \beta_0 r_o = 10 \times 200 \times 10^5\ \Omega = 200\ \text{M}\Omega$$

The equivalent circuit with these parameter values is shown in Fig. 1.21.

1.4.8 Specification of Transistor Frequency Response

The high-frequency gain of the transistor is controlled by the capacitive elements in the equivalent circuit of Fig. 1.20. The frequency capability of the transistor is

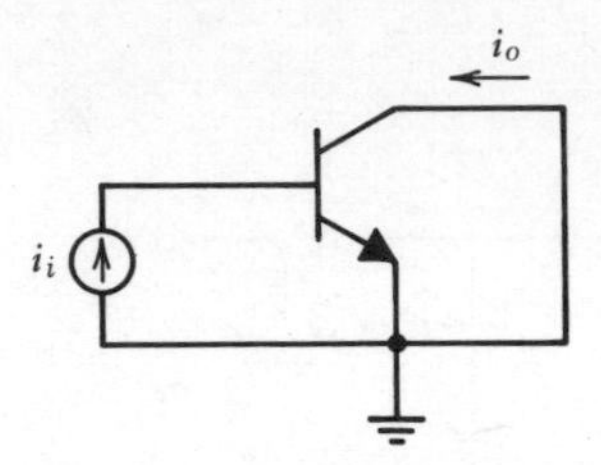

Fig. 1.22 ac schematic for measurement of f_T.

most often specified in practice by determining the frequency where the magnitude of the short-circuit, common-emitter current gain falls to unity. This is called the *transition frequency*, f_T, and is a measure of the maximum useful frequency of the transistor when it is used as an amplifier. The value of f_T can be measured as well as calculated, using the ac circuit of Fig. 1.22. A small-signal current, i_i, is applied to the base and the output current, i_o, is measured with the collector short circuited for ac signals. A small-signal equivalent circuit can be formed for this situation by using the equivalent circuit of Fig. 1.20 as shown in Fig. 1.23 where r_{ex} and r_μ have been neglected. If r_c is assumed small, then r_o and C_{cs} have no influence, and we have

$$v_1 \simeq \frac{r_\pi}{1 + r_\pi(C_\pi + C_\mu)s} i_i \tag{1.119}$$

If the current fed forward through C_μ is neglected,

$$i_o \simeq g_m v_1 \tag{1.120}$$

Substitution of (1.119) in (1.120) gives

$$i_o \simeq i_i \frac{g_m r_\pi}{1 + r_\pi(C_\pi + C_\mu)s}$$

and thus

$$\frac{i_o}{i_i}(j\omega) = \frac{\beta_0}{1 + \beta_0 \dfrac{C_\pi + C_\mu}{g_m} j\omega} \tag{1.121}$$

using (1.110).

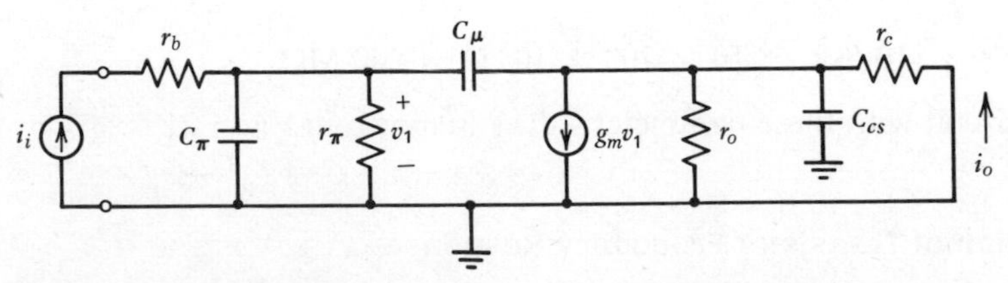

Fig. 1.23 Small-signal equivalent circuit for the calculation of f_T.

Now if $i_o/i_i(j\omega)$ is written as $\beta(j\omega)$ (the high-frequency, small-signal current gain) then

$$\beta(j\omega) = \frac{\beta_0}{1 + \beta_0 \dfrac{C_\pi + C_\mu}{g_m} j\omega} \tag{1.122}$$

At high frequencies the imaginary part of the denominator of (1.122) is dominant and we can write

$$\beta(j\omega) \simeq \frac{g_m}{j\omega(C_\pi + C_\mu)} \tag{1.123}$$

From (1.123), $|\beta(j\omega)| = 1$ when

$$\omega = \omega_T = \frac{g_m}{C_\pi + C_\mu} \tag{1.124}$$

and thus

$$f_T = \frac{1}{2\pi} \frac{g_m}{C_\pi + C_\mu} \tag{1.125}$$

The transistor behavior can be illustrated by plotting $|\beta(j\omega)|$ using (1.122) as shown in Fig. 1.24. The frequency ω_β is defined as the frequency where $|\beta(j\omega)|$ is equal to $\beta/\sqrt{2}$ (3 dB down from the low-frequency value). From (1.122) we have

$$\omega_\beta = \frac{1}{\beta_0} \frac{g_m}{C_\pi + C_\mu} = \frac{\omega_T}{\beta_0} \tag{1.126}$$

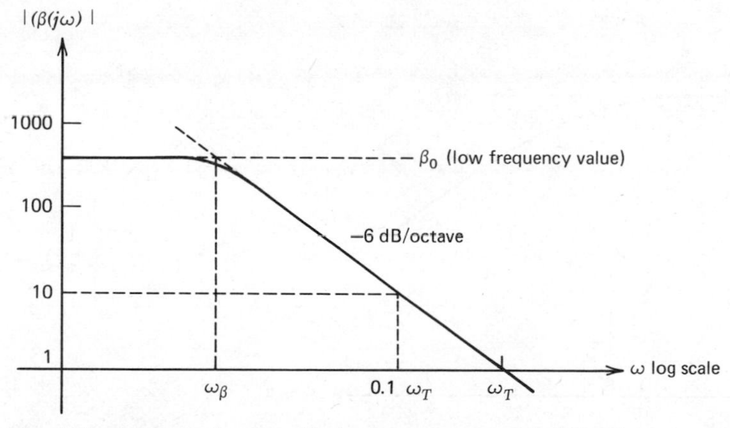

Fig. 1.24 Magnitude of small-signal ac current gain $|\beta(j\omega)|$ versus frequency for a typical bipolar transistor.

From Fig. 1.24 it can be seen that ω_T can be determined by measuring $|\beta(j\omega)|$ at some frequency ω_x where $|\beta(j\omega)|$ is falling at 6 dB/octave and using

$$\omega_T = \omega_x |\beta(j\omega_x)| \tag{1.127}$$

This is the method used in practice, since deviations from ideal behavior tend to occur as $|\beta(j\omega)|$ approaches unity. Thus $|\beta(j\omega)|$ is typically measured at some frequency where its magnitude is about 5 or 10, and (1.127) is used to determine ω_T.

It is interesting to examine the time constant, τ_T, associated with ω_T. This is defined as

$$\tau_T = \frac{1}{\omega_T} \tag{1.128}$$

and use of (1.124) in (1.128) gives

$$\tau_T = \frac{C_\pi}{g_m} + \frac{C_\mu}{g_m} \tag{1.129}$$

Substitution of (1.118) and (1.104) in (1.129) gives

$$\tau_T = \frac{C_b}{g_m} + \frac{C_{je}}{g_m} + \frac{C_\mu}{g_m} = \tau_F + \frac{C_{je}}{g_m} + \frac{C_\mu}{g_m} \tag{1.130}$$

Equation 1.130 indicates that τ_T is dependent on I_C (through g_m) and approaches a constant value of τ_F at high collector bias currents. At low values of I_C, the terms involving C_{je} and C_μ dominate and they cause τ_T to *rise* and f_T to *fall* as I_C is decreased. This behavior is illustrated in Fig. 1.25, which is a typical plot

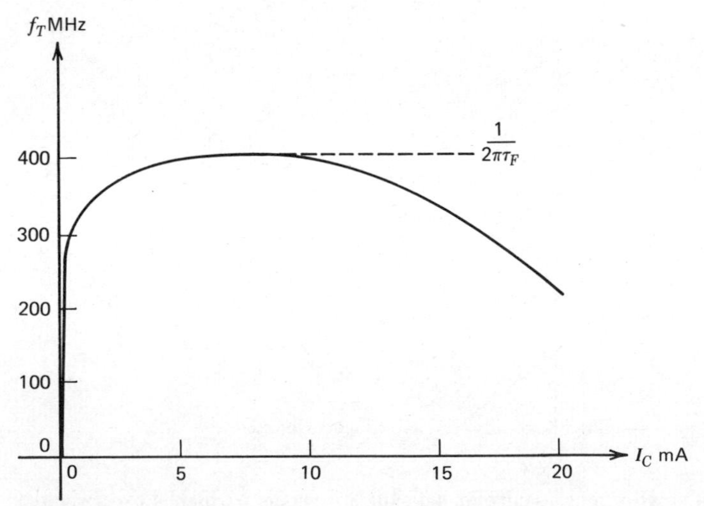

Fig. 1.25 Typical curve of f_T versus I_C for an *npn* bipolar transistor.

of f_T versus I_C for an integrated-circuit *npn* transistor. The decline in f_T at high collector currents is not predicted by this simple theory and is due to an increase in τ_F caused by high-level injection and Kirk effect at high currents. These are the same mechanisms that cause a decrease in β_F at high currents as described in Section 1.3.5.

EXAMPLE

A bipolar transistor has a short-circuit, common-emitter current gain at 100 MHz of 4 with $I_C = 1$ mA and 4.5 with $I_C = 4$ mA. Assuming that high-level injection effects are negligible, calculate C_{je} and τ_F, assuming both are constant. The measured value of C_μ is 0.3 pF.

From the data, values of f_T are

$$f_{T1} = 4 \times 100 = 400 \text{ MHz} \quad \text{at } I_C = 1 \text{ mA}$$
$$f_{T2} = 4.5 \times 100 = 450 \text{ MHz} \quad \text{at } I_C = 4 \text{ mA}$$

Corresponding values of τ_T are

$$\tau_{T1} = \frac{1}{2\pi f_{T1}} = 398 \text{ psec}$$

$$\tau_{T2} = \frac{1}{2\pi f_{T2}} = 354 \text{ psec}$$

Using this data in (1.130) we have

$$398 \times 10^{-12} = \tau_F + 26(C_\mu + C_{je}) \tag{1.131}$$

at $I_C = 1$ mA. At $I_C = 4$ mA we have

$$354 \times 10^{-12} = \tau_F + 6.5(C_\mu + C_{je}) \tag{1.132}$$

Subtraction of (1.132) from (1.131) yields

$$C_\mu + C_{je} = 2.26 \text{ pF}$$

Since C_μ was measured as 0.3 pF, the value of C_{je} is given by

$$C_{je} \simeq 2.0 \text{ pF}$$

Substitution in (1.131) gives

$$\tau_F = 339 \text{ psec}$$

This is an example of how basic device parameters can be determined from high-frequency current-gain measurements. Note that the assumption that C_{je} is constant is a useful approximation in practice because V_{BE} changes by only 36 mV as I_C increases from 1 mA to 4 mA.

1.5 LARGE-SIGNAL BEHAVIOR OF FIELD-EFFECT TRANSISTORS

The most common type of field-effect transistor (FET) used in analog integrated circuits is the junction field-effect transistor (JFET), and this device will now be considered in detail. As described in Chapter 2, most JFET structures require additional processing steps if they are fabricated on the same chip as bipolar devices. Thus circuits using JFETs as active elements tend to be more costly than standard circuits using only bipolar devices. However, in some applications JFETs can offer substantial performance improvement over that obtainable from bipolar devices, and the additional processing is often justified by the results obtained. For example JFETs can be used as input devices in linear monolithic amplifiers where they realize much larger input impedances than bipolar transistors.

1.5.1 Transfer Characteristics of the JFET

A cross section of a typical diffused JFET structure as fabricated in an integrated circuit is shown in Fig. 1.26*a*. These devices can be made simultaneously with bipolar transistors and using the same set of diffusions or by additional ion implantation steps. In the former case an additional n^+ diffusion is usually added to fabricate the top gate structure in order to narrow the *p*-type region connecting the source and drain. The reason for this will become apparent later. The gate is the control electrode and the gate voltage modulates the current flowing between source and drain. Note that the gate exists on both sides of the *p*-type channel, which is the conducting region connecting source and drain. This is called a *p*-channel JFET and its circuit symbol is shown in Fig. 1.26*b* together with the sign conventions used for currents and voltages.

The transfer characteristics of the JFET can be derived from the idealized structure of Fig. 1.27 where a uniformly doped *p*-type channel is assumed with symmetrical *n*-type gates on either side. If all the terminals of Fig. 1.27*a* are grounded, depletion regions exist at the junctions as explained in Section 1.2. The voltage across each junction is just the built-in potential ψ_0 and the depletion regions will be uniform in width in both *p*-type and *n*-type regions. If a negative voltage V_{DS} is now applied to the drain as shown in Fig. 1.27*a*, current will flow from source to drain. This will produce a voltage gradient in the channel and the gate-channel junction will be more reverse biased at the drain end. Since the reverse bias on the junction now varies down the channel, the depletion layer width also varies and is narrowest at the drain end. As the channel width decreases in this manner with the applications of V_{DS}, the channel resistance increases. As the magnitude of V_{DS} is increased in the limit, the channel width at the drain end approaches zero as shown in Fig. 1.27*b*; this is called "*pinch-off*." The voltage from gate to channel required to cause pinch-off is defined as the pinch-off voltage V_P, and this is a positive voltage for a *p*-channel JFET.

Consider now the situation in Fig. 1.27*b* where a positive voltage, V_{GS}, is applied to the gate and a negative voltage, V_{DS}, is applied to the drain while the

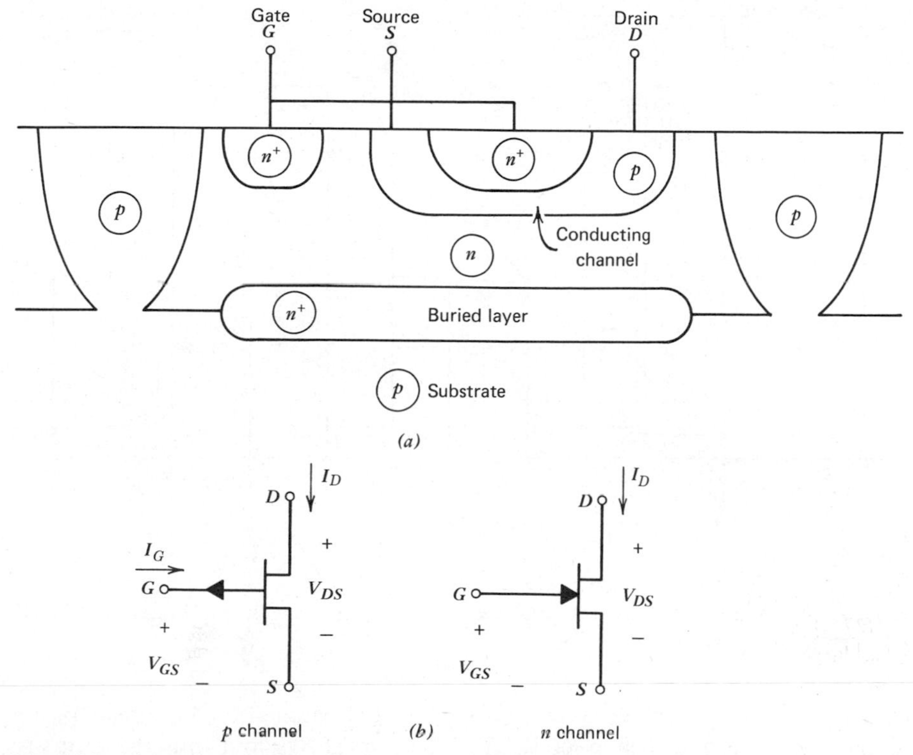

Fig. 1.26 (*a*) Typical monolithic *p*-channel JFET structure (not to scale). (*b*) JFET symbols and sign convention.

source is grounded. The applied gate voltage will cause the depletion regions to penetrate further into the channel along the whole channel length. When V_{GS} equals the pinch-off voltage, V_P, the whole channel is depleted and ideally zero current will flow between source and drain. The device is then said to be *cutoff*.

Taking the general situation shown in Fig. 1.27*b* we can calculate the width, $W(y)$, of the depletion region in the channel at distance y from the source contact. From (1.14) this is

$$W(y) = K_1\sqrt{\psi_0 + V_R(y)} \tag{1.133}$$

where

$$K_1 = \sqrt{\frac{2\epsilon}{qN_A\left(1 + \dfrac{N_A}{N_D}\right)}} \tag{1.134}$$

In these equations, N_A and N_D are the doping densities in the channel and gate, respectively, and $V_R(y)$ is reverse bias across the gate-channel junction at a distance y into the channel. The voltage, $V_R(y)$, can be written as

$$V_R(y) = V_{GS} - V(y) \tag{1.135}$$

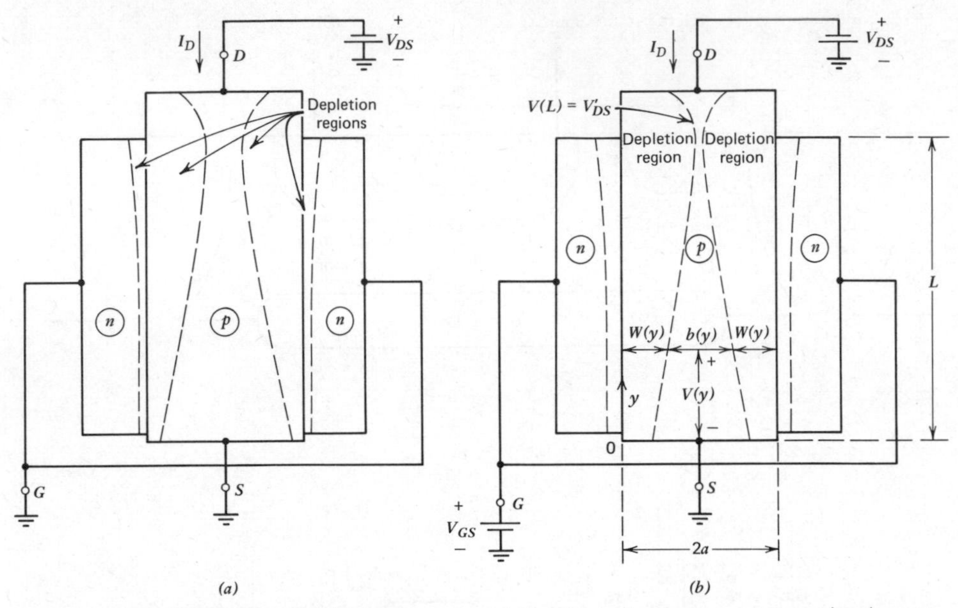

Fig. 1.27 Idealized p-channel JFET structure. (a) Source and gate grounded and $|V_{DS}|$ small. (b) $|V_{DS}|$ large, and positive V_{GS} applied.

where $V(y)$ is the voltage between the point in the channel under consideration and the source. For a p-channel JFET, $V(y)$ is negative and thus the terms in (1.135) add. Substitution of (1.135) in (1.133) gives

$$W(y) = K_1\sqrt{\psi_0 + V_{GS} - V(y)} \tag{1.136}$$

The channel width, $b(y)$, a distance y into the channel, can be calculated from Fig. 1.27b as

$$b(y) = 2a - 2W(y) \tag{1.137}$$

and substitution of (1.136) in (1.137) gives

$$b(y) = 2a - 2K_1\sqrt{\psi_0 + V_{GS} - V(y)} \tag{1.138}$$

where $2a$ is the channel depth. Ohm's law for conduction in the channel is

$$J_y = \sigma\mathscr{E}_y \tag{1.139}$$

where J_y is the current density in the y-direction, $\mathscr{E}_y$ is the electric field in the y-direction, and σ is the conductivity of the p-type channel region. Equation 1.139 can be written as

$$\frac{-I_D}{Zb(y)} = -\sigma\frac{dV(y)}{dy} \tag{1.140}$$

where Z is the width of the device perpendicular to the plane of Fig. 1.27.

Since I_D is constant in the channel, integration of (1.140) along the channel gives

$$I_D \int_0^L dy = \sigma Z \int_0^{V(L)} b\, dV \tag{1.141}$$

where $V(L)$ is the voltage in the channel a distance L from the source. This is the point where pinch-off occurs and may be considered the *internal* drain node. The external drain-source voltage, V_{DS}, will have a slightly larger magnitude because of the additional voltage drop from the pinch-off point to the drain contact. If we designate $V(L) = V'_{DS}$, substitute (1.138) in (1.141) and integrate we obtain

$$I_D = G_0 \left[V'_{DS} + \tfrac{2}{3} \frac{K_1}{a} (\psi_0 + V_{GS} - V'_{DS})^{3/2} - \tfrac{2}{3} \frac{K_1}{a} (\psi_0 + V_{GS})^{3/2} \right] \tag{1.142}$$

where

$$G_0 = \frac{2a\sigma Z}{L} \tag{1.143}$$

Quantity K_1/a, which appears in (1.142), can be evaluated using (1.133) by noting that the depletion layer width, $W(y)$, equals half the channel depth when the reverse voltage, V_R, equals the pinch-off voltage, V_P. Thus

$$a = K_1 \sqrt{\psi_0 + V_P}$$

and we have

$$\frac{K_1}{a} = \frac{1}{\sqrt{\psi_0 + V_P}} \tag{1.144}$$

Substitution of (1.144) in (1.142) gives

$$I_D = G_0 \left[V'_{DS} + \tfrac{2}{3} \frac{(\psi_0 + V_{GS} - V'_{DS})^{3/2} - (\psi_0 + V_{GS})^{3/2}}{(\psi_0 + V_P)^{1/2}} \right] \tag{1.145}$$

Equation 1.145 can be used to plot the transfer characteristics of the JFET (see Fig. 1.28*a*). The device operates with a negative voltage, V'_{DS}, and the drain current, I_D, flows *out* of the drain terminal and is also negative. For small values of V_{DS}, the device behaves as a near-linear resistor whose value is controlled by V_{GS}. For larger values of V_{DS} the depletion regions narrow the channel near the drain causing the channel resistance to increase and the characteristic to bend over. For $V_{GS} = 0$, the device approaches pinch-off as $V'_{DS} \to -V_P$ and the slope of the characteristic approaches zero. In the actual device, further increases in external V_{DS} beyond this point ideally produce no further change in V'_{DS} because the channel is pinched off. Any additional applied external drain voltage is absorbed by further constriction of the portion of the channel near the drain contact, and the drain current remains approximately constant. Note that in the pinch-off region, the channel thickness at the drain end is not zero in practice, but instead it approaches some small limiting value and the drain current I_D flows through this narrow neck.

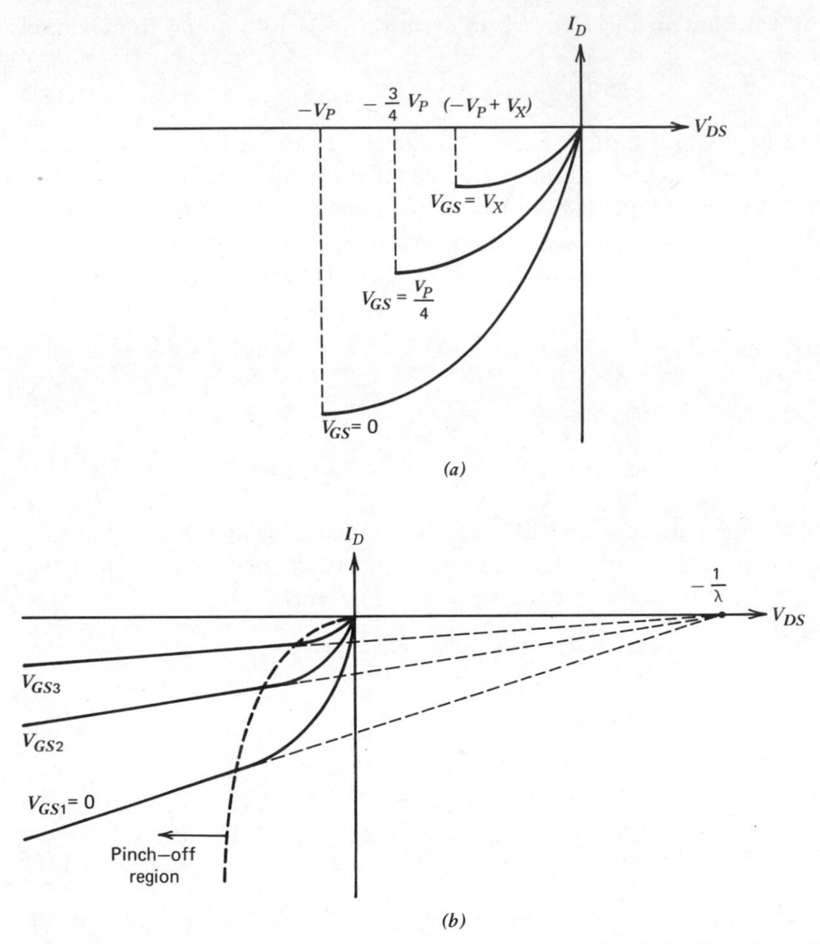

Fig. 1.28 *p*-channel JFET characteristics. (*a*) Below pinch-off. (*b*) Complete characteristics.

Consider now the situation for nonzero values of V_{GS}. As V_{GS} is made positive the depletion regions in Fig. 1.27 widen along the length of the gate and constrict the channel. It is apparent that for $V_{GS} = V_P$ the depletion regions extend completely across the channel and no drain current can flow. For values of V_{GS} less than V_P a constricted channel exists and the device functions as described previously except that the magnitude of voltage V_{DS} required to cause pinch-off is reduced by the value of V_{GS}. To cause pinch-off at the drain we require that the voltage from gate to drain be V_P and thus

$$V_{GS} - V'_{DS} = V_P \tag{1.146}$$

It follows from (1.146) that

$$V'_{DS} = -V_P + V_{GS} \tag{1.147}$$

at the point of pinch-off. In addition, since the channel is now initially constricted, the drain-current magnitude at the pinch-off point is reduced as shown in Fig. 1.28*a*.

The characteristics in Fig. 1.28*a* are plotted as a function of the internal drain-source voltage, V'_{DS}. Typical characteristics plotted as a function of external drain-source voltage V_{DS} are shown in Fig. 1.28*b*. These are quite similar to those of Fig. 1.28*a* for values of drain-source voltage below pinch-off. However V_{DS} can be increased beyond the pinch-off point and I_D remains approximately constant. The curves do have a finite slope in this region however, and experiment shows that like the bipolar transistor, the JFET characteristics can be extrapolated to a single point on the V_{DS} axis.

In the pinch-off region, I_D is approximately independent of V_{DS} and its value can be determined from (1.145) by substituting for V'_{DS} from (1.147) to obtain

$$I_D = G_0\left[-V_P + V_{GS} + \frac{2}{3}\frac{(\psi_0 + V_P)^{3/2} - (\psi_0 + V_{GS})^{3/2}}{(\psi_0 + V_P)^{1/2}}\right] \tag{1.148}$$

The maximum value of I_D (designated I_{DSS}) occurs for $V_{GS} = 0$ and can be calculated from (1.148) as

$$I_{DSS} = G_0\left[-V_P + \frac{2}{3}\frac{(\psi_0 + V_P)^{3/2} - \psi_0^{3/2}}{(\psi_0 + V_P)^{1/2}}\right] \tag{1.149}$$

If I_D is normalized to I_{DSS} and plotted as a function of V_{GS}/V_P using (1.148) and (1.149), the curve shown in Fig. 1.29 results (where a typical value of $\psi_0 \simeq \frac{1}{3}V_P$ has been assumed). Also plotted in Fig. 1.29 is a square-law transfer characteristic given by

$$I_D = I_{DSS}\left(1 - \frac{V_{GS}}{V_P}\right)^2 \tag{1.150}$$

The two curves agree quite closely and (1.150) is commonly used as an approximation to the JFET characteristic in the pinch-off region where the device is normally biased for use as an amplifier. The square-law characteristic of (1.150) is found to be[16] an even closer approximation to the characteristics of JFETs which have graded doping profiles across the channel. Devices of the type of Fig. 1.26*a* fit this description.

As described above, the JFET is usually operated in the pinch-off region of its characteristics for use in analog circuit applications. It should be pointed out at this stage that this region of the characteristics of a JFET is often referred to as the "saturation region." Thus the word "saturation" applied to a JFET usually means something quite different from what it does when applied to a bipolar

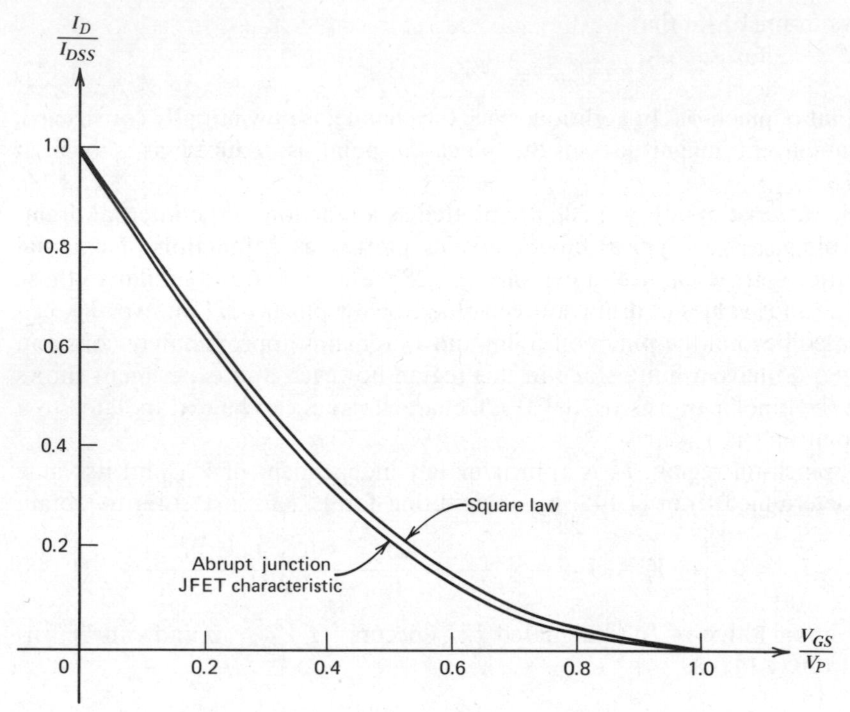

Fig. 1.29 Normalized transfer characteristics of an abrupt junction JFET compared with a square-law characteristic.

transistor. In order to avoid confusion, the term "pinch-off region" will be used in this book to describe the flat region of the JFET characteristics.

Let us return now to our examination of JFET characteristics and consider (1.150), which applies in the pinch-off region. The parameter, I_{DSS}, is the drain current for zero gate-source voltage and is negative for a *p*-channel device. This parameter depends on a number of device characteristics including device size, but typical values are in the region of several milliamperes. The pinch-off voltage, V_P, is another important device parameter, and it can be related to basic device parameters using (1.134) and (1.144) to give

$$V_P = a^2 \frac{qN_A\left(1 + \frac{N_A}{N_D}\right)}{2\epsilon} - \psi_0 \tag{1.151}$$

The pinch-off voltage thus depends on the square of the channel thickness and is positive for a *p*-channel JFET. In order that the JFET may be operated in the pinch-off region with reasonably small values of V_{DS}, it is necessary that V_P be small. Typical values of V_P for devices fabricated as in Fig. 1.26*a* are in the range

1 to 3 V. This requires a small value of channel thickness and is the reason for the separate n^+ diffusion used in the structure of Fig. 1.26*a*.

In the characteristics of Figs. 1.28 and 1.29, positive values of V_{GS} only were considered. Under these conditions the gate-source junction is reverse biased and the gate current is the reverse leakage of a *pn* junction. The gate-source leakage current is designated I_{GSS} and is measured with $V_{DS} = 0$. Since this has a typical value of 10^{-10}–10^{-12} A, the gate electrode appears as an extremely high impedance. The device can be operated with a small forward bias on the gate (small negative gate-source voltage for a *p*-channel device) as long as significant minority-carrier injection does not occur at the gate. Since this limits operation to several tenths of a volt of forward bias, the device is not often used in this way.

The above analysis was performed on a *p*-channel JFET. However the results apply equally to an *n*-channel JFET if the current and voltage polarities are reversed. Thus V_{DS} and I_D are normally positive using the convention of Fig. 1.26*b*, and V_{GS} is normally negative. The parameter I_{DSS} is positive and V_P is negative for an *n*-channel device.

EXAMPLE

Calculate the pinch-off voltage for a *p*-channel JFET with parameters $N_A = 10^{15}\text{ cm}^{-3}$, $N_D = 10^{16}\text{ cm}^{-3}$, $a = 2\ \mu\ (2 \times 10^{-4}\text{ cm})$.

From (1.151)

$$V_P = 4 \times 10^{-8}\,\frac{1.6 \times 10^{-19} \times 10^{15} \times 1.1}{2 \times 1.04 \times 10^{-12}} - \psi_0 = 3.38 - \psi_0$$

From (1.1)

$$\psi_0 = 26 \ln \frac{10^{15} \times 10^{16}}{2.25 \times 10^{20}}\text{ mV} = 0.638\text{ V}$$

Thus $\quad V_P = 2.74\text{ V}$

1.5.2 Large-Signal Model of the JFET

The results derived above can be used to form a large-signal model of the JFET in the pinch-off region for use in bias calculations. This is shown in Fig. 1.30 where the device is assumed to have an infinite output resistance.

The characteristics of Fig. 1.28*b* show that the JFET does have a finite output resistance in the pinch-off region, and the characteristics all extrapolate to the same point on the V_{DS} axis. This behavior can be modeled in (1.150) by adding a term to give

$$I_D = I_{DSS}\left(1 - \frac{V_{GS}}{V_P}\right)^2 (1 + \lambda V_{DS}) \tag{1.152}$$

where the intercept on the V_{DS} axis is $-(1/\lambda)$. The parameter, λ, is negative for a *p*-channel device and positive for an *n*-channel device with a typical magnitude

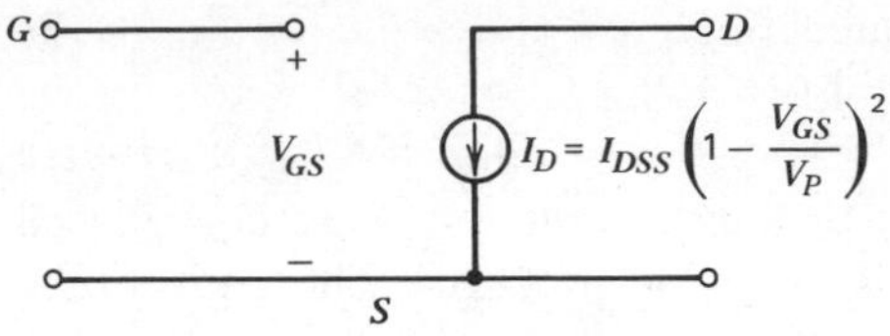

$$I_D = I_{DSS}\left(1 - \frac{V_{GS}}{V_P}\right)^2$$

Fig. 1.30 Large-signal model for the JFET. The signs of the variables are as follows.

	V_P	I_{DSS}	V_{GS}
p channel	positive	negative	normally positive
n channel	negative	positive	normally negative

of 10^{-2} V^{-1}. Note that λ is analogous to the reciprocal of the Early voltage V_A for bipolar transistors.

1.5.3 JFET Breakdown Voltages

The *pn* junctions between gate and channel of a JFET are subject to avalanche breakdown as described in Section 1.2.2. Since the JFET is a majority-carrier device whose operation does not depend on minority-carrier concentrations, the breakdown process is quite straightforward. Breakdown occurs when the voltage between gate and channel exceeds a critical value and the breakdown characteristic is sharp, as in the case of a Zener diode. There is no amplification of the avalanche process as in the case of the bipolar transistor in the common-emitter connection.

A typical breakdown characteristic for a JFET of the type shown in Fig. 1.26*a* is shown in Fig. 1.31. For $V_{GS} = 0$ breakdown occurs for $V_{DS} \simeq -7$ V, at which point avalanche breakdown occurs between the drain and gate. This is the same breakdown that occurs with reverse bias on the base-emitter junction of *npn* transistors. As V_{GS} is made positive, breakdown occurs for the same voltage from gate to drain but this now requires a smaller value of V_{DS} because of the finite value of V_{GS}. The magnitude of V_{DS} required for breakdown is reduced by the value of V_{GS}. The breakdown characteristic of the JFET can be described by specifying BV_{DGO}, which is the breakdown voltage from drain to gate with the source open.

The rather low breakdown voltage of the devices described above limits their usefulness. However as described in Chapter 2, ion-implanted *p*-channel JFETs can be fabricated with much larger breakdown voltages.

1.6 SMALL-SIGNAL MODEL OF THE JFET

A small-signal model of the JFET can be derived using methods similar to those employed for the bipolar transistor. Since the JFET is almost always operated in the pinch-off region for small-signal applications, the equivalent circuit is

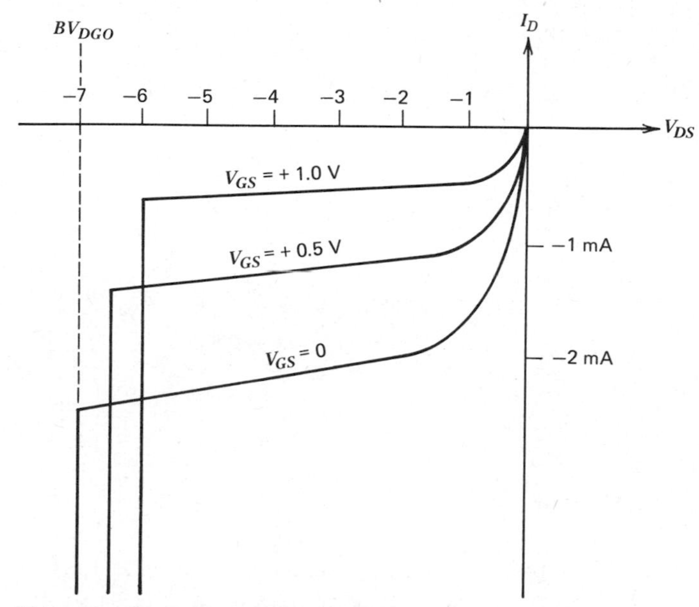

Fig. 1.31 Typical breakdown characteristics for a diffused monolithic JFET.

derived for operation in this region. The small-signal transconductance g_m can be derived from (1.150) as

$$g_m = \frac{dI_D}{dV_{GS}} = -\frac{2I_{DSS}}{V_P}\left(1 - \frac{V_{GS}}{V_P}\right) \tag{1.153}$$

$$= g_{m0}\left(1 - \frac{V_{GS}}{V_P}\right) \tag{1.154}$$

where

$$g_{m0} = -\frac{2I_{DSS}}{V_P} \tag{1.155}$$

The quantity g_{m0} is always positive, and for typical values of $I_{DSS} = -1$ mA and $V_P = 2$ V the value of g_{m0} is 1 mA/V. Equation 1.154 indicates that g_m varies linearly with V_{GS}, as illustrated in Fig. 1.32. A useful experimental method to determine V_P is to plot g_m versus V_{GS} as in Fig. 1.32 and extrapolate to the V_{GS} axis, since this gives a much sharper intersection than plotting I_D versus V_{GS}.

The small-signal output resistance r_o of the JFET can be determined from (1.152) as

$$\frac{1}{r_o} = \frac{\partial I_D}{\partial V_{DS}} = \lambda I_{DSS}\left(1 - \frac{V_{GS}}{V_P}\right)^2 \tag{1.156}$$

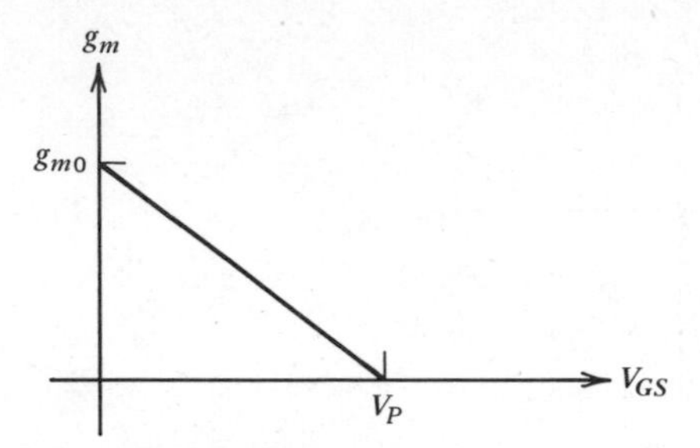

Fig. 1.32 Variation of g_m with V_{GS} for a JFET in the pinch-off region assuming a square-law transfer characteristic.

Substitution of (1.152) in (1.156) gives

$$r_o \simeq \frac{1}{\lambda I_D} \tag{1.157}$$

where it is assumed in (1.152) that $\lambda V_{DS} \ll 1$. A typical value of r_o is 100 kΩ at $I_D = 1$ mA.

It can be seen from Figs. 1.26*a* and 1.27 that the small-signal equivalent circuit of the JFET will also contain depletion layer capacitances from gate to source, gate to drain, and gate to substrate. Since the doping density at the gate-channel junction is usually graded, the gate-source capacitance, C_{gs}, and the gate-drain capacitance, C_{gd}, can be expressed as follows, using (1.22) for a *p*-channel JFET in the pinch-off region.

$$C_{gs} = \frac{C_{gs0}}{\left(1 + \dfrac{V_{GS}}{\psi_0}\right)^{1/3}} \tag{1.158}$$

$$C_{gd} = \frac{C_{gd0}}{\left(1 + \dfrac{V_{GD}}{\psi_0}\right)^{1/3}} \tag{1.159}$$

The capacitance C_{gss} from gate to substrate corresponds to the collector-substrate capacitance in an *npn* transistor, and, using (1.21), this can be expressed as

$$C_{gss} = \frac{C_{gss0}}{\left(1 + \dfrac{V_{GSS}}{\psi_0}\right)^{1/2}} \tag{1.160}$$

where V_{GSS} is the voltage from gate to substrate.

Finally, parasitic resistance exists in the JFET between the source and drain contacts and the active-channel region. The resistance in series with the source will affect the measured transfer characteristics of the device in that I_{DSS} and g_{m0} will be lower than expected. However, since measured data is generally used for I_{DSS}, V_P, and g_{m0}, the effect of series source resistance is conveniently included

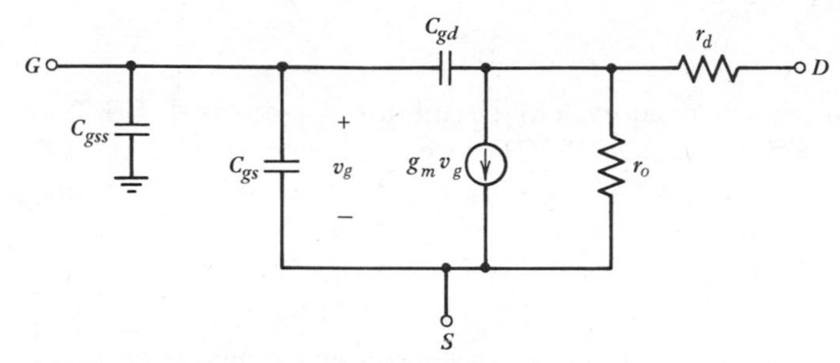

Fig. 1.33 Small-signal JFET equivalent circuit.

in these parameters and need not be included in the equivalent circuit. In most cases the presence of series source resistance does not affect the validity of the square-law approximation of (1.150).

The complete small-signal equivalent circuit of the monolithic JFET is shown in Fig. 1.33. This equivalent circuit is valid for both *p*-channel and *n*-channel JFETs although the position of the parasitic capacitance to the substrate may be different for different JFET structures. The parasitic drain resistance, r_d, included in series with the drain contact has a typical value of 50 to 100 Ω. Typical zero-bias values of the capacitances are $C_{gs0} = 1$ to 4 pF, $C_{gd0} = 0.3$ to 1 pF, and $C_{gss0} = 4$ to 8 pF. Typical parameters for various monolithic JFET structures are summarized in Chapter 2.

The JFET small-signal equivalent circuit of Fig. 1.33 is quite similar to that of the bipolar transistor of Fig. 1.20, and much of the theory developed for bipolar transistor circuits applies equally to JFET circuits if the appropriate parameter values are used. One major difference is the absence of any input shunt resistance in the JFET because of the reverse-biased *pn* junction in series with the gate electrode. This makes the JFET very attractive for circuits requiring high input impedance. An additional difference between the JFET and the bipolar transistor is the absence of any resistance in the JFET, which is equivalent to base resistance r_b in the bipolar transistor. Note also that at a bias current of $I_D = 1$ mA, a typical value of g_m for a JFET is 1 mA/V compared with 38 mA/V for a bipolar transistor with $I_C = 1$ mA. Thus, in most circuits, a bipolar transistor will give a higher gain than a JFET, and JFETs are only employed when their unique characteristics can be used to advantage.

The frequency response of the JFET can be specified in a fashion similar to that used for the bipolar transistor. If the analysis performed in Section 1.4.8 is repeated for the JFET, the frequency of unity current gain is found to be

$$f_c = \frac{1}{2\pi} \frac{g_m}{C_{gs} + C_{gd} + C_{gss}} \tag{1.161}$$

and this is similar in form to (1.125) for the bipolar transistor. A JFET with $g_m =$ 1 mA/V and $(C_{gs} + C_{gd} + C_{gss}) = 5$ pF has a frequency f_c of about 30 MHz.

EXAMPLE

Derive the complete small-signal equivalent circuit for a *p*-channel JFET at $I_D = -0.5$ mA, $V_{DS} = -5$ V, and $V_{GSS} = 10$ V. Device parameters are $C_{gs0} = 2$ pF, $C_{gd0} = 0.5$ pF, $C_{gss0} = 4$ pF, $I_{DSS} = -1$ mA, $V_P = 2$ V, $\lambda = -10^{-2}$ V^{-1}, and $r_d = 50\ \Omega$

The value of ψ_0 is 0.7 V for C_{gs} and C_{gd} and 0.52 V for C_{gss}.

From (1.150) the gate bias voltage is

$$V_{GS} = V_P\left(1 - \sqrt{\frac{I_D}{I_{DSS}}}\right) = 2\left(1 - \sqrt{\frac{0.5}{1}}\right) \text{V} = 0.586 \text{ V}$$

From (1.153) the small-signal transconductance is

$$g_m = -\frac{2(-1)}{2}\left(1 - \frac{0.586}{2}\right) \text{mA/V} = 0.707 \text{ mA/V}$$

The small-signal output resistance can be calculated from (1.157)

$$r_o = \frac{1}{10^{-2} \times 0.5} \text{k}\Omega = 200 \text{ k}\Omega$$

Finally the device capacitances can be calculated as follows. Using $V_{GS} = 0.586$ V in (1.158) gives

$$C_{gs} = \frac{2}{\left(1 + \dfrac{0.586}{0.7}\right)^{1/3}} \text{pF} = 1.63 \text{ pF}$$

The voltage from gate to drain is

$$V_{GD} = V_{GS} - V_{DS} = 5.586 \text{ V}$$

and substitution in (1.159) gives

$$C_{gd} = \frac{0.5}{\left(1 + \dfrac{5.586}{0.7}\right)^{1/3}} \text{pF} = 0.24 \text{ pF}$$

From (1.160) the capacitance from gate to substrate is

$$C_{gss} = \frac{4}{\left(1 + \dfrac{10}{0.52}\right)^{1/2}} \text{pF} = 0.89 \text{ pF}$$

The complete small-signal equivalent circuit is shown in Fig. 1.34.

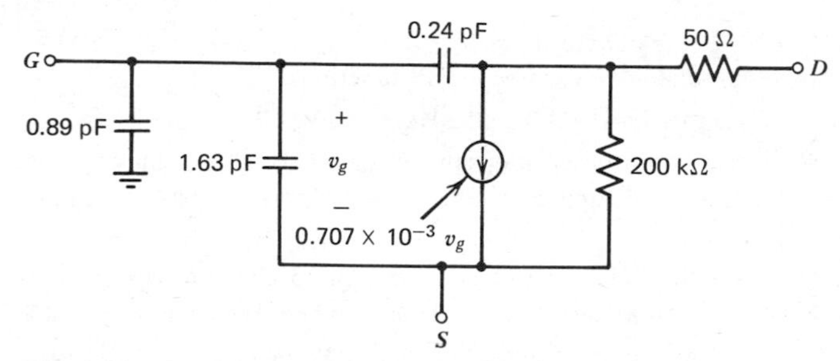

Fig. 1.34 Complete small-signal equivalent circuit for a diffused monolithic p-channel JFET at $I_D = -0.5$ mA, $V_{DS} = -5$ V, and $V_{GSS} = 10$ V. Device parameters are $C_{gs0} = 2$ pF, $C_{gd0} = 0.5$ pF, $C_{gss0} = 4$ pF, $I_{DSS} = -1$ mA, $V_P = 2$ V, $\lambda = -10^{-2}$ V^{-1}, $r_d = 50\ \Omega$.

PROBLEMS

1.1 (a) Calculate the built-in potential, depletion-layer depths, and maximum field in a plane-abrupt pn junction in silicon with doping densities $N_A - 8 \times 10^{15}$ atoms/cm^3 and $N_D = 10^{17}$ atoms/cm^3. Assume a reverse bias of 5 V.
(b) Repeat (a) for zero external bias and 0.3 V forward bias.

1.2 Calculate the zero-bias junction capacitance for the example in Problem 1.1 and also calculate the value at 5 V reverse bias and 0.3 V forward bias. Assume a junction area of 2×10^{-5} cm^2.

1.3 Calculate the breakdown voltage for the junction of Problem 1.1 if the critical field is $\mathscr{E}_{crit} = 4 \times 10^5$ V/cm.

1.4 If junction curvature causes the maximum field at a practical junction to be 1.5 times the theoretical value, calculate the doping density required to give a breakdown voltage of 150 V with an abrupt pn junction in silicon. Assume one side of the junction much more heavily doped than the other and $\mathscr{E}_{crit} = 3 \times 10^5$ V/cm.

1.5 If the collector doping density in a transistor is 6×10^{15} atoms/cm^3, and is much less than the base doping, find BV_{CEO} for $\beta = 200$ and $n = 4$. Use $\mathscr{E}_{crit} = 3 \times 10^5$ V/cm.

1.6 Repeat Problem 1.5 for doping of 10^{15} atoms/cm^3 and $\beta = 400$.

1.7 (a) Sketch the I_C–V_{CE} characteristics in the forward-active region for an npn transistor with $\beta_F = 100$ (measured at low V_{CE}), $V_A = 50$ V, $BV_{CBO} = 120$ V, and $n = 4$. Use

$$I_C = \left(1 + \frac{V_{CE}}{V_A}\right)\frac{M\alpha_F}{1 - M\alpha_F} I_B$$

where M is given by (1.78).
Plot I_C from 0 to 10 mA and V_{CE} from 0 to 50 V. Use $I_B = 1\ \mu$A, 10 μA, 30 μA, and 60 μA.
(b) Repeat (a) but sketch V_{CE} from 0 to 10 V.

1.8 Derive and sketch the complete small-signal equivalent circuit for a bipolar transistor at $I_C = 0.2$ mA, $V_{CB} = 10$ V, $V_{CS} = 15$ V. Device parameters are $C_{je0} = 0.6$ pF, $C_{\mu 0} =$

0.25 pF, $C_{cs0} = 1.5$ pF, $\beta_0 = 200$, $\tau_F = 0.3$ ns, $\eta = 2 \times 10^{-4}$, $r_b = 200\ \Omega$, $r_c = 100\ \Omega$, $r_{ex} = 1\ \Omega$, and $r_\mu = 5\ \beta_0 r_o$. Assume $\psi_0 = 0.55$ V for all junctions.

1.9 Repeat Problem 1.8 for $I_C = 5$ mA, $V_{CB} = 5$ V, and $V_{CS} = 10$ V.

1.10 Sketch the graph of small-signal common-emitter current gain versus frequency on log scales from 0.1 to 1000 MHz for the examples of Problems 1.8 and 1.9. Thus calculate the f_T of the device in each case.

1.11 An integrated-circuit *npn* transistor has the following measured characteristics: $r_b = 100\ \Omega$, $r_c = 100\ \Omega$, $\beta_0 = 100$, $r_o = 50$ kΩ at $I_C = 1$ mA, $f_T = 600$ MHz with $I_C = 1$ mA and $V_{CB} = 10$ V, $f_T = 1$ GHz with $I_C = 10$ mA and $V_{CB} = 10$ V, $C_\mu = 0.15$ pF with $V_{CB} = 10$ V, and $C_{cs} = 1$ pF with $V_{CS} = 10$ V. Assume $\psi_0 = 0.5$ V for all junctions and assume C_{je} is constant in the forward-bias region. Use $r_\mu = 5\ \beta_0 r_o$.
(a) Form the *complete* small-signal equivalent circuit for this transistor at $I_C = 0.1$, 1, and 5 mA with $V_{CB} = 2$ V and $V_{CS} = 15$ V.
(b) Sketch the graph of f_T versus I_C for this transistor on log scales from 1 μA to 10 mA with $V_{CB} = 2$ V.

1.12 A lateral *pnp* transistor has an effective base width of 10 μ (1 $\mu = 10^{-4}$ cm).
(a) If the emitter-base depletion capacitance is 2 pF in the forward-bias region and is constant, calculate the device f_T at $I_C = -0.5$ mA. (Neglect C_μ). Also calculate the minority-carrier charge stored in the base of the transistor at this current level.
Data: $D_P = 13$ cm^2/sec in silicon.
(b) If the collector-base depletion layer width changes 0.1 μ per volt of V_{CE}, calculate r_o for this transistor at $I_C = -0.5$ mA.

1.13 If the area of the transistor in Problem 1.11 was effectively doubled by connecting two transistors in parallel, which model parameters in the small-signal equivalent circuit of the composite transistor would differ from those of the original device if the *total* collector current was unchanged? What is the relationship between the parameters of the composite and original devices?

1.14 An integrated *npn* transistor has the following characteristics: $\tau_F = 0.25$ ns, small-signal short-circuit current gain is 9 with $I_C = 1$ mA at $f = 50$ MHz, $V_A = 40$ V, $\beta_0 = 100$, $r_b = 150\ \Omega$, $r_c = 150\ \Omega$, $C_\mu = 0.6$ pF, and $C_{cs} = 2$ pF at the bias voltage used. Determine all elements in the small-signal equivalent circuit at $I_C = 2$ mA and sketch the circuit.

1.15 Calculate the pinch-off voltage for a *p*-channel JFET with parameters $N_A = 5 \times 10^{15}$ atoms/cm^3, $N_D = 10^{17}$ atoms/cm^3, and $a = 1.2\ \mu$.

1.16 Calculate the ratio of channel width to channel length Z/L required for the JFET of Problem 1.15 to realize a value of $I_{DSS} = -2$ mA. Assume a channel resistivity of $\rho = 3\ \Omega$-cm.

1.17 Use (1.152) to sketch the I_D–V_{DS} characteristics of a *p*-channel JFET with $I_{DSS} = -5$ mA, $V_P = 3$ V, $\lambda = -10^{-2}$ V^{-1}, and $BV_{DGO} = -40$ V. Plot I_D from 0 to -10 mA and V_{DS} from 0 to -40 V. Use $V_{GS} = 0$, $+0.5$ V, $+1$ V, $+2$ V, and $+3$ V.

1.18 Derive the complete small-signal equivalent circuit for a *p*-channel JFET at $I_D = -1$ mA, $V_{DS} = -10$ V, and $V_{GSS} = 15$ V. Device parameters are $C_{gs0} = 3$ pF, $C_{gd0} = 0.4$ pF, $C_{gss0} = 6$ pF, $I_{DSS} = -5$ mA, $V_P = 3$ V, $\lambda = -2 \times 10^{-2}$ V^{-1}, and $r_d = 40\ \Omega$. Assume $\psi_0 = 0.6$ V for all junctions.

1.19 Calculate the frequency of unity current gain f_c for the JFET of Problem 1.18 at I_D values of -0.1 mA, -1 mA, and -5 mA. Sketch the curve of f_c versus I_D. Assume that C_{gd} and C_{gss} are independent of bias point.

REFERENCES

1. P. E. Gray, D. DeWitt, A. R. Boothroyd, and J. F. Gibbons. *Physical Electronics and Circuit Models of Transistors.* Wiley, New York, 1964, p. 20.
2. H. C. Poon and H. K. Gummel. "Modeling of Emitter Capacitance," *Proc. IEEE*, Vol. 57, pp. 2181–2182, December 1969.
3. B. R. Chawla and H. K. Gummel. "Transition Region Capacitance of Diffused pn Junctions," *IEEE Trans. Electron Devices*, Vol. ED-18, pp. 178–195, March 1971.
4. S. L. Miller. "Avalanche Breakdown in Germanium," *Phys. Rev.*, Vol. 99, p. 1234, 1955.
5. A. S. Grove. *Physics and Technology of Semiconductor Devices.* Wiley, New York, 1967, Ch. 6.
6. A. S. Grove. Op. cit., Ch. 4.
7. A. S. Grove. Op. cit., Ch. 7.
8. P. E. Gray et al. Op. cit., p. 10.
9. P. E. Gray et al. Op. cit., p. 129.
10. P. E. Gray et al. Op. cit., p. 180.
11. B. A. McDonald. "Avalanche degradation of h_{FE}," *IEEE Trans. Electron Devices*, Vol. ED-17, pp. 871–878, October 1970.
12. H. DeMan. "The influence of Heavy Doping on the Emitter Efficiency of a Bipolar Transistor," Vol. ED-18, *IEEE Trans. Electron Devices*, pp. 833–835, October 1971.
13. R. J. Whittier and D. A. Tremere. "Current Gain and Cutoff Frequency Falloff at High Currents," *IEEE Trans. Electron Devices*, Vol. ED-16, pp. 39–57, January 1969.
14. J. L. Moll and I. M. Ross. "The Dependence of Transistor Parameters on the Distribution of Base Layer Resistivity," *Proc. IRE*, Vol. 44, p. 72, 1956.
15. P. E. Gray et al. Op. cit., Ch. 8.
16. R. D. Middlebrook and I. Richer. "Limits on the Power-Law Exponent for Field-Effect Transistor Transfer Characteristics," *Solid-State Electronics*, Vol. 6, pp. 542–544, September–October 1963.

GENERAL REFERENCES

(a) P. E. Gray and C. L. Searle. *Electronic Principles.* Wiley, New York, 1969.
(b) T. Kamins and R. S. Muller. *Semiconductor Devices.* Wiley, New York, 1977.
(c) I. Getreu. *Modelling the Bipolar Transistor.* Tektronix Inc., 1976.

CHAPTER 2

ANALOG INTEGRATED-CIRCUIT TECHNOLOGY

2.1 INTRODUCTION

For the designer and user of integrated circuits, a knowledge of the details of the fabrication process is important for two reasons. First, the reason for the pervasiveness of IC technology has been the economic advantage of the planar process for fabricating complex circuitry at low cost through batch processing. Thus a knowledge of the factors influencing the cost of fabrication of integrated circuits is essential to either the selection of a circuit approach to a given design problem by the designer or the selection of a particular circuit for fabrication as a custom integrated circuit by the user. Second, the bipolar integrated-circuit technology presents a completely different set of cost constraints to the circuit designer from those encountered with discrete components. The optimum choice of a circuit approach to realize a specified circuit function requires an understanding of the degrees of freedom available with the technology, and the nature of the devices that are most easily fabricated on the integrated-circuit chip.

In this chapter we first enumerate the basic processes that are fundamental to the planar process: solid-state diffusion, optical photolithography, and epitaxial growth. Next the sequence of steps used to fabricate a bipolar integrated circuit is described, and the properties of the passive and active devices that result in the circuit are discussed. In the next section the economic factors in integrated-circuit fabrication are discussed. Next a number of modifications to the basic process are enumerated, which provide improved performance in the resulting circuits at increased fabrication cost. Finally, the problem of circuit encapsulation is discussed.

2.2 BASIC PROCESSES IN INTEGRATED-CIRCUIT FABRICATION

The fabrication of integrated circuits, as well as most modern discrete component transistors, is based on a sequence of photomasking, diffusion, and epitaxial growth steps applied to a slice of silicon starting material called a wafer.[1,2] Before beginning a discussion of the basic process steps, we will first review the effects produced on the electrical properties of silicon by the addition of impurity atoms.

2.2.1 Electrical Resistivity of Silicon

The addition of small concentrations of n-type or p-type impurities to a crystalline silicon sample has the effect of increasing the number of majority carriers (electrons for n-type, holes for p-type) and decreasing the number of minority carriers. The addition of impurities is called "doping" the sample. For practical concentrations of impurities, the density of majority carriers is approximately equal to the density of the impurity atoms in the crystal. Thus, for n-type material,

$$n_n \approx N_D \tag{2.1}$$

where $n_n(\text{cm}^{-3})$ is the equilibrium concentration of electrons and $N_D(\text{cm}^{-3})$ is the concentration of n-type donor impurity atoms. For p-type material,

$$p_p \approx N_A \tag{2.2}$$

where $p_p(\text{cm}^{-3})$ is the equilibrium concentration of holes and $N_A(\text{cm}^{-3})$ is the concentration of p-type acceptor impurities. Any increase in the equilibrium concentration of one type of carrier in the crystal must result in a decrease in the equilibrium concentration of the other. This occurs because the holes and electrons recombine with each other at a rate that is proportional to the product of the concentration of holes and the concentration of electrons. Thus the number of recombinations per second, R, is given by

$$R = \gamma np \tag{2.3}$$

where γ is a constant and n and p are electron and hole concentrations, respectively, in the silicon sample. The generation of the hole-electron pairs is a thermal process that depends only on temperature; the rate of generation, G, is not dependent on impurity concentration. In equilibrium, R and G must be equal, so that

$$G = \text{constant} = R = \gamma np \tag{2.4}$$

If no impurities are present, then

$$n = p = n_i(T) \tag{2.5}$$

where $n_i(\text{cm}^{-3})$ is the *intrinsic* concentration of carriers in a pure sample of silicon. Equations 2.4 and 2.5 establish that, for any impurity concentration, it must be true that $\gamma np = \text{constant} = \gamma n_i^2$ and thus

$$np = n_i^2(T) \tag{2.6}$$

Equation 2.6 shows that as the majority carrier concentration is increased by impurity doping, the minority carrier concentration is decreased by the same factor so that product np is constant in equilibrium. For impurity concentrations of practical interest, the majority carriers outnumber the minority carriers by many orders of magnitude.

The importance of minority and majority-carrier concentrations in the operation of the transistor was discussed in Chapter 1. Another important effect of the addition of impurities is an increase in the ohmic conductivity of the material itself. This conductivity is given by:

$$\sigma = q(\mu_n n + \mu_p p) \tag{2.7}$$

where μ_n(cm^2/V-sec) is the electron mobility, μ_p(cm^2/V-sec) is the hole mobility, and $\sigma(\Omega\ \text{cm})^{-1}$ is the electrical conductivity. For an n-type sample, substitution of (2.1) and (2.6) in (2.7) gives

$$\sigma = q\left(\mu_n N_D + \mu_p \frac{n_i^2}{N_D}\right) \approx q\mu_n N_D \tag{2.8}$$

For a p-type sample, substitution of (2.2) and (2.6) in (2.7) gives

$$\sigma = q\left(\mu_n \frac{n_i^2}{N_A} + \mu_p N_A\right) \approx q\mu_p N_A \tag{2.9}$$

The mobility μ is different for holes and electrons and is also a function of the impurity concentration in the crystal for high impurity concentrations. Measured values of mobility in silicon as a function of impurity concentration are shown in Fig. 2.1. The resistivity $\rho(\Omega\text{-cm})$ is usually specified in preference to the conductivity, and the resistivity of n- and p-type silicon as a function of impurity concentration is shown in Fig. 2.2. The conductivity and resistivity are related by the simple expression $\rho = 1/\sigma$.

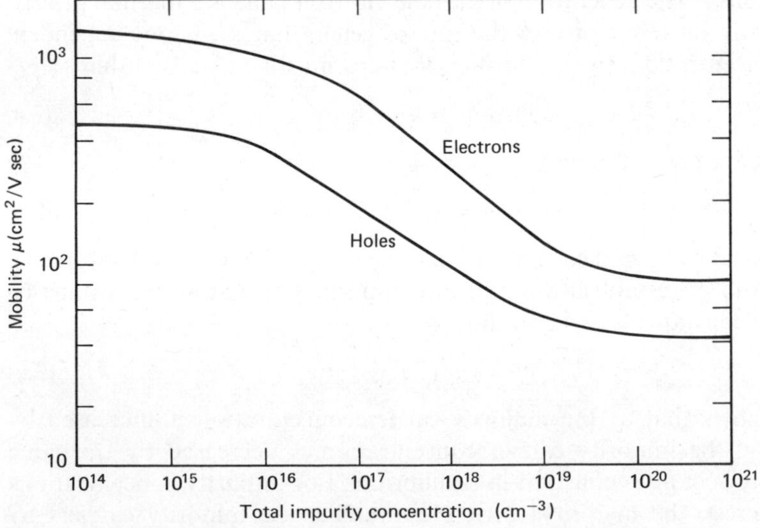

Fig. 2.1 Hole and electron mobility as a function of doping in silicon.[12]

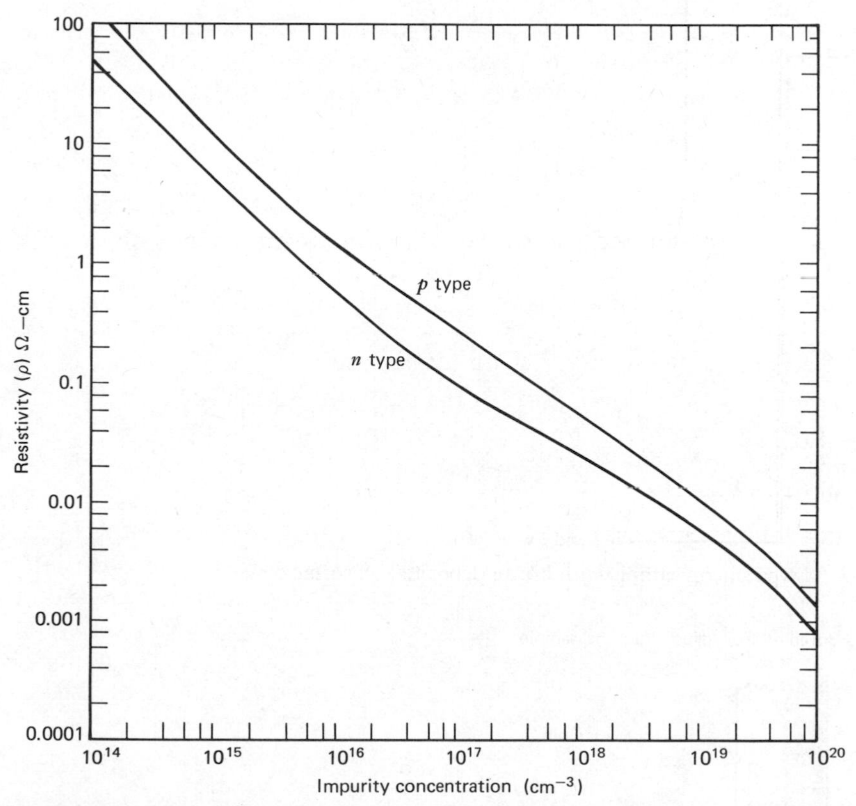

Fig. 2.2 Resistivity of *p*- and *n*-type silicon as a function of impurity concentration.[13]

2.2.2 Solid-State Diffusion

Solid-state diffusion of impurities in silicon is the movement, usually at high temperature, of impurity atoms from the surface of the silicon sample into the bulk material. During this high-temperature process, the impurity atoms replace silicon atoms in the lattice and are termed substitutional impurities. Since the doped silicon behaves electrically as *p*-type or *n*-type material depending on the type of impurity present, regions of *p*-type and *n*-type material can be formed by solid-state diffusion.

The nature of the diffusion process is illustrated by the conceptual example shown in Figs. 2.3 and 2.4. We assume that the silicon sample initially contains a uniform concentration of *n*-type impurity of 10^{15} atoms per cubic centimeter. Commonly used *n*-type impurities in silicon are phosphorous, arsenic, and antimony. We further assume that by some means we deposit atoms of *p*-type impurity on the top surface of the silicon sample. The most commonly used

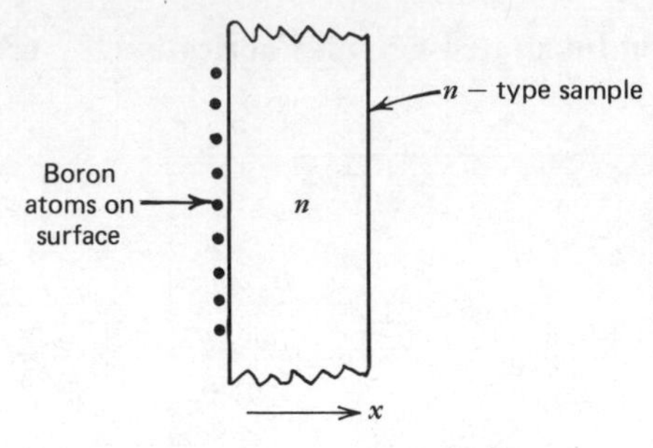

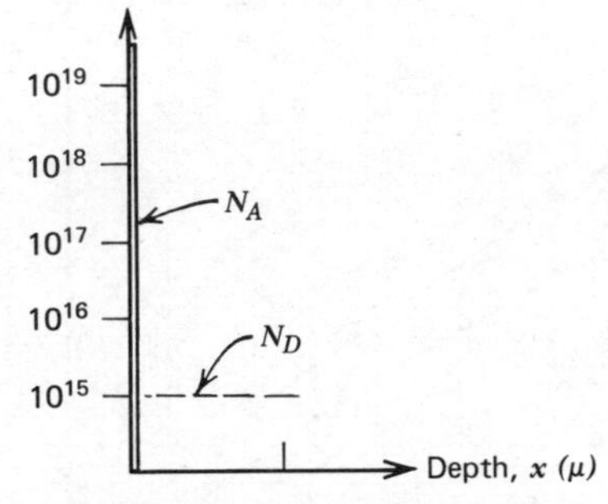

Fig. 2.3 *n*-type silicon sample with boron deposited on surface.

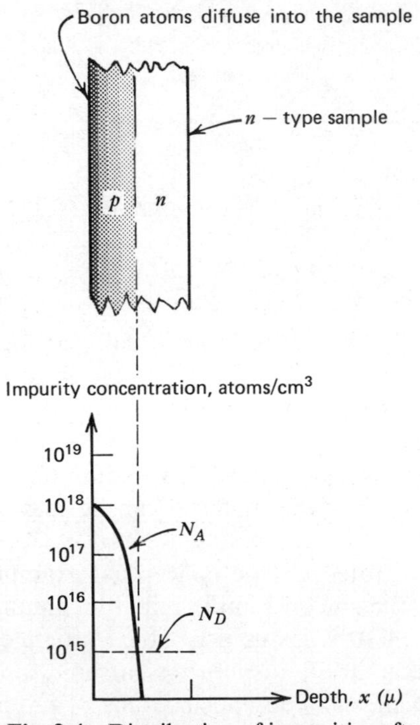

Fig. 2.4 Distribution of impurities after diffusion.

p-type impurity in silicon device fabrication is boron. The distribution of impurities prior to the diffusion step is illustrated in Fig. 2.3. The initial placement of the impurity atoms on the surface of the silicon is called the predeposition step and can be accomplished by a number of different techniques.

If the sample is now subjected to a high temperature on the order of 1100 degrees centigrade for a time on the order of one hour, the impurities *diffuse* into the sample as illustrated in Fig. 2.4. Within the silicon, the regions in which the p-type impurities outnumber the original n-type impurities show p-type electrical behavior, while the regions in which the n-type impurities are more numerous show n-type electrical behavior. The diffusion process has allowed the formation of a pn junction within the continuous crystal of silicon material. The depth of this junction from the surface varies from 0.1 μ to 20 μ for silicon integrated-circuit diffusions (where 1 μ = 1 micron = 10^{-6} m).

2.2.3 Electrical Properties of Diffused Layers

The result of the diffusion process is often a thin layer near the surface of the silicon sample that has been converted from one impurity type to another. Silicon devices and integrated circuits are constructed primarily from these layers. From an electrical standpoint, if the pn junction formed by this diffusion is reverse biased, then the layer is electrically isolated from the underlying material by the reverse-biased junction, and the electrical properties of the layer itself can be measured. The electrical parameter most often used to characterize such layers is the *sheet resistance*. In order to define this quantity, consider the resistance of a uniformly doped sample of length L, width W, thickness T, and n-type doping concentration N_D, as shown in Fig. 2.5. The resistance is:

$$R = \frac{\rho L}{WT} = \frac{1}{\sigma}\frac{L}{WT}$$

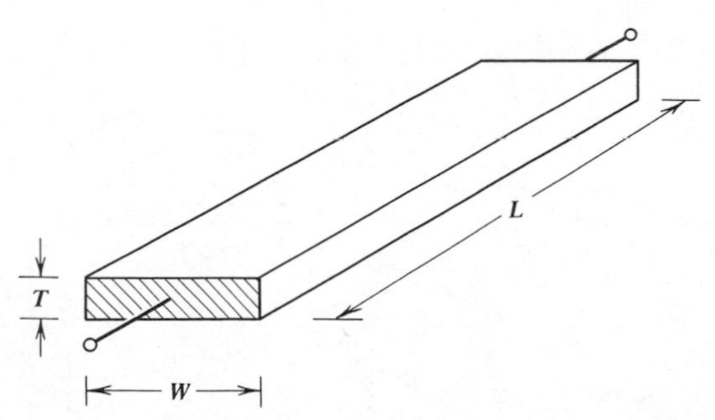

Fig. 2.5 Rectangular sample for calculation of sheet resistance.

Substitution of the expression for conductivity σ from (2.8) gives:

$$R = \left(\frac{1}{q\mu_n N_D}\right)\frac{L}{WT} = \frac{L}{W}\left(\frac{1}{q\mu_n N_D T}\right) = \frac{L}{W} R_\square \tag{2.10}$$

Quantity $R_\square$ is the *sheet resistance* of the layer and has units of ohms. Since the sheet resistance is the resistance of any *square* sheet of material with thickness T, its units are often given as *ohms per square* ($\Omega/\square$) rather than simply ohms. The sheet resistance can be written in terms of the resistivity of the material, using (2.8), as:

$$R_\square = \frac{1}{q\mu_n N_D T} = \frac{\rho}{T} \tag{2.11}$$

The diffused layer illustrated in Fig. 2.6 is similar to the case considered above except that the impurity concentration is not uniform. However, we can consider the layer to be made up of a parallel combination of many thin conducting sheets. The conducting sheet of thickness, dx, at depth x has a conductance:

$$dG = q\left(\frac{W}{L}\right)\mu_n N_D(x)\, dx \tag{2.12}$$

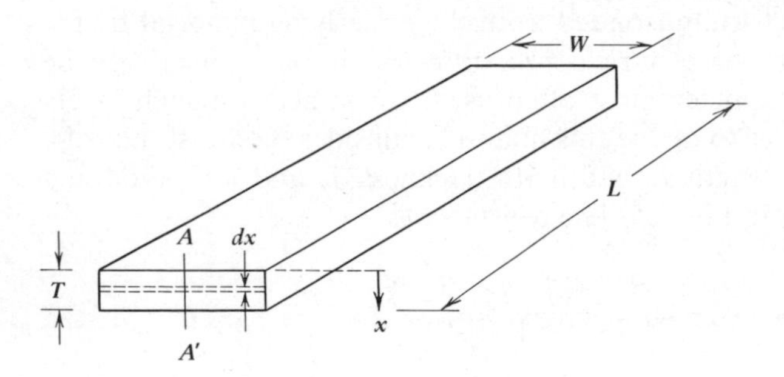

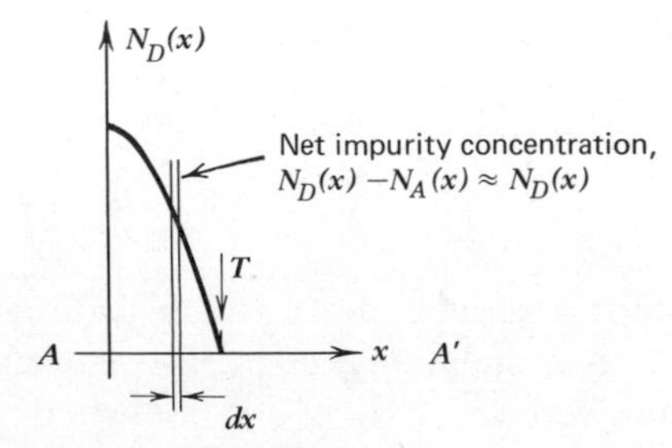

Fig. 2.6 Calculation of the resistance of a diffused layer.

To find the total conductance, we sum all the contributions.

$$G = \int_0^{x_j} q \frac{W}{L} \mu_n N_D(x)\, dx = \frac{W}{L} \int_0^{x_j} q\mu_n N_D(x)\, dx \tag{2.13}$$

Inverting (2.13) we obtain

$$R = \frac{L}{W}\left[\frac{1}{\int_0^{x_j} q\mu_n N_D(x)\, dx}\right] \tag{2.14}$$

Comparison of (2.10) and (2.14) gives

$$R_\square = \left[\int_0^{x_j} q\mu_n N_D(x)\, dx\right]^{-1} \approx \left[q\bar{\mu}_n \int_0^{x_j} N_D(x)\, dx\right]^{-1} \tag{2.15}$$

where $\bar{\mu}_n$ is the average mobility. Thus (2.10) can be used for diffused layers if the appropriate value of $R_\square$ is used. Equation 2.15 shows that the sheet resistance of the diffused layer depends on the total number of impurity atoms in the layer per unit area. The depth x_j in (2.13), (2.14), and (2.15) is actually the distance from the surface to the edge of the junction depletion layer, since the donor atoms within the depletion layer do not contribute to conduction. Sheet resistance is a very useful parameter for the electrical characterization of diffusion processes and is a key parameter in the design of integrated resistors. The sheet resistance of a diffused layer is easily measured in the laboratory; the actual evaluation of (2.15) is seldom necessary.

EXAMPLE

Calculate the resistance of a layer with length 50 μ and width 5 μ in material of sheet resistance 200 $\Omega/\square$.

From (2.10)

$$R = \frac{50}{5} \times 200\ \Omega = 2\ \text{k}\Omega$$

Note that this region constitutes 10 squares in series and R is thus 10 times the sheet resistance.

In order to use the diffusion process steps described above to fabricate useful devices, the diffusion must be restricted to a small region on the surface of the sample rather than the entire planar surface. This is accomplished with photolithography.

2.2.4 Photolithography

When a sample of crystalline silicon is placed in an oxidizing environment, a layer of silicon dioxide will form at the surface. This layer acts as a barrier to the diffusion of impurities, so that impurities separated from the surface of the silicon by a layer of oxide do not diffuse into the silicon during high-temperature processing. A *pn* junction can thus be formed in a selected location on the sample by first covering the sample with a layer of oxide (called an oxidation step), removing the oxide in the selected region, and then performing a predeposition and diffusion step. The selective removal of the oxide in the desired areas is accomplished with photolithography. This process is illustrated by the conceptual example of Fig. 2.7. Again we assume the starting material is a sample of *n*-type silicon. We first perform an oxidation step in which a layer of silicon dioxide (SiO_2) is thermally grown on the top surface, usually of thickness of 0.2 to 1 μ. The wafer following this step is shown in Fig. 2.7*a*. Then the sample is coated with a thin layer of photosensitive material called photoresist. When this material is exposed to a particular wavelength of light it undergoes a chemical change and, in the case of positive photoresist becomes soluble in certain chemicals in which the unexposed photoresist is insoluble. The sample at this stage is illustrated in Fig. 2.7*b*. To define the desired diffusion areas on the silicon sample, a photomask is placed over the surface of the sample; this photomask is opaque except for clear areas where the diffusion is to take place. Light of the appropriate wavelength is directed at the sample, as shown in Fig. 2.7*c*, and falls on the photoresist only in the clear areas of the

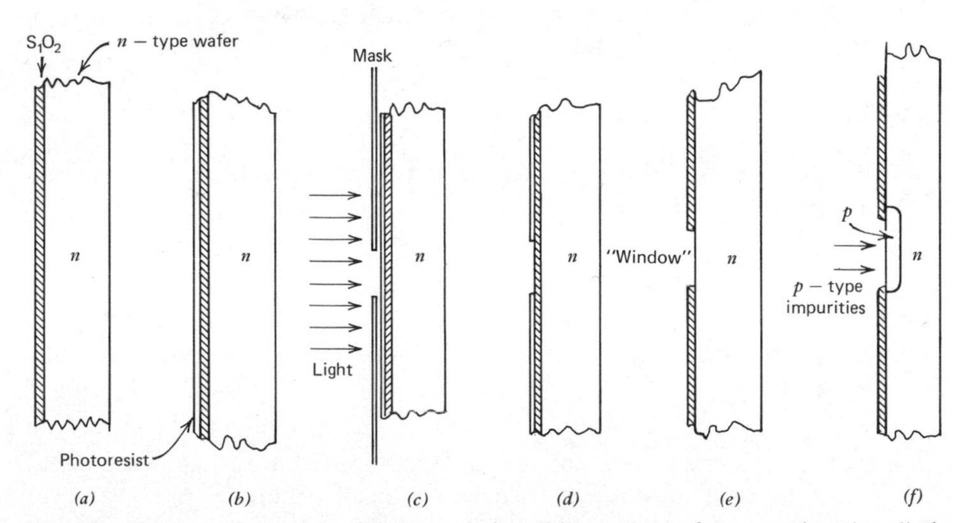

Fig. 2.7 Conceptual example of the use of photolithography to form a *pn* junction diode. (*a*) Grow SiO_2. (*b*) Apply photoresist. (*c*) Expose through mask. (*d*) Develop photoresist. (*e*) Etch SiO_2 and remove photoresist. (*f*) Predeposit and diffuse impurities.

mask. These areas of the resist are then chemically dissolved, as shown in Fig. 2.7*d*. This is the development step. The unexposed areas of the photoresist are impervious to the developer.

Since the objective is the formation of a region clear of SiO_2, the next step is the etching of the oxide. The sample is dipped in an etching solution, usually containing hydrofluoric acid, and, in the regions where the photoresist has been removed, the oxide is etched away leaving the bare silicon surface.

The remaining photoresist is next removed by a chemical stripping operation, leaving the sample with holes, or "windows" in the oxide at the desired locations as shown in Fig. 2.7*e*. The sample now undergoes a predeposition and diffusion step, resulting in the formation of *p*-type regions where the oxide had been removed, as shown in Fig. 2.7*f*. The minimum dimension of the diffused region, which can be routinely formed with this technique in device production, has decreased with time, and at present is approximately 4 μ by 4 μ. The number of such regions that can be fabricated simultaneously can be calculated by noting that the silicon sample used in the production of integrated circuits is a round slice, typically 3 to 5 inches in diameter and 250 μ thick. Thus the number of electrically independent *pn* junctions of dimension 4×4 μ, spaced 4 μ apart (which can be formed on one such wafer), is on the order of 10^8. Of course, in actual integrated circuits a number of masking and diffusion steps are used to form more complex structures such as transistors, but the key point is that photolithography is capable of defining a large number of devices on the surface of the sample, and that all of these devices are batch fabricated at the same time. Thus the cost of the photomasking and diffusion steps applied to the wafer during the process are divided among the devices or circuits on the wafer. This ability to fabricate hundreds or thousands of devices at once is the key to the economic advantage of IC technology.

2.2.5 Epitaxial Growth

Early planar transistors and the first integrated circuits used only photomasking and diffusion steps in the fabrication process. However, it became evident that all-diffused integrated circuits had severe limitations compared with discrete component circuits. In a triple-diffused transistor as illustrated in Fig. 2.8, the collector region is formed by an *n*-type diffusion into the *p*-type wafer. The drawbacks of this structure are that the series collector resistance is high and the collector-to-emitter breakdown voltage is low. The former occurs because the impurity concentration in the portion of the collector diffusion below the collector-base junction is low, giving the region high resistivity. The latter occurs because near the surface of the collector the concentration of impurities is relatively high, resulting in a low breakdown voltage between the collector and base diffusions at the surface as described in Chapter 1. Thus the concentration profile provided by the diffused collector is very disadvantageous; what is required is a low impurity

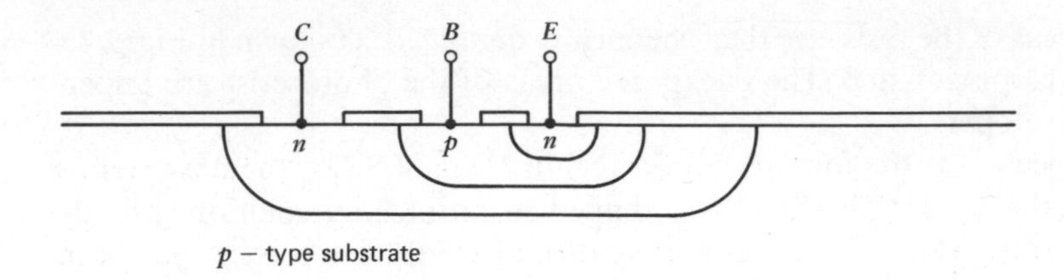

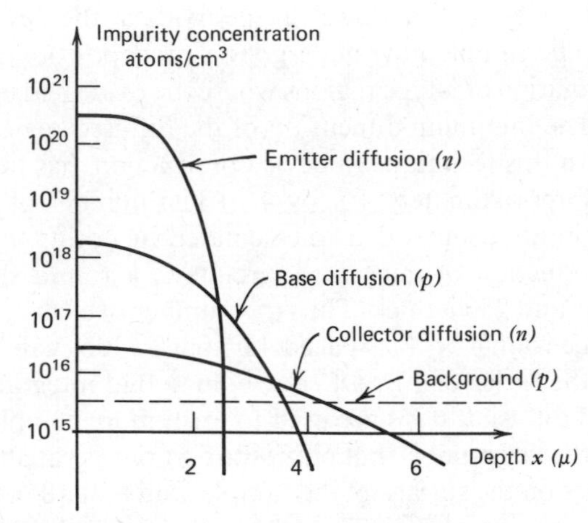

Fig. 2.8 Triple-diffused transistor and resulting impurity profile.

concentration at the collector-base junction for high breakdown voltage and a high concentration below the junction for low collector resistance. Such a concentration profile cannot be realized with diffusions alone, and the epitaxial growth technique was adopted as a result.

Epitaxial (epi) growth consists of formation of a layer of single-crystal silicon on the surface of the silicon sample so that the crystal structure of the silicon is continuous across the interface. The impurity concentration in the epi layer can be controlled independently and can be greater or smaller than in the substrate material. In addition, the epi layer is often of opposite impurity type from the substrate on which it is grown. The thickness of epi layers used in integrated-circuit fabrication vary from 1 to 20 μ, and the growth of thc laycr is accomplished by placing the wafer in an ambient atmosphere containing silicon tetrachloride ($SiCl_4$) or silane (SiH_4) at an elevated temperature. A chemical reaction takes place in which elemental silicon is deposited on the surface of the wafer, and if the conditions are carefully controlled the resulting surface layer of silicon is crystalline in structure with few defects. Such a layer is suitable as starting material for the fabrication of bipolar transistors.

2.2.6 Ion Implantation

Ion implantation is a technique for directly inserting impurity atoms into a silicon wafer.[3,4] The wafer is placed in an evacuated chamber and ions of the desired impurity species are directed at the sample at high velocity. These ions penetrate the surface of the silicon wafer to an average depth of from less than 0.1 μ to about 0.6 μ, depending on the velocity with which they strike the sample. The wafer is then held at a moderate temperature for a period of time (e.g., 800°C for 10 minutes) in order to allow the ions to become mobile and fit into the crystal lattice. This is called an *anneal step* and is essential to allow repair of any crystal damage caused by the implantation. The principal advantages of ion implantation over conventional diffusion are (1) that small amounts of impurities can be reproducibly deposited and (2) that the amount of impurity deposited per unit area can be very precisely controlled. In addition the deposition can be made to be very uniform across the wafer. An additional useful property of ion-implanted layers is that, unlike diffused layers, the peak of the impurity concentration profile can be made to occur below the surface of the silicon. This allows the fabrication of implanted bipolar and JFET structures with properties that are significantly better than those of diffused devices.

2.3 BIPOLAR INTEGRATED-CIRCUIT FABRICATION

The fabrication of a complete bipolar integrated circuit involves a sequence of from five to eight masking and diffusion steps of the type just discussed. The starting material is a wafer of *p*-type silicon, usually 250 μ thick and with an impurity concentration of approximately 10^{16} atoms/cm^3. We will consider the sequence of diffusion steps required to form an *npn* integrated-circuit transistor. The first mask and diffusion step, illustrated in Fig. 2.9 forms a low-resistance *n*-type layer that will eventually become a low-resistance path for the collector current of the transistor. This step is called the buried-layer diffusion, and the layer itself is called the buried layer. The sheet resistance of the layer is in the range of 20 to 50 $\Omega/\square$ and the impurity used is usually arsenic or antimony because these impurities diffuse slowly and thus do not greatly redistribute during subsequent processing.

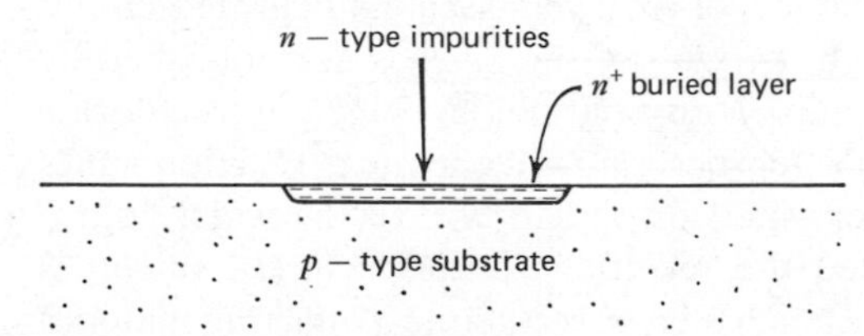

Fig. 2.9 Buried-layer diffusion.

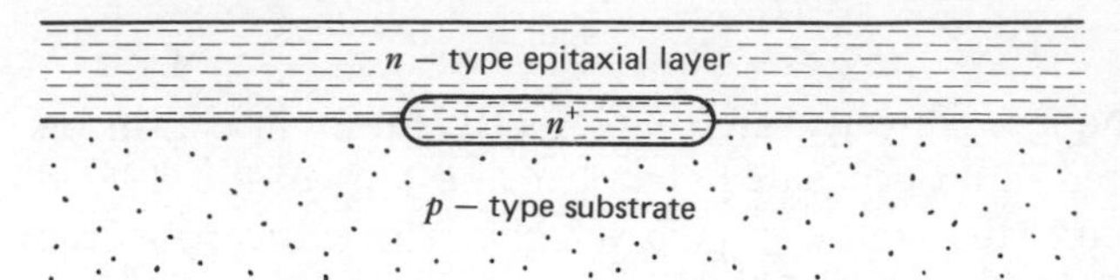

Fig. 2.10 Bipolar integrated-circuit wafer following epitaxial growth.

After the buried-layer step, the wafer is stripped of all oxide and an epi layer is grown as shown in Fig. 2.10. The thickness of the layer and its n-type impurity concentration determine the collector-base breakdown voltage of the transistors in the circuit since this material forms the collector region of the transistor. For example, if the circuit is to operate at a power supply voltage of 36 V, the devices generally are required to have BV_{CEO} breakdown voltages above this value. As discussed in Chapter 1, this implies that the plane breakdown voltage in the collector-base junction must be several times this value because of the effects of collector avalanche multiplication. For $BV_{CEO} = 36$ V, a collector-base plane breakdown voltage of approximately 90 V is required, which implies an impurity concentration in the collector of approximately 10^{15} atoms/cm^3 and a resistivity of 5 Ω-cm. The thickness of the epitaxial layer then must be large enough to accommodate the depletion layer associated with the collector-base junction. At 36 V, the results of Chapter 1 can be used to show that the depletion-layer thickness is approximately 6 μ. Since the buried layer diffuses outward approximately 8 μ during subsequent processing, and the base diffusion will be approximately 3 μ deep, a total epitaxial layer thickness of 17 μ is required for a 36-V circuit. For circuits with lower operating voltages, thinner and more heavily doped epitaxial layers are used, because this reduces the transistor collector series resistance as will be shown later.

Following the epitaxial growth, an oxide layer is grown on the top surface of the epitaxial layer. A mask step and boron (p-type) predeposition and diffusion are performed, resulting in the structure shown in Fig. 2.11. The function of this diffusion is to isolate the collectors of the transistors from each other with reverse-

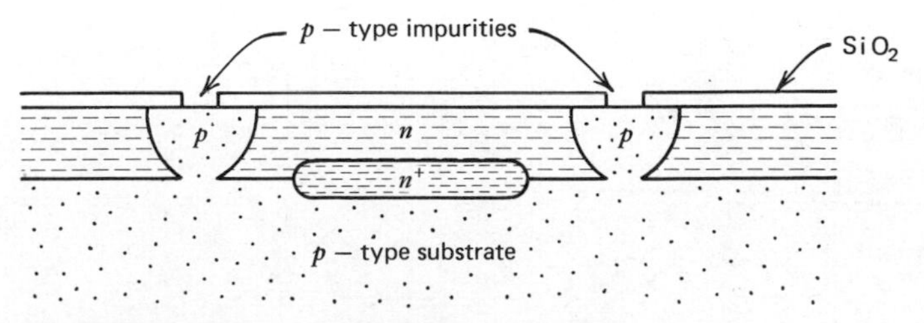

Fig. 2.11 Structure following isolation diffusion.

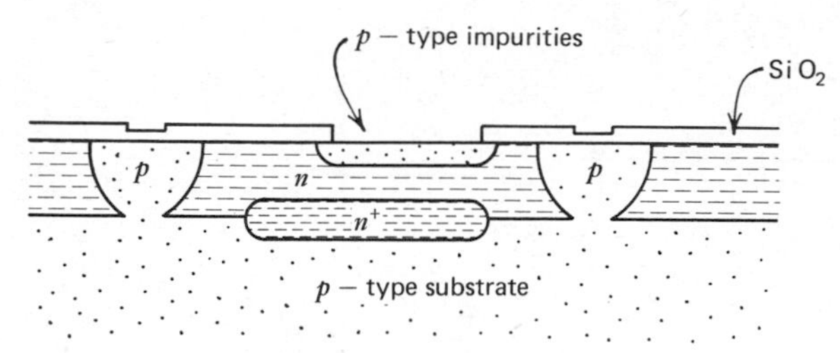

Fig. 2.12 Structure following base diffusion.

biased *pn* junctions, and it is termed the isolation diffusion. Because of the depth to which the diffusion must penetrate, this diffusion requires several hours in a diffusion furnace at temperatures of about 1200°C. The isolation diffused layer has a sheet resistance of from 20 to 40 Ω/□.

The next steps are the base mask, base predeposition, and base diffusion as shown in Fig. 2.12. The latter is usually a boron diffusion, and the resulting layer has a sheet resistance of from 100 to 300 Ω/□, and a depth of 1 to 3 μ at the end of the process. This diffusion forms not only the bases of the transistors, but also many of the resistors in the circuit, so that control of the sheet resistance is important.

Following the base diffusion, the emitters of the transistors are formed by a mask step, *n*-type predeposition, and diffusion as shown in Fig. 2.13. The sheet resistance is between 2 and 10 Ω/□, and the depth is 0.5 to 2.5 μ after the diffusion. This diffusion step is also used to form a low-resistance region, which serves as the contact to the collector region. This is necessary because ohmic contact is difficult to accomplish between aluminum metallization and the high-resistivity epitaxial material directly. The next masking step, the contact mask, is used to open holes in the oxide over the emitter, base, and collector of the transistors so that electrical contact can be made to them. Contact windows are also opened for the passive components on the chip. The entire wafer is then coated with a thin (about 1 μ) layer of aluminum that will interconnect the circuit elements. The actual interconnect pattern is defined by the last mask step in which the

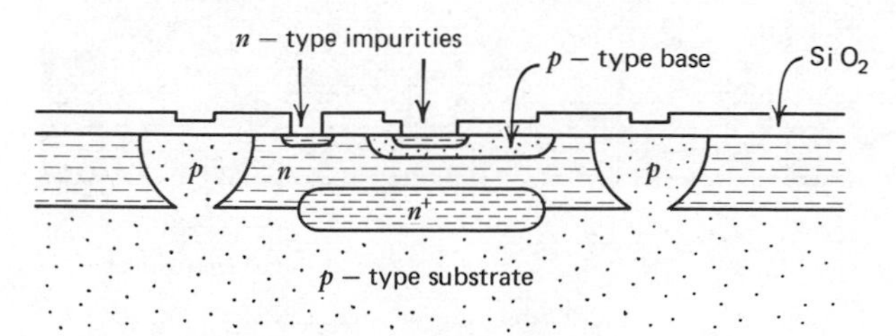

Fig. 2.13 Structure following emitter diffusion.

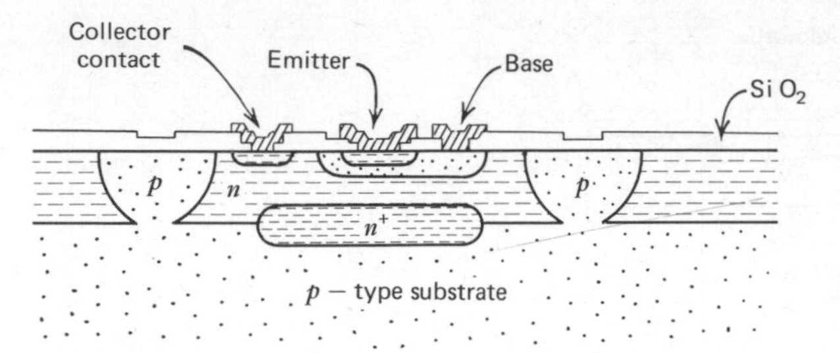

Fig. 2.14 Final structure following contact mask and metallization.

aluminum is etched away in the areas in which the photoresist is removed in the develop step. The final structure is shown in Fig. 2.14. A microscope photograph of an actual structure of the same type is shown in Fig. 2.15. The terraced effect on the surface of the device results from the fact that additional oxide is grown during each diffusion cycle, so that the oxide thickness is greatest over the epitaxial region where no oxide has been removed, is less thick over the base and isolation regions, which are both opened at the base mask step, and is smallest over the emitter diffusion. A typical diffusion profile for an analog integrated circuit is shown in Fig. 2.16.

The sequence described above allows simultaneous fabrication of a large number (often thousands) of complex circuits on a single wafer. The wafer is then placed

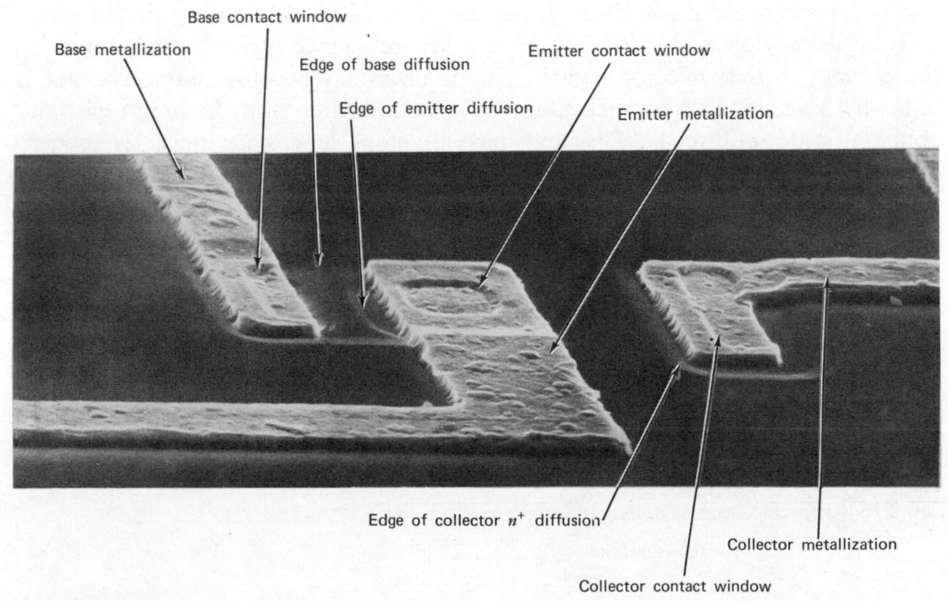

Fig. 2.15 Scanning electron microscope photograph of *npn* transistor structure.

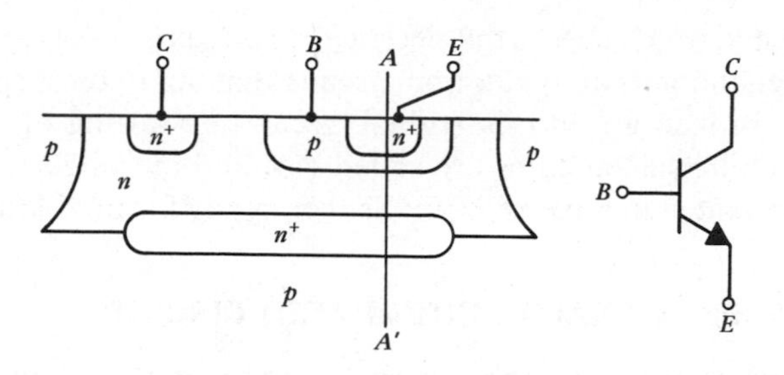

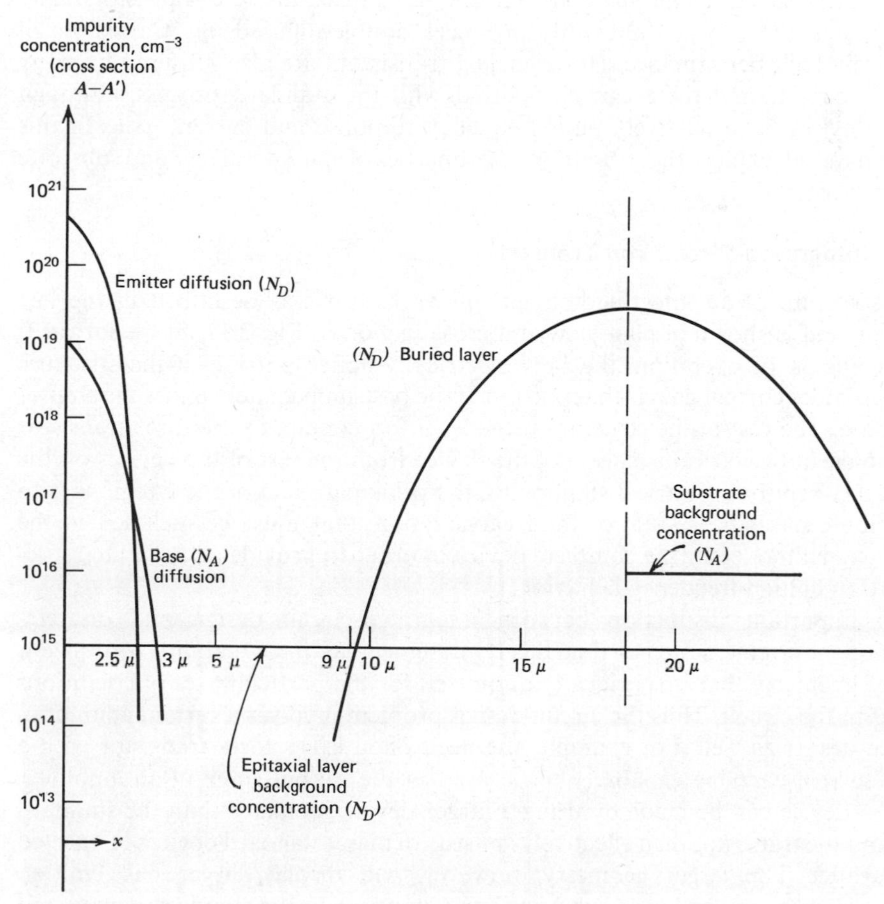

Fig. 2.16 Typical impurity concentration in a monolithic *npn* transistor.

in an automatic tester, which checks the electrical characteristics of each individual circuit on the wafer and puts an ink dot on circuits that fail to meet specifications. The wafer is then broken up into individual circuits by sawing or scribing and breaking. The resulting silicon chips are called *dice* and the singular is *die*. Each good die is then mounted in a package and is then ready for final testing.

2.4 ACTIVE DEVICES IN ANALOG INTEGRATED CIRCUITS

The IC fabrication process described in the previous section is an outgrowth of the one used to make *npn* epitaxial double-diffused discrete transistors, and, as a result, the process inherently produces double-diffused *npn* transistors of relatively high performance. However *pnp* transistors are also required in many analog circuits and these can be realized with the standard process, although these devices have relatively poor frequency response and current gain. In this section we will explore the structure and properties of *npn*, lateral *pnp*, and substrate *pnp* transistors.

2.4.1 Integrated-Circuit *npn* Transistor

The structure of an integrated-circuit *npn* transistor was described in the last section, and is shown in plan view and cross section in Fig. 2.17. In the forward-active region of operation, the only electrically active portion of the structure that provides current gain is that portion of the base immediately under the emitter diffusion. The rest of the structure provides a top contact to the three transistor terminals and electrical isolation of the device from the rest of the devices on the same die. From an electrical standpoint, the principal effect of these regions is to contribute parasitic resistances and capacitances that must be included in the small-signal model for the complete device in order to provide an accurate representation of high-frequency behavior.

An important distinction between integrated-circuit design and discrete-component circuit design is that the IC designer has the capability to utilize a device geometry that is specifically optimized for the particular set of conditions found in the circuit. Thus the circuit design problem involves a certain amount of device design as well. For example, the need often exists for a transistor with a high current-carrying capability to be used in the output stage of an amplifier. Such a device can be made by using a larger device geometry than the standard one and the transistor then effectively consists of many standard devices connected in parallel. The larger geometry, however, will display larger base-emitter, collector-base, and collector-substrate capacitance than the standard device, and this must be taken into account in analyzing the frequency response of the circuit. The circuit designer, then, must be able to determine the effect of changes in device geometry on device characteristics and to estimate the important device parameters when the device structure and doping levels are known. To illustrate this procedure,

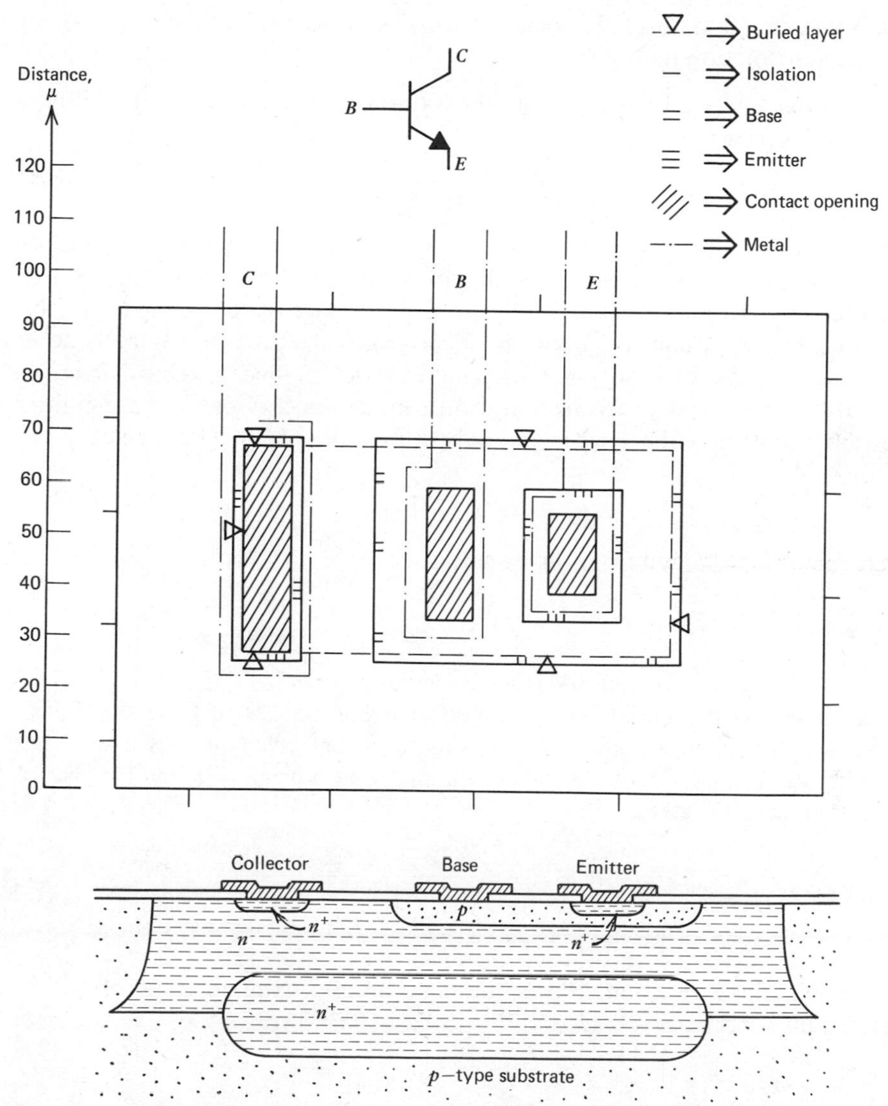

Fig. 2.17 Integrated-circuit *npn* transistor. The mask layers are coded as shown.

we will calculate the model parameters of the *npn* device shown in Fig. 2.17. This structure is typical of the minimum-geometry devices used in circuits with a 5 Ω-cm, 17 μ epitaxial layer. The emitter diffusion is 20 μ × 25 μ, the base diffusion is 45 μ × 60 μ, and the base-isolation spacing is 25 μ. The overall device dimensions are 140 μ × 95 μ. Device geometries intended for lower epi resistivity and thickness can be much smaller; the base-isolation spacing is dictated by the side diffusion of the isolation region plus the depletion layers associated with the base-collector and collector-isolation junctions.

Saturation Current, I_S. In Chapter 1, the saturation current of a graded-base transistor was shown to be:

$$I_S = \frac{qA\bar{D}_n n_i^2}{Q_B} \tag{2.16}$$

where A is the emitter-base junction area, Q_B is the total number of impurity atoms in the base per unit area, n_i is the intrinsic carrier concentration, and $\bar{D}_n$ is the effective diffusion constant for electrons in the base region of the transistor. From Fig. 2.16, the quantity Q_B can be identified as the area under the concentration curve in the base region. This could be determined graphically, but is most easily determined experimentally from measurements of the base-emitter voltage at a constant collector current. Substitution of (2.16) in (1.35) gives

$$\frac{Q_B}{\bar{D}_n} = A\frac{qn_i^2}{I_C}\exp\frac{V_{BE}}{V_T} \tag{2.17}$$

and Q_B can be determined from this equation.

EXAMPLE

A base-emitter voltage of 550 mV is measured at a collector current of 10 μA on a test transistor with a 4 mil $\times$ 4 mil emitter area. Estimate Q_B if T = 300°K.

First note that device surface dimensions are often given in mils (thousands of an inch) and 1 mil = 25.4 μ. Also from Chapter 1 we have $n_i = 1.5 \times 10^{10}\text{ cm}^{-3}$. Substitution in (2.17) gives

$$\frac{Q_B}{\bar{D}_n} = (4 \times 25.4 \times 10^{-4})^2\,\frac{1.6 \times 10^{-19} \times 2.25 \times 10^{20}}{10^{-5}}\,e^{550/26}$$

$$= 5.72 \times 10^{11}\text{ cm}^{-4}\text{ sec}$$

At the doping levels encountered in the base, an approximate value of $\bar{D}_n$, the electron diffusivity, is:

$$\bar{D}_n = 13\text{ cm}^2\text{ sec}^{-1}$$

Thus for this example,

$$Q_B = 5.72 \times 10^{11} \times 13\text{ cm}^{-2} = 7.4 \times 10^{12}\text{ atom/cm}^2$$

Note that Q_B depends on the diffusion profiles and will be different for different types of processes. Generally speaking, fabrication processes intended for lower-voltage operation use thinner base regions and display lower values of Q_B. Within one nominally fixed process, Q_B can vary by a factor of two or three to one because of diffusion process variations. The principal significance of the numerical value for Q_B is that it allows the calculation of the saturation current I_S for any device structure once the emitter-base junction area is known.

Series Base Resistance r_b. Because the base contact is physically removed from the active base region, a significant series ohmic resistance is observed between the contact and the active base. This resistance can have a significant effect on the high-frequency gain and the noise performance of the device. As illustrated in Fig. 2.18*a*, this resistance consists of two parts. The first is the resistance, r_{b1}, of the path between the base contact and the edge of the emitter diffusion. The second part, r_{b2}, is that resistance between the edge of the emitter and the site within the base region at which the current is actually flowing. The former component can be estimated by neglecting fringing and assuming that this component of the resistance is that of a rectangle of material as shown in Fig. 2.18*b*. For a

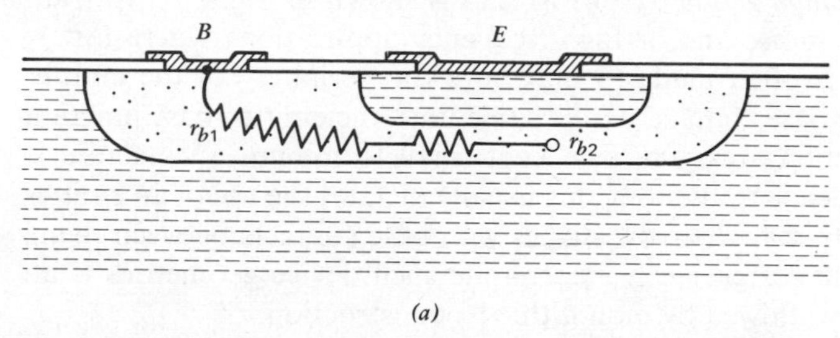

Fig. 2.18*a* Base resistance components for the *npn* transistor.

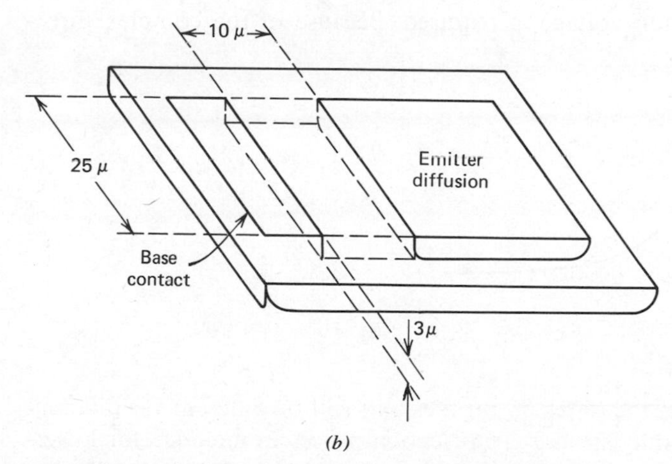

Fig. 2.18*b* Calculation of r_{b1}. The r_{b1} component of base resistance can be estimated by calculating the resistance of the rectangular block above.

base sheet resistance of 100 Ω/□, and typical dimensions as shown in Fig. 2.18*b* this would give a resistance of

$$r_{b1} = \frac{10\ \mu}{25\ \mu} 100\ \Omega = 40\ \Omega$$

The calculation of r_{b2} is complicated by several factors. First, the current flow in this region is not well modeled by a single resistor because the base resistance is distributed throughout the base region and two-dimensional effects are important. Second, at even moderate current levels, the effect of current crowding[5] in the base causes most of the carrier injection from the emitter into the base to occur near the periphery of the emitter diffusion. At higher current levels, essentially all of the injection takes place at the periphery and the effective value of r_b approaches r_{b1}. In this situation the portion of the base directly beneath the emitter is not involved in transistor action. A typically observed variation of r_b with collector current for the *npn* geometry of Fig. 2.17 is shown in Fig. 2.19. In transistors designed for low-noise and/or high-frequency applications where low r_b is important, an effort is often made to maximize the periphery of the emitter which is adjacent to the base contact. At the same time, the emitter-base junction and collector-base junction areas must be kept small to minimize capacitance. In the case of high-frequency transistors, this usually dictates the use of an emitter geometry that consists of many narrow stripes with base contacts between them. The ease with which the designer can incorporate such device geometries is an example of the flexibility allowed by monolithic IC construction.

Series Collector Resistance r_c. The series collector resistance is important both in high-frequency circuits as well as in low-frequency applications where low collector-emitter saturation voltage is required. Because of the complex three-

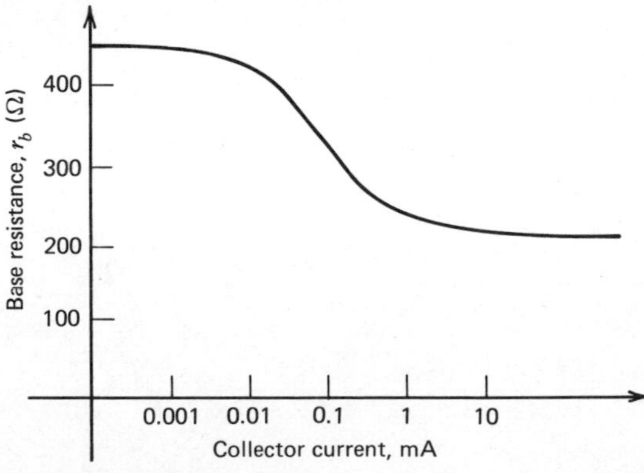

Fig. 2.19 Typical variation of effective small-signal base resistance with collector current for integrated-circuit *npn* transistor.

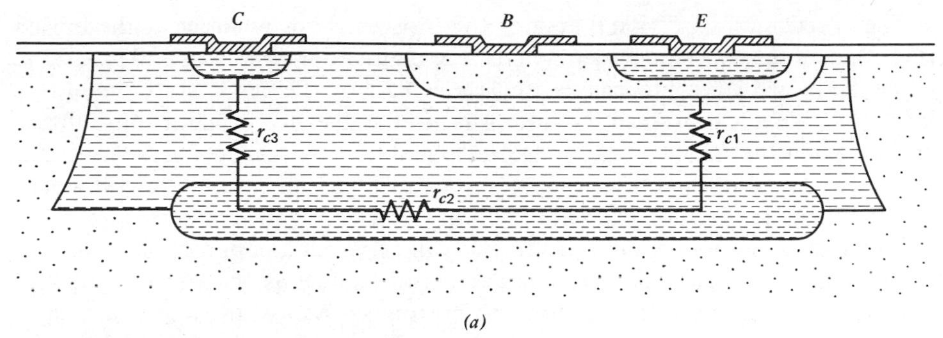

Fig. 2.20a Components of collector resistance r_c.

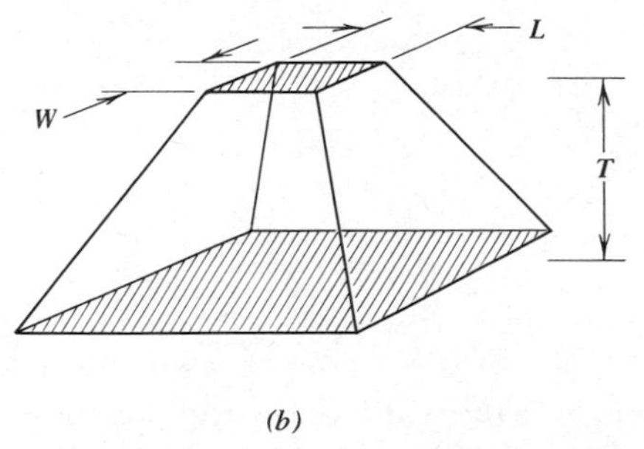

Fig. 2.20b Model for calculation of collector resistance.

dimensional shape of the collector region itself, only an approximate value for the collector resistance can be obtained by hand analysis. From Fig. 2.20 we see that the resistance consists of three parts: that from the collector-base junction under the emitter down to the buried layer, r_{c1}; that of the buried layer from the region under the emitter over to the region under the collector contact, r_{c2}; and finally that portion from the buried layer up to the collector contact, r_{c3}. The small-signal series collector resistance in the forward-active region can be estimated by adding the resistance of these three paths.

EXAMPLE

Estimate the collector resistance of the transistor of Fig. 2.17, assuming the doping profile of Fig. 2.16. We first calculate the r_{c1} component. The thickness of the lightly doped epi layer between the collector-base junction and the buried layer is 6 μ. Assuming that the collector-base junction is at zero bias, the results of Chapter 1 can be used to show that the depletion layer is about 1 μ thick. Thus the undepleted epi material under the base is 5 μ thick.

The effective cross-sectional area of the resistance, r_{c1}, is larger at the buried layer than at the collector-base junction. The emitter dimensions are 20 $\mu \times 25\ \mu$, while the buried layer dimensions are 41 $\mu \times 85\ \mu$ on the mask. Since the buried layer side-diffuses a distance roughly equal to the distance which it out-diffuses, about 8 μ must be added on each edge, giving an effective size of 57 $\mu \times 101\ \mu$. An exact calculation of the ohmic resistance of this three-dimensional region would require a solution of Laplaces's equation in the region with a rather complex set of boundary conditions. Consequently, we will carry out an approximate analysis by modeling the region as a rectangular parallelepiped as shown in Fig. 2.20*b*. Under the assumptions that the top and bottom surfaces of the region are equipotential surfaces, and that the current flow in the region takes place only in the vertical direction, the resistance of the structure can be shown to be:

$$R = \frac{\rho T}{WL} \frac{\ln \dfrac{a}{b}}{a - b} \tag{2.18}$$

where:

T = thickness of the region

ρ = resistivity of the material

W,L = width, length of the top rectangle

a = ratio of the width of the bottom rectangle to the width of the top rectangle

b = ratio of the length of the bottom rectangle to the length of the top rectangle

Direct application of this expression to the case at hand would give an unrealistically low value of resistance, because the assumption of one-dimensional flow is seriously violated when the dimensions of the lower rectangle are very much larger than those of the top rectangle. Equation 2.18 gives realistic results when the sides of the region form an angle of about 60° or less with the vertical. When the angle of the sides is increased beyond this point, the resistance does not increase very much because of the long path for current flow between the top electrode and the remote regions of the bottom electrode. Thus the dimensions of the bottom electrode should be determined by either the edge of the buried layer, or by the dimension of the emitter plus a distance equal to about twice the vertical thickness T of the region, whichever is smaller. For the case of r_{c1},

$$T = 5\ \mu = 5 \times 10^{-4}\ \text{cm}$$
$$\rho = 5\ \Omega\text{-cm}$$

We assume that the effective emitter dimensions are those defined by the mask plus approximately 2 μ of side diffusion on each edge. Thus:

$$W = 20\ \mu + 4\ \mu = 24 \times 10^{-4}\ \text{cm}$$
$$L = 25\ \mu + 4\ \mu = 29 \times 10^{-4}\ \text{cm}$$

For this case, the buried-layer edges are further away from the emitter edge than twice the thickness, T, on all four sides when side diffusion is taken into account. Thus the *effective* buried-layer dimensions that we use in (2.18) are:

$$W_{BL} = W + 2T = 24\ \mu + 10\ \mu = 34\ \mu$$
$$L_{BL} = L + 2T = 29\ \mu + 10\ \mu = 39\ \mu$$

and

$$a = \frac{34\ \mu}{24\ \mu} = 1.42$$

$$b = \frac{39\ \mu}{29\ \mu} = 1.34$$

Thus from (2.18)

$$r_{c1} = \frac{(5)(5 \times 10^{-4})}{(24 \times 10^{-4})(29 \times 10^{-4})}(0.72)\ \Omega = 258\ \Omega$$

We will now calculate r_{c2}, assuming a buried-layer sheet resistance of 20 $\Omega/\square$. The distance from the center of the emitter to the center of the collector-contact diffusion is 62 μ, and the width of the buried layer is 41 μ. The r_{c2} component is thus, approximately,

$$r_{c2} = (20\ \Omega/\square)\left(\frac{L}{W}\right) = 20\ \Omega/\square\left(\frac{62\ \mu}{41\ \mu}\right) = 30\ \Omega$$

Here the buried-layer side diffusion was not taken into account because the ohmic resistance of the buried layer is determined entirely by the number of impurity atoms actually diffused (see Equation 2.15) into the silicon, which is determined by the mask dimensions and the sheet resistance of the buried layer.

For the calculation of r_{c3}, the dimensions of the collector-contact n^+ diffusion are 18 $\mu \times$ 49 μ, including side diffusion. The distance from the buried layer to the bottom of the n^+ diffusion is seen on Fig. 2.16 to be 6.5 μ and thus $T = 6.5\ \mu$ in this case. On the three sides of the collector n^+ diffusion that do not face the base region, the out-diffused buried layer extends only 4 μ outside the n^+ diffusion, and thus the effective dimension of the buried layer is determined by the actual buried-layer edge on these sides. On the side facing the base region, the effective edge of the buried layer is a distance $2T$, or 13 μ, away from the edge of the n^+ diffusion. The effective buried-layer dimensions for the calculation of r_{c3} are thus 35 $\mu \times$ 57 μ. Using (2.18)

$$r_{c3} = \frac{(5)(6.5 \times 10^{-4})}{(18 \times 10^{-4})(49 \times 10^{-4})}0.66 = 243\ \Omega$$

The total collector resistance is thus:

$$r_c = r_{c1} + r_{c2} + r_{c3} = 531\ \Omega$$

The value actually observed in such devices is somewhat lower than this for three reasons. First, we have approximated the flow as one dimensional, and it is actually three dimensional. Second, for larger collector-base voltages, the collector-base depletion layer extends further into the epi, decreasing r_{c1}. Finally, the value of r_c which is important, is often that for a saturated device. In saturation, holes are injected into the epi region under the emitter by the forward-biased collector-base junction and they modulate the conductivity of the region even at moderate current levels.[6] Thus the collector resistance one measures when the device is in saturation is closer to $(r_{c2} + r_{c3})$, or about 250 to 300 Ω. Thus r_c is smaller in saturation than in the forward-active region.

Collector-Base Capacitance. The collector-base capacitance is simply the capacitance of the collector-base junction including both the flat bottom portion of the junction as well as the sidewalls. This junction is formed by the diffusion of boron into an *n*-type epitaxial material that we will assume has a resistivity of 5 Ω-cm, corresponding to an impurity concentration of 10^{15} atoms/cm^3. The uniformly doped epi layer is much more lightly doped than the *p*-diffused region, and, as a result, this junction is well approximated by a step junction in which the depletion layer lies almost entirely in the epitaxial material. Under this assumption, the results of Chapter 1 regarding step junctions can be applied, and for convenience this relationship has been plotted in nomograph form in Fig. 2.21. This nomograph is a graphical representation of the relation:

$$C_j = \sqrt{\frac{q\epsilon N_B}{2(\psi_0 + V_R)}} \tag{2.19}$$

where N_B is the doping density in the epi material and V_R is the reverse bias on the junction.

Note that the horizontal axis in Fig. 2.21 is the *total* junction potential, which is the applied potential plus the built-in voltage ψ_0. In order to use the curve, then, the built-in potential must be calculated. While this would be an involved calculation for a diffused junction, the built-in potential is actually only weakly dependent on the details of the diffusion profile, and can be assumed to be about 0.55 V for the collector-base, 0.52 V for the collector-substrate junctions, and about 0.7 V for the emitter-base junction.

EXAMPLE

Calculate the collector-base capacitance of the device of Fig. 2.17. The zero-bias capacitance per unit area of the collector-base junction can be found from Fig. 2.21 to be approximately 10^{-4} pF/μ^2. The total area of the collector-base junction is the sum of the area of the bottom of the base diffusion plus the base sidewall area.

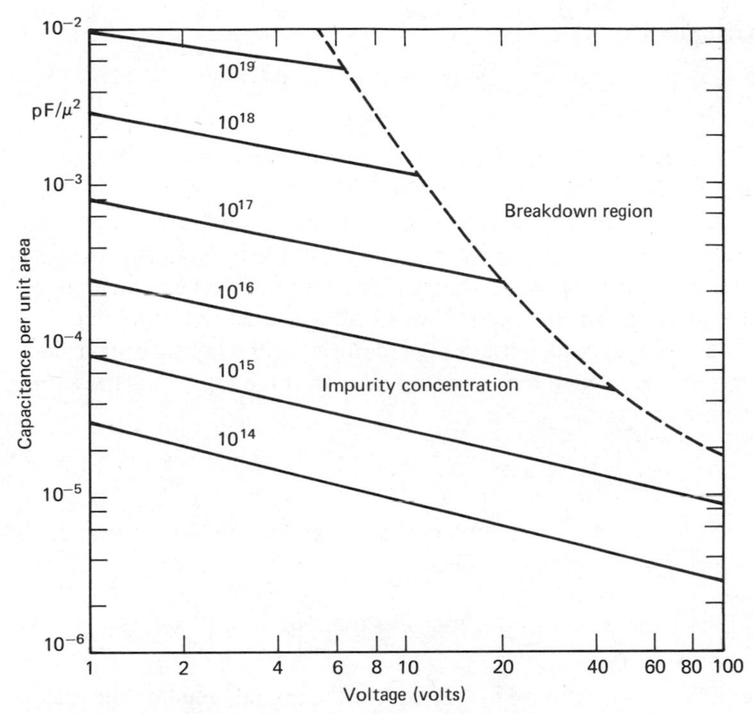

Fig. 2.21 Capacitance of an abrupt *pn* junction as a function of applied voltage and concentration on the lightly doped side of the junction.[14]

From Fig. 2.17, the bottom area is:

$$A_{\text{bottom}} = 60\ \mu \times 45\ \mu = 2700\ \mu^2$$

The edges of the base region can be seen from Fig. 2.16 to have the shape similar to one-quarter of a cylinder. We will assume that the region is cylindrical in shape, which yields a sidewall area of:

$$A_{\text{sidewall}} = P \times d \times \frac{\pi}{2}$$

where

$$P = \text{base region periphery}$$
$$d = \text{base diffusion depth}$$

Thus we have

$$A_{\text{sidewall}} = 3\ \mu \times (60 + 60 + 45 + 45) \times \frac{\pi}{2} = 989\ \mu^2$$

and the total capacitance is:

$$C_{\mu 0} = (A_{\text{bottom}} + A_{\text{sidewall}})(10^{-4}\ \text{pF}/\mu^2) = 0.36\ \text{pF}$$

Collector-substrate Capacitance. The collector-substrate capacitance consists of three portions: that of the junction between the buried layer and the substrate, that of the sidewall of the isolation diffusion, and that between the epitaxial material and the substrate. Since the substrate has an impurity concentration of about $10^{16}\ \text{cm}^{-3}$ it is more heavily doped than the epi material, and we can analyze both the sidewall and epi-substrate capacitance under the assumption that the junction is a one-sided step junction with the epi material as the lightly doped side. Under this assumption, the capacitance per unit area in these regions is the same as in the collector-base junction.

EXAMPLE

Calculate the collector-substrate capacitance of the standard device of Fig. 2.17. The area of the collector-substrate sidewall is:

$$A_{\text{sidewall}} = (17\ \mu)(140\ \mu + 140\ \mu + 95\ \mu + 95\ \mu)\left(\frac{\pi}{2}\right) = 12{,}550\ \mu^2$$

We will assume that the actual buried layer covers the area defined by the mask, indicated on Fig. 2.17 as an area of 41 $\mu \times 85\ \mu$, plus 8 μ of side-diffusion on each edge. This gives a total area of 57 $\mu \times 101\ \mu$. The area of the junction between the epi material and the substrate is the total area of the isolated region, minus that of the buried layer.

$$\begin{aligned} A_{\text{epi-substrate}} &= (140\ \mu \times 95\ \mu) - (57\ \mu \times 101\ \mu) \\ &= 7543\ \mu^2 \end{aligned}$$

The capacitances of the sidewall and epi-substrate junctions are, using a capacitance per unit area of $10^{-4}\ \text{pF}/\mu^2$

$$\begin{aligned} C_{cs0}(\text{sidewall}) &= (12{,}550\ \mu^2)(10^{-4}\ \text{pF}/\mu^2) = 1.2\ \text{pF} \\ C_{cs0}(\text{epi-substrate}) &= (7543\ \mu^2)(10^{-4}\ \text{pF}/\mu^2) = 0.75\ \text{pF} \end{aligned}$$

For the junction between the buried layer and substrate, the lightly-doped side of the junction is the substrate. Assuming a substrate doping level of 10^{16} atoms/cm^3, and a built-in voltage of 0.52 V, we can calculate the zero-bias capacitance per unit area as $3.3 \times 10^{-4}\ \text{pF}/\mu^2$. The area of the buried layer is:

$$A_{BL} = 57\ \mu \times 101\ \mu = 5757\ \mu^2$$

and the zero-bias capacitance from the buried layer to the substrate is thus

$$C_{cs0}(BL) = (5757\ \mu^2)(3.3 \times 10^{-4}\ \text{pF}/\mu^2) = 1.89\ \text{pF}$$

The total zero-bias collector-substrate capacitance is thus:

$$C_{cs0} = 1.2 \text{ pF} + 0.75 \text{ pF} + 1.89 \text{ pF} = 3.84 \text{ pF}$$

Emitter-Base Capacitance. The emitter-base junction of the transistor has a doping profile that is not well approximated by a step junction as the impurity concentration on both sides of the junction varies with distance in a rather complicated way. Furthermore, the sidewall capacitance per unit area is not constant but varies with distance from the surface because the base impurity concentration varies with distance. A precise evaluation of this capacitance can be carried out numerically, but a first-order *estimate* of the capacitance can be obtained by calculating the capacitance of an abrupt junction with an impurity concentration on the lightly doped side which is equal to the concentration in the base at the edge of the junction. The sidewall contribution is neglected.

EXAMPLE

Calculate the zero-bias, emitter-base junction capacitance of the standard device of Fig. 2.17.

We first estimate the impurity concentration at the emitter edge of the base region. From Fig. 2.16 it can be seen that this concentration is approximately 10^{17} atoms/cm^3. From the nomograph of Fig. 2.21, this abrupt junction would have a zero-bias capacitance per unit area of 10^{-3} pF/μ^2. Since the area of the bottom portion of the emitter-base junction is $25 \times 20\ \mu^2$, the capacitance of the bottom portion is:

$$C_{\text{bottom}} = (500\ \mu^2)(10^{-3}\ \text{pF}/\mu^2) = 0.5 \text{ pF}$$

Again assuming a cylindrical cross section, the sidewall area is given by

$$A_{\text{sidewall}} = 2(25\ \mu + 20\ \mu)(\pi/2)(2.5\ \mu) = 353\ \mu^2$$

Assuming that the capacitance per unit area of the sidewall is approximately the same as the bottom,

$$C_{\text{sidewall}} = (353\ \mu^2)(10^{-3}\ \text{pF}/\mu^2) = 0.35 \text{ pF}$$

The total emitter-base capacitance is

$$C_{je0} = 0.85 \text{ pF}$$

Current Gain. As discussed in Chapter 1, the current gain of the transistor depends on minority-carrier lifetime in the base, which affects the base transport factor, and on the diffusion length in the emitter, which affects the emitter efficiency. In analog IC processing, the base minority-carrier lifetime is sufficiently long that the base transport factor is not a limiting factor in the forward current gain in *npn* transistors. Because the emitter region is heavily doped with phosphorous, the minority-carrier lifetime is degraded in this region, and current gain is limited primarily by emitter efficiency.[7] Because the doping level, and hence lifetime, vary with distance in the emitter, the calculation of emitter efficiency for the *npn* transistor is difficult, and measured parameters must be used. The room-temperature current gain typically lies between 200 and 1000 for these devices. The current gain falls with decreasing temperature, usually to a value of from 0.5 to 0.75 of the room temperature value at -55 degrees centigrade.

Summary of npn Device Parameters. A typical set of device parameters for the device of Fig. 2.17 is shown in Fig. 2.22. This transistor geometry is typical of that

	Parameter	Typical Value, 5 Ω-cm, 17 μ epi 44-V Device	Typical Value, 1 Ω-cm, 10 μ epi 20-V Device
	β_F	200	200
	β_R	2	2
	V_A	130 V	90 V
	η	2×10^{-4}	2.8×10^{-4}
	I_S	5×10^{-15} A	1.5×10^{-15} A
	I_{CO}	10^{-10} A	10^{-10} A
	BV_{CEO}	50 V	25 V
	BV_{CBO}	90 V	50 V
	BV_{EBO}	7 V	7 V
	τ_F	0.35 ns	0.25 ns
	β_0	200	150
	r_b	200 Ω	200 Ω
	r_c (saturation)	200 Ω	75 Ω
	r_{ex}	2 Ω	2 Ω
Base-emitter junction	C_{je0}	1 pF	1.3 pF
	ψ_0	0.7 V	0.7 V
	n	0.33	0.33
Base-collector junction	$C_{\mu 0}$	0.3 pF	0.6 pF
	ψ_0	0.55 V	0.6 V
	n	0.5	0.5
Collector-substrate junction	C_{cs0}	3 pF	3 pF
	ψ_0	0.52 V	0.58 V
	n	0.5	0.5

Fig. 2.22 Typical parameters for integrated *npn* transistors. The thick epi device is typical of those used in circuits operating at up to 44 V power-supply voltage, while the thinner device can operate up to about 20 V. While the geometry of the thin epi device is smaller, the collector-base capacitance is larger because of the heavier epi doping. The emitter-base capacitance is higher because the base is shallower, and the doping level in the base at the emitter-base junction is higher.

used for circuits that must operate at power supply voltages up to 40 V. For lower operating voltages, thinner epitaxial layers can be used, and smaller device geometries can be used as a result. Also shown in Fig. 2.22 are typical parameters for a device made with 1 Ω-cm epi material, which is 10 μ thick. Such a device is physically smaller and has a collector-emitter breakdown voltage of about 25 V.

2.4.2 Integrated-Circuit *pnp* Transistors

As mentioned previously, the integrated-circuit bipolar fabrication process is an outgrowth of that used to build double-diffused epitaxial *npn* transistors, and the technology inherently produces *npn* transistors of high performance. However, *pnp* transistors of comparable performance are not easily produced in the same process, and the earliest analog integrated circuits used no *pnp* transistors. The lack of a complementary device for use in biasing, level shifting, and as load devices in amplifier stages proved to be a severe limitation on the performance attainable in analog circuits, and this led to the development of several *pnp* transistor structures that are compatible with the standard IC fabrication process. Because these devices utilize the lightly doped *n*-type epitaxial material as the base of the transistor, they are generally inferior to the *npn* devices in frequency response and high-current behavior, but are very useful nonetheless. In this section we will discuss the lateral *pnp* and substrate *pnp* structures.

Lateral pnp Transistors. A typical lateral *pnp* transistor structure is illustrated in Fig. 2.23*a*.[8] The emitter and collector are formed with the same diffusion that forms the base of the *npn* transistors. The collector is a *p*-type ring around the emitter and the base contact is made in the *n*-type epi material *outside* the collector ring. The flow of minority carriers across the base is illustrated in Fig. 2.23*b*. Holes are injected from the emitter, flow parallel to the surface across the *n*-type base region, and ideally are collected by the *p*-type collector before reaching the base contact. Thus the transistor action is *lateral* rather than *vertical* as in the case for *npn* transistors. The principal drawback of the structure is the fact that the base region is more lightly doped than the collector. As a result, the collector-base depletion layer extends almost entirely into the base. The base region must then be made wide enough so that the depletion layer does not reach the emitter when the maximum collector-emitter voltage is applied. In a typical analog IC process, the width of this depletion layer is 6 to 8 μ when the collector-emitter voltage is in the 40-V range. Thus the minimum base width for such a device is about 8 μ and the minimum base transit time can be estimated from (1.99) as

$$\tau_F = \frac{{W_B}^2}{2D_p} \tag{2.20}$$

Use of $W_B = 8\ \mu$ and $D_p = 10\ \text{cm}^2/\text{sec}$ (for holes) in (2.20) gives

$$\tau_F = 32\ \text{ns}$$

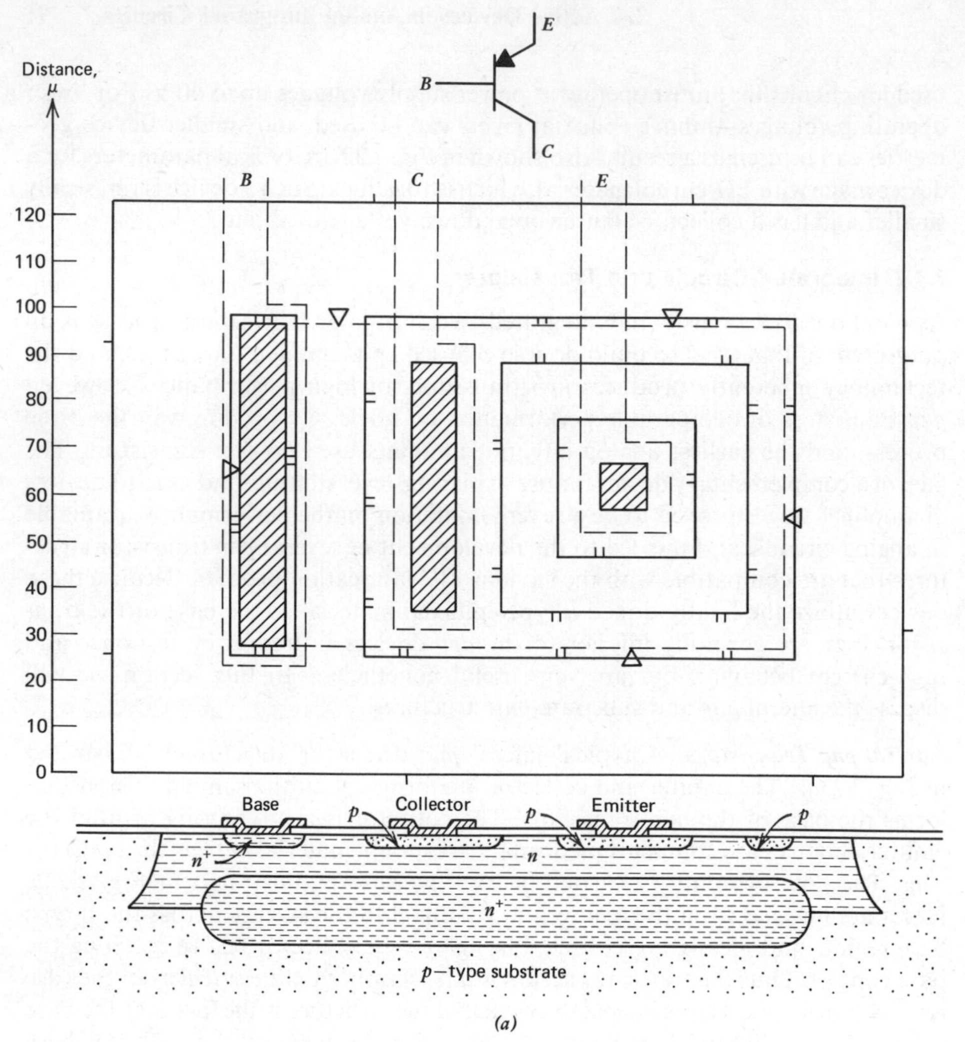

Fig. 2.23*a* Lateral *pnp* structure.

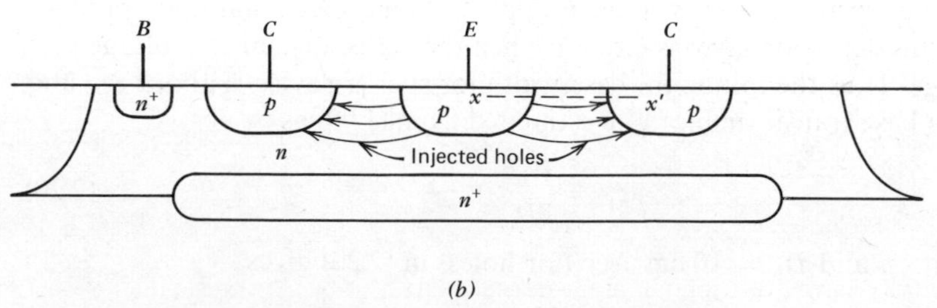

Fig. 2.23*b* Minority-carrier flow in the lateral *pnp* transistor.

This corresponds to a peak f_T of 5 Mhz, which is a factor of 100 lower than a typical *npn* transistor.

The current gain of lateral pnp transistors tends to be low for several reasons. First, minority carriers (holes) in the base are injected downward from the emitter as well as laterally, and some of these are collected by the substrate, which acts as the collector of a parasitic vertical *pnp* transistor. The buried layer sets up a retarding field that tends to inhibit this process, but it still produces a measurable degradation of β_F. Second, the emitter of the *pnp* is not as heavily doped as is the case for the *npn* devices, and thus the emitter injection efficiency given by (1.51b) is not optimized for the *pnp* devices. Finally, the wide base of the lateral *pnp* results in both a low emitter injection efficiency and also a low base transport factor as given by (1.51a).

Another drawback resulting from the use of a lightly doped base region is that the current gain of the device falls very rapidly with increasing collector current due to high-level injection. The minority-carrier distribution in the base of a lateral *pnp* transistor in the forward-active region is shown in Fig. 2.24. The collector current per unit of cross-sectional area can be obtained from (1.32) as

$$J_p = qD_p \frac{p_n(0)}{W_B} \tag{2.21}$$

Inverting this relationship, we can calculate the minority-carrier density at the emitter edge of the base as

$$p_n(0) = \frac{J_p W_B}{qD_p} \tag{2.22}$$

As long as this concentration is much less than the majority-carrier density in the base, low-level injection conditions exist and the base minority-carrier lifetime

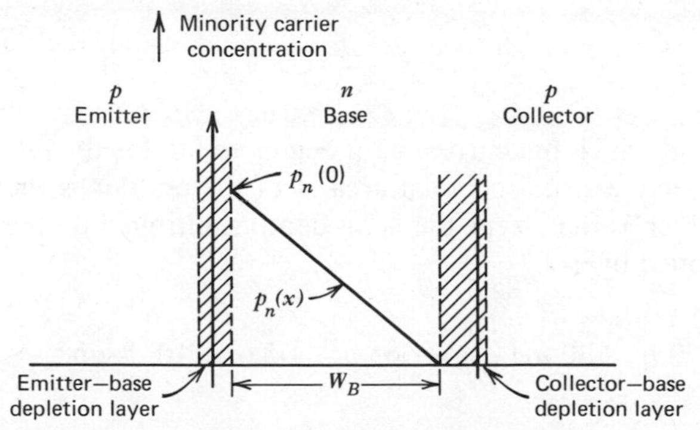

Fig. 2.24 Minority-carrier distribution in the base of a lateral *pnp* transistor in the forward-active region. This distribution is that observed through section *x–x′* in Fig. 2.23*b*.

remains constant. However, when the minority-carrier density becomes comparable with the majority-carrier density, the majority-carrier density must increase to maintain charge neutrality in the base. This causes a decrease in β_F for two reasons. First, there is a decrease in the effective lifetime of minority carriers in the base, since there is an increased number of majority carriers with which recombination can occur. Thus the base transport factor given by (1.51a) decreases. Second, the increase in the majority-carrier density represents an effective increase in base doping density. This causes a decrease in emitter injection efficiency given by (1.51b). Both these mechanisms are also present in *npn* transistors, but occur at much higher current levels due to the higher doping density in the base of the *npn* transistor.

The collector current at which these effects become significant can be calculated for a lateral *pnp* transistor by equating the minority-carrier concentration given by (2.22) to the equilibrium majority-carrier concentration. Thus

$$\frac{J_p W_B}{qD_p} = n_n \simeq N_D \tag{2.23}$$

where (2.1) has been substituted for n_n, and N_D is the donor density in the *pnp* base (*npn* collector). We can calculate from (2.23) the collector current for the onset of high-level injection in a *pnp* transistor as

$$I_C = \frac{qAN_D D_p}{W_B} \tag{2.24}$$

where A is the effective area of the emitter-base junction. Note that this current depends directly on the base doping density in the transistor, and, since this is quite low in a lateral *pnp* transistor, the current density at which this falloff begins is quite low.

EXAMPLE

Calculate the collector current at which the current gain begins to fall for the *pnp* structure of Fig. 2.23*a*. The effective cross-sectional area A of the emitter is the sidewall area of the emitter, which is the *p*-type diffusion depth multiplied by the periphery of the emitter multiplied by $\pi/2$.

$$A = (3\ \mu)(30\ \mu + 30\ \mu + 30\ \mu + 30\ \mu)\left(\frac{\pi}{2}\right) = 565\ \mu^2 = 5.6 \times 10^{-6}\ \text{cm}^2$$

The majority-carrier density is 10^{15} atoms/cm^3 for an epi-layer resistivity of 5 Ω-cm. In addition we can assume $W_B = 8\ \mu$ and $D_p = 10$ cm^2/sec. Substitution of this data in (2.24) gives

$$I_C = 5.6 \times 10^{-6} \times 1.6 \times 10^{-19} \times 10^{15} \times 10 \frac{1}{8 \times 10^{-4}} \text{ A} = 11.2\ \mu\text{A}$$

The typical lateral *pnp* structure of Fig. 2.23*a* shows a low-current beta of approximately 30 to 50, which begins to decrease at a collector current of a few tens of microamperes, and has fallen to less than 10 at a collector current of 1 mA. A typical set of parameters for a structure of this type is shown in Fig. 2.25. Note that in the lateral *pnp* transistor the substrate junction capacitance appears between the *base* and the substrate.

Substrate pnp Transistors. One reason for the poor high-current performance of the lateral *pnp* is the relatively small effective cross-sectional area of the emitter,

	Parameter	Typical Value, 5 Ω-cm, 17 μ epi, 44 V Device	Typical Value, 1 Ω-cm, 10 μ epi, 20 V Device
	β_F	50	20
	β_R	4	2
	V_A	50 V	50 V
	η	5×10^{-4}	5×10^{-4}
	I_S	2×10^{-15} A	2×10^{-15} A
	I_{CO}	10^{-10} A	5×10^{-9} A
	BV_{CEO}	60 V	30 V
	BV_{CBO}	90 V	50 V
	BV_{EBO}	90 V	50 V
	τ_F	30 ns	20 ns
	β_0	50	20
	r_b	300 Ω	150 Ω
	r_c	100 Ω	75 Ω
	r_{ex}	10 Ω	10 Ω
Base-emitter junction	C_{je0}	0.3 pF	0.6 pF
	ψ_0	0.55 V	0.6 V
	n	0.5	0.5
Base-collector junction	$C_{\mu 0}$	1 pF	2 pF
	ψ_0	0.55 V	0.6 V
	n	0.5	0.5
Base-substrate junction	C_{bs0}	3 pF	3.5 pF
	ψ_0	0.52 V	0.58 V
	n	0.5	0.5

Fig. 2.25 Typical parameters for a lateral *pnp* transistor.

which results from the lateral nature of the injection. A common application for a *pnp* transistor is in a class B output stage where the device is called on to operate at collector currents in the 10-mA range. A lateral *pnp* designed to do this would require a large amount of die area, and in this application a different structure is usually used in which the substrate itself is used as the collector instead of a diffused *p*-type region. Such a substrate *pnp* transistor is shown in Fig. 2.26*a*. The *p*-type

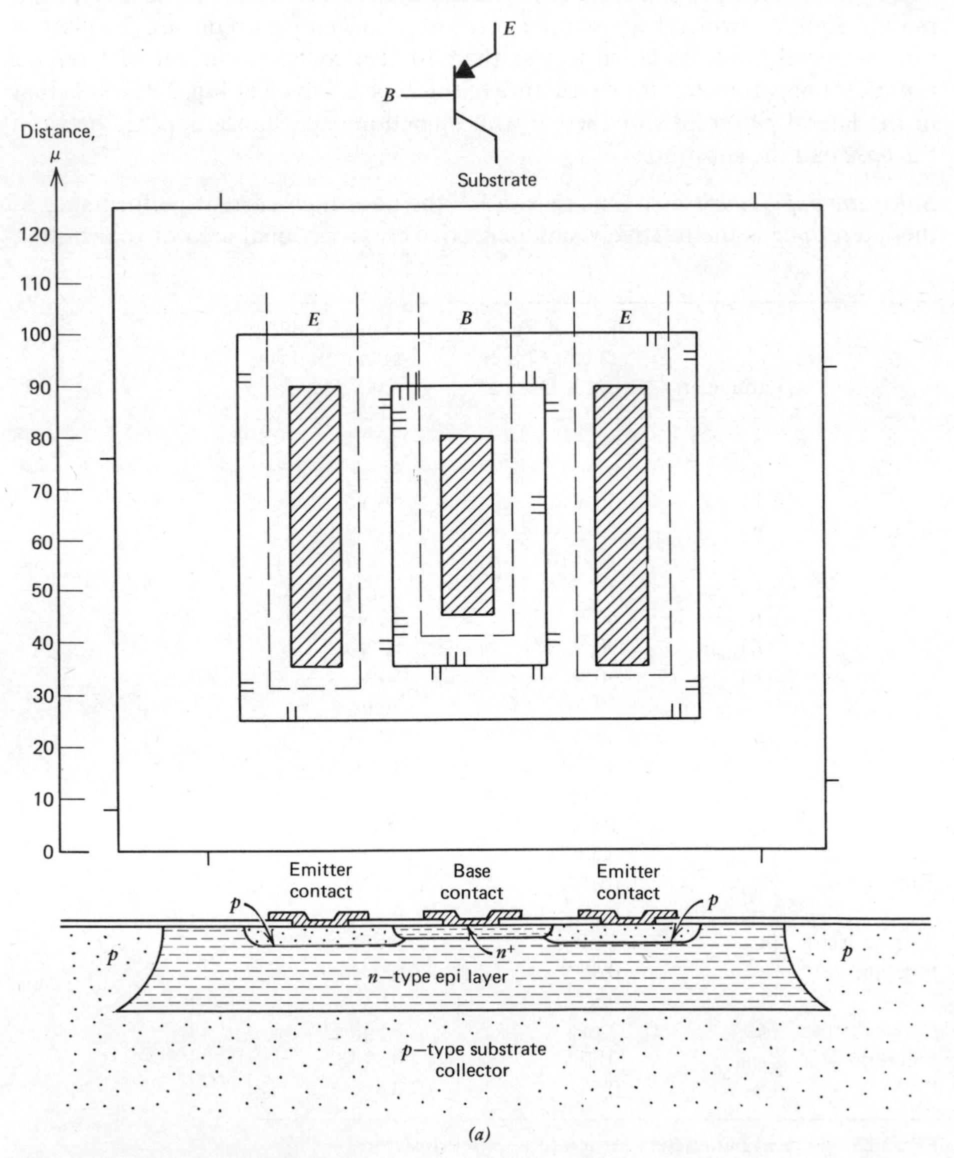

Fig. 2.26*a* Substrate *pnp* structure.

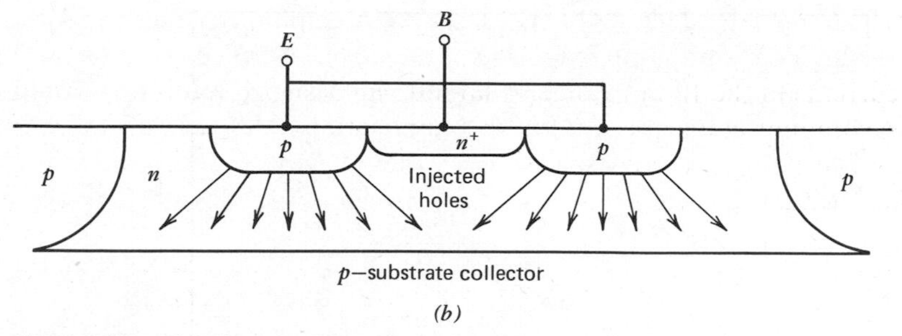

Fig. 2.26*b* Minority-carrier flow in the substrate *pnp* transistor.

emitter diffusion for this particular substrate *pnp* geometry is rectangular with a rectangular hole in the middle. In this hole an n^+ region is formed with the *npn* emitter diffusion to provide a contact for the *n*-type base. Because of the lightly doped base material, the series base resistance can become quite large if the base contact is far removed from the active base region. In this particular structure, the n^+ base contact diffusion is actually allowed to come in contact with the *p*-type

	Parameter	Typical Value, 5 Ω-cm, 17 μ epi, 44 V Device 5100 μ^2 Emitter area	Typical Value, 1 Ω-cm, 10 μ epi, 20 V Device 5100 μ^2 Emitter area
	β_F	50	30
	β_R	4	2
	V_A	50 V	30 V
	η	5×10^{-4}	9×10^{-4}
	I_S	10^{-14} A	10^{-14} A
	I_{CO}	2×10^{-10} A	2×10^{-10} A
	BV_{CEO}	60 V	30 V
	BV_{CBO}	90 V	50 V
	BV_{EBO}	7 V or 90 V	7 V or 50 V
	τ_F	20 ns	14 ns
	β_0	50	30
	r_b	150 Ω	50 Ω
	r_c	50 Ω	50 Ω
	r_{ex}	2 Ω	2 Ω
Base-emitter junction	C_{je0}	0.5 pF	1 pF
	ψ_0	0.55 V	0.58 V
	n	0.5	0.5
Base-collector junction	$C_{\mu 0}$	2 pF	3 pF
	ψ_0	0.52 V	0.58 V
	n	0.5	0.5

Fig. 2.27 Typical device parameters for a substrate *pnp* with 5100 μ^2 emitter area.

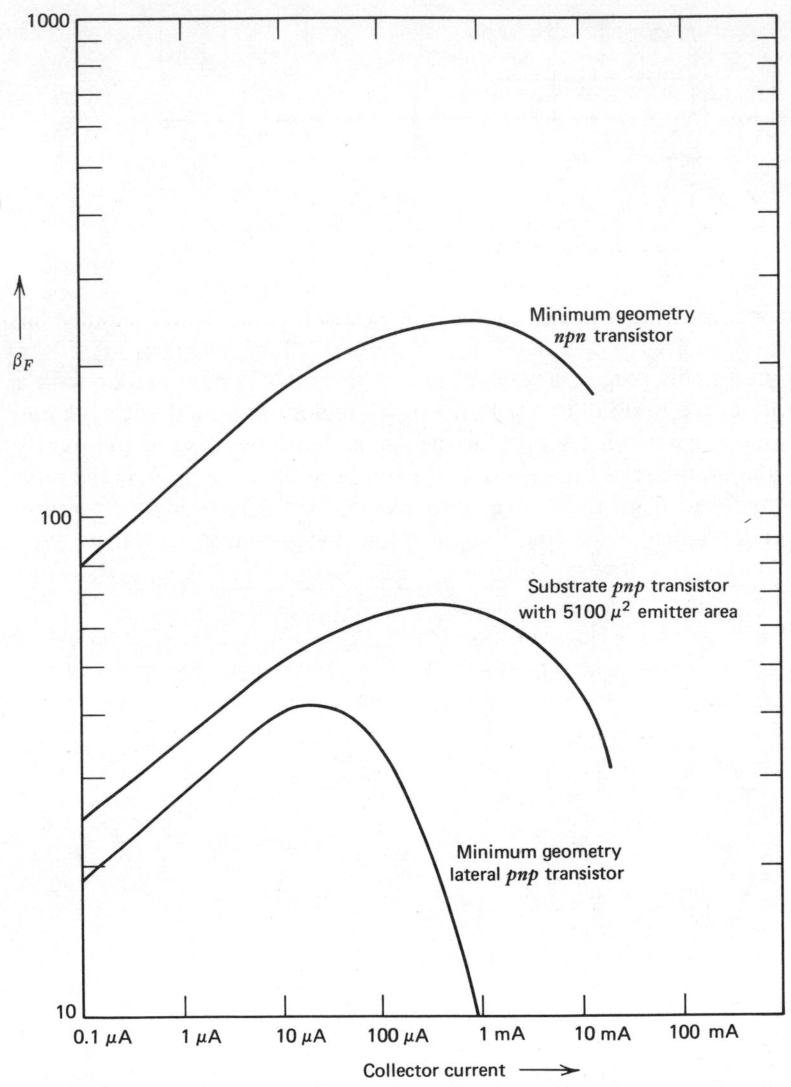

Fig. 2.28 Current gain as a function of collector current for typical lateral *pnp*, substrate *pnp*, and *npn* transistor geometries.

emitter diffusion, in order to get the low-resistance base contact diffusion as close as possible to the active base. The only drawback of this, in a substrate *pnp* structure, is that the emitter-base breakdown voltage is reduced to approximately 7 V. If larger emitter-base breakdown is required, then the *p*-emitter diffusion must be separated from the n^+ base contact diffusion by a distance on the order of 10 to 15 μ. Many variations exist on the substrate *pnp* geometry shown in Fig. 2.26*a*.

The minority-carrier flow in the forward-active region is illustrated in Fig. 2.26*b*. The principal advantage of this device is that the current flow is vertical and the effective cross-sectional area of the emitter is much larger than in the case of the lateral *pnp* for the same overall device size. The device is however restricted to use in emitter-follower configurations, since the collector is electrically identical with the substrate that must be tied to the most negative circuit potential. Other than the better current-handling capability, the properties of substrate *pnp* transistors are similar to those for lateral *pnp* transistors since the base width is similar in both cases. An important consideration in the design of substrate *pnp* structures is that the collector current flows in the *p*-substrate region, which usually has relatively high resistivity. Thus, unless care is taken to provide an adequate low-resistance path for the collector current, a high series collector resistance can result. This resistance can degrade device performance in two ways. First, large collector currents in the *pnp* can cause enough voltage drop in the substrate region itself that other substrate-epitaxial layer junctions within the circuit can become forward biased. This usually has a catastrophic effect on circuit performance. Second, the effects of the collector-base junction capacitance on the *pnp* are multiplied by the Miller effect resulting from the large series collector resistance, as discussed further in Chapter 7. In order to minimize these effects, the collector contact is usually made by contacting the isolation diffusion immediately adjacent to the substrate *pnp* itself with metallization. For high-current devices, this isolation diffusion contact is made to surround the device to as great an extent as possible. The properties of a typical substrate *pnp* transistor are summarized in Fig. 2.27. The dependence of current gain on collector current for a typical *npn*, lateral *pnp*, and substrate *pnp* transistor are shown in Fig. 2.28. The low-current reduction in β, which is apparent for all three devices, is due to recombination in the base-emitter depletion region, described in Section 1.3.5.

2.5 PASSIVE COMPONENTS IN ANALOG INTEGRATED CIRCUITS

In this section we describe the structures available to the integrated-circuit designer for realization of resistance and capacitance. Resistor structures include base-diffused, emitter-diffused, pinch, epitaxial, and pinched epitaxial resistors. Capacitance structures include MOS and junction capacitors. Inductors have not proven to be feasible in monolithic technology. Other resistor technologies, such as thin film resistors, are discussed in a later section.

2.5.1 Diffused Resistors

In an earlier section of this chapter the sheet resistance of a diffused layer was calculated. Integrated-circuit resistors are generally fabricated using one of the diffused layers formed during the fabrication process, or in some cases a combination of two diffused layers. The layers available for use as resistors include the base diffusion, emitter diffusion, epitaxial layer, the active-base region layer of a transistor and the epitaxial layer pinched between the base diffusion and the *p*-type substrate. The choice of layer generally depends on the value, tolerance, and temperature coefficient of the resistor required.

Base and Emitter Diffused Resistors. The structure of a typical base-diffused resistor is shown in Fig. 2.29. The resistor is formed from the *p*-type base diffusion for the *npn* transistors and is situated in a separate isolation region. The epitaxial region into which the resistor structure is diffused must be biased in such a way that the *pn* junction between the resistor and the epi layer is always reverse biased. For this reason a contact is made to the *n*-type epi region as shown in Fig. 2.29 and this is connected either to that end of the resistor that is most positive, or to a potential that is more positive than either end of the resistor. The junction between these two regions contributes a parasitic capacitance between the resistor and the epi layer, and this capacitance is distributed along the length of the resistor.

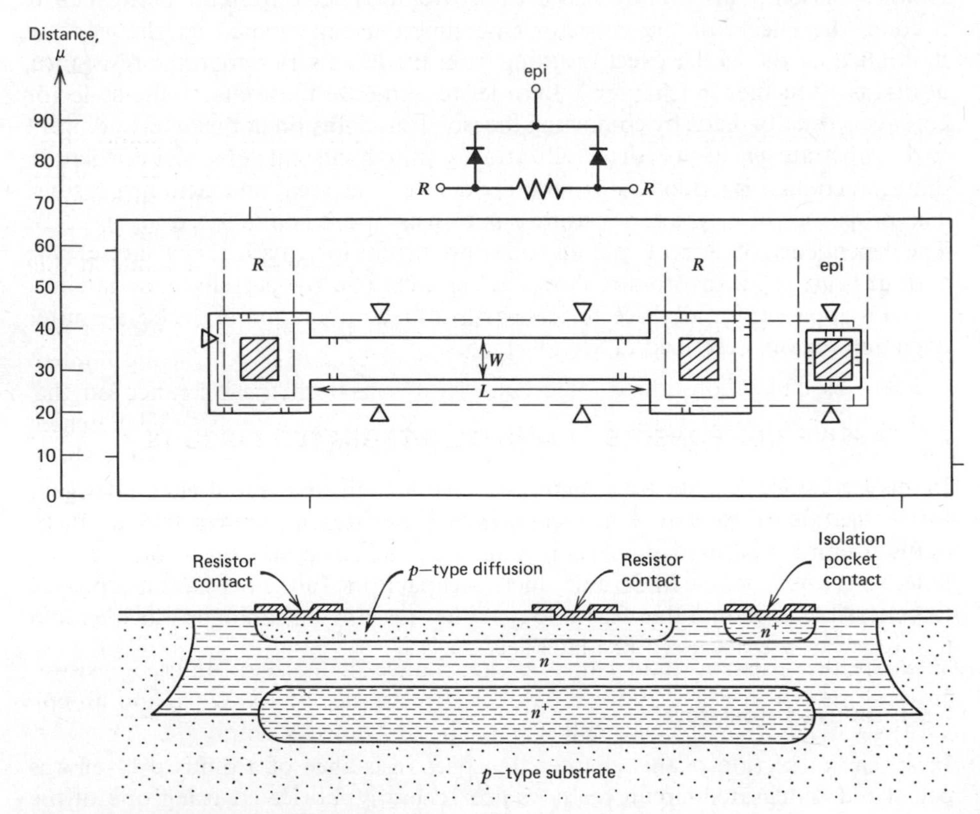

Fig. 2.29 Base-diffused resistor structure.

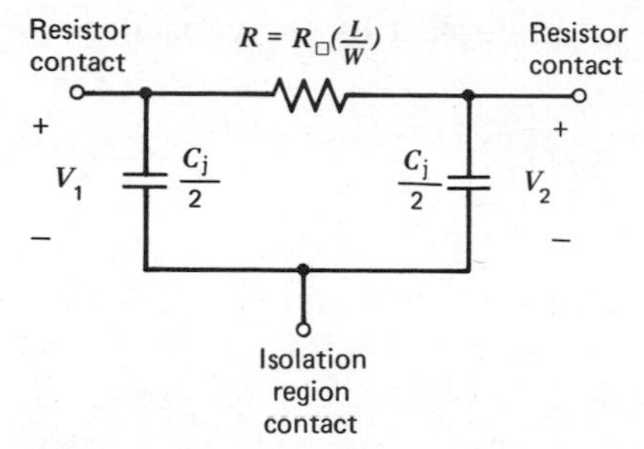

Fig. 2.30 Lumped model for the base-diffused resistor.

For most applications this parasitic capacitance can be adequately modeled by separating it into two lumped portions and placing one lump at each end of the resistor as illustrated in Fig. 2.30.

The resistance of a structure as shown in Fig. 2.29 is given by

$$R = \frac{L}{W} R_{\square} \tag{2.10}$$

where L is the resistor length and W is the width. The base sheet resistance, $R_{\square}$, lies in the range 100 to 200 $\Omega/\square$, and thus resistances in the range 50 Ω to 50 kΩ are practical using the base diffusion. The resistance contributed by the "clubheads" at each end of the resistor can be significant, particularly for small values of L/W. The clubheads are required to allow space for ohmic contact to be made at the ends of the resistor.

Since minimization of die area is an important objective, the width of the resistor is kept as small as possible, the minimum practical width being limited to about 5 μ by photolithographic considerations. Both the tolerance on the resistor value and the precision with which two identical resistors can be matched can be improved by the use of wider geometries. However for a given base sheet resistance and a given resistor value, the area occupied by the resistor increases as the *square* of its width. This can be seen from (2.10) since the ratio L/W is constant.

EXAMPLE

Calculate the resistance and parasitic capacitance of the base diffused resistor structure shown in Fig. 2.29 for a base sheet resistance of 100 $\Omega/\square$, and an epi resistivity of 2.5 Ω-cm. Neglect end effects. The resistance is simply:

$$R = 100\ \Omega/\square \left(\frac{100\ \mu}{10\ \mu}\right) = 1\ \text{k}\Omega$$

The capacitance is the total area of the resistor multiplied by the capacitance per unit area. The area of the resistor body is:

$$A_1 = (10\ \mu)(100\ \mu) = 1000\ \mu^2$$

The area of the clubheads is:

$$A_2 = 2(30\ \mu \times 30\ \mu) = 1800\ \mu^2$$

The total zero-bias capacitance is, from Fig. 2.21,

$$C_{j0} = (10^{-4}\ \text{pF}/\mu^2)(2800\ \mu^2) = 0.28\ \text{pF}$$

As a first-order approximation this can be divided into two parts, one placed at each end. Note that this capacitance will vary depending on the voltage at the clubhead with respect to the epitaxial pocket.

Emitter-diffused resistors are fabricated using geometries similar to the base resistor, but the emitter diffusion is used to form the actual resistor. Since the sheet resistance of this diffusion is in the 2- to 10-Ω/□ range, these resistors can be used to advantage where very low resistance values are required. In fact, they are widely used simply to provide a crossunder beneath an aluminum metallization interconnection. The parasitic capacitance can be calculated in a way similar to that for the base diffusion. However these resistors have different temperature dependence from base-diffused resistors and the two types do not track with temperature.

Base Pinch Resistors. A third layer available for use as a resistor is the layer that in the *npn* transistor forms the active base region. This layer is "pinched" between the n^+ emitter and the *n*-type collector regions, giving rise to the term pinch resistor. The layer can be electrically isolated by reverse biasing the emitter-base and collector-base junctions, which is usually accomplished by connecting the *n*-type regions to the most positive end of the resistor. The structure of a typical pinch resistor is shown in Fig. 2.31; the n^+ diffusion overlaps the *p*-diffusion so that the n^+ region is electrically connected to the *n*-type epi region. The sheet resistance is in the 5 to 15 kΩ/□ range, so that this resistor allows the fabrication of very large values of resistance. Unfortunately, the sheet resistance undergoes the same process-related variations as does the Q_B of the transistor, which is approximately ± 50 percent. Also, because the material making up the resistor itself is relatively lightly doped, the resistance displays a relatively large variation with temperature. Another significant drawback is that the maximum voltage that can be applied across the resistor is limited to around 6 V because of the breakdown voltage between the emitter-diffused top layer and the base diffusion. Nonetheless, this type of resistor has found wide application where the large tolerance and low breakdown voltage are not significant drawbacks.

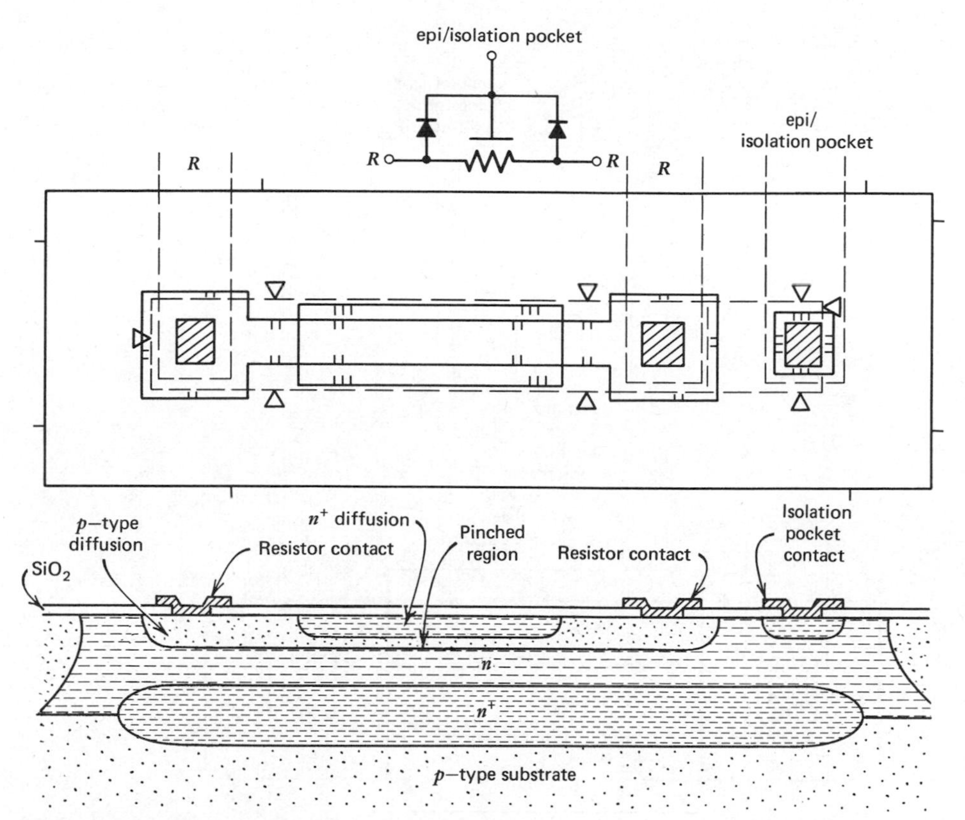

Fig. 2.31 Pinch resistor structure.

2.5.2 Epitaxial and Epitaxial Pinch Resistors

The limitation of the pinch resistor to low operating voltages does not permit its use in circuits where a small bias current is to be derived directly from the power supply voltage using a large-value resistor. The epitaxial layer itself has a sheet resistance much larger than the base diffusion, and for this application the epi layer is often used as a resistor. For example, the sheet resistance of a 17-μ thick, 5 Ω-cm epi layer can be calculated from (2.11) as

$$R_{\square} = \frac{\rho_{\text{EPI}}}{T} = \frac{5\ \Omega\text{-cm}}{(17\ \mu) \times (10^{-4}\ \text{cm}/\mu)} = 2.9\ \text{k}\Omega/\square \tag{2.25}$$

Large values of resistance can be realized in a small area using structures of the type shown in Fig. 2.32. Again, because of the light doping in the resistor body, these resistors display a rather large temperature coefficient. A still-larger sheet resistance can be obtained by putting a p-type base diffusion over the top of an epitaxial resistor as shown in Fig. 2.32. Such a structure actually behaves as a

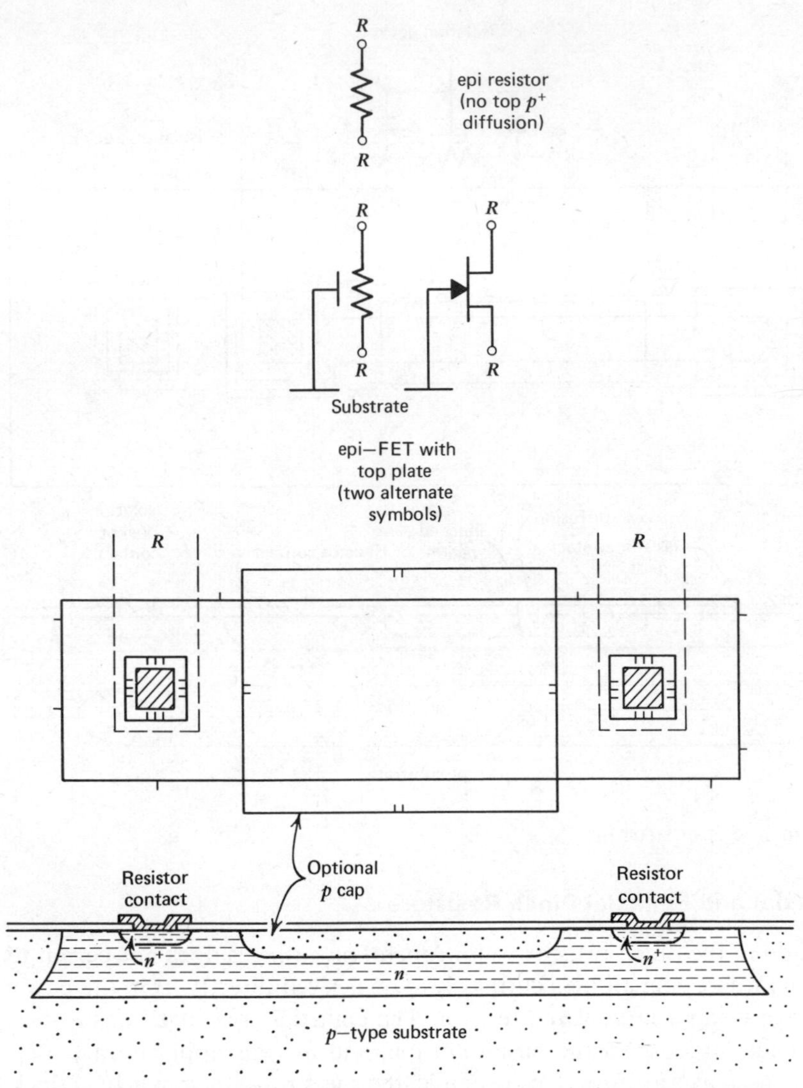

Fig. 2.32 Epitaxial resistor structure. The p-cap diffusion is optional, and forms an epitaxial pinch resistor.

junction FET, in which the p-type gate is tied to the substrate. The pinch-off voltage varies depending on epi thickness and doping, but for the case of 17 μ thick, 5 Ω-cm epi, and a 3-μ-deep base, the pinch-off voltage is [from (1.151) interchanging N_D and N_A and assuming $N_D \ll N_A$]

$$V_P = \left(a^2 \frac{qN_D}{2\epsilon} - \psi_0 \right) = 38 \text{ V} \tag{2.26}$$

Resistor Type	Sheet ρ $\Omega/\square$	Absolute Tolerance (%)	Matching Tolerance (%)	Temperature Coefficient
Base diffused	100–200	± 20	± 2 (5 μ wide) ± 0.2 (50 μ wide)	$+1500$ to $+2000$ ppm/°C
Emitter diffused	2–10	± 20	± 2	$+600$ ppm/°C
Ion implanted	100–1000	± 3	± 2 (5 μ wide) ± 0.15 (50 μ wide)	Controllable to $\pm$ 100 ppm/°C
Base pinch	2k–10k	± 50	± 10	$+2500$ ppm/°C
Epitaxial	2k–5k	± 30	± 5	$+3000$ ppm/°C
Epitaxial pinch	4k–10k	± 50	± 7	$+3000$ ppm/°C
Thin film	0.1k–2k	$\pm 5 - \pm 20$	$\pm 0.2 - \pm 2$	± 10 to ± 200 ppm/°C

Fig. 2.33 Summary of resistor properties for different types of IC resistors.

When used as a biasing element, this structure provides a measure of independence of bias current to power-supply current if the supply is larger than this pinch-off voltage. However, the I_{DSS} of the JFET is sensitive to process variations.

The properties of the various diffused and pinch-resistor structures are summarized in Fig. 2.33.

2.5.3 Integrated-Circuit Capacitors

Early analog integrated circuits were designed on the assumption that capacitors of usable value were impractical to integrate on the chip itself because they would take too much area, and external capacitors were used where required. It is still true that monolithic capacitors of value larger than a few tens of picofarads are expensive in terms of die area, but design approaches have evolved for monolithic circuits that allow small values of capacitance to be used to perform functions that previously required large capacitance values. The compensation of operational amplifiers is perhaps the best example of this trend, and monolithic capacitors are now widely used in all types of analog integrated circuits. These capacitors fall into two categories. First, *pn* junctions under reverse bias inherently display depletion capacitance, and in certain circumstances this capacitance can be effectively utilized. The drawbacks of junction capacitance are that the junction must always be kept reverse biased, that the capacitance varies with reverse voltage, and that, for the emitter-base junction, the breakdown voltage is only about 7 V. For the collector-base junction the breakdown voltage is higher but the capacitance per unit area is quite low.

By far the most commonly used monolithic capacitor is the MOS capacitor structure shown in Fig. 2.34. In the fabrication sequence an additional mask step is inserted to define a region over an emitter diffusion on which a thin layer of silicon dioxide is grown. Aluminum metallization is then placed over this thin oxide, producing a capacitor between the aluminum and the emitter diffusion, which has a capacitance of 0.2 to 0.3 pF/mil^2 and a breakdown voltage of 60 to

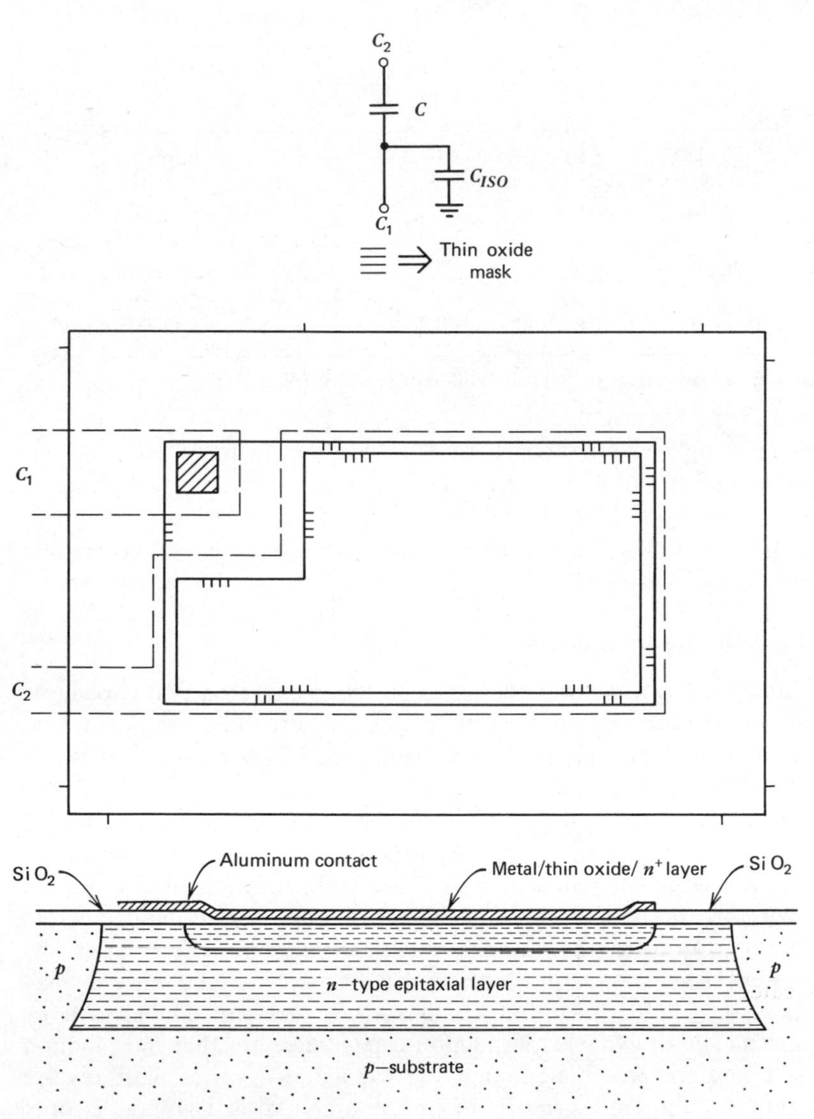

Fig. 2.34 MOS capacitor structure.

100 V. This capacitor is extremely linear and has a low temperature coefficient. A sizable parasitic capacitance is present between the *n*-type bottom plate and the substrate because of the depletion capacitance of the epi-substrate junction, but this parasitic is unimportant in many applications.

2.5.4 Zener Diodes

As described in Chapter 1, the emitter-base junction of the *npn* transistor structure displays a reverse breakdown voltage of between 6 and 8 V, depending on processing details. This voltage is of a low enough value to be useful as a voltage reference for the stabilization of bias reference circuits, and for such functions as level shifting. The reverse bias I-V characteristic of a typical emitter-base junction is illustrated in Fig. 2.35*a*.

An important aspect of the behavior of this device is the temperature sensitivity of the breakdown voltage. The actual breakdown mechanism is dominated by quantum mechanical tunneling through the depletion layer when the breakdown voltage is below about 6 V, and is dominated by avalanche multiplication in the depletion layer at the larger breakdown voltages. Because these two mechanisms have opposite temperature coefficients of breakdown voltage, the actually observed

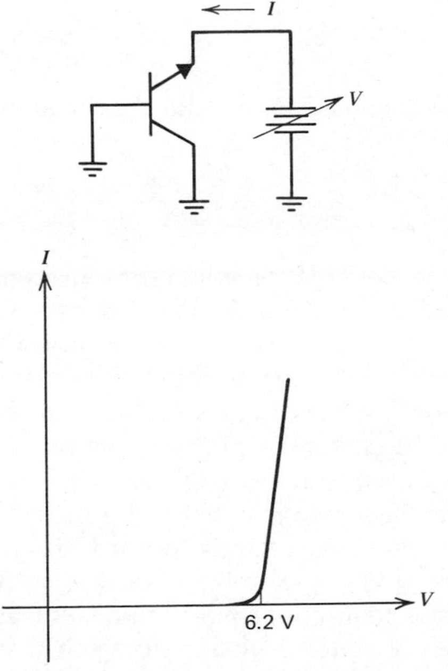

Fig. 2.35*a* Current-voltage characteristic of a typical emitter-base Zener diode.

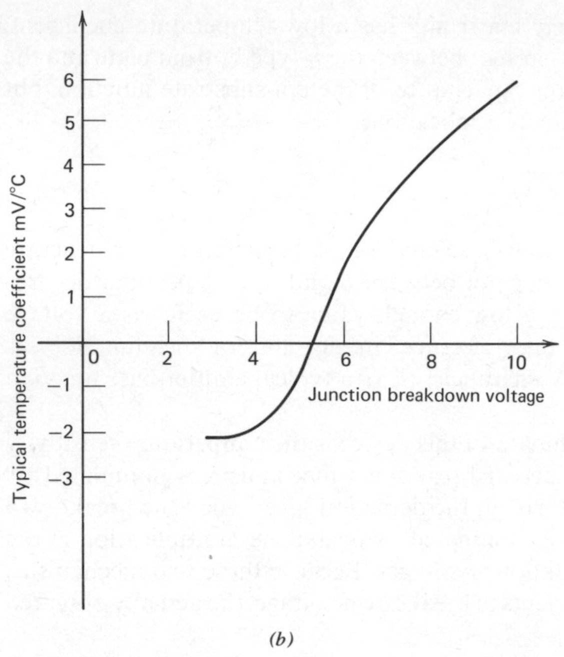

Fig. 2.35*b* Temperature coefficient of junction breakdown voltage as a function of breakdown voltage.

breakdown voltage has a temperature coefficient that varies with the value of breakdown voltage itself. The value of this is shown in Fig. 2.35*b*.

2.5.5 Junction Diodes

Junction diodes can be formed by various connections of the *npn* and *pnp* transistor structures, as illustrated in Fig. 2.36. When the diode is forward-biased in the diode connections *a*, *b*, and *d* of Fig. 2.36, the collector-base junction becomes forward biased as well. When this occurs, the collector-base junction injects holes into the epi region that can be collected by the reverse-biased epi-isolation junction or by other devices in the same isolation region. A similar phenomenon occurs when a transistor enters saturation. As a result, substrate currents can flow that can cause voltage drops in the high-resistivity substrate material and other epi-isolation junctions within the circuit can become inadvertantly forward biased. Thus the diode connections of Fig. 2.36*c* are usually preferable since they keep the base-collector junction at zero bias. These connections have the additional advantage of resulting in the smallest amount of minority charge storage within the diode under forward-bias conditions.

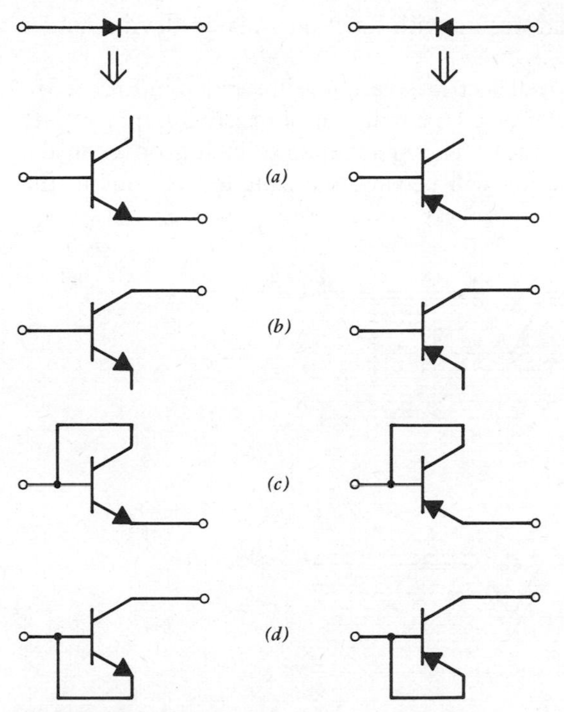

Fig. 2.36 Diode connections for *npn* and *pnp* transistors.

2.6 MODIFICATIONS TO THE BASIC PROCESS

The basic fabrication process consisting of six or seven mask steps and four diffusions is used for the vast majority of the analog integrated circuits produced. However, where special performance requirements exist, certain circuit performance parameters can be greatly improved by the addition of extra processing steps to produce special devices or characteristics.

2.6.1 Dielectric Isolation

We first consider a special isolation technique—dielectric isolation—that has been used in digital and analog integrated circuits that must operate at very high speed and/or must operate in the presence of large amounts of radiation. The objective of the isolation technique is to isolate electrically the collectors of the devices from each other with a layer of silicon dioxide rather than with a *pn* junction. This layer has much lower capacitance per unit area than a *pn* junction, and as a result the collector-substrate capacitance of the transistors is greatly reduced.

Also, the reverse photocurrent that occurs with junction isolated devices under intense radiation is eliminated.

The fabrication sequence used for dielectric isolation is illustrated in Figs. 2.37*a* to *d*. The starting material is a wafer of *n*-type material of resistivity appropriate for the collector region of the transistor. The first step is to etch grooves in the back side of the starting wafer, which will become the isolation regions in the

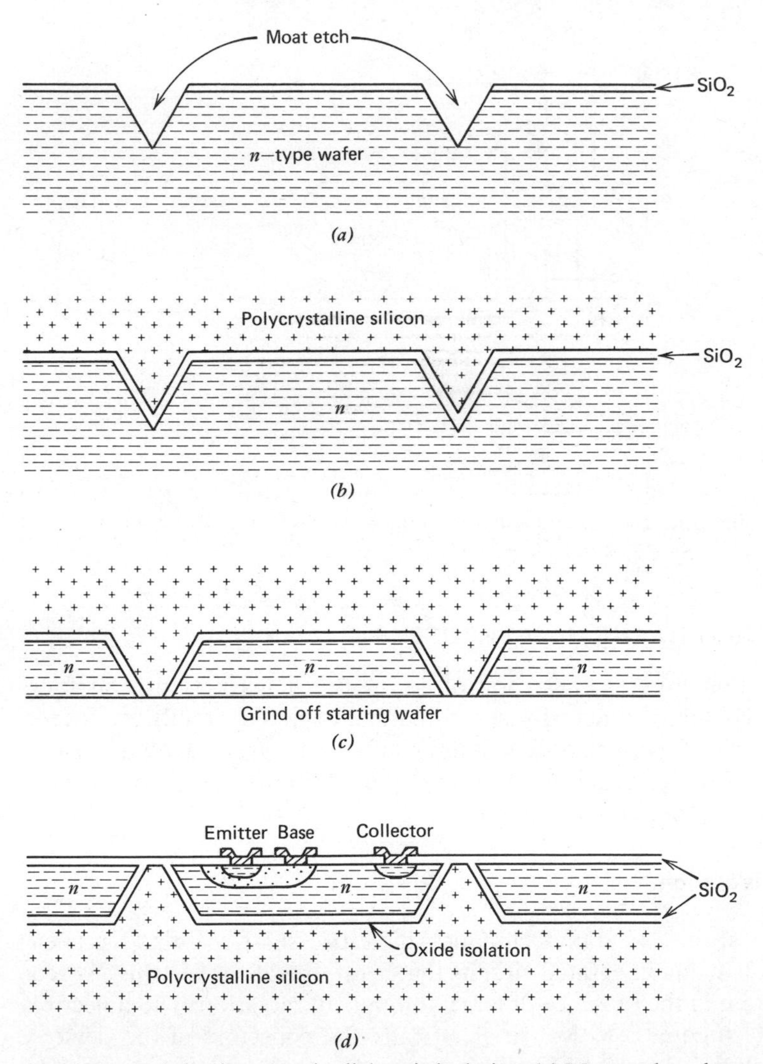

Fig. 2.37 Fabrication steps in dielectric isolation. (*a*) Moat etch on bottom of starting wafer. (*b*) Deposit polycrystalline silicon support layer. (*c*) Grind off starting wafer and polish. (*d*) Carry out standard process, starting with base mask.

finished circuit. These grooves are about 20 μ deep for typical analog circuit processing. This step, called *moat etch*, can be accomplished with a variety of techniques, including a preferential etch that allows precise definition of the depth of the moats. Next an oxide is grown on the surface and a thick layer of polycrystalline silicon is deposited on the surface. This layer will be the mechanical support for the finished wafer and thus must be on the order of 200 μ thick. Next the starting wafer is etched or ground from the top side until it is entirely removed except for the material left in the isolated islands between the moats, as illustrated in Fig. 2.37*c*. After the growth of an oxide, the wafer is ready for the rest of the standard process sequence. Note that the isolation of each device is accomplished by means of an oxide layer.

2.6.2 Compatible Processing for High-Performance Active Devices

Many specialized circuit applications require a particular type of active device other than the *npn* and *pnp* transistors which result from the standard process schedule. These include JFETs for analog switching and low-input-current amplifiers, MOSFETs for the same applications, high-beta ("superbeta") *npn* transistors for low-input-current amplifiers, and high-speed *pnp* transistors for fast analog circuits. The fabrication of these devices generally requires the addition of one or more mask steps to the basic fabrication process. We now describe the more important of these special structures.

Double-Diffused JFETs. Junction FETs can be fabricated on the same die with bipolar devices by a number of different techniques. If the base region of a standard bipolar integrated-circuit transistor is used as the channel of the JFET, as in the pinch resistor of Fig. 2.31, the resulting pinch-off voltage is much too high because of the relatively wide channel width of the resulting structure. A narrower-channel double-diffused device can be fabricated on the same die, however, by inserting an additional diffusion in the process as illustrated in Fig. 2.38*a*. In this process a separate n^+ diffusion is used to form the top gate of the JFET. The predeposition for this diffusion is made after the base diffusion. A short diffusion is then performed to partially redistribute these impurities into the wafer, and the standard emitter predeposition and diffusion are then performed. Since the gate predeposition was performed earlier, those impurities diffuse further, resulting in a pinched channel region in the JFET which is narrower than the base of the *npn* transistor. In physical structure, the JFET is conceptually identical to the pinch resistor shown in Fig. 2.31, but usually has different lateral dimensions. Similar structures can be formed by using two separate *p*-base diffusions and a single n^+ diffusion.

The resulting JFET structure has a typical pinch-off voltage of from 2 to 5 V, and a gate leakage current in the 10- to 100-pA range. This technique has been widely used to fabricate FET-input operational amplifiers. As an input device in such applications, the structure has several drawbacks. The channel width of the device is quite narrow, and a relatively small variation in conditions during

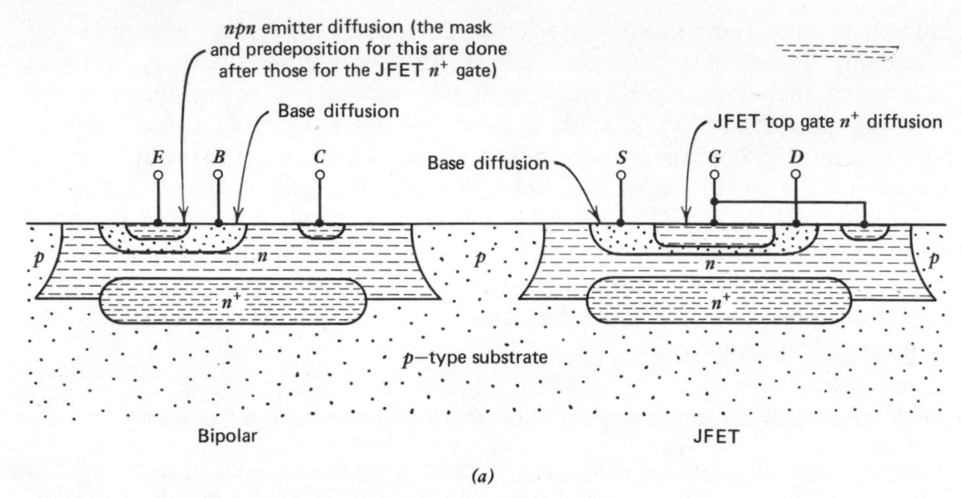

Fig. 2.38*a* Compatible double-diffused JFET structure.

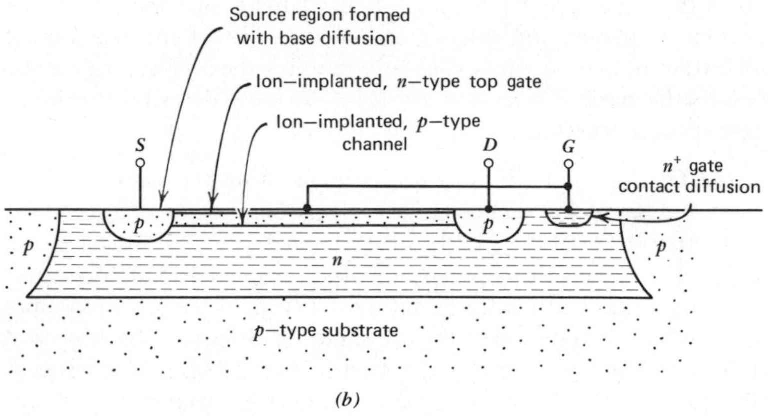

Fig. 2.38*b* Ion-implanted JFET structure.

the predeposition and diffusion will result in large variations in the channel width and the pinchoff voltage. Also, since the bottom gate is more lightly doped than the channel, most of the depletion layer associated with this junction occurs in the lightly doped material, the gate material. As a result, the bottom gate contributes little to the transconductance of the device. Another drawback is the fact that the breakdown voltage from drain to gate is limited to that of an emitter-base junction of a bipolar transistor, or about seven volts. Therefore these devices must be used in an input stage that limits the source-drain voltage applied to the JFETs. The application of these devices in operational amplifier circuits is discussed in Chapter 6.

Ion-Implanted JFETs. The drawbacks of the double-diffused structure can be overcome by fabricating the JFET with a sequence of ion implantation steps of the type discussed in Section 2.2.6.[3] A typical ion-implanted JFET structure is shown in Fig. 2.38*b*. One *p*-type implantation is used to form the channel, and since the implantation process is capable of very uniform and precise implacement of impurities, the number of impurity atoms in the channel profile is precisely controlled. This allows much more precise pinch-off voltage control and matching than is possible in a double-diffused JFET. Also, since the peak doping density in the channel region is only approximately 10^{16} atoms/cm^3, the drain-gate breakdown voltage can be kept high. The top gate is formed by implantation of a thin layer of *n*-type impurities near the surface, and since this layer does not extend very far into the channel region, the precise depth and doping density in this implantation does not strongly affect the pinch-off voltage or the I_{DSS}. As a result, pairs of these devices can be made with very closely matched characteristics. The significance of this fact in FET-input op amp design is discussed in Chapter 6. Typical parameter values for integrated-circuit JFETs are shown in Fig. 2.39.

Superbeta Transistors. Another approach to decreasing the input bias current in amplifiers is to increase the current gain of the input stage transistors.[9] Since a decrease in the base width of a transistor improves both the base transport factor and the emitter efficiency (see Section 1.3.1), the current gain increases as the

	Parameter	Diffused *p*-channel	Ion-implanted *p*-channel
	I_{DSS}	$-500\ \mu A\left(\frac{Z}{L}=25\right)$	$-300\ \mu A\left(\frac{Z}{L}=25\right)$
	V_P	2 V	1 V
	λ	$-10^{-2}\ V^{-1}$	$-10^{-2}\ V^{-1}$
	I_{GSS}	10^{-10} A	10^{-10} A
	BV_{GDO}	7 V	60 V
	g_{m0}	1 mA/V	0.6 mA/V
	r_d	100 Ω	50 Ω
Gate-source junction	C_{gs0}	2 pF	2 pF
	ψ_0	0.7 V	0.5 V
	n	0.33	0.33
Gate-drain junction	C_{gd0}	0.5 pF	0.5 pF
	ψ_0	0.7 V	0.5 V
	n	0.33	0.33
Gate-substrate junction	C_{gss0}	4 pF	4 pF
	ψ_0	0.52 V	0.52 V
	n	0.5	0.5

Fig. 2.39 Typical parameters for integrated-circuit JFETs.

base width is made smaller. Thus the current gain of the devices in the circuit can be increased by simply increasing the emitter diffusion time and narrowing the base width in the resulting devices. However, any increase in the current gain also causes a reduction in the breakdown voltage, BV_{CEO}, of the transistors. It was shown in Section 1.3.4 that

$$BV_{CEO} = \frac{BV_{CBO}}{\sqrt[n]{\beta}} \tag{2.27}$$

where BV_{CBO} is the plane breakdown voltage of the collector-base junction. Thus for a given epitaxial layer resistivity and corresponding collector-base breakdown voltage, an increase in the beta gives a decrease in BV_{CEO}. Because of this, it is not possible to simply increase the beta of all the transistors in the operational amplifier circuit by process modifications, because the circuit could not withstand the required operating voltage.

The problem of the tradeoff between current gain and breakdown voltage can be avoided by fabricating two different types of devices on the same die. One, the standard device, is similar to conventional transistors in structure. By inserting a second diffusion, however, high-beta devices are formed that are used as the input transistors of the ampliffer. A structure typical of such devices is shown in Fig. 2.40. As in the case of the double-diffused JFET, these devices may be made by utilizing the same base diffusion for both devices and using separate emitter diffusions, or by using two different base diffusions and the same emitter diffusion. Both techniques are used. Since the superbeta devices are not required to have a breakdown voltage of more than about one volt, they can be diffused to extremely narrow basewidths, giving current gain on the order of 2000 to 5000. At these basewidths, the actual breakdown mechanism is often no longer collector multiplication at all but is due to the depletion layer of the collector-base junction depleting the whole base region and reaching the emitter-base depletion layer.

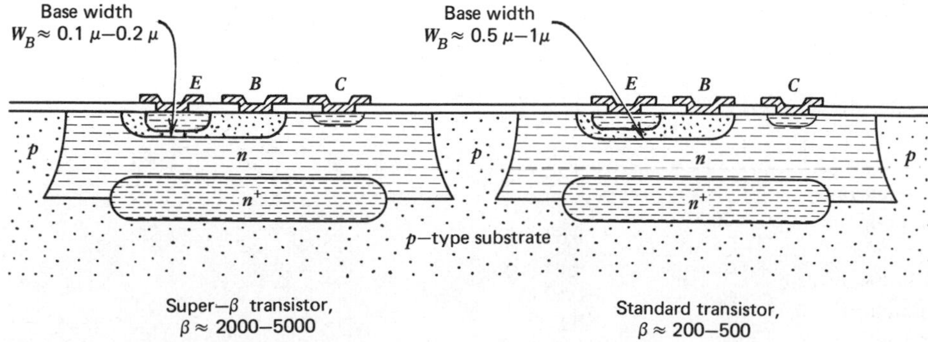

Fig. 2.40 Superbeta device structure.

This breakdown mechanism is called punchthrough. The application of these devices in op amp design is discussed in Chapter 6.

MOS Transistors. MOS transistors are useful in integrated-circuit design because they provide high performance analog switches, low-input-current amplifiers, and particularly because complex digital logic can be realized in a small area using MOS technology. The latter consideration is important since the partitioning of subsystems into analog and digital chips becomes more and more cumbersome as the complexity of the individual chips becomes greater.

Metal-gate p-channel MOS transistors can be formed with a standard analog IC process with one extra mask step.[10] If a capacitor mask was included in the original sequence then no extra mask steps are required. As illustrated in Fig. 2.41, the source and drain are formed in the epi material using the base diffusion. The capacitor mask is used to define the oxide region over the channel and the aluminum metallization forms the metal gate.

Although the process complexity of such a sequence is not great, the dependence of the MOS transistor properties on surface conditions requires a more careful control of surface conditions on the wafer than is typically required in standard bipolar IC processing. Thus the inclusion of these devices is more difficult than it at first appears.

Double-Diffused pnp Transistors. The limited frequency response of the lateral *pnp* transistor places a limitation on the high-frequency performance attainable with certain types of analog circuits. While it has generally proven to be possible to circumvent this problem in many cases by clever circuit design, the resulting circuit is often quite complex and costly. An alternate approach is to use a more complex process that produces a high-speed double-diffused *pnp* transistor with properties comparable with the *npn* transistor.[11] The process usually utilizes three additional mask steps and diffusions: one to form a lightly doped *p*-type region, which will be the collector of the *pnp*, one *n*-type diffusion to form the base of the *pnp*, and one *p*-type diffusion to form the emitter of the *pnp*. A typical resulting

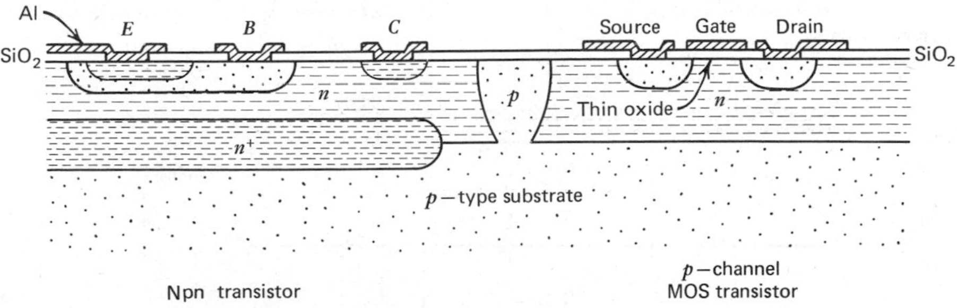

Fig. 2.41 Compatible *p*-channel MOS transistor.

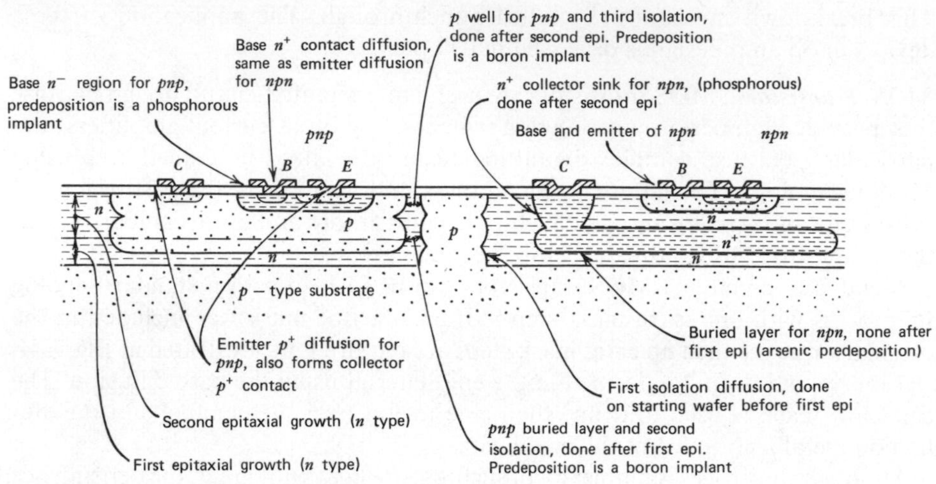

Fig. 2.42 Compatible double-diffused *pnp* process.

structure is shown in Fig. 2.42. This process requires 10 masking steps and two epitaxial growth steps.

2.6.3 High-Performance Passive Components

Diffused resistors have the drawback that they have high temperature coefficient, poor tolerance, and are junction isolated. The latter means that a parasitic capacitance is associated with each resistor, and exposure to radiation causes photocurrents to flow across the isolating junction. These drawbacks can be overcome by the use of thin-film resistors deposited on the top surface of the die over an

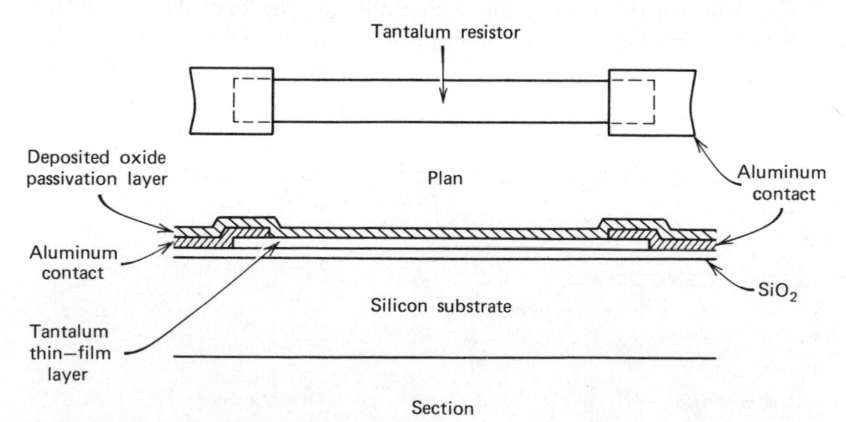

Fig. 2.43 Typical thin-film resistor structure.

	Nichrome	Tantalum	Cermet (Cr-SiO)
Range of sheet resistance Ω/□	10–1000	10–1000	30–2500
Temperature coefficient ppm/°C	±10 to ±150	±5 to ±200	±50 to ±150

Fig. 2.44 Properties of monolithic thin-film resistors.

insulating layer of oxide. After the resistor material itself is deposited, the individual resistors are defined in a conventional way using a masking step. They are then interconnected with the rest of the circuit using the standard aluminum interconnect process. The most common materials for the resistors are nichrome and tantalum, and a typical structure is shown in Fig. 2.43. The properties of the resulting resistors using these materials are summarized in Fig. 2.44.

2.7 ECONOMICS OF INTEGRATED-CIRCUIT FABRICATION

The principal reason for the growing pervasiveness of integrated circuits in systems of all types is the reduction in cost attainable through integrated-circuit fabrication. Proper utilization of the technology to achieve this cost reduction requires an understanding of the factors influencing the cost of an integrated circuit in completed, packaged form. In this section we discuss these factors.

2.7.1 Yield Considerations in Integrated-Circuit Fabrication

As discussed earlier in this chapter, integrated circuits are batch-fabricated on single wafers, each containing up to several thousand separate but identical circuits. At the end of the processing sequence, the individual circuits on the wafer are probed and tested prior to the breaking up of the wafer into individual dice. The percentage of the circuits that are electrically functional and within specification at this point is termed the wafer-sort yield, Y_{ws}, and is usually in the range of 10 to 90 percent. The nonfunctional units can result from a number of factors, but one major source of yield loss is point defects of various kinds that occur during the photoresist and diffusion operations. These defects can result from mask defects, pinholes in the photoresist, airborne particles that fall on the surface of the wafer, crystalline defects in the epitaxial layer, and so on. If such a defect occurs in the active region on one of the transistors or resistors making up the circuit, a nonfunctional unit usually results. The frequency of occurrence per unit of wafer area of such defects is usually dependent primarily on the particular fabrication process used and not on the particular circuit being fabricated. Generally speaking, the more mask steps and diffusion operations that the wafer is subjected to, the higher will be the density of defects on the surface of the finished wafer.

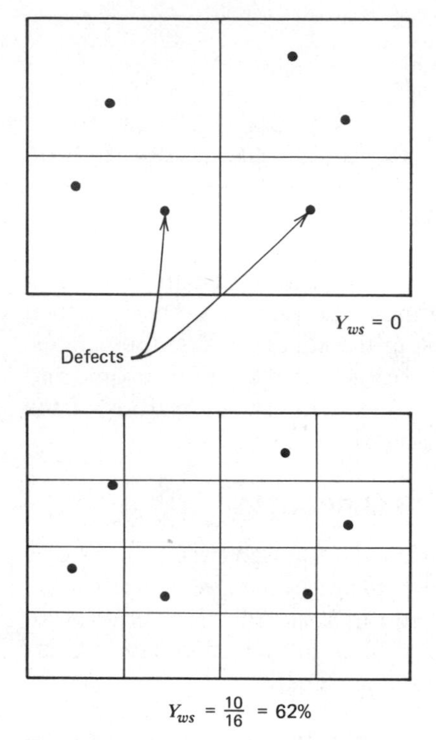

Fig. 2.45 Conceptual example of the effect of die size on yield.

The existence of these defects limits the size of the circuit that can be economically fabricated on a single die. Consider the two cases illustrated in Fig. 2.45 where two identical wafers with the same defect locations have been used to fabricate circuits of different area. The defect locations in both cases are the same, but for the case of the large die the wafer-sort yield would be zero. When the die size is cut to one-fourth of the original size, the wafer sort yield is 62 percent. This conceptual example illustrates the effect of die size on wafer-sort yield; a typical measured wafer-sort yield (Y_{ws}) versus die size is illustrated in Fig. 2.46. This particular curve is derived from yield data on a typical six-mask analog integrated-circuit process, and it should be emphasized that other types of processes with fewer mask steps will generally have higher yield at the larger die sizes. Also, the curve can be raised or lowered by more conservative design rules, more careful processing, and other factors. Uncontrolled factors such as testing problems, design problems in the circuit, and so on can cause the results for a particular circuit to deviate widely from the curve, but such a trend line is still very useful.

In addition to affecting the yield, the die size also affects the total number of dice that can be fabricated on a wafer of a given size. For the case of a three-inch

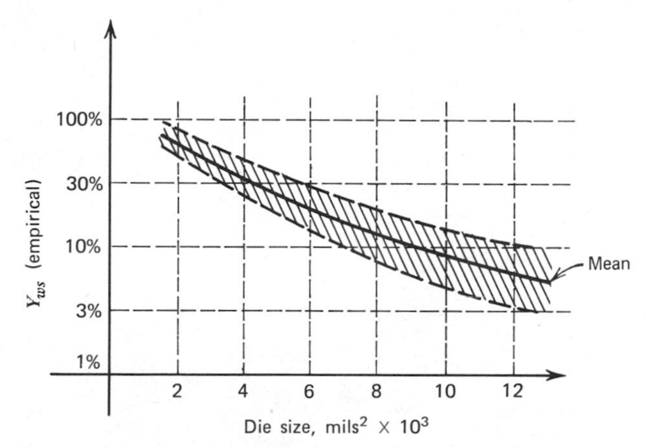

Fig. 2.46 Typically observed yield versus die size, six-mask bipolar process.

wafer, the total number of usable dice on the wafer, called the gross die per wafer, N, is plotted in Fig. 2.47 as a function of die size. The product of the gross die per wafer and the wafer sort yield gives the net good die per wafer, plotted in Fig. 2.48 for the yield curve of Fig. 2.46.

Once the wafer has undergone the wafer-probe test, it is separated into individual dice by sawing or scribing and breaking. The dice are visually inspected, sorted, and readied for assembly into packages. This step is termed die fab, and there is some loss of good dice in the process. Of the original electrically good dice on the wafer, some will be lost in the die fab process due to breakage and scratching of the surface. The ratio of the electrically good dice following die fab to the number of electrically good dice on the wafer before die fab is called the die fab yield, Y_{df}. The good dice are then inserted in a package and the electrical connections to the die are made with bonding wires to the pins on the package. The packaged circuits then undergo a final test, and there is generally some loss of functional

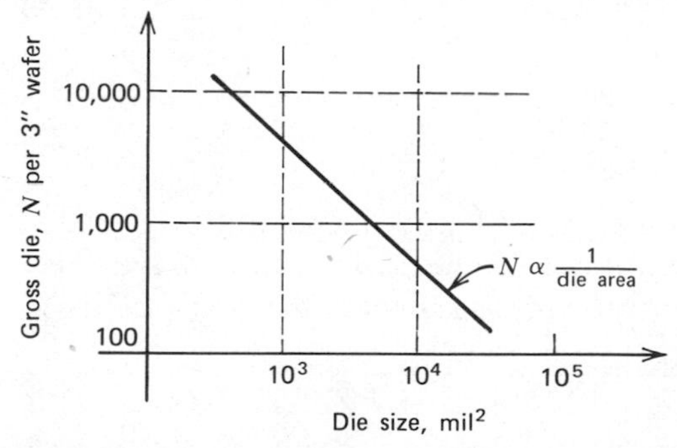

Fig. 2.47 Gross die/wafer versus die size, 3 in. wafer.

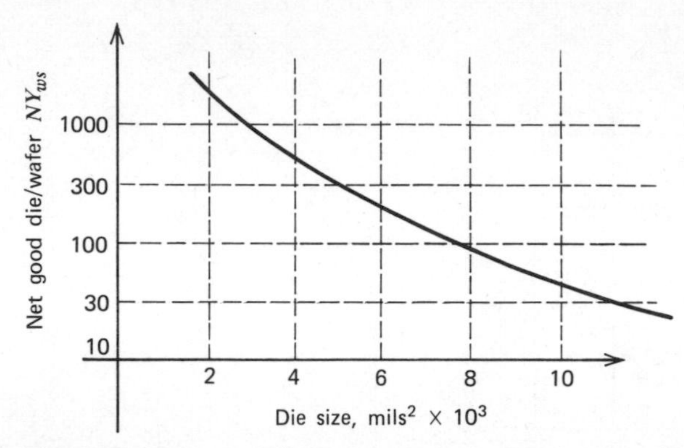

Fig. 2.48 Net good die/wafer versus die size for the yield curve of Fig. 2.46.

units because of improper bonding and handling losses. The ratio of the number of good units at final test to the number of good dice into assembly is called the final test yield, Y_{ft}.

2.7.2 Cost Considerations in Integrated-Circuit Fabrication

The principal direct costs to the manufacturer can be divided into two categories: those associated with fabricating and testing the wafer, called the wafer fab cost, C_w, and those associated with packaging and final testing the individual dice, called the packaging cost, C_p. If we consider the costs incurred by the complete fabrication of one wafer of dice, we first have the wafer cost itself, C_w. The number of electrically good dice that are packaged from the wafer is $NY_{ws}Y_{df}$. The total cost, C_t, incurred once these units have been packaged and tested, is:

$$C_t = C_w + C_p N Y_{ws} Y_{df} \tag{2.28}$$

The total number of good finished units, N_g, is

$$N_g = N Y_{ws} Y_{df} Y_{ft} \tag{2.29}$$

Thus the cost per unit is:

$$C = \frac{C_t}{N_g} = \frac{C_w}{N Y_{ws} Y_{df} Y_{ft}} + \frac{C_p}{Y_{ft}} \tag{2.30}$$

The first term in the cost expression is wafer fab cost, while the second is associated with assembly and final testing. This expression can be used to calculate the direct cost of the finished product to the manufacturer as shown in the example below.

EXAMPLE

Plot the direct fabrication cost as a function of die size for the yield curve of Fig. 2.46. Assume a wafer fab cost of \$20, a package and testing cost of \$0.06, a

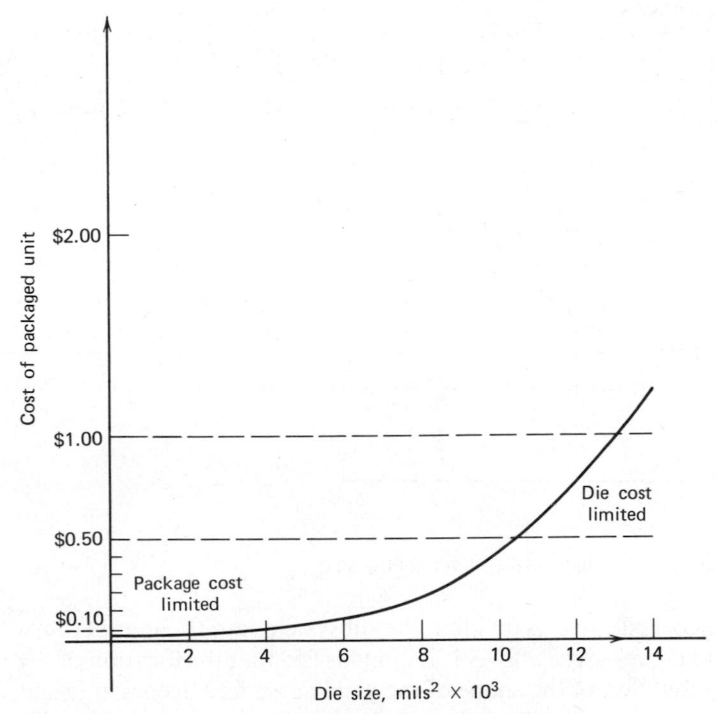

Fig. 2.49 Example cost of packaged unit versus die size.

die fab yield of 0.9 and a final test yield of 0.9. From (2.30)

$$C = \frac{\$20}{(NY_{ws})(0.81)} + \frac{0.06}{0.9} = \frac{\$24.69}{NY_{ws}} + 0.066$$

This cost is plotted in Fig. 2.49 using values of NY_{ws} from Fig. 2.48.

It is clear from this example that for small die sizes most of the direct cost comes from the assembly and packaging cost, while for large die sizes most of the cost comes from the wafer fab cost. This relationship is made clearer by considering the cost of the integrated circuit in terms of cost per unit area of silicon, as illustrated in Fig. 2.50. This curve is a plot of the ratio of packaged unit cost to die area, as a function of die area. The lowest cost per unit area results midway between the package-cost and die-cost limited regions. Thus the fabrication of excessively small or excessively large die is uneconomical in terms of utilization of the silicon die area at minimum cost. The significance of this curve is that if, for example, an analog subsystem containing a fixed amount of circuitry is to be partitioned into a number of separate chips, then the partitioning should be done so that the die size lies in the range where minimum cost results. If the subsystem requires, for example, 20,000 mil^2 of die area, then putting the entire subsystem on one

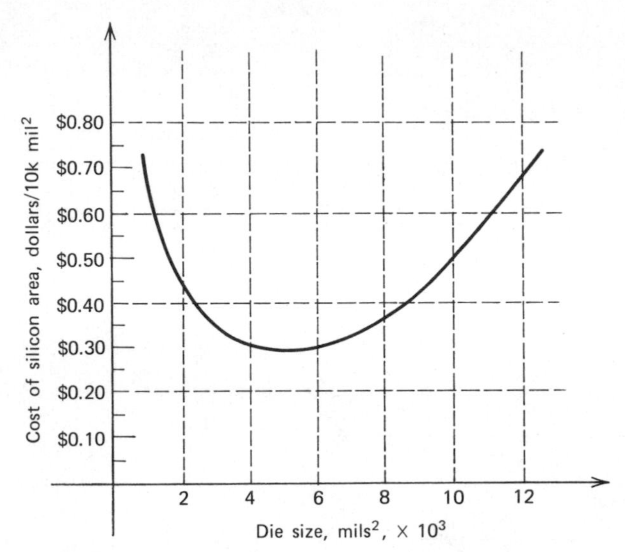

Fig. 2.50 Example cost per unit area of silicon versus die size.

chip would not be as cost effective as dividing the subsystem into three or perhaps two chips for the yield curves given above. It is emphasized that the location of the minimum is a strong function of the shape of the yield curve and occurs at larger die sizes for simpler processes. The minimum is also higher when more expensive packages are used; in the example, a $0.06 assembly cost was assumed, which reflects the cost of a plastic dual-in-line package.

The preceding discussion concerned only the direct costs to the manufacturer of the fabrication of the finished product; the actual selling price is much higher and reflects additional research and development, engineering, and selling costs. Many of these costs are fixed, however, so that the selling price of a particular integrated circuit tends to vary inversely with the quantity of the circuits sold by the manufacturer.

2.8 PACKAGING CONSIDERATIONS FOR INTEGRATED CIRCUITS

As illustrated in the last section, the finished cost of the integrated circuit is heavily dependent on the cost of the package in which it is encapsulated. In addition to the cost, the package also strongly affects two other important parameters. The first is the maximum allowable power dissipation in the circuit, and the second is the reliability of the circuit. We will discuss these limitations individually.

2.8.1 Maximum Power Dissipation

When power is dissipated within a device on the surface of the integrated circuit die, two distinct changes occur. First, the dissipated heat must flow away from the individual device through the silicon material, which gives rise to temperature gradients across the top surface of the chip. These gradients can strongly affect circuit performance, and their effect is discussed further in Chapter 6. Second, the heat must then flow out of the silicon material into the package structure, and then out of the package and to the ambient atmosphere. The flow of heat from the package to the ambient atmosphere can occur primarily by radiation and convection or, if the package is attached to a heat sink, can occur primarily by conduction. This flow of heat to the ambient environment causes the die as a whole to experience an increase in temperature, and in the steady state the average die temperature will be higher than the ambient temperature by an amount proportional to the power dissipation on the chip and the *thermal resistance* of the package.

The steady-state thermal behavior of the die/package structure can be analyzed approximately using the electrical analog shown in Fig. 2.51. In this analog, current is analogous to a flow of heat, and voltage is analogous to temperature. The current source represents the power dissipation on the integrated circuit die. The voltage drop across the resistance θ_{jc} represents the temperature drop between the surface of the chip and the outside of the case of the package. Finally, the drop across the resistor θ_{ca} represents the temperature drop between the outside of the case and the ambient atmosphere. This representation is only approximate since in reality the structure is distributed and neither the top surface of the die nor the outside of the case is isothermal. However, this equivalent circuit is useful for approximate analysis.

The resistance θ_{jc} is termed the *junction-to-case thermal resistance* of the package. This resistance varies from about 30°C/W for the TO-99 metal package to about 4°C/W for the TO-3 metal power package. These packages are shown in Fig. 2.52 along with the plastic dual-in-line package (DIP). The resistance θ_{ca} is termed the *case-to-ambient* thermal resistance. For the situation in which no heat

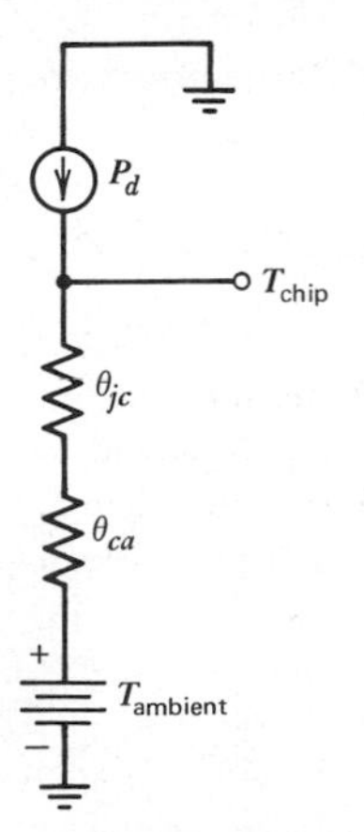

Fig. 2.51 Electrical analog for the thermal behavior of the die-package structure.

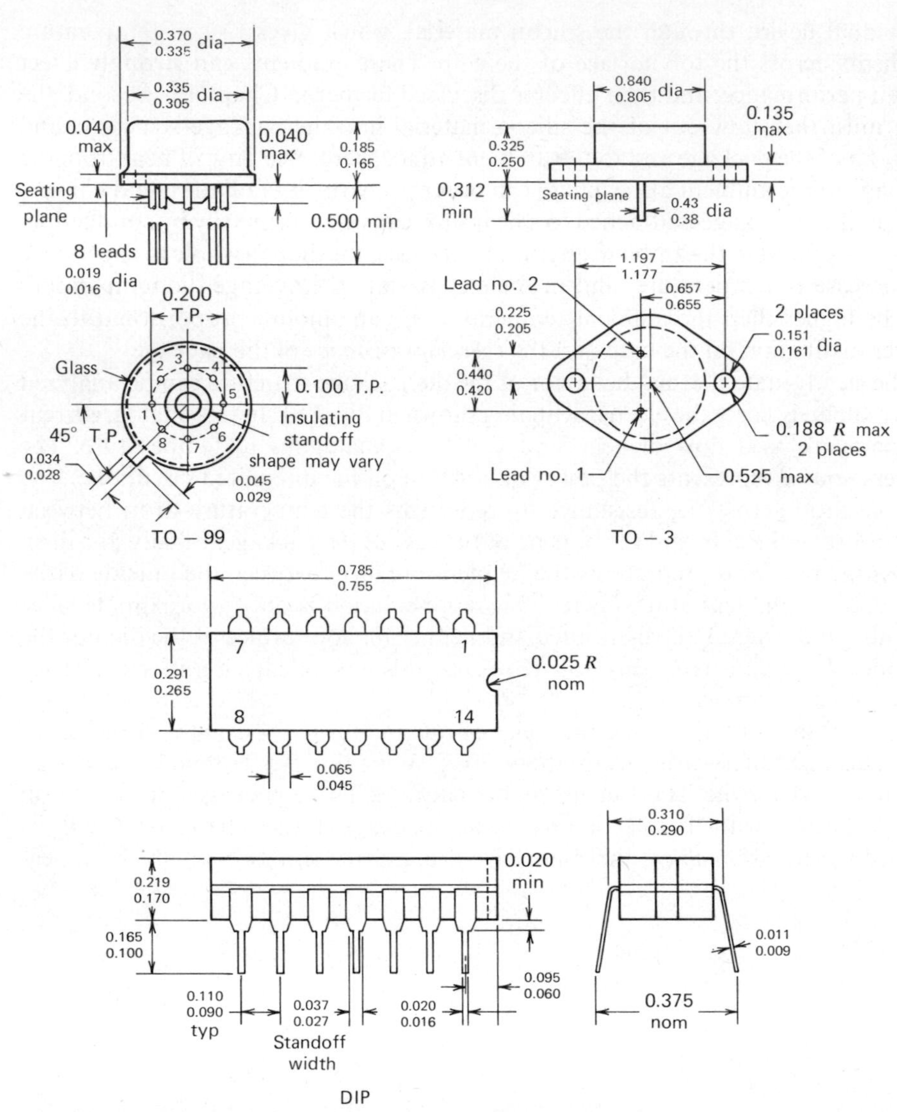

Fig. 2.52 TO-99, TO-3, dual-in-line (DIP) integrated-circuit packages. Dimensions are in inches. The TO-3 is used as shown for three-terminal ICs such as voltage regulators, and in versions with up to 10 leads when required. The basic TO-99 package shape is available in 3, 4, 6, 8, 10, and 12 lead versions. The DIP package is available in 8, 14, 16, 18, and 22 lead versions. It is available in both hermetically sealed ceramic and plastic versions.

sink is used, this resistance is determined primarily by the rate at which heat can be transferred from the outside surface of the package to the surrounding air. This is dependent on package size and on the rate of airflow around the package, if any. Because thermal radiation effects are present, the rate of heat transfer is not a linear function of case temperature, but the approximation is usually made that this thermal resistance is linear. For the case in which the surrounding air is still and no heat sink is used, the resistance θ_{ca} varies from about 100°C/W for the TO-99 to about 40°C/W for the TO-3.

For integrated circuits that dissipate large amounts of power, the use of a heat sink is often necessary to prevent excessive die temperatures. For this situation the case-to-ambient thermal resistance is determined by the heat sink. Heat sinks for use with integrated circuit packages vary from small finned structures having a thermal resistance of about 30°C/W, to massive structures achieving thermal resistances in the range of 2°C/W. Effective utilization of low-thermal-resistance heat sinks requires that the package and heat sink be in intimate thermal contact with each other; for TO-3 packages, special mica washers and heat sink grease are used to attach the package to the heat sink while maintaining electrical isolation.

The choice of package and heat sink for a particular circuit is dependent on the power to be dissipated in the circuit, the range of ambient temperatures to be encountered, and the maximum allowable chip temperature. These three quantities are related under steady-state conditions by the following expression:

$$T_{\text{chip}} = T_{\text{ambient}} + (\theta_{jc} + \theta_{ca})\, P_d$$

where T_{ambient} is the ambient temperature, T_{chip} is the chip temperature, and P_d is the power dissipation on the chip. For silicon integrated circuits, reliability considerations dictate that the chip temperature be kept below about 150°C, and this temperature is normally taken as the maximum allowable chip temperature. Thus once the maximum ambient temperature is known, the temperature drop across the series combination of θ_{jc} and θ_{ca} is specified. Once the power dissipation is known, the maximum allowable thermal resistance of the package and heat sink can be calculated.

EXAMPLE

What is the maximum permissible power dissipation in a circuit in a TO-99 package in still air when the ambient temperature is 70°C? 125°C?

Solution:

For the TO-99 in still air, $(\theta_{jc} + \theta_{ca}) = 30°\text{C/W} + 100°\text{C/W} = 130°\text{C/W}$

$$T_{\text{chip}} = T_{\text{ambient}} + (130°\text{C/W})\,(P_d)$$

For $T_{\text{ambient}} = 70°\text{C}$, we have

$$150°\text{C} = 70°\text{C} + (130°\text{C/Watt})\, P_{d\max}$$

and thus

$$P_{d\max} = 620 \text{ mW}$$

For $T_{\text{ambient}} = 125°\text{C}$, we have

$$150°\text{C} = 125°\text{C} + (130°\text{C/W})\, P_{d\max}$$

and thus

$$P_{d\max} = 190 \text{ mW}$$

2.8.2 Reliability Considerations in Integrated-Circuit Packaging

In applications where field servicing is difficult or impossible, or where device failure has catastrophic consequences, circuit reliability becomes a primary concern. The primary parameter describing circuit reliability is the mean time to failure of a sample of integrated circuits under a specified set of worst-case environmental conditions. The study of the various failure modes that can occur in integrated circuits under such conditions and the means to avoid such failures has evolved into a separate discipline, which is beyond the scope of this book. However, integrated circuit packages can be divided into two distinct groups from a reliability standpoint: those in which the die is in a hermetically sealed cavity and those in which the cavity is not hermetically sealed. The former group includes most of the metal can packages and the ceramic dual-in-line and flat packages. The latter group includes the plastic packages. The plastic packages are less expensive to produce and are as reliable as the hermetic packages under mild environmental conditions. The hermetic packages are generally more expensive to produce, but are more reliable under adverse environmental conditions, particularly in the case of high temperature/high humidity conditions.

PROBLEMS

2.1 What impurity concentration corresponds to 1 Ω-cm resistivity in *p*-type silicon? In *n*-type silicon?

2.2 What is the sheet resistance of a layer of 1 Ω-cm material that is 5 μ thick?

2.3 Consider a hypothetical layer of silicon that has an *n*-type impurity concentration of 10^{17} cm^{-3} at the top surface, and in which the impurity concentration decreases exponentially with distance into the silicon. Assume the concentration has decreased to $1/e$ of its surface value at a depth of 0.5 μ, and that the impurity concentration in the sample before the insertion of the *n*-type impurities was 10^{15} cm^{-3} *p*-type. Determine the depth below the surface of the *pn* junction that results and determine the sheet resistance of the *n*-type layer. Assume a constant electron mobility of 800 cm^2/V-sec. Assume that the width of the depletion layer is negligible.

2.4 A diffused resistor has a length of 200 μ and a width of 5 μ. The sheet resistance of the base diffusion is 100 Ω/□ and the emitter diffusion is 5 Ω/□. The base pinched layer has a sheet resistance of 5 kΩ/□. Determine the resistance of the resistor if it is an emitter diffused, base diffused, or pinch resistor.

2.5 A base-emitter voltage of from 520 to 580 mV is measured on a test *npn* transistor structure with 10 μA collector current. The emitter dimensions on the test transistor are 100 $\mu \times$ 100 μ. Determine the range of values of Q_B that this data implies. Use this information to calculate the range of values of sheet resistance that will be observed in the pinch resistors in the circuit. Assume a constant electron diffusivity, $\bar{D}_n$, of 13 cm^2/sec, and a constant hole mobility of 150 cm^2/V-sec. Assume that the width of the depletion layer is negligible.

2.6 Estimate the series base resistance, series collector resistance, I_S, base-emitter capacitance, base-collector capacitance, and collector-substrate capacitance of the high-current *npn* transistor structure shown in Fig. 2.53. This structure is typical of those used as the

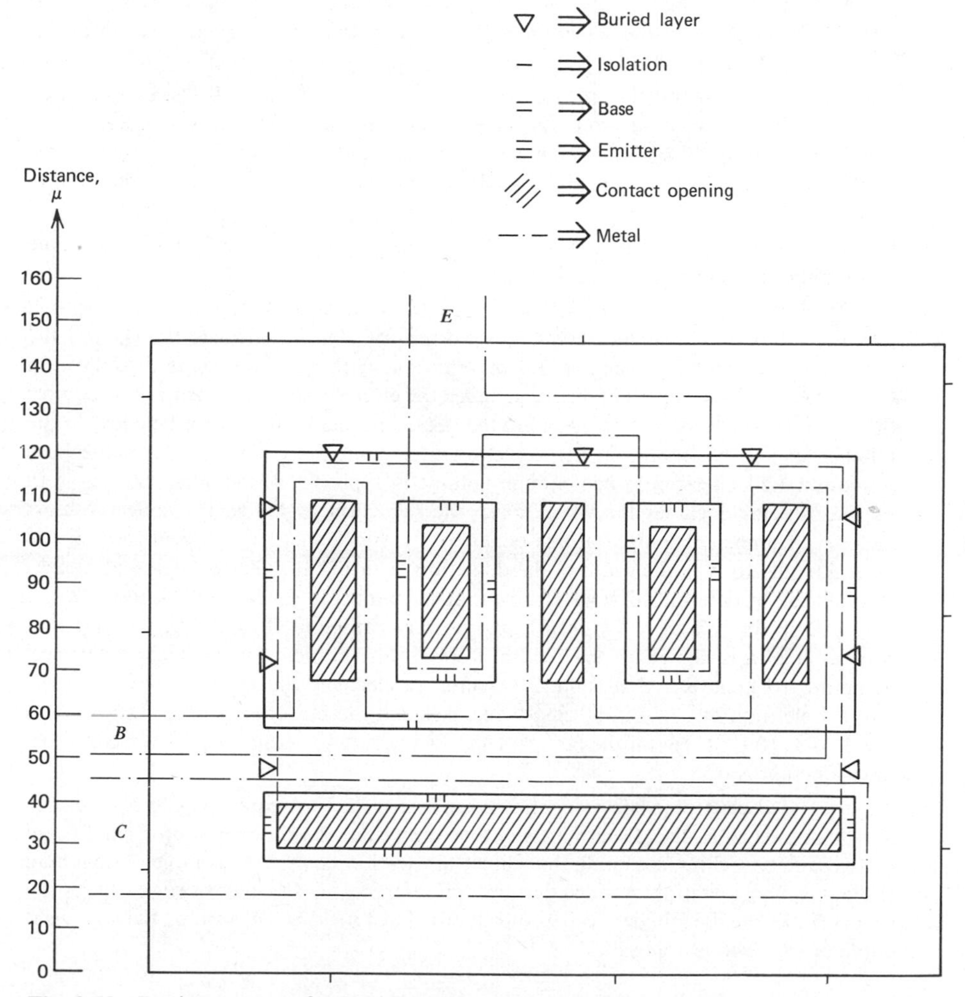

Fig. 2.53 Device structure for Problem 2.6.

output transistor in operational amplifiers that must supply up to about 20 mA. Assume a doping profile as shown in Fig. 2.16.

2.7 If the lateral *pnp* structure of Fig. 2.23*a* is fabricated with an epi layer resistivity of 0.5 Ω-cm, determine the value of collector current at which the current gain begins to fall off. Assume a diffusivity for holes of: $\bar{D}_P = 10\ \text{cm}^2/\text{sec}$. Assume a base width of 8 μ.

2.8 The substrate *pnp* of Fig. 2.26*a* is to be used as a test device to monitor epitaxial layer thickness. Assume that the flow of minority carriers across the base is vertical, and that the width of the emitter-base and collector-base depletion layers are negligible. Assume that by independent measurement the epi layer resistivity is known to be 2 Ω-cm. The base-emitter voltage is observed to vary from 525 to 560 mV over several wafers at a collector current of 10 μA. What range of epitaxial layer thickness does this imply? What is the corresponding range of sheet resistance that will be observed in the epitaxial pinch resistors? Assume a hole diffusivity of 10 cm^2/sec, and an electron mobility of 800 cm^2/V-sec. Neglect the depletion layer thickness. Assume a junction depth of 3 μ for the base diffusion.

2.9 Calculate the total parasitic junction capacitance associated with a 10-kΩ base-diffused resistor if the base sheet resistance is 100 Ω/□ and the resistor width is 6 μ. Repeat for a resistor width of 12 μ. Assume the doping profiles are as shown in Fig. 2.16. Assume the clubheads are 26 μ × 26 μ, and that the junction depth is 3 μ. Account for sidewall effects.

2.10 For the substrate *pnp* structure shown in Fig. 2.26*a*, calculate I_S, C_{je}, C_μ, and τ_F. Assume the doping profiles are as shown in Fig. 2.16.

2.11 A base-emitter voltage of 480 mV is measured on a super-β test transistor with a 100 μ × 100 μ emitter area at a collector current of 10 μA. Calculate the Q_B and the sheet resistance of the base region. Estimate the punchthrough voltage in the following way. When the base depletion region includes the entire base, charge neutrality requires that the number of ionized acceptors in the depletion region in the base be equal to the number of ionized donors in the depletion region on the collector side of the base (Equation 1.2). Therefore, when enough voltage is applied that the depletion region in the base region includes the whole base, the depletion region in the collector must include a number of ionized atoms equal to Q_B. Since the density of these atoms is known (equal to N_D), the width of the depletion layer in the collector region at punch-through can be determined. If we assume that the doping in the base, N_A, is much larger than that in the collector, N_D, then Equation 1.14 can be used to find the voltage that will result in this depletion layer width. Repeat this problem for the standard device, assuming a V_{BE} measured at 560 mV. Assume an electron diffusivity $\bar{D}_n$ of 13 cm^2/sec, and a hole mobility, $\bar{\mu}_p$ of 150 cm^2/V-sec. Assume the epi doping is $10^{15}\ \text{cm}^{-3}$. Use $\epsilon = 1.04 \times 10^{-12}$ F/cm for the permittivity of silicon. Also, assume ψ_0 for the collector-base junction is 0.55 V.

2.12 An integrated electronic subsystem has been determined to require 12,000 square mils of silicon area to realize. Determine whether the system should be put on a single chip or two different chips, assuming that direct fabrication cost of the complete subsystem is the only consideration. Assume that the wafer fab cost is $30, the package and testing cost is $0.06, the die fab yield is 0.9, and the final test yield is 0.9. Assume that the yield curve of Fig. 2.46 is applicable.

2.13 Determine the direct fabrication cost of an integrated circuit that is 80 mils × 80 mils in size. Assume a wafer fab cost of $40, packaging cost of $0.09, final test yield of 0.9, and die fab yield of 0.9. Assume that the yield curve of Fig. 2.46 is applicable.

2.14 Figure 2.50 illustrates the cost of silicon per unit area as a function of die size for a typical set of package and wafer-fab cost constraints. Construct additional curves of the same type for the following cases:
   (a) Device must be packaged in a multilead hermetic power package. Package cost increased from $0.06 to $0.50.
   (b) Manufacturing changed from 3 to 4 in. wafers. Wafer-fab cost stays the same, but NY_{ws} is increased by a factor of 1.8 for any die size. Determine the optimum die size in each case.

2.15 A TO-99 package is used with a heat sink, providing a thermal resistance from case to ambient of 20°C/W. The junction-to-case thermal resistance of the TO-99 is 30°C/W. Plot the maximum allowable power dissipation in this package as a function of ambient temperature, assuming a maximum die temperature of 150°C.

REFERENCES

1. A. S. Grove. *Physics and Technology of Semiconductor Devices.* Wiley, New York, 1967.
2. T. Kamins and R. S. Muller. *Semiconductor Devices.* Wiley, New York, 1977.
3. R. W. Russell and D. D. Culmer. "Ion-Implanted JFET-Bipolar Monolithic Analog Circuits," Digest of Technical Papers, 1974 International Solid-State Circuits Conference, Philadelphia, Pennsylvania, pp. 140–141, February 1974.
4. D. J. Hamilton and W. G. Howard. *Basic Integrated Circuit Engineering.* McGraw-Hill, New York, 1975.
5. R. M. Burger and R. P. Donovan. *Fundamentals of Silicon Integrated Device Technology.* Vol. 2, pp. 134–136, Prentice-Hall, New Jersey, 1968.
6. R. J. Whittier and D. A. Tremere. "Current Gain and Cutoff Frequency Falloff at High Currents," *IEEE Trans. Electron Devices*, Vol. ED-16, pp. 39–57, January 1969.
7. H. J. DeMan. "The influence of Heavy Doping on the Emitter Efficiency of a Bipolar Transistor," *IEEE Transactions on Electron Devices*, Vol. ED-18, pp. 833–835, October 1971.
8. H. C. Lin. *Integrated Electronics.* Holden-Day, San Francisco, 1967.
9. R. J. Widlar, "Design Techniques for Monolithic Operational Amplifiers," *IEEE Journal of Solid-State Circuits*, Vol. SC-4, pp. 184–191, August 1969.
10. K. R. Stafford, P. R. Gray, and R. A. Blanchard. "A Complete Monolithic Sample/Hold Amplifier," *IEEE Journal of Solid-State Circuits*, Vol. SC-9, pp. 381–387, December 1974.
11. P. C. Davis, S. F. Moyer, and V. R. Saari. "High Slew Rate Monolithic Operational Amplifier Using Compatible Complementary *pnp*'s," *IEEE Journal of Solid-State Circuits*, Vol. SC-9, pp. 340–346, December 1974.
12. E. M. Conwell. "Properties of Silicon and Germanium," *Proc. IRE*, Vol. 46, pp. 1281–1300, June 1958.
13. J. C. Irvin. "Resistivity of Bulk Silicon and of Diffused Layers in Silicon," *Bell System Tech. Journal*, Vol. 41, pp. 387–410, March 1962.
14. H. R. Camenzind. *Electronic Integrated Systems Design.* Van Nostrand Reinhold, New York, 1972. Copyright © 1972 Litton Educational Publishing, Inc. Reprinted by permission of Van Nostrand Reinhold Company.

CHAPTER 3

SINGLE-TRANSISTOR AND TWO-TRANSISTOR AMPLIFIERS

The technology used to fabricate integrated circuits presents a unique set of component cost constraints to the circuit designer. The most cost-effective circuit approach to accomplish a given function may be quite different when the realization of the circuit is to be in monolithic form as opposed to discrete transistors and passive elements.[1] As an illustration, consider the two realizations of a three-stage audio amplifier shown in Figs. 3.1 and 3.2. The first reflects a cost-effective solution in the context of discrete-component circuits, since passive components such as resistors and capacitors are less expensive than the active components, the transistors. Hence, the circuit contains a minimum number of transistors and the interstage coupling is accomplished with capacitors. However, for the case of monolithic construction, the principal determining factor in cost is the die area used. Capacitors of the values used in the discrete-component circuit are not feasible and would have to be external to the chip. This would increase the pin count of the package, which increases cost; a high premium is then placed on

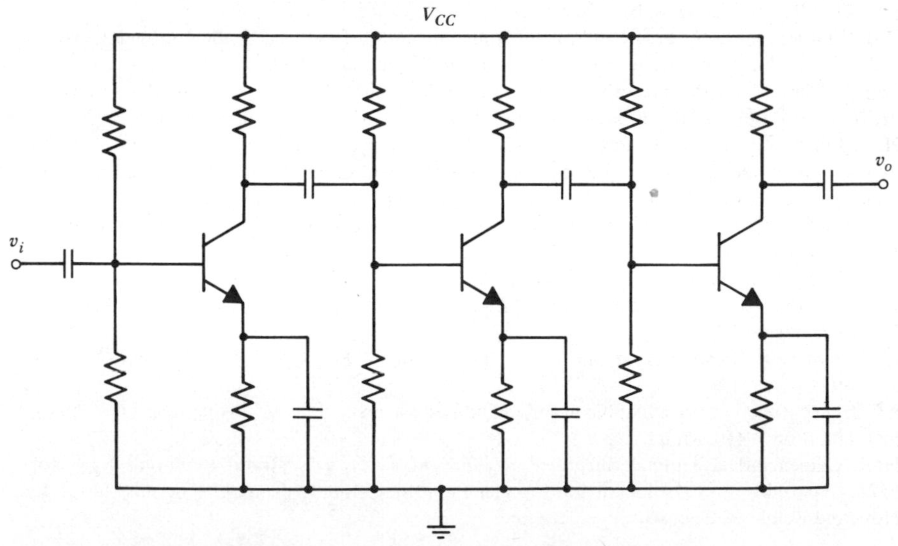

Fig. 3.1 Typical discrete-component realization of an audio amplifier.

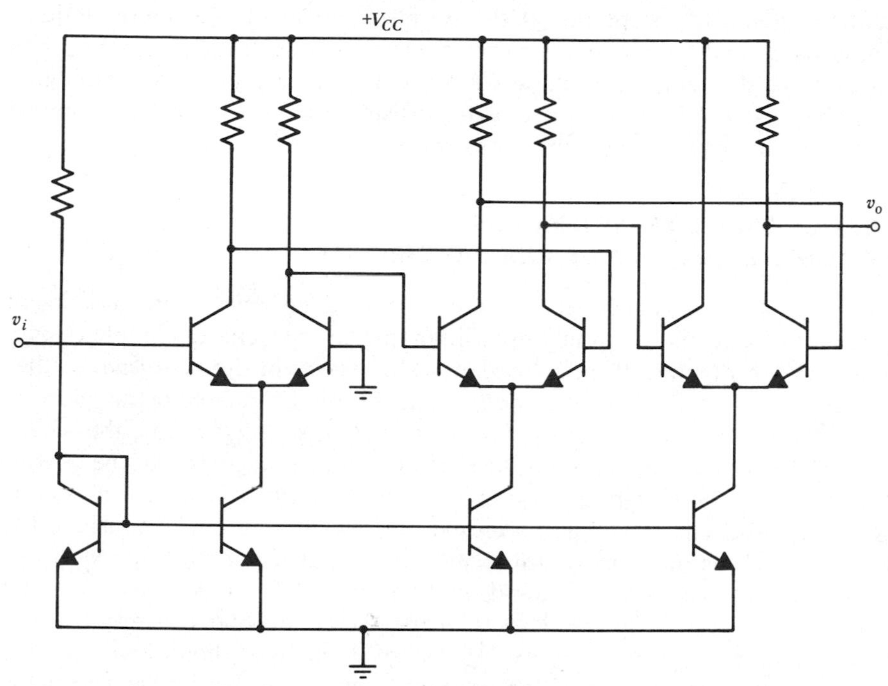

Fig. 3.2 Typical integrated-circuit realization of an audio amplifier.

eliminating large capacitors. A dc-coupled circuit realization is therefore very desirable. A second constraint is that the *cheapest* component that can be fabricated in the integrated circuit is one that takes up the least area, usually a transistor. Thus a circuit realization that contains the minimum possible total resistance while using more active components may be optimum. The circuit of Fig. 3.2 reflects these constraints[2,3]. It uses many more transistors, has less total resistance, and has no coupling capacitors. Emitter-coupled pairs are used to allow direct coupling between stages, while transistor current sources provide biasing without large amounts of resistance.

The next three chapters analyze the various circuit configurations encountered in linear integrated circuits. In discrete component circuits, the number of transistors is usually minimized, and in the analysis of such circuits it is usually best to proceed by regarding each individual transistor as a "stage," and to analyze the circuit as a collection of single transistor stages. A typical monolithic circuit, however, contains a large number of transistors that perform many functions, both passive and active. Thus it is more useful to regard the monolithic circuit as a collection of "subcircuits," which perform specific functions, rather than as a collection of individual transistor stages. In this chapter we first consider the dc and low-frequency properties of the simplest subcircuits—common-emitter, common-base, and common-collector single-transistor amplifiers. We then consider three different two-transistor subcircuits that are useful as amplifying stages. The most widely used of these two-transistor circuits is the emitter-coupled pair, which is analyzed extensively in this chapter.

3.1 DEVICE MODEL SELECTION FOR APPROXIMATE ANALYSIS OF ANALOG CIRCUITS

Much of this book is concerned with the salient performance characteristics of a variety of subcircuits commonly used in analog circuits, and of complete functional blocks made up of these subcircuits. The aspects of the performance that are of interest are the dc currents and voltages within the circuit, the effect of mismatches in device characteristics on these voltages and currents, the small-signal, low-frequency input and output resistance, and voltage gain of the circuit. In later chapters the high-frequency small-signal behavior of the circuit is considered. The subcircuit or circuit under investigation is often one of considerable complexity, and the most important single principle that must be adhered to in order to achieve success in the hand analysis of such circuits is selecting the simplest possible model for the devices within the circuit that will result in the required accuracy. For example, in the case of dc analysis, hand analysis of a complex circuit is greatly simplified by neglect of certain aspects of transistor behavior, such as the output resistance, which may result in a 10- to 20-percent error in the dc currents calculated. The principal objective of hand analysis, however, is to obtain the intuitive understanding of factors affecting circuit behavior so that an iterative design procedure resulting in improved performance can be carried out. The performance of the circuit can at any point in this cycle be determined precisely by computer simulation, but this does not yield the intuitive understanding necessary for design.

Unfortunately, no specific rules can be formulated regarding the selection of the simplest device model for analysis. For example, in the dc analysis of biasing circuits, it is often sufficient to assume a constant base-emitter voltage and neglect the output resistance of the transistor. However, certain bias circuits depend on the nonlinear relation between the collector current and base-emitter voltage to control the bias current, and in the analysis of these circuits the assumption of a constant V_{be} will result in gross errors in the results. In analyzing the dc transfer characteristic of the active-load stage in Chapter 4, the output resistance must be considered in order to obtain meaningful results. A similar situation exists for the small-signal case. A key step in every analysis, then, is to inspect the circuit to determine what aspects of the behavior of the transistor strongly affect the performance of the circuit, and then simplify the model to include only those aspects. This step in the procedure is emphasized in the following chapters.

3.2 BASIC SINGLE-TRANSISTOR AMPLIFIER STAGES

The bipolar transistor is capable of providing useful amplification of signals when used in three different configurations. In the common-emitter configuration, the signal is applied to the base of the transistor and the amplified output is taken from the collector. In the common-base configuration, the signal is applied to the emitter and the output is taken from the collector. In the common-collector, or emitter-follower, configuration, the signal is applied to the base and the output is taken from the emitter. Each of these configurations provides a unique combination of input resistance, output resistance, voltage gain, and current gain. The analysis of complex multistage amplifiers can, in many instances, be reduced to the analysis of a number of single-transistor stages of these types. In this section we consider the dc and small-signal ac behavior of the resistively loaded common-emitter amplifier, and then consider the small-signal behavior of the common-base and common-collector configurations.

3.2.1 Common-Emitter Configuration

The resistively loaded common-emitter (CE) amplifier configuration is shown in Fig. 3.3. The resistor, R_C, represents the collector load resistance. The short horizontal line labeled V_{CC} at the top end of R_C implies that a voltage source of value V_{CC} is connected between that point and ground. This symbology will be used throughout the book. We first calculate the dc transfer characteristic of the amplifier as the input voltage is increased in the positive direction from zero. We assume that the base of the transistor is driven by a voltage source of value V_i. When V_i is zero, the transistor is in the cutoff state and no collector current flows other than the leakage current, I_{CO}. As the input voltage is increased, the

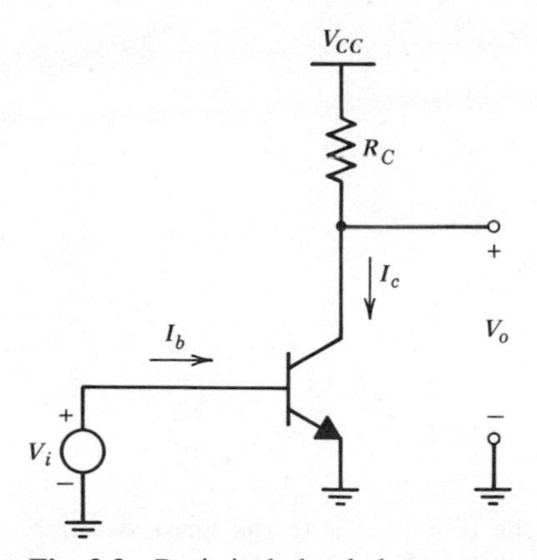

Fig. 3.3 Resistively loaded common-emitter amplifier.

transistor enters the forward-active region, and the collector current is given by:

$$I_c = I_S \exp \frac{V_i}{V_T} \tag{3.1}$$

The equivalent circuit for the amplifier when the transistor is in the forward-active region was derived in Chapter 1 and is repeated in Fig. 3.4. Because of the exponential relationship between I_c and V_{be}, the value of the collector current is very small until the input voltage reaches approximately 0.5 V. As long as the transistor is in the forward-active region, the base current is equal to the collector current divided by β_F, or:

$$I_b = \frac{I_c}{\beta_F} = \frac{I_S}{\beta_F} \exp \frac{V_i}{V_T} \tag{3.2}$$

The output voltage is equal to the supply voltage, V_{CC}, minus the voltage drop across the collector resistor.

$$V_o = V_{CC} - I_c R_C = V_{CC} - R_C I_S \exp \frac{V_i}{V_T} \tag{3.3}$$

When the output voltage approaches zero, the collector-base junction of the transistor becomes forward biased and the device enters saturation. Once this occurs, the output voltage takes on a nearly constant value and decreases no

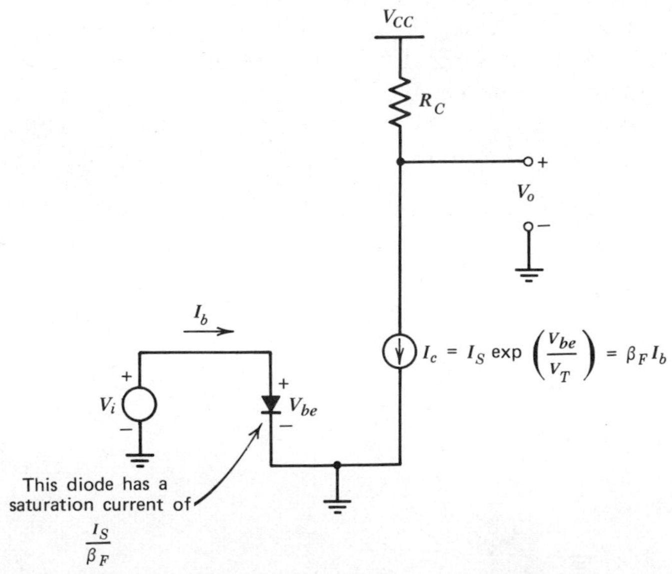

Fig. 3.4 Large-signal equivalent circuit valid when the transistor is in the forward-active region. The saturation current of the equivalent base-emitter diode is I_S/β_F.

further. The base current, however, continues to increase with further increases in V_i beyond saturation. In a practical application, the limited current available from the signal source driving the amplifier would prevent any further increases in V_i because of the large base current that would be required. The output voltage and the base current are plotted as a function of the input voltage in Fig. 3.5. Note that when the device is in the forward-active region, small changes in the input voltage can give rise to large changes in the output voltage. The circuit thus provides *voltage gain.* We now proceed to calculate the voltage gain in the forward-active region.

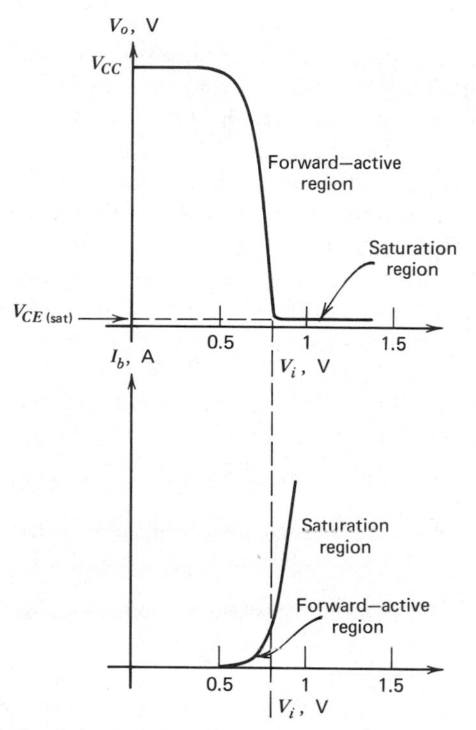

Fig. 3.5 Output voltage and base current as a function of V_i for the common-emitter circuit.

Whereas in calculating the dc transfer characteristic the large-signal model of the transistor was appropriate, calculation of incremental performance parameters such as voltage gain are most efficiently carried out using the small-signal hybrid-π model for the transistor developed in Chapter 1. The small-signal equivalent circuit for the common-emitter amplifier is shown in Fig. 3.6. Here we have neglected r_b, assuming that it is much smaller than r_π. We have also neglected r_μ. This equivalent circuit does not include the resistance associated with the

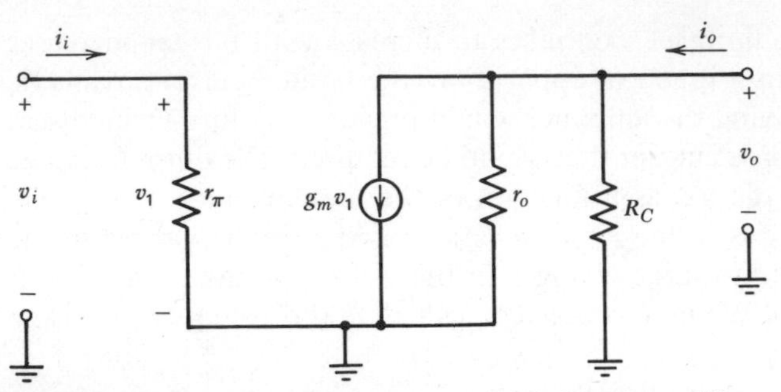

Fig. 3.6 Small-signal equivalent circuit for the CE amplifier.

source from which the amplifier is driven and does not include the resistance of the load connected to the amplifier output. The collector resistor, R_C, is included because it is usually present in some form as a biasing element. Our objective is the characterization of the amplifier alone so that the voltage gain can then be calculated under arbitrary conditions of loading at the input and output. In order to accomplish this, we will calculate the small-signal input resistance, transconductance, and output resistance of the circuit. If the circuit is approximately unilateral as is the case for the common-emitter amplifier, then the performance of the circuit when used with a particular combination of load resistance and source resistance can be evaluated using the admittance-parameter two-port equivalent network for the circuit illustrated in Fig. 3.7.

The transconductance, G_m, is the change in short-circuit output current per unit change of input voltage, and for the CE amplifier is given by

$$G_m = g_m \tag{3.4}$$

The input resistance is the Thévenin equivalent resistance seen looking into the input, or:

$$R_i = r_\pi \tag{3.5}$$

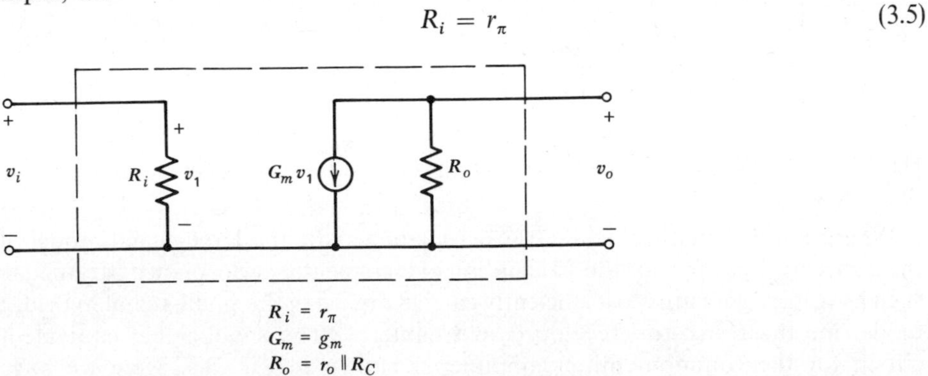

Fig. 3.7 Two-port representation of the small-signal properties of the common-emitter amplifier.

The output resistance is the Thévenin equivalent resistance seen looking into the output with the input shorted, or:

$$R_o = R_C \| r_o \tag{3.6}$$

The amplifier can thus be represented by the two-port network shown in Fig. 3.7. If no load is attached to the output, the output voltage is given by:

$$v_o = -G_m v_i R_o = -g_m (r_o \| R_C) v_i \tag{3.7}$$

Thus the *open circuit*, or *unloaded*, *voltage gain* is:

$$a_v = \frac{v_o}{v_i} = -g_m (r_o \| R_C) \tag{3.8}$$

Note that if the collector load resistor, R_C, is made very large, then the voltage gain, a_v, becomes:

$$\lim_{R_C \to \infty} a_v = -g_m r_o = -\frac{1}{\eta} \tag{3.9}$$

where η is defined in (1.113). For typical *npn* devices this value of gain is approximately 5×10^3 and represents the maximum low-frequency voltage gain obtainable from the transistor. A second parameter of interest, which can be directly calculated from the two-port representation of the amplifier, is the *short-circuit current gain*, a_i. This is the ratio of i_o to i_i when the output is shorted. For the common-emitter amplifier,

$$a_i = \frac{i_o}{i_i} = \frac{G_m v_i}{\dfrac{v_i}{R_i}} = g_m r_\pi = \beta_0 \tag{3.10}$$

EXAMPLE

(a) Find the input resistance, output resistance, voltage gain, and current gain of the common-emitter amplifier in Fig. 3.8*a*. Assume that $I_C = 100\ \mu\text{A}$, $\beta_0 = 100$, $r_b = 0$, and $r_o = \infty$.

Solution:

$$R_{\text{in}} = r_\pi = \frac{\beta_0}{g_m} = \frac{100}{100\ \mu\text{A}/26\ \text{mV}} = 26\ \text{k}\Omega$$

$$R_{\text{out}} = R_C = 5\ \text{k}\Omega$$

$$a_v = -g_m R_C = -\left(\frac{100\ \mu\text{A}}{26\ \text{mV}}\right)(5\ \text{k}\Omega) = (-19.2)$$

$$a_i = \beta_0 = 100$$

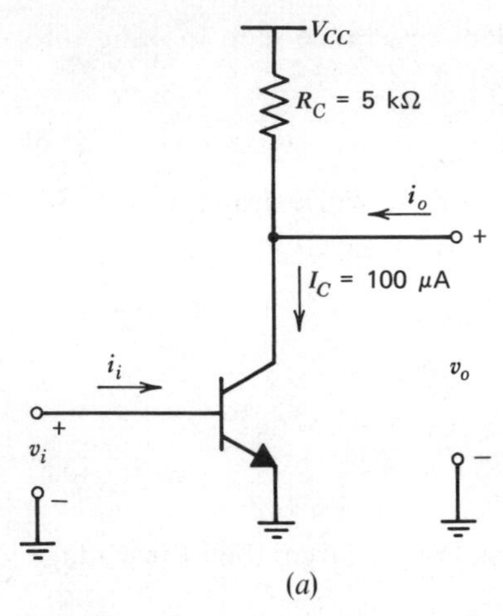

Fig. 3.8*a* Example amplifier circuit.

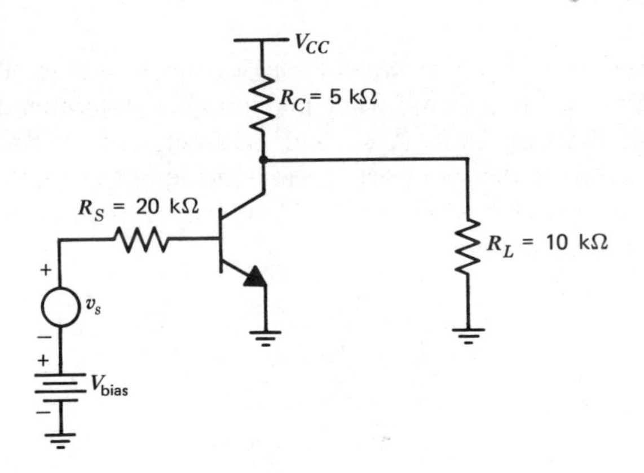

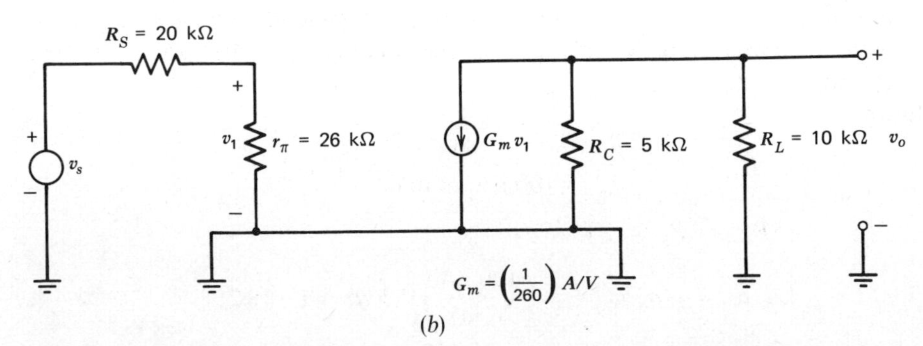

Fig. 3.8*b* Circuit for calculation of voltage gain with typical source and load resistance values.

(b) Calculate the voltage gain of the circuit of Fig. 3.8*b*. Assume that V_{bias} is adjusted so that the collector current is maintained at 100 μA.

$$v_1 = v_s\left(\frac{R_i}{R_S + R_i}\right)$$

$$v_o = -G_m v_1(R_o \| R_L) = -G_m\left(\frac{R_i}{R_S + R_i}\right)(R_o \| R_L)v_s$$

$$\frac{v_o}{v_s} = -\left(\frac{1}{260\ \Omega}\right)\left(\frac{26\ \text{k}\Omega}{26\ \text{k}\Omega + 20\ \text{k}\Omega}\right)\left[\frac{(10\ \text{k}\Omega)(5\ \text{k}\Omega)}{10\ \text{k}\Omega + 5\ \text{k}\Omega}\right]$$

$$= -7.25$$

3.2.2 Common-Base Configuration

In the common-base (CB) configuration,[4] the signal is applied to the emitter of the transistor, and the output is taken from the collector. The base is tied to ac ground. The common-base connection is shown in Fig. 3.9. While the configuration is not as widely used as the common emitter, it has properties that make it useful in certain circumstances. In this section we calculate the small-signal gain, input resistance, and output resistance of the common-base stage.

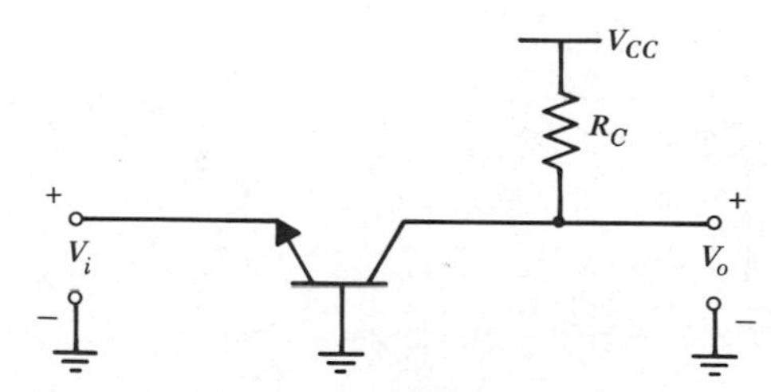

Fig. 3.9 Typical common-base amplifier.

The hybrid-π model provides an accurate representation of the small-signal behavior of the transistor independent of the circuit configuration, but for the common-base stage it is somewhat cumbersome because the collector-current-dependent generator is connected between the input and output terminals.[4] The analysis of common-base stages can be carried out more conveniently if the model is modified as shown in Fig. 3.10. The small-signal hybrid-π model is shown in Fig. 3.10*a*. In order to transform this model to an emitter-current-controlled model, we first note that the controlled current source flows from the collector terminal to the emitter terminal. The circuit behavior is unchanged if we replace this single current source with two current sources of the same value, one going from the collector to the base, and one going from the base to the emitter as shown

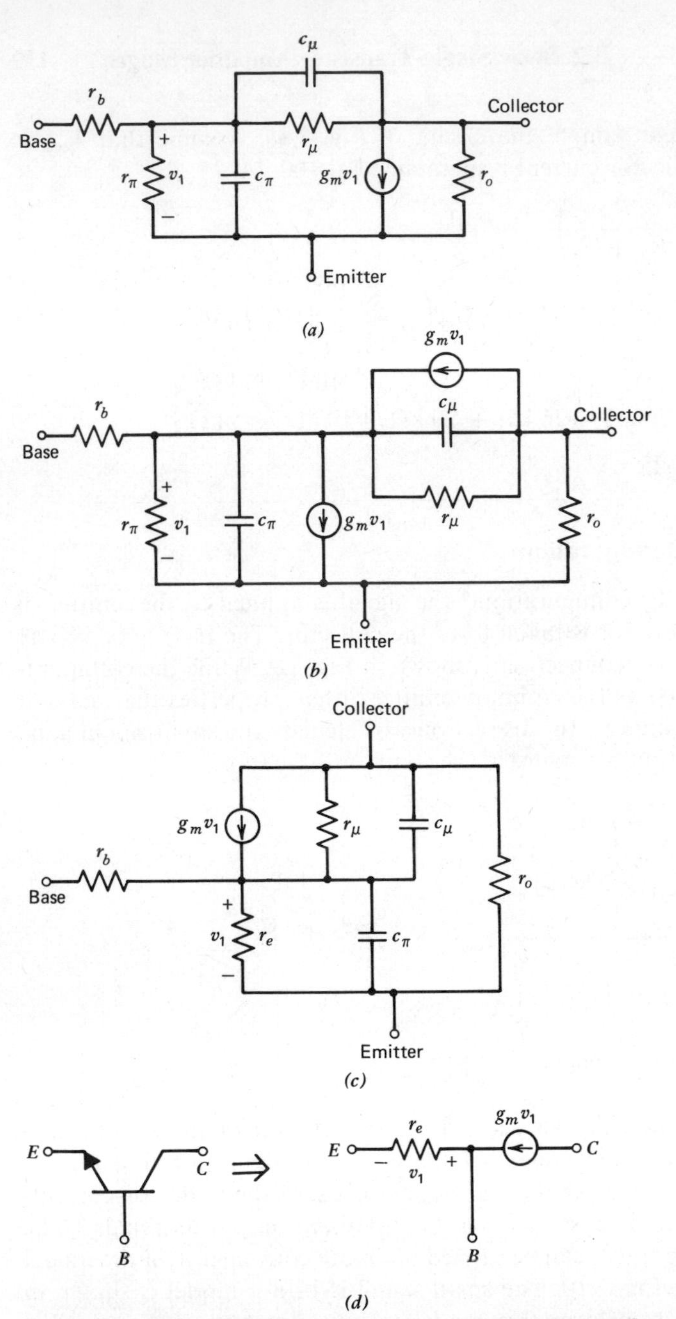

Fig. 3.10 Generation of emitter-current-controlled T-model from the hybrid-π. (*a*) Hybrid-π model. (*b*) The collector current source $g_m v_1$ is changed to two current sources in series, and the point between them attached to the base. This results in no net change in current flowing in the base. (*c*) The current source between base and emitter is converted to a resistor of value $1/g_m$. (*d*) "T" model for low frequencies, neglecting r_o, r_μ, and the charge-storage elements.

in Fig. 3.10*b*. Note that the current fed into the base is equal to the current taken out by the current sources so the circuit is indistinguishable from the original hybrid-π model. We next note that the controlled current source connecting the base and emitter is controlled by the voltage across its own terminals. Thus it can be replaced by a resistor of value $1/g_m$. This resistance appears in parallel with r_π, and the parallel combination of the two is called the emitter resistance, r_e.

$$r_e = \frac{1}{g_m + \dfrac{1}{r_\pi}} = \frac{1}{g_m\left(1 + \dfrac{1}{\beta_0}\right)} = \frac{\alpha_0}{g_m} \tag{3.11}$$

The new equivalent circuit, called the T model, and shown in Fig. 3.10*c*, has terminal properties exactly equivalent to the hybrid-π, but is more convenient to use in common-base calculations, particularly where r_o can be neglected. For the case where r_o, r_μ, and r_b are negligible, the model reduces to the simple form shown in Fig. 3.10*d*. This model does not include any frequency-dependent effects. Utilizing the T-model, the small-signal equivalent circuit of the common-base stage becomes that of Fig. 3.11. The output resistance, r_o, has been assumed large compared to R_C, and we assume for the moment that r_b is small. By inspection of Fig. 3.11, the short-circuit transconductance is:

$$G_m = g_m \tag{3.12}$$

The input resistance is just the resistance r_e:

$$R_i = r_e \tag{3.13}$$

The output resistance is given by

$$R_o = R_C \tag{3.14}$$

Using these parameters, the open-circuit voltage gain and short-circuit current gain are:

$$a_v = G_m R_o = g_m R_C \tag{3.15}$$

$$a_i = G_m R_i = g_m r_e = \alpha_0 \tag{3.16}$$

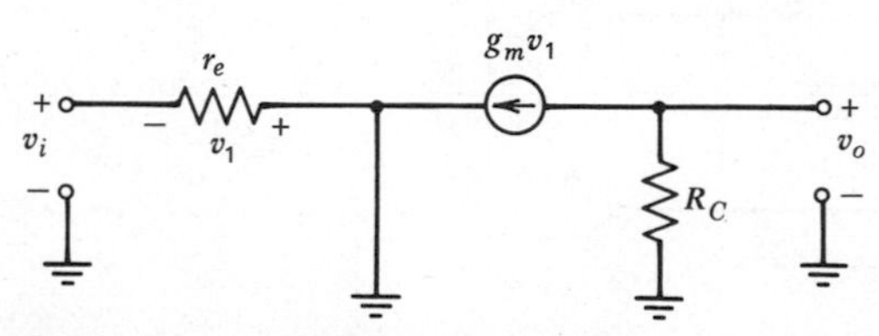

Fig. 3.11 Small-signal equivalent circuit of the common-base stage; r_o, r_b, r_μ are assumed negligible.

The two-port equivalent, small-signal circuit for the common-base stage is shown in Fig. 3.12*a*. We have assumed thus far that r_b is negligible; actually, if the common-base stage is operated at sufficiently high current levels, the base resistance has a significant effect on the transconductance and the input resistance. If these parameters are recalculated including the effects of r_b, the following modified expressions for G_m and R_i are obtained:

$$G_m = g_m \left(\frac{1}{1 + \dfrac{r_b}{r_\pi}} \right) \tag{3.17}$$

$$R_i = \frac{\alpha_0}{g_m} \left(1 + \frac{r_b}{r_\pi} \right) \tag{3.18}$$

Thus if the dc collector current is large enough that r_π is comparable with r_b, then the effects of base resistance must be included. If, for example, $r_b = 100\ \Omega$ and $\beta = 100$, then a collector current of 26 mA is required to make r_b and r_π equal. In calculating the above expressions for G_m, R_i, and R_o, we have neglected the effects of r_o. When R_C becomes large enough that it is comparable with r_o, r_o must be included in the small-signal model to predict the output resistance of the amplifier accurately. Since r_o is connected from the amplifier output back to the input, it causes the circuit to be nonunilateral, and the feedback contributed by r_o modifies the input resistance as well as the output resistance. Perhaps the most convenient

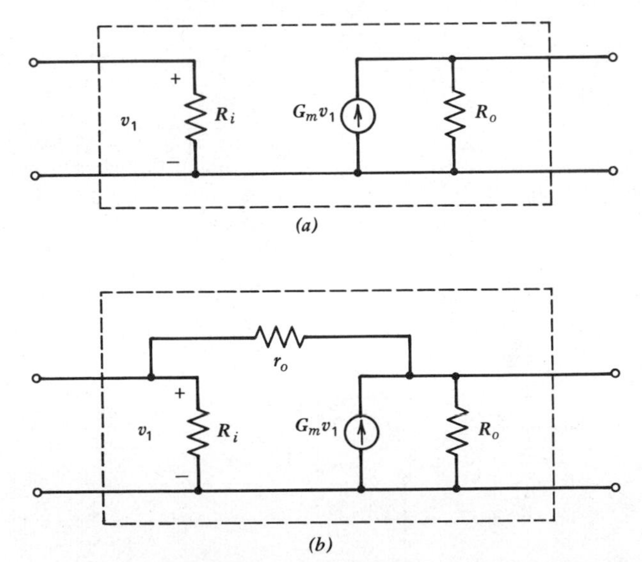

Fig. 3.12 (*a*) Two-port, small-signal equivalent circuit for CB stage. $R_i = r_e$, $G_m = g_m$, $R_o = R_C$. The effects of r_o are neglected. (*b*) Inclusion of the transistor output resistance in the common-base equivalent circuit.

way of including the effects of r_o when they are important is to use the two-port equivalent circuit of Fig. 3.12*b*, in which the output resistance has been added to the equivalent circuit. Compared with the common-emitter configuration, the common-base stage displays an input impedance, which is lower by a factor of β_0, and a current gain from input to output, which is less than one. The latter makes the cascading of many common-base stages impractical unless interstage transformers are used, since the current gain of the entire cascade would be less than one. An alternate way of viewing this fact is to note that in cascades of common-base stages a severe impedance mismatch exists between the output of one stage and the input of the next. That is, the input impedance is very low and the output impedance high. The motivation for using common-base stages is twofold. First, the collector-base capacitance does not cause high-frequency feedback from output to input as in the common-emitter amplifier. As discussed in Chapter 7, this can be important in the design of high-frequency amplifiers. Second, as discussed in Chapter 4, in the limiting case in which $R_C \to \infty$, the common-base amplifier can achieve much larger output resistance than the common-emitter stage. As a result, the common-base configuration can be used as a current source whose current is nearly independent of the voltage across it.

EXAMPLE

Calculate the input resistance, output resistance, and voltage gain of the common-base amplifier shown in Fig. 3.13. Assume that $\beta_0 = 100$, $r_b = 0$, $r_o = \infty$, and $I_C = 100\ \mu\text{A}$.

Solution:

$$R_{\text{in}} = \frac{\alpha_0}{g_m} = \frac{\dfrac{\beta_0}{1+\beta_0}}{g_m} = \frac{0.99}{100\ \mu\text{A}/26\ \text{mV}} \simeq 260\ \Omega$$

$$R_{\text{out}} = R_C = 5\ \text{k}\Omega$$

$$a_v = g_m R_C = 19.2$$

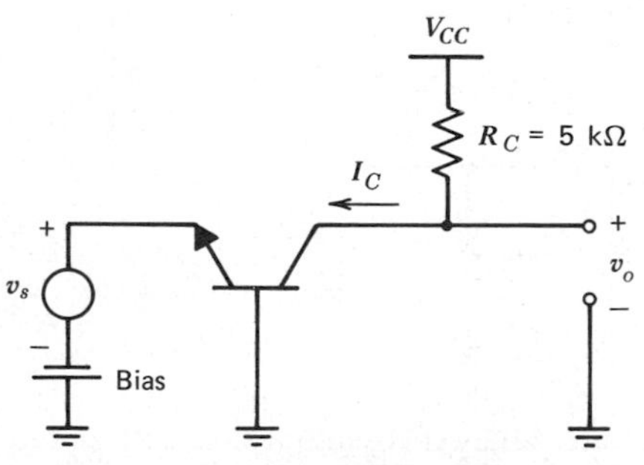

Fig. 3.13 Example common-base amplifier.

3.2.3 Common-Collector Configuration (Emitter Follower)

The common-collector connection is shown in Fig. 3.14*a*. The distinguishing feature of this configuration is that the signal is applied to the base and the output is taken from the emitter.[4] The appropriate small-signal transistor model is the hybrid-π, and the small-signal equivalent circuit is shown in Fig. 3.14*b*. Note that in this configuration the transistor is not unilateral, and, as a result, the input resistance depends on the load resistor, R_L, and the output resistance depends on the source resistance, R_S. Because of this, the characterization of the emitter follower by the corresponding equivalent two-port network is not particularly useful for intuitive understanding and first-order analysis. Instead, we simply analyze the emitter-follower circuit of Fig. 3.14*b*, including both the source resistance R_S and the load resistor R_L. Summing currents at the output node,

$$\frac{v_s - v_o}{R_S + r_\pi} + \beta_0\left(\frac{v_s - v_o}{R_S + r_\pi}\right) - \frac{v_o}{R_L} = 0 \tag{3.19}$$

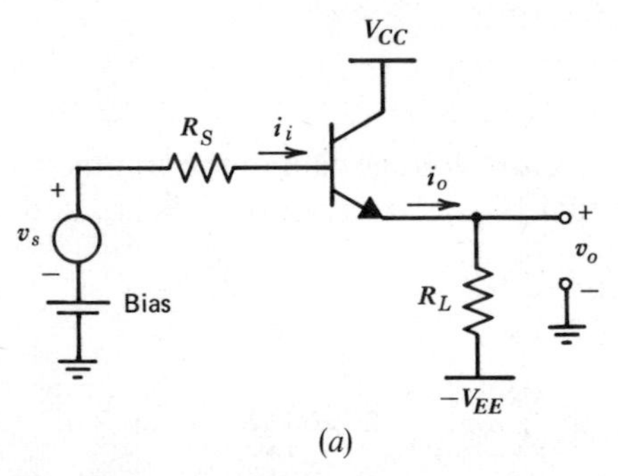

(*a*)

Fig. 3.14*a* Common-collector configuration.

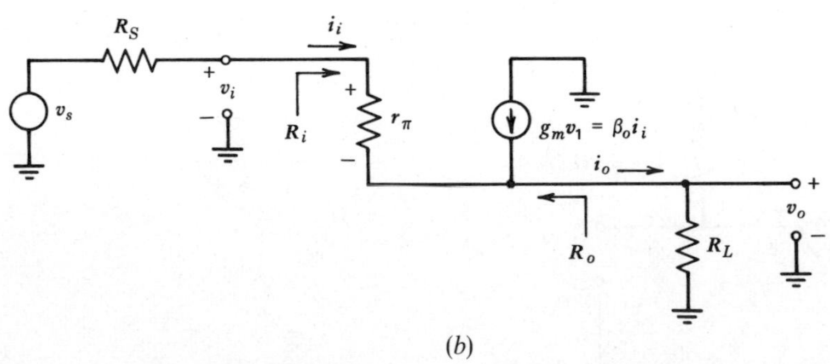

(*b*)

Fig. 3.14*b* Small-signal equivalent circuit of the emitter-follower circuit including R_L and R_S.

from which we find

$$\frac{v_o}{v_s} = \frac{1}{1 + \dfrac{R_S + r_\pi}{(\beta_0 + 1)R_L}} \tag{3.20}$$

If the base resistance, r_b, is significant, it can simply be added to R_S in these expressions. The voltage gain is always less than unity and will be close to unity if $\beta_0 R_L \gg (R_S + r_\pi)$. Normally this is the case in practical circuits. Note that the value of v_o/v_s we have calculated is not analogous to a_v for the CE and CB stages, since we have included the source resistance in this calculation.

We calculate R_i by removing the source, placing a test current source i_x across the input terminals, and calculating the resulting voltage, v_x, across the terminals. The circuit used to perform this calculation is shown in Fig. 3.15*a*. The current flowing in the load resistor, R_L, is:

$$i_o = i_x + \beta_0 i_x \tag{3.21}$$

So that the voltage, v_x, is:

$$v_x = i_x r_\pi + R_L(i_x + \beta_0 i_x) \tag{3.22}$$

and thus

$$R_i = \frac{v_x}{i_x} = r_\pi + R_L(\beta_0 + 1) \tag{3.23}$$

A general property of emitter followers is that the resistance seen looking into the base is equal to r_π plus the incremental resistance at the emitter multiplied by $(\beta_0 + 1)$.

We now calculate the output resistance by removing the load resistance, R_L, and finding the Thévenin equivalent resistance looking into the output terminals. We do this by either inserting a test current and calculating the resulting voltage or applying a test voltage and calculating the current. In this case, the calculation

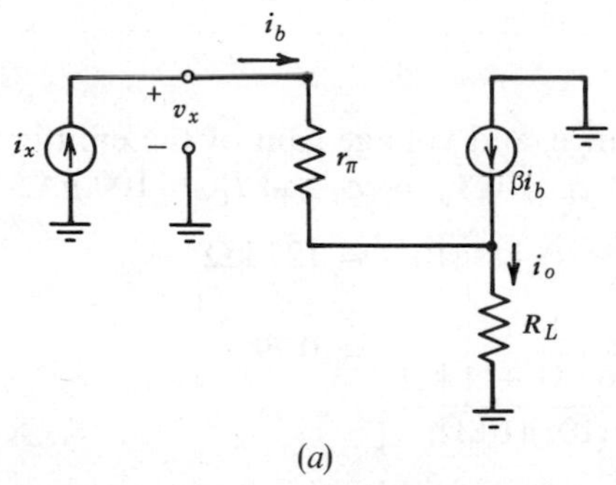

Fig. 3.15*a* Circuit for calculation of the input resistance of the emitter follower.

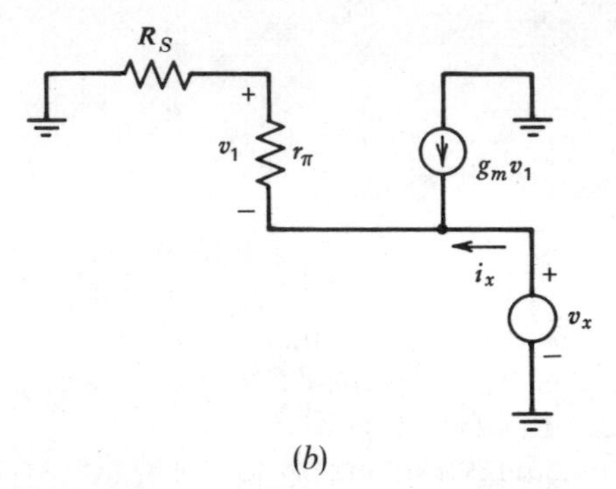

Fig. 3.15*b* Circuit for calculation of the output resistance of the emitter follower.

is simpler if a test voltage, v_x, is applied as shown in Fig. 3.15*b*. The voltage, v_1, is given by:

$$v_1 = -v_x\left(\frac{r_\pi}{r_\pi + R_S}\right) \tag{3.24}$$

The total output current, i_x, is thus:

$$i_x = \frac{v_x}{r_\pi + R_S} + g_m v_x\left(\frac{r_\pi}{r_\pi + R_S}\right) \tag{3.25}$$

and thus

$$R_o = \frac{v_x}{i_x} = \frac{r_\pi + R_S}{1 + \beta_0} \approx \left(\frac{1}{g_m} + \frac{R_S}{1 + \beta_0}\right) \tag{3.26}$$

The resistance seen at the output is thus the resistance in the base lead, divided by $(\beta_0 + 1)$, plus $1/g_m$. The emitter follower thus has high input resistance, low output resistance, and near-unity voltage gain. It is most widely used as an impedance transformer, to prevent loading of a preceding signal source by the low input impedance of a following stage.

EXAMPLE

Calculate the input resistance, output resistance, and voltage gain of the emitter follower of Fig. 3.15*c*. Assume that $\beta_0 = 100$, $r_b = 0$, $r_o = \infty$, and $I_C = 100\ \mu\text{A}$.

$$R_i = r_\pi + R_L(1 + \beta_0) = 26\text{ k}\Omega + (1\text{ k}\Omega)(101) = 127\text{ k}\Omega$$

$$\frac{v_o}{v_s} = \frac{1}{1 + \dfrac{r_\pi + R_S}{(\beta_0 + 1)R_L}} = \frac{1}{1 + \left[\dfrac{26\text{ k}\Omega + 1\text{ k}\Omega}{(101)(1\text{ k}\Omega)}\right]} = 0.79$$

$$R_o = \frac{R_S + r_\pi}{1 + \beta_0} = \frac{1\text{ k}\Omega + 26\text{ k}\Omega}{101} = 270\ \Omega$$

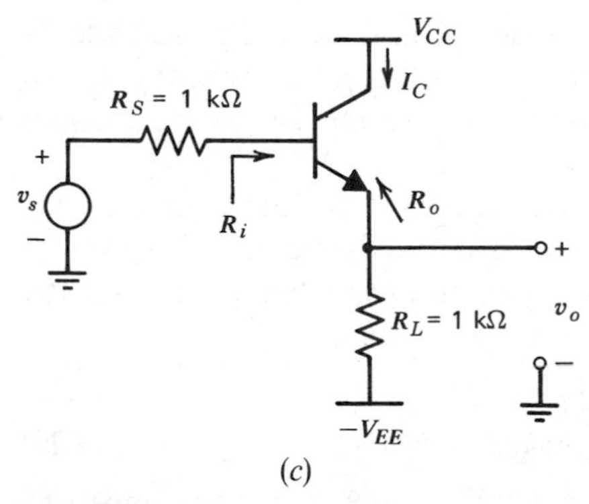

(c)

Fig. 3.15c Example emitter follower.

In summary, of the three transistor configurations, the common-emitter is most widely used because it provides both current and voltage gain, and many stages can be cascaded without interstage transformers. The common-base stage can provide voltage gain but no current gain, and the emitter follower provides current gain but no voltage gain.

3.2.4 Common-Emitter Amplifier with Emitter Degeneration

In the common-emitter amplifier considered earlier, the signal is applied to the base, the output is taken from the collector, and the emitter is attached to ac ground. Often, however, the common-emitter circuit is used with a resistance in series with the emitter as shown in Fig. 3.16*a*. The resistance has several effects, including a reduction in transconductance, an increase in output resistance, and an increase in input resistancc. Since the gain is decreased, the presence of an

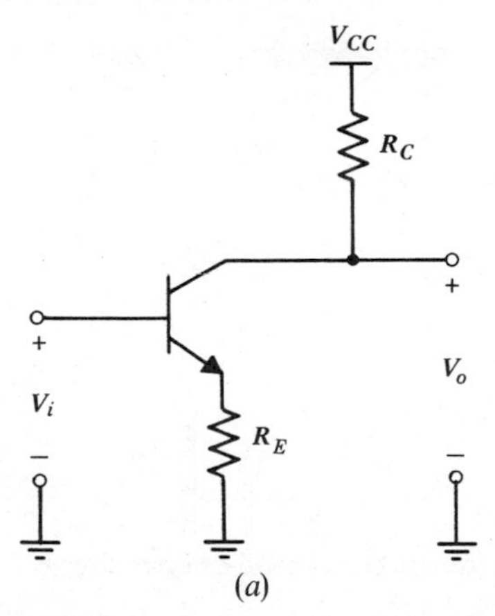

(a)

Fig. 3.16a Common-emitter amplifier with emitter degeneration.

emitter resistor is termed emitter degeneration. The resistor actually causes feedback to occur since the voltage across the emitter resistor, which is proportional to output current, subtracts directly from the input voltage. This circuit is examined from a feedback standpoint in Chapter 8.

In this section we calculate the input resistance, output resistance, and transconductance of the emitter-degenerated common-emitter amplifier. The small-signal equivalent circuit is shown in Fig. 3.16*b*. The input resistance is exactly the same as the emitter follower with a load resistor equal to R_E, or:

$$R_i = r_\pi + R_E(\beta_0 + 1) \tag{3.27}$$

$$\approx r_\pi(1 + g_m R_E) \tag{3.28}$$

Thus the input resistance is increased by a factor $(1 + g_m R_E)$ over the undegenerated case. The transconductance can be calculated by first calculating the small-signal voltage across r_π.

$$v_i = i_b r_\pi + (i_b + i_c)R_E = \frac{i_c}{\beta_0} r_\pi + i_c\left(1 + \frac{1}{\beta_0}\right) R_E \tag{3.29}$$

and thus,

$$v_i = i_c\left[\frac{1}{g_m} + R_E\left(1 + \frac{1}{\beta_0}\right)\right] \tag{3.30}$$

from which we obtain

$$G_m = \frac{i_c}{v_i} \approx \frac{g_m}{1 + g_m R_E} \tag{3.31}$$

Thus the transconductance is *decreased* by a factor $(1 + g_m R_E)$ compared with the case for $R_E = 0$.

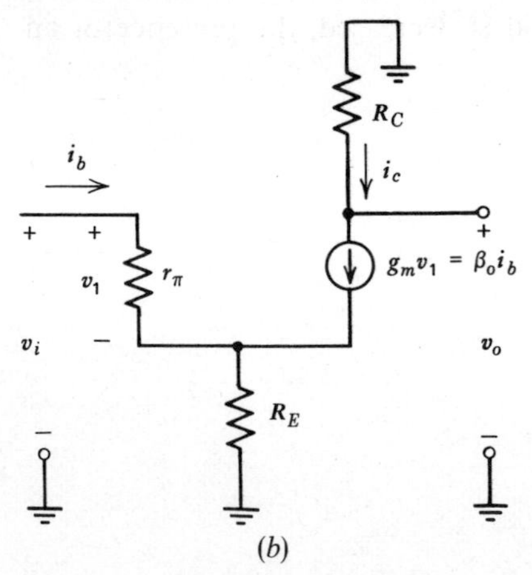

Fig. 3.16*b* Small-signal equivalent circuit for emitter-degenerated, common-emitter amplifier.

The output resistance is calculated using the equivalent circuit of Fig. 3.16*c*. For the time being we assume hypothetically that R_C is very large and can be neglected. The test current i_x flows in the parallel combination of r_π and R_E, so that

$$v_1 = -i_x(r_\pi \| R_E) \tag{3.32}$$

The current i_1 is thus

$$i_1 = i_x - g_m v_1 \tag{3.33}$$
$$= i_x + i_x g_m(r_\pi \| R_E) \tag{3.34}$$

And the voltage, v_x, is:

$$v_x = -v_1 + i_1 r_o \tag{3.35}$$
$$= i_x\{(r_\pi \| R_E) + r_o[1 + g_m(r_\pi \| R_E)]\} \tag{3.36}$$

using (3.32) and (3.34).
Thus

$$R_o = \frac{v_x}{i_x} = (r_\pi \| R_E) + r_o[1 + g_m(r_\pi \| R_E)] \tag{3.37}$$

The first term above is much smaller than the second in this equation, and if that term is neglected we obtain:

$$R_o \approx r_o\left(\frac{1 + g_m R_E}{1 + \dfrac{g_m R_E}{\beta_0}}\right) \tag{3.38}$$

If $g_m R_E \ll \beta_0$, then

$$R_o \approx r_o(1 + g_m R_E) \tag{3.39}$$

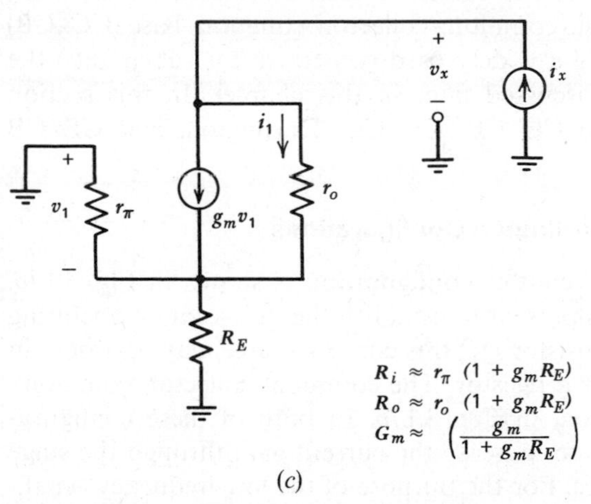

Fig. 3.16*c* Circuit for calculation of output resistance.

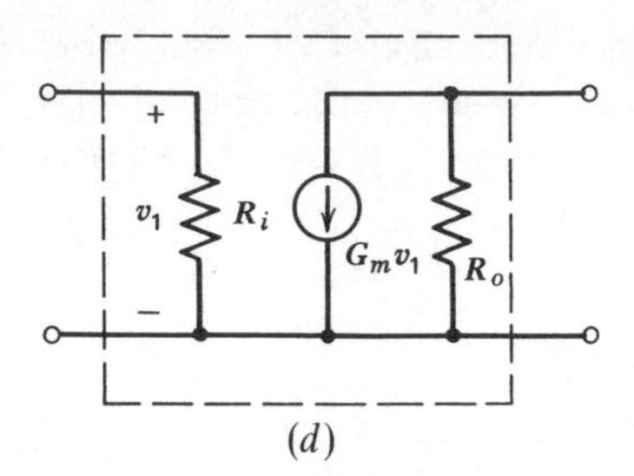

Fig. 3.16*d* Small-signal, two-port equivalent of emitter-degenerated CE amplifier.

Thus the output resistance is increased by a factor $(1 + g_m R_E)$. This fact makes the use of emitter degeneration desirable in transistor current sources. A small-signal equivalent circuit, neglecting the collector load resistor, R_C, is shown in Fig. 3.16*d*.

3.3 TWO-TRANSISTOR AMPLIFIER STAGES

Most integrated-circuit amplifiers consist of a number of stages, each of which provides voltage gain, current gain, and/or impedance level transformation from input to output. Such circuits can of course, be analyzed by considering each transistor to be a "stage," and analyzing the circuit as a collection of individual transistors. However, certain combinations of two transistors occur so frequently in such circuits that it is convenient to characterize these as two-transistor "subcircuits" and to regard them together as a single stage when they occur in amplifiers.[5] These subcircuits include the common–collector–common–emitter (CC–CE) connection, the common–collector–common–collector (CC–CC) connection, the Darlington connection, the common–emitter–common–base (CE–CB) or cascode, connection, and the common–collector–common–base (CC–CB) connection. The latter is a form of a widely used two-transistor subcircuit—the emitter-coupled pair—which is discussed later in this chapter. In this section we analyze the properties of the CC–CE, CC–CC, Darlington, and CE–CB connections.

3.3.1 The CC-CE, CC-CC, and Darlington Configurations

The common–collector–common–emitter configuration is shown in Fig. 3.17*a*. The biasing current source, I_{bias}, is present to establish the quiescent dc operating current in the emitter-follower transistor Q_1; this current source may be absent in some cases or may be replaced by a resistor. The common–collector–common–collector configuration is illustrated in Fig. 3.17*b*. In both of these configurations, the effect of transistor Q_1 is to increase the current gain through the stage and to increase the input resistance. For the purpose of the low-frequency, small-signal analysis of circuits, the two transistors, Q_1 and Q_2, can be thought of as a

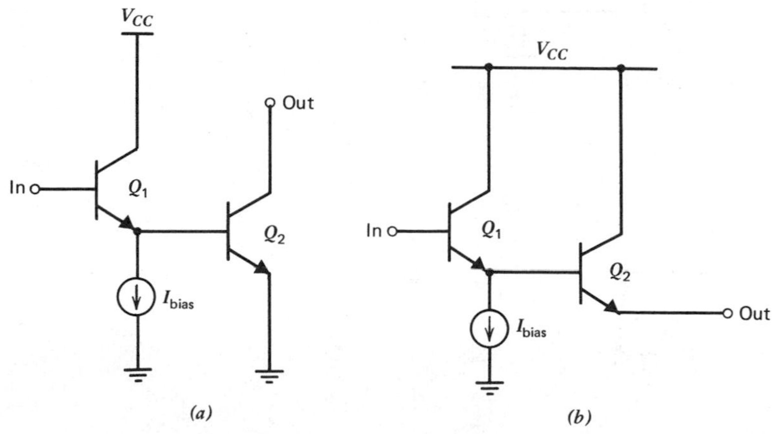

Fig. 3.17 (*a*) Common-collector-common-emitter cascade. (*b*) Common-collector-common-collector cascadc.

single composite transistor as illustrated in Fig. 3.18. Provided the effects of the r_o of Q_1 are negligible, the small-signal equivalent circuit for this composite device is shown in Fig. 3.19. We will now calculate effective values for the r_π, g_m, β_0, and r_o of the composite device, and we will designate these composite parameters with a superscript *c*. We will also denote the terminal voltages and currents of the composite device with a superscript *c*. We assume that β_0 is constant.

The effective value of r_π, r_π^c, is the resistance seen looking into the base with the emitter grounded. Referring to Fig. 3.19, we see that the resistance between the emitter of Q_1 and ground is simply $r_{\pi 2}$. Thus (3.23) for the input resistance of the emitter follower can be used. Substituting $r_{\pi 2}$ for R_E,

$$r_\pi^c = r_{\pi 1} + (\beta_0 + 1) r_{\pi 2} \tag{3.40}$$

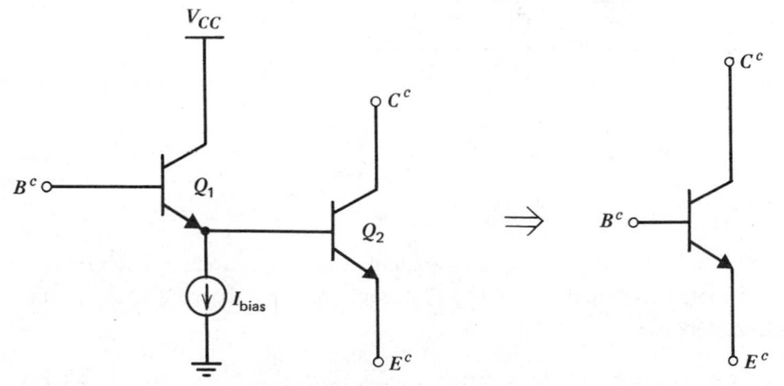

Fig. 3.18 The composite transistor representation of the CC–CE and CC–CC connections.

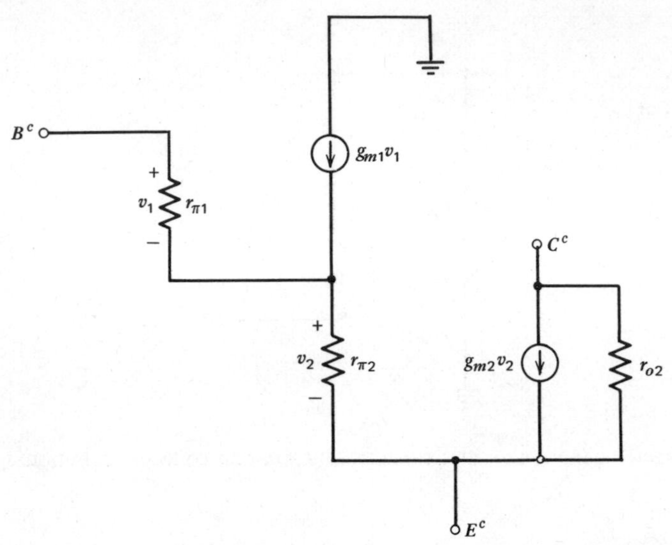

Fig. 3.19 Small-signal equivalent circuit for the CC–CE and CC–CC connected transistors.

The effective transconductance of the configuration, $g_m{}^c$, is the change in collector current of Q_2, $i_c{}^c$, for a unit change in $v_{be}{}^c$. To calculate this, we must find the change in v_2 that occurs for a unit change in $v_{be}{}^c$. Equation 3.20 can be used directly, giving:

$$\frac{v_2}{v_{be}{}^c} = \frac{1}{1 + \left[\dfrac{r_{\pi 1}}{(\beta_0 + 1)r_{\pi 2}}\right]} \tag{3.41}$$

and

$$i_c{}^c = g_m{}^c v_{be}{}^c = g_{m2} v_2 \tag{3.42}$$

$$= \frac{g_{m2} v_{be}{}^c}{1 + \left[\dfrac{r_{\pi 1}}{(\beta_0 + 1)r_{\pi 2}}\right]} \tag{3.43}$$

Thus

$$g_m{}^c = \left[\frac{g_{m2}}{1 + \dfrac{r_{\pi 1}}{(\beta_0 + 1)r_{\pi 2}}}\right] \tag{3.44}$$

For the special case in which the biasing current source, I_{bias}, is zero, the emitter current of Q_1 is equal to the base current of Q_2. Thus $r_{\pi 1}$ and $r_{\pi 2}$ have a ratio equal to β_0, and (3.44) reduces to:

$$g_m{}^c = \frac{g_{m2}}{2} \tag{3.45}$$

The effective current gain is the ratio:

$$\beta^c = \frac{i_c^{\ c}}{i_b^{\ c}} = \frac{i_{c2}}{i_{b1}} \tag{3.46}$$

The emitter current of Q_1 is given by:

$$i_{e1} = (\beta_0 + 1)i_{b1} \tag{3.47}$$

This is equal to i_{b2}, so that:

$$i_{c2} = i_c^{\ c} = \beta_0 i_{b2} = \beta_0(\beta_0 + 1)i_{b1} \tag{3.48}$$

$$= \beta_0(\beta_0 + 1)i_b^{\ c} \tag{3.49}$$

Therefore:

$$\beta^c = \beta_0(\beta_0 + 1) \tag{3.50}$$

The current gain of the composite transistor is approximately equal to $\beta_0^{\,2}$. Also, by inspection of Fig. 3.19, assuming r_μ is negligible,

$$r_o^{\ c} = r_{o2} \tag{3.51}$$

The small-signal, two-port network equivalent for the CC–CE connection is shown in Fig. 3.20. The collector resistor has not been included. This small-signal equivalent can be used to represent the performance of the composite device, thus simplifying the analysis of circuits containing composite transistors.

The Darlington configuration, illustrated in Fig. 3.21, is a composite two-transistor device in which the collectors are tied together and the emitter of the first device drives the base of the second. A biasing element of some sort is required to control the emitter current of Q_1. The result is a three-terminal composite transistor that can, in principle, be used in place of a single transistor in common–emitter, common–base, and in common-collector configurations. When used as an emitter follower, the device is identical to the CC–CC connection already

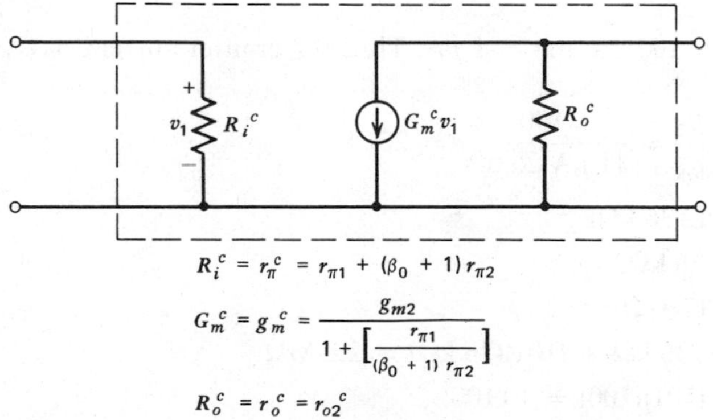

Fig. 3.20 Two-port representation, CC–CE connection.

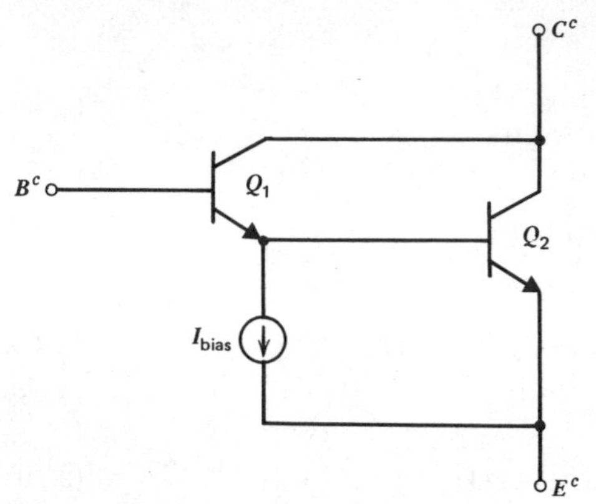

Fig. 3.21 The Darlington configuration.

discussed. When used as a common-emitter amplifier, the device is very similar to the CC–CE connection, except that the collector of Q_1 is connected to the output instead of to the power supply. The effect of this change is to reduce the effective output resistance of the device because of feedback through r_o of Q_1, and to cause the input capacitance to increase because of the connection of the collector-base capacitance of Q_1 from the input to the output. Because of these drawbacks, the CC–CE connection is normally preferable in integrated small-signal amplifiers. The term Darlington is often used to refer to both the CC–CE and CC–CC connections.

EXAMPLE

Find the effective $r_\pi^{\,c}$, β^c, and $g_m^{\,c}$ for the composite transistor shown in Fig. 3.18. Assume, for both devices, that $\beta_0 = 100$, $r_b = 0$, and $r_o = \infty$. Assume for Q_2 that $I_C = 100\ \mu\text{A}$ and that $I_{\text{bias}} = 10\ \mu\text{A}$.

Solution:

The base current of Q_2 is $100\ \mu\text{A}/100 = 1\ \mu\text{A}$. Thus the emitter current of Q_1 is $11\ \mu\text{A}$.

$$r_{\pi 1} = \frac{\beta_0}{g_m} = \frac{100}{11\ \mu\text{A}/26\ \text{mV}} = 236\ \text{k}\Omega$$

$$g_{m1} = (2.36\ \text{k}\Omega)^{-1}$$

$$r_{\pi 2} = 26\ \text{k}\Omega$$

$$g_{m2} = (260\ \Omega)^{-1}$$

$$r_\pi^{\,c} = 236\ \text{k}\Omega + (101)(26\ \text{k}\Omega) = 2.8\ \text{M}\Omega$$

$$\beta^c = (101)(100) = 10{,}100$$

$$g_m^{\,c} = g_{m2}(0.916) = (283\ \Omega)^{-1}$$

Thus the composite transistor has much higher input resistance and current gain than a single transistor.

3.3.2 The CE–CB, or Cascode, Configuration

The cascode two-transistor subcircuit is shown in Fig. 3.22. The principal attributes of this configuration are that the output resistance is very high and that no high-frequency feedback occurs from the output back to the input through C_μ as occurs in the common-emitter configuration.[4] The high output impedance attainable is particularly useful in attaining power supply desensitization in bias reference supplies and achieving large amounts of voltage gain in a single amplifying stage with an active *pnp* load. These applications are discussed further in Chapter 4. In this section we calculate the low-frequency, small-signal properties of the CE–CB connection. We will assume that r_b in both devices is zero; in considering the high-frequency performance of this combination we must account for the effects of r_b on the frequency response as discussed in Chapter 7. The base resistances have a negligible effect on the low-frequency performance.

The small-signal equivalent circuit for the cascode circuit is shown in Fig. 3.23. Since we are considering only the low-frequency performance, we neglect the capacitive energy storage elements. We will determine the input resistance, output resistance, and transconductance of the cascode circuit. By inspection of Fig. 3.23, the input resistance is simply r_π of Q_1. Since the current gain from emitter to collector of Q_2 is nearly unity, the transconductance of the circuit from input to output is approximately equal to the transconductance of Q_1. The output resistance can be calculated by applying a test voltage source, v_x, to the collector of Q_2 and calculating the current, i_x, which results as shown in Fig. 3.24. We first note that r_{o1} appears in parallel with r_{e2}, and is thus negligible since it is larger by approximately 10^4. The current i_{x1} is given by:

$$i_{x1} = \frac{v_x}{r_{o2} + r_{e2}} \approx \frac{v_x}{r_{o2}} \tag{3.52}$$

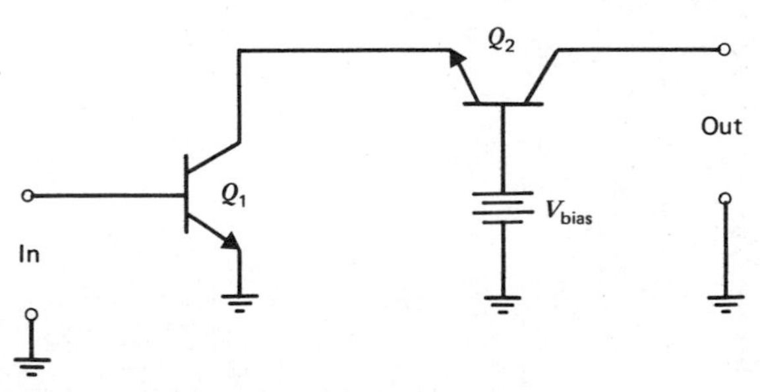

Fig. 3.22 The cascode amplifier.

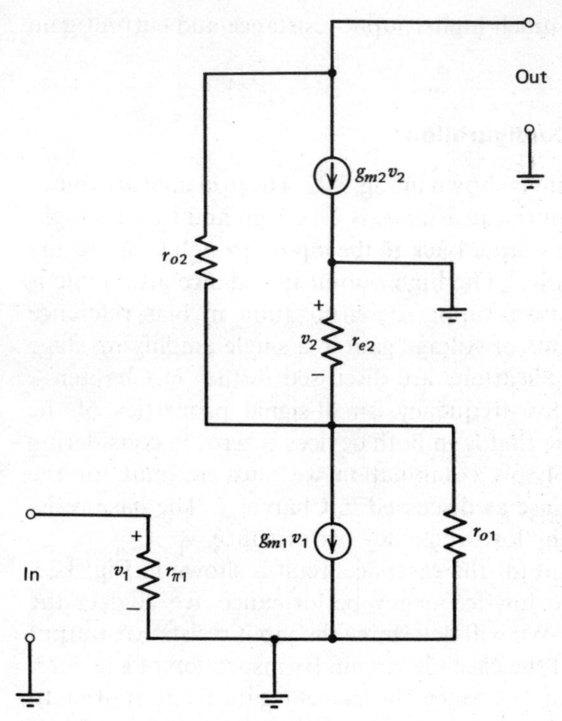

Fig. 3.23 Small-signal equivalent circuit, cascode amplifier.

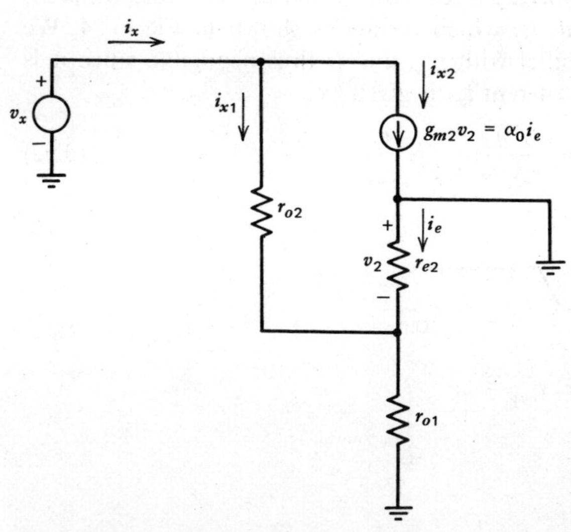

Fig. 3.24 Equivalent circuit for calculation of output resistance of the cascode.

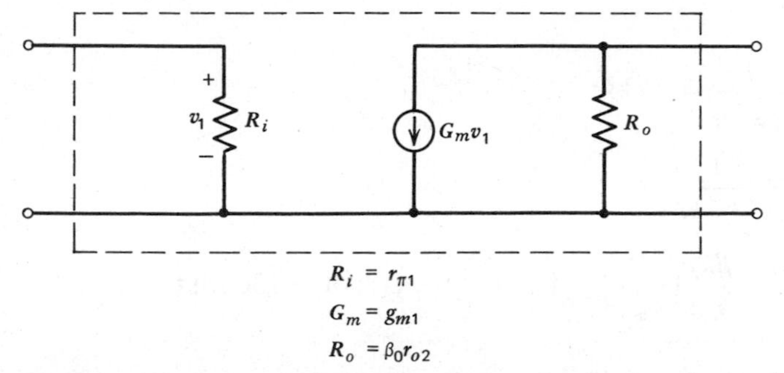

Fig. 3.25 Two-port small-signal equivalent for the cascode circuit.

This current is also equal to $-i_e$. Thus i_{x2} is:

$$i_{x2} = \alpha_0 i_e = -\alpha_0 \frac{v_x}{r_{o2}} \tag{3.53}$$

The total output current, i_x, is thus:

$$i_x = i_{x1} + i_{x2} = \frac{v_x}{r_{o2}}(1 - \alpha_0) \approx \frac{v_x}{\beta_0 r_{o2}} \tag{3.54}$$

and thus,

$$R_o = \beta_0 r_{o2} \tag{3.55}$$

The CE–CB cascade thus displays an output resistance that is larger by factor β_0 than the CE stage alone. The small-signal two-port equivalent circuit is shown in Fig. 3.25. Note that if this circuit is operated with a hypothetical collector load that has infinite incremental resistance, that the voltage gain is:

$$A_v = G_m R_o = g_m r_{o2} \beta_0 = \frac{\beta_0}{\eta} \tag{3.56}$$

For a typical *npn* transistor this ratio is approximately 2×10^5. Thus the maximum available voltage gain is higher by factor β_0 than for the case of a single transistor. In this analysis we have neglected r_μ. As discussed in Chapter 1, the value of r_μ for integrated circuit *npn* transistors is usually much larger than $\beta_0 r_o$, and r_μ will have little effect on R_o. However, for lateral *pnp* transistors, r_μ is comparable with $\beta_0 r_o$ and the effect of r_μ will be to decrease R_o somewhat.

EXAMPLE

Calculate the input resistance, transconductance, and output resistance of the cascode circuit of Fig. 3.22. Assume that $I_C = 100\ \mu\text{A}$, $\beta_0 = 100$, $r_b = 0$, and $\eta = 2 \times 10^{-4}$.

Solution:

$$R_i = r_{\pi 1} = \frac{\beta_0}{g_{m1}} = 26\ \text{k}\Omega$$

$$G_m = g_{m1} = \frac{1}{260\ \Omega}$$

$$R_o = \beta_0 r_{o2} = \frac{\beta_{02}}{\eta_2}\left(\frac{1}{g_{m2}}\right) = \left(\frac{100}{2 \times 10^{-4}}\right)(260) = 130\ \text{M}\Omega$$

3.4 EMITTER-COUPLED PAIRS

The emitter-coupled pair is perhaps the most widely used two-transistor subcircuit in monolithic analog circuits. The usefulness of this circuit stems from the fact that cascades of emitter-coupled pairs can be directly coupled to one another without interstage coupling capacitors, and from the fact that the differential input characteristics provided by the emitter-coupled pair are required in many types of analog circuits[6,7]. The simplest form of emitter-coupled pair is shown in Fig. 3.26. The biasing circuit in the common emitter lead can be either a simple

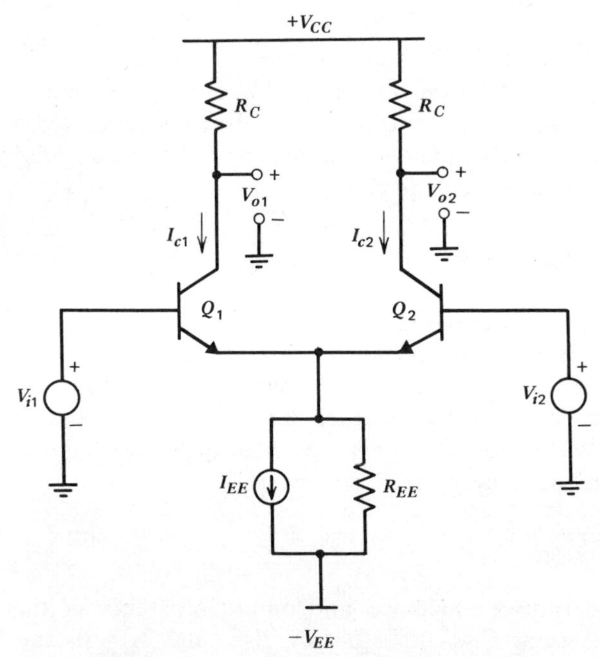

Fig. 3.26 Emitter-coupled pair circuit diagram.

resistor, in which case the equivalent current source will be zero, or a transistor current source, in which case the circuit is the Norton equivalent of the transistor.

3.4.1 dc Transfer Characteristic

The large-signal behavior of the emitter-coupled pair is important both because it illustrates the limited range of input voltages over which the circuit behaves linearly and because it illustrates an important aspect of the circuit behavior, that of nonsaturating limiting of analog waveforms. For simplicity in the analysis we assume that the emitter current source output resistance R_{EE} is infinite, that the base resistance of each transistor is negligible, and that the output resistance of each transistor is infinite. These assumptions do not strongly affect the low-frequency, large-signal behavior of the circuit, although the effect on the small-signal behavior can be significant. We first sum the voltages around the loop consisting of the two voltage sources and the two base-emitter junctions:

$$V_{i1} - V_{be1} + V_{be2} - V_{i2} = 0 \tag{3.57}$$

From the Ebers-Moll equations, assuming $V_{be1}, V_{be2} \gg V_T$,

$$V_{be1} = V_T \ln \frac{I_{c1}}{I_{S1}} \tag{3.58}$$

$$V_{be2} = V_T \ln \frac{I_{c2}}{I_{S2}} \tag{3.59}$$

Combining (3.57), (3.58), and (3.59), and assuming that $I_{S1} = I_{S2}$, we find

$$\frac{I_{c1}}{I_{c2}} = \exp\left(\frac{V_{i1} - V_{i2}}{V_T}\right) = \exp\left(\frac{V_{id}}{V_T}\right) \tag{3.60}$$

Here we have defined V_{id} as the difference between V_{i1} and V_{i2}. Summing currents at the emitters of the transistors,

$$-(I_{e1} + I_{e2}) = I_{EE} = \frac{1}{\alpha_0}(I_{c1} + I_{c2}) \tag{3.61}$$

Combining Equations 3.60 and 3.61:

$$I_{c1} = \frac{\alpha_0 I_{EE}}{1 + \exp\left(-\dfrac{V_{id}}{V_T}\right)} \tag{3.62}$$

$$I_{c2} = \frac{\alpha_0 I_{EE}}{1 + \exp\left(\dfrac{V_{id}}{V_T}\right)} \tag{3.63}$$

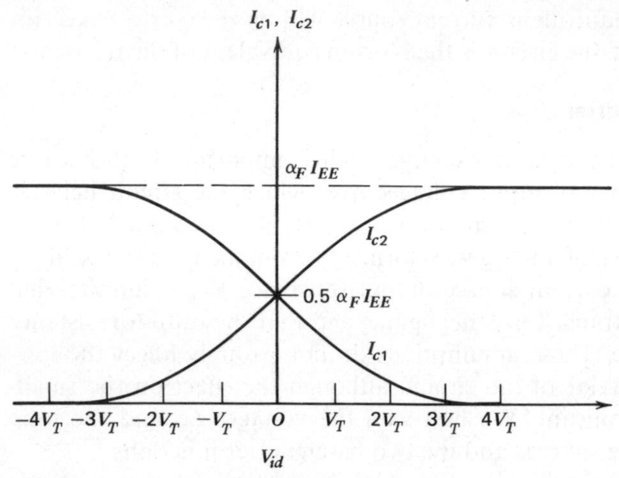

Fig. 3.27 Emitter-coupled pair collector currents as a function of differential input voltage.

These two currents are shown as a function of V_{id} in Fig. 3.27. Note that for input voltage differences of greater than several hundred millivolts, the collector currents become independent of V_{id}, since all the current is flowing in one of the transistors. Only for difference voltages less than approximately 50 mV does the circuit behave in an approximately linear fashion. We can now compute the output voltages, since

$$V_{o1} = V_{CC} - I_{c1}R_C \tag{3.64}$$

$$V_{o2} = V_{CC} - I_{c2}R_C \tag{3.65}$$

The output signal of interest is often the difference between V_{o1} and V_{o2}, which we term V_{od}.

$$V_{od} = V_{o1} - V_{o2} = \alpha_0 I_{EE} R_C \tanh\left(\frac{-V_{id}}{2V_T}\right) \tag{3.66}$$

This function is plotted in Fig. 3.28. Here a significant advantage of differential amplifiers is apparent: When V_{id} is zero, V_{od} is zero, which allows direct coupling of cascaded stages without introducing dc offsets.

3.4.2 Emitter Degeneration

In order to increase the range of input voltage over which the emitter coupled pair behaves approximately as a linear amplifier, emitter-degeneration resistors are frequently included in series with the emitters of the transistors as shown in Fig. 3.29. The analysis of this circuit proceeds in the same manner as before, except the voltage drop across these resistors must be included in the voltage summation corresponding to (3.57). A closed form result like that of (3.66) does not result from the analysis, but the effect of the resistors may be understood

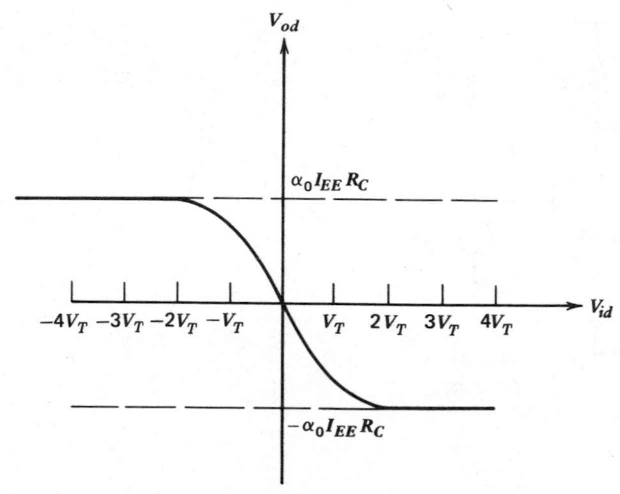

Fig. 3.28 Emitter-coupled pair, differential output voltage as a function of differential input voltage.

intuitively from the examples plotted on Fig. 3.29. For large values of emitter degeneration, the linear range of operation is extended by an amount approximately equal to $I_{EE}R_E$. The voltage gain is reduced by approximately the same factor that the input range is increased. The emitter resistor actually introduces local feedback in the pair and is discussed further in Chapter 8.

3.4.3 Small-Signal Analysis

In many cases, the features of interest in the performance of the emitter-coupled pair are the small-signal properties of the circuit for dc differential input voltages near zero volts. In the following analysis we assume that the dc differential input voltage is zero and calculate the small-signal gain and input resistance for signals that are small enough that they do not violate the assumption of linear operation. We also assume that the Norton equivalent resistance of the emitter biasing element is finite since this resistance has a considerable effect on the small-signal behavior of the circuit. We also assume that $r_b = 0$, $r_o = \infty$, and $r_\mu = \infty$ for the transistors. The small-signal equivalent circuit for the emitter-coupled pair under these assumptions is shown in Fig. 3.30.

The emitter-coupled pair is unlike the circuits considered in earlier sections in that there are two input terminals and two output terminals. Since the small-signal circuit is linear, superposition may be applied and we expect to obtain four constants that describe the behavior of the circuit instead of one:

$$v_{o1} = A_{11}v_{i1} + A_{12}v_{i2} \tag{3.67}$$

$$v_{o2} = A_{21}v_{i1} + A_{22}v_{i2} \tag{3.68}$$

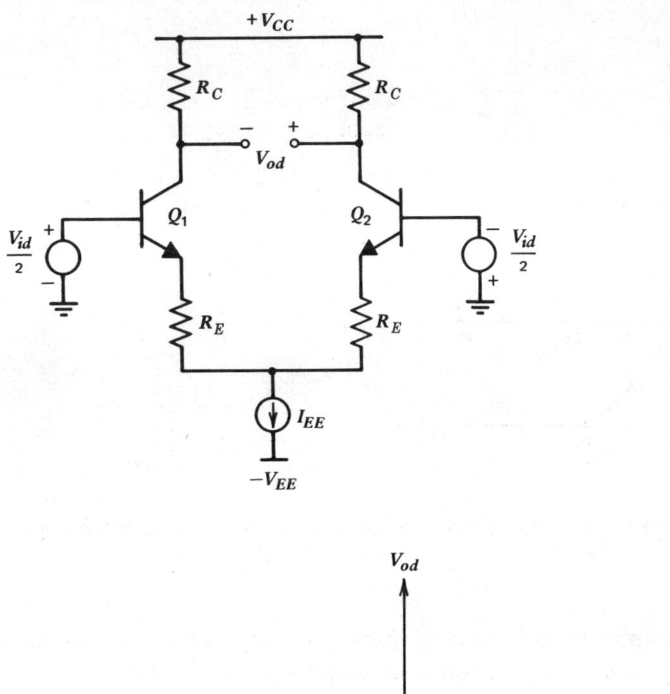

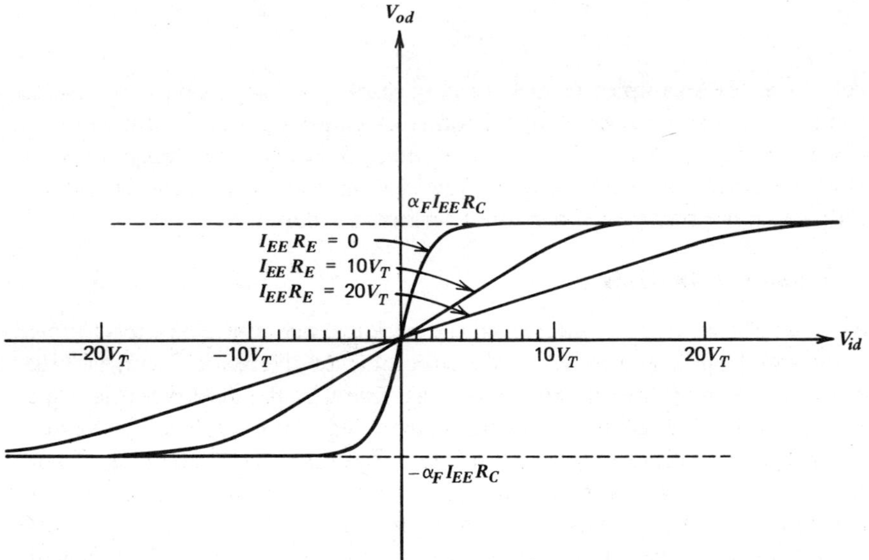

Fig. 3.29 Output voltage as a function of input voltage, emitter-coupled pair with emitter degeneration.

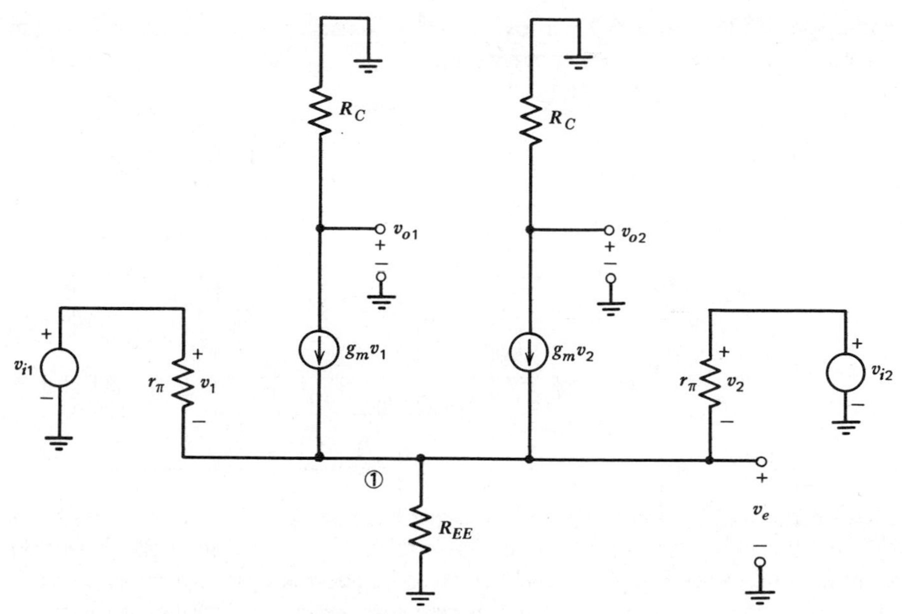

Fig. 3.30 Small-signal equivalent circuit, emitter-coupled pair.

We now calculate the values for A_{11} and A_{21} by summing currents at the emitter node.

$$\frac{v_{i1} - v_e}{r_\pi} + g_m(v_{i1} - v_e) + g_m(v_{i2} - v_e) + \frac{v_{i2} - v_e}{r_\pi} - \frac{v_e}{R_{EE}} = 0 \tag{3.69}$$

Solving for v_e, we obtain

$$v_e = \frac{v_{i1} + v_{i2}}{2 + \left[\dfrac{1}{g_m R_{EE}\left(1 + \dfrac{1}{\beta_0}\right)}\right]} \tag{3.70}$$

The output voltage v_{o1} is given by:

$$v_{o1} = -g_m(v_{i1} - v_e)R_C \tag{3.71}$$

Substituting (3.70) above in (3.71) we obtain

$$v_{o1} = -\frac{g_m R_C}{2}\left\{v_{i1}\left[\frac{1 + \dfrac{1}{g_m R_{EE}\left(1 + \dfrac{1}{\beta_0}\right)}}{1 + \dfrac{1}{2g_m R_{EE}\left(1 + \dfrac{1}{\beta_0}\right)}}\right] - v_{i2}\left[\frac{1}{1 + \dfrac{1}{2g_m R_{EE}\left(1 + \dfrac{1}{\beta_0}\right)}}\right]\right\} \tag{3.72}$$

Comparison of (3.67) and (3.72) allows direct calculation of A_{11} and A_{12}. From the symmetry of the circuit it is apparent that $A_{11} = A_{22}$ and $A_{21} = A_{12}$. Thus we find

$$A_{22} = A_{11} = -\frac{g_m R_C}{2}\left[\frac{1 + \dfrac{1}{g_m R_{EE}\left(1 + \dfrac{1}{\beta_0}\right)}}{1 + \dfrac{1}{2g_m R_{EE}\left(1 + \dfrac{1}{\beta_0}\right)}}\right] \tag{3.73}$$

$$A_{21} = A_{12} = +\frac{g_m R_C}{2}\left[\frac{1}{1 + \dfrac{1}{2g_m R_{EE}\left(1 + \dfrac{1}{\beta_0}\right)}}\right] \tag{3.74}$$

These expressions, however, do not contribute insight into actual behavior of the circuit as most commonly used. The circuit is balanced in such a way that it amplifies differential signals, and tends to reject voltages common to both inputs. This aspect of circuit behavior is much more clearly evident if the input and output voltages are redefined as follows. The differential-mode input voltage is defined as:

$$v_{id} = v_{i1} - v_{i2} \tag{3.75}$$

and the common-mode input voltage is defined as:

$$v_{ic} = \frac{v_{i1} + v_{i2}}{2} \tag{3.76}$$

These equations can be inverted to give the actual inputs v_{i1} and v_{i2} in terms of these two new input variables

$$v_{i1} = \frac{v_{id}}{2} + v_{ic} \tag{3.77}$$

$$v_{i2} = -\frac{v_{id}}{2} + v_{ic} \tag{3.78}$$

The physical significance of these new variables can be understood by redrawing the equivalent circuit as shown in Fig. 3.31. Note that v_{id} is the *difference* between the two inputs, while v_{ic} is the *average* of the two inputs. New output variables are defined in the same way:

$$v_{od} = v_{o1} - v_{o2} \tag{3.79}$$

$$v_{oc} = \frac{v_{o1} + v_{o2}}{2} \tag{3.80}$$

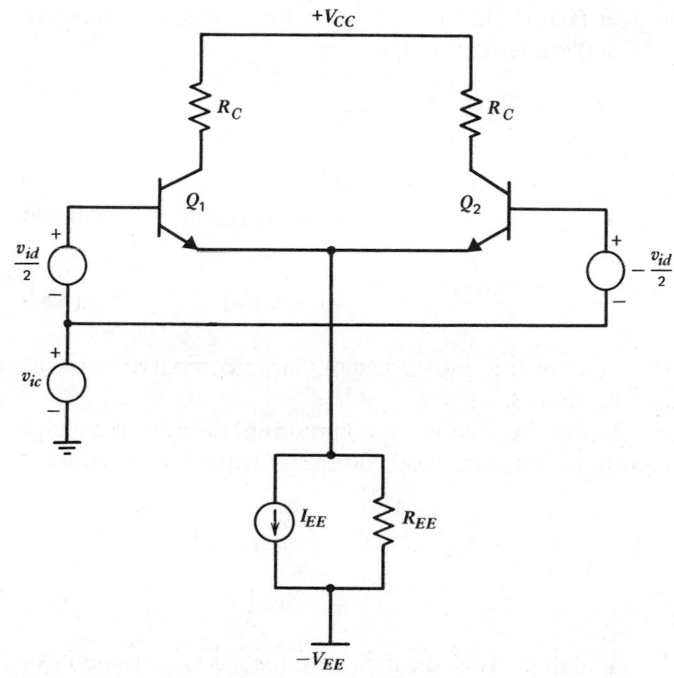

Fig. 3.31 Emitter-coupled pair with source decomposed into differential and common-mode components.

Solving these equations for v_{o1} and v_{o2}, we obtain

$$v_{o1} = \frac{v_{od}}{2} + v_{oc} \tag{3.81}$$

$$v_{o2} = -\frac{v_{od}}{2} + v_{oc} \tag{3.82}$$

We have now defined two new input variables and two new output variables. By substituting the expressions for v_{o1}, v_{o2}, v_{i1}, v_{i2} in terms of the new variables back into (3.67) and (3.68), we obtain the new relations:

$$v_{od} = \left(\frac{A_{11} - A_{12} - A_{21} + A_{22}}{2}\right) v_{id} + (A_{11} + A_{12} - A_{21} - A_{22}) v_{ic} \tag{3.83}$$

$$v_{oc} = \left(\frac{A_{11} - A_{12} + A_{21} - A_{22}}{4}\right) v_{id} + \left(\frac{A_{11} + A_{12} + A_{21} + A_{22}}{2}\right) v_{ic} \tag{3.84}$$

We now define four new gain factors that are equal to the coefficients in these equations. Thus (3.83) and (3.84) are written in the form

$$v_{od} = A_{dm}v_{id} + A_{cm-dm}v_{ic} \tag{3.85}$$

$$v_{oc} = A_{dm-cm}v_{id} + A_{cm}v_{ic} \tag{3.86}$$

The *differential-mode gain*, A_{dm}, is the change in differential output $(v_{o1} - v_{o2})$ per unit change in differential input $v_{i1} - v_{i2}$. Its value for the emitter-coupled pair is:

$$A_{dm} = \frac{A_{11} - A_{12} - A_{21} + A_{22}}{2} = (-g_m R_C) \tag{3.87}$$

using (3.73) and (3.74). The value of this gain is usually large compared with the other three coefficients in (3.85) and (3.86).

The *common-mode gain*, A_{cm}, is the change in common-mode output voltage $[(v_{o1} + v_{o2})/2]$ per unit change in common-mode input. Its value for the emitter-coupled pair is:

$$A_{cm} = \frac{A_{11} + A_{12} + A_{21} + A_{22}}{2} = \frac{-g_m R_C}{1 + 2g_m R_{EE}\left(1 + \dfrac{1}{\beta_0}\right)} \tag{3.88}$$

using (3.73) and (3.74). The remaining terms, the *differential-mode-to-common-mode gain*, A_{dm-cm}, and the *common-mode-to-differential-mode gain*, A_{cm-dm}, are the changes in common-mode and differential-mode outputs for a unit change in differential and common-mode inputs, respectively. These two quantities are given by:

$$A_{cm-dm} = A_{11} + A_{12} - A_{21} - A_{22} = 0 \tag{3.89}$$

$$A_{dm-cm} = \frac{A_{11} - A_{12} + A_{21} - A_{22}}{4} = 0 \tag{3.90}$$

The symmetry of the circuit causes these quantities to be zero for identical transistors and load resistors. The behavior of the circuit in this case is thus completely specified by the two parameters A_{cm} and A_{dm} as opposed to the four parameters used originally. These terms are not precisely zero in actual circuits because of device mismatches; this important second-order effect is discussed in Chapter 6.

EXAMPLE

In the circuit of Fig. 3.32, find the voltage gain v_o/v_i. Assume for bias calculations that $V_{BE} = 0.7$ V.

1. dc analysis: $-(I_{E1} + I_{E2}) = \dfrac{15\text{ V} - 0.7\text{ V}}{14.3\text{ k}\Omega} = 1\text{ mA}$

$$I_{C1} = I_{C2} \approx -I_{E1} = 0.5\text{ mA}$$

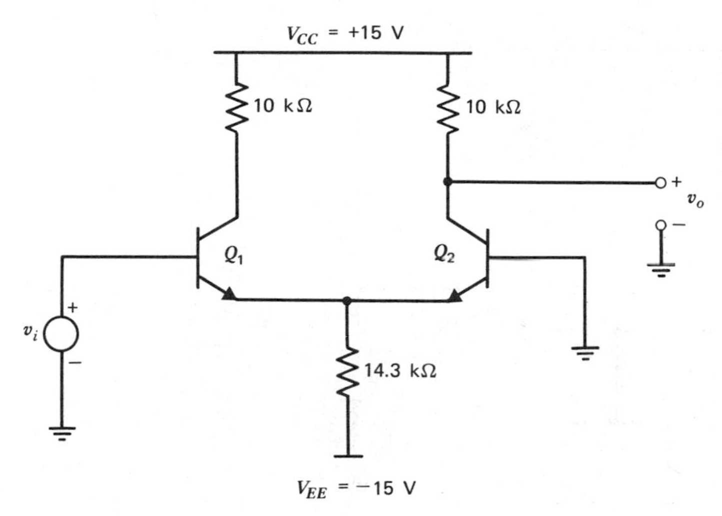

Fig. 3.32 Example circuit for calculation of voltage gain.

2. A_{dm}, A_{cm}:

$$A_{dm} = -g_m R_C = -\frac{I_C}{V_T} R_C = -\frac{10 \text{ k}\Omega}{52\Omega} = -192.3$$

$$A_{cm} = \frac{-g_m R_C}{1 + 2g_m R_E\left(1 + \dfrac{1}{\beta_0}\right)} = -0.35$$

3. Gain calculation

$$v_o = v_{o2} = -\frac{v_{od}}{2} + v_{oc} = -A_{dm}\frac{v_{id}}{2} + A_{cm}v_{ic} = \frac{-A_{dm}}{2}(v_i) + A_{cm}\left(\frac{v_i}{2}\right)$$

$$= \left(\frac{+192.3}{2} - \frac{0.35}{2}\right) v_i = 95.98\, v_i$$

3.4.4 Differential and Common-Mode Gain Using the Half-Circuit Concept

In the previous section the common-mode and differential-mode gain were calculated by direct analysis of the small-signal circuit. The same results can be obtained by determining the response of the circuit to pure differential-mode and pure common-mode inputs separately, and superposing the results. This approach yields additional insight into the function of the emitter-coupled pair. The response of the circuit to pure differential-mode signals is first calculated.

The circuit of Fig. 3.31 is redrawn in Fig. 3.33 with the common-mode input voltage set to zero. The small-signal equivalent circuit is shown in Fig. 3.34. Note first that equal and opposite small-signal voltages are applied to the two base terminals. Because the circuit is balanced, the voltage at the emitters of the

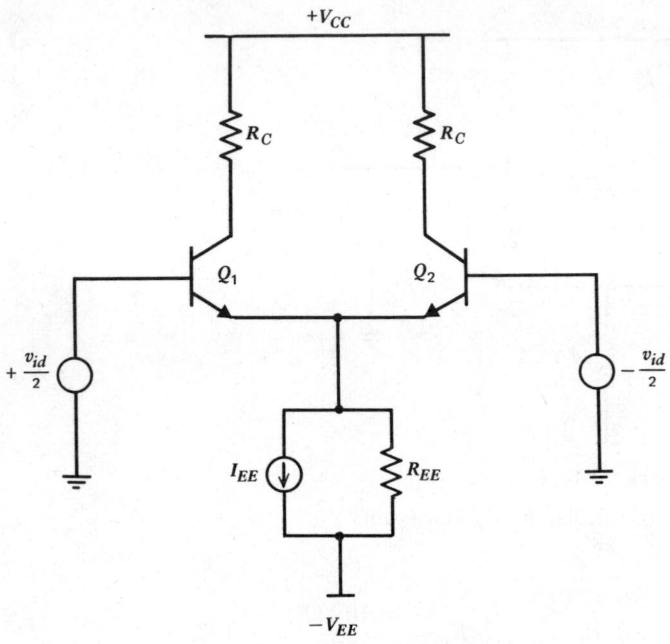

Fig. 3.33 Emitter-coupled pair with pure differential input.

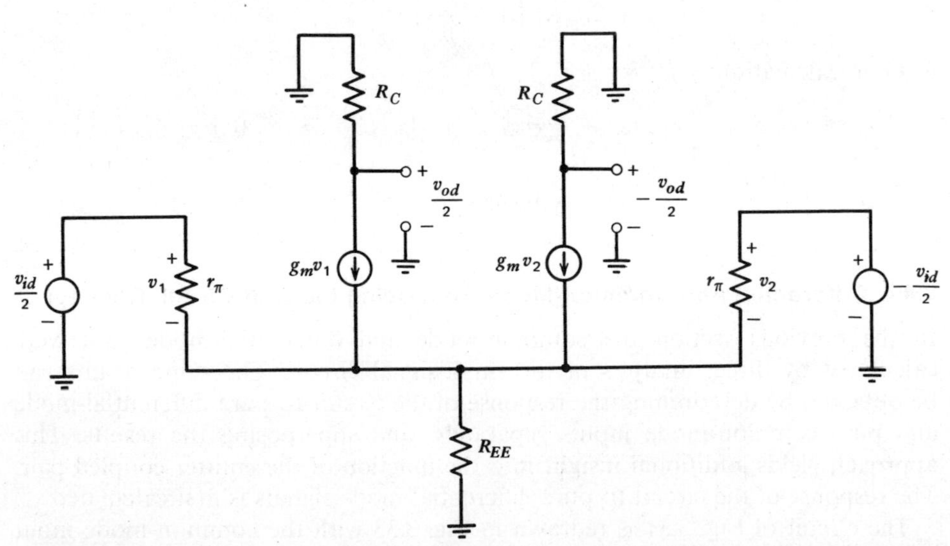

Fig. 3.34 Small-signal equivalent circuit for emitter-coupled pair with pure differential-mode input.

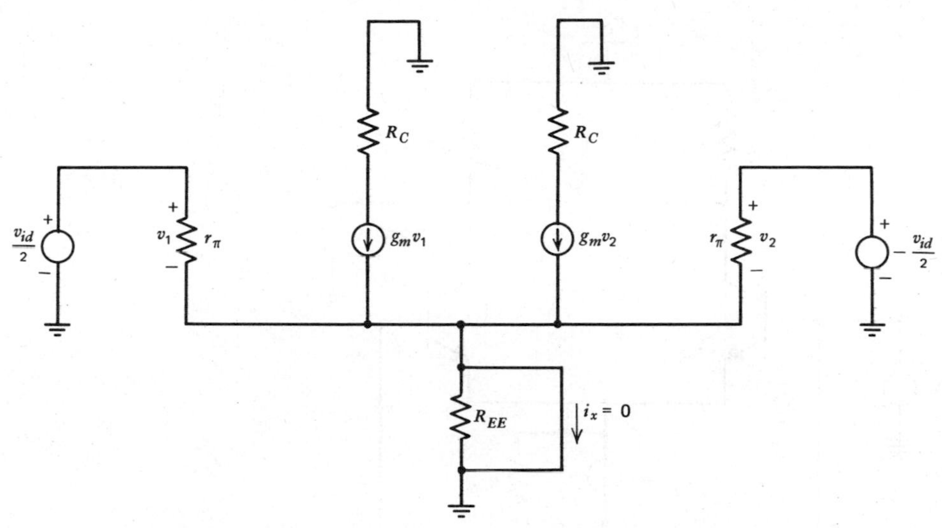

Fig. 3.35 Differential-mode circuit with emitter node grounded. Because of the symmetry of the circuit, $i_x = 0$.

transistors does not vary at all. Since this point experiences no voltage variation, the behavior of the circuit is unaffected by the placement of a short circuit from that point to ground as shown in Fig. 3.35. The gain of the circuit can now be calculated by analyzing only one side of the balanced amplifier. This simplified circuit, shown in Fig. 3.36, is termed the differential-mode half-circuit, and is useful for analysis of both the low- and high-frequency performance of differential amplifiers of all types. By inspection of Fig. 3.36,

$$\frac{v_{od}}{2} = -g_m R_C \frac{v_{id}}{2} \tag{3.91}$$

and

$$A_{dm} = \frac{v_{od}}{v_{id}} = -g_m R_C \tag{3.92}$$

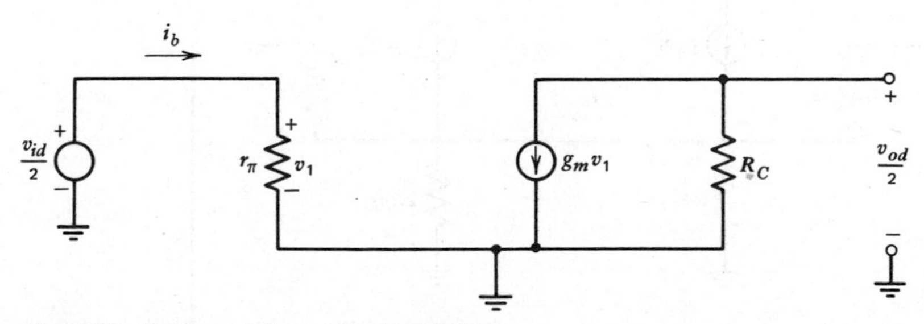

Fig. 3.36 Differential-mode half circuit.

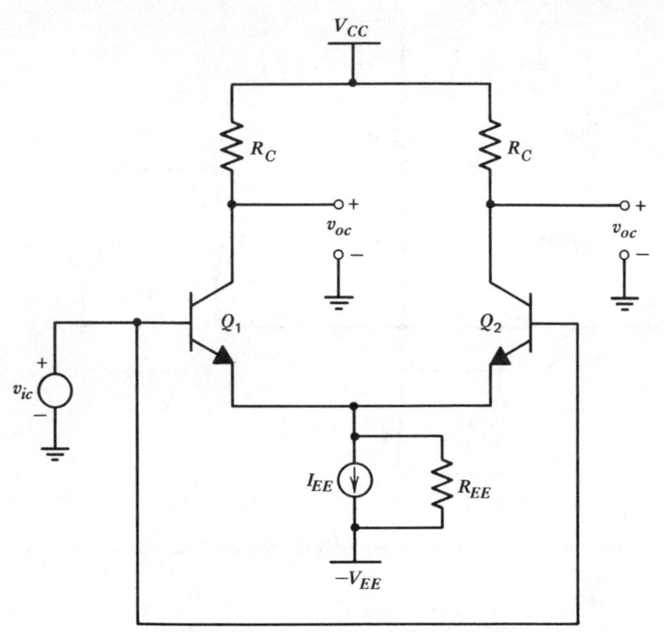

Fig. 3.37 Emitter-coupled pair with pure common-mode input.

The circuit of Fig. 3.31 is now redrawn in Fig. 3.37 with the differential input set to zero. The small-signal equivalent circuit is shown in Fig. 3.38, but with the modification that the resistor R_{EE} has been split into two parallel resistors, each of value twice the original. Now observe that the two transistors have exactly the same voltage applied across the base-emitter junction, which implies that the collector currents must be identical. Note also that because of the symmetry of

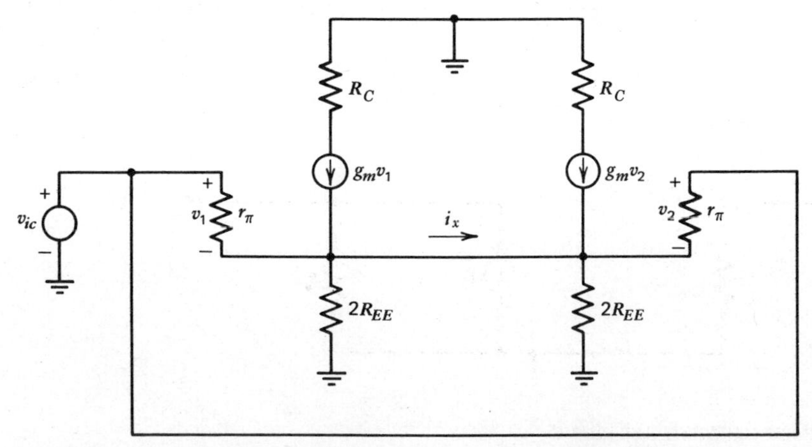

Fig. 3.38 Small-signal equivalent circuit, pure common-mode input.

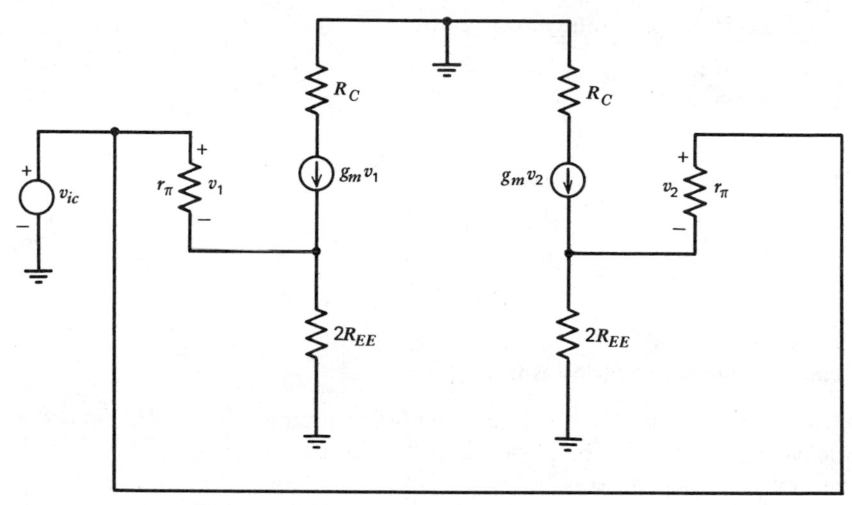

Fig. 3.39 Modified common-mode equivalent circuit.

the circuit, no current i_x flows in the lead connecting the two emitters. The circuit behavior is thus unchanged if this lead is removed as shown in Fig. 3.39, and when this is done the circuit reduces to two half circuits that are completely independent. Only one of the half circuits need be analyzed to predict circuit behavior, and the common-mode half-circuit is shown in Fig. 3.40. Summing voltages around the loop containing the source voltage, base-emitter junction, and $2R_{EE}$, we obtain

$$v_{ic} - i_b r_\pi - i_b(\beta_0 + 1)2R_{EE} = 0 \tag{3.93}$$

or

$$i_b = \frac{v_{ic}}{r_\pi + 2R_{EE}(\beta_0 + 1)} \tag{3.94}$$

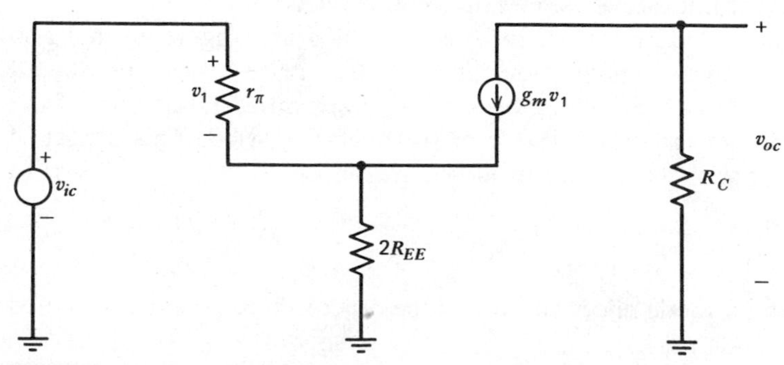

Fig. 3.40 Common-mode half circuit.

The common-mode output voltage is given by:

$$v_{oc} = -R_C i_c = -R_C \beta_0 i_b = v_{ic}\left[\frac{-\beta_0 R_C}{r_\pi + 2R_{EE}(\beta_0 + 1)}\right] \tag{3.95}$$

Rearranging,

$$A_{cm} = \frac{v_{oc}}{v_{ic}} = -\frac{g_m R_C}{1 + 2g_m R_{EE}\left(1 + \dfrac{1}{\beta_0}\right)} \tag{3.96}$$

3.4.5 Common-Mode Rejection Ratio

Most circuit applications of differential amplifiers require the amplification of differential voltages, often in the presence of fluctuating common-mode voltages. Since the desired signal is usually the differential voltage, the response to the common-mode signal produces an error at the output that is indistinguishable from the signal. The minimization of both the common-mode gain and the common-mode to differential-mode gain are thus important objectives in differential amplifier design.

The common-mode rejection ratio (CMRR) is conventionally defined as the magnitude of the ratio of differential-mode to common-mode gain.

$$\text{CMRR} \triangleq \left|\frac{A_{dm}}{A_{cm}}\right| \tag{3.97}$$

For the particular case of the single-stage emitter-coupled pair, using (3.92) and (3.96), we obtain

$$\text{CMRR} = 1 + 2g_m R_{EE}\left(1 + \frac{1}{\beta_0}\right) \tag{3.98}$$

From this expression it is clear that increasing the output resistance of the biasing current source, R_{EE}, will improve the common-mode rejection ratio.

The term common-mode rejection ratio is also used as a figure of merit to describe the performance of multistage differential amplifiers and operational amplifiers. In such circuits, the common-mode-to-differential-mode gain of the first stage is usually an important factor in the overall CMRR. This aspect of differential amplifier performance is considered in Chapter 6.

EXAMPLE

Calculate the common-mode rejection ratio of the resistor-biased emitter-coupled pair shown in Fig. 3.41.

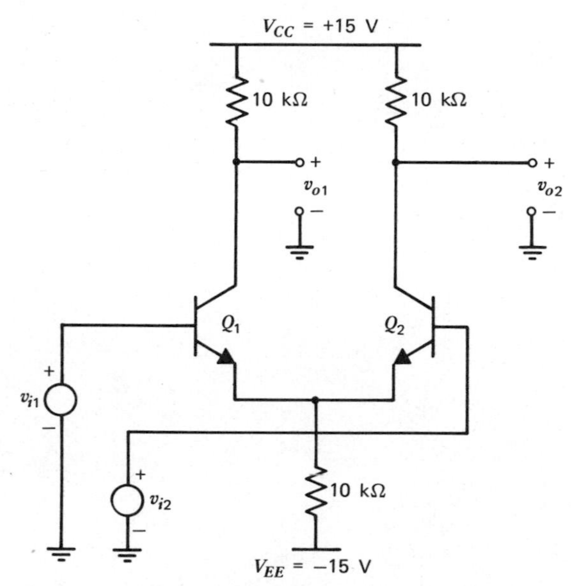

Fig. 3.41 Example circuit for calculation of CMRR.

$$\text{CMRR} = \left[\frac{1}{1 + 2g_m R_{EE}\left(1 + \dfrac{1}{\beta_0}\right)}\right]^{-1}$$

$$\approx \left(\frac{1}{2\dfrac{I_C R_{EE}}{V_T}}\right)^{-1} = \frac{2I_C R_{EE}}{V_T} = \frac{I_{EE} R_{EE}}{V_T}$$

For this case,

$$\text{CMRR} = \frac{14.3\ \text{V}}{26\ \text{mV}} = 550 = 54\ \text{dB}$$

Note that given the resistive biasing it is not possible to improve the CMRR of this basic circuit without increasing the voltage drop across the resistor R_{EE}. In practice this would require increasing the supply voltage, which is often not possible.

3.4.6 Differential and Common-Mode Input Resistance

Since differential amplifiers are often used as the input stage of instrumentation circuits, input resistance is an important design consideration. The differential input resistance, R_{id}, is defined as the ratio of the small-signal differential input voltage, v_{id}, to the small-signal input current, i_b, when a pure differential input

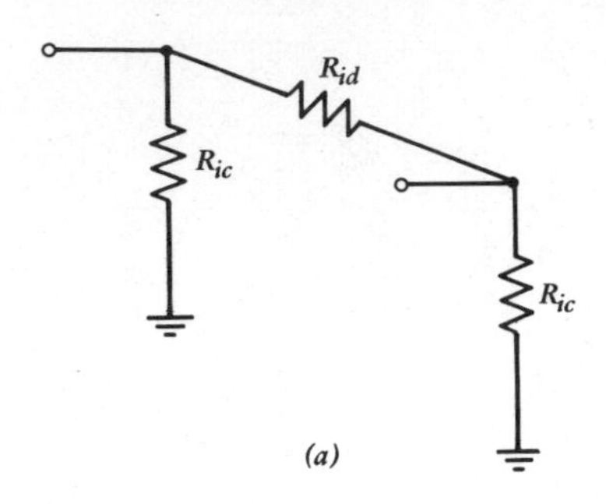

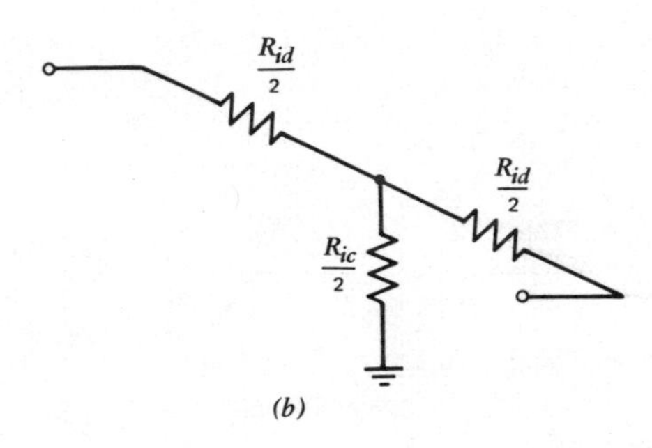

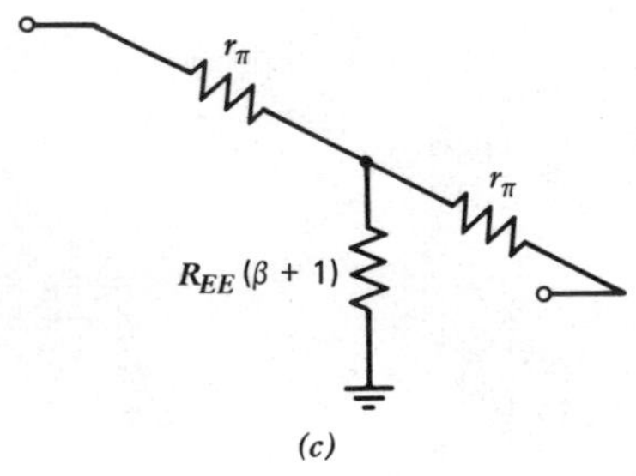

Fig. 3.42 (*a*) General low-frequency, small-signal, equivalent input circuit for the differential amplifier. (*b*) "T" equivalent input circuit. (*c*) Equivalent input circuit for the emitter-coupled pair (with no emitter degeneration resistors).

voltage is applied. By inspecting Fig. 3.36, the input resistance of the emitter-coupled pair is:

$$\frac{v_{id}}{2} = i_b r_\pi \tag{3.99}$$

and thus

$$R_{id} = \frac{v_{id}}{i_b} = 2r_\pi \tag{3.100}$$

Thus the differential input resistance depends on the r_π of the transistor, which increases with increasing beta and decreasing collector current. High input resistance is thus obtained when the emitter-coupled pair is operated at low bias current levels.

The common-mode input resistance is defined as the ratio of the small-signal, common-mode input voltage to the small-signal input current, i_b, in one input terminal when a pure common-mode input is applied. For the emitter-coupled pair, this can be seen by inspection of Fig. 3.39 to be:

$$R_{ic} = \frac{v_{ic}}{i_b} = [r_\pi + 2R_{EE}(1 + \beta_0)] \tag{3.101}$$

using (3.94).

The small-signal input current, which will flow when mixed common-mode and differential-mode input voltages are applied, can be found by superposition and is given by:

$$i_{b1} = \frac{v_{id}}{R_{id}} + \frac{v_{ic}}{R_{ic}}$$

$$i_{b2} = -\frac{v_{id}}{R_{id}} + \frac{v_{ic}}{R_{ic}}$$

The input resistance can thus be represented by the π equivalent circuit of Fig. 3.42*a* assuming that R_{ic} is much larger than R_{id}. Alternatively, this circuit can be transformed to a T-equivalent circuit as shown in Fig. 3.42*b*, again assuming that R_{ic} is much greater than R_{id}. The values of the resistances in the T-equivalent circuit for the emitter-coupled pair are shown in Fig. 3.42*c*.

3.5 SOURCE-COUPLED JFET PAIRS

An important objective in differential amplifier design is minimization of the dc bias current flowing into the input leads of the circuit and maximization of the differential input resistance. Since JFETs have the property that their input resistance and bias current are merely the incremental resistance and leakage current, respectively, of a reverse-biased junction, they are often used in differential amplifiers to improve input resistance and input bias current. In the context of analog ICs, using JFETs involves the addition of considerable complexity to the standard fabrication process, as described in Chapter 2. Thus they are used only when a very high input resistance is required. In this section we analyze both the dc and small-signal behavior of the source-coupled pair.

3.5.1 dc Large-Signal Analysis

The *n*-channel JFET source-coupled pair[8] is shown in Fig. 3.43*a*. The analysis applies equally well to the *p*-channel source-coupled circuit with appropriate sign

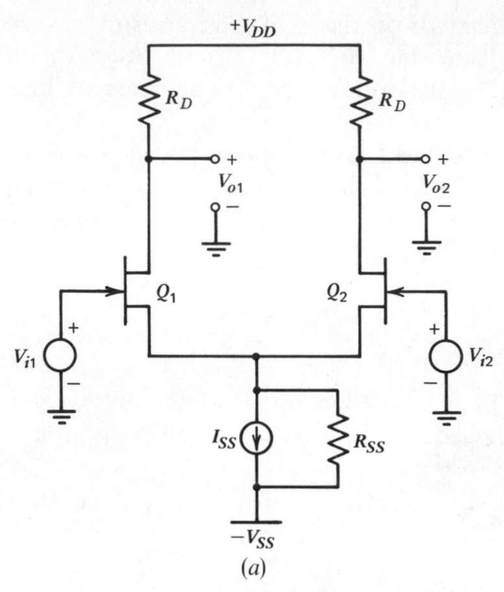

Fig. 3.43*a* *n*-channel JFET source-coupled pair.

changes. For this large-signal analysis, we assume that the incremental output resistance of the biasing current source is infinite, and we neglect the output resistance of the JFETs as well. We assume that the JFETs and drain resistors are respectively identical. We first sum the voltages around the source loop:

$$V_{i1} - V_{gs1} + V_{gs2} - V_{i2} = 0 \tag{3.102}$$

For each JFET, we assume that the drain current is related to V_{gs} by the approximate square-law relationship discussed in Chapter 1, and given in (1.150),

$$I_d = I_{DSS}\left(1 - \frac{V_{gs}}{V_P}\right)^2 \tag{3.103}$$

or

$$\frac{V_{gs}}{V_P} = \left(1 - \sqrt{\frac{I_d}{I_{DSS}}}\right) \tag{3.104}$$

Combining (3.102) and (3.104)

$$\frac{V_{i1} - V_{i2}}{V_P} = -\sqrt{\frac{I_{d1}}{I_{DSS}}} + \sqrt{\frac{I_{d2}}{I_{DSS}}} \tag{3.105}$$

Summing currents at the source node,

$$I_{d1} + I_{d2} = I_{SS} \tag{3.106}$$

Combining (3.105) and (3.106), and solving the resulting quadratic, yields

$$I_{d1} = \frac{I_{SS}}{2}\left[1 + \frac{V_{id}}{V_P}\sqrt{2\left(\frac{I_{DSS}}{I_{SS}}\right) - \left(\frac{V_{id}}{V_P}\right)^2\left(\frac{I_{DSS}}{I_{SS}}\right)^2}\right] \tag{3.107}$$

$$I_{d2} = \frac{I_{SS}}{2}\left[1 - \frac{V_{id}}{V_P}\sqrt{2\left(\frac{I_{DSS}}{I_{SS}}\right) - \left(\frac{V_{id}}{V_P}\right)^2\left(\frac{I_{DSS}}{I_{SS}}\right)^2}\right] \tag{3.108}$$

where $V_{id} = V_{i1} - V_{i2}$. Note that when a large dc differential input voltage is applied, all of the bias current I_{SS} must flow in one of the JFETs. Thus if I_{SS} were *larger* than I_{DSS}, the gate-channel junctions of the JFETs would become forward biased when a large differential input signal were applied. Thus, in the analysis above, the implicit assumption is made that $I_{SS} \leq I_{DSS}$. Also, the range of differential input voltages for which both transistors conduct current is given by:

$$\left|\frac{V_{id}}{V_P}\right| < \sqrt{\frac{2I_{SS}}{I_{DSS}}} \tag{3.109}$$

Outside this range of V_{id}, the output currents take on a constant value of either zero or I_{SS}. Equations 3.108 and 3.107 are valid only for the range of V_{id} given by (3.109). The two currents are plotted for several typical cases in Fig. 3.43*b*. When compared with the corresponding curves for the bipolar case (Fig. 3.27), the JFET circuit is seen to have a much higher range of V_{id} over which the circuit behaves in a near-linear way. This range is a fraction of the pinch-off voltage V_P of the JFET (typically 2 to 5 V) rather than a comparable fraction of V_T (26 mV) as in the bipolar case. The differential output voltage, $V_{od} = (V_{o1} - V_{o2})$, is given by:

$$V_{od} = -(I_{d1}R_D - I_{d2}R_D) \tag{3.110}$$

Substitution of (3.107) and (3.108) in (3.110) gives

$$V_{od} = -\frac{I_{SS}R_D}{V_P}V_{id} \times \sqrt{2\left(\frac{I_{DSS}}{I_{SS}}\right) - \left(\frac{V_{id}}{V_P}\right)^2\left(\frac{I_{DSS}}{I_{SS}}\right)^2} \tag{3.111}$$

3.5.2 Small-Signal Analysis

Since the small-signal model for the JFET is similar to that for the bipolar transistor, the results from the emitter-coupled pair can be used with slight modification. For the JFET, r_π is essentially infinite, and the transconductance is given by (1.153), which we repeat here.

$$g_m = -\frac{2I_{DSS}}{V_P}\left(1 - \frac{V_{GS}}{V_P}\right) = \frac{2}{|V_P|}\sqrt{|I_D|\,|I_{DSS}|} \tag{3.112}$$

The differential and common-mode gains for the source-coupled pair are given by (3.87) and (3.88), using the value of g_m given by (3.112).

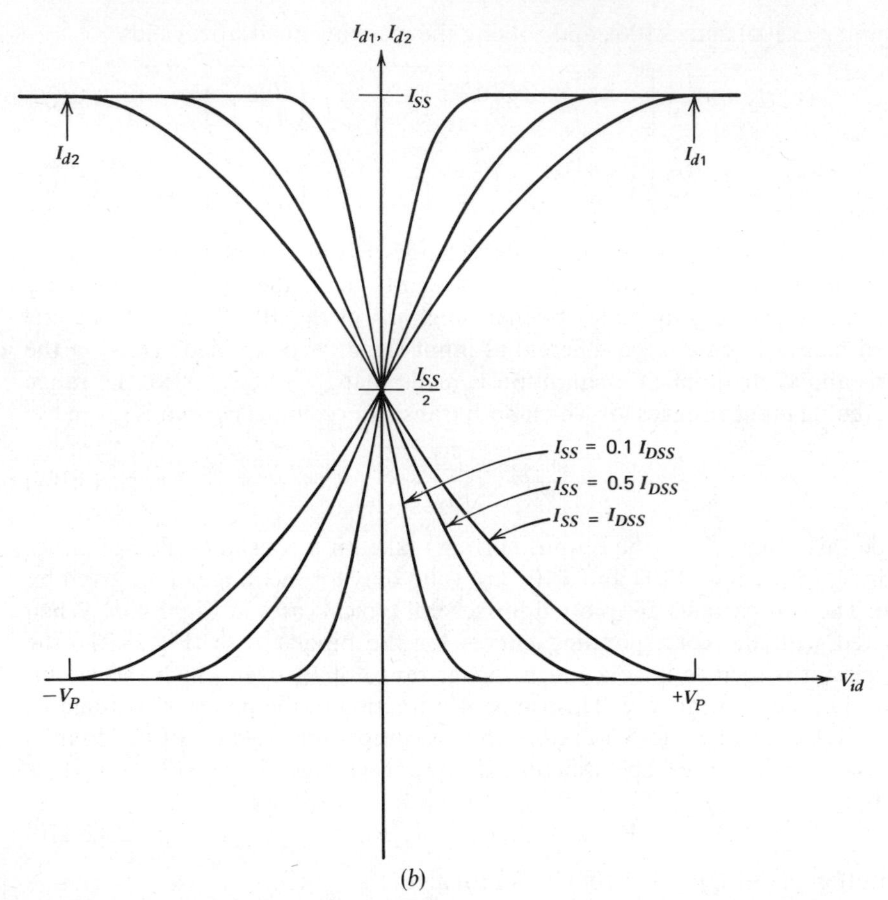

(*b*)

Fig. 3.43*b* dc transfer characteristic, JFET source-coupled pair.

In comparison with the bipolar transistor operated at the same bias current, the transconductance of the JFET is much lower, typically by a factor of about 40. Thus, for the same value of load resistor, the JFET differential amplifier will have a lower differential voltage gain than the bipolar circuit. However, the common-mode gain is not significantly lower, so the JFET circuit has a much worse common-mode rejection ratio.

EXAMPLE

Find the differential-mode and common-mode gain and the CMRR of the source-coupled pair shown in Fig. 3.44. Assume that $V_P = -2$ V and that $I_{DSS} = 2$ mA.

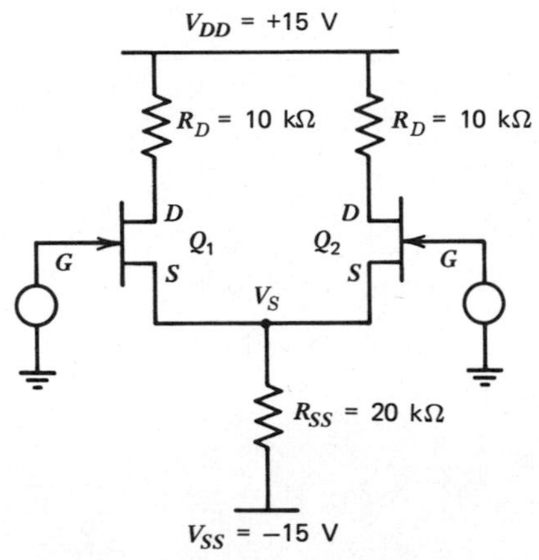

Fig. 3.44 Example source-coupled circuit for calculation of CM and DM gain.

The first task is to determine the dc drain current and gate-source voltage of the JFETs. Consider the dc common-mode half circuit of Fig. 3.45 for this circuit,

$$V_{GS} + 2I_D R_{SS} - 15 \text{ V} = 0$$

But

$$V_{GS} = V_P\left(1 - \sqrt{\frac{I_D}{I_{DSS}}}\right)$$

Thus

$$V_P\left(1 - \sqrt{\frac{I_D}{I_{DSS}}}\right) + 2I_D R_{SS} = 15 \text{ V}$$

This quadratic equation yields the solution:

$$I_{D1,2} = I_{DSS}\left\{\frac{V_P}{4I_{DSS}R_{SS}}\left[1 - \sqrt{1 - \frac{8I_{DSS}R_{SS}}{V_P}\left(1 - \frac{15 \text{ V}}{V_P}\right)}\right]\right\}^2$$

$$= 0.40 \text{ mA}$$

and the gate-source voltage is:

$$V_{GS} = V_P\left(1 - \sqrt{\frac{I_D}{I_{DSS}}}\right) = -1.1 \text{ V}$$

Therefore,

$$V_S = +1.1 \text{ V with respect to ground}$$

and:

$$I_{SS} = \frac{16.1 \text{ V}}{20 \text{ k}\Omega} \cong 0.80 \text{ mA}$$

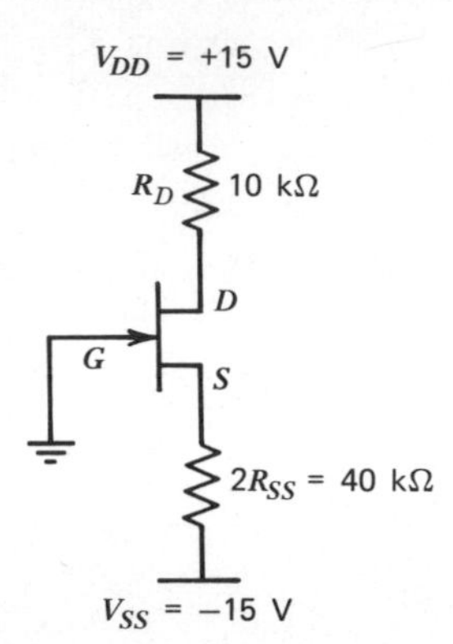

Fig. 3.45 Circuit for calculation of the dc drain current, I_D.

We can now calculate g_m:

$$g_m = \frac{2}{|V_P|}\sqrt{|I_D \cdot I_{DSS}|} = \left(\frac{2}{2\text{ V}}\right)[\sqrt{(0.4)(2) \times 10^{-6}}] = 0.89 \times 10^{-3}\text{ mho}$$

and

$$A_{dm} = -g_m R_D = (0.89 \times 10^{-3})(10\text{ k}\Omega) = \underline{-8.9}$$

$$A_{cm} = \frac{-g_m R_D}{1 + 2g_m R_{SS}} = \frac{-8.9}{1 + 2(0.89 \times 10^{-3})(20\text{ k}\Omega)} = -0.\underline{24}$$

$$\text{CMRR} = \frac{A_{dm}}{A_{cm}} = 37 = \underline{31\text{ dB}}$$

3.6 DEVICE MISMATCH EFFECTS IN DIFFERENTIAL AMPLIFIERS

An important aspect of the performance of differential amplifiers is the minimum dc differential voltage that can be detected. The presence of component mismatches within the amplifier itself and drifts of component values with temperature produce differential voltages at the output that are indistinguishable from the signal being amplified. In many analog systems, this type of dc error is the basic limitation on the resolution of the system, and hence consideration of mismatch-induced offsets is often central to the design of analog circuits.

For transistor differential amplifiers, the effect of mismatches on dc performance is most conveniently represented by two quantities, the input offset voltage and the input offset current. These quantities represent the effect of all of the component mismatches within the amplifier referred to the input[6,7]. As illustrated in Fig. 3.46, the dc behavior of the amplifier containing the mismatches is identical to an ideal amplifier with no mismatches but with the input offset voltage source added in

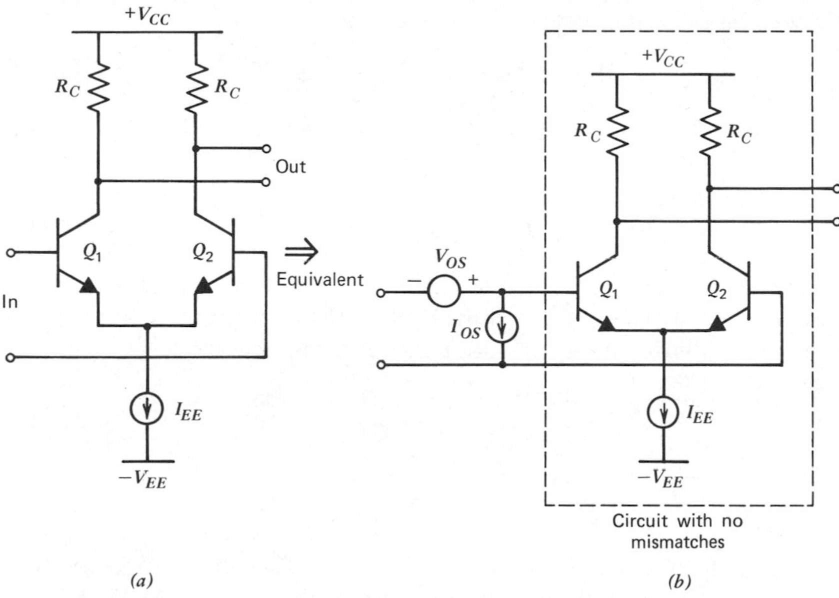

Fig. 3.46 Equivalent input offset voltage (V_{OS}) and current (I_{OS}) for an emitter-coupled pair. (*a*) Actual circuit containing mismatches. (*b*) Equivalent circuit with identically matched devices and the offset voltage and current referred to the input.

series with the input and the input offset current source in shunt across the input terminals. These quantities are usually a function of both temperature and common-mode input voltage. In this section we calculate the input offset voltage and current of both the emitter-coupled pair and the source-coupled JFET pair.

3.6.1 Input Offset Voltage of the Emitter-Coupled Pair

The predominant sources of offset error in the emitter-coupled pair shown in Fig. 3.46*a* are the mismatches in the base width, base doping level, and collector doping level of the transistors, mismatches in the effective emitter area of the transistors, and mismatches in the collector load resistors. In order to provide analytical results simple enough for intuitive interpretation, the analysis will be carried out assuming a uniform-base transistor. The results are similar for the nonuniform case, although the analytical procedure is more tedious. In most instances the dc base current is low enough that the dc voltage drop in r_b is negligible, so we neglect r_b.

Referring to Fig. 3.46*b*, we note that the input offset voltage V_{OS} is equal to the value of V_{id} which must be applied to the input to drive the differential output voltage to zero. Assuming that V_{id} has been adjusted to drive V_{od} to zero in

Fig. 3.46*a*, we sum the voltages around the source loop

$$V_{OS} - V_{BE1} + V_{BE2} = 0 \tag{3.113}$$

$$V_{OS} = V_T \ln \frac{I_{C1}}{I_{S1}} - V_T \ln \frac{I_{C2}}{I_{S2}} \tag{3.114}$$

$$= V_T \ln \frac{I_{C1}}{I_{C2}} \frac{I_{S2}}{I_{S1}} \tag{3.115}$$

The factors determining the saturation current, I_S, of a bipolar transistor were discussed in Chapter 1. There it was shown that if the impurity concentration in the base region is uniform, these saturation currents can be written

$$I_{S1} = \frac{qn_i^2\bar{D}_n}{N_A W_{B1}(V_{CB})} A_1 = \frac{qn_i^2\bar{D}_n}{Q_{B1}(V_{CB})} A_1 \tag{3.116}$$

$$I_{S2} = \frac{qn_i^2\bar{D}_n}{N_A W_{B2}(V_{CB})} A_2 = \frac{qn_i^2\bar{D}_n}{Q_{B2}(V_{CB})} A_2 \tag{3.117}$$

where $W_B(V_{CB})$ is the base width as a function of V_{CB}, N_A is the acceptor density in the base, and A is the emitter area. We denote the product $N_A W_B(V_{CB})$ as $Q_B(V_{CB})$, the total base impurity doping per unit area. In order for V_{od} to be zero,

$$I_{C1}R_{C1} = I_{C2}R_{C2}, \qquad \frac{I_{C1}}{I_{C2}} = \frac{R_{C2}}{R_{C1}} \tag{3.118}$$

Combining these relationships,

$$V_{OS} = V_T \ln \left[\left(\frac{R_{C2}}{R_{C1}} \right) \left(\frac{A_2}{A_1} \right) \left(\frac{Q_{B1}(V_{CB})}{Q_{B2}(V_{CB})} \right) \right] \tag{3.119}$$

This expression relates the input offset voltage to the device parameters and R_C mismatch. Usually, however, the argument of the log function is very close to unity and the equation can be interpreted in a more intuitively satisfying way. In the following section we perform an approximate analysis, valid if the mismatches are small.

3.6.2 Offset Voltage of the Emitter-Coupled Pair-Approximate Analysis

In cases of practical interest involving offset voltages and currents, the mismatch between any two nominally matched circuit parameters is usually small compared to the absolute value of that parameter. This leads to a procedure by which the individual contributions to offset voltage can be considered separately and summed. Rewriting (3.119),

$$V_{OS} = V_T \ln \left[\left(\frac{R_{C2}}{R_{C1}} \right) \left(\frac{A_2}{A_1} \right) \left(\frac{Q_{B1}(V_{CB})}{Q_{B2}(V_{CB})} \right) \right] \tag{3.120}$$

Let us define new parameters to describe the mismatch in the components, using the relations

$$\Delta X = X_1 - X_2 \tag{3.121}$$

$$X = \frac{X_1 + X_2}{2} \tag{3.122}$$

Thus ΔX is the difference between two parameters and X is the average of the two nominally matched parameters. Note that ΔX can be positive or negative. For example,

$$\Delta R_C = R_{C1} - R_{C2}$$

$$R_C = \frac{R_{C1} + R_{C2}}{2} \tag{3.123}$$

We first invert (3.121) and (3.122) to give:

$$X_1 = X + \frac{\Delta X}{2} \tag{3.124}$$

$$X_2 = X - \frac{\Delta X}{2} \tag{3.125}$$

These relations can be used with (3.120) to give:

$$V_{OS} = V_T \ln \left[\left(\frac{R_C - \dfrac{\Delta R_C}{2}}{R_C + \dfrac{\Delta R_C}{2}} \right) \left(\frac{A - \dfrac{\Delta A}{2}}{A + \dfrac{\Delta A}{2}} \right) \left(\frac{Q_B + \dfrac{\Delta Q_B}{2}}{Q_B - \dfrac{\Delta Q_B}{2}} \right) \right] \tag{3.126}$$

With the assumptions that:

$$\Delta R_C \ll R_C \qquad \Delta A \ll A \qquad \Delta Q_B \ll Q_B \tag{3.127}$$

this expression can be simplified to:

$$V_{OS} = V_T \ln \left[\left(1 - \frac{\Delta R_C}{R_C} \right) \left(1 - \frac{\Delta A}{A} \right) \left(1 + \frac{\Delta Q_B}{Q_B} \right) \right] \tag{3.128}$$

The Taylor series for the logarithm function can be used to expand this expression. If the higher-order terms in the expansion are neglected, we obtain:

$$V_{OS} = V_T \left(-\frac{\Delta R_C}{R_C} - \frac{\Delta A}{A} + \frac{\Delta Q_B}{Q_B} \right) \tag{3.129}$$

Thus, under the assumptions made, we have obtained an expression for the input offset voltage, which is the linear superposition of the effects of the different components. It can be shown that this can always be done for small component mismatches. Note that the signs of the individual terms of (3.129) are not particularly significant, since the mismatch factors can be positive or negative depending on the direction of the random parameter variation. The worst-case offset occurs when the terms have signs such that the individual contributions add.

Equation 3.129 relates the offset voltage to mismatches in the resistors and in the structural parameters, A and Q_B, of the transistors. For the purpose of predicting the offset voltage from device parameters that are directly measurable electrically, it is convenient to recombine (3.129) to express the offset in terms of the resistor mismatch and the mismatch in the saturation currents of the transistors.

$$V_{OS} = V_T \left(-\frac{\Delta R_C}{R_C} - \frac{\Delta I_S}{I_S} \right) \tag{3.130}$$

where

$$\frac{\Delta I_S}{I_S} = \frac{\Delta A}{A} - \frac{\Delta Q_B}{Q_B}$$

is the offset voltage contribution from the transistors themselves, as reflected in the mismatch in saturation current. Mismatch factors $\Delta R_C/R_C$ and $\Delta I_S/I_S$ are actually random parameters that take on a different value for each circuit fabricated, and the distribution of the observed values is described by a probability distribution. For large samples the distribution tends toward a normal, or Gaussian, distribution with zero mean. Typically observed standard deviations for the mismatch parameters above are:

$$\sigma_{\Delta R/R} = .01 \qquad \sigma_{\Delta I_S/I_S} = .05 \tag{3.131}$$

In the Gaussian distribution, 68 percent of the samples have a value within $\pm\,\sigma$ of the mean value. If we assume the mean value of the distribution is zero, then 68 percent of the resistor pairs in a large sample will match within 1 percent, and 68 percent of the transistor pairs will have saturation currents that match within 5 percent for the distributions described by (3.131). These values can be heavily influenced by device geometry and processing as discussed further in Chapter 6. If, from the distributions, we pick one sample with parameters equal to the standard deviation, and the mismatch factors are in the direction that they add, the resulting offset from (3.130) would be:

$$V_{OS} = (26\text{ mV})(.06) \cong 1.5\text{ mV} \tag{3.132}$$

A parameter of more interest to the circuit designer is the standard deviation of the total offset voltage. Since the offset is the sum of two uncorrelated random parameters, the standard deviation of the sum is equal to the square root of the sum of the squares of the standard deviation of the two mismatch contributions, or

$$\sigma_{V_{OS}} = V_T \sqrt{(\sigma_{\Delta R/R})^2 + (\sigma_{\Delta I_S/I_S})^2} \tag{3.133}$$

The properties of the Gaussian distribution are summarized in Appendix A3.1.

3.6.3 Offset Voltage Drift in the Emitter-Coupled Pair

When emitter-coupled pairs are used as low-level dc amplifiers where the offset voltage is critical, provision is often made to manually adjust the input offset voltage to zero with an external potentiometer. When this is done, the important parameter becomes not the offset voltage itself, but the variation of this offset voltage with temperature, often referred to as drift. For most practical circuits, the sensitivity of the input offset voltage to temperature is not zero, and the wider the excursion of temperature experienced by the circuit, the more error the offset voltage drift will contribute. This parameter is easily calculated for the emitter-coupled pair by differentiating (3.129)

$$\frac{dV_{OS}}{dT} = \frac{V_{OS}}{T} \tag{3.134}$$

using $V_T = kT/q$ and assuming the other terms in (3.129) are independent of temperature. Thus, the drift and offset are proportional for the emitter-coupled pair. Under the assumptions we have made, an emitter-coupled pair with a measured offset voltage of, for example, 2mV, would display a drift of 2mV/300°K, or 6.6 μV/°C. This relationship is observed experimentally.

It would appear from (3.134) that the drift would also be nulled by externally adjusting the offset to zero. This is only approximately true because of the way in which the nulling is accomplished.[9] Usually an external potentiometer is placed in parallel with a portion of one of the collector load resistors in the pair. The temperature coefficient of the nulling potentiometer generally does not match that of the diffused resistors, so a resistor-mismatch temperature coefficient is introduced that can make the drift worse than it was with no nulling. Voltage drifts in the 1 μV/°C range can be obtained with careful design.

3.6.4 Input Offset Current of the Emitter-Coupled Pair

The input offset current is equal to the difference between the two base currents of the transistors. Since the base current is equal to the collector current divided by beta, the offset current is:

$$I_{OS} = \frac{I_{C1}}{\beta_{F1}} - \frac{I_{C2}}{\beta_{F2}} \tag{3.135}$$

As before, we can write:

$$I_{C1} = I_C + \frac{\Delta I_C}{2} \qquad I_{C2} = I_C - \frac{\Delta I_C}{2} \tag{3.136}$$

$$\beta_{F1} = \beta_F + \frac{\Delta \beta_F}{2} \qquad \beta_{F2} = \beta_F - \frac{\Delta \beta_F}{2} \tag{3.137}$$

The offset current becomes, inserting (3.136) and (3.137) into (3.135),

$$I_{OS} = \left(\frac{I_C + \frac{\Delta I_C}{2}}{\beta_F + \frac{\Delta \beta_F}{2}} - \frac{I_C - \frac{\Delta I_C}{2}}{\beta_F - \frac{\Delta \beta_F}{2}} \right) \tag{3.138}$$

Neglecting higher-order terms, this becomes:

$$I_{OS} = \frac{I_C}{\beta_F}\left(\frac{\Delta I_C}{I_C} - \frac{\Delta \beta_F}{\beta_F}\right) \tag{3.139}$$

The mismatch in collector currents is, from (3.118),

$$\frac{\Delta I_C}{I_C} = -\frac{\Delta R_C}{R_C} \tag{3.140}$$

so that:

$$I_{OS} = -\frac{I_C}{\beta_F}\frac{\Delta R_C}{R_C} - \frac{I_C}{\beta_F}\frac{\Delta \beta_F}{\beta_F} \tag{3.141}$$

A typically observed beta mismatch distribution displays a standard deviation of about 10 percent. Assuming a beta mismatch of 10 percent and a mismatch in collector resistors of 1 percent,

$$I_{OS} = -\frac{I_C}{\beta_F}\left(\frac{\Delta R_C}{R_C} + \frac{\Delta \beta_F}{\beta_F}\right) = -\frac{I_C}{\beta_F}(.11) = -.11 I_B \tag{3.142}$$

The input offset current as well as the input current itself must be minimized in many applications, a good example being the input stage of operational amplifiers. Various circuit and technological approaches to their minimization are discussed in Chapter 6.

3.6.5 Input Offset Voltage and Drift of the Source-Coupled JFET Pair

As mentioned earlier in the chapter, field-effect transistors inherently provide higher input resistance and lower input bias current than bipolar transistors when used as source-coupled pair amplifiers. However, the lower transconductance of the devices generally results in poorer input offset voltage and common-mode rejection ratio than in the case of bipolar transistors.[10] In this section we calculate these parameters for the source-coupled pair.

The circuit to be analyzed is shown in Fig. 3.47. Referring to Fig. 1.27, we can see that the predominant sources of offset are mismatches in channel width Z, channel length L, thickness $2a$, and doping N_D. The analysis is again carried out assuming a uniform channel, but similar results are obtained for nonuniform channels.

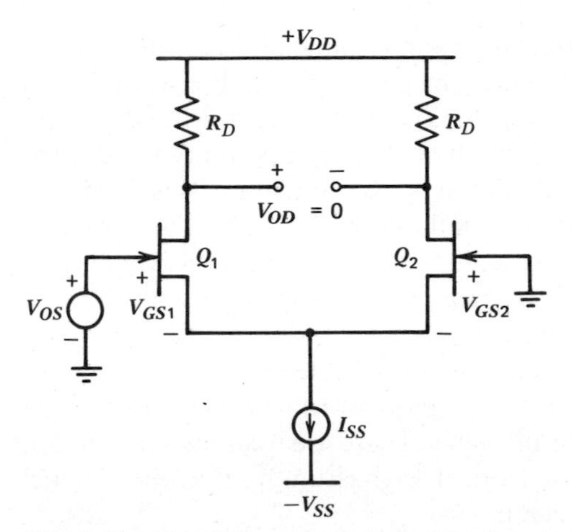

Fig. 3.47 Circuit for calculation of the input offset voltage of the source-coupled pair.

Assuming again that V_{id} has been adjusted to drive V_{od} to zero, we sum voltages around the source loop.

$$V_{OS} - V_{GS1} + V_{GS2} = 0 \tag{3.143}$$

Using (3.104), this becomes:

$$V_{OS} = V_{P1}\left(1 - \sqrt{\frac{I_{D1}}{I_{DSS1}}}\right) - V_{P2}\left(1 - \sqrt{\frac{I_{D2}}{I_{DSS2}}}\right) \tag{3.144}$$

We now use the relationships

$$I_{D1} = I_D + \frac{\Delta I_D}{2} \qquad I_{D2} = I_D - \frac{\Delta I_D}{2}$$

$$I_{DSS1} = I_{DSS} + \frac{\Delta I_{DDS}}{2} \qquad I_{DSS2} = I_{DSS} - \frac{\Delta I_{DSS}}{2}$$

$$V_{P1} = V_P + \frac{\Delta V_P}{2} \qquad V_{P2} = V_P - \frac{\Delta V_P}{2}$$

Substituting these into (3.144), and neglecting higher-order terms we obtain

$$V_{OS} = \Delta V_P - V_P\sqrt{\frac{I_D}{I_{DSS}}}\left(\frac{\Delta I_D}{2I_D} - \frac{\Delta I_{DSS}}{2I_{DSS}} + \frac{\Delta V_P}{V_P}\right) \tag{3.145}$$

assuming $\Delta V_P \ll V_P$, $\Delta I_D \ll I_D$, $\Delta I_{DSS} \ll I_{DSS}$. The offset voltage thus consists of two parts, one equal to pinch-off voltage mismatch and one that contains mismatches in load resistance (through I_D), I_{DSS}, and the pinch-off voltage. The pinch-off voltage depends on channel thickness and doping, and I_{DSS} depends on channel width, length, thickness, and doping. The evaluation of the second term of structural parameters is tedious. However, this term is current dependent, and can generally be made small compared to the first term by operating the devices at very low currents compared to I_{DSS}. In fact, it is experimentally observed that mean offset over a large sample of circuits is reduced by reducing the operating current to the lowest practicable value. Note, however, that in an individual pair the two terms can be opposite in sign, giving an offset voltage that goes through zero at some particular value of bias current.

Assuming that the second term has been made negligible by operating at very low current, the offset can be calculated in a straightforward way. Here we assume the uniform channel case. The results for the general case are qualitatively similar. In terms of the channel half-thickness, a, from (1.151) assuming that the channel is lightly doped compared to the gate region,

$$V_P = \left(\frac{qN_D}{2\epsilon}a^2 - \psi_0\right)$$

Assuming the channel doping, N_D, is the same for both devices, and thus that ψ_0 is the same,

$$\Delta V_P = 2\frac{qN_D}{2\epsilon}a^2\left(\frac{\Delta a}{a}\right) = 2(V_P + \psi_0)\left(\frac{\Delta a}{a}\right) \tag{3.146}$$

Assuming a typical thickness mismatch of .025, and a value of $(V_P + \psi_0)$ of 1 V, we obtain

$$\Delta V_P = 50 \text{ mV} \tag{3.147}$$

Note that this is much larger than the offset voltage of the emitter-coupled pair for comparable percentage mismatch. In practice, careful device design and optimized processing can yield channel thickness matching much better than 2.5 percent. However, all else being equal, the sensitivity of V_P to channel width will result in higher offset for the JFET pair than for the bipolar pair.

Offset voltage drift in JFET source-coupled pairs does not show the good correlation with offset voltage observed in bipolar pairs. The offset consists of several terms that have different temperature coefficients. Both V_P and I_{DSS} have a strong temperature dependence, which affects V_{GS} in opposite directions. The temperature dependence of I_{DSS} stems primarily from mobility variation, which gives a negative temperature coefficient to the drain current, while the pinch-off voltage depends on the built-in potential of the gate-channel junction. The latter decreases with temperature and contributes a positive temperature coefficient to the drain current. These two effects can be made to cancel at one value of I_D, which is a useful phenomenon for temperature stable biasing of single-ended amplifiers. It is not greatly useful, however, in *differential* configurations where first-order cancellation of V_{GS} temperature variations is already achieved.

APPENDIX

A3.1 ELEMENTARY STATISTICS AND THE GAUSSIAN DISTRIBUTION

In this chapter reference was made to the fact that, from the viewpoint of the circuit designer, many circuit parameters are best regarded as random variables whose behavior is described by a probability distribution. This is particularly important in the case of a parameter such as offset voltage. Even though the design value of the offset may be zero, random variations in resistors and transistors cause a spread of offset voltage around the mean value, and the size of this spread determines the fraction of circuits that will meet the required offset specification.

There are several factors causing the parameters of an integrated circuit to show random variations as described above. One of these is the randomness of the edge definition when windows are cut in the oxide to form resistors and transistor emitters. In addition, random variations across the wafer in the diffusion of impurities can be a significant factor. These processes usually give rise to a *Gaussian* distribution (sometimes called a *normal* distribution) of the parameters. A Gaussian distribution of a parameter x is specified by a probability density function $p(x)$ given by

$$p(x) = \frac{1}{\sqrt{2\pi}\,\sigma} \exp\left[-\frac{(x-m)^2}{2\sigma^2}\right] \tag{3.148}$$

where σ is the standard deviation of the distribution and m is the mean or average value of x. The significance of this function is that for one particular circuit chosen at random from a large collection of circuits the *probability* of the parameter having values between x and $(x + dx)$ is given by $p(x)\,dx$, which is the *area under the curve* $p(x)$ in the range x to $(x + dx)$. For example, the probability that x has a value less than X is obtained by integrating (3.148) to give

$$P(x < X) = \int_{-\infty}^{X} p(x)\,dx \tag{3.149}$$

$$= \int_{-\infty}^{X} \frac{1}{\sqrt{2\pi}\,\sigma} \exp\left[-\frac{(x-m)^2}{2\sigma^2}\right] dx \tag{3.150}$$

In a large sample, the *fraction of circuits* where x is less than X will be given by the *probability*, $P(x < X)$, and thus this quantity has real practical significance. The probability density function, $p(x)$ in (3.148), is sketched in Fig. 3.48 and shows a characteristic bell shape. The peak value of the distribution occurs when $x = m$ where m is the mean value of x. The standard deviation, σ, is a measure of the *spread* of the distribution, and large values of σ give rise to a broad distribution. The distribution extends to $x = \pm\infty$, as shown by (3.148), but most of the area under the curve is found in the range $x = m \pm 3\sigma$ as will be seen below.

The above development has shown that the probability of the parameter x having values in a certain range is just equal to the *area* under the curve of Fig. 3.48

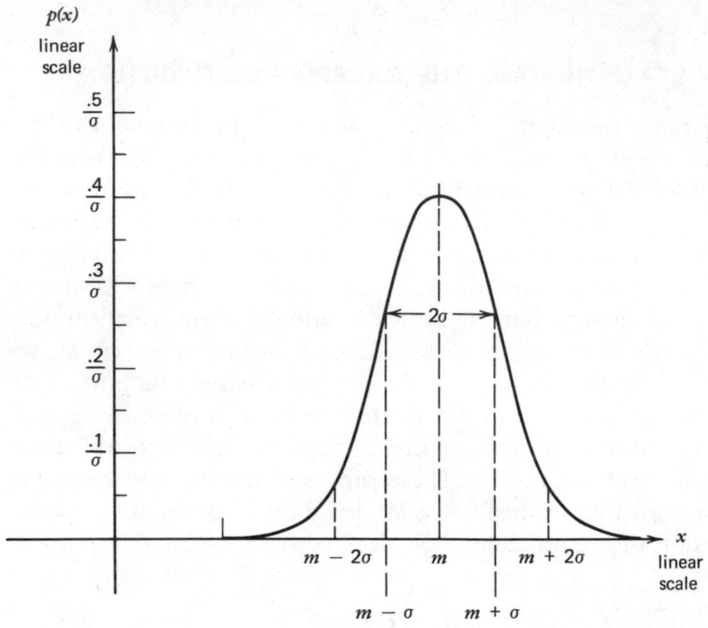

Fig. 3.48 Probability density function $p(x)$ for a Gaussian distribution with mean value m and standard deviation σ. $p(x) = 1/\sqrt{2\pi}\,\sigma\{\exp[-(x - m)^2/2\sigma^2]\}$.

in that range. Since x must lie somewhere in the range $\pm\infty$, the total area under the curve must be unity, and integration of (3.148) will show that this is so. The most common specification of interest to circuit designers is the fraction of a large sample of circuits (the production run) that lies *inside* a band about the mean. For example, if a circuit has a gain x that has a Gaussian distribution with mean value 100, what fraction of circuits have gain values in the range 90 to 110? This fraction can be found by evaluating the probability that x takes on values in the range $x = m \pm 10$ where $m = 100$. This could be found from (3.148) if σ is known by integrating as follows:

$$P(m - 10 < x < m + 10) = \int_{m-10}^{m+10} \frac{1}{\sqrt{2\pi}\,\sigma} \exp\left[-\frac{(x - m)^2}{2\sigma^2}\right] dx \tag{3.151}$$

This is the area under the Gaussian curve in the range $x = m \pm 10$.

In order to simplify calculations of the kind described above, values of the integral in (3.151) have been calculated and tabulated. In order to make the tables general, the range of integration is normalized to σ to give

$$P(m - k\sigma < x < m + k\sigma) = \int_{m-k\sigma}^{m+k\sigma} \frac{1}{\sqrt{2\pi}\,\sigma} \exp\left[-\frac{(x - m)^2}{2\sigma^2}\right] dx \tag{3.152}$$

Values of this integral for various values of k are tabulated in Fig. 3.49. There it is shown that $P = .683$ for $k = 1$ and thus 68.3 percent of a large sample of a Gaussian distribution lies within a range $x = m \pm \sigma$. For $k = 3$, the value of $P = .997$ and thus 99.7 percent of a large sample lies within a range $x = m \pm 3\sigma$.

Circuit parameters such as offset or gain often can be expressed as a linear combination of other parameters as shown in (3.129) for offset voltage. If all the parameters are independent random variables with Gaussian distributions, the standard deviations and means can be related as follows. Assume that the random variable x can be expressed in terms of random variables a, b, and c using

$$x = a + b - c \tag{3.153}$$

Then it can be shown that

$$m_x = m_a + m_b - m_c \tag{3.154}$$

$$\sigma_x^2 = \sigma_a^2 + \sigma_b^2 + \sigma_c^2 \tag{3.155}$$

where m_x is the mean value of x and σ_x is the standard deviation of x. Note that (3.155) shows that the square of the standard deviation of x is the *sum* of the squares of the standard deviations of a, b, and c. This result extends to any number of variables.

The results described above were treated in the context of the random variations found in circuit parameters. However the Gaussian distribution is also useful in the treatment of random noise in Chapter 11.

k	Area under the Gaussian curve in the range $m \pm k\sigma$
0.2	0.159
0.4	0.311
0.6	0.451
0.8	0.576
1	0.683
1.2	0.766
1.4	0.838
1.6	0.890
1.8	0.928
2	0.954
2.2	0.972
2.4	0.984
2.6	0.991
2.8	0.995
3	0.997

Fig. 3.49 Values of the integral in Equation 3.152 for various values of k. This is the area under the Gaussian curve of Fig. 3.48 in the range $x = m \pm k\sigma$.

EXAMPLE

The offset voltage of a circuit has a mean value of $m = 0$ and a standard deviation of $\sigma = 2$ mV. What fraction of circuits will have offsets with magnitudes less than 4 mV?

A range of offset of ± 4 mV corresponds to $\pm 2\sigma$. From Fig. 3.49 we find that the area under the Gaussian curve in this range is .954 and thus 95.4 percent of circuits will have offsets with magnitudes less than 4 mV.

PROBLEMS

For these problems, use the device parameters given in Fig. 2.22 for the high-voltage devices.

3.1 Determine the input resistance, transconductance, and output resistance of the CE amplifier of Fig. 3.3 if $R_C = 20$ kΩ and $I_C = 250$ μA. Assume that $r_b = 0$.

3.2 A CE transistor is to be used in the amplifier of Fig. 3.50 with a source resistance R_S and collector resistor R_C. First find the overall small-signal gain, v_o/v_i, as a function of R_S, R_C, β_0, V_A, and collector current I_C. Next determine the value of dc collector bias current I_C, which maximizes the small-signal voltage gain. Explain qualitatively why the gain falls at very high and very low collector currents. Do not neglect r_o in this problem. What is the voltage gain at the optimum I_C? Assume that $r_b = 0$.

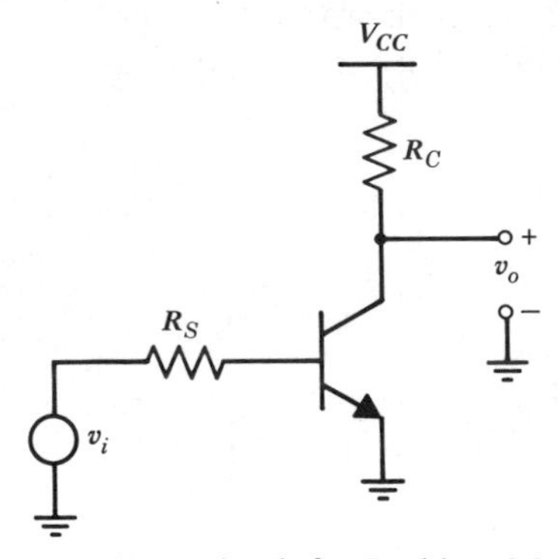

Fig. 3.50 Circuit for Problem 3.2.

3.3 Assume that $R_S = 50$ kΩ, $R_C = 50$ kΩ in Problem 3.2, and calculate the optimum I_C. What is the dc voltage drop across R_C? What is the voltage gain?

3.4 Determine the input resistance, transconductance, and output resistance of the CB amplifier of Fig. 3.9 if $I_C = 250$ μA, $R_C = 10$ kΩ. Neglect r_b and r_o.

3.5 Assume that R_C is made large compared with r_o in the CB amplifier of Fig. 3.9. Use the equivalent circuit of Fig. 3.12*b* and calculate the output resistance for the case of:
(a) The amplifier driven by an ideal current source.
(b) The amplifier driven by an ideal voltage source. Neglect r_b.

3.6 Determine the input resistance, voltage gain v_o/v_s, and output resistance of the CC amplifier of Fig. 3.14*a* if $R_S = 5$ kΩ, $R_L = 500$ Ω, $I_C = 1$ mA. Neglect r_π and r_o. Do not include R_S in calculating the input resistance. Include R_L, however, in calculating the output resistance. Include both R_S and R_L in the gain calculation.

3.7 Determine the dc collector currents in Q_1 and Q_2, and then the input resistance and voltage gain for the Darlington emitter follower of Fig. 3.51. Neglect r_μ and r_o. Assume that $V_{BE(\text{on})} = 0.7$ V.

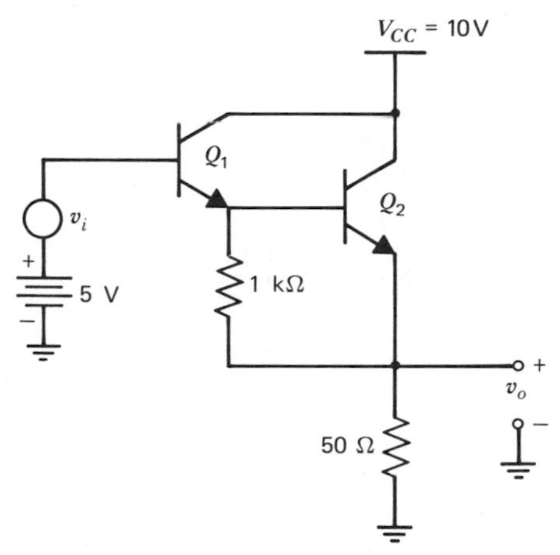

Fig. 3.51 Circuit for Problem 3.7.

3.8 Calculate the output resistance, r_o^c, of the common-emitter Darlington transistor of Fig. 3.52 as a function of I_{bias}. Do not neglect either r_{o1} or r_{o2} in this calculation, but you may neglect r_b and r_μ. If $I_{C2} = 1$ mA, what is r_o^c for $I_{\text{bias}} = 1$ mA? For $I_{\text{bias}} = 0$?

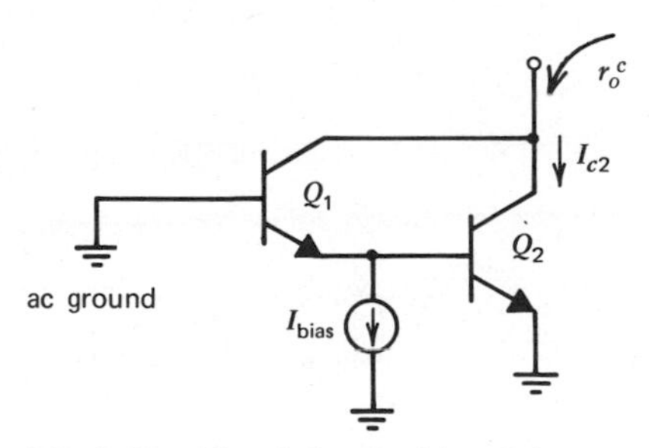

Fig. 3.52 Circuit for Problem 3.8.

3.9 Determine the input resistance, transconductance, output resistance, and maximum open-circuit voltage gain for the CE-CB circuit of Fig. 3.22 if $I_{C1} = I_{C2} = 250$ μA. Neglect r_b and r_μ.

3.10 Determine the differential-mode gain, common-mode gain, differential-mode input resistance, and common-mode input resistance for the circuit of Fig. 3.26 with $I_{EE} = 20$ μA, $R_{EE} = 10$ MΩ, $R_C = 100$ kΩ, $V_{EE} = -5$ V. Neglect r_b, r_o, and r_μ. Calculate the CMRR.

3.11 Repeat Problem 3.10, but with the addition of emitter degeneration resistors of value 4 kΩ each.

3.12 Determine the overall input resistance, voltage gain, and output resistance of the CC–CB connection of Fig. 3.53. Neglect r_o, r_μ, and r_b. Note that the addition of a 10 kΩ resistor in the collector of Q_1 would not change the results, so that the results of the emitter-coupled pair analysis can be used.

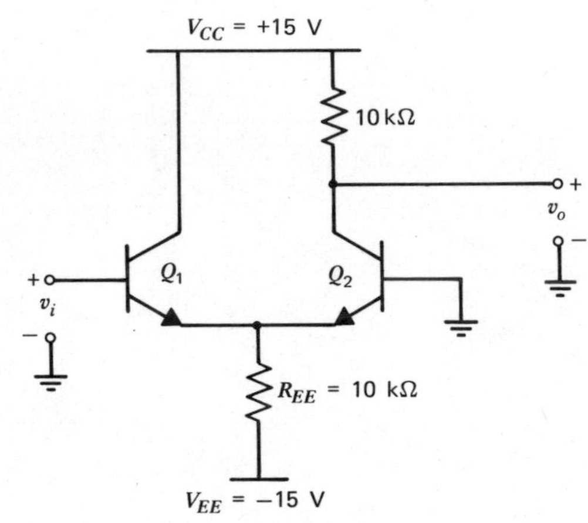

Fig. 3.53 Circuit for Problem 3.12.

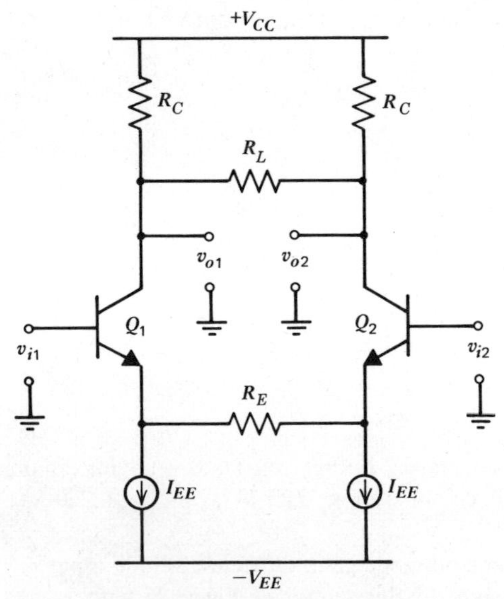

Fig. 3.54 Circuit for Problem 3.13.

3.13 Use half-circuit concepts to determine the differential-mode and common-mode gain of the circuit shown in Fig. 3.54. Neglect r_b, r_o, and r_μ. Calculate the differential-mode and common-mode input resistance.

3.14 Design an emitter-coupled pair of the type shown in Fig. 3.41, selecting new values of R_C and R_{EE} to give a differential input resistance of 2 MΩ, a DM voltage gain of 500, and a CMRR of 500. What are the minimum values of V_{CC} and V_{EE} that will yield this performance while keeping the transistors biased in the forward-active region under zero-signal conditions? Assume the dc common-mode input voltage is zero. Neglect r_b, r_μ, and r_o.

3.15 Determine the dc gate-source voltage of the JFETs, and the differential mode voltage gain for the JFET circuit of Fig. 3.55. Neglect the output resistance of the JFETs. Assume for the JFETs that $I_{DSS} = 1$ mA, $V_P = -2$ V.

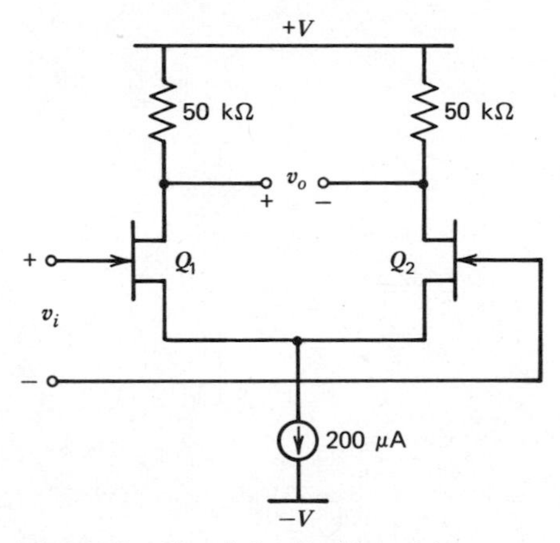

Fig. 3.55 Circuit for Problem 3.15.

3.16 For the circuit of Fig. 3.46*a*, determine the input offset voltage if the transistor base widths mismatch by 10 percent but otherwise the circuit is balanced.

3.17 For the circuit of Fig. 3.55, determine the input offset voltage if the JFET channel widths mismatch by 10 percent but otherwise the circuit is balanced. Assume that $I_{DSS} = 1$ mA, $V_P = -2$ V. The bias current is much lower than I_{DSS} so the dominant offset contribution is ΔV_P.

REFERENCES

1. R. J. Widlar. "Some Circuit Design Techniques for Linear Integrated Circuits," *IEEE Transactions on Circuit Theory*, Vol. CT-12, pp. 586–590, December 1965.
2. H. R. Camenzind and A. B. Grebene. "An Outline of Design Techniques for Linear Integrated Circuits," *IEEE Journal of Solid State Circuits*, Vol. SC-4, pp. 110–122, June 1969.
3. J. Giles. *Fairchild Semiconductor Linear Integrated Circuits Applications Handbook.* Fairchild Semiconductor, 1967.

4. C. L. Searle, A. R. Boothroyd, E. J. Angelo, P. E. Gray, and D. O. Pederson. *Elementary Circuit Properties of Transistors.* Wiley, New York, 1964, Chapter 7.
5. R. D. Thornton, C. L. Searle, D. O. Pederson, R. B. Adler, and E. J. Angelo. *Multistage Transistor Circuits*, Wiley, New York, 1965, Chapter 6.
6. R. D. Middlebrook. *Differential Amplifiers.* Wiley, New York, 1963.
7. L. J. Giacoletto. *Differential Amplifiers.* Wiley, New York, 1970.
8. J. Wallmark and H. Johnson (Eds.). *Field Effect Transistors: Physics, Technology, and Applications.* Prentice-Hall, Inc., Englewood Cliffs, N.J. 1966, Chapter 5.
9. G. Erdi. "A Low-Drift, Low-Noise Monolithic Operational Amplifier for Low Level Signal Processing," *Fairchild Semiconductor Applications Brief*, 136, July 1969.
10. H. C. Lin. "Comparison of Input Offset Voltage of Differential Amplifiers Using Bipolar Transistor and Field-Effect Transistors," *IEEE Journal of Solid-State Circuits*, Vol. SC-5, pp. 126–129, June 1970.

CHAPTER 4

TRANSISTOR CURRENT SOURCES AND ACTIVE LOADS

4.1 INTRODUCTION

Transistor current sources have come to be widely used in analog integrated circuits both as biasing elements and as load devices for amplifier stages. The use of current sources in biasing can result in superior insensitivity of circuit performance to power-supply variations and to temperature. Current sources are frequently more economical than resistors in terms of the die area required to provide bias current of a certain value, particularly when the value of bias current required is small. When used as a load element in transistor amplifiers, the high incremental resistance of the current source results in high voltage gain at low power-supply voltages.

The first section of this chapter analyzes the basic types of current-source circuits that are commonly used. The output current and output resistance of each type is calculated, and the effect of device mismatches in current-source circuits is considered. The second section of the chapter deals with the design of bias circuits for integrated circuits with the objective of obtaining insensitivity of biasing current to temperature and power-supply voltage. Finally, the use of the current source as an active load is considered.

4.2 CURRENT SOURCES

4.2.1 Simple Current Source

The simplest form of current source consists of a resistor and two transistors[1] as shown in Fig. 4.1. Transistor Q_1 is diode connected, forcing the collector-base voltage to zero. In this mode, no injection takes place at the collector-base junction since it is at zero bias, and the transistor behaves as if it were in the forward-active region. We will neglect junction leakage currents, assume identical transistors, and assume that the output resistance of Q_2 is infinite. Since Q_1 and Q_2 have the same base-emitter voltage, their collector currents are equal:

$$I_{C1} = I_{C2} \tag{4.1}$$

Summing currents at the collector of Q_1,

$$I_{\text{ref}} - I_{C1} - 2\frac{I_{C1}}{\beta_F} = 0$$

and thus

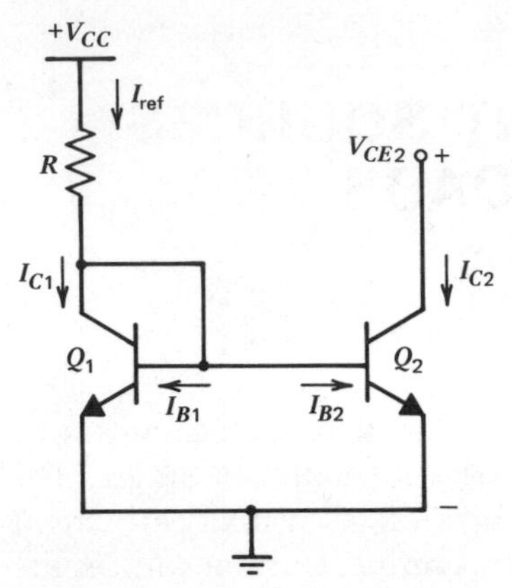

Fig. 4.1 A simple two-transistor current source.

$$I_{C1} = \frac{I_{\text{ref}}}{1 + \dfrac{2}{\beta_F}} = I_{C2} \tag{4.2}$$

If β_F is large, the collector current of Q_2 is nearly equal to the reference current.

$$I_{C2} \cong I_{\text{ref}} = \frac{V_{CC} - V_{BE(\text{on})}}{R} \tag{4.3}$$

Thus for the case of identical devices Q_1 and Q_2, the output and reference current are equal. Actually, the devices need not be identical; the emitter areas of Q_1 and Q_2 can be made different, which will cause the I_S values for the two transistors to be different. The two collector currents, I_{C1} and I_{C2}, will then have a constant ratio rather than being equal as shown by (4.1). This ratio can be either less than or greater than unity, and thus any desired output current I_{C2} can be derived from a fixed reference current. However, area ratios greater than about five to one consume a large die area because of the area of the larger of the two devices. Thus for the generation of large current ratios the Widlar current source, discussed in the next section, is usually preferable.

One of the most important aspects of current-source performance is the variation of the current-source current with changes in voltage at the output terminal. This is characterized by the small-signal output resistance of the current source. For example, the common-mode rejection ratio of the differential amplifier depends directly on this resistance, as does the gain of the active-load circuit. We assumed, in writing (4.1), that the collector currents of the transistors are independent of their collector-emitter voltages. Actually, the collector current increases slowly with increasing collector-emitter voltage as illustrated in Fig. 4.2. As dis-

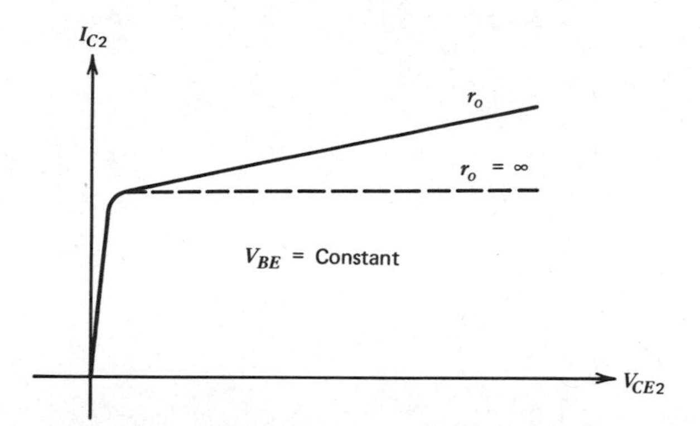

Fig. 4.2 Collector characteristics for an *npn* transistor for the hypothetical case of $r_o = \infty$, and the actual case for which r_o is finite.

cussed in Chapter 1, this base-width modulation effect can be represented for large-signal conditions by the expression:

$$I_C = I_S\left(\exp\frac{V_{BE}}{V_T}\right)\left(1 + \frac{V_{CE}}{V_A}\right)$$

where V_A is the Early voltage. A typical value of the Early voltage for *npn* transistors is 130 V. Thus, for example, if the collector-emitter voltage of Q_1 is held at $V_{BE(\text{on})}$, and if the collector voltage of Q_2 is at 30 V, then the ratio of I_{C2} to I_{C1} would be:

$$\frac{I_{C2}}{I_{C1}} = \frac{1 + \dfrac{V_{CE2}}{V_A}}{1 + \dfrac{V_{CE1}}{V_A}} = \frac{1 + \dfrac{30}{130}}{1 + \dfrac{0.6}{130}} \approx 1.25 \tag{4.4}$$

Thus for a circuit operating at a power supply voltage of 30 V, the current-source currents can differ by as much as 25 percent from those values calculated by assuming that the transistor output resistance is negligible.

A useful figure of merit for transistor current sources is the equivalent open circuit voltage, V_{Thev}. As long as the transistors in the current source are in the forward-active region, each current source configuration can be characterized by an output resistance, R_o, and an output current, I_o, as shown in the Norton equivalent circuit of Fig. 4.3*a*. Generally speaking, the output resistance will decrease in practical circuits when the output current is increased. The Thévenin equivalent is shown in Fig. 4.3*b*, for which the Thévenin voltage generator is equal to:

$$V_{\text{Thev}} = I_o R_o \tag{4.5}$$

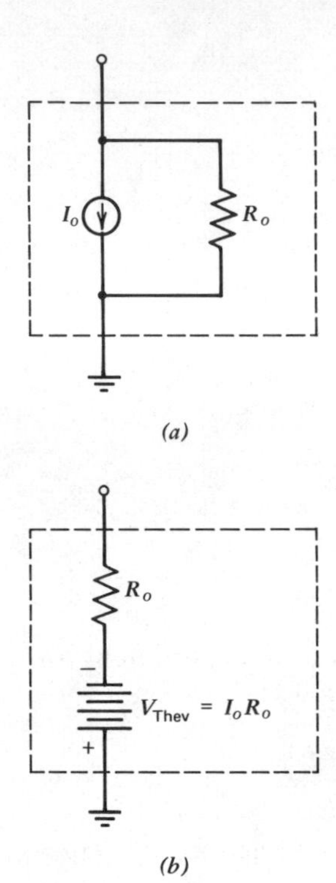

Fig. 4.3 (*a*) Norton equivalent representation of a transistor current source. (*b*) Thévenin equivalent representation of a transistor current source.

The Thévenin voltage generally remains constant for a given current-source configuration, independent of the particular design value of I_o. For example, for the simple current source,

$$V_{\text{Thev}} = I_o R_o = I_{C2} r_{o2} = I_{C2} \frac{V_A}{I_{C2}} = V_A \tag{4.6}$$

where we have used the fact that $r_o = V_A/I_C$, as shown in Chapter 1. Thus for the simple current source the equivalent open-circuit voltage is equal to V_A, the Early voltage. More complex current-source circuits display values of V_{Thev} that are higher than V_A. If, for example, the output current of a simple current source is set at 1 mA, then

$$R_o = \frac{V_{\text{Thev}}}{I_o} = \frac{V_A}{I_o} = \frac{130 \text{ V}}{1 \text{ mA}} = 130 \text{ k}\Omega \tag{4.7}$$

Note that the equivalent open-circuit voltage, V_{Thev}, is never actually observed at the output of the current source. If the output of the current source is open circuited, the current-source equivalent circuit of Fig. 4.3*b* indicates that an output voltage of $-V_{\text{Thev}}$ will be observed. This does not actually occur because the current-source transistor will saturate when the voltage across the current source approaches zero, since this voltage is equal to the collector-emitter voltage of the transistor. Thus the Thévenin and Norton representations are only valid for values of voltage at the current-source output for which the devices in the circuit are in the forward-active region.

In addition to the variation in output current due to finite output resistance, the collector current, I_{C2}, is different from the reference current by a factor $[1 + (2/\beta_F)]$. When the current source is constructed with low-gain *pnp* transistors, the value of β_F may be low enough that this factor is significant. In order to reduce this source of error, an additional transistor can be added as shown in Fig. 4.4. The emitter current of transistor Q_3 is equal to:

$$-I_{E3} = \frac{I_{C1}}{\beta_F} + \frac{I_{C2}}{\beta_F} = \frac{2}{\beta_F} I_{C2} \tag{4.8}$$

where I_E, I_C, and I_B are defined as positive when flowing into the transistor, and where we have neglected the effects of finite output resistance. The base current of transistor Q_3 is equal to:

$$I_{B3} = \frac{-I_{E3}}{\beta_F + 1} = \frac{2}{\beta_F(\beta_F + 1)} I_{C2} \tag{4.9}$$

Finally, summing currents at the collector of Q_1,

$$I_{\text{ref}} - I_{C1} - \frac{2}{\beta_F(\beta_F + 1)} I_{C2} = 0 \tag{4.10}$$

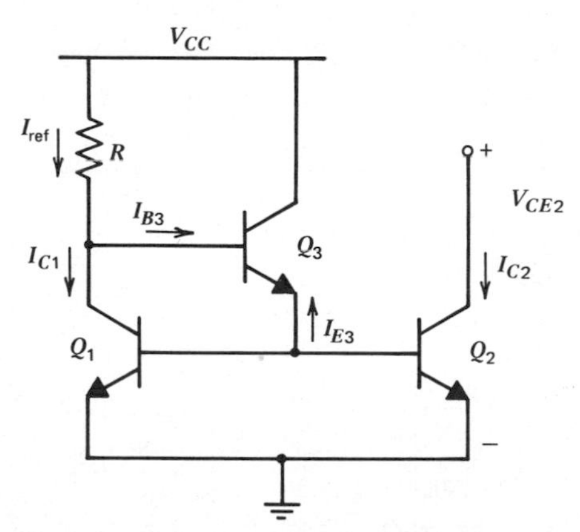

Fig. 4.4 Simple current source with current gain.

Since I_{C1} and I_{C2} are equal,

$$I_o = I_{C2} = \frac{I_{\text{ref}}}{1 + \dfrac{2}{\beta_F^2 + \beta_F}} \tag{4.11}$$

Thus the reference current and output current differ only by a factor containing $1/\beta_F^2$, neglecting the effects of output resistance.

4.2.2 Widlar Current Source

In operational amplifiers, the low input current required dictates that the input emitter-coupled pair be biased at very low current, typically at a collector current of the order of 5 μA. Bias currents of this magnitude are also required in a variety of other applications. Using a simple current source and assuming a maximum practical emitter area ratio of ten-to-one in Q_2 and Q_1, we would need a reference current of 50 μA for an output current of 5 μA. This would require a resistor of 600 kΩ if the reference current were developed by connecting the resistor across a 30-V power supply, and resistors of this magnitude are very costly in terms of die area. Currents of such low magnitude can be obtained with moderate values of resistance, however, by modifying the simple current source so that the two devices, Q_1 and Q_2, operate at different values of V_{BE}. In the Widlar[2,3] current source of Fig. 4.5, this is accomplished by inserting a resistor in series with the emitter of Q_2. We will now calculate the output current and equivalent open-circuit voltage for this current source.

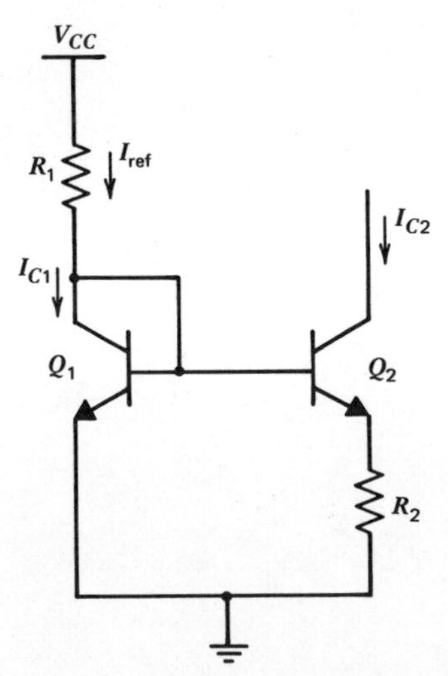

Fig. 4.5 Widlar current source.

Summing voltages around the emitter-base loop, assuming that V_A is infinite, and neglecting base currents,

$$V_{BE1} - V_{BE2} - I_{C2}R_2 = 0 \tag{4.12}$$

and thus

$$V_T \ln \frac{I_{C1}}{I_{S1}} - V_T \ln \frac{I_{C2}}{I_{S2}} - I_{C2}R_2 = 0 \tag{4.13}$$

For identical transistors, I_{S1} and I_{S2} are equal, and (4.13) becomes

$$V_T \ln \frac{I_{C1}}{I_{C2}} = I_{C2}R_2 \tag{4.14}$$

This transcendental equation must be solved by trial and error if R_2 and I_{C1} are known and I_{C2} must be found. For design purposes, I_{C1} and I_{C2} are usually known and (4.14) provides the value of R_2 required to achieve the desired value of I_{C2}.

EXAMPLE

In the circuit of Fig. 4.5, determine the proper value of R_2 to give $I_{C2} = 10\ \mu\text{A}$. Assume that $V_{CC} = 30$ V, $R_1 = 29.3\ \text{k}\Omega$, $V_{BE(\text{on})} = 0.7$ V. Neglecting base current,

$$I_{C1} = \frac{30\ \text{V} - 0.7\ \text{V}}{29.3\ \text{k}\Omega} = 1\ \text{mA}$$

$$V_T \ln \frac{I_{C1}}{I_{C2}} = 26\ \text{mV} \ln \left(\frac{1\ \text{mA}}{10\ \mu\text{A}}\right) = 119\ \text{mV}$$

Thus, from (4.14),

$$I_{C2}R_2 = 119\ \text{mV}$$

and

$$R_2 = \frac{119\ \text{mV}}{10\ \mu\text{A}} = 11.9\ \text{k}\Omega$$

The total resistance in the circuit is 41.2 kΩ.

EXAMPLE

For the circuit of Fig. 4.5 assume that $I_{\text{ref}} = 1$ mA and $R_2 = 5\ \text{k}\Omega$, neglect base current, and find I_{C2}.

From (4.14),

$$V_T \ln \frac{1\ \text{mA}}{I_{C2}} - 5\ \text{k}\Omega(I_{C2}) = 0$$

Try
$$I_{C2} = 15\ \mu\text{A}$$
$$108\text{ mV} - 75\text{ mV} \neq 0$$

The logarithm term is too large, thus a larger current should be tried.

Try
$$I_{C2} = 20\ \mu\text{A}$$
$$101.7\text{ mV} - 100\text{ mV} \approx 0$$

The current is very close to 20 μA.

The output resistance of this current source can be calculated using the small-signal equivalent circuit shown in Fig. 4.6*a*. Since I_{C2} is smaller than I_{C1}, $r_{\pi 2}$ will be larger than $(1/g_{m1})$ by a factor of at least β_0. The circuit can thus be simplified

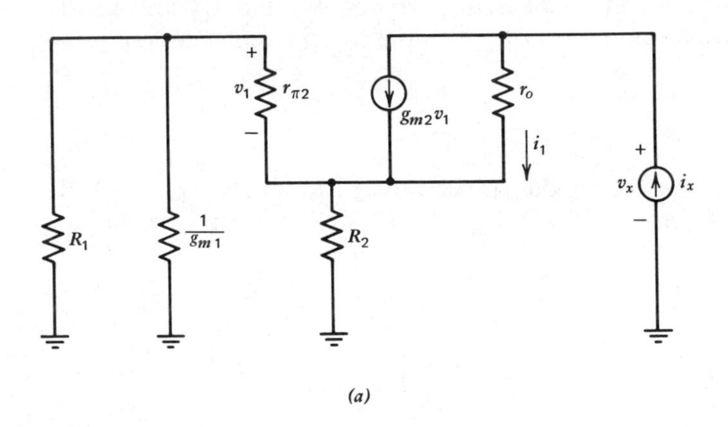

(a)

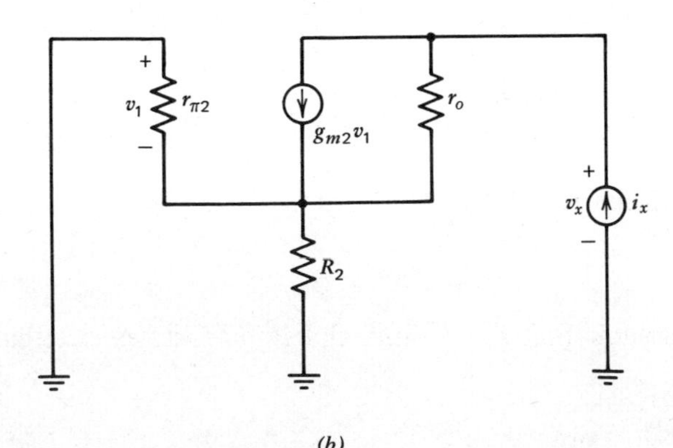

(b)

Fig. 4.6 Small-signal equivalent circuits for the Widlar source, used to calculate R_o.

to that of Fig. 4.6*b*. The test current source i_x is applied to calculate the output resistance. Since i_x flows in the parallel combination of $r_{\pi 2}$ and R_2,

$$v_1 = -i_x(r_{\pi 2} \| R_2) \tag{4.15}$$

The current i_1 is then

$$\begin{aligned} i_1 &= i_x - g_{m2} v_1 \\ &= i_x + i_x g_{m2}(r_{\pi 2} \| R_2) \end{aligned} \tag{4.16}$$

using (4.15). The voltage, v_x, is simply

$$v_x = -v_1 + i_1 r_o \tag{4.17}$$

and substitution of (4.15) and (4.16) gives

$$v_x = i_x(r_{\pi 2} \| R_2) + i_x r_o[1 + g_{m2}(r_{\pi 2} \| R_2)] \tag{4.18}$$

Thus the output resistance of the current source is

$$R_o = \frac{v_x}{i_x} = (r_{\pi 2} \| R_2) + r_o[1 + g_{m2}(r_{\pi 2} \| R_2)] \tag{4.19}$$

The first term of (4.19) is negligible compared with the second, which can be rearranged to give:

$$R_o \approx r_o \left(\frac{1 + g_{m2} R_2}{1 + \dfrac{g_{m2} R_2}{\beta_0}} \right) = \frac{V_A}{I_o} \left(\frac{1 + g_{m2} R_2}{1 + \dfrac{g_{m2} R_2}{\beta_0}} \right) \tag{4.20}$$

where $I_o = I_{C2}$ is the Norton equivalent current source. The equivalent open-circuit voltage of the source is thus:

$$V_{\text{Thev}} = I_o R_o = V_A \left(\frac{1 + g_{m2} R_2}{1 + \dfrac{g_{m2} R_2}{\beta_0}} \right) \tag{4.21}$$

The output resistance is increased by the presence of R_2. Equation 4.20 for R_o can be rewritten under the assumption that $g_{m2} R_2 \ll \beta_0$, and using $g_{m2} = I_{C2}/V_T$ we obtain:

$$R_o = r_o \left(1 + \frac{I_{C2} R_2}{V_T} \right) \tag{4.22}$$

Thus R_o depends on $I_{C2} R_2$, which is the dc voltage drop across R_2. The larger this voltage drop is made, the higher the output resistance. For the Widlar source, $I_{C2} R_2$ is limited to several hundred millivolts for practical current ratios, and the corresponding value of V_{Thev} is limited to about $10 V_A$.

In a similar fashion, the output resistance of the simple current source can be increased by the addition of resistances in the emitters of both of the current source transistors. This modification is illustrated in Fig. 4.7*a* for the simple current source. Summing voltages around the loop composed of R_2, Q_2, Q_1, and R_3, and neglecting base currents, we obtain:

$$I_{\text{ref}}R_3 - I_{C2}R_2 + V_T \ln\left(\frac{I_{S2}}{I_{S1}}\right)\left(\frac{I_{\text{ref}}}{I_{C2}}\right) = 0 \tag{4.23a}$$

Note that if the two resistors are equal and the transistors are identical, I_{C2} is equal to I_{ref}. The resistor R_2 appears in the emitter of Q_2, and the output resistance will be increased according to (4.20) as long as R_3 is small compared to $r_{\pi 2}$, as assumed in the analysis. The addition of the resistors also improves the sensitivity of the output current to transistor mismatches as discussed in Appendix A4.1.

In the current source of Fig. 4.7*a*, if a value of I_{C2} is desired that is proportional to I_{ref} but not equal to it, then the value of the ratio R_2/R_3 and the ratio I_{S2}/I_{S1} must be scaled accordingly. However, note that by making R_3 and R_2 unequal, the output current can be made unequal to the reference current even for identical transistors. Using (4.23a),

$$I_{C2} = \frac{1}{R_2}\left[I_{\text{ref}}R_3 + V_T \ln\left(\frac{I_{S2}}{I_{S1}}\right)\left(\frac{I_{\text{ref}}}{I_{C2}}\right)\right] \tag{4.23b}$$

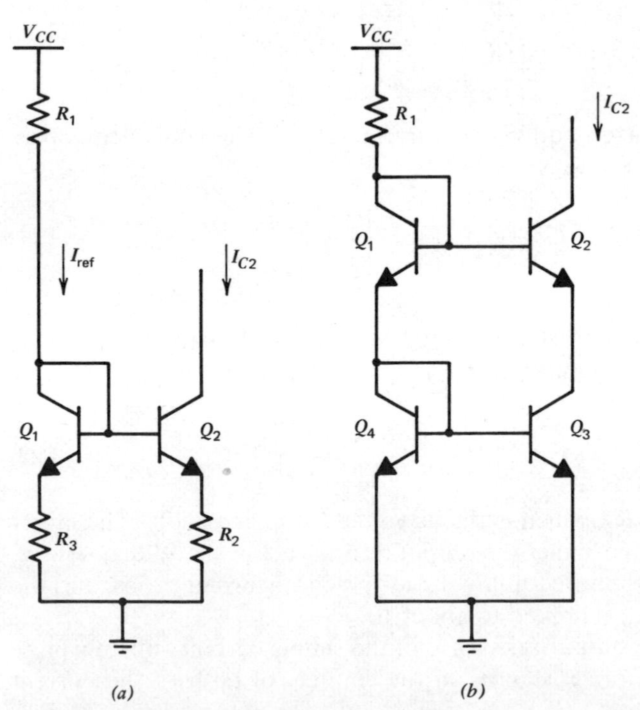

Fig. 4.7 (*a*) Simple current source with emitter resistors. (*b*) Cascode current source.

The rightmost term represents the difference between the base-emitter voltages, and has a magnitude of only 10 to 150 mV even for current ratios as large as 100. Thus if $I_{ref}R_3$ is large compared to this value,

$$I_{C2} \approx I_{ref}\left(\frac{R_3}{R_2}\right) \tag{4.23c}$$

Thus if enough voltage is dropped across R_3, the current ratio becomes dependent on the resistor ratio rather than on transistor area ratios.

Another approach to achieving high output resistance is the use of the cascode configuration as shown in Fig. 4.7*b*. Since utilization of a large value of R_2 gives high output resistance, we want to replace R_2 with a transistor current source so that the effective value of R_2 becomes the output resistance of that current source. The effective value of R_2 for transistor Q_2 in Fig. 4.7*b* is actually r_{o3}, the output resistance of transistor Q_3. Since transistors Q_2 and Q_3 operate at the same current, (4.20) becomes:

$$R_o = \frac{V_A}{I_o}\left(\frac{1 + g_{m2}r_{o3}}{1 + \dfrac{g_{m2}r_{o3}}{\beta_0}}\right) = \frac{V_A}{I_o}\left(\frac{1 + \dfrac{1}{\eta}}{1 + \dfrac{1}{\eta\beta_0}}\right) \cong \beta_0\frac{V_A}{I_o} = \beta_0 r_o \tag{4.24}$$

since $1/\eta \gg 1$. Thus the output resistance and Thévenin voltage can be increased by a factor β_0 in this way, and:

$$V_{\text{Thev}} = \beta_0 r_o I_o = \beta_0 V_A = 12{,}000 \text{ V} \tag{4.25}$$

for $\beta_0 = 100$, $V_A = 120$ V. For an output current of 1 mA, for example,

$$R_o = \frac{12{,}000 \text{ V}}{1 \text{ mA}} - \underline{12 \text{ M}\Omega}$$

In the analysis of the Widlar and cascode current sources we have again neglected the effects of r_μ. Whereas this assumption was easy to justify in the case of the simple current source, it must be reexamined for these higher impedance sources. The collector-base resistance, r_μ, results from modulation of the base recombination current as a consequence of the Early effect, as discussed in Chapter 1. For the case of a transistor whose base current is composed entirely of base recombination current, the percentage change in base current when V_{CE} is changed at a constant V_{BE} would equal that of the collector current, and r_μ would be related to r_o by a factor β_0. If this were the case, the effect of r_μ would be to reduce the output resistance of the cascode current source by a factor of 2 and to affect the Widlar source much less significantly.

In an actual integrated-circuit *npn* transistor, however, only a very small percentage of the base current results from recombination in the base. Since only this component is modulated by the Early effect, the observed values of r_μ are a factor of 10 or more larger than $\beta_0 r_o$. This means that the feedback resistor has a negligible effect in the *npn* current sources considered so far. For the case of lateral *pnp* transistors, the feedback resistance, r_μ, is much smaller than for *npn* transistors because most of the base current results from base-region recombination. The actual value of this resistance depends on a number of processes and device geometry variables but observed values range from twice to five times $\beta_0 r_o$. Thus for *pnp* transistors, the effect of the feedback resistance can be significant when the base of the transistor is connected to a high incremental resistance.

In considering current sources that give output resistances higher than $\beta_0 r_o$, the effects of r_μ must be considered, of course. Also, for the cases we have considered so far, the base of the current-source transistor has been returned to a low-impedance point, so that current variations in r_μ did not change V_{BE}. If this is not the case, the variations in the current flowing in r_μ flow into the base and are amplified by the beta of the transistor, resulting in much larger variations in output current than in the cases we have considered.

4.2.3 Wilson Current Source

As is apparent from Fig. 4.7*b*, current sources that achieve Thévenin equivalent voltages in the thousand-volt range require large numbers of devices. One relatively simple circuit configuration that achieves a Thévenin voltage much larger than the Early voltage is the Wilson configuration[4] shown in Fig. 4.8. This source uses the negative feedback provided by Q_3 to raise the output resistance. Furthermore, first-order cancellation of the base currents is achieved with this configuration, making the ratio of the output current to reference current less sensitive to β_F. The reference current is developed by the resistor, R.

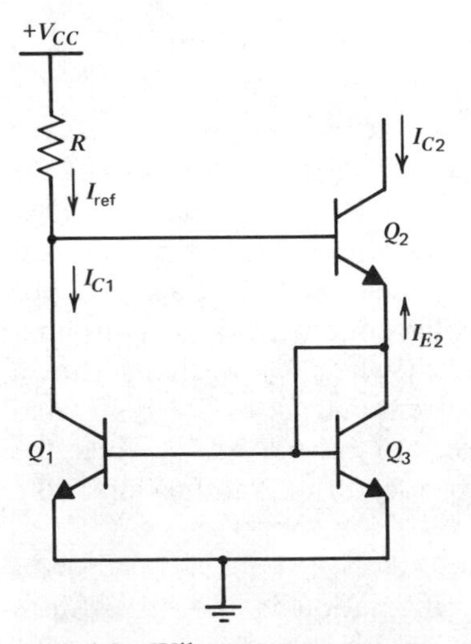

Fig. 4.8 Wilson current source.

From a qualitative standpoint, the difference between the reference current and I_{C1} will flow into the base of Q_2. This base current is multiplied by $(\beta_F + 1)$ and flows in the diode-connected transistor, Q_3, which causes current of the same magnitude to flow in Q_1. A feedback path is thus formed that regulates I_{C1} so that it is nearly equal to the reference current. Note that transistors Q_1 and Q_3 operate at collector-emitter voltages that are different by only one diode drop, and that transistor Q_3 does not experience a variation in collector-emitter voltage when the voltage across the current source is varied. Thus the collector current of Q_3 remains very nearly equal to that in Q_1, independent of the voltage on the collector of Q_2. This implies that the collector current of Q_2 remains nearly constant, giving high output impedance.

For the purposes of dc analysis, we assume that $V_A = \infty$ and that the transistors are identical. The emitter current of Q_2 is equal to the collector current of Q_3 plus the base currents of Q_1 and Q_3.

$$-I_{E2} = I_{C3} + I_{B3} + I_{B1} = I_{C3}\left(1 + \frac{1}{\beta_F}\right) + \frac{I_{C1}}{\beta_F} \tag{4.26}$$

$$= I_{C3}\left(1 + \frac{2}{\beta_F}\right) \tag{4.27}$$

The collector current of Q_2 is then:

$$I_{C2} = -I_{E2}\left(\frac{\beta_F}{1 + \beta_F}\right) = I_{C3}\left(1 + \frac{2}{\beta_F}\right)\left(\frac{\beta_F}{1 + \beta_F}\right) \tag{4.28}$$

using (4.27). Rearranging (4.28) we obtain

$$I_{C3} = I_{C2}\left[\frac{1}{\left(1 + \frac{2}{\beta_F}\right)\left(\frac{\beta_F}{1 + \beta_F}\right)}\right] \tag{4.29}$$

Summing currents at the base of Q_2,

$$I_{C1} = I_{\text{ref}} - \frac{I_{C2}}{\beta_F} \tag{4.30}$$

We assumed in writing (4.26) that the transistors are identical, so that:

$$I_{C1} = I_{C3} \tag{4.31}$$

Inserting (4.29) and (4.30) into (4.31), we find

$$I_{C2} = I_{\text{ref}}\left(1 - \frac{2}{\beta_F^2 + 2\beta_F + 2}\right) \tag{4.32}$$

Thus the output and reference currents differ by only a factor on the order of $2/\beta_F^2$. A small-signal analysis of the circuit, carried out in the same way as in the Widlar source, and neglecting r_μ, gives an output resistance and Thévenin voltage of:

$$R_o \approx \beta_0 r_{o2}/2 \tag{4.33}$$

$$V_{\text{Thev}} \approx \beta_0 V_A/2 \tag{4.34}$$

In the discussion of the simple current source and the Widlar and Wilson circuits, we have concentrated on the problem of obtaining the desired value of output current, which may be small, and on the problem of obtaining high output resistance in the current source. The Widlar circuit is useful for obtaining small output currents, and the Wilson circuit for obtaining high output resistance and low sensitivity to transistor base currents.

Particular circuit design problems frequently require careful control of other aspects of the performance of transistor current sources. For example, the need often arises for the generation of two currents that are equal to as high a degree of accuracy as possible. The ability to produce such matched current sources is usually limited by mismatches in the devices which make up the circuit. This problem is discussed more thoroughly in Appendix A4.1 at the end of this chapter.

A second frequently encountered requirement is that the performance of the integrated circuit be invariant when the power-supply voltage and/or the temperature varies between specified limits. This generally requires that the biasing current sources within the circuit have output currents that are insensitive to variations in power-supply voltage and temperature. The design of bias circuits to achieve these objectives is discussed in Appendices A4.2 and A4.3.

4.3 CURRENT SOURCES AS ACTIVE LOADS

In conventional differential amplifiers of the type discussed in Chapter 3, resistors are used as the collector load element as shown in Fig. 3.26. For this circuit, the voltage gain is:

$$A_{dm} = -g_m R_C = -\frac{I_C R_C}{V_T} \tag{4.35}$$

In order to achieve large voltage gain, the $I_C R_C$ product must be made large. This requires both a large power-supply voltage and large values of resistance. For example, a voltage gain of 500 would require that $I_C R_C = 13$ V and, if $I_C = 100\ \mu$A, R_C would have to be 130 kΩ. For a 15-V power supply, the range of dc common-mode input voltage for which the transistors would remain unsaturated would be very restricted, and the two resistors required would take a very large amount of die area. As will be discussed in Chapter 9, an important objective in feedback amplifier design is to obtain the required voltage gain in as few stages as possible. The use of the r_o of a *pnp* transistor as a load element allows high voltage gain without requiring large power supply voltages.[5] Since the load element in such a circuit is a *pnp* transistor instead of a resistor, the collector load element is said to be *active.*

4.3.1 Common-Emitter Amplifier with Active Load

A common-emitter amplifier with *pnp* active load is shown in Fig. 4.9. First consider the dc large-signal behavior of the circuit. The I-V characteristic of the *pnp* current-source transistor is shown in Fig. 4.10*a*. This curve we designate:

$$I_{c2} = I_{c2}(V_{ce2}, V_{BE2})$$

where V_{BE2} is the dc base-emitter voltage of Q_2. The collector curves of the *npn* transistor with V_{BE} as a parameter are shown in Fig. 4.10*b*. For the active-load circuit,

$$I_{c1} = -I_{c2} \tag{4.36}$$

$$V_{ce2} = -V_{CC} + V_{ce1} \tag{4.37a}$$

Using these relations, we can construct the *pnp* I-V characteristic on the *npn* collector characteristics. The I-V characteristic of the load device requires that I_{c1}, the collector current of Q_1, and V_{ce1}, the collector-emitter voltage of Q_1, be related in the following way:

$$I_{c1} = -I_{c2}(V_{ce2}, V_{BE2}) \tag{4.37b}$$

Using (4.37a)

$$I_{c1} = -I_{c2}[(-V_{CC} + V_{ce1}), V_{BE2}] \tag{4.37c}$$

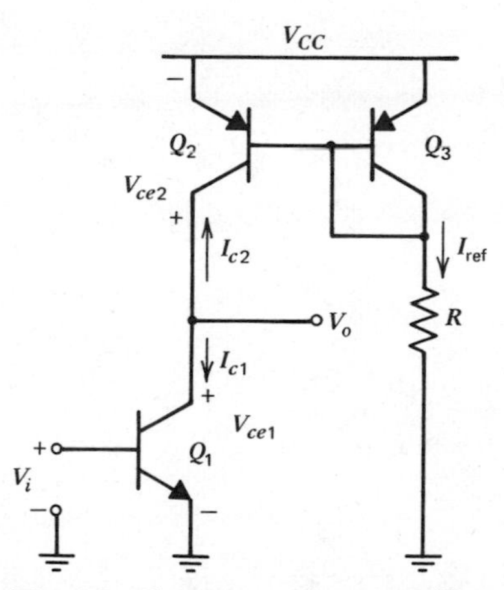

Fig. 4.9 Common-emitter amplifier with active load.

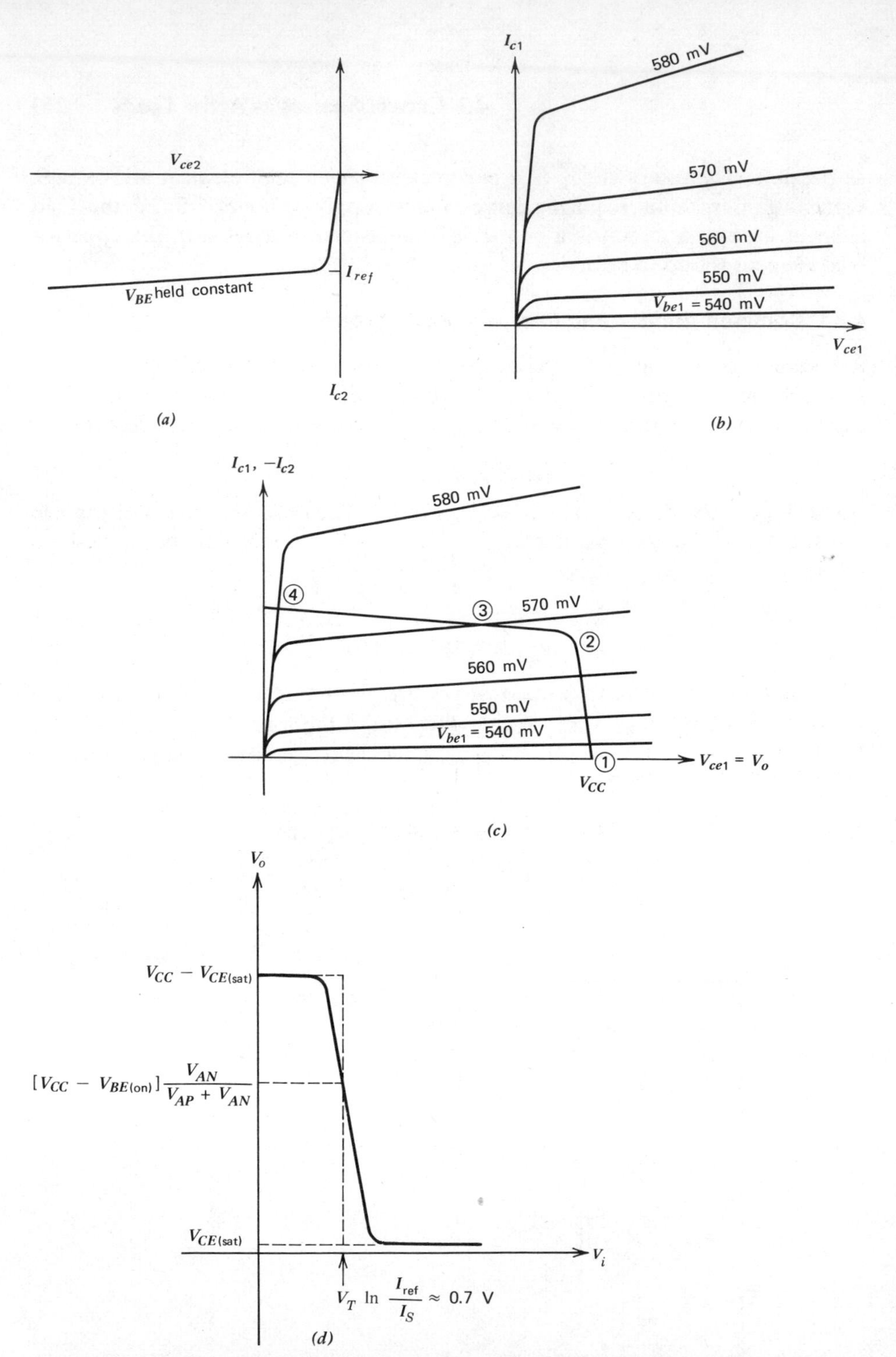

Fig. 4.10 (*a*) I-V characteristic of *pnp* transistor active load. (*b*) Collector characteristics of the *npn* transistor. (*c*) *npn* collector characteristic with *pnp* load line superimposed. (*d*) dc transfer characteristic of common-emitter amplifier with active load.

Thus, according to (4.37c), the load line has the shape of the *pnp* collector current curve of Fig. 4.10*a*, but the curve is mirrored about the horizontal axis and shifted to the right by an amount equal to V_{CC}. The resulting load line is shown in Fig. 4.10*c*. We now consider the dc transfer characteristic of the circuit.

Initially, assume that $V_i = 0$. The *npn* transistor is turned off, and the *pnp* is saturated; this corresponds to point ① on Fig. 4.10*c*. As V_i is increased, the *npn* begins to conduct current but the *pnp* remains saturated until point ② is reached. Here the *pnp* comes out of saturation, and a small further increase in V_i moves the operating point rapidly through point ③ to point ④, at which the *npn* transistor is saturated. The change in V_i required to move from ② to ④ is only a few millivolts because of the small slope of the collector characteristics. The transfer curve (V_o as a function of V_i) is sketched in Fig. 4.10*d*.

For the region of the transfer characteristic between points ② and ④, for which both transistors are in the forward-active region, we can determine the transfer curve analytically using the large-signal model for the transistor discussed in Chapter 1. For the *npn* transistor, Q_1, in Fig. 4.9 we have

$$I_{c1} = I_{S1}\left(\exp\frac{V_i}{V_T}\right)\left(1 + \frac{V_{ce1}}{V_{AN}}\right) \tag{4.38}$$

where V_{AN} is the Early voltage for the *npn* transistor. For the *pnp* transistor, Q_2,

$$I_{c2} = -I_{S2}\left(\exp\frac{|V_{BE2}|}{V_T}\right)\left(1 + \frac{|V_{ce2}|}{V_{AP}}\right) \tag{4.39}$$

where V_{AP} is the Early voltage for the *pnp* transistors. Since transistor Q_3 is diode connected, its collector-emitter voltage V_{CE} is equal to $V_{BE(\text{on})}$, and

$$I_{\text{ref}} \simeq -I_{c3} = I_{S2}\left(\exp\frac{|V_{BE3}|}{V_T}\right)\left[1 + \frac{V_{BE(\text{on})}}{V_{AP}}\right] \tag{4.40}$$

assuming $\beta_F \gg 1$.

Since $V_{BE2} = V_{BE3}$, we can combine (4.39) and (4.40) to yield:

$$I_{c2} = -I_{\text{ref}}\left[\frac{1 + \dfrac{|V_{ce2}|}{V_{AP}}}{1 + \dfrac{V_{BE(\text{on})}}{V_{AP}}}\right] \tag{4.41}$$

Since the two collector currents, I_{c1} and I_{c2}, must be equal in magnitude, the combination of (4.38) and (4.41) yields:

$$I_{S1}\left(\exp\frac{V_i}{V_T}\right)\left(1 + \frac{V_{ce1}}{V_{AN}}\right) = I_{\text{ref}}\left(\frac{1 + \dfrac{|V_{ce2}|}{V_{AP}}}{1 + \dfrac{V_{BE(\text{on})}}{V_{AP}}}\right) \tag{4.42}$$

The output voltage, V_o, is related to the collector-emitter voltages by:

$$V_{ce1} = V_o \tag{4.43}$$

$$|V_{ce2}| = V_{CC} - V_o \tag{4.44}$$

Thus

$$I_{S1}\left(\exp\frac{V_i}{V_T}\right)\left(1+\frac{V_o}{V_{AN}}\right) = I_{\text{ref}}\left(\frac{1+\dfrac{V_{CC}-V_o}{V_{AP}}}{1+\dfrac{V_{BE(\text{on})}}{V_{AP}}}\right) \tag{4.45}$$

and we obtain

$$I_{S1}\exp\frac{V_i}{V_T} = I_{\text{ref}}\left[\frac{1+\dfrac{V_{CC}-V_o}{V_{AP}}}{\left(1+\dfrac{V_{BE(\text{on})}}{V_{AP}}\right)\left(1+\dfrac{V_o}{V_{AN}}\right)}\right] \tag{4.46}$$

We now make the assumption that the quantities $(V_{CC} - V_o)/V_{AP}$, $V_{BE(\text{on})}/V_{AP}$, and V_o/V_{AN} are much less than unity. Using the approximations:

$$\frac{1}{1+x} \approx 1 - x \qquad (x \ll 1) \tag{4.47}$$

and

$$(1+x)(1+y) \approx 1 + x + y \qquad (x, y \ll 1) \tag{4.48}$$

(4.46) can be simplified to:

$$I_{S1}\exp\frac{V_i}{V_T} = I_{\text{ref}}\left[1+\frac{V_{CC}}{V_{AP}} - V_o\left(\frac{1}{V_{AP}}+\frac{1}{V_{AN}}\right) - \frac{V_{BE(\text{on})}}{V_{AP}}\right] \tag{4.49}$$

This expression can be solved for V_o to yield:

$$V_o = [V_{CC} - V_{BE(\text{on})}]\left(\frac{V_{AN}}{V_{AN}+V_{AP}}\right) + V_{A(\text{eff})}\left[1 - \frac{I_{S1}\exp\dfrac{V_i}{V_T}}{I_{\text{ref}}}\right] \tag{4.50}$$

where

$$V_{A(\text{eff})} = \left(\frac{V_{AN}V_{AP}}{V_{AN}+V_{AP}}\right) \tag{4.51}$$

is the effective Early voltage.

The first term of (4.50) is constant and the second is a function of V_i. Equation 4.50 is valid only for

$$[V_{CC} - V_{CE(\text{sat})}] > V_o > [V_{CE(\text{sat})}] \tag{4.52}$$

Thus the transfer curve between points ② and ④ is actually a small portion of an exponential characteristic. Note that when

$$\frac{I_{S1} \exp \dfrac{V_i}{V_T}}{I_{\text{ref}}} = 1 \tag{4.53}$$

in (4.50), the output voltage is

$$V_o = [V_{CC} - V_{BE(\text{on})}]\left(\frac{V_{AN}}{V_{AN} + V_{AP}}\right) \tag{4.54}$$

as shown in Fig. 4.10*d*. Under this condition, the output voltage lies at a point between V_{CC} and ground, dictated by the relative values of V_{AN} and V_{AP}. If it were true that V_{AN} and V_{AP} were equal, then the output voltage would be approximately one-half of V_{CC}. From (4.53), the value of V_i at this point is $V_T \ln (I_{\text{ref}}/I_{S1})$, which is the value that would produce a current I_{ref} in the collector of Q_1 if V_{CE1} were small.

We will perform a separate small-signal analysis to determine the voltage gain and output resistance of this circuit, but the small-signal voltage gain can be obtained directly from (4.50) by differentiation.

$$A_v = \frac{dV_o}{dV_i} = -\frac{V_{A(\text{eff})}}{V_T}\left(\frac{I_{S1} \exp \dfrac{V_i}{V_T}}{I_{\text{ref}}}\right) \tag{4.55}$$

If we take the case in which the term in brackets is unity, then, the small-signal gain is

$$A_v = -\frac{V_{A(\text{eff})}}{V_T} = -\left(\frac{V_{AP}V_{AN}}{V_{AN} + V_{AP}}\right)\frac{1}{V_T} \tag{4.56}$$

This can be manipulated to give:

$$A_v = \frac{-1}{\dfrac{V_T}{V_{AN}} + \dfrac{V_T}{V_{AP}}} = \frac{-1}{\eta_{npn} + \eta_{pnp}} \tag{4.57}$$

where the relation $\eta = V_T/V_A$ has been used from (1.114). The voltage gain is thus the inverse of the sum of the two Early factors. We now calculate the same parameter using a small-signal analysis.

The primary feature of interest in the small-signal performance of this circuit is the voltage gain and output resistance when both devices are in the forward-active region. The small-signal equivalent circuit is shown in Fig. 4.11, and it is clear by inspection that, with no load attached to the output,

$$A_v = -g_{m1}(r_{o1}\|r_{o2}) = \frac{-g_{m1}}{\dfrac{1}{r_{o1}} + \dfrac{1}{r_{o2}}} = \frac{-g_{m1}}{\eta_{npn}g_{m1} + \eta_{pnp}g_{m2}} \tag{4.58}$$

Since the two transistors operate at the same collector current, $g_{m1} = g_{m2}$ and thus

$$A_v = \frac{-1}{\eta_{npn} + \eta_{pnp}} \tag{4.59}$$

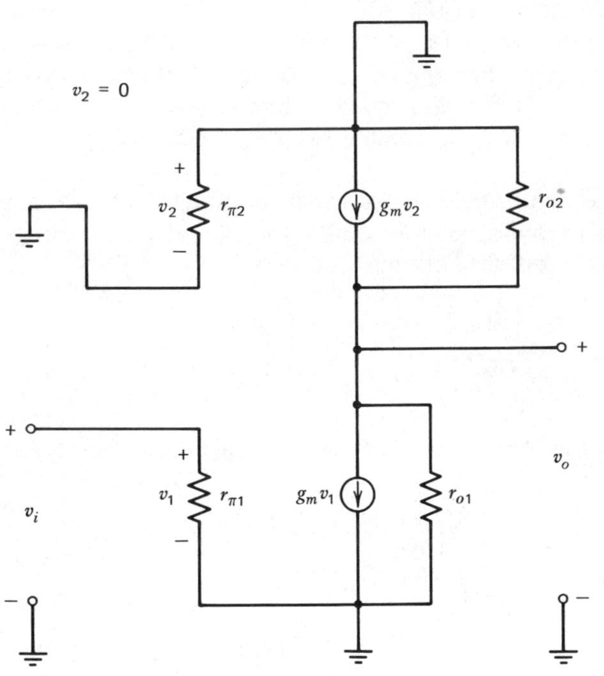

Fig. 4.11 Small-signal equivalent circuit for common-emitter amplifier with active load.

This is the same expression for voltage gain that was found by differentiating (4.50). Typical values for this voltage gain are in the 1000 to 2000 range, so that the actively loaded stage provides very high voltage gain. By inspection of the small-signal circuit, it is apparent that the resistance seen looking into the output is just the parallel combination of the two transistor output resistances.

$$R_o = (r_{onpn}\|r_{opnp}) \tag{4.60}$$

EXAMPLE

Find the voltage gain of the circuit in Fig. 4.11, when both devices are active. Assume $\eta_{pnp} = 5 \times 10^{-4}$, $\eta_{npn} = 2 \times 10^{-4}$

$$A_v = \frac{-1}{5 \times 10^{-4} + 2 \times 10^{-4}} = -1.4 \times 10^3$$

4.3.2 Differential Amplifier with Active Load

A straightforward application of the active-load concept to the emitter-coupled pair would yield the circuit shown in Fig. 4.12*a*. The differential-mode half circuit for this emitter-coupled pair is identical to the common-emitter amplifier of Fig. 4.9. Thus the differential-mode voltage gain is indeed very large. The circuit

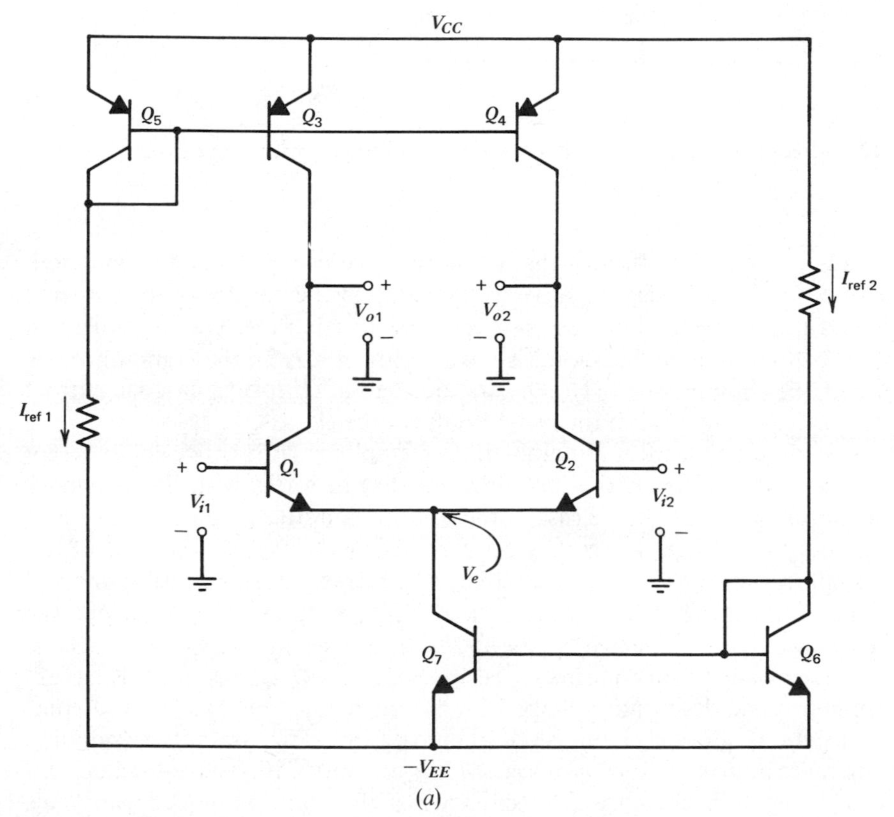

Fig. 4.12*a* Emitter-coupled pair with active load.

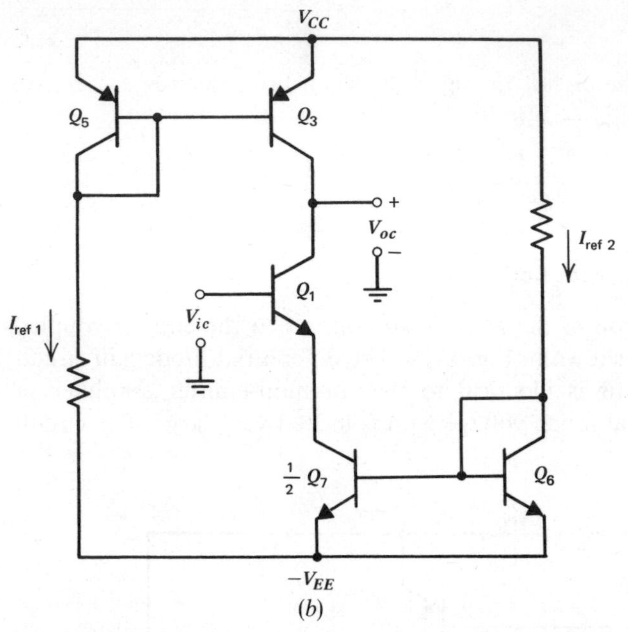

Fig. 4.12*b* Common-mode half circuit for emitter-coupled pair with active load.

as it stands, however, has the drawback that the quiescent value of the common-mode output voltage is very sensitive to the value of the emitter-biasing current source and the active-load current sources. This fact is more clearly illustrated by the dc common-mode half-circuit shown in Fig. 4.12*b*. In the common-mode half circuit, the combination of Q_1, Q_6, and one half of Q_7 form a cascode current source, which is connected to the *pnp* current source Q_3 to Q_5.

This circuit resembles the common-emitter amplifier with active load, except that the cascode transistor, Q_1, has been inserted in series with the common-emitter transistor, Q_7. As in the case of the common-emitter amplifier with active load, the output voltage, V_{oc}, is very sensitive to the voltage at the base of Q_6, which is influenced by the reference current, I_{ref2}. For example, a detailed analysis shows that if the reference currents, I_{ref1} and I_{ref2}, are different by 4 percent, the output voltage, V_{oc}, will change by about 2 V from its nominal value. The same change results from a 1-mV mismatch between Q_6 and Q_7. Thus, for this circuit, the common-mode dc output voltage is very sensitive to mismatches and component variations, and the circuit is not practical from a bias stability standpoint.

An alternate approach, shown in Fig. 4.13, is to control the active-load current source with the collector current of one side of the emitter-coupled pair. This circuit not only eliminates the common-mode problem but provides a single output with much better rejection of common-mode input signals than a standard

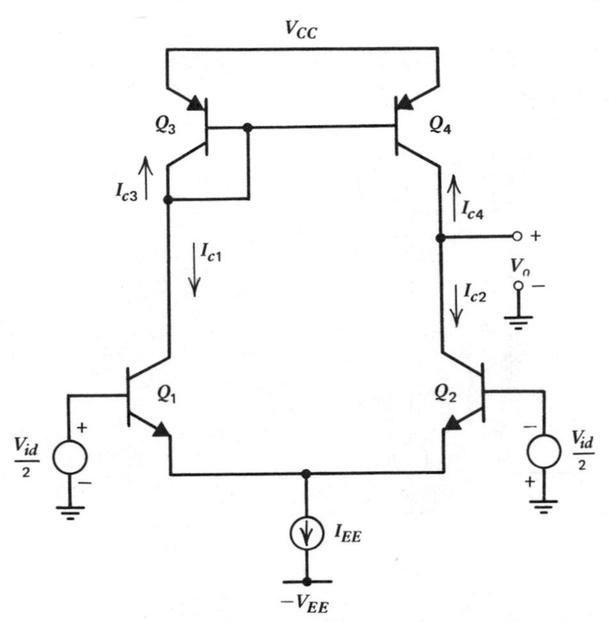

Fig. 4.13 Modified active load for emitter-coupled pair.

resistively loaded emitter-coupled pair with the output taken off one collector. We will analyze both the large-signal and small-signal behavior of this circuit.

In order to obtain an intuitive insight into the behavior of the circuit, we will first analyze its large-signal behavior under the unrealistic condition that the output terminal is attached to a voltage source as shown in Fig. 4.14, so that the load resistance is zero. Since the circuit is used primarily to obtain large voltage gain, it is not actually used with a dc short-circuit load, but an analysis of the short-circuit output current as a function of input voltage provides a good starting point for understanding the operation of the circuit. In Fig. 4.14, the voltage source V_{bias} is adjusted to any convenient value in order to keep Q_2 and Q_4 in the forward-active region. Since the load resistance is zero, the output resistances of Q_2 and Q_4 have little effect on the output current, and we will initially neglect them as well as r_{o1} and r_{o3}. We also neglect *pnp* base current and assume that the transistors are identical. Using the results of Chapter 3,

$$I_{c1} = \frac{\alpha_F I_{EE}}{1 + \exp\left(-\dfrac{V_{id}}{V_T}\right)} \tag{4.61}$$

$$I_{c2} = \frac{\alpha_F I_{EE}}{1 + \exp\left(\dfrac{V_{id}}{V_T}\right)} \tag{4.62}$$

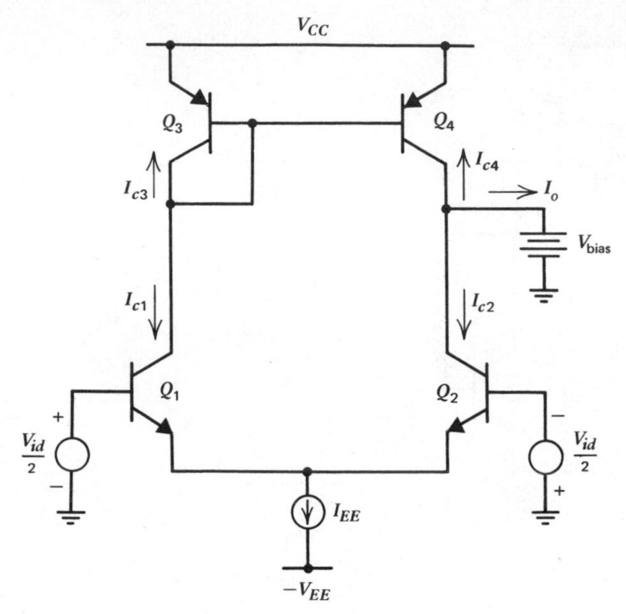

Fig. 4.14 Active-load stage with output connected to a voltage source.

Since Q_3 and Q_4 are identical,

$$I_{c3} = I_{c4} = -I_{c1} \tag{4.63}$$

The output current is:

$$I_o = -I_{c4} - I_{c2} = I_{c1} - I_{c2} \tag{4.64}$$

Using (4.61) and (4.62) in (4.64) we obtain

$$I_o = \alpha_F I_{EE} \tanh\left(\frac{V_{id}}{2V_T}\right) \tag{4.65}$$

This transfer curve is shown in Fig. 4.15. Note that $I_o = 0$ when $V_{id} = 0$, so that when current is taken as the output the circuit has no inherent input offset voltage. The circuit is capable of both absorbing current from and supplying current to the load attached to the output. When the output is shorted, the output current varies between $+I_{EE}$ and $-I_{EE}$ as V_{id} is varied over a range of a few hundred millivolts.

We will perform a separate small-signal analysis, but (4.65) can be used to calculate the short-circuit transconductance of the circuit by differentiation:

$$G_m = \frac{dI_o}{dV_{id}} = \frac{\alpha_F I_{EE}}{2V_T} \operatorname{sech}^2 \frac{V_{id}}{2V_T} \tag{4.66}$$

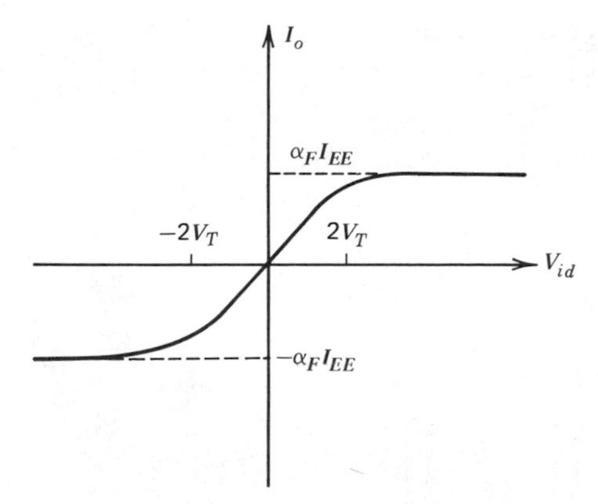

Fig. 4.15 Short-circuit dc transfer characteristic of actively loaded emitter-coupled pair.

For the case of $V_{id} = 0$ we find that

$$G_m = \frac{\alpha_F I_{EE}}{2V_T} = \frac{\alpha_F I_C}{V_T} \approx g_m \tag{4.67}$$

Thus the transconductance of the circuit is the same as that of the individual transistors.

We now consider the large-signal behavior of the circuit in the more useful configuration of Fig. 4.13 in which no load is attached to the output. We will calculate the output voltage as a function of V_{id}. Qualitatively, we expect that the voltage gain will be large when all four devices are in the forward-active region. We neglect base currents in this analysis, but the finite output resistance of the devices must be considered. For the four devices, we have,

$$I_{c1} = I_S\left(\exp\frac{V_{be1}}{V_T}\right)\left(1 + \frac{V_{ce1}}{V_{AN}}\right) \tag{4.68}$$

$$I_{c2} = I_S\left(\exp\frac{V_{be2}}{V_T}\right)\left(1 + \frac{V_{ce2}}{V_{AN}}\right) \tag{4.69}$$

$$I_{c3} = -I_S\left(\exp\left|\frac{V_{be3}}{V_T}\right|\right)\left(1 + \left|\frac{V_{ce3}}{V_{AP}}\right|\right) \tag{4.70}$$

$$I_{c4} = -I_S\left(\exp\left|\frac{V_{be4}}{V_T}\right|\right)\left(1 + \left|\frac{V_{ce4}}{V_{AP}}\right|\right) \tag{4.71}$$

where V_{AN} is the Early voltage of the *npn* devices and V_{AP} is that of the *pnp* devices. We assume that the dc potential at the bases of Q_1 and Q_2 is approximately ground, so that, for Q_1,

$$V_{ce1} = V_{CC} + V_{be3} + V_{be1} \approx V_{CC} \tag{4.72}$$

since $V_{be3} \simeq -V_{be1}$. For Q_3, which is diode connected,

$$V_{ce3} = -V_{BE(\text{on})} \tag{4.73}$$

Summing the voltages around the loop containing the base-emitter junctions of Q_1 and Q_2, we have

$$V_{id} = V_{be1} - V_{be2} \tag{4.74}$$

Inverting (4.68) and (4.69), and substituting in (4.74) we obtain

$$V_{id} = V_T \ln \left[\frac{I_{c1}\left(1 + \dfrac{V_{ce2}}{V_{AN}}\right)}{I_{c2}\left(1 + \dfrac{V_{CC}}{V_{AN}}\right)} \right] \tag{4.75}$$

where (4.72) has been used for V_{ce1}. The base-emitter voltages of Q_3 and Q_4 are the same, so that (4.70) and (4.71) can be combined to yield:

$$\frac{I_{c4}}{I_{c3}} = \frac{1 + \left|\dfrac{V_{ce4}}{V_{AP}}\right|}{1 + \left|\dfrac{V_{BE(\text{on})}}{V_{AP}}\right|} \tag{4.76}$$

where (4.73) has been used for V_{ce3}. If we neglect base currents, then I_{c1} is equal in magnitude to I_{c3}, and I_{c2} is equal in magnitude to I_{c4}. Thus (4.76) can be written as

$$\frac{I_{c2}}{I_{c1}} = \frac{1 + \left|\dfrac{V_{ce4}}{V_{AP}}\right|}{1 + \left|\dfrac{V_{BE(\text{on})}}{V_{AP}}\right|} \tag{4.77}$$

Inserting this ratio in (4.75), we obtain:

$$V_{id} = V_T \ln \left[\frac{1 + \dfrac{V_{BE(\text{on})}}{V_{AP}}}{1 + \dfrac{|V_{ce4}|}{V_{AP}}} \right] \left(\frac{1 + \dfrac{V_{ce2}}{V_{AN}}}{1 + \dfrac{V_{CC}}{V_{AN}}} \right) \tag{4.78}$$

We now must relate voltages V_{ce4} and V_{ce2} to the output voltage, V_o. We have

$$V_{ce2} = V_o + V_{BE2(\text{on})} \tag{4.79}$$

$$V_{ce4} = V_o - V_{CC} \tag{4.80}$$

and

$$|V_{ce4}| = V_{CC} - V_o \tag{4.81}$$

Substituting (4.79) and (4.81) in (4.78), we obtain:

$$V_{id} = V_T \ln \frac{\left[1 + \frac{V_{BE(\text{on})}}{V_{AP}}\right]\left[1 + \frac{V_{BE(\text{on})}}{V_{AN}} + \frac{V_o}{V_{AN}}\right]}{\left(1 + \frac{V_{CC}}{V_{AN}}\right)\left(1 + \frac{V_{CC}}{V_{AP}} - \frac{V_o}{V_{AP}}\right)} \tag{4.82}$$

We now make the assumption that quantities $V_{BE(\text{on})}/V_{AP}$, $V_{BE(\text{on})}/V_{AN}$, V_{CC}/V_{AP}, V_{CC}/V_{AN}, V_o/V_{AN}, and V_o/V_{AP} are much less than unity. If this is the case, we can again use the approximations:

$$\frac{1}{1+x} \approx 1 - x \qquad (x \ll 1) \tag{4.83}$$

and

$$(1 + x)(1 + y) \approx 1 + x + y \qquad (x, y \ll 1) \tag{4.84}$$

to simplify (4.82) to:

$$V_{id} = V_T \ln \frac{1 + \left[\frac{V_o - V_{CC} + V_{BE(\text{on})}}{V_{AN}}\right]}{1 - \left[\frac{V_o - V_{CC} + V_{BE(\text{on})}}{V_{AP}}\right]} \tag{4.85}$$

Exponentiating both sides, we obtain:

$$\left[1 - \frac{V_o - V_{CC} + V_{BE(\text{on})}}{V_{AP}}\right]\left(\exp \frac{V_{id}}{V_T}\right) = 1 + \frac{V_o - V_{CC} + V_{BE(\text{on})}}{V_{AN}} \tag{4.86}$$

Solving for $[V_o - V_{CC} + V_{BE(\text{on})}]$,

$$V_o - V_{CC} + V_{BE(\text{on})} = \frac{\exp \frac{V_{id}}{V_T} - 1}{\frac{1}{V_{AP}} \exp \frac{V_{id}}{V_T} + \frac{1}{V_{AN}}} \tag{4.87}$$

We next multiply numerator and denominator by $(V_{AP}V_{AN} \exp -V_{id}/2V_T)$, giving

$$V_o - V_{CC} + V_{BE(\text{on})} = \frac{V_{AN}V_{AP}\left(\exp \frac{V_{id}}{2V_T} - \exp \frac{-V_{id}}{2V_T}\right)}{V_{AN} \exp \frac{V_{id}}{2V_T} + V_{AP} \exp \frac{-V_{id}}{2V_T}} \tag{4.88}$$

This expression can be simplified using the identities:

$$\exp x - \exp(-x) = 2 \sinh x$$
$$a \exp x + b \exp(-x) = 2(a - b) \sinh x + 2(a + b) \cosh x$$

to give:

$$V_o - V_{CC} + V_{BE(\text{on})} = \frac{V_{AN}V_{AP}\sinh\dfrac{V_{id}}{2V_T}}{(V_{AN} + V_{AP})\cosh\dfrac{V_{id}}{2V_T} + (V_{AN} - V_{AP})\sinh\dfrac{V_{id}}{2V_T}} \tag{4.89}$$

and thus

$$V_o = V_{CC} - V_{BE(\text{on})} + \left(\frac{2V_{A(\text{eff})}\tanh\dfrac{V_{id}}{2V_T}}{1 + \dfrac{V_{AN} - V_{AP}}{V_{AN} + V_{AP}}\tanh\dfrac{V_{id}}{2V_T}} \right) \tag{4.90}$$

where

$$V_{A(\text{eff})} = \frac{V_{AP}V_{AN}}{V_{AN} + V_{AP}} \tag{4.91}$$

This transfer characteristic is illustrated in Fig. 4.16 and was derived assuming all four devices remain in the forward-active region. Actually, when V_o approaches V_{CC}, Q_4 saturates, and when V_o approaches -0.6 V, Q_2 saturates. Thus (4.90) is valid only for a small range of V_{id} about zero. This fact can be used to simplify

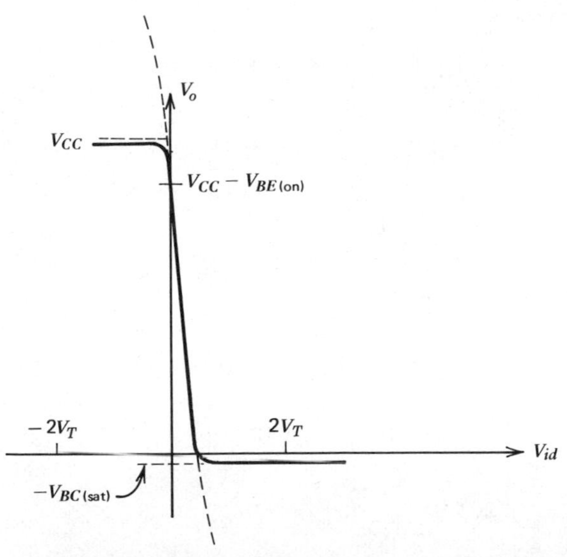

Fig. 4.16 Output voltage as a function of input voltage for the emitter-coupled pair with active load. The actually observed portion (solid line) is only a small section of the calculated tanh transfer characteristic. For large positive V_{id}, the output voltage is equal to the collector-base voltage of Q_2 in saturation, assuming that the base of Q_2 is at ground.

the result further, since the quantity

$$\left| \frac{V_{AN} - V_{AP}}{V_{AN} + V_{AP}} \tanh \frac{V_{id}}{2V_T} \right| \ll 1 \tag{4.92}$$

if

$$\frac{V_{id}}{2V_T} \ll 1$$

Thus (4.90) can be simplified to:

$$V_o \approx V_{CC} - V_{BE(\text{on})} + 2V_{A(\text{eff})} \tanh\left(\frac{V_{id}}{2V_T}\right) \tag{4.93}$$

for

$$V_{CC} > V_o > [-V_{BE(\text{on})} + V_{CE(\text{sat})}]$$

The output voltage in this case thus has the same tanh dependence on V_{id} as did the output current with a short-circuit load. Note that when the differential input voltage, V_{id}, is zero, the output voltage is equal to $[V_{CC} - V_{BE(\text{on})}]$. This must be the case since this value of output voltage is the only one for which $V_{ce2} = V_{ce1}$, $V_{ce4} = V_{ce3}$, $|I_{c1}| = |I_{c3}|$, and $|I_{c2}| = |I_{c4}|$ in Fig. 4.13. We will perform a separate small-signal analysis, but (4.93) can also be used to give the voltage gain of the circuit by differentiation.

$$A_{vd} = \frac{dV_o}{dV_{id}} = \frac{d}{dV_{id}}\left[V_{CC} - V_{BE(\text{on})} + 2V_{A(\text{eff})} \tanh \frac{V_{id}}{2V_T}\right] \tag{4.94}$$

$$= \frac{V_{A(\text{eff})}}{V_T} \operatorname{sech}^2\left(\frac{V_{id}}{2V_T}\right) \tag{4.95}$$

For the condition that the quiescent value of $V_{id} = 0$,

$$A_{vd} = \frac{V_{A(\text{eff})}}{V_T} = \frac{1}{V_T}\left(\frac{V_{AN}V_{AP}}{V_{AN} + V_{AP}}\right) = \frac{1}{\dfrac{V_T}{V_{AN}} + \dfrac{V_T}{V_{AP}}} \tag{4.96}$$

In terms of the Early factors,

$$A_{vd} = \frac{1}{\eta_{npn} + \eta_{pnp}} = g_m(r_{onpn} \| r_{opnp}) \tag{4.97}$$

where

$$\eta = \text{Early factor} = \frac{V_T}{V_A}$$

Since the Early factor is on the order of 2×10^{-4}, the circuit has very high voltage gain.

EXAMPLE

For a power supply voltage V_{CC} of 15 V, determine how much differential input voltage must be applied to the active-load circuit of Fig. 4.13 to drive the output voltage from its zero-input value of $[V_{CC} - V_{BE(on)}]$ to ground potential. Assume $V_{BE(on)} = 0.7$ V, $V_{AN} = 130$ V, $V_{AP} = 50$ V. From (4.93)

$$V_o = V_{CC} - V_{BE(on)} + 2V_{A(eff)} \tanh \frac{V_{id}}{2V_T}$$

and substitution of data gives

$$0 = 15 - 0.7 + 2\left[\frac{(130)(50)}{(130) + (50)}\right] \tanh \frac{V_{id}}{2V_T}$$

This gives

$$\tanh \frac{V_{id}}{2V_T} = \frac{14.3}{72.0} = 0.20$$

and thus

$$V_{id} = 2V_T \tanh^{-1}(0.20)$$

Since

$$\tanh^{-1} x = x + \frac{x^3}{3} \cdots \approx x \qquad \text{for } x \ll 1$$

we find

$$V_{id} \cong (2)(0.20)V_T = 10.4 \text{ mV}$$

EXAMPLE

In the circuit of Fig. 4.13, assume that Q_1 has an I_S that is 10 percent smaller than that of Q_2. Determine the dc output voltage if the two inputs are grounded so that $V_{id} = 0$. Assume $V_{CC} = 15$ V, $V_{BE(on)} = 0.7$, $V_{AN} = 130$ V, $V_{AP} = 50$ V. We first calculate the input offset voltage, as defined in Section 3.6.1,

$$V_{OS} = V_T \ln \frac{I_{S2}}{I_{S1}} = 2.6 \text{ mV}$$

The offset voltage may be regarded as a dc voltage source in series with the input so that, under the given conditions, the effective differential input voltage is

$$V_{id(eff)} = -2.6 \text{ mV}$$

Thus

$$V_o = V_{CC} - V_{BE(on)} + 2V_{A(eff)} \tanh \frac{V_{id(eff)}}{2V_T}$$

$$= 15 - 0.7 - 2\left(\frac{(130)(50)}{(130) + (50)}\right) \tanh \frac{2.6 \text{ mV}}{2(26 \text{ mV})}$$

$$= 14.3 - 72.0 \tanh(0.05)$$

Since

$$\tanh x = x - \frac{x^3}{3} + \cdots \approx x \qquad \text{for } x \ll 1$$

we have

$$V_o \approx 14.3 - (72.0)(0.05) = 14.3 - 3.60$$

and thus

$$V_o = 10.70 \text{ V}$$

While the dc analysis provides an intuitive feel for currents flowing in the circuit, the primary feature of interest in this circuit is the small-signal voltage gain with no load attached, and also the output resistance. Since the circuit is not symmetrical, a half-circuit approach is not useful. We will analyze the low-frequency, small-signal circuit directly assuming that $r_\mu = \infty$, $r_b = 0$, and $R_{EE} = \infty$, where R_{EE} is the output resistance of the biasing current source. This circuit is shown in Fig. 4.17.

First summing currents at node ①,

$$\frac{v_i - v_1}{r_{\pi 1}}(1 + \beta_0) + \frac{v_2 - v_1}{r_{o1}} + \frac{v_3 - v_1}{r_{o2}} - \frac{v_1}{r_{\pi 2}}(1 + \beta_0) = 0 \qquad (4.98)$$

where v_1, v_2, and v_3 are the voltages at the nodes labeled ①, ②, and ③. Rearranging,

$$-v_1\left(\frac{1 + \beta_0}{r_{\pi 1}} + \frac{1 + \beta_0}{r_{\pi 2}} + \frac{1}{r_{o1}} + \frac{1}{r_{o2}}\right) + \left(\frac{v_2}{r_{o1}}\right) + \left(\frac{v_3}{r_{o2}}\right) = -\frac{v_i}{r_{\pi 1}}(\beta_0 + 1) \qquad (4.99)$$

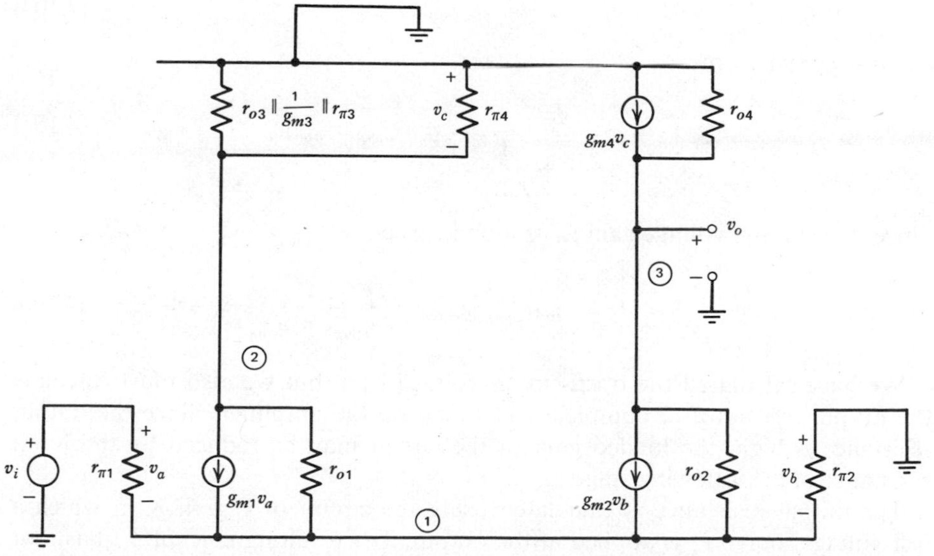

Fig. 4.17 Small-signal equivalent circuit, emitter-coupled pair with active load.

We now sum currents at node ②. Note that the total resistance connected between node ② and V_{CC} is the parallel combination of $r_{\pi 3}$, r_{o3}, $1/g_{m3}$, and $r_{\pi 4}$. We make the simplifying assumption that β_0 and r_o are large, and this parallel combination is approximately equal to $1/g_{m3}$. Thus,

$$\frac{v_2}{\dfrac{1}{g_{m3}}} + \frac{v_2 - v_1}{r_{o1}} + (v_i - v_1)g_{m1} = 0 \tag{4.100}$$

Rearranging,

$$-v_1\left(\frac{1}{r_{o1}} + g_{m1}\right) + v_2\left(\frac{1}{r_{o1}} + \frac{1}{\dfrac{1}{g_{m3}}}\right) = -g_{m1}v_i \tag{4.101}$$

Summing currents at node ③,

$$\frac{v_3}{r_{o4}} + v_2 g_{m4} + \frac{v_3 - v_1}{r_{o2}} - g_{m2}v_1 = 0 \tag{4.102}$$

Rearranging,

$$-v_1\left(g_{m2} + \frac{1}{r_{o2}}\right) + v_2 g_{m4} + v_3\left(\frac{1}{r_{o4}} + \frac{1}{r_{o2}}\right) = 0 \tag{4.103}$$

Equations 4.101 and 4.103 can be used to eliminate v_1 and v_2 from (4.99). Assuming that $g_{m1} = g_{m2} = g_{m3} = g_m$ and that

$$\frac{1}{r_o} \ll g_m \tag{4.104}$$

equation 4.99 becomes:

$$v_3 = \frac{g_m v_i}{\dfrac{1}{r_{onpn}} + \dfrac{1}{r_{opnp}}} = g_m(r_{opnp} \| r_{onpn})v_i \tag{4.105}$$

Since $v_o = v_3$, the voltage gain is, as found earlier,

$$\frac{v_o}{v_i} = A_v = g_m(r_{opnp} \| r_{onpn}) = \frac{1}{\eta_{pnp} + \eta_{npn}} \tag{4.106}$$

We have calculated the open-circuit voltage gain, but we also must calculate the output resistance to completely characterize the amplifier. Since the output resistance is high, the loaded gain of the circuit may be reduced by the input resistance of the following stage.

The output resistance is calculated using the circuit of Fig. 4.18, in which a test voltage source v_x is applied at the output. The resulting current, i_x, has four

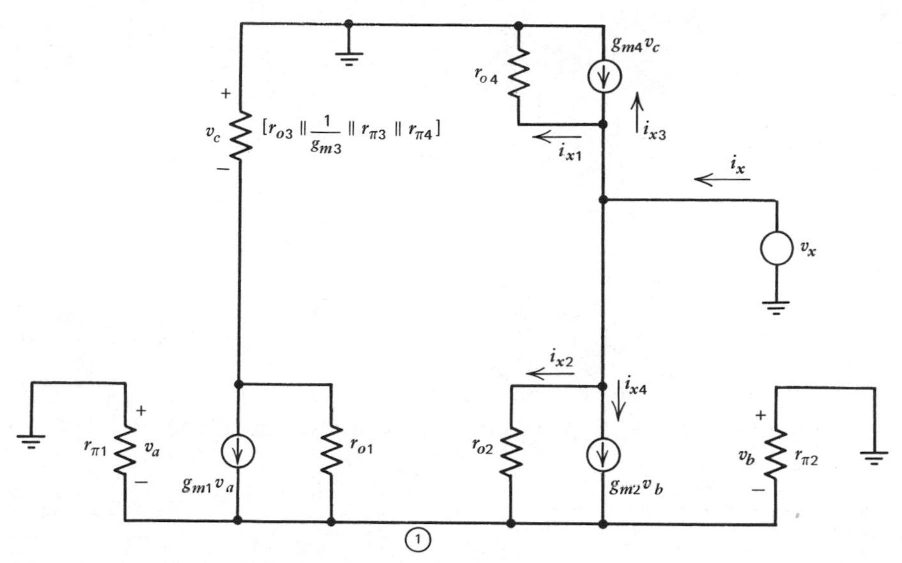

Fig. 4.18 Circuit for calculation of the output resistance of the emitter-coupled pair with active load.

components. The current in r_{o4} is:

$$i_{x1} = \frac{v_x}{r_{o4}} \tag{4.107}$$

The resistance in the emitter lead of Q_2 is that seen looking into the emitter of Q_1, which is approximately $1/g_{m1}$. Thus, using (3.39) for a transistor with emitter degeneration, we find that the effective output resistance of Q_2 is

$$R_{o2} = r_{o2}\left(1 + g_{m2}\frac{1}{g_{m1}}\right) = 2r_{o2} \tag{4.108}$$

and hence

$$i_{x2} + i_{x4} = \frac{v_x}{2r_{o2}} \tag{4.109}$$

This current flows into the emitter of Q_1, and passes through Q_3 and Q_4 with unity gain to produce

$$\begin{aligned} i_{x3} &= i_{x2} + i_{x4} \\ &= \frac{v_x}{2r_{o2}} \end{aligned} \tag{4.110}$$

Thus

$$i_x = i_{x1} + i_{x2} + i_{x3} + i_{x4} \tag{4.111}$$

$$= v_x\left(\frac{1}{r_{o4}} + \frac{1}{r_{o2}}\right) \tag{4.112}$$

and thus

$$r_o = \frac{v_x}{i_x}$$

$$= \frac{1}{\dfrac{1}{r_{onpn}} + \dfrac{1}{r_{opnp}}} = r_{onpn} \| r_{opnp} \tag{4.113}$$

Thus the output resistance of the circuit is simply that of the *npn* and *pnp* transistors in parallel.

EXAMPLE

Calculate the voltage gain and output resistance of the circuit of Fig. 4.13 if $\eta_{npn} = 2 \times 10^{-4}$, $\eta_{pnp} = 5 \times 10^{-4}$, $I_{EE} = 100\ \mu\text{A}$. Using (4.106)

$$A_v = \frac{1}{2 \times 10^{-4} + 5 \times 10^{-4}} = 1400$$

The output resistance is:

$$R_o = r_{opnp} \| r_{onpn}$$

$$r_{onpn} = \frac{1}{\eta_{npn} g_m} = \frac{1}{(2 \times 10^{-4})\left(\dfrac{50\ \mu\text{A}}{26\ \text{mV}}\right)} = 2.6\ \text{M}\Omega$$

$$r_{opnp} = \frac{1}{\eta_{pnp} g_m} = \frac{1}{(5 \times 10^{-4})\left(\dfrac{50\ \mu\text{A}}{26\ \text{mV}}\right)} = 1.04\ \text{M}\Omega$$

and thus

$$R_o = 2.6\ \text{M}\Omega \| 1.04\ \text{M}\Omega = 740\ \text{k}\Omega$$

If the effects of the r_o of Q_2 and Q_4 are neglected, the differential input resistance of the actively loaded emitter-coupled pair is simply $2r_\pi$ as in the resistively loaded case. Actually, the asymmetry of the circuit together with the high voltage gain cause feedback to occur through the output resistance of Q_4 to node 1 . This causes the input resistance as seen from the two inputs to be slightly different from that calculated neglecting these effects.

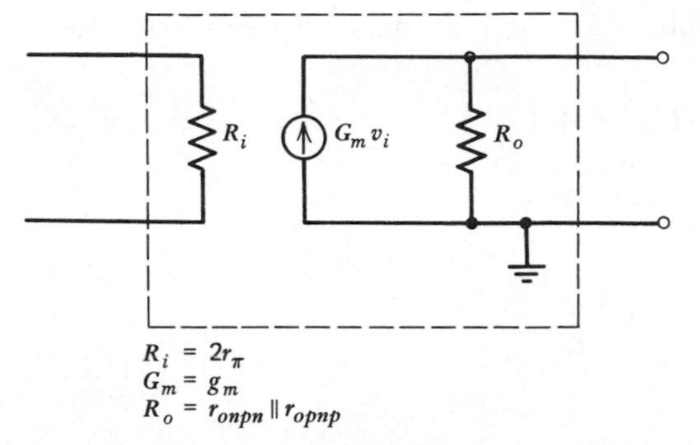

Fig. 4.19 Two-port representation of small-signal properties of active-load, emitter-coupled pair. The effects of assymetrical input resistance have been neglected.

In summary, the actively loaded differential pair is thus capable of providing differential to single-ended conversion (i.e., the conversion from a differential voltage to a voltage referenced to the ground potential). The high output impedance of the circuit requires that care must be taken to use a following stage with high input resistance if the large gain is to be realized. A small-signal two-port equivalent circuit for the stage is shown in Fig. 4.19.

4.3.3 Input Offset Voltage of the Emitter-Coupled Pair with Active Load

In the case of the resistively loaded emitter-coupled pair, it was found in Chapter 3 that the input offset voltage arose primarily from mismatches in I_S in the input transistors and from mismatches in the collector load resistors. In the active-load case, the input offset voltage results from mismatches in the input transistors and load devices and from the base current of the load devices. Referring to Fig. 4.13, we begin by assuming that V_{id} has been adjusted in order to drive the output to the value $[V_{CC} - V_{BE(\text{on})}]$. This is the value of output voltage that results when the devices are identical and the inputs are grounded. At this output voltage,

$$V_{CE3} = V_{CE4} \qquad \text{and} \qquad V_{CE1} = V_{CE2}$$

The collector current I_{C4} is related to I_{C3} by:

$$I_{C4} = I_{C3}\left(\frac{I_{S4}}{I_{S3}}\right) \tag{4.114}$$

The collector current of Q_2 is equal to $(-I_{C4})$ and thus (4.114) can be written as

$$I_{C2} = -I_{C3}\left(\frac{I_{S4}}{I_{S3}}\right) \tag{4.115}$$

The current I_{C1} is equal to $(-I_{C3})$, plus the base currents in the *pnp* transistors:

$$I_{C1} = -I_{C3}\left[1 + \left(\frac{2}{\beta_F}\right)\right] \tag{4.116}$$

The input offset voltage is then given by:

$$V_{OS} = V_{BE1} - V_{BE2} \tag{4.117}$$

$$= V_T \ln \frac{I_{C1}}{I_{C2}} \frac{I_{S2}}{I_{S1}} \tag{4.118}$$

Using (4.115) and (4.116), in (4.118)

$$V_{OS} = V_T \ln \left[\frac{I_{S3}}{I_{S4}} \frac{I_{S2}}{I_{S1}} \left(1 + \frac{2}{\beta_F}\right)\right] \tag{4.119}$$

If the mismatches are small, this expression can be approximated as:

$$V_{OS} = V_T \left(\frac{\Delta I_{SP}}{I_{SP}} - \frac{\Delta I_{SN}}{I_{SN}} + \frac{2}{\beta_F}\right) \tag{4.120}$$

using the technique described in Section 3.6.2, where

$$\Delta I_{SP} = I_{S3} - I_{S4} \tag{4.121}$$

$$I_{SP} = \frac{I_{S3} + I_{S4}}{2} \tag{4.122}$$

$$\Delta I_{SN} = I_{S1} - I_{S2} \tag{4.123}$$

$$I_{SN} = \frac{I_{S1} + I_{S2}}{2} \tag{4.124}$$

Assuming a worst-case value for $\Delta I_S/I_S$ of ± 5 percent and a *pnp* beta of 20, we have, for the worst-case offset voltage,

$$V_{OS} = V_T(0.05 + 0.05 + 0.1) \tag{4.125}$$

$$= 0.2V_T \simeq 5 \text{ mV} \tag{4.126}$$

Thus these circuits have significantly higher offset than the resistively loaded case. The offset can be reduced by inserting resistors in the emitters of Q_3 and Q_4 in order to reduce their offset contribution as discussed in Section A4.1. The base-current contribution to offset can be reduced by including a transistor in the active load as described in Section 4.2.1. These modifications are shown in Fig. 4.20.

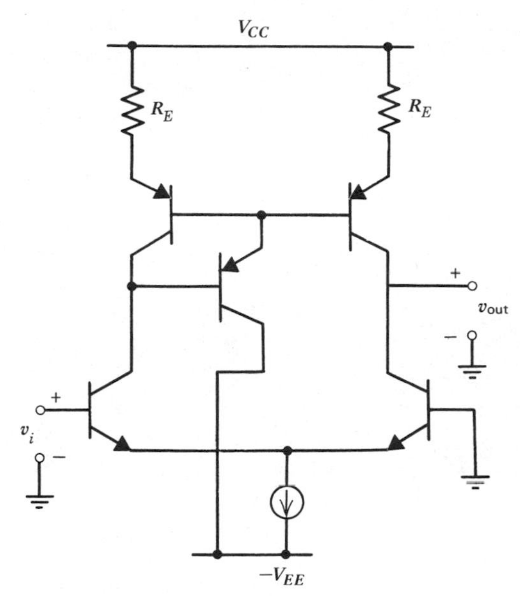

Fig. 4.20 Actively loaded emitter-coupled pair with modifications for improved offset.

4.3.4 Common-Mode Rejection Ratio of the Emitter-Coupled Pair with Active Load

In addition to providing high voltage gain, the circuit of Fig. 4.13 provides conversion from a differential signal to a signal that is referenced to ground. Such a conversion is required in all differential input, single-ended output amplifiers, a good example being the operational amplifier.

The simplest differential to single-ended converter is the resistively loaded emitter-coupled pair shown in Fig. 4.21*a*. Here the output is taken from only one of the collectors. The output, however, is given by:

$$v_o = -\frac{v_{od}}{2} + v_{oc} = -\frac{A_{dm}v_{id}}{2} + A_{cm}v_{ic} \tag{4.127}$$

$$= -\frac{A_{dm}}{2}\left(v_{id} + \frac{2A_{cm}}{A_{dm}}v_{ic}\right) \tag{4.128}$$

$$= -\frac{A_{dm}}{2}\left(v_{id} + \frac{2v_{ic}}{\text{CMRR}}\right) \tag{4.129}$$

Thus, common-mode signals at the input will cause changes in the output voltage. In Chapter 3, the CMRR of this stage was shown to be:

$$\text{CMRR} = 2g_m R_{EE} = 2g_m r_{o3} \tag{4.130}$$

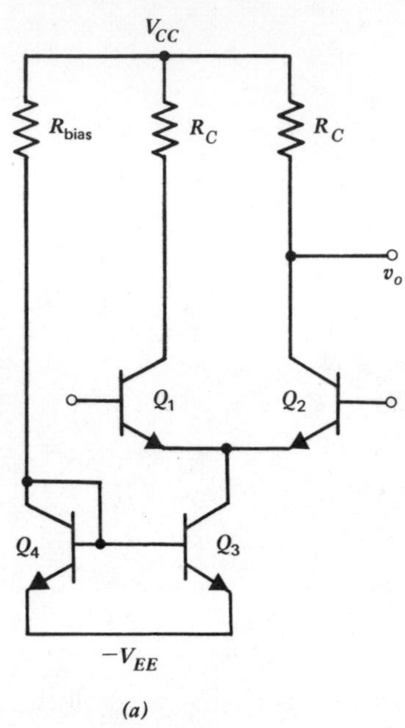

Fig. 4.21*a* Differential-to-single-ended conversion using the resistively loaded emitter-coupled pair.

The current source transistor, Q_3, operates at twice the current of Q_1 and Q_2, so that, for the particular circuit of Fig. 4.21*a*,

$$\text{CMRR} = g_{m3}r_{o3} = \frac{1}{\eta_3} \tag{4.131}$$

Thus the CMRR of the resistively loaded stage is the inverse of the η of the current-source transistor when a simple current source is used as the biasing element.

On the other hand, the active-load stage shown in Fig. 4.21*b* has a common-mode rejection ratio that is much superior to that of the circuit of Fig. 4.21*a*. As in the resistively loaded case, changes in the common-mode input will cause changes in the bias current, I_{EE}, because of the finite output resistance of the biasing current source. This will cause a change in I_{C2} and an identical change in I_{C1}. Because of the behavior of the active load, the change in I_{C1} will produce a change in the currents flowing in the *pnp* load transistors, which produces a change in the collector current of Q_4 that will precisely cancel the original change in the collector current of Q_2. As a result, the output does not change at all in response to common-mode inputs.

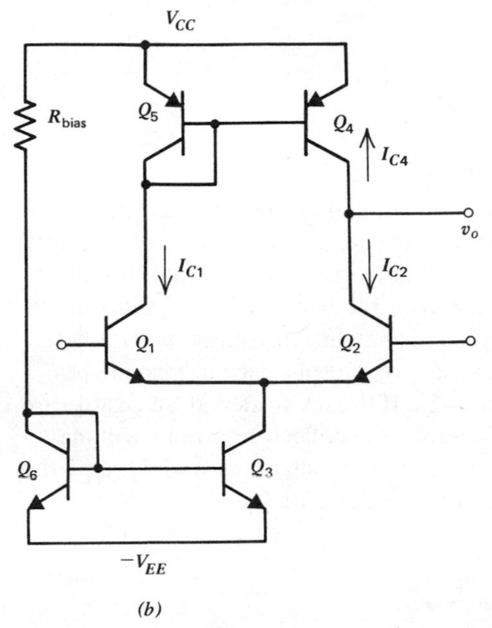

Fig. 4.21*b* Differential-to-single-ended conversion using the actively loaded emitter-coupled pair.

The superior common-mode rejection ratio of this stage can be predicted analytically using a relationship to be developed in Chapter 6. There it is shown that, for circuits with differential inputs and a single-ended output, the CMRR is given by:

$$\text{CMRR} = \left|\frac{A_{dm}}{A_{cm}}\right| = \left(\frac{\partial V_{OS}}{\partial V_{ic}}\right)^{-1} \tag{4.132}$$

Here, V_{OS} is the value of differential input voltage required to keep the output voltage at a constant value. V_{ic} is the common-mode input voltage. If, for example, the offset voltage does not change at all when V_{ic} is changed, then it must be true that if the differential input is held constant while V_{ic} is changed, the output voltage does not change. This is equivalent to saying that the common-mode gain is zero, and the CMRR is infinite, as predicted by (4.132).

For the emitter-coupled pair with active load, the offset-voltage expression, Equation 4.120, does not depend on the bias current, I_{EE}, at all. Thus, when (4.132) is applied, the derivative of input offset voltage with respect to common-mode input voltage is zero. Thus the CMRR is infinite if the transistors are identical, even for the case of finite current gain in the *pnp* transistors.

APPENDIX

A4.1 MATCHING CONSIDERATIONS IN TRANSISTOR CURRENT SOURCES

In many types of circuits, an objective of current-source design is generation of two or more current sources whose values are identical. This is particularly important in the design of digital-to-analog converters, operational amplifiers, and instrumentation amplifiers. We discussed in the last section means of minimizing the effect of base current. However, mismatches in transistor parameters I_S and α_F, and resistor mismatches, can cause large output current mismatches.

Consider the dual current source of Fig. 4.22. If the resistors and transistors are identical and the collector voltages are the same, the collector currents will match precisely. However, mismatch in the transistor parameters, α_F and I_S, and the emitter resistors will cause the currents to be unequal. For Q_3,

$$V_T \ln \frac{I_{C3}}{I_{S3}} + \frac{I_{C3}}{\alpha_{F3}} R_3 = V_B \tag{4.133}$$

For Q_4,

$$V_T \ln \frac{I_{C4}}{I_{S4}} + \frac{I_{C4}}{\alpha_{F4}} R_4 = V_B \tag{4.134}$$

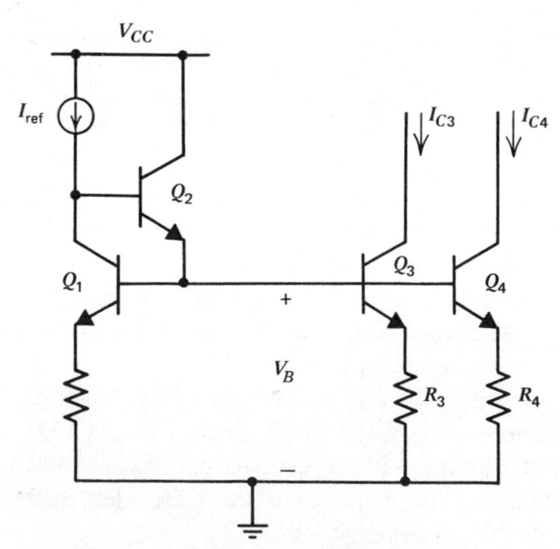

Fig. 4.22 Matched current sources.

Subtraction of these two equations gives:

$$V_T \ln \frac{I_{C3}}{I_{C4}} - V_T \ln \frac{I_{S3}}{I_{S4}} + \frac{I_{C3}}{\alpha_{F3}} R_3 - \frac{I_{C4}}{\alpha_{F4}} R_4 = 0 \tag{4.135}$$

We now define new "average" and "mismatch" parameters as follows:

$$\Delta I_C = I_{C3} - I_{C4} \tag{4.136}$$

$$I_C = \frac{I_{C3} + I_{C4}}{2} \tag{4.137}$$

$$\Delta I_S = I_{S3} - I_{S4} \tag{4.138}$$

$$I_S = \frac{I_{S3} + I_{S4}}{2} \tag{4.139}$$

$$\Delta R = R_3 - R_4 \tag{4.140}$$

$$R = \frac{R_3 + R_4}{2} \tag{4.141}$$

$$\Delta \alpha_F = \alpha_{F3} - \alpha_{F4} \tag{4.142}$$

$$\alpha_F = \frac{\alpha_{F3} + \alpha_{F4}}{2} \tag{4.143}$$

These relations can be inverted to give the original parameters in terms of the average and mismatch parameters. For example,

$$I_{C3} = I_C + \frac{\Delta I_C}{2} \tag{4.144}$$

$$I_{C4} = I_C - \frac{\Delta I_C}{2} \tag{4.145}$$

This set of equations for the various parameters is now substituted into (4.135). The result is:

$$V_T \ln \left(\frac{I_C + \dfrac{\Delta I_C}{2}}{I_C - \dfrac{\Delta I_C}{2}} \right) - V_T \ln \left(\frac{I_S + \dfrac{\Delta I_S}{2}}{I_S - \dfrac{\Delta I_S}{2}} \right) + \frac{\left(I_C + \dfrac{\Delta I_C}{2}\right)\left(R + \dfrac{\Delta R}{2}\right)}{\alpha_F + \dfrac{\Delta \alpha_F}{2}} - \frac{\left(I_C - \dfrac{\Delta I_C}{2}\right)\left(R - \dfrac{\Delta R}{2}\right)}{\alpha_F - \dfrac{\Delta \alpha_F}{2}} = 0 \tag{4.146}$$

If it is true that the mismatches are small, then for example, the first term above can be reduced as follows:

$$V_T \ln\left(\frac{I_C + \dfrac{\Delta I_C}{2}}{I_C - \dfrac{\Delta I_C}{2}}\right) = V_T \ln\left(\frac{1 + \dfrac{\Delta I_C}{2I_C}}{1 - \dfrac{\Delta I_C}{2I_C}}\right) \tag{4.147}$$

If $\Delta I_C/2I_C \ll 1$, this term becomes

$$V_T \ln\left(\frac{I_C + \dfrac{\Delta I_C}{2}}{I_C - \dfrac{\Delta I_C}{2}}\right) \simeq V_T \ln\left[\left(1 + \frac{\Delta I_C}{2I_C}\right)\left(1 + \frac{\Delta I_C}{2I_C}\right)\right] \tag{4.148}$$

$$\simeq V_T \ln\left[1 + \frac{\Delta I_C}{I_C} + \left(\frac{\Delta I_C}{2I_C}\right)^2\right] \tag{4.149}$$

$$\simeq V_T \ln\left(1 + \frac{\Delta I_C}{I_C}\right) \tag{4.150}$$

where the squared term is neglected. The logarithm function has the infinite series:

$$\ln(1 + x) = x - \frac{x^2}{2} + \cdots \tag{4.151}$$

If x is much less than unity, then

$$\ln(1 + x) \simeq x \tag{4.152}$$

and, from (4.150),

$$V_T \ln\left(\frac{I_C + \dfrac{\Delta I_C}{2}}{I_C - \dfrac{\Delta I_C}{2}}\right) \simeq V_T \frac{\Delta I_C}{I_C} \tag{4.153}$$

Applying the same approximation to the other terms in (4.146) we obtain

$$\left|\frac{\Delta I_C}{I_C}\right| \simeq \left(\frac{1}{1 + \dfrac{g_m R}{\alpha_F}}\right)\frac{\Delta I_S}{I_S} + \frac{\dfrac{g_m R}{\alpha_F}}{1 + \dfrac{g_m R}{\alpha_F}}\left(-\frac{\Delta R}{R} + \frac{\Delta \alpha_F}{\alpha_F}\right) \tag{4.154}$$

Note that for $g_m R \ll 1$, $[(I_C R/V_T) \ll 1]$, the mismatch is entirely determined by transistor I_S mismatch. Typically observed mismatches in I_S range from ± 10 to ± 1 percent depending on geometry.

In the other case in which $g_m R \gg 1$, $[(I_C R/V_T) \gg 1]$, the mismatch is determined by the second term containing resistor mismatch and transistor α_F mismatch. Resistor mismatch ranges from ± 2 to ± 0.1 percent depending on geometry, and alpha matching is in the ± 0.1 percent range for *npn* transistors (± 1 percent for *pnp*s). Thus, for *npn* current sources, the use of emitter resistors offers significantly improved current matching. For *pnp* current sources, the alpha mismatch is larger due to the lower beta, and the advantage is less significant.

A4.2 SUPPLY-INDEPENDENT BIASING

As a biasing source, the simple current source of Fig. 4.1 has the drawback that the output current is proportional to the power-supply voltage. If, for example, it were used in an operational amplifier circuit that had to function with power-supply voltages ranging from 10 to 30 V, the bias current would vary over a three-to-one range, and the power dissipation would vary over a nine-to-one range. One measure of this aspect of bias-circuit performance is the fractional change in bias current that results from a given fractional change in supply voltage. For the simple current source, this figure of merit is unity. The Widlar current source of Fig. 4.5 is somewhat better in power-supply sensitivity. The presence of the emitter resistor gives an approximately logarithmic dependence of output current on supply voltage, as indicated implicitly by (4.14) and illustrated in Fig. 4.23.

The most useful parameter for describing the variation of output current with power-supply voltage is the sensitivity, S. This is the percentage, or fractional, change in output current per percentage or fractional change in power-supply voltage. For small variations,

$$S_{V_{CC}}^{I_{C2}} = \frac{\dfrac{\Delta I_{C2}}{I_{C2}}}{\dfrac{\Delta V_{CC}}{V_{CC}}} = \frac{V_{CC}}{I_{C2}} \frac{\Delta I_{C2}}{\Delta V_{CC}} = \frac{V_{CC}}{I_{C2}} \frac{\partial I_{C2}}{\partial V_{CC}} \tag{4.155}$$

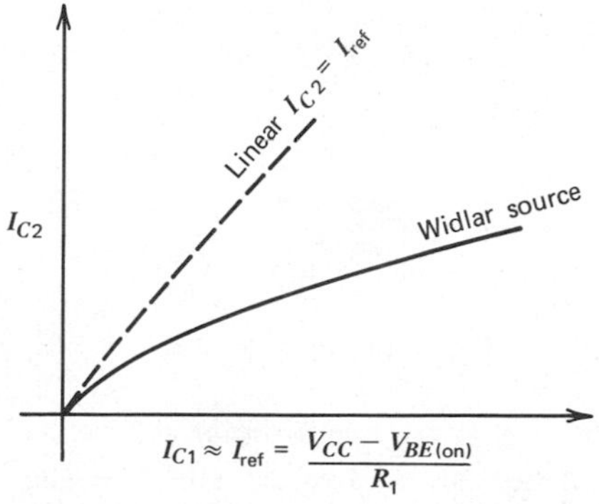

Fig. 4.23 Dependence of I_{C2} on supply voltage for the Widlar current source.

The sensitivity of any parameter y to a second one x is defined similarly:

$$S_x^y = \frac{x}{y}\frac{\partial y}{\partial x} \tag{4.156}$$

For the case of the Widlar source, the output current, I_{C2}, is given implicitly by the equation:

$$V_T \ln \frac{I_{\text{ref}}}{I_{C2}} = I_{C2}R_2 \tag{4.157}$$

To determine the sensitivity of I_{C2} to the power-supply voltage, this equation is differentiated with respect to V_{CC}.

$$V_T \frac{\partial}{\partial V_{CC}} \ln \frac{I_{\text{ref}}}{I_{C2}} = R_2 \frac{\partial I_{C2}}{\partial V_{CC}} \tag{4.158}$$

Differentiating

$$V_T\left(\frac{I_{C2}}{I_{\text{ref}}}\right)\left(\frac{1}{I_{C2}}\frac{\partial I_{\text{ref}}}{\partial V_{CC}} - \frac{I_{\text{ref}}}{I_{C2}{}^2}\frac{\partial I_{C2}}{\partial V_{CC}}\right) = R_2\left(\frac{\partial I_{C2}}{\partial V_{CC}}\right) \tag{4.159}$$

Solving this equation for $\partial I_{C2}/\partial V_{CC}$, we obtain:

$$S_{V_{CC}}^{I_{C2}} = \frac{V_{CC}}{I_{C2}}\frac{\partial I_{C2}}{\partial V_{CC}} = \left(\frac{1}{1 + \dfrac{I_{C2}R_2}{V_T}}\right)\frac{V_{CC}}{I_{\text{ref}}}\frac{\partial I_{\text{ref}}}{\partial V_{CC}} \tag{4.160}$$

and thus

$$S_{V_{CC}}^{I_{C2}} = \left(\frac{1}{1 + \dfrac{I_{C2}R_2}{V_T}}\right)S_{V_{CC}}^{I_{\text{ref}}} \tag{4.161}$$

If V_{CC} is much larger than one diode drop, then the sensitivity of V_{CC} to I_{ref} is approximately unity, since:

$$I_{\text{ref}} \approx \frac{V_{CC}}{R} \tag{4.162}$$

For the previous example in which $I_{\text{ref}} = 1$ mA, $I_{C2} = 10\ \mu$A, and $R_2 = 11.9$ kΩ, (4.161) can be used to give:

$$S_{V_{CC}}^{I_{C2}} = \frac{V_{CC}}{I_{C2}}\frac{\partial I_{C2}}{\partial V_{CC}} = \frac{1}{1 + \dfrac{119\text{ mV}}{26\text{ mV}}} = 0.13 \tag{4.163}$$

Thus for this case a 10-percent power-supply voltage change results in only 1.3 percent change in I_{C2}.

This level of power-supply independence is not adequate for many types of analog circuits. Much better independence can be obtained by causing the bias currents in the circuit to depend on a voltage standard other than the supply voltage. Bias reference circuits can be classified by the source of the voltage standard by which the bias currents are established. The most convenient standards are the V_{BE} of a transistor, the thermal voltage, V_T, and the breakdown voltage of a reverse-biased emitter-base junction of a transistor. Each of these can be used to establish supply independence, but the drawback of the first two is that the reference voltage is quite temperature dependent; the V_{BE} has a negative temperature coefficient and the thermal voltage a positive one. The Zener diode has the disadvantage that at least 7 or 8 V of supply voltage are required since standard integrated-circuit technology produces *npn* transistors with an emitter-base breakdown voltage of about 6 V, and, furthermore, that *pn* junctions produce large amounts of voltage noise under the reverse-breakdown conditions encountered in a bias reference circuit. Noise in avalanche breakdown is discussed further in Chapter 11.

We first consider bias reference circuits based on the base-emitter junction voltage. The circuit in simplest form is shown in Fig. 4.24*a*. This circuit can be seen to be similar to the Wilson circuit with the diode-connected transistor replaced with a resistor. The reference current again is forced to flow in Q_1, and, in order for this to occur, the transistor, Q_2, must supply enough current into

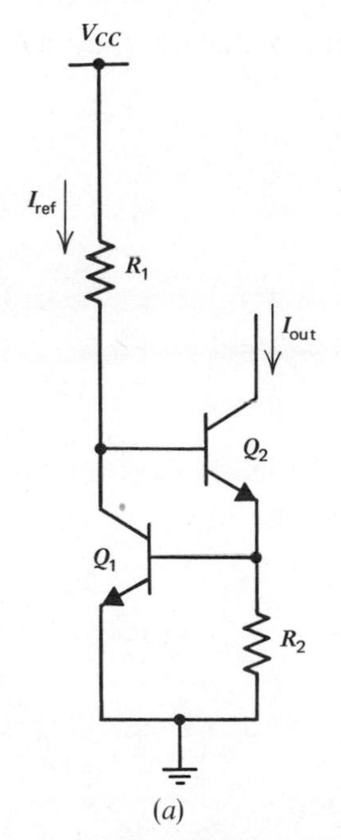

Fig. 4.24*a* Supply-independent bias using V_{BE} as a reference.

R_2 so that the base-emitter voltage of Q_1 is equal to $V_{BE}(I_{ref})$. If we neglect base currents, the current I_{out} is equal to the emitter current of Q_2, which is the current flowing through R_2. Since R_2 has a voltage equal to one base-emitter drop across it, the output current is proportional to this base-emitter voltage. Thus, neglecting base currents,

$$I_{out} = \frac{V_{BE1}}{R_2} = \frac{V_T}{R_2} \ln \frac{I_{ref}}{I_{S1}} \tag{4.164}$$

This circuit is not fully supply independent since the base-emitter voltage of Q_1 will change slightly with power supply voltage. This occurs because the collector current of Q_1 is proportional to V_{CC}. This is often a problem with bias circuits whose output current is derived from a resistor connected to the supply terminal, since the currents in some portion of the circuit will change with supply voltage. Power-supply independence can be greatly improved by the use of the so-called bootstrap bias technique, also referred to as self-biasing. Instead of developing the reference current by connecting a resistor to the supply, the reference current is made to depend directly on the output current of the current source itself. Assuming that the feedback loop formed by this connection has a stable operating point, the currents flowing in the circuit are much more independent of power supply voltage than in the resistively biased case. The application of this technique to the V_{BE}-referenced current source is illustrated in Fig. 4.24*b*. We assume for simplicity that $V_A = \infty$. The circuit composed of Q_1, Q_2, and R dictates that the

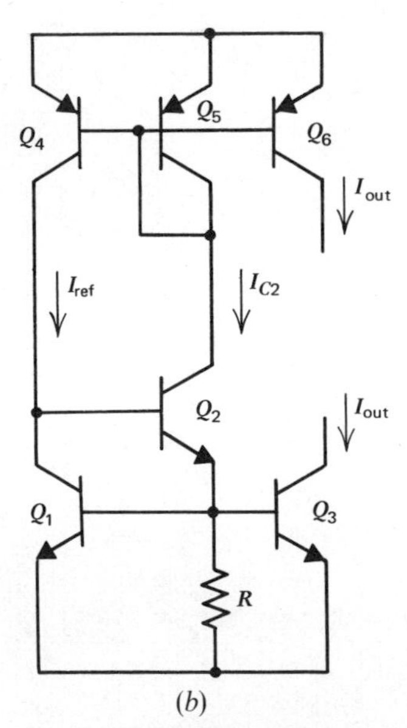

Fig. 4.24*b* Self-biasing V_{BE} reference.

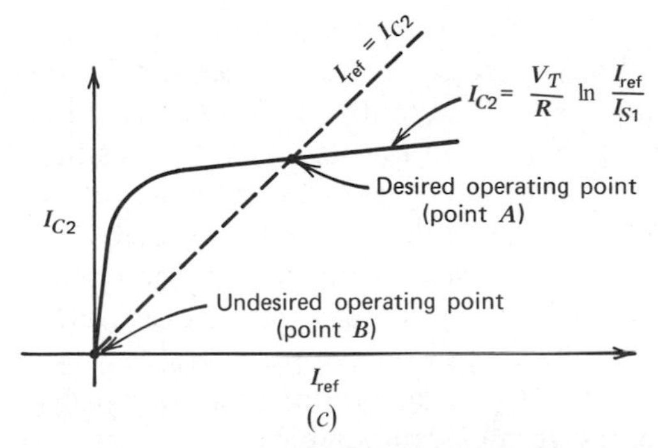

Fig. 4.24c Determination of operating point.

current I_{C2} have a logarithmic dependence on I_{ref} as shown in Fig. 4.24*c* and indicated by (4.164). Second, the current source composed of equal-area transistors Q_4 and Q_5 dictates that I_{ref} and I_{C2} be equal, as indicated by the dotted line in Fig. 4.24*c*. The operating point of the circuit must satisfy both constraints and hence is at the intersection of the two lines. Except for the effects of finite output resistance of the transistors, the bias currents are independent of supply voltage. If required, the output impedance of the current sources could be made larger by the use of cascode or Wilson sources in the circuit. The actual output current is supplied by Q_6 and Q_3, whose emitter areas equal those of Q_5 and Q_1, respectively.

An important aspect of self-biased circuits of this type is that they often have a stable state in which zero current flows in the circuit even when the power supply voltage is nonzero. A separate start-up circuit is usually required to prevent the circuit from remaining in this state. For example, the circuit of Fig. 4.24*b* has two possible operating points at the intersections of the transfer characteristics shown in Fig. 4.24*c*. Point *B* would normally be an unstable operating point; further detailed analysis of the circuit shows that the feedback in the circuit at this point is positive and that the circuit tends to drive itself out of this state. The behavior of circuits containing positive and negative feedback is discussed further in Chapters 8 and 9.

In actual circuits of this type, point *B* is frequently a stable point because the currents in the transistors at this point are very small, often in the picoampere range. As discussed in Chapter 1, the current gain of the transistors at very low current levels falls and is often less than unity at collector currents in the picoamp range. As a result the circuit is usually unable to drive itself out of the zero-current state. Thus, unless precautions are taken, the circuit may operate in the zero-current condition.

The zero-current state can be avoided by insuring that some current always flows in the transistors in the circuit so that their current gain does not fall to a very low value. An additional requirement is that the circuitry added in order to do this must not interfere with the normal operation of the reference once the circuit has reached the desired operating point. The additional devices are called the start-up circuit; a typical example is illustrated in Fig. 4.24*d*. We first assume that the circuit is in the undesired zero-current state. If this were true, the voltage at the base of Q_1 would be at ground. The voltage at the base of Q_2 would be tens of millivolts above ground, determined by the leakage currents in the circuit. However, the voltage on the left-hand end of D_1 is four diode drops above ground, so that a voltage of approximately three diode drops would appear across R_x, and a current would flow through R_x into the Q_1–Q_2 combination. This would cause current to flow in Q_4 and Q_5, avoiding the zero-current state.

The bias reference circuit then drives itself toward the desired stable state, and we require that the start-up circuit not affect the steady-state current values. This can be accomplished by causing R_x to be large enough that when the steady-state current is established in Q_1, the voltage drop across R_x is large enough to reverse bias D_1. In the steady state the voltage at the collector of Q_1 is two diode drops above ground, and the left-hand end of D_1 is four diode drops above ground. Thus if we make $I_{C1}R_x$ equal to two diode drops, D_1 will have zero voltage across it in the steady state. As a result, the start-up circuit composed of R_s, D_2–D_5, and D_1 is,

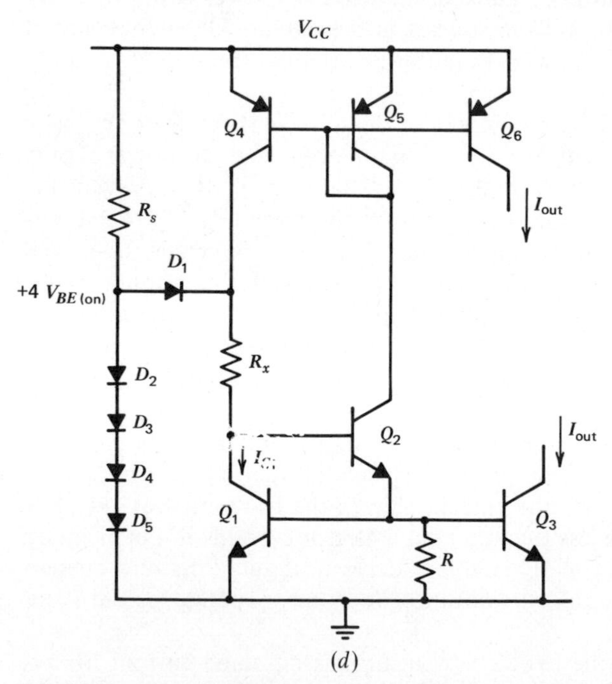

Fig. 4.24*d* Start-up circuitry to avoid zero-current state.

in effect, disconnected from the circuit for steady-state operation. Since the value of R_s is not critical, an epitaxial resistor, or FET of the type described in Chapter 2, is often used to achieve a large value of resistance in a small area.

A second important aspect of the performance of such biasing circuits is their dependence on temperature. This variation is most conveniently expressed in terms of the fractional change in output current per degree centigrade of temperature variation, which we term the fractional temperature coefficient, TC_F.

$$TC_F = \frac{1}{I_{\text{out}}} \frac{\partial I_{\text{out}}}{\partial T} \tag{4.165}$$

For the V_{BE}-referenced circuit of Fig. 4.24*b*,

$$I_{\text{out}} = \frac{V_{BE1}}{R} \tag{4.166}$$

$$\frac{\partial I_{\text{out}}}{\partial T} = \frac{1}{R} \frac{\partial V_{BE1}}{\partial T} - \frac{V_{BE1}}{R^2} \frac{\partial R}{\partial T} \tag{4.167}$$

$$= I_{\text{out}} \left(\frac{1}{V_{BE1}} \frac{\partial V_{BE1}}{\partial T} - \frac{1}{R} \frac{\partial R}{\partial T} \right) \tag{4.168}$$

Therefore,

$$TC_F = \frac{1}{V_{BE1}} \frac{\partial V_{BE1}}{\partial T} - \frac{1}{R} \frac{\partial R}{\partial T} \tag{4.169}$$

Thus the temperature dependence of I_{out} is related to the difference between the resistor temperature coefficient and that of the base-emitter junction. Since the former has a positive and the latter a negative coefficient, the net TC_F is quite large.

EXAMPLE

Design a bias reference as shown in Fig. 4.24*b* to produce 100 μA output current. Find the TC_F. Assume that, for Q_1, $I_S = 10^{-14}$ A. Assume that $\partial V_{BE}/\partial T = -2$ mV/°C and that $(1/R)(\partial R/\partial T) = +1500$ ppm/°C.

The current in Q_1 will be equal to I_{out}, so that

$$V_{BE1} = V_T \ln \frac{100\ \mu\text{A}}{10^{-14}\ \text{A}} = 598\ \text{mV}$$

Thus, from (4.166),

$$R = \frac{598\ \text{mV}}{0.1\ \text{mA}} = 5.98\ \text{k}\Omega$$

From (4.169),

$$TC_F = \frac{-2 \text{ mV/}^\circ\text{C}}{598 \text{ mV}} - 1.5 \times 10^{-3} = -3.3 \times 10^{-3} - 1.5 \times 10^{-3}$$

and thus

$$TC_F = -4.8 \times 10^{-3}/^\circ\text{C} = -4800 \text{ ppm/}^\circ\text{C}$$

The term ppm is an abbreviation for parts per million, and implies a multiplier of 10^{-6}.

An alternate source for the voltage reference is the thermal voltage, V_T. The difference in junction potential between two junctions operated at different current densities can be shown to be proportional to V_T. This voltage difference must be converted to a current to provide the bias current. For the Widlar source shown in Fig. 4.5, it was shown in (4.14) that the voltage across the resistor R_2 was:

$$V_X = I_{C2}R_2 = V_T \ln \frac{I_{C1}I_{S2}}{I_{C2}I_{S1}} \tag{4.170}$$

Thus if the ratio of the two collector currents is held constant, the voltage across R_2 is indeed proportional to V_T. This fact is utilized in the self-biased circuit of Fig. 4.25*a*. Here the collector currents, I_{C1} and I_{C2}, are constrained to be equal by the current source formed by Q_3 and Q_4, which have equal areas. Let us assume, for example, that transistor Q_2 has twice the area and thus twice the I_S of Q_1, so that the voltage across R_2 is:

$$V_X = V_T \ln \frac{I_{C1}}{I_{C2}} \frac{I_{S2}}{I_{S1}} = V_T \ln 2 \tag{4.171}$$

since $I_{C1} = I_{C2}$. This voltage appears across R_2 and produces a current of value:

$$I_{C2} = \frac{V_T}{R_2} \ln 2 \tag{4.172}$$

The self-biased feedback circuit used in this case to constrain the currents in Q_1 and Q_2 to be equal consists of a *pnp* current source formed by Q_3 and Q_4. As illustrated in Fig. 4.25*b*, only two possible operating points exist for the circuit-one at zero and one at the desired operating current. This circuit also requires start-up circuitry to avoid occurrence of the zero-current state.

The temperature variation of the output current can be calculated as follows. From (4.172)

$$\frac{1}{I_{C2}} \frac{\partial I_{C2}}{\partial T} = \frac{1}{I_{C2}} \frac{\partial}{\partial T} \left(\frac{V_T}{R_2} \ln 2 \right) \tag{4.173}$$

$$= \frac{1}{I_{C2}} \left[\frac{V_T}{R_2} (\ln 2) \left(\frac{1}{V_T} \frac{\partial V_T}{\partial T} - \frac{1}{R_2} \frac{\partial R_2}{\partial T} \right) \right] \tag{4.174}$$

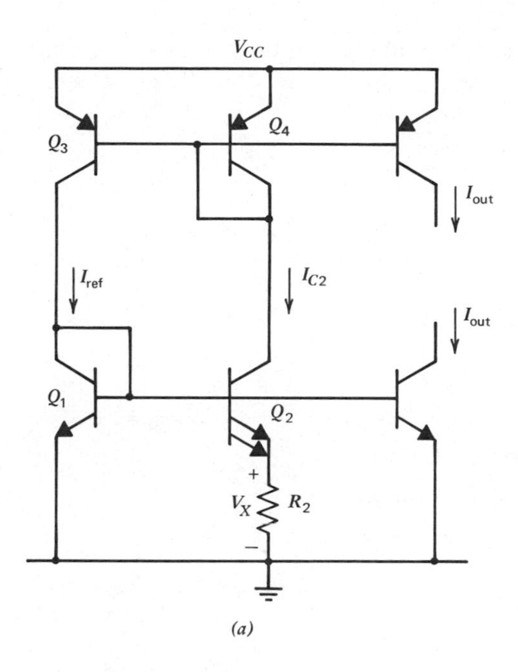

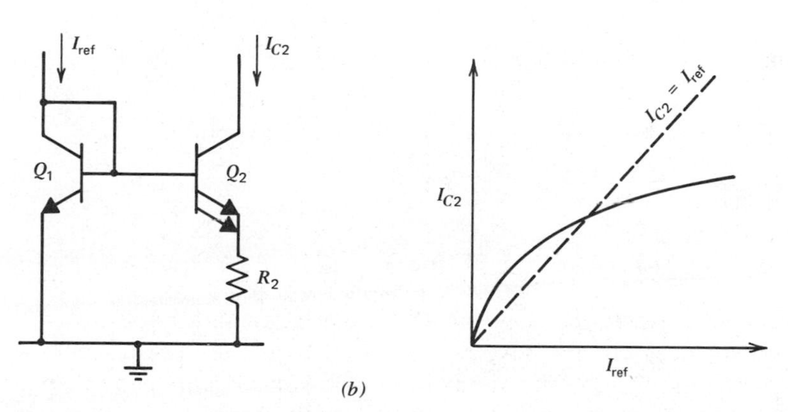

Fig. 4.25 (*a*) Bias source utilizing the thermal voltage. (*b*) Determination of operating point.

and using (4.172) again we find

$$\frac{1}{I_{C2}}\frac{\partial I_{C2}}{\partial T} = \left(\frac{1}{V_T}\frac{\partial V_T}{\partial T} - \frac{1}{R_2}\frac{\partial R_2}{\partial T}\right) \tag{4.175}$$

We have chosen a transistor area ratio of two to one as an example. Actually, this ratio is chosen to minimize the total area required for the transistors and for resistor R_2. Compared with the case of the V_{BE} reference, this circuit produces a much smaller temperature coefficient of the output current because the fractional sensitivities of both V_T and that of a diffused resistor R_2 are positive and tend to cancel.

EXAMPLE

Design a bias reference of the type shown in Fig. 4.25 to produce an output current of 100 μA. Find the TC_F of I_{out}. Assume base-diffused resistors, $(1/R)(\partial R/\partial T) = +1500$ ppm/°C.

From (4.172)

$$I_{out} = \frac{V_T}{R_2} \ln 2$$

which gives

$$R_2 = \frac{(26 \text{ mV})(\ln 2)}{100 \ \mu\text{A}} = 180 \ \Omega$$

From (4.175)

$$TC_F = \frac{1}{I_{out}} \frac{\partial I_{out}}{\partial T} = \frac{1}{V_T} \frac{\partial}{\partial T}(V_T) - 1500 \times 10^{-6} = \frac{1}{V_T}\left(\frac{V_T}{T}\right) - 1500 \times 10^{-6}$$

$$= \frac{1}{T} - 1500 \times 10^{-6}$$

Assuming room temperature, $T = 300$°K,

$$\frac{1}{I_{out}} \frac{\partial I_{out}}{\partial T} = 3300 \times 10^{-6} - 1500 \times 10^{-6} = 1800 \text{ ppm/°C}.$$

A4.3 TEMPERATURE-INDEPENDENT BIASING

As illustrated by the examples in Appendix A4.2, the base-emitter voltage and thermal-voltage-referenced circuits have rather high temperature coefficients of output current, although the thermal voltage circuit is considerably better. This difficulty can be overcome by two different approaches: the Zener diode reference and the band-gap reference. We consider the Zener-diode reference first.

A4.3.1 Zener-Referenced Bias Circuits

A typical bias circuit using a Zener reference is shown in Fig. 4.26. When conducting in the reverse direction, the emitter-base junction of an *npn* transistor displays a Zener or avalanche breakdown of between 6 and 8 V, depending on processing details. In this region, the incremental resistance is small and the

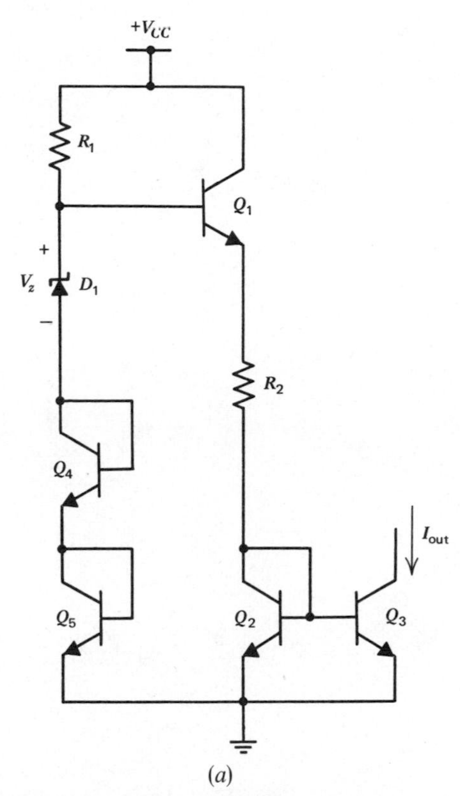

Fig. 4.26*a* Zener diode bias reference.

device resembles a voltage source with a temperature coefficient that is near zero for a breakdown voltage of about 5 V and increases to about 4 mV/°C for Zener voltages of approximately 8 V. The temperature coefficient of this voltage is shown as a function of the breakdown voltage itself in Fig. 2.35. In the circuit shown in Fig. 4.26*a*, the resistor R_1 provides the dc current necessary to bias D_1, Q_4, and Q_5. The voltage at the base of Q_1, V_{B1}, is equal to V_Z plus two diode drops. The voltage across R_2 is equal to V_{B1} minus the diode drops across Q_2 and the base-emitter junction of Q_1. Thus the voltage across R_2 is nearly equal to V_Z, and

$$I_{out} \approx \frac{V_Z}{R_2} \tag{4.176}$$

This circuit displays a slight supply dependence because the current in R_1 varies when the power supply changes. Thus the current in Q_3, Q_4, and D_1 varies. These devices have finite incremental resistance, so that the voltage at the base of Q_1

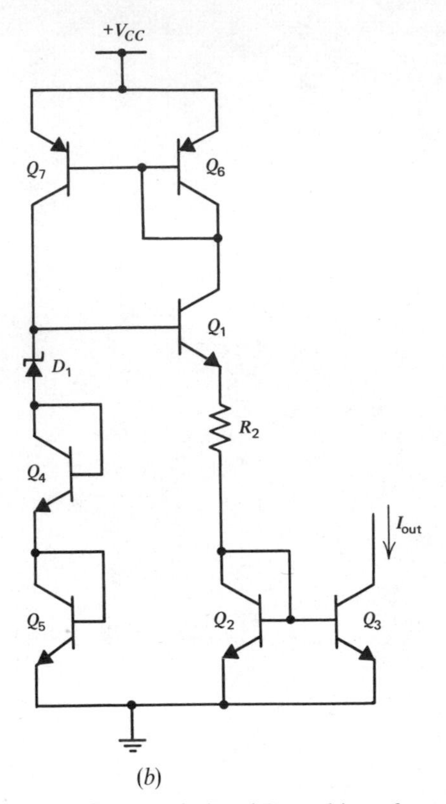

Fig. 4.26*b* Self-biased Zener bias reference.

varies. The circuit can be self-biased as in the previous cases to avoid this problem, as illustrated in Fig. 4.26*b*.

As they stand, the circuits of Fig. 4.26*a* and *b* produce an output *current*, which has a low TC_F if the TC_F of R_2 is low, but which has significant temperature variation when R_2 is a diffused resistor with high TC_F. If zero TC_F of I_{out} is the objective, diodes can be added in series with R_2 so as to compensate for the temperature variations of R_2 and V_Z as shown in Fig. 4.26*c*. Here,

$$V_Z = I_{out} R_2 + (n + 2) V_{BE(on)} \tag{4.177}$$

where n is the number of temperature compensating diodes. Differentiating with respect to temperature,

$$\frac{\partial}{\partial T} V_Z = R_2 \frac{\partial I_{out}}{\partial T} + I_{out} \frac{\partial R_2}{\partial T} + (n + 2) \frac{\partial V_{BE(on)}}{\partial T} \tag{4.178}$$

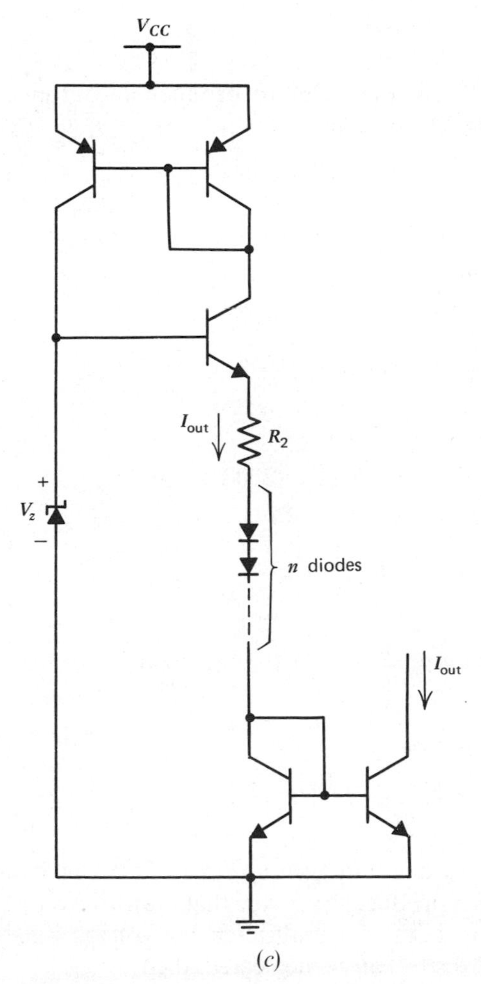

Fig. 4.26*c* Temperature-compensated Zener reference source.

Dividing by $R_2 I_{out}$ we obtain

$$\frac{1}{R_2 I_{out}} \frac{\partial V_Z}{\partial T} = \frac{1}{I_{out}} \frac{\partial I_{out}}{\partial T} + \frac{1}{R_2} \frac{\partial R_2}{\partial T} + \frac{(n+2)}{I_{out} R_2} \frac{\partial V_{BE(on)}}{\partial T} \tag{4.179}$$

Thus, in order to obtain zero TC_F for I_{out}, we require

$$\frac{1}{I_{out}} \frac{\partial I_{out}}{\partial T} = 0 = \frac{1}{R_2 I_{out}} \left[\frac{\partial V_Z}{\partial T} - (n+2) \frac{\partial V_{BE(on)}}{\partial T} \right] - \frac{1}{R_2} \frac{\partial R_2}{\partial T} \tag{4.180}$$

EXAMPLE

Determine the required values of n and R_2 in Fig. 4.26c to produce an output current of 100 μA with the lowest possible TC. Assume

$$\frac{\partial V_Z}{\partial T} \approx +2.5 \text{ mV/°C}; \qquad V_Z = 6.2 \text{ V}$$

$$\frac{1}{R}\frac{\partial R}{\partial T} = +2000 \text{ ppm/°C}$$

$$\frac{\partial V_{BE(\text{on})}}{\partial T} = -2 \text{ mV/°C}, \qquad V_{BE(\text{on})} = 0.6 \text{ V}$$

From (4.180) we have

$$0 = 2.5 \text{ mV/°C} - (n + 2)(-2 \text{ mV/°C}) - R_2(I_{\text{out}})(2000 \times 10^{-6})$$

We also have the auxiliary condition from (4.177)

$$V_Z = I_{\text{out}} R_2 + (n + 2)V_{BE(\text{on})}$$

Combining these relations,

$$0 = 2.5 \text{ mV/°C} - (n + 2)(-2 \text{ mV/°C}) - [V_Z - (\text{n} + 2)V_{BE(\text{on})}](2000 \times 10^{-6})$$

Solving this equation,

$$(n + 2) = 3.09$$
$$n = 1.09$$

Utilizing (4.177) $R_2 = 43.4 \text{ k}\Omega$

Thus the optimum number of diodes is a noninteger number. This can be realized using the circuit of Fig. 4.27*a*, which produces a voltage that is an arbitrary fraction of $V_{BE(\text{on})}$. This circuit is referred to as the V_{BE} multiplier. If a voltage V is applied, and if the base current of the transistor can be neglected, then:

$$V_{BE} = V\left(\frac{R_2}{R_1 + R_2}\right)$$

The collector current is given by:

$$I_C = I_S \exp\left(\frac{V_{BE}}{V_T}\right) = I_S \exp\left(\frac{VR_2}{R_1 + R_2}\right)\left(\frac{1}{V_T}\right)$$

and the total current, I, is equal to $(I_1 + I_C)$ such that

$$I = I_1 + I_C = \frac{V}{R_1 + R_2} + I_S \exp\left(\frac{VR_2}{R_1 + R_2}\right)\left(\frac{1}{V_T}\right)$$

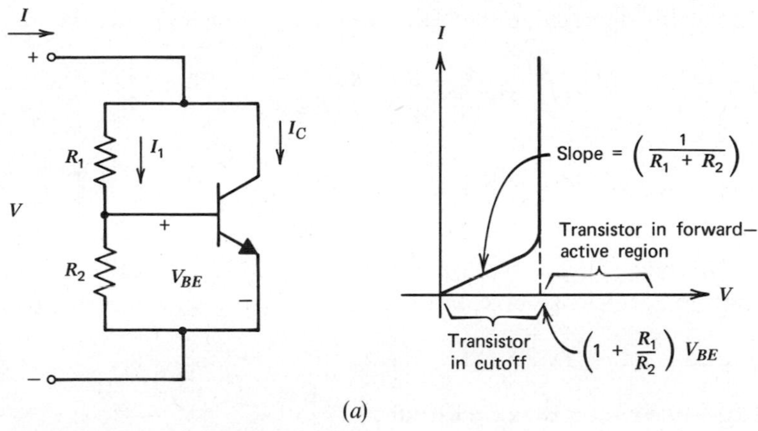

Fig. 4.27*a* The V_{BE} multiplier circuit.

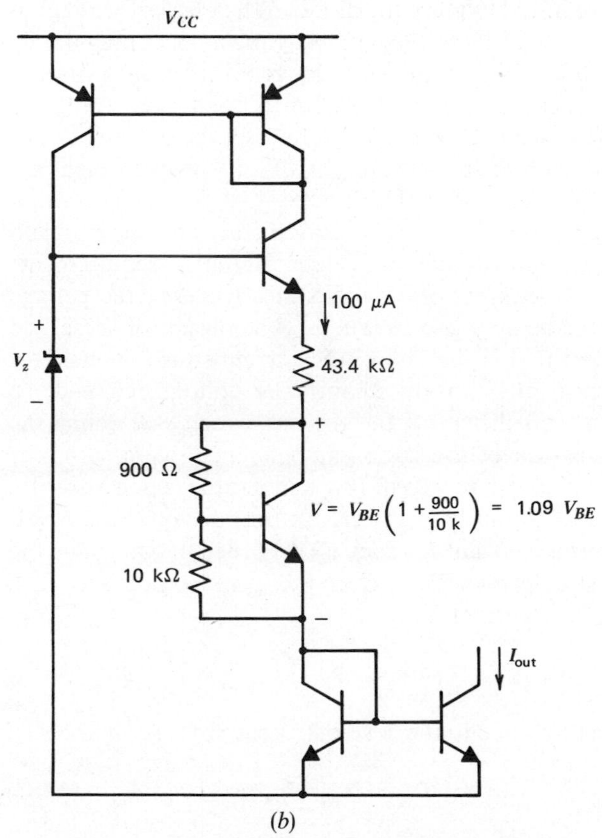

Fig. 4.27*b* Low TC_F, 100 μA, Zener-referenced bias current source.

The circuit is normally operated in the region where I_C is much greater than I_1. In this region,

$$I \approx I_S \exp\left(\frac{VR_2}{R_1 + R_2}\right)\frac{1}{V_T}$$

and, inverting,

$$V = \left(1 + \frac{R_1}{R_2}\right) V_T \ln \frac{I}{I_S} = \left(1 + \frac{R_1}{R_2}\right) V_{BE}(I)$$

Thus the terminal voltage is a constant times the transistor V_{BE}. A realization of the zero temperature-coefficient current source using this circuit is shown in Fig. 4.27*b*.

A4.3.2 Band-Gap-Referenced Biasing Circuits

The drawbacks of the Zener-referenced bias circuit are that a power supply voltage of at least 7 to 10 V is required to place the diode in the breakdown region and that substantial noise is introduced into the circuit by the avalanching diode. Thus we are led to examine other possibilities for the realization of a biasing circuit with low temperature coefficient. Since the sources referenced to $V_{BE(\text{on})}$ and V_T have opposite TC_F, the possibility exists for referencing the output current to a composite voltage that is a weighted sum of $V_{BE(\text{on})}$ and V_T. By proper weighting, zero temperature coefficient should be attainable.

In the biasing sources discussed thus far, we have concentrated on the problem of obtaining a *current* with low temperature coefficient. Actually, requirements often arise for low-temperature-coefficient bias or reference *voltages*, the voltage reference for a voltage regulator being a good example. The design of these two types of circuits is similar, except that in the case of the current source a temperature coefficient must be intentionally introduced into the voltage reference to compensate for the temperature coefficient of the resistor which will define the current. In the following discussion of the band-gap reference, we assume for simplicity that the objective is a *voltage* source of low temperature coefficient.

First consider the hypothetical circuit of Fig. 4.28. An output voltage is developed that is equal to $V_{BE(\text{on})}$ plus a constant K times V_T. In order to determine the required value for K, we must determine the TC of $V_{BE(\text{on})}$ more precisely. The V_{BE} can be written, neglecting base current,

$$V_{BE(\text{on})} = V_T \ln \frac{I_1}{I_S} \tag{4.181}$$

The saturation current, I_S, can be related to the device structure by (see Chapter 1):

$$I_S = \frac{qA{n_i}^2\bar{D}_n}{Q_B} = B{n_i}^2\bar{D}_n = B'{n_i}^2T\bar{\mu}_n \tag{4.182}$$

where n_i is the intrinsic minority-carrier concentration, Q_B is the total base doping per unit area, $\bar{\mu}_n$ is the average electron mobility in the base, A is the emitter-base

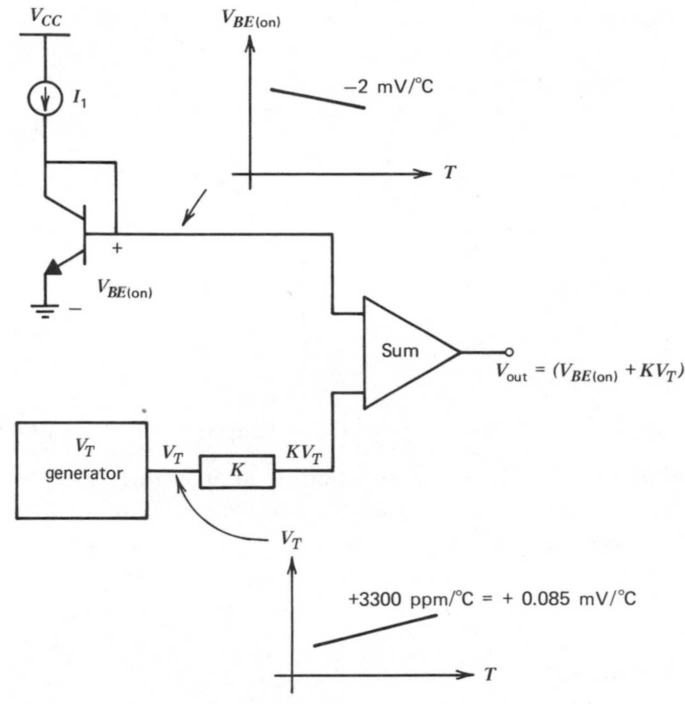

Fig. 4.28 Hypothetical band-gap reference circuit.

junction area, and T is the temperature. Here, the constants B and B' involve only temperature-independent quantities. The Einstein relation $\mu_n = (q/kT)D_n$ was used to write I_S in terms of $\bar{\mu}_n$ and n_i^2. The quantities in (4.182), which are temperature dependent, are given by:[7]

$$\bar{\mu}_n = CT^{-n} \tag{4.183}$$

$$n_i^2 = DT^3 \exp\left(-\frac{V_{GO}}{V_T}\right) \tag{4.184}$$

where V_{GO} is the band-gap voltage of silicon extrapolated to zero degrees Kelvin. Here again, D and C are temperature-independent quantities whose exact values are unimportant in the analysis. The exponent, n, in the expression for base region electron mobility $\bar{\mu}_n$ is dependent on doping level in the base. Combining (4.181), (4.182), (4.183), and (4.184),

$$V_{BE(on)} = V_T \ln\left(I_1 T^{-\gamma} E \exp\frac{V_{GO}}{V_T}\right) \tag{4.185}$$

where E is another constant and

$$\gamma = 4 - n \tag{4.186}$$

In actual band-gap circuits, the current, I_1, is not constant but varies with temperature. We assume for the time being that this temperature variation is known, and that it can be written in the form:

$$I_1 = GT^{\alpha} \tag{4.187}$$

where G is another temperature-independent constant. Combining (4.185) and (4.187)

$$V_{BE(\text{on})} = V_{GO} - V_T[(\gamma - \alpha) \ln T - \ln EG] \tag{4.188}$$

From Fig. 4.28, the output voltage is:

$$V_{\text{out}} = V_{BE(\text{on})} + KV_T \tag{4.189}$$

Substitution of (4.188) in (4.189) gives

$$V_{\text{out}} = V_{GO} - V_T(\gamma - \alpha) \ln T + V_T(K + \ln EG) \tag{4.190}$$

This expression gives the output voltage as a function of temperature in terms of the circuit parameters, G, α, and K, and the device parameters, E and γ. Our objective is to make V_{out} independent of temperature, and to this end we take the derivative of V_{out} with respect to temperature to find the required values of G, γ, and K to give zero TC_F. Differentiating (4.190),

$$0 = \left.\frac{dV_{\text{out}}}{dT}\right|_{T=T_0} = \frac{V_{TO}}{T_0}(K + \ln EG) - \frac{V_{TO}}{T_0}(\gamma - \alpha) \ln T_0 - \frac{V_{TO}}{T_0}(\gamma - \alpha) \tag{4.191}$$

where V_{TO} is the thermal voltage, V_T, evaluated at T_0. Equation 4.191 can be rearranged to give:

$$(K + \ln EG) = (\gamma - \alpha) \ln T_0 + (\gamma - \alpha) \tag{4.192}$$

This equation gives the required values of circuit parameters, K, α, and G, in terms of the device parameters, E and γ. In principle, these values could be calculated directly from (4.192). However, further insight is gained by back-substituting (4.192) into (4.190). The result is:

$$V_{\text{out}}(T) = V_{GO} + V_T(\gamma - \alpha)\left(1 + \ln \frac{T_0}{T}\right) \tag{4.193}$$

Thus the temperature dependence of the output voltage is entirely described by the single parameter, T_0, which in turn is determined by the constants K, E, and G. A typical family of output voltage variation characteristics is shown in Fig. 4.29 for different values of T_0, for the special case in which $\alpha = 0$ and I_1 is temperature independent.

Using (4.193), the output voltage at the zero TC_F temperature ($T = T_0$) is given by:

$$V_{\text{out}}(T)|_{T=T_0} = V_{GO} + V_{TO}(\gamma - \alpha) \tag{4.194}$$

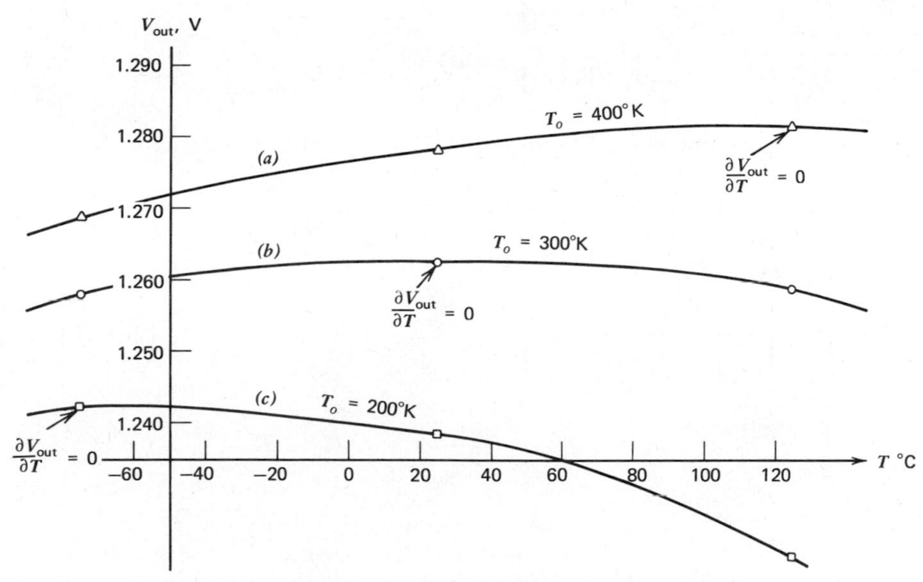

Fig. 4.29 Variation of band-gap reference output voltage with temperature.

For example, in order to achieve zero TC_F at 25°C, assuming that $\gamma = 3.2$ and $\alpha = 1$,

$$V_{out}(T)|_{T_0=25^\circ C} = V_{GO} + 2.2V_{T0} \tag{4.195}$$

The band-gap voltage of silicon is $V_{GO} = 1.205$ V, so that

$$V_{out}(T_0)|_{T_0=25^\circ C} = 1.205 \text{ V} + (2.2)(26 \text{ mV}) \tag{4.196}$$

$$= 1.262 \text{ V}$$

EXAMPLE

A band-gap reference is designed to give a norminal output voltage of 1.262 V, which gives zero TC_F at 25°C. Because of component variations, the actual room temperature output voltage is 1.280 V. Find the temperature of actual zero TC_F of V_{out}, find $V_{out}(T)$, and calculate the TC_F at room temperature. Assume that $\gamma = 3.2$ and $\alpha = 1$. Since the behavior of the output voltage is characterized by T_0, we first calculate the value of T_0. From (4.193),

$$V_{out}(T) = V_{GO} + V_T(\gamma - \alpha)\left(1 + \ln \frac{T_0}{T}\right)$$

At 25°C,

$$1.280 \text{ V} = 1.205 \text{ V} + (26 \text{ mV})(2.2)\left(1 + \ln \frac{T_0}{300^\circ\text{K}}\right)$$

and thus

$$T_0 = 300^\circ\text{K}\left(\exp\frac{18 \text{ mV}}{57 \text{ mV}}\right) = 411^\circ\text{K}.$$

This is the temperature where TC_F of V_{out} will be zero. Thus we can express V_{out} as

$$V_{out}(T) = 1.205 + 57 \text{ mV}\left(1 + \ln\frac{411^\circ\text{K}}{T}\right)$$

Differentiating (4.193),

$$\frac{dV_{out}}{dT} = \frac{1}{T}\left[V_T(\gamma - \alpha)\left(1 + \ln\frac{T_0}{T}\right)\right] - \frac{V_T}{T}(\gamma - \alpha) = (\gamma - \alpha)\frac{V_T}{T}\left(\ln\frac{T_0}{T}\right)$$

If T is near T_0, then

$$\ln\frac{T_0}{T} = \ln\left(1 + \frac{T_0 - T}{T}\right) \simeq \frac{T_0 - T}{T}$$

and we have

$$\frac{dV_{out}}{dT} = \frac{V_T}{T}\left(\frac{T_0 - T}{T}\right)(\gamma - \alpha)$$

For $T_0 = 411^\circ\text{K}$, $T = 300^\circ\text{K}$,

$$\frac{dV_{out}}{dT} = \frac{26 \text{ mV}}{300^\circ\text{K}}\left(\frac{411 - 300}{300}\right)(2.2) = 70\ \mu\text{V}/^\circ\text{C}$$

This is the TC_F of V_{out} at room temperature.

The parameter of interest in reference voltage sources is the variation in the output voltage that will be encountered over the entire temperature range. Since the TC_F expresses the temperature sensitivity only at one temperature, a different parameter must be used to characterize the behavior of the circuit over a broad temperature range. An effective TC_F can be defined for a voltage reference as:

$$TC_F = \frac{1}{V_{out}}\left(\frac{V_{max} - V_{min}}{T_{max} - T_{min}}\right) \tag{4.197}$$

where V_{max} and V_{min} are the largest and smallest output voltages observed over the temperature range, and $T_{max} - T_{min}$ is the temperature excursion. V_{out} is the nominal output voltage. By this standard, the effective TC_F over the -55 to $+125^\circ$C range, for example (*b*) of Fig. 4.29, is 44 ppm/°C. If the temperature range is restricted to 0 to 70°C, the TC_F improves to 17 ppm/°C. Thus over a restricted temperature range this reference is comparable with the standard cell in temperature stability once the zero TC_F temperature has been set at room temperature. Saturated standard cells have a TC_F of about ± 30 ppm/°C.

Practical realizations of band-gap references take on several forms.[6,7,8] One such circuit is illustrated in Fig. 4.30*a*. The circuit of Fig. 4.30*a* utilizes a feedback loop to establish an operating point in the circuit such that the output voltage is equal to a $V_{BE(on)}$ plus a voltage proportional to the difference between two base-emitter voltages. The operation of the feedback loop is best understood by reference to Fig. 4.30*b*, in which a portion of the circuit is shown. We first consider the variation of the output voltage, V_2, as the input voltage, V_1, is varied from zero in the positive direction. Initially, with V_1 set at zero, devices Q_1 and Q_2 are not conducting and $V_2 = 0$. As V_1 is increased, Q_1 and Q_2 do not conduct significant current until the input voltage reaches about 0.6 V. During this time, output voltage V_2 is equal to V_1 since there is no voltage drop in R_2. When V_1 exceeds 0.6 V, however, Q_1 begins to conduct current. This corresponds to point 1 on Fig. 4.30*b*. The magnitude of the current in Q_1 is roughly equal to $(V_1 - 0.6\text{ V})/R_1$. For small values of this current, Q_1 and Q_2 carry the same current since the drop across R_3 will be negligible at low currents. Since the resistor, R_2, is much larger than R_1, the voltage drop across it is much larger than $(V_1 - 0.6\text{ V})$, and transistor Q_2 saturates. This corresponds to point ② on Fig. 4.30*b*. Because of the presence of R_3, the collector current that *would* flow in Q_2 if it were in the forward-active region has an approximately logarithmic dependence on V_1, exactly as in the Widlar source. Thus as V_1 is further increased, a point is reached at which Q_2 comes out of saturation. This occurs because V_1 increases faster than the voltage drop across R_2. This is labeled point ③ on Fig. 4.30*b*.

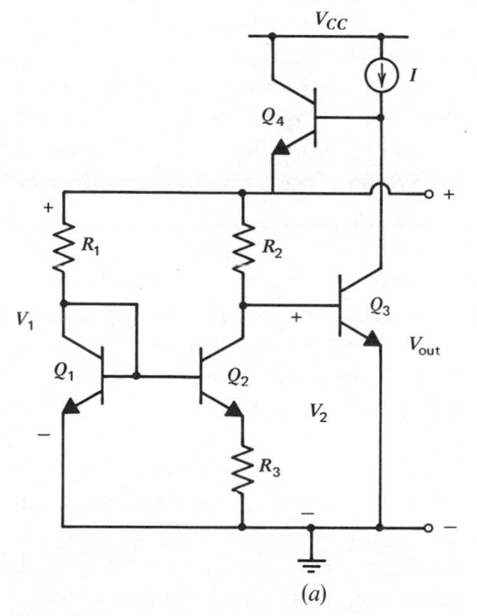

Fig. 4.30*a* Widlar band-gap reference.

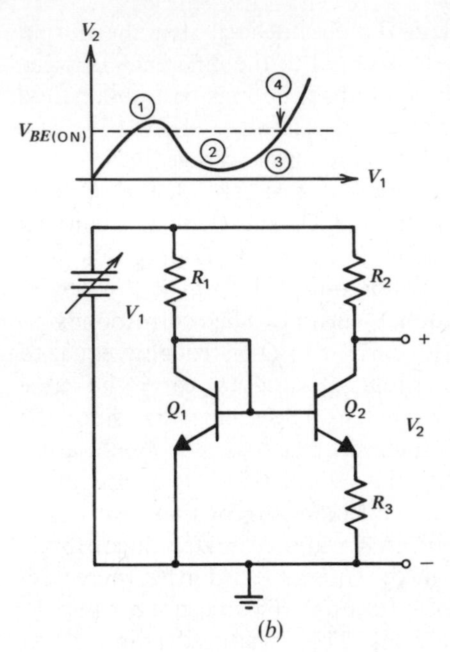

Fig. 4.30*b* Band-gap subcircuit.

Now consider the complete circuit of Fig. 4.30*a*. If transistor Q_3 is initially turned off, transistor Q_4 will drive V_1 in the positive direction. This will continue until enough voltage is developed at the base of Q_3 to produce a collector current in Q_3 approximately equal to I. Thus the circuit stabilizes with voltage V_2 equal to one diode drop, the base-emitter voltage of Q_3. Note that this can occur at point ① and at point 4 in Fig. 4.30*b*. Appropriate start-up circuitry must be included to insure operation at point ④.

Assuming that the circuit has reached a stable operating point at point 4, it can be seen that output voltage V_{out} is the sum of the base-emitter voltage of Q_3 and the voltage drop across R_2. The drop across R_2 is equal to the voltage drop across R_3 multiplied by (R_2/R_3) since the collector current of Q_2 is approximately equal to the emitter current. The voltage drop across R_3 is equal to the difference in base-emitter voltages of Q_1 and Q_2. The ratio of currents in Q_1 and Q_2 is set by the ratio of R_2 to R_1.

A drawback of this reference is that the current, I_1, is derived from the power supply and may vary with power-supply variations. An improved reference circuit is shown in Fig. 4.30*c*. If we assume a priori that a stable operating point exists for this circuit, then the differential input voltage of the op amp must be zero and resistors R_1 and R_2 have equal voltages across them. Thus the two currents I_1 and I_2 must have a ratio determined by the ratio of R_1 to R_2. Note that these two currents are the collector currents of the two diode-connected transistors Q_2

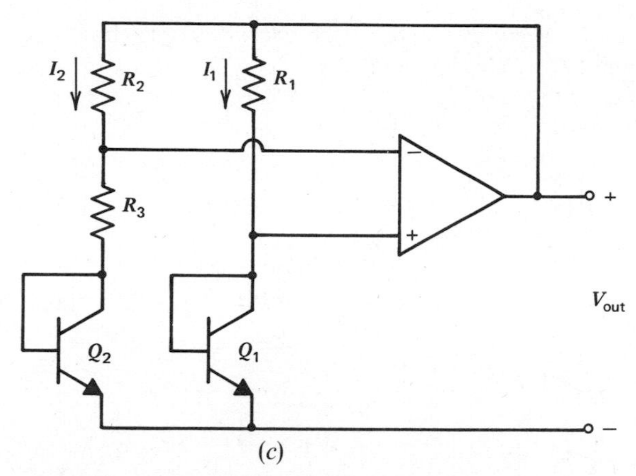

Fig. 4.30c Improved band-gap reference.

and Q_1, assuming base currents are negligible. Thus the difference between their base-emitter voltages, again assuming that base currents are negligible, is:

$$\Delta V_{BE} = V_T \ln \frac{I_1}{I_2}\frac{I_{S2}}{I_{S1}} = V_T \ln \frac{R_2 I_{S2}}{R_1 I_{S1}} \tag{4.198}$$

This voltage appears across resistor R_3. The same current that flows in R_3 also flows in R_2, so that the voltage across R_2 must be:

$$V_{R2} = \frac{R_2}{R_3}\Delta V_{BE} = \frac{R_2}{R_3} V_T \ln \frac{R_2 I_{S2}}{R_1 I_{S1}} \tag{4.199}$$

Note that this implies that the currents I_1 and I_2 are both proportional to temperature if the resistors have zero temperature coefficient. Thus for this reference, $\alpha = 1$.

The output voltage is the sum of the voltage across R_1 and the voltage across Q_1. The voltage across R_1 is equal to that across R_2 indicated above. The output voltage is thus:

$$V_{\text{out}} = V_{BE1} + \frac{R_2}{R_3} V_T \ln \frac{R_2 I_{S2}}{R_1 I_{S1}} = V_{BE1} + K V_T \tag{4.200}$$

The circuit thus behaves as a band-gap reference, with the value of K set by the ratios of (R_2/R_1), (R_2/R_3), and I_{S2}/I_{S1}.

PROBLEMS

NOTE: For these problems use the high-voltage device parameters given in Fig. 2.22 and Fig. 2.25. Assume that $r_b = 0$ and $r_\mu = \infty$ in all problems. Assume all transistors are in the forward-active region and nelgect base currents in bias calculations.

4.1 Determine the output current and output resistance of the current source shown in Fig. 4.31. Find the output current if $V_o = 1$ V, 5 V, and 30 V.

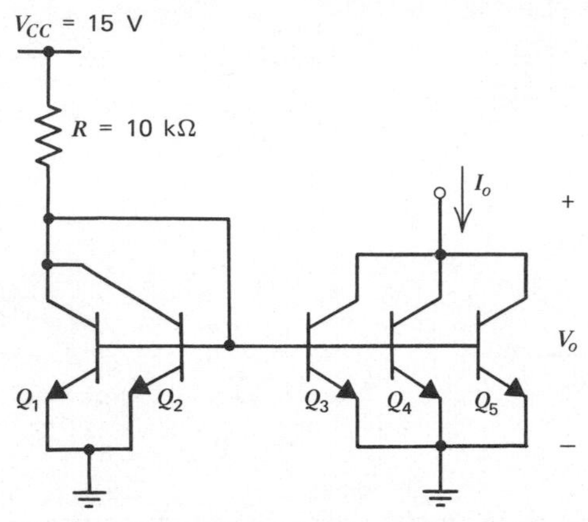

Fig. 4.31 Circuit for Problem 4.1.

4.2 Design a Widlar current source using *npn* transistors that produces a 5-μA output current. Assume a V_{CC} of 30 V and a reference current resistor of 30 kΩ. Find the output resistance.

4.3 In the design of a Widlar source to produce a specified output current, two resistors must be selected. One resistor, R_1, sets I_{ref} and then the emitter resistor, R_2, sets I_o. Assuming a supply voltage of V_{CC} and a desired output current I_o, determine the values of the two resistors R_1 and R_2 so that the total resistance in the circuit is minimized. Your answer should be expressions for R_2 and R_1 in terms of V_{CC} and I_o. What values would this give for Problem 4.2? Are these practical values?

4.4 Determine the output current in the circuit of Fig. 4.32.

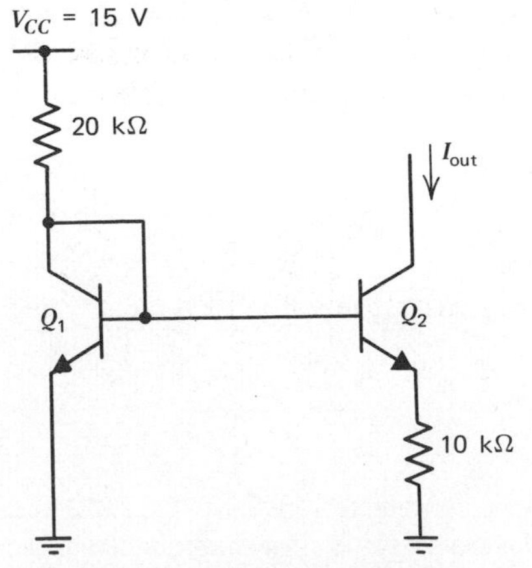

Fig. 4.32 Circuit for Problem 4.4.

4.5 Determine the output current and output resistance of the circuit shown in Fig. 4.33.

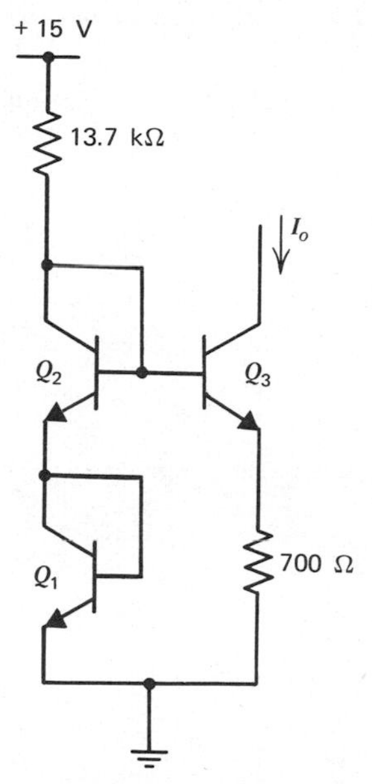

Fig. 4.33 Circuit for Problem 4.5.

4.6 For the Wilson current source shown in Fig. 4.34, calculate the output resistance using the small-signal equivalent circuit. What is the percentage change in I_o for a 5 V change in V_o?

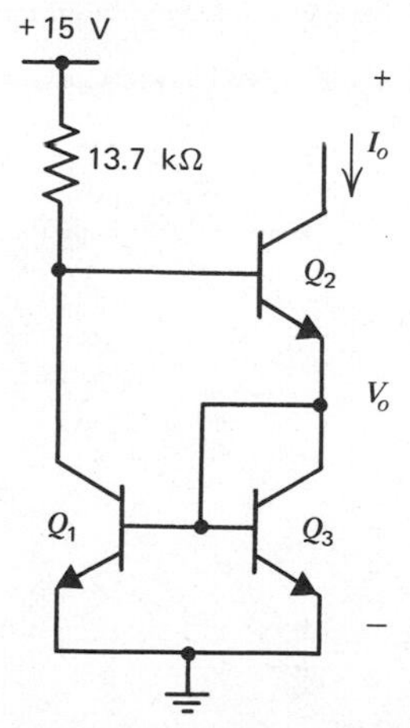

Fig. 4.34 Circuit for Problem 4.6.

4.7 A pair of current sources is to be designed to produce output currents that match within ± 1 percent. If resistors display a worst-case mismatch of ± 0.5 percent, and transistors a worst-case V_{BE} mismatch of 2 mV, how much voltage must be dropped across the emitter resistors?

4.8 Determine the value of sensitivity S of output current to supply voltage for the circuit of Fig. 4.35, where $S = (V_{CC}/I_o)(\partial I_o/\partial V_{CC})$.

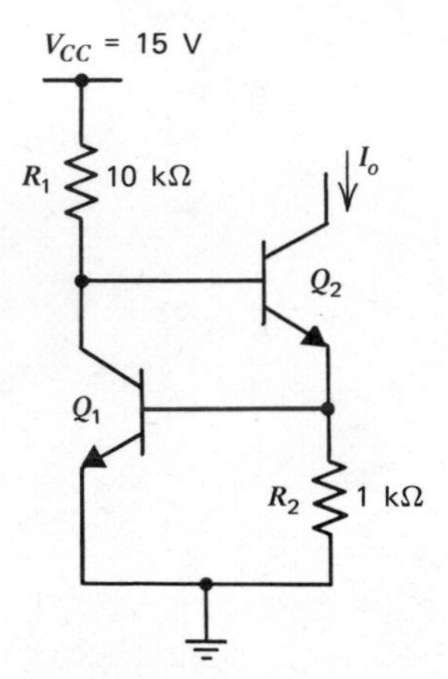

Fig. 4.35 Circuit for Problem 4.8.

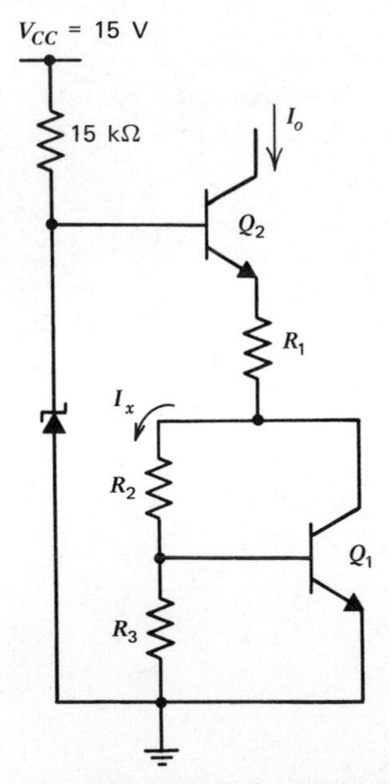

Fig. 4.36 Circuit for Problem 4.9.

4.9 In the circuit of Fig. 4.36, select values of R_1, R_2, and R_3 such that the *current*, I_o, has zero TC_F, and has a value of 1 mA. Choose R_2 and R_3 so that I_x is equal to 50 μA and is thus a negligible part of the output current. Neglect base currents. Assume the Zener voltage is 6.2 V, and $\partial V_Z/\partial T = +2.5$ mV/°C. Also assume that $(1/R)(\partial R/\partial T) = +2000$ ppm/°C.

4.10 The circuit of Fig. 4.30*c* is to be used as a band-gap reference. If the op amp is ideal, its differential input voltage and current are both zero and

$$V_{\text{out}} = (V_{BE1} + I_1 R_1) = (V_{BE1} + I_2 R_2)$$

$$V_{\text{out}} = V_{BE1} + R_2 \left(\frac{V_{BE1} - V_{BE2}}{R_3} \right)$$

Assume that I_1 is to be made equal to 200 μA, and that $(V_{BE1} - V_{BE2})$ is to be made equal to 100 mV. Determine R_1, R_2, and R_3 to realize zero TC_F of V_{out}. Neglect base currents.

4.11 In the analysis of the hypothetical reference of Fig. 4.28, the current, I_1, was assumed constant. Assume instead that this current is derived from a diffused resistor, and thus has a TC_F of -1500 ppm/°C. Determine the new value of V_{out} required to achieve zero TC_F at 25°C. Neglect base current.

4.12 A band-gap reference like that of Fig. 4.30*c* is designed to have nominally zero TC_F at 25°C. Due to process variations, the saturation current, I_S, of the transistors is actually twice the nominal value. What is the output TC_F at 25°C?

4.13 Repeat Problem 4.12 assuming that the value of I_S, R_2, and R_1 are nominal but that R_3 is 1 percent low. Assume $V_{BE(\text{on})} = 0.6$ V.

4.14 Determine the unloaded voltage gain and output resistance for the circuit of Fig. 4.37.

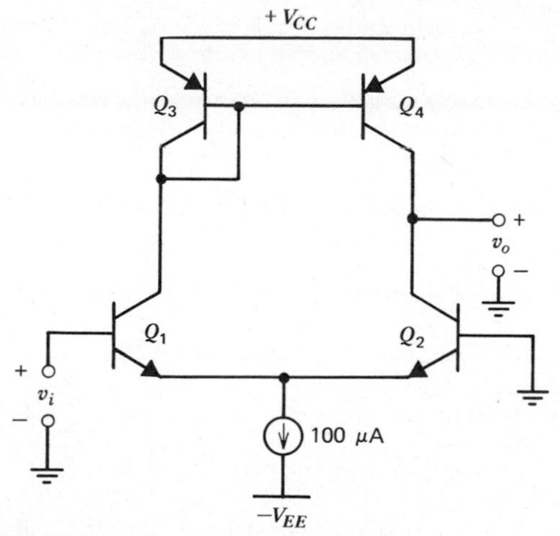

Fig. 4.37 Circuit for Problem 4.14.

4.15 Repeat Problem 4.14, but now assuming that 2 kΩ resistors are inserted in series with the emitters of Q_3 and Q_4.

4.16 Determine the unloaded voltage gain of the circuit shown in Fig. 4.38. Neglect r_μ.

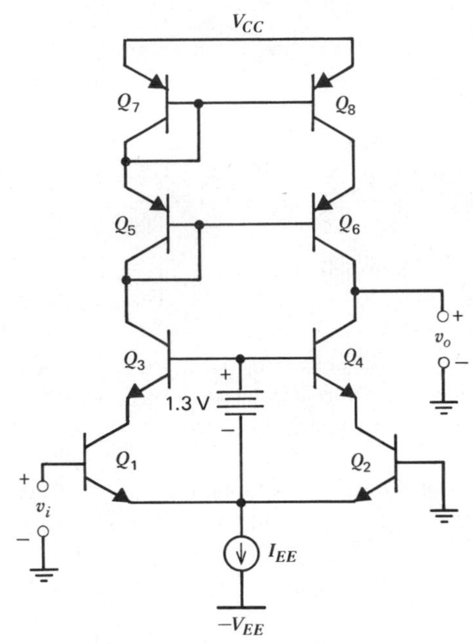

Fig. 4.38 Cascode active-load circuit for Problem 4.16.

4.17 Determine the worst case input offset voltage for the circuit of Fig. 4.37 assuming worst case I_S mismatches in the transistors of ± 5 percent and a *pnp* β of 15. Assume $V_o = V_{CC} - V_{BE(on)}$.

4.18 Repeat Problem 4.17, but assume that 2 kΩ resistors are placed in series with the emitters of Q_3 and Q_4. Assume the resistor mismatch is ± 0.5 percent and the *pnp* β mismatch is 10 percent.

REFERENCES

1. J. Giles. *Fairchild Semiconductor Linear Integrated Circuits Applications Handbook.* Fairchild Semiconductor, 1967.
2. R. J. Widlar. "Some Circuit Design Techniques for Linear Integrated Circuits," *IEEE Transactions on Circuit Theory*, Vol. CT-12, pp. 586–590, December 1965.
3. R. J. Widlar. "Design Techniques for Monolithic Operational Amplifiers," *IEEE Journal of Solid-State Circuits*, Vol. SC-4, pp. 184–191, August 1969.
4. G. R. Wilson. "A Monolithic Junction FET-NPN Operational Amplifier," *IEEE Journal of Solid-State Circuits*, Vol. SC-3, pp. 341–348, December 1968.
5. D. Fullagar. "A New High-Performance Monolithic Operational Amplifier," *Fairchild Semiconductor Applications Brief*, May 1968.

6. R. J. Widlar. "New Developments in IC Voltage Regulators," *IEEE Journal of Solid-State Circuits*, Vol. SC-6, pp. 2–7, February 1971.
7. K. E. Kujik. "A Precision Reference Voltage Source," *IEEE Journal of Solid-State Circuits*, Vol. SC-8, pp. 222–226, June 1973.
8. A. P. Brokaw. "A Simple Three-Terminal IC Bandgap Reference," *IEEE Journal of Solid-State Circuits*, Vol. SC-9, pp. 388–393, December 1974.

General Reference

D. J. Hamilton and W. G. Howard. *Basic Integrated-Circuit Engineering*. McGraw-Hill, New York, 1975.

CHAPTER 5

OUTPUT STAGES

5.1 INTRODUCTION

The output stage of an amplifier must satisfy a number of special requirements. One of the most important of these is to deliver a specified amount of signal power to a load with acceptably low levels of signal distortion. Another common objective of output-stage design is to minimize the output impedance so that the voltage gain is relatively unaffected by the value of load impedance. A well-designed output stage should achieve these performance specifications while consuming low quiescent power and, in addition, should not be a major limitation on the frequency response of the amplifier.

In this chapter, several output-stage configurations will be considered to realize the above specifications. The simplest output-stage configuration is the emitter follower, considered first, followed by an examination of the common-emitter and common-base output stages. Finally, more complex output stages employing multiple output devices are treated, and comparisons are made of power-output capability and efficiency.

5.2 THE EMITTER FOLLOWER AS AN OUTPUT STAGE

An emitter-follower output stage is shown in Fig. 5.1. In order to simplify the analysis, positive and negative bias supplies of equal magnitude V_{CC} are assumed, although in practice these may have different values. When output voltage V_o is zero, output current I_o is also zero. The emitter-follower output device, Q_1, is biased to a quiescent current I_Q by current source Q_2. The output stage is driven by voltage V_i, which has a quiescent dc value of V_{be1} for $V_o = 0$ V. The bias components R_1, R_3, and Q_3 can be those used to bias other stages in the circuit. Since the quiescent current, I_Q in Q_2, will usually be larger than the reference current, I_R, resistor R_2 is usually smaller than R_1 to accommodate this difference.

5.2.1 Transfer Characteristic of the Emitter Follower

In analyzing the circuit of Fig. 5.1, consider the fact that, as an output stage, it must handle large signal amplitudes. That is, the current and voltage swings resulting from the presence of signals may be a large fraction of the bias values. This means that the small-signal analyses that have been used extensively up to this point must be used with care in this situation. For this reason we first determine

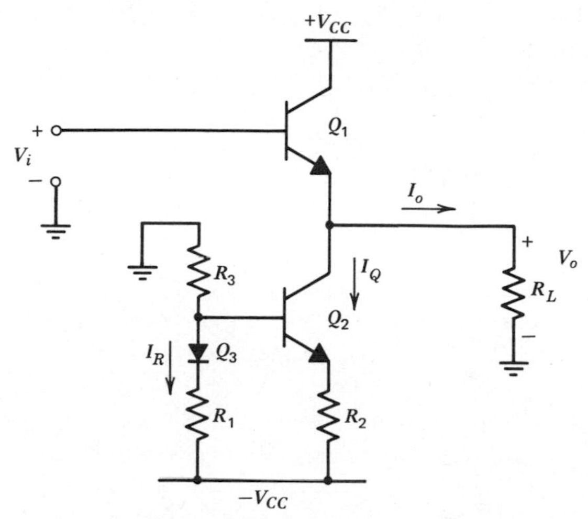

Fig. 5.1 Emitter-follower output stage with current-source bias.

the dc transfer characteristic of the emitter follower. This allows calculation of the gain of the circuit and also gives important information on the *linearity* of the transfer characteristic and thus on the *distortion* performance of the stage.

Consider the circuit of Fig. 5.1. The large-signal transfer characteristic can be derived as follows.

$$V_i = V_{be1} + V_o \tag{5.1}$$

In this case, the base-emitter voltage V_{be1} of Q_1 cannot be assumed constant but must be expressed in terms of the collector current I_{c1} of Q_1 and the saturation current, I_S. Thus

$$V_{be1} = \frac{kT}{q} \ln \frac{I_{c1}}{I_S} \tag{5.2}$$

if Q_1 is in the forward-active region. Also

$$I_{c1} = I_Q + \frac{V_o}{R_L} \tag{5.3}$$

if Q_2 is in the forward-active region and β is assumed large. Substitution of (5.3) and (5.2) in (5.1) gives

$$V_i = \frac{kT}{q} \ln \frac{I_Q + \dfrac{V_o}{R_L}}{I_S} + V_o \tag{5.4}$$

Equation 5.4 is a nonlinear equation relating V_o and V_i if both Q_1 and Q_2 are in the forward-active region. The load resistance, R_L, is assumed small compared with the output resistance of the transistors.

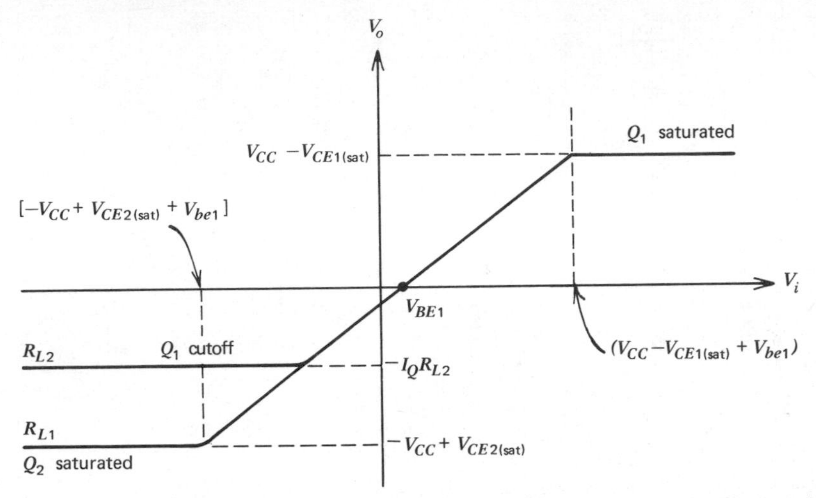

Fig. 5.2 Transfer characteristic of the circuit of Fig. 5.1 for a low (R_{L2}) and a high (R_{L1}) value of load resistance.

The transfer characteristic can be plotted from (5.4) and this has been done in Fig. 5.2. Consider first the case where R_L is large. This is labeled R_{L1}, and in this case the first term on the right-hand side of (5.4) (which represents the base-emitter voltage V_{be1} of Q_1) is relatively constant as V_o changes. This is due to the fact that for a large R_L, the current in the load is small and thus the current in Q_1 is relatively constant as V_o changes and V_{be1} is also relatively constant. As a result the transfer characteristic for $R_L = R_{L1}$ is a relatively straight line with unity slope that is offset on the V_i axis by V_{BE1}, the quiescent value of V_{be1}. This near-linear region depends on both Q_1 and Q_2 being in the forward-active region. However as V_i is made large positive or negative, one or other of these devices *saturates* and the transfer characteristic abruptly changes slope.

Consider V_i made large and positive. Output voltage V_o follows V_i until $V_o = V_{CC} - V_{CE1(sat)}$ at which point Q_1 saturates. The collector-base junction of Q_1 is then forward biased and large currents will flow from base to collector. In practice, the transistor base resistance (and any source resistance present) will limit the current in the forward-biased, collector-base junction and prevent the voltage at the internal transistor base from rising appreciably higher. Further increases in V_i thus produce little change in V_o and the characteristic flattens out as shown in Fig. 5.2. Note that the value of V_i required to cause this is slightly larger than the supply voltage because V_{be1} is larger than the saturation voltage $V_{CE(sat)}$. Consequently, in a practical circuit the preceding stage often limits the maximum positive output voltage because a voltage larger than V_{CC} usually cannot be generated at the base of the output stage.

Consider V_i now made large and negative. The output voltage follows until $V_o = -V_{CC} + V_{CE2(sat)}$, at which point Q_2 saturates. [The voltage drop across

R_2 is assumed small and neglected. If necessary it could be lumped in with the saturation voltage $V_{CE2(\text{sat})}$ of Q_2.] When Q_2 saturates there is another discontinuity in the transfer curve and the slope abruptly decreases. For acceptable distortion performance in the circuit, the voltage swing must be limited to the region between these two break points. As mentioned above, the driver stage supplying V_i usually cannot produce values of V_i that have a magnitude exceeding V_{CC} (if it is connected to the same supply voltages) and the driver itself then limits the amplitude.

Consider now the case where R_L in Fig. 5.1 has a relatively small value. Then when V_o is made large and negative, the first term in (5.4) can become large. In particular this term approaches minus infinity when V_o approaches the critical value

$$V_o = -I_Q R_L \tag{5.5}$$

In this situation, the current drawn from the load $[-(V_o/R_L)]$ is equal to the current, I_Q, and device Q_1 cuts off, leaving Q_2 to draw the current I_Q from the load. Further decreases in V_i produce no change in V_o and the transfer characteristic is the one labeled R_{L2} in Fig. 5.2. The transfer characteristic for positive V_i is similar for both cases.

For the case $R_L = R_{L2}$ the stage will produce severe waveform distortion if V_i is a sinusoid with amplitude exceeding $I_Q R_{L2}$. Consider the two sinusoidal waveforms in Fig. 5.3*a*. Waveform ① has an amplitude $V_1 < I_Q R_{L2}$ and waveform ② has an amplitude $V_2 > I_Q R_{L2}$. If these signals are applied as inputs at V_i in Fig. 5.1 (together with a bias voltage) the output waveforms that result are shown in Fig. 5.3*b* for $R_L = R_{L2}$. For the smaller input signal, the circuit acts as a near-linear amplifier and the output is near sinusoidal. The output waveform distortion, which is apparent for the larger input, is termed "clipping" and must be avoided in normal operation of the circuit as a linear output stage. For a given I_Q and R_L, the onset of clipping limits the maximum signal that can be handled. Note that if $I_Q R_L$ is larger than V_{CC}, the situation shown for $R_L = R_{L1}$ in Fig. 5.2 holds and the output voltage can swing almost to the positive and negative supply voltages before excessive distortion occurs.

5.2.2 Power Output and Efficiency

Further insight into the operation of the circuit of Fig. 5.1 can be obtained from Fig. 5.4 where three different load lines are drawn on the $I_c - V_{ce}$ characteristics of Q_1. The equation for the load lines can be written from Fig. 5.1 and is

$$V_{ce1} = V_{CC} - (I_{c1} - I_Q)R_L \tag{5.6}$$

when both Q_1 and Q_2 are in the forward-active region. The values of V_{ce1} and I_{c1} are related by (5.6) for any value of V_i and the line includes the quiescent point, Q, where $I_{c1} = I_Q$ and $V_{ce1} = V_{CC}$. Equation 5.6 is plotted in Fig. 5.4 for load resistances R_{L1}, R_{L2}, and R_{L3} and the device operating point moves up and down these lines as V_i varies. As V_i increases and V_{ce1} decreases, Q_1 eventually saturates,

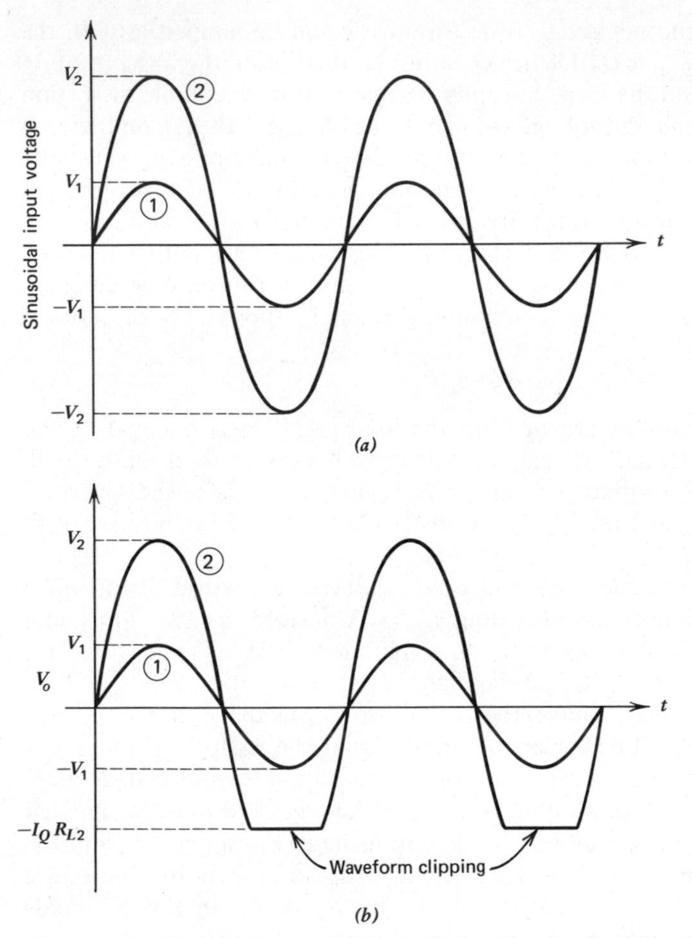

Fig. 5.3 (*a*) Sinusoidal input signals applied to the circuit of Fig. 5.1. (*b*) Output waveforms corresponding to the inputs in (*a*) with $R_L = R_{L2}$.

as was illustrated in Fig. 5.2. As V_i decreases and V_{ce1} increases, there are two possibilities as described above. If R_L is large (R_{L1}), V_o decreases and V_{ce1} increases until Q_2 saturates. Thus the maximum possible value that V_{ce1} can attain is $[2V_{CC} - V_{CE2(\text{sat})}]$ and this is marked on Fig. 5.4. If, however, R_L is small (R_{L2}), the maximum negative value of V_o as illustrated in Fig. 5.2 is $-I_Q R_{L2}$ and the maximum possible value of V_{ce1} is $(V_{CC} + I_Q R_{L2})$.

So far no mention has been made of the maximum voltage limitations of the output stage. As described in Chapter 1, avalanche breakdown of a transistor occurs for $V_{ce} = BV_{CEO}$ in the common-emitter configuration, and this is the worst case for breakdown voltage. In a conservative design the value of V_{ce} in the circuit

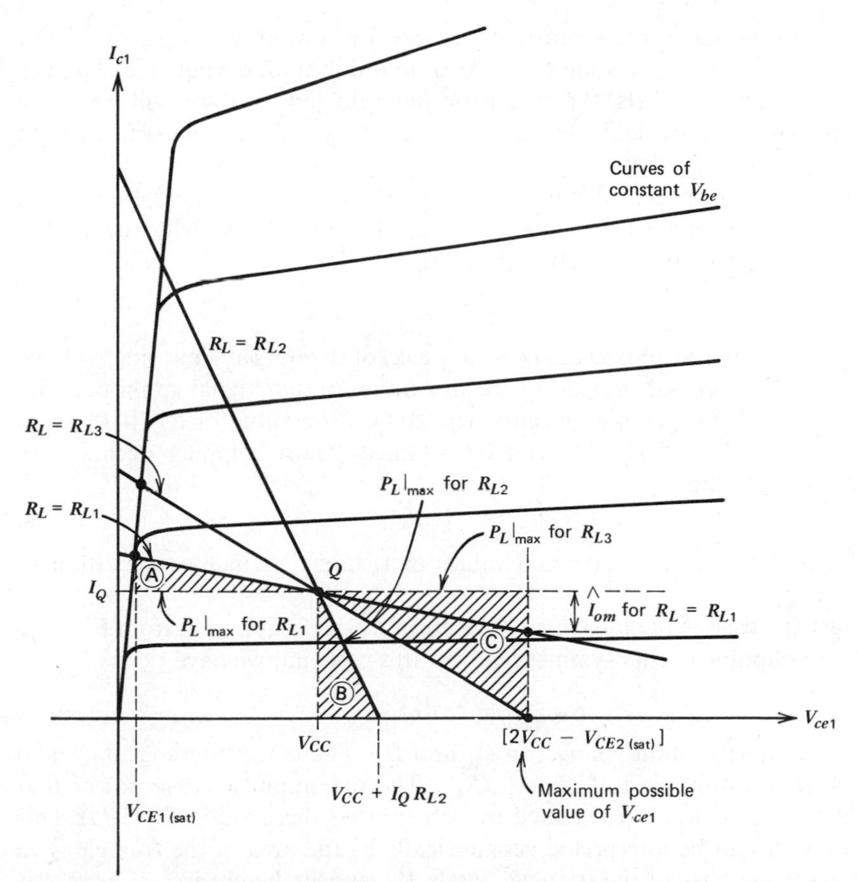

Fig. 5.4 Load lines in the I_{c1}–V_{ce1} plane for emitter follower Q_1 of Fig. 5.1.

of Fig. 5.1 should always be less than BV_{CEO} by an appropriate safety margin. In the preceding analysis it was shown that the maximum value that V_{ce1} can attain in this circuit for *any* load resistance is approximately $2V_{CC}$, and thus BV_{CEO} must be greater than this value.

Consider now the power relationships in the circuit. When sinusoidal signals are present, the power dissipated in various elements will vary with time. We are concerned both with the *instantaneous* power dissipated and with the *average* power dissipated. Instantaneous power is important when considering transistor dissipation with low-frequency or dc signals. The junction temperature of the transistor will tend to rise and fall with the instantaneous power dissipated in the device, and there will be a maximum allowable instantaneous power dissipation for safe operation of any device.

Average power levels are important because the power delivered to a load is usually specified as an *average* value. Also note that if an output stage handles only high-frequency signals, the transistor junction temperature will not vary appreciably over a cycle and the *average* device power dissipation will then be the limiting quantity.

Consider the output signal power that can be delivered to load R_L when a *sinusoidal* input is applied at V_i. The *average* output power delivered to R_L, assuming that V_o is approximately sinusoidal, is

$$P_L = \tfrac{1}{2}\hat{V}_o\hat{I}_o \tag{5.7}$$

where $\hat{V}_o$ and $\hat{I}_o$ are the amplitudes (zero to peak) of the output sinusoidal voltage and current. As described previously, the maximum output signal amplitude that can be attained before clipping occurs depends on the value of R_L. If $P_L|_{\max}$ is the maximum value of P_L that can be attained before clipping occurs with sinusoidal signals, then

$$P_L|_{\max} = \tfrac{1}{2}\hat{V}_{om}\hat{I}_{om} \tag{5.7a}$$

where $\hat{V}_{om}$ and $\hat{I}_{om}$ are the maximum values of $\hat{V}_o$ and $\hat{I}_o$, which can be attained before clipping.

Consider the case of the large load resistance, R_{L1}. It is apparent from Figs. 5.2 and 5.4 that clipping occurs symmetrically in this case and we have

$$\hat{V}_{om} = V_{CC} - V_{CE(\text{sat})} \tag{5.8}$$

assuming equal saturation voltages in Q_1 and Q_2. The corresponding sinusoidal output current amplitude is $\hat{I}_{om} = \hat{V}_{om}/R_{L1}$. The maximum average power that can be delivered to R_{L1} is calculated by substituting these values in (5.7a). This value of power can be interpreted geometrically as the area of the triangle A in Fig. 5.4 since the base of the triangle equals $\hat{V}_{om}$ and its height is $\hat{I}_{om}$. As R_{L1} is made larger, the maximum average output power that can be delivered diminishes because the triangle becomes smaller. The maximum output voltage amplitude remains essentially the same but the current amplitude decreases as R_{L1} increases.

If $R_L = R_{L2}$ in Fig. 5.4, the maximum output voltage swing before clipping occurs is

$$\hat{V}_{om} = I_Q R_{L2} \tag{5.9}$$

The corresponding current amplitude is $\hat{I}_{om} = I_Q$, and, using (5.7a), the maximum average output power $P_L|_{\max}$ that can be delivered is given by the area of triangle B, shown in Fig. 5.4. As R_{L2} is made smaller it is apparent that the maximum average power that can be delivered diminishes.

An examination of Fig. 5.4 shows that the power-output capability of the stage is maximized for $R_L = R_{L3}$, and this can be calculated from (5.6) and Fig. 5.4 as

$$R_{L3} = \frac{V_{CC} - V_{CE(\text{sat})}}{I_Q} \tag{5.10}$$

This load line gives the triangle of largest area (C) and thus the largest average output power. In this case $\hat{V}_{om} = [V_{CC} - V_{CE(\text{sat})}]$, $\hat{I}_{om} = I_Q$ and using (5.7a) we have

$$P_L\big|_{\max} = \tfrac{1}{2}\hat{V}_{om}\hat{I}_{om}$$
$$= \tfrac{1}{2}[V_{CC} - V_{CE(\text{sat})}]I_Q \tag{5.11}$$

In order to calculate the *efficiency* of the circuit, the power drawn from the supply voltages must now be calculated. The current drawn from the positive supply is the collector current of Q_1, which is assumed sinusoidal with an average value I_Q. The current flowing in the negative supply is constant and equal to I_Q (neglecting bias current I_R). Since the supply voltages are constant, the *average power* drawn from the supplies is constant and *independent* of the presence of sinusoidal signals in the circuit. The total power drawn from the two supplies is thus

$$P_{\text{supply}} = 2V_{CC}I_Q \tag{5.12}$$

The *power conversion efficiency* (η_C) of the circuit at an arbitrary output power level is defined as the ratio of the average power delivered to the load to the power drawn from the supplies. Since the power drawn from the supplies is constant in this circuit, the efficiency obviously increases as the output power increases. Also since it has been shown that the power-output capability of the circuit depends on the value of R_L, the efficiency also depends on R_L. The best efficiency occurs for $R_L = R_{L3}$ since this gives maximum average power output. We have, in general, that

$$\eta_C = \frac{P_L}{P_{\text{supply}}} \tag{5.13}$$

If $R_L = R_{L3}$ and $\hat{V}_o = \hat{V}_{om}$ then substitution of (5.11) and (5.12) in (5.13) gives for the maximum possible efficiency

$$\eta_{\max} = \tfrac{1}{4}\left(1 - \frac{V_{CE(\text{sat})}}{V_{CC}}\right) \tag{5.14}$$

Thus if $V_{CE(\text{sat})} \ll V_{CC}$, the maximum efficiency of the stage is $\frac{1}{4}$ or 25 percent.

Another aspect of circuit performance that is important is the power dissipated in the active device. The current and voltage waveforms in Q_1 at maximum signal swing and with $R_L = R_{L3}$ are shown in Fig. 5.5 [assuming $V_{CE(\text{sat})} \simeq 0$ for simplicity] together with their product which is the *instantaneous* power dissipation in the transistor. The curve of instantaneous power dissipation in Q_1 as a function of time varies at twice the signal frequency and has an average value half the quiescent value. This can be shown analytically as follows. The instantaneous power dissipation in Q_1 is

$$P_{c1} = V_{ce1}I_{c1} \tag{5.15}$$

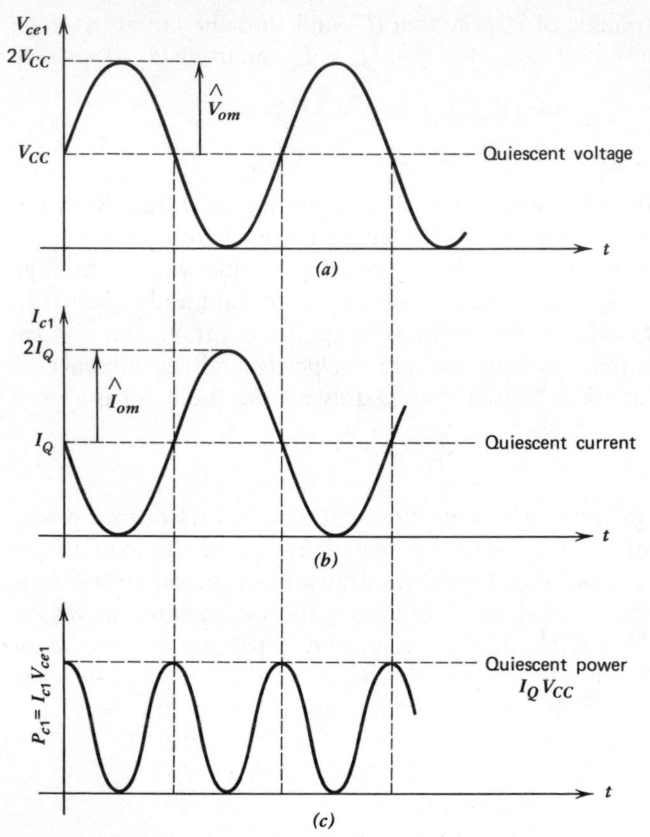

Fig. 5.5 Waveforms for the transistor Q_1 of Fig. 5.1 at full output with $R_L = R_{L3}$. (*a*) Collector-emitter voltage waveform. (*b*) Collector current waveform. (*c*) Collector power dissipation waveform.

At maximum signal swing with a sinusoidal signal, P_{c1} can be expressed as (from Fig. 5.5)

$$\begin{aligned} P_{c1} &= V_{CC}(1 + \sin \omega t)I_Q(1 - \sin \omega t) \\ &= V_{CC}I_Q(1 - \sin^2 \omega t) \\ &= \frac{V_{CC}I_Q}{2}(1 + \cos 2\,\omega t) \end{aligned} \tag{5.15a}$$

The average value of P_{c1} from (5.15a) is $V_{CC}I_Q/2$. Thus at maximum output the *average* power dissipated in Q_1 is half its quiescent value and the average device temperature when delivering power with $R_L = R_{L3}$ is *less than* its quiescent value.

Further information on the power dissipated in Q_1 can be obtained by plotting curves of constant device dissipation in the I_c–V_{ce} plane. Equation 5.15 indicates

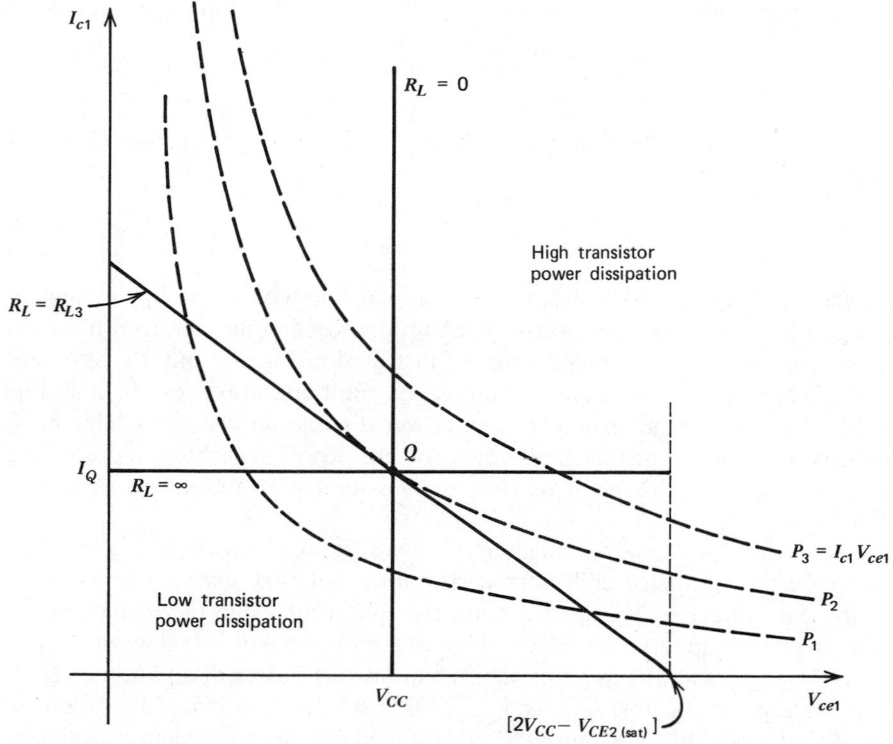

Fig. 5.6 Hyperbolas of constant instantaneous transistor power dissipation P_1, P_2, and P_3 in the I_{c1}–V_{ce1} plane for emitter follower Q_1 of Fig. 5.1. Load lines are included for $R_L = R_{L3}$, 0, and ∞. Note that $P_1 < P_2 < P_3$.

that such curves are hyperbolas and these are plotted in Fig. 5.6 for constant transistor instantaneous power dissipation values P_1, P_2, and P_3 (where $P_1 < P_2 < P_3$). The power hyperbola of value P_2 passes through the quiescent point, Q_1, and the equation to this curve can be calculated from (5.15) as

$$I_{c1} = \frac{P_2}{V_{ce1}} \tag{5.16}$$

The slope of the curve is

$$\frac{dI_{c1}}{dV_{ce1}} = -\frac{P_2}{V_{ce1}^2}$$

and substitution of (5.16) in this equation gives

$$\frac{dI_{c1}}{dV_{ce1}} = -\frac{I_{c1}}{V_{ce1}} \tag{5.17}$$

At the quiescent point, Q, we have $I_{c1} = I_Q$ and $V_{ce1} = V_{CC}$. Thus the slope is

$$\left.\frac{dI_{c1}}{dV_{ce1}}\right|_Q = -\frac{I_Q}{V_{CC}} \tag{5.18}$$

However, the slope of the load line with $R_L = R_{L3}$ is $-(1/R_{L3})$ from (5.6) and from (5.10)

$$-\frac{1}{R_{L3}} \simeq -\frac{I_Q}{V_{CC}} \tag{5.19}$$

Comparing (5.18) with (5.19) shows that the load line with $R_L = R_{L3}$ is tangent to the power hyperbola passing through the quiescent point, since both curves have the same slope at that point. This is illustrated in Fig. 5.6 and it is apparent that as the operating point leaves the quiescent point and moves on the load line with $R_L = R_{L3}$ the instantaneous device power dissipation *decreases*. This must be so because the load line then intersects constant-power hyperbolas representing lower power values. This point of view is consistent with the power waveform shown in Fig. 5.5.

The load line for $R_L \to \infty$ (open-circuit load) is also shown in Fig. 5.6 and in that case the transistor collector current does not vary over a period but is constant. For values of V_{ce1} greater than the quiescent value the instantaneous device power dissipation increases. The maximum possible value of V_{ce1} is $[2V_{CC} - V_{CE2(\text{sat})}]$ and at this value the instantaneous power dissipation in Q_1 is approximately $2V_{CC}I_Q$ if $V_{CE2(\text{sat})} \ll V_{CC}$. This dissipation is twice the quiescent value of $V_{CC}I_Q$, and this possibility should be taken into account when considering the power handling requirements of Q_1. Note that at the other extreme of the swing where $V_{ce1} \simeq 0$, the power dissipation in Q_2 is also $2V_{CC}I_Q$.

A situation that is potentially even more damaging can occur if the load is short circuited. In that case the load line is vertical through the quiescent point as shown in Fig. 5.6, and with large input signals the collector current (and thus the device power dissipation) of Q_1 can become quite large. The limit on collector current is set by the ability of the driver to supply the base current to Q_1, and also by the high-current falloff in β_F of Q_1, described in Chapter 1. In practice these limits may be sufficient to prevent burnout of Q_1, but current-limiting circuitry may be necessary. An example of such protection is given in Section 5.5.6.

A useful general result can be derived from the calculations above involving load lines and constant-power hyperbolas. It is apparent by inspecting Fig. 5.6 that the maximum instantaneous device power dissipation for $R_L = R_{L3}$ occurs at the quiescent point, Q (since $P_1 < P_2 < P_3$), and this is the *midpoint* of the load line if $V_{CE2(\text{sat})} \ll V_{CC}$. (The midpoint of the load line is assumed to be midway between its intersections with the I_c and V_{ce} axes.) It can be seen from (5.17) that *any* load line tangent to a power hyperbola makes contact with the hyperbola at the midpoint of the load line. Consequently the midpoint is the point of maximum instantaneous device power dissipation with *any* load line. For example in Fig. 5.4 with $R_L = R_{L2}$ the maximum instantaneous device power dissipation occurs at the midpoint of the load line where $V_{ce1} = \frac{1}{2}(V_{CC} + I_Q R_{L2})$.

An output stage of the type described in this section where the output device is always conducting appreciable current is called a *Class A* output stage. This type of operation can be realized with different transistor configurations but always has a maximum efficiency of 25 percent.

Finally it should be pointed out that in this development the emitter follower was assumed to have a current source I_Q in its emitter as a bias element. In practice this may be approximated by simply using a resistor connected to the negative supply, and this will result in some deviations from the above calculations. In particular the output power available from the circuit will be reduced.

EXAMPLE

An output stage such as shown in Fig. 5.1 has the following parameters: $V_{CC} = 10$ V, $R_3 = 5$ kΩ, $R_1 = R_2 = 0$, $V_{CE(\text{sat})} = 0.2$ V, $R_L = 1$ kΩ.
(a) Calculate the maximum average output power that can be delivered to R_L before clipping occurs and the corresponding efficiency. What is the maximum possible efficiency with this stage and what value of R_L is required to achieve this? Assume the signals are sinusoidal.
(b) Calculate the maximum possible instantaneous power dissipation in Q_1. Also calculate the average power dissipation in Q_1 when $\hat{V}_o = 1.5$ V and the output voltage is sinusoidal.

The solution proceeds as follows.
(a) The bias current, I_Q, is first calculated.

$$I_Q = I_R = \frac{V_{CC} - V_{BE3}}{R_3} = \frac{10 - 0.7}{5}\ \text{mA} = 1.86\ \text{mA}$$

The product, $I_Q R_L$, is given by

$$I_Q R_L = 1.86 \times 1 = 1.86\ \text{V}$$

Since $I_Q R_L$ is less than V_{CC}, the maximum sinusoidal output voltage swing is limited to 1.86 V by clipping on negative voltage swings and the situation corresponds to $R_L = R_{L2}$ in Fig. 5.4. The maximum output voltage and current swings are thus $\hat{V}_{om} = 1.86$ V and $\hat{I}_{om} = I_Q = 1.86$ mA. The maximum average output power available from the circuit for sinusoidal signals can be calculated from (5.7a) as

$$P_L\big|_{\max} = \tfrac{1}{2}\hat{V}_{om}\hat{I}_{om} = \tfrac{1}{2} \times 1.86 \times 1.86\ \text{mW} = 1.73\ \text{mW}$$

The power drawn from the supplies is calculated from (5.12) as

$$P_{\text{supply}} = 2V_{CC}I_Q = 2 \times 10 \times 1.86 \text{ mW} = 37.2 \text{ mW}$$

The efficiency of the circuit at the output power level calculated above can be determined from (5.13)

$$\eta_C = \frac{P_L|_{\max}}{P_{\text{supply}}} = \frac{1.73}{37.2} = 0.047$$

The efficiency of 4.7 percent is quite low and is due to the limitation on the negative voltage swing.

The maximum possible efficiency with this stage occurs for $R_L = R_{L3}$ in Fig. 5.4 and R_{L3} is given by (5.10)

$$R_{L3} = \frac{V_{CC} - V_{CE(\text{sat})}}{I_Q} = \frac{10 - 0.2}{1.86} \text{ k}\Omega = 5.27 \text{ k}\Omega$$

In this case the maximum average power that can be delivered to the load before clipping occurs is found from (5.11) as

$$P_L|_{\max} = \tfrac{1}{2}[V_{CC} - V_{CE(\text{sat})}]I_Q = \tfrac{1}{2}(10 - 0.2)1.86 \text{ mW} = 9.11 \text{ mW}$$

The corresponding efficiency using (5.14) is

$$\eta_{\max} = \tfrac{1}{4}\left[1 - \frac{V_{CE(\text{sat})}}{V_{CC}}\right] = \tfrac{1}{4}\left(1 - \frac{0.2}{10}\right) = 0.245$$

This is close to the theoretical maximum of 25 percent.

(b) The maximum possible instantaneous power dissipation in Q_1 occurs at the midpoint of the load line. Reference to Fig. 5.4 and the load line, $R_L = R_{L2}$, shows that this occurs for

$$V_{ce1} = \tfrac{1}{2}(V_{CC} + I_Q R_L) = \tfrac{1}{2}(10 + 1.86) = 5.93 \text{ V}$$

The corresponding collector current in Q_1 is

$$I_{c1} = \frac{5.93}{R_L} = 5.93 \text{ mA}$$

since

$$R_L = 1 \text{ k}\Omega$$

Thus the maximum possible instantaneous power dissipation in Q_1 is

$$P_{c1} = I_{c1}V_{ce1} = 35.2 \text{ mW}$$

Note that this occurs for $V_{ce1} = 5.93$ V, which represents a signal swing beyond the linear limits of the circuit [clipping occurs on negative output voltage swings when the output voltage amplitude reaches 1.86 V as calculated in (a)]. However, this condition could easily occur if the circuit is overdriven by a large input signal.

The *average* power dissipation in Q_1 can be calculated by noting that for sinusoidal signals, the average power drawn from the two supplies is constant and independent of the presence of signals. Since the power input to the circuit from the supplies is constant, the *total* average power dissipated in Q_1, Q_2, and R_L must be constant and independent of the presence of sinusoidal signals. The average power dissipated in Q_2 is constant because I_Q is constant, and thus the average power dissipated in Q_1 and R_L together is constant. Thus as $\hat{V}_o$ is increased, the average power dissipated in Q_1 *decreases* by the same amount as the average power in R_L *increases.* With no input signal, the quiescent power dissipated in Q_1 is

$$P_{CQ} = V_{CC}I_Q = 10 \times 1.86 \text{ mW} = 18.6 \text{ mW}$$

For $\hat{V}_o = 1.5$ V, the average power delivered to the load is

$$P_L = \frac{1}{2}\frac{\hat{V}_o^2}{R_L} = \frac{1}{2}\frac{2.25}{1} \text{ mW} = 1.13 \text{ mW}$$

Thus the *average* power dissipated in Q_1 when $\hat{V}_o = 1.5$ V with a sinusoidal signal is

$$P_{av} = P_{CQ} - P_L = 17.5 \text{ mW}$$

5.2.3 Emitter-Follower Drive Requirements

The calculations above have been concerned with the performance of the emitter-follower output stage when driven by a sinusoidal input voltage. The stage preceding the output stage is called the driver stage and in practice it may introduce additional limitations on the circuit performance. For example, it was shown that to drive the output voltage, V_o, of the emitter follower to its maximum positive value required a voltage slightly greater than the supply voltage. Since, in most cases, the driver stage is connected to the same supplies as the output stage, the driver stage generally cannot produce voltages greater than the supply and this further reduces the possible output voltage swing.

The above limitations arise because the emitter follower has a voltage gain of unity and thus the driver stage must handle the same voltage swing as the output. The driver can however be a much *lower power* stage than the output stage because the current it must deliver is the base current of the emitter follower that is $1/\beta$ times the output current. Consequently the driver bias current can be much lower than the output-stage bias current, and a smaller geometry can be used for the driver device. Although it has only unity voltage gain, the emitter follower has substantial *power gain,* which is a requirement of any output stage.

5.2.4 Small-Signal Properties of the Emitter Follower

A simplified low-frequency, small-signal equivalent circuit of the emitter follower of Fig. 5.1 is shown in Fig. 5.7. As described in Chapter 7, the emitter follower is

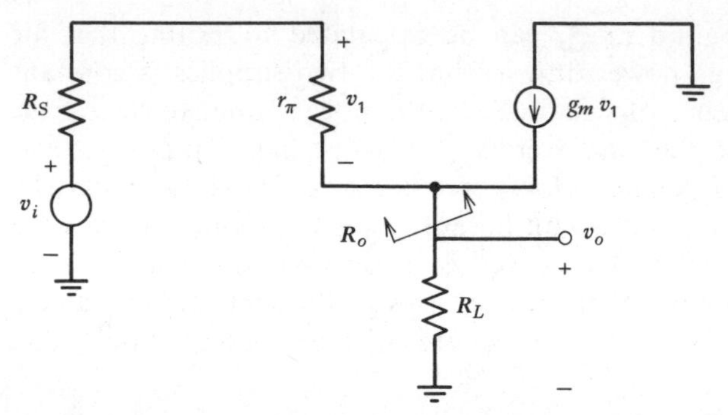

Fig. 5.7 Low-frequency, small-signal equivalent circuit for the emitter follower of Fig. 5.1.

an extremely wideband circuit and rarely is a source of frequency limitation in the small-signal gain of an amplifier. Thus the equivalent circuit of Fig. 5.7 is useful over a wide frequency range and an analysis of this circuit shows that the voltage gain, A_v, and the output resistance, R_o, can be expressed approximately as

$$A_v = \frac{v_o}{v_i} \simeq \frac{R_L}{R_L + \dfrac{1}{g_m} + \dfrac{R_S}{\beta_0}} \tag{5.20}$$

$$R_o = \frac{1}{g_m} + \frac{R_S}{\beta_0} \tag{5.21}$$

for

$$\beta_0 \gg 1$$

These quantities are small-signal quantities, and, since $g_m = qI_C/kT$ is a function of bias point, both A_v and R_o are functions of I_C. Since the emitter follower is being considered here for use as an output stage where the signal swing may be large, (5.20) and (5.21) must be applied with caution. However for small-to-moderate signal swings, these equations may be used to estimate the average gain and output resistance of the stage if quiescent bias values are used for transistor parameters in the equations. Equation 5.20 can also be used as a means of estimating the nonlinearity[1] in the stage by recognizing that it gives the *incremental slope* of the large-signal characteristic of Fig. 5.2 at any point. If this is evaluated at the extremes of the signal swing, an estimate of the curvature of the characteristic is obtained, as illustrated in the following example.

EXAMPLE

Calculate the incremental slope of the transfer characteristic of the circuit of Fig. 5.1 at the quiescent point and at the extremes of the signal swing with a peak sinusoidal output of 0.6 V. Use data as in the previous example and assume that $R_S = 0$.

From (5.20) the small-signal gain with $R_S = 0$ is

$$A_v = \frac{R_L}{R_L + \dfrac{1}{g_m}} \tag{5.22}$$

Since $I_Q = 1.86$ mA, $1/g_m = 14\ \Omega$ at the quiescent point and the quiescent gain is

$$A_{vQ} = \frac{1000}{1000 + 14} = 0.9862$$

Since the output voltage swing is 0.6 V, the output current swing is

$$\hat{I}_o = \frac{\hat{V}_o}{R_L} = \frac{0.6}{1000} = 0.6 \text{ mA}$$

Thus at the positive signal peak, the transistor collector current is

$$I_Q + \hat{I}_o = 1.86 + 0.6 = 2.46 \text{ mA}$$

At this current, $1/g_m = 10.6\ \Omega$ and use of (5.22) gives the small-signal gain as

$$A_v' = \frac{1000}{1010.6} = 0.9895$$

This represents an increase of 0.3 percent over the quiescent value. At the negative signal peak, the transistor collector current is

$$I_Q - \hat{I}_o = 1.86 - 0.6 = 1.26 \text{ mA}$$

At this current $1/g_m = 20.6\ \Omega$ and use of (5.22) gives the small-signal gain as

$$A_v'' = \frac{1000}{1020.6} = 0.9798$$

This represents a decrease of 0.7 percent compared with the quiescent value. Note the extremely small variation in the small-signal gain of the circuit even though the collector-current signal amplitude is one third of the bias current. This circuit thus has a high degree of linearity.

5.3 THE COMMON-EMITTER OUTPUT STAGE

Common-emitter output stages are not as widely used as emitter followers in integrated-circuit design because of the generally superior characteristics (lower output resistance and less distortion) available from emitter-follower circuits. However, because of their high gain, common-emitter stages are often used as output-stage drivers and thus their performance is of some interest.

5.3.1 Transfer Characteristic of the Common-Emitter Stage

A common-emitter output stage is shown in Fig. 5.8. For convenience, the input voltage, V_i, is taken between the base of Q_1 and the negative supply. Current-source transistor Q_2 establishes a current I_Q, and when V_i is adjusted for $V_o = I_o = 0$ at the quiescent point, the bias current in Q_1 is also I_Q. Note that the load, R_L, is fed from a high source impedance (the collector of Q_1). Consequently the voltage gain depends directly on R_L assuming that R_L is much less than the r_o of Q_1 and Q_2.

The large-signal transfer characteristic of the circuit can be derived from Fig. 5.8.

$$I_o = I_Q - I_{c1} \tag{5.23}$$

where I_{c1} is the collector current of Q_1. Also

$$I_o = \frac{V_o}{R_L} \tag{5.24}$$

Substitution of (5.24) and (5.2) in (5.23) gives

$$V_o = -R_L\left(I_S \exp\frac{V_i}{V_T} - I_Q\right) \tag{5.25}$$

where

$$V_T = \frac{kT}{q} \tag{5.26}$$

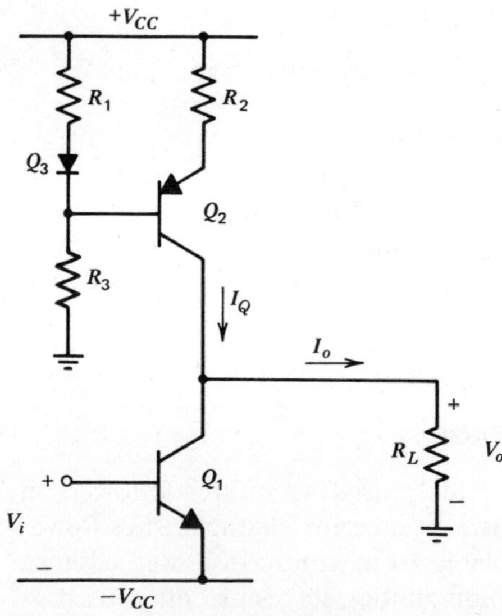

Fig. 5.8 Common-emitter Class A output stage.

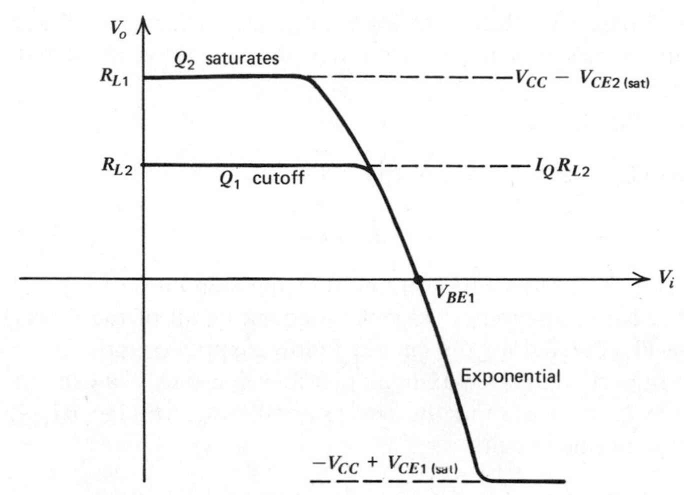

Fig. 5.9 Transfer characteristic of the circuit of Fig. 5.8 for a low (R_{L2}) and a high (R_{L1}) value of load resistance.

The transfer characteristic from V_i to V_o can be plotted from (5.25) and this is shown in Fig. 5.9. There are a number of important differences between this result and that derived in the previous section for the emitter follower. First, Equation 5.25 describing the transfer characteristic shows that it is basically a simple exponential. As a consequence, for given voltage swings and given R_L, the characteristic will show more *curvature* and thus produce more *distortion* than the emitter follower. However, the input voltage swing required to produce maximum output voltage swing in the common-emitter circuit is typically a few tens of millivolts or less, and this is much smaller than in the case of the emitter follower.

As in the case of the emitter follower, the limits of the transfer characteristic depend on the value of R_L. The case of large R_L is labeled R_{L1} in Fig. 5.9 and in this case as V_i is reduced, V_o increases until the current source, Q_2, saturates. If R_L is small as shown by R_{L2}, the maximum value V_o can attain is obtained from (5.25) for V_i large negative and is

$$V_o = I_Q R_{L2} \tag{5.27}$$

If this value is less than $[V_{CC} - V_{CE(\text{sat})}]$ the positive output voltage swing is reduced as shown. With Q_1 cut off, Q_2 then supplies I_Q to the load.

For both values of R_L, as V_i is increased the current in Q_1 increases and causes V_o to go negative until Q_1 saturates for $V_o = -V_{CC} + V_{CE(\text{sat})}$. The possibility of clipping of the output signal for low R_L obviously exists and is similar to that shown in Fig. 5.3 for the emitter follower. In this case it tends to occur first on the positive output voltage excursion. Note that for a large enough input voltage swing, clipping eventually occurs on both positive and negative peaks of the signal.

5.3.2 Power Output and Efficiency

The equation for the load line of the circuit of Fig. 5.8 is

$$V_{ce1} = V_{CC} - (I_{c1} - I_Q)R_L \tag{5.28}$$

This is identical to (5.6) for the emitter-follower and thus the load lines of Fig. 5.4 apply equally well to the common-emitter stage. Consequently, all of the power output and efficiency results derived for the emitter follower apply equally to the common-emitter stage. In particular the maximum possible efficiency of a Class A common-emitter stage is 25 percent, and the device breakdown voltage BV_{CEO} must be greater than $2V_{CC}$ in this circuit.

5.3.3 Distortion in the Common-Emitter Stage

The transfer function of the common-emitter stage was calculated in (5.25) where V_o is expressed as a function of V_i. This makes the direct calculation of distortion in this circuit much simpler than in the case of the emitter follower where the transfer characteristic given by (5.4) did not give V_o as a function of V_i. The calculation of signal distortion from a nonlinear transfer function will now be illustrated using the common-emitter stage as an example.

If the input signal voltage applied is V_s then

$$V_s = V_i - V_{BE1} \tag{5.29}$$

where V_{BE1} is the bias voltage. Substitution of (5.29) in (5.25) gives

$$V_o = -R_L\left(I_S \exp\frac{V_s + V_{BE1}}{V_T} - I_Q\right)$$

and thus

$$V_o = -R_L\left(I_S \exp\frac{V_{BE1}}{V_T} \exp\frac{V_s}{V_T} - I_Q\right) \tag{5.30}$$

But

$$I_Q = I_S \exp\frac{V_{BE1}}{V_T} \tag{5.31}$$

and substitution of (5.31) in (5.30) gives

$$V_o = -R_L I_Q\left(\exp\frac{V_s}{V_T} - 1\right) \tag{5.32}$$

Expansion of the exponential term in (5.32) in a power series gives

$$V_o = -R_L I_Q\left[\frac{V_s}{V_T} + \frac{1}{2}\left(\frac{V_s}{V_T}\right)^2 + \frac{1}{6}\left(\frac{V_s}{V_T}\right)^3 + \cdots\right] \tag{5.33}$$

$$= a_1 V_s + a_2 V_s^2 + a_3 V_s^3 + \cdots \tag{5.34}$$

where

$$a_1 = -\frac{R_L I_Q}{V_T} \tag{5.35}$$

$$a_2 = -\frac{R_L I_Q}{2V_T^2} \tag{5.36}$$

$$a_3 = -\frac{R_L I_Q}{6V_T^3} \tag{5.37}$$

Equation 5.33 allows calculation of distortion in the common-emitter stage. For values of $V_s/V_T \ll 1$ the first term in parentheses dominates and the circuit is essentially linear. However, as V_s becomes comparable to V_T, the terms involving V_s^2 and V_s^3 become significant and distortion products are generated as will now be illustrated. A common method of describing the nonlinearity of an amplifier is the specification of *harmonic distortion.* This is defined for a single sinusoidal input applied to the amplifier. Thus let

$$V_s = \hat{V}_s \sin \omega t \tag{5.38}$$

and substitution of (5.38) in (5.34) gives

$$V_o = a_1 \hat{V}_s \sin \omega t + a_2 \hat{V}_s^2 \sin^2 \omega t + a_3 \hat{V}_s^3 \sin^3 \omega t + \cdots$$

$$= a_1 \hat{V}_s \sin \omega t + \frac{a_2 \hat{V}_s^2}{2}(1 - \cos 2\,\omega t)$$

$$+ \frac{a_3 \hat{V}_s^3}{4}(3 \sin \omega t - \sin 3\,\omega t) + \cdots \tag{5.39}$$

Equation 5.39 shows that the output voltage contains frequency components at the fundamental frequency, ω (the input frequency), and also at harmonic frequencies 2ω, 3ω, and so on. The latter terms represent *distortion products* that are not present in the input signal. Second-harmonic distortion HD_2 is defined as the ratio of the amplitude of the output-signal component at frequency 2ω to the amplitude of the first harmonic (or fundamental) at frequency ω. For small distortion, the term $\frac{3}{4} a_3 \hat{V}_s^3 \sin \omega t$ in (5.39) is small compared to $a_1 \hat{V}_s \sin \omega t$ and the amplitude of the fundamental is approximately $a_1 \hat{V}_s$. Again for small distortion, higher-order terms in (5.39) may be neglected and

$$HD_2 = \frac{a_2 \hat{V}_s^2}{2}\frac{1}{a_1 \hat{V}_s} = \frac{1}{2}\frac{a_2}{a_1}\hat{V}_s \tag{5.40}$$

Note that with the assumptions made, HD_2 varies *linearly* with signal level $\hat{V}_s$. The value of HD_2 can be expressed in terms of known parameters by substituting (5.35) and (5.36) in (5.40) to give

$$HD_2 = \tfrac{1}{4}\frac{\hat{V}_s}{V_T} \tag{5.41}$$

This important result shows that second-harmonic distortion in *any* voltage-driven bipolar transistor at low frequencies depends only on the normalized input voltage. This result shows that 10 percent of second-harmonic distortion ($HD_2 = 0.1$) occurs for $\hat{V}_s = 0.1 \times 4 \times 26$ mV $\simeq 10$ mV.

Third-harmonic distortion HD_3 is defined as the ratio of the output signal component at frequency 3ω to the first harmonic. From (5.39) and assuming small distortion

$$HD_3 = \frac{a_3\hat{V}_s^3}{4}\frac{1}{a_1\hat{V}_s} = \tfrac{1}{4}\frac{a_3}{a_1}\hat{V}_s^2 \tag{5.42}$$

With the assumptions made, HD_3 varies as the *square* of the signal amplitude. The value of HD_3 can be expressed in terms of known parameters by substituting (5.35) and (5.37) in (5.40) to give

$$HD_3 = \tfrac{1}{24}\left(\frac{\hat{V}_s}{V_T}\right)^2 \tag{5.43}$$

Thus HD_3 also depends on the normalized input voltage amplitude. For $\hat{V}_s = 10$ mV, $HD_3 = 0.62$ percent.

Equations 5.41 and 5.43 can be used to calculate harmonic distortion in *any voltage-driven*, common-emitter transistor stage (assuming saturation does not occur). However the presence of finite source resistance will change the situation, and in the extreme case of a current drive to Q_1 the distortion will be that due to variation of β_F with I_C, which is usually much less than distortion created by the exponential.

Note that in this derivation, the effect of the output resistance of Q_1 and Q_2 has been neglected and this is valid for low to moderate values of R_L.

EXAMPLE

Calculate second and third harmonic distortion in the circuit of Fig. 5.8 for a peak sinusoidal output voltage $\hat{V}_o = 0.6$ V. Assume that $I_Q = 1.86$ mA and $R_L = 1$ kΩ as in the previous example using the emitter follower.

Since (5.41) and (5.43) are expressed in terms of input voltage, the value of $\hat{V}_s$ corresponding to $\hat{V}_o = 0.6$ V must be estimated. This can be done by recalling that the analysis is only valid for small distortion, and thus the value of $\hat{V}_s$ can be estimated using a small-signal calculation.

The small-signal voltage gain of the circuit of Fig. 5.8 is

$$A_v = g_{m1} R_L = \frac{qI_C}{kT} R_L \tag{5.44}$$

At the quiescent point the gain is

$$A_{vQ} = \frac{1.86}{26} \times 1000 = 70.6$$

Thus

$$\hat{V}_s = \frac{\hat{V}_o}{70.6} = \frac{600}{70.6} \text{ mV} = 8.5 \text{ mV}$$

Note that this is much smaller than the input drive required by the emitter follower for the same output.

Using the above value of $\hat{V}_s$ in (5.41) and (5.43) gives

$$HD_2 = \tfrac{1}{4} \frac{8.5}{26} = 0.082$$

$$HD_3 = \tfrac{1}{24} \left(\frac{8.5}{26}\right)^2 = 0.0045$$

Thus the second-harmonic distortion is 8.2 percent and the third-harmonic distortion is 0.45 percent.

Comparison with the emitter follower can be made by calculating the variation of the small-signal gain of the circuit over the extremes of the signal swing. At the quiescent point the gain was calculated above as 70.6. Since the peak output voltage is 0.6 V and R_L is 1 kΩ, this gives a peak output current swing of $\hat{I}_o =$ 0.6 mA. Thus at the positive signal peak the collector current is

$$I_Q + \hat{I}_o = (1.86 + 0.6) \text{ mA} = 2.46 \text{ mA}$$

At this current, (5.44) gives for the small-signal gain

$$A_v' = \frac{2.46}{26} \times 1000 = 94.6$$

This represents an increase of 34 percent over the quiescent value.

At the negative signal peak the collector current is

$$I_Q - \hat{I}_o = 1.86 - 0.6 = 1.26 \text{ mA}$$

At this current (5.44) gives for the small-signal gain

$$A_v'' = \frac{1.26}{26} \times 1000 = 48.5$$

This represents a decrease of 31 percent compared with the quiescent value.

Note that the variation of the small-signal gain in the case of the emitter follower was about $\frac{1}{60}$ of the variation in the common-emitter stage for the same output signal amplitude. Further calculation shows that corresponding to this fact, the distortion in the emitter follower is about $\frac{1}{60}$ of the distortion in the common-emitter circuit.

5.4 THE COMMON-BASE OUTPUT STAGE

The common-base configuration is only used as an output stage in special circumstances. Its disadvantage is that its current gain is close to unity and thus its driver stage must supply the same current as the output stage. However it does have voltage gain (and thus power gain since the current gain is unity) and the driver experiences less voltage swing than the output stage. The main advantage of the common-base stage is that the breakdown voltage of the transistor in this configuration is BV_{CBO} and this is higher than the breakdown voltage of the other configurations (typically by a factor of about 2). A second advantage (described further in Chapter 7) is that the common-base stage has good high-frequency performance when driving large load impedances.

An application that requires the characteristics listed above is in an oscilloscope vertical amplifier where high-frequency performance is important and voltage

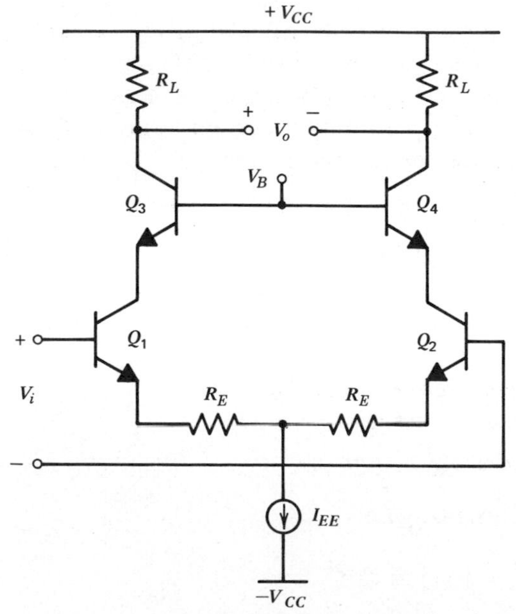

Fig. 5.10 Class A output stage using common-base output transistors.

swings are large. A simplified circuit of a monolithic vertical amplifier output stage is shown in Fig. 5.10. The common-base output devices, Q_3 and Q_4, drive balanced loads R_L. The driver stage is a common-emitter pair Q_1 and Q_2, which handles the same current as the output stage. The high breakdown voltage of the common-base output means that a larger collector-base bias voltage can be used in this circuit compared with previously considered output stages. Thus a larger output voltage swing can be realized before clipping occurs.

In this configuration the common-base stages are current fed and thus contribute little distortion since their current gain α_F is close to unity and almost independent of bias. The major source of distortion is now the common-emitter drivers, and emitter-degeneration resistors R_E are included to help linearize the driver transfer characteristic as described in Chapter 3.

If power dissipated in the driver stage is neglected, it can be shown that the maximum efficiency of the Class A common-base output stage is 25 percent as in the case of the common-emitter and emitter-follower configurations. The common-base stage, like the common-emitter stage, has a high output impedance unless modified by the use of negative feedback.

5.5 CLASS B (PUSH-PULL) OUTPUT STAGES[2,3]

The major disadvantage of Class A output stages is that large power dissipation occurs even for no ac input. In many applications of power amplifiers the circuit may spend long periods of time in a "standby" condition with no input signal, or with intermittent inputs as in the case of voice signals. Power dissipated in these standby periods is wasted, and this is important for two reasons. First, in battery-operated equipment, supply power must be conserved to extend the battery life. Second, any power wasted in the circuit is dissipated in the active devices, which means that they operate at higher temperatures and thus have a greater chance of failure. The power dissipated in the devices affects the physical size of device required, and larger devices are more expensive in terms of silicon area.

A Class B output stage alleviates this problem by having essentially zero power dissipation with zero input signal. *Two* active devices are used to deliver the power instead of one, and each device conducts for alternate half cycles. This is the origin of the name *push-pull*. Another advantage of Class B output stages is that the efficiency is much higher than for a Class A output stage (ideally 78.6 percent at full output power).

A typical integrated-circuit realization of the Class B output stage is shown in Fig. 5.11. This uses both *pnp* and *npn* devices and is known as a *complementary* output stage. The *pnp* transistor is usually a substrate *pnp*. Note that the load resistance, R_L, is connected to the emitters of the active devices and thus the devices act as emitter followers.

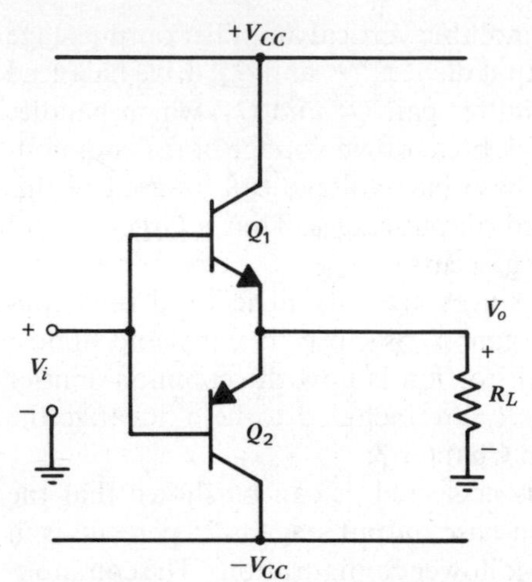

Fig. 5.11 Simple integrated-circuit Class B output stage.

5.5.1 Transfer Characteristic of the Class B Stage

The transfer characteristic of the circuit of Fig. 5.11 is shown in Fig. 5.12. For V_i equal to zero, V_o is also zero and both devices are off with $V_{be} = 0$. As V_i is made positive, the base-emitter voltage of Q_1 increases until it reaches the value $V_{BE(\text{on})}$

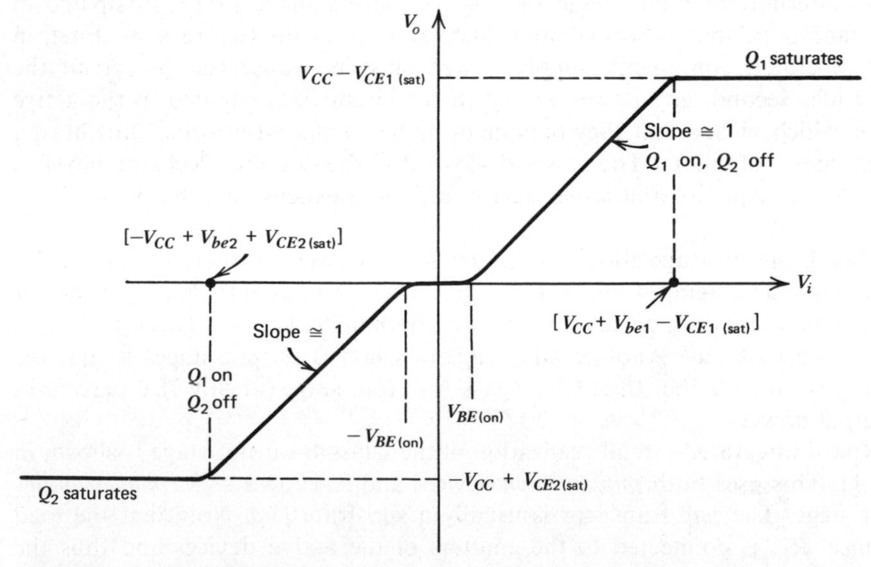

Fig. 5.12 Transfer characteristic of the Class B output stage of Fig. 5.11.

when appreciable current will start to flow in Q_1. At this point V_o is still approximately zero, but further increases in V_i will cause similar increases in V_o as Q_1 acts as an emitter follower. Note that Q_2 is off with a *reverse* bias of $V_{BE(on)}$ across its base-emitter junction. As V_i is made even more positive, Q_1 eventually saturates [for $V_i = V_{CC} + V_{be1} - V_{CE1(sat)}$] and the characteristic flattens out as for the conventional emitter follower considered earlier.

As V_i is taken negative from $V_i = 0$, a similar characteristic is obtained except that Q_2 now acts as an emitter follower for V_i more negative than $-V_{BE(on)}$. In this region Q_1 is held in the off condition with a reverse bias of $V_{BE(on)}$ across its base-emitter junction.

The characteristic of Fig. 5.12 shows a notch (or "deadband") of $2V_{BE(on)}$ in V_i centered around $V_i = 0$. This is common in Class B output stages and gives rise to *crossover* distortion. This is illustrated in Fig. 5.13 where the output waveform from the circuit is shown for various amplitude input sinusoidal signals. It is apparent that in this circuit the distortion is high for small input signals with amplitudes somewhat larger than $V_{BE(on)}$. This source of distortion diminishes as the input signal becomes larger and the deadband represents a smaller fraction of the signal amplitude. Eventually, for very large signals, saturation of Q_1 and Q_2 occurs and distortion rises sharply again due to clipping. This behavior is characteristic of Class B output stages and is why distortion figures are often quoted for both low and high output power operation.

The crosssover distortion described above can be reduced by using *Class AB* operation of the circuit. In this scheme the active devices are biased so that each conducts a small quiescent current for $V_i = 0$. This can be achieved as shown in Fig. 5.14 where the current source, I_Q, forces bias current in diodes Q_3 and Q_4. Since the diodes are connected in parallel with the base-emitter junctions of Q_1 and Q_2, the output transistors are biased on with a current that is dependent on the area ratios of Q_1, Q_2, Q_3, and Q_4. A typical transfer characteristic for this circuit is shown in Fig. 5.15 and the deadband has been effectively eliminated. The remaining nonlinearities due to crossover in conduction from Q_1 to Q_2 can be reduced by using negative feedback as described in Chapter 8.

The operation of the circuit of Fig. 5.14 is quite similar to that of Fig. 5.11. As V_i is taken negative from its quiescent value, emitter follower Q_2 forces V_o to follow. The load current flows through Q_2 whose base-emitter voltage will increase slightly. Since the diodes maintain a constant total bias voltage across the base-emitter junctions of Q_1 and Q_2, the base-emitter voltage of Q_1 will decrease by the same amount that Q_2 increased. Thus during the negative output voltage excursion, Q_1 stays on but conducts little current and plays no part in delivering output power. For V_i taken positive, the opposite occurs and Q_1 acts as the emitter follower delivering current to R_L with Q_2 conducting only a very small current. In this case, the current source, I_Q, supplies the base-current drive to Q_1.

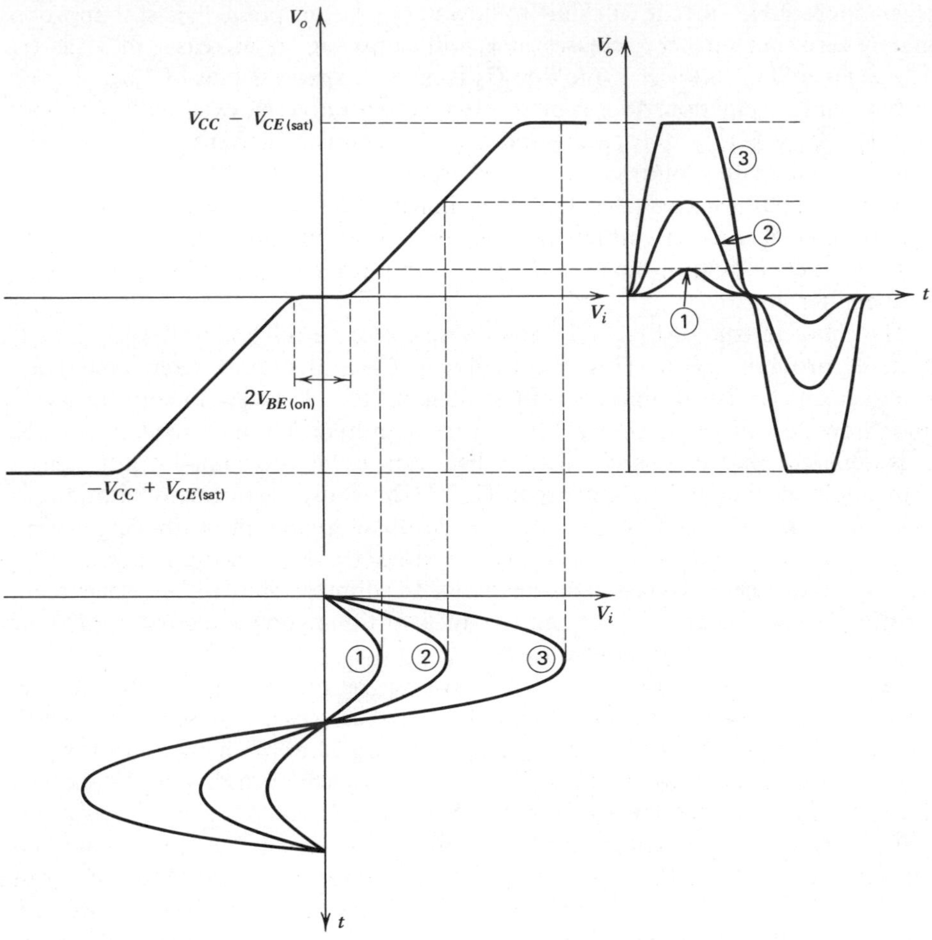

Fig. 5.13 Output waveforms for various amplitude input signals applied to the Class B circuit of Fig. 5.11.

In the derivation of the characteristics of the circuits of Fig. 5.11 and Fig. 5.14 we assumed that there was no limitation on the magnitude of the input voltage, V_i. In the characteristic of Fig. 5.12 the magnitude of V_i required to cause saturation of Q_1 or Q_2 exceeds the supply voltage, V_{CC}. However, as in the case of the single emitter follower described earlier, practical driver stages generally cannot produce values of V_i exceeding V_{CC} if they are connected to the same supply voltages as the output stage. For example, the current source, I_Q, in Fig. 5.14 is usually realized with a *pnp* transistor and thus the voltage at the base of Q_1 cannot exceed $[V_{CC} - V_{CE(\text{sat})}]$ at which point saturation of the current source occurs. Conse-

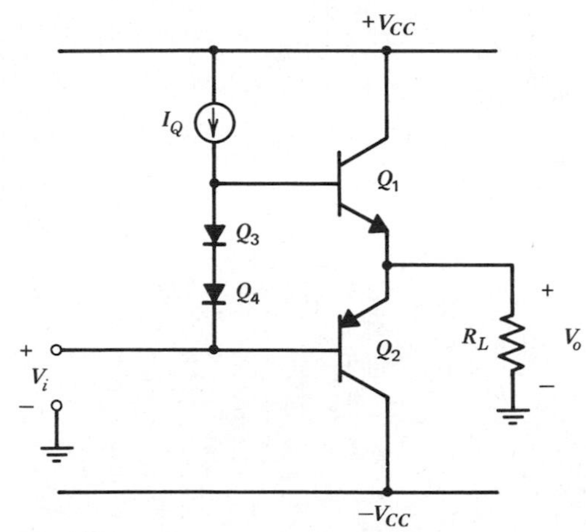

Fig. 5.14 Class B output stage with diodes to reduce crossover distortion.

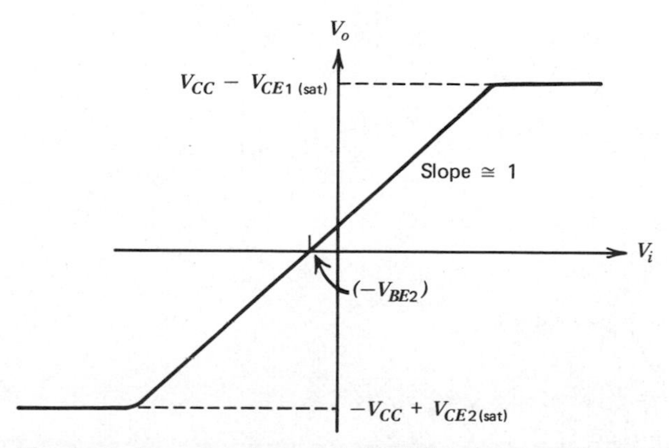

Fig. 5.15 Transfer characteristic of the circuit of Fig. 5.14.

quently, the positive and negative limits of V_o where clipping occurs are generally somewhat less than shown in Fig. 5.12 and Fig. 5.15, and the limitation usually occurs in the driver stage. This point will be investigated further when practical output stages are considered in later sections.

5.5.2 Power Output and Efficiency of the Class B Stage

The method of operation of a Class B stage can be further appreciated by plotting the collector current waveforms in the two devices, and this is done in Fig. 5.16. Note that each transistor conducts current to R_L for half a cycle.

The collector current waveforms of Fig. 5.16 also represent the waveforms of the current drawn from the two supplies. If the waveforms are assumed to be

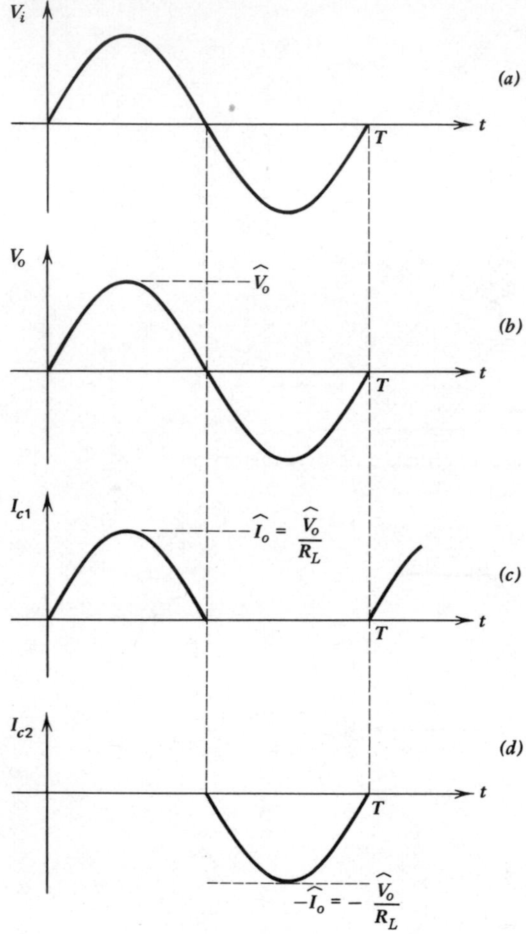

Fig. 5.16 Voltage and current waveforms for a Class B output stage. (*a*) Input voltage. (*b*) Output voltage. (*c*) First output device collector current. (*d*) Second output device collector current.

half-sinusoids then the average current drawn from each supply is

$$I_{\text{supply}} = \frac{1}{T}\int_0^T I_c(t)\,dt$$

where T is the period of the input signal. Integration of the waveform of Fig. 5.16*c* gives

$$I_{\text{supply}} = \frac{1}{\pi}\hat{I}_o\Big|_{\text{peak}} = \frac{1}{\pi}\frac{V_o}{R_L} \tag{5.45}$$

where $\hat{V}_o$ and $\hat{I}_o$ are the amplitudes (zero to peak) of the output sinusoidal voltage and current. Since each supply delivers the same current magnitude, the total average power drawn from the two supplies is

$$P_{\text{supply}} = 2V_{CC}I_{\text{supply}} \tag{5.46}$$

$$= \frac{2}{\pi}\frac{V_{CC}}{R_L}\hat{V}_o \tag{5.47}$$

where (5.45) has been substituted. Note that in the case of the Class B stage, the average power drawn from the supplies *does vary* with signal level, and is directly proportional to $\hat{V}_o$.

The average power delivered to R_L is given by

$$P_L = \frac{1}{2}\frac{\hat{V}_o^2}{R_L} \tag{5.48}$$

and, from the definition of circuit efficiency, we have

$$\eta_C = \frac{P_L}{P_{\text{supply}}} = \frac{\pi}{4}\frac{\hat{V}_o}{V_{CC}} \tag{5.49}$$

where (5.47) and (5.48) have been substituted. Equation 5.49 shows that η for a Class B stage is independent of R_L but increases linearly as the output voltage amplitude, $\hat{V}_o$, increases.

The maximum value that $\hat{V}_o$ can attain before clipping occurs with the characteristic of Fig. 5.15 is $\hat{V}_{om} = [V_{CC} - V_{CE(\text{sat})}]$ and thus the maximum average signal power that can be delivered to R_L for sinusoidal signals can be calculated from (5.48) as

$$P_L\big|_{\max} = \frac{1}{2}\frac{[V_{CC} - V_{CE(\text{sat})}]^2}{R_L} \tag{5.50}$$

The corresponding maximum efficiency from (5.49) is

$$\eta_{\max} = \frac{\pi}{4}\frac{V_{CC} - V_{CE(\text{sat})}}{V_{CC}} \tag{5.51}$$

If $V_{CE(\text{sat})}$ is small compared with V_{CC}, the circuit has a maximum efficiency of 0.786 or 78.6 percent. This maximum efficiency is much higher than the value of 25 percent achieved in Class A circuits and in addition the standby power dissipation is essentially zero in the Class B circuit. These are the reasons for the widespread use of Class B output stages.

The load line for one device in a Class B stage is shown in Fig. 5.17. For values of V_{ce} less than the quiescent value (which is V_{CC}), the load line has a slope of $-(1/R_L)$. For values of V_{ce} greater than V_{CC}, the load line lies along the V_{ce} axis.

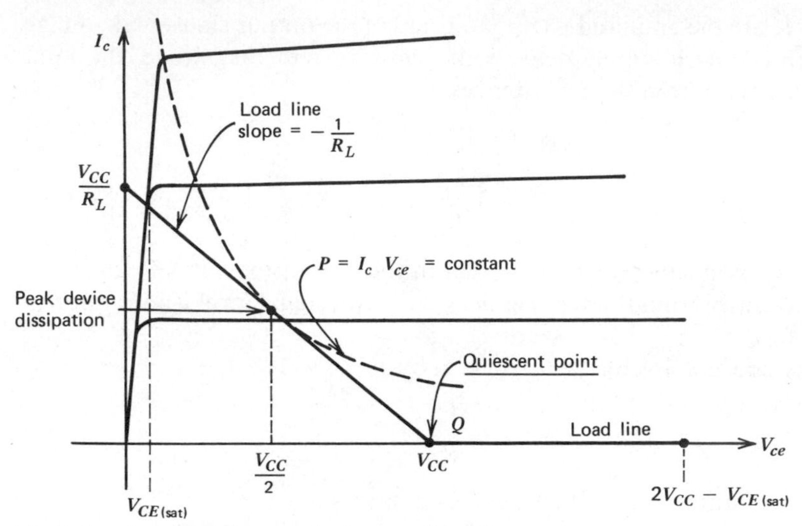

Fig. 5.17 Load line for one device in a Class B stage.

This is due to the fact that the other device is then conducting and the V_{ce} of the device under consideration increases while its collector current is zero. The maximum value of V_{ce} is $[2V_{CC} - V_{CE(sat)}]$. As in the case of a Class A stage, a geometrical interpretation of the average power, P_L, delivered to R_L can be obtained by noting that $P_L = \frac{1}{2}\hat{I}_o\hat{V}_o$ where $\hat{I}_o$ and $\hat{V}_o$ are the peak sinusoidal current and voltage delivered to R_L. Thus P_L is the area of the triangle in Fig. 5.17 between the V_{ce} axis and the portion of the load line traversed by the operating point.

Consider the instantaneous power dissipated in one device. This is

$$P_c = V_{ce}I_c \tag{5.52}$$

But

$$V_{ce} = V_{CC} - I_cR_L \tag{5.53}$$

Substitution of (5.53) in (5.52) gives

$$P_c = I_c(V_{CC} - I_cR_L) = I_cV_{CC} - I_c^2R_L \tag{5.54}$$

Differentiation of (5.54) shows that P_c reaches a peak for

$$I_c = \frac{V_{CC}}{2R_L} \tag{5.55}$$

This on the load line midway between the I_c and V_{ce} axis intercepts and agrees with the result derived earlier for the Class A stage. As in that case, the load line in Fig. 5.17 is tangent to a power hyperbola at the point of peak dissipation. Thus, in a Class B stage, maximum instantaneous device dissipation occurs for a voltage equal to about half the maximum swing. Since the quiescent device power dissipation is zero, a Class B device always operates at a *higher* temperature when signal is applied.

The instantaneous device power dissipation as a function of time is shown in Fig. 5.18 where collector current, collector-emitter voltage and their product is displayed for one device in a Class B stage at maximum output [$V_{CE(\text{sat})}$ is assumed zero]. When the device is conducting, the power dissipation varies at twice the signal frequency. The device power dissipation is zero for the half cycle when the device is cut off. As in the case of the Class A stage, the load line in Fig. 5.17 becomes vertical through the quiescent point for a short-circuited load and the instantaneous device power dissipation can then become excessive. Methods of protection against such a possibility are described in Section 5.5.6. Note that for an open-circuited load the load line in Fig. 5.17 lies along the V_{ce} axis and the device has zero dissipation.

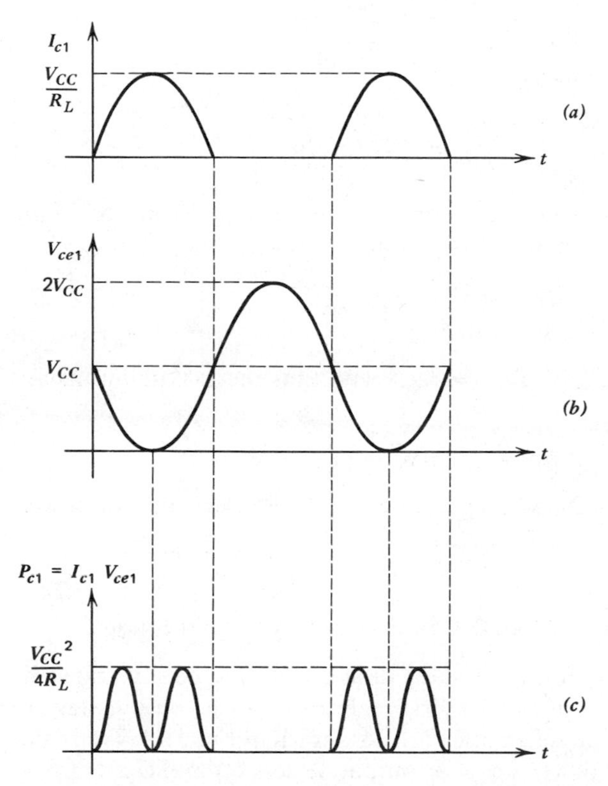

Fig. 5.18 Waveforms at maximum output for one device in a Class B stage. (*a*) Collector current waveform. (*b*) Collector voltage waveform. (*c*) Collector power dissipation waveform.

EXAMPLE

A Class B stage of the type shown in Fig. 5.11 drives a load $R_L = 500\ \Omega$. If the positive and negative supplies have magnitudes of 15 V, calculate the maximum average power that is delivered to R_L for $\hat{V}_o = 14.4$ V, the corresponding efficiency and the maximum instantaneous device dissipation. Assume that V_o is sinusoidal.

From (5.45) the average supply current is

$$I_{\text{supply}} = \frac{1}{\pi}\frac{\hat{V}_o}{R_L} = \frac{1}{\pi}\frac{14.4}{500} = 9.17 \text{ mA}$$

Use of (5.46) gives the average power drawn from the supplies as

$$P_{\text{supply}} = I_{\text{supply}} \times 2V_{CC} = 9.17 \times 30 \text{ mW} = 275 \text{ mW}$$

The average power delivered to R_L is

$$P_L = \tfrac{1}{2}\frac{\hat{V}_o^2}{R_L} = \tfrac{1}{2}\frac{14.4^2}{500} = 207 \text{ mW}$$

The corresponding efficiency is (from (5.13))

$$\eta_C = \frac{P_L}{P_{\text{supply}}} = \frac{207}{275} = 75.3 \text{ percent}$$

and this is close to the theoretical maximum of 78.6 percent. From (5.55) the maximum instantaneous device power dissipation occurs when

$$I_c = \frac{V_{CC}}{2R_L} = \frac{15}{1000} = 15 \text{ mA}$$

The corresponding value of V_{ce} is $V_{CC}/2 = 7.5$ V and thus the maximum instantaneous device dissipation is

$$P_c = I_c V_{ce} = 15 \times 7.5 \text{ mW} = 112.5 \text{ mW}$$

Note that the *average* power dissipated per device is (by conservation of power)

$$P_{\text{av}} = \tfrac{1}{2}(P_{\text{supply}} - P_L) = \tfrac{1}{2}(275 - 207) \text{ mW} = 34 \text{ mW}$$

5.5.3 Practical Realizations of Class B Complementary Output Stages[4]

The practical aspects of Class B output stage design will now be illustrated by considering two examples. One of the simplest realizations is the output stage of the 709 op amp, and a simplified schematic of this is shown in Fig. 5.19. Transistor Q_3 acts as a common-emitter driver stage for output devices Q_1 and Q_2.

The transfer characteristic of this stage can be calculated as follows. In the quiescent condition, $V_o = 0$ and $V_1 = 0$. Since Q_1 and Q_2 are then off, there is

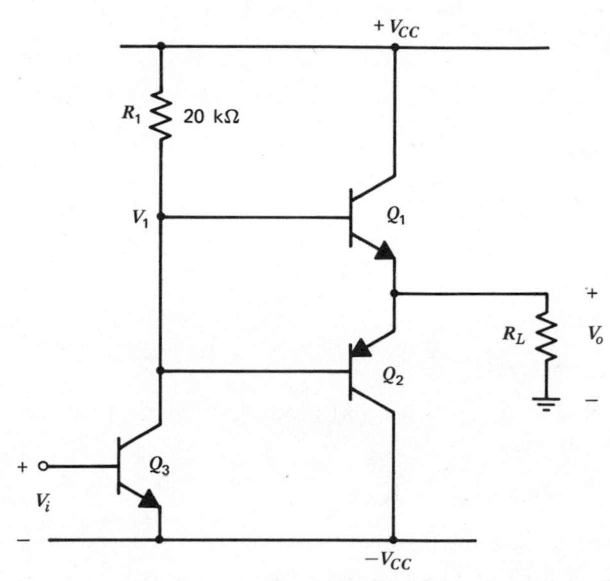

Fig. 5.19 Simplified schematic of the output stage of the 709 op amp.

no base current in these devices and the bias current in Q_3 is

$$I_{C3} = \frac{V_{CC}}{R_1} = \frac{V_{CC}}{20{,}000}$$

If $V_{CC} = 10$ V, this gives

$$I_{C3} = 0.50 \text{ mA}$$

The maximum values that V_o can take are determined by the *driver stage.* When V_i is taken large positive, V_1 decreases until Q_3 saturates, at which point the negative voltage limit V_o^- is reached.

$$V_o^- = -V_{CC} + V_{CE3(\text{sat})} - V_{be2} \tag{5.56}$$

For values of V_1 between $[-V_{BE(\text{on})}]$ and $[-V_{CC} + V_{CE3(\text{sat})}]$ both Q_3 and Q_2 are in the forward-active region and V_o follows V_1 with Q_2 acting as an emitter follower.

As V_i is taken negative, the current in Q_3 decreases and V_1 rises, turning Q_1 on. The positive voltage limit V_o^+ limit is reached when Q_3 cuts off and the base of Q_1 is simply fed from the positive supply via R_1. Then

$$V_{CC} = I_{b1}R_1 + V_{be1} + V_o^+ \tag{5.57}$$

If β_1 is large then

$$V_o^+ = I_{c1}R_L = \beta_1 I_{b1} R_L$$

where β_1 is the current gain of Q_1. Thus

$$I_{b1} = \frac{V_o^+}{\beta_1 R_L} \tag{5.58}$$

Substitution of (5.58) in (5.57) gives

$$V_o^+ = \frac{V_{CC} - V_{be1}}{1 + \dfrac{R_1}{\beta_1 R_L}} \tag{5.59}$$

For $R_L = 10\ \text{k}\Omega$ and $\beta_1 = 100$, (5.59) gives

$$V_o^+ = 0.98(V_{CC} - V_{be1})$$

In this case the limit on V_o is similar for positive and negative swings. However if $R_L = 1\ \text{k}\Omega$ and $\beta_1 = 100$, (5.59) gives

$$V_o^+ = 0.83(V_{CC} - V_{be1})$$

For this lower value of R_L, the maximum positive value of V_o is reduced and clipping on a sine wave will occur first for V_o going positive.

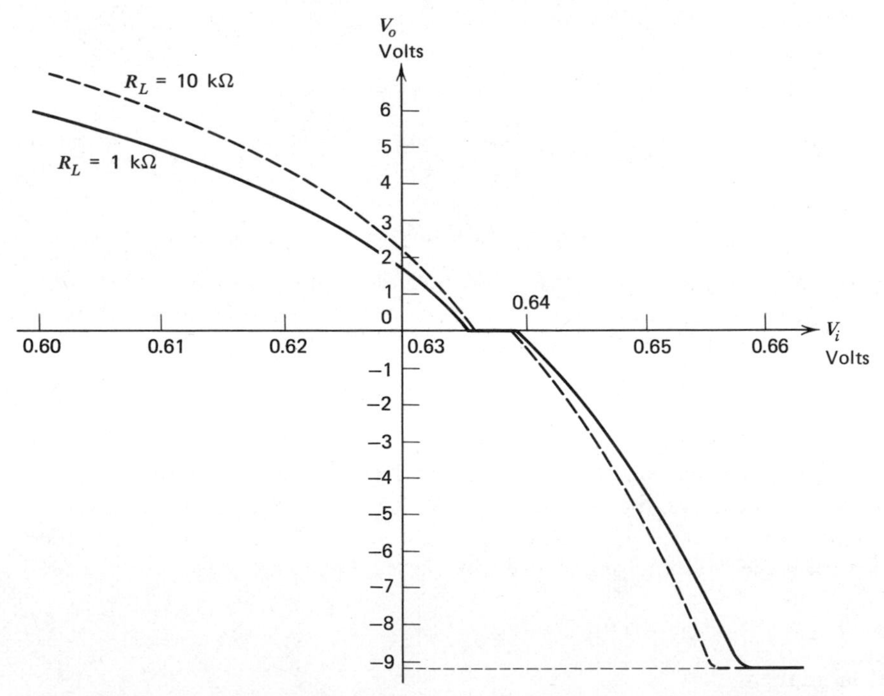

Fig. 5.20 Computer-generated transfer characteristic for the circuit of Fig. 5.19 with $V_{CC} = 10$ V and $R_L = 1\ \text{k}\Omega$ and $10\ \text{k}\Omega$.

Computer-generated transfer curves for this circuit with $V_{CC} = 10$ V are shown in Fig. 5.20 for $R_L = 1$ kΩ and $R_L = 10$ kΩ. ($\beta = 100$ is assumed for all devices.) The reduced positive voltage capability for $R_L = 1$ kΩ is apparent, as is the deadband present in the transfer characteristic. The curvature in the characteristic is due to the exponential nonlinearity of the driver, Q_3. In practice, the transfer characteristic may be even more nonlinear since β for the *npn* transistor, Q_1, is generally larger than β for the *pnp* transistor, Q_2, causing the positive and negative sections of the characteristic to differ significantly. This can be seen by calculating the small-signal gain $\Delta V_o/\Delta V_i$ for positive and negative V_o. In the actual 709 integrated circuit, negative feedback is applied around this output stage to reduce these nonlinearities in the transfer characteristic.

A second example of a practical Class B output stage is shown in Fig. 5.21 where computer-calculated bias currents are included. This is a simplified schematic of the 741 op amp output circuitry. The output devices, Q_{14} and Q_{20}, are biased to a collector current of about 0.17 mA by the diodes, Q_{18} and Q_{19}. The value of the bias current in Q_{14} and Q_{20} depends on the effective area ratio between

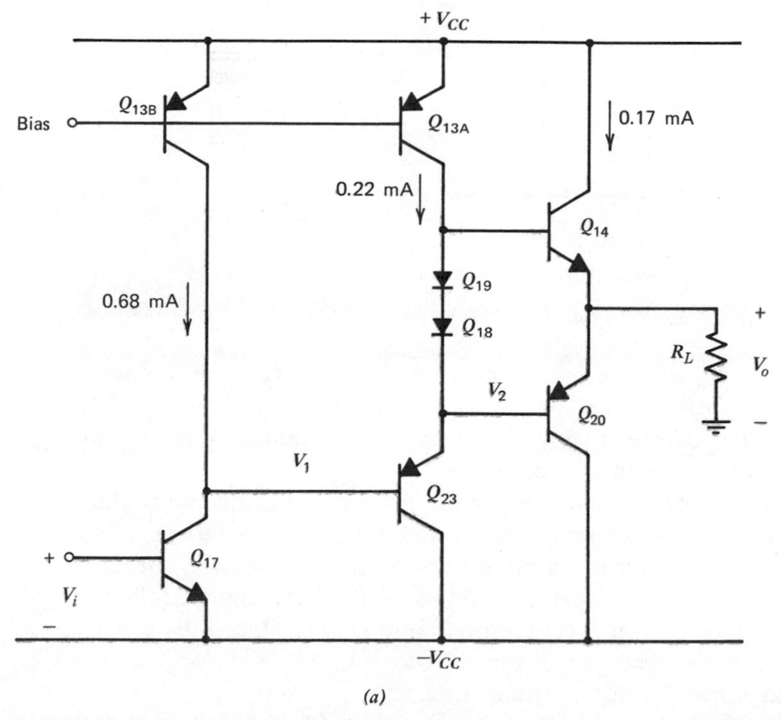

Fig. 5.21 (*a*) Simplified schematic of the 741 op amp output stage. (*b*) Schematic of the 741 output stage showing the detail of Q_{18} and Q_{19}.

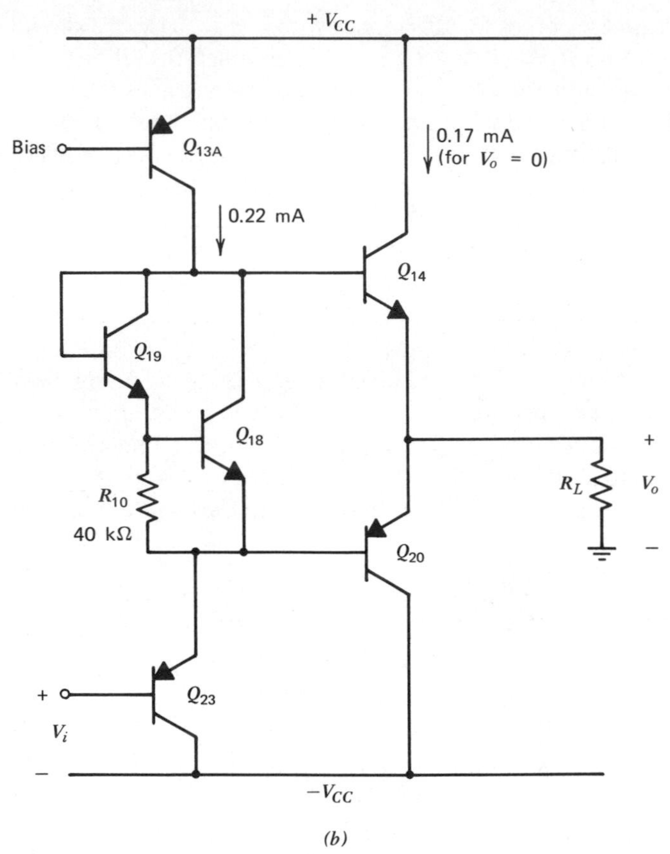

Fig. 5.21 (*Continued*)

diodes Q_{18} and Q_{19} and the output devices. (Q_{18} and Q_{19} are, in fact, diode-connected transistors and are considered later.) The output stage is driven by lateral *pnp* emitter-follower Q_{23}, which is driven by common-emitter stage Q_{17} biased to 0.68 mA by current source Q_{13B}.

The diodes in Fig. 5.21 essentially eliminate crossover distortion in the circuit and this can be seen in the computer-generated transfer characteristic of Fig. 5.22. The linearity of this stage is further improved by the fact that the output devices are driven from a low-impedance source provided by the emitter follower, Q_{23}. Consequently, differences in β between Q_{14} and Q_{20} produce little effect on the transfer characteristic since small-signal gain $\Delta V_o/\Delta V_1 \simeq 1$ for *any* practical value of β with either Q_{14} or Q_{20} conducting.

The limits on the output voltage swing shown in Fig. 5.22 can be determined as follows. As V_i is taken positive, the voltage V_1 at the base of Q_{23} goes negative

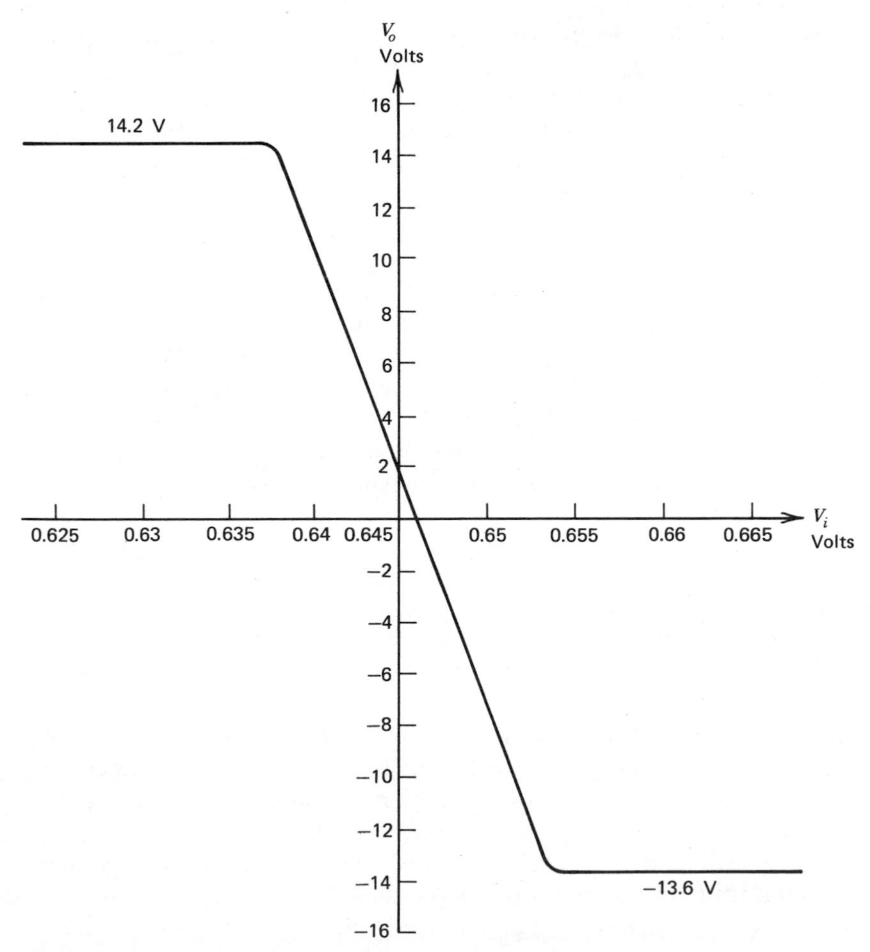

Fig. 5.22 Computer-generated transfer curve for the circuit of Fig. 5.21*a* with $V_{CC} = 15$ V and $R_L = 1$ kΩ.

and voltages V_2 and V_o follow with Q_{20} drawing current from R_L. When Q_{17} saturates, the output voltage limit for negative excursions is reached at

$$V_o^- = -V_{CC} + V_{CE17(\text{sat})} - V_{be23} - V_{be20} \tag{5.60}$$

This is about 1.4 V more positive than the negative supply and V_o^- is thus limited by saturation in Q_{17}, which is the stage *preceding* driver stage Q_{23}.

As V_i is taken negative from its quiescent value (where $V_o = 0$) voltage V_1 rises and voltages V_2 and V_o follow with Q_{14} delivering current to the load. The positive output voltage limit V_o^+ is reached when current source Q_{13A} saturates and

$$V_o^+ = V_{CC} - V_{CE13A(\text{sat})} - V_{be14} \tag{5.61}$$

This is about 0.8 V below the positive supply and V_o^+ is limited by the driver stage.

The power requirements of the driver circuits in a configuration such as shown in Fig. 5.21 require some consideration. The basic requirement of the driver is to supply sufficient drive to the output stage so that it can supply the desired power to R_L. As V_o is taken negative, Q_{23} draws current from the base of Q_{20} with essentially no limit. In fact, the circuit must be protected in case of a short-circuited load, in which case a large input signal could cause Q_{23} and Q_{20} to conduct such heavy currents that they burn out. As explained before, the negative voltage limit is reached when Q_{17} saturates and can no longer drive the base of Q_{23} negative.

As V_o is taken positive (by V_i going negative and V_1 going positive) Q_{23} conducts less and current source Q_{13A} supplies base current to Q_{14}. The maximum output current is limited by the current of 0.22 mA available for driving Q_{14}. As V_1, V_2 and V_o go positive, the current in Q_{14} increases, and the current in Q_{13A} is progressively diverted to the base of Q_{14}. The maximum possible output current delivered by Q_{14} is thus

$$I_o = \beta_{14} \times 0.22 \text{ mA}$$

If $\beta_{14} = 100$ this gives a limit of 22 mA. The driver stage may thus limit the maximum positive current available from the output stage. However this output current level is only reached if R_L is small enough so that Q_{13A} does not saturate on the positive voltage excursion.

The stage preceding the driver in this circuit is Q_{17} and it was mentioned above that the negative voltage limit of V_o is reached when Q_{17} saturates. The bias current of 0.68 mA in Q_{17} is much greater than the base current of Q_{23}, and Q_{23} thus produces very little loading on Q_{17}. Consequently, voltage V_1 at the base of Q_{23} can be driven to within $V_{CE(\text{sat})}$ of either supply voltage with only a very small fractional change in the collector current of Q_{17}.

Finally, it is of some interest to examine the detail of the fabrication of diodes Q_{18} and Q_{19} in the 741. The actual circuit is shown in Fig. 5.21*b* with the output protection circuitry omitted. It can be seen that diode Q_{19} only conducts a current equal to the base current of Q_{18} plus the bleed current in pinch resistor R_{10}. Transistor Q_{18} thus conducts most of the bias current of current source Q_{13A}. There are two reasons for using this arrangement. First, the basic aim of achieving a voltage drop equal to two base-emitter voltages is achieved, but since Q_{18} and Q_{19} have common collectors they can be placed in the same isolation region. Second, since Q_{19} conducts only a small current, the bias voltage produced by Q_{18} and Q_{19} across the bases of Q_{14} and Q_{20} is less than would result from a connection as shown in Fig. 5.21*a*. This is important because output transistors Q_{14} and Q_{20} generally have emitter areas larger than the standard device geometry (typically four times larger or more) so that they can maintain high β_F while conducting large output currents. Thus in the circuit of Fig. 5.21*a* the bias current in Q_{14} and Q_{20} would be about four times the current in Q_{18} and Q_{19} and this would be excessive in a 741-type circuit. The circuit of Fig. 5.21*b* however can be designed to bias Q_{14} and Q_{20} to a current comparable to the current in the diodes, even though the output devices have a large area. The basic reason for this is that the small bias current in Q_{19} in Fig. 5.21*b* gives it a smaller base-emitter voltage than the same device in Fig. 5.21*a*, and thus the total bias voltage between the bases of Q_{14} and Q_{20} is reduced.

The results described above can be illustrated quantitatively by calculating the bias currents in Q_{14} and Q_{20} of Fig. 5.21*b*. We have

$$V_{BE19} + V_{BE18} = V_{BE14} + |V_{BE20}|$$

and thus

$$V_T \ln \frac{I_{C19}}{I_{S19}} + V_T \ln \frac{I_{C18}}{I_{S18}} = V_T \ln \frac{I_{C14}}{I_{S14}} + V_T \ln \left|\frac{I_{C20}}{I_{S20}}\right| \tag{5.62}$$

If we assume that the circuit is biased for $V_o = 0$ V and also that $\beta_{14} \gg 1$ and $\beta_{20} \gg 1$, then $|I_{C14}| = |I_{C20}|$ and (5.62) becomes

$$\frac{I_{C19} I_{C18}}{I_{S18} I_{S19}} = \frac{I_{C14}^2}{I_{S14} I_{S20}}$$

from which

$$I_{C14} = -I_{C20} = \sqrt{I_{C19} I_{C18}} \sqrt{\frac{I_{S14} I_{S20}}{I_{S18} I_{S19}}} \tag{5.63}$$

Equation 5.63 may be used to calculate the output bias current in circuits of the type shown in Fig. 5.21*b*. Note that the output-stage bias current from (5.63) varies as $\sqrt{I_{C19}}$. For this specific example, the collector current in Q_{19} is approximately equal to the current in R_{10} if β_F is large and thus

$$I_{C19} \simeq \frac{V_{BE18}}{R_{10}} \simeq \frac{0.6}{40} \text{ mA} = 15\ \mu\text{A}$$

If the base currents of Q_{14} and Q_{20} are neglected, the collector current of Q_{18} is

$$I_{C18} \simeq |I_{C13A}| - I_{C19} = (220 - 15)\ \mu\text{A} = 205\ \mu\text{A}$$

In order to calculate the output-stage bias currents from (5.63), values for the various reverse saturation currents are required. These values depend on the particular IC process used, but typical values are $I_{S18} = I_{S19} = 2 \times 10^{-15}$ A, $I_{S14} = 4I_{S18} = 8 \times 10^{-15}$ A, $I_{S20} = 4 \times 10^{-15}$ A. Substitution of this data in (5.63) gives $I_{C14} = -I_{C20} = 0.16$ mA, which is close to the computed value.

EXAMPLE

For the output stage of Fig. 5.21*a* calculate bias currents in all devices for $V_o = +10$ V. Assume that $V_{CC} = 15$ V, $R_L = 2$ kΩ, and $\beta = 100$. For simplicity assume all devices of equal area and for each device

$$|I_c| = 10^{-14} \exp \left| \frac{V_{be}}{V_T} \right| \tag{5.64}$$

Assuming Q_{14} supplies the load current for positive output voltages, we have

$$I_{c14} = \frac{10}{2} = 5 \text{ mA}$$

Substitution in (5.64) gives

$$V_{be14} = 700 \text{ mV}$$

also

$$I_{b14} = \frac{I_{c14}}{\beta_{14}} = \frac{5}{100} = 0.05 \text{ mA}$$

Thus

$$I_{c19} \simeq I_{c18} \simeq -I_{c23} = 0.22 - 0.05 = 0.17 \text{ mA}$$

Substitution in (5.64) gives

$$V_{be19} = V_{be18} = -V_{be23} = 613 \text{ mV}$$

Thus

$$V_{be20} = -(V_{be19} + V_{be18} - V_{be14}) = -525 \text{ mV}$$

Use of (5.64) gives

$$I_{c20} = -5.9\ \mu\text{A}$$

and the collector current in Q_{20} is quite small as predicted. Finally

$$I_{c17} = 0.68 \text{ mA} - \frac{I_{c23}}{\beta_{23}} = \left(0.68 - \frac{0.17}{100}\right) \text{mA} = 0.68 \text{ mA}$$

also

$$V_2 = V_o - V_{be20} = 5 - 0.525 = 4.475 \text{ V}$$

and

$$V_1 = V_2 - V_{be23} = 4.475 - 0.613 = 3.862 \text{ V}$$

5.5.4 All-*npn* Class B Output Stage[5,6]

The Class B circuits described above are adequate for many integrated-circuit applications where the output power to be delivered to the load is of the order of several hundred milliwatts or less. However, if output-power levels of several watts or more are required, these circuits are inadequate because the substrate *pnp* transistors used in the output stage have a limited current-carrying capability. This is due to the fact that the doping levels in the emitter, base, and collector of these devices are not optimized for *pnp* structures because the *npn* devices in the circuit have conflicting requirements.

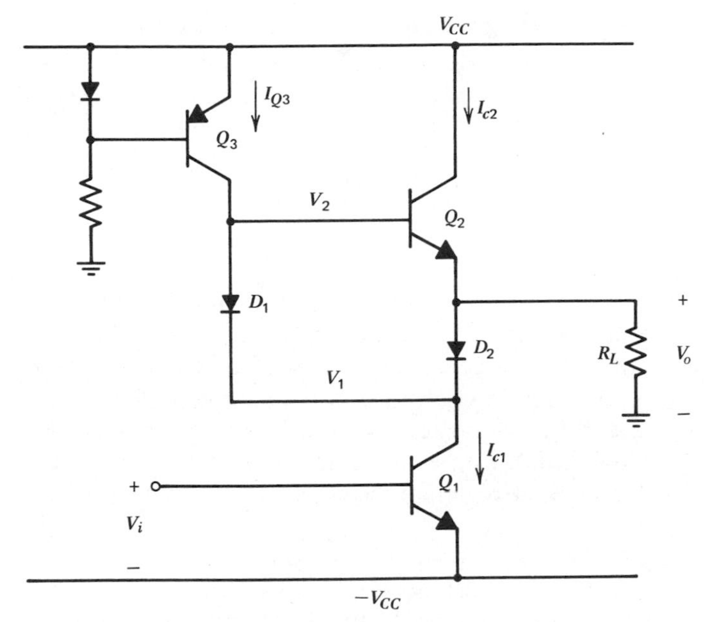

Fig. 5.23 All-*npn* Class B output stage.

A circuit design that uses high-power *npn* transistors in both halves of a Class B configuration is shown in Fig. 5.23. In this circuit, common-emitter transistor Q_1 delivers power to the load during the negative half-cycle, and emitter follower Q_2 delivers power during the positive half-cycle.

In order to examine the operation of this circuit, consider V_i taken negative from its quiescent value so that Q_1 is off and $I_{c1} = 0$. Then diodes D_1 and D_2 must both be off and all of the collector current of Q_3 is delivered to the base of Q_2. The output voltage then has its maximum positive value V_o^+. In general, Q_3 will then be saturated and

$$V_o^+ = V_{CC} - V_{CE3(\text{sat})} - V_{be2} \tag{5.65}$$

If V_o is to be able to attain the maximum positive value given by (5.65), transistor Q_3 must be saturated in this extreme condition. Note that Q_2 in this circuit cannot saturate, as there is no way for the base voltage of Q_2 to exceed the positive supply voltage where the collector of Q_2 is connected. The condition for Q_3 to be saturated is that the nominal collector bias current I_{Q3} in Q_3 (when Q_3 is *not* saturated) should be larger than the base current of Q_2 when $V_o = V_o^+$. Thus we require

$$I_{Q3} > I_{b2} \tag{5.66}$$

Now since Q_2 supplies the current to R_L for V_o positive we have

$$V_o^+ = -I_{e2}R_L = (\beta_2 + 1)I_{b2}R_L \tag{5.67}$$

Substitution of (5.67) and (5.65) in (5.66) gives the requirement on the bias current of Q_3 as

$$I_{Q3} > \frac{V_{CC} - V_{CE3(\text{sat})} - V_{be2}}{(\beta_2 + 1)R_L} \tag{5.68}$$

Equation 5.68 also applies to the circuit of Fig. 5.21*a*. It gives limits on I_{C3}, β_2, and R_L for V_o to be able to swing close to the positive supply. If I_{Q3} is less than the value given by (5.68), V_o will begin clipping at a positive value *less than* that given by (5.65) and Q_3 will never saturate.

Consider V_i now made positive to turn Q_1 on and produce finite I_{c1}. Since the base of Q_2 is more positive than its emitter, diode D_1 will turn on in preference to D_2 and D_2 will be off with zero volts across its junction. The current I_{c1} will flow through D_1 and will be drawn from Q_3, which is assumed saturated. Transistor Q_3 will eventually come out of saturation as I_{c1} increases, and voltage V_2 at the base of Q_2 will then be pulled down. Since Q_2 acts as an emitter follower, V_o will follow V_2 down. This is the positive half of the cycle and Q_1 acts as a driver with Q_2 as the output device.

When V_o is reduced to 0 V, there is no load current and $I_{c2} = 0$. This corresponds to $I_{c1} = I_{C3}$ and all of the bias current in Q_3 passes through D_1 to Q_1. If I_{c1} is increased further, V_o stays constant at 0 V while V_2 is reduced to 0 V also. This means that V_1 is negative by an amount equal to the diode voltage drop of D_1, and thus power diode D_2 is turned on. Since the current in D_1 is essentially fixed by Q_3, further increases in I_{c1} cause current to flow through D_2 and the negative half of the cycle consists of Q_1 acting as the output device and feeding R_L through D_2. The maximum negative voltage occurs when Q_1 saturates and is

$$V_o^- = -V_{CC} + V_{CE1(\text{sat})} + V_{d2} \tag{5.69}$$

where V_{d2} is the forward voltage drop across D_2.

The sequence just described gives rise to a highly nonlinear transfer characteristic as shown in Fig. 5.24 where V_o is plotted as a function of I_{c1} for convenience. When V_o is positive, the current I_{c1} feeds into the base of Q_2 and the small-signal gain is

$$\frac{\Delta V_o}{\Delta I_{c1}} \simeq \frac{\Delta V_2}{\Delta I_{c1}} = r_{o1} \| r_{o3} \| (r_{\pi 2} + (\beta_2 + 1)R_L)$$

where the impedance of D_1 is assumed negligible. That is, the impedance at the base of Q_2 is equal to the parallel combination of the output resistances of Q_1 and Q_3 and the input resistance of emitter follower, Q_2.

When V_o in Fig. 5.23 is negative, I_{c1} feeds R_L directly and the small-signal gain is

$$\frac{\Delta V_o}{\Delta I_{c1}} \simeq r_{o1} \| R_L$$

where the impedance of D_2 is assumed negligible.

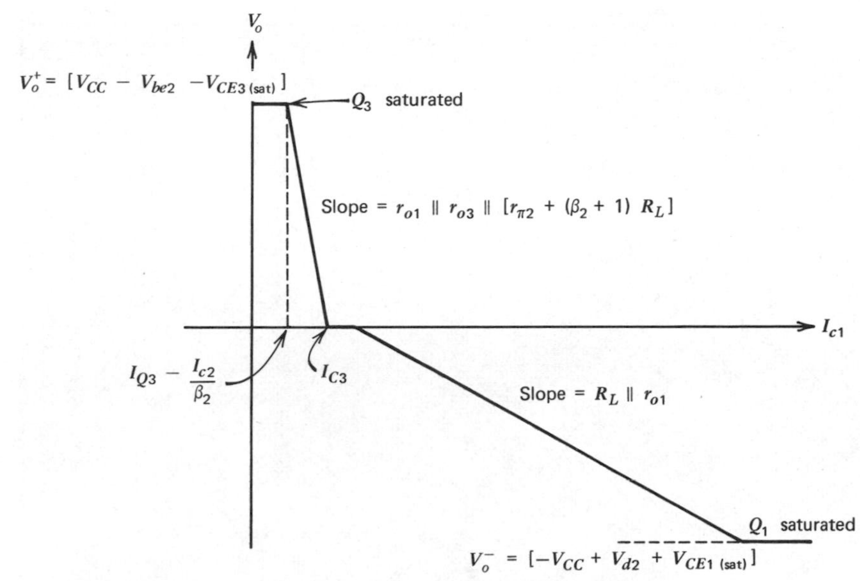

Fig. 5.24 Transfer characteristic of the circuit of Fig. 5.23 from I_{c1} to V_o.

Note the small deadband in Fig. 5.24 where diode D_2 turns on. This can be eliminated by adding a second diode in series with D_1. In practice, negative feedback must be used around this circuit to linearize the transfer characteristic and this will reduce any crossover effects. The transfer characteristic of the circuit from V_i to V_o is even more nonlinear because it includes the exponential nonlinearity of Q_1.

In integrated-circuit fabrication of the circuit of Fig. 5.23, devices Q_1 and Q_2 are identical large power transistors, and in high-power (delivering several watts or more) they may occupy 50 percent of the whole die. Diode D_2 is a large power diode that also occupies considerable area. These features are illustrated in Fig. 5.25, which is a die photo of the 791 high-power op amp. This circuit can dissipate 10 W of power and can deliver 15 W of output power into an 8-Ω load. The large power transistors in the output stage can be seen on the right-hand side of the die.

Finally, the power and efficiency results derived previously for the complementary Class B stage apply equally to the all-*npn* Class B stage if allowance is made for the voltage drop in D_2. Thus the ideal maximum efficiency is 79 percent.

5.5.5 Quasi-Complementary Output Stages[7]

The all-*npn* stage described above is one solution to the problem of the limited power-handling capability of the substrate, *pnp*. Another solution is shown in

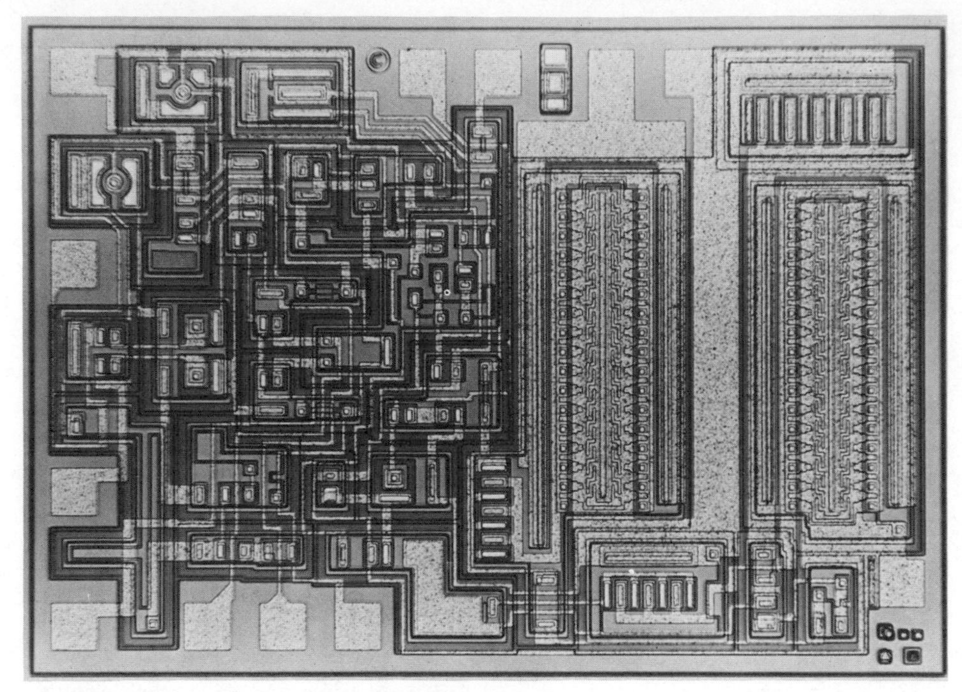

Fig. 5.25 Die photo of the 791 high-power op amp.

Fig. 5.26 where a *composite pnp* has been made from a lateral *pnp* Q_3 and a high-power *npn* transistor Q_4. This is called a quasi-complementary output stage.

The operation of the circuit of Fig. 5.26 is almost identical to that of Fig. 5.21. The pair Q_3–Q_4 is equivalent to a *pnp* transistor as shown in Fig. 5.27, and the collector current of Q_3 is

$$I_{C3} = -I_S \exp\left(-\frac{V_{BE}}{V_T}\right) \tag{5.70}$$

The composite collector current I_C is the emitter current of Q_4, which is

$$I_C = (\beta_4 + 1)I_{C3} = -(\beta_4 + 1)I_S \exp\left(-\frac{V_{BE}}{V_T}\right) \tag{5.71}$$

The composite device thus shows the standard relationship between I_C and V_{BE} for a *pnp* transistor. However most of the current is carried by the high-power *npn* transistor. Note that the saturation voltage of the composite device is $[V_{CE3(\text{sat})} + V_{BE4}]$ and is higher than normal. This is because saturation occurs when Q_3 saturates and V_{BE4} must be added to this voltage.

The major problem with the configuration of Fig. 5.26 is potential instability of the local feedback loop formed by Q_3 and Q_4, particularly with capacitive loads on the amplifier. These problems are considered in Chapter 9.

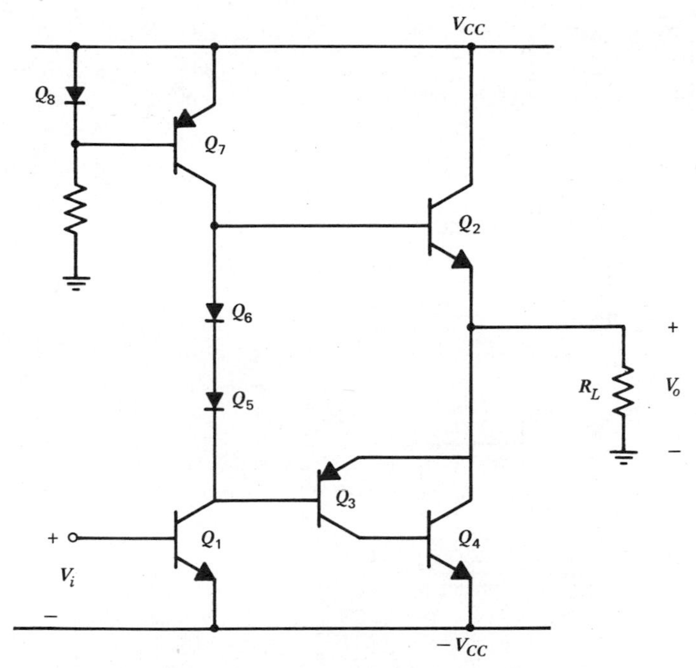

Fig. 5.26 Quasi-complementary Class B output stage.

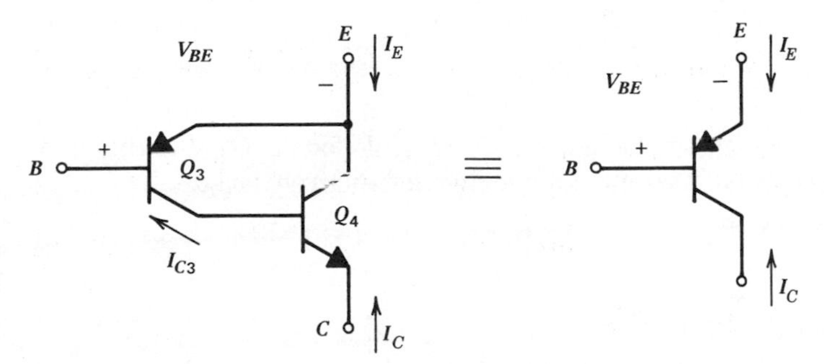

Fig. 5.27 Equivalence of the composite connection and a *pnp* transistor.

5.5.6 Overload Protection

The most common type of overload protection in integrated-circuit output stages is short-circuit current protection. As an example, consider the 741 output stage shown in Fig. 5.28 with partial short-circuit protection included. Initially assume that $R_6 = 0$ and ignore Q_{15}. The maximum positive drive delivered to the output stage occurs for V_i large positive. If $R_L = 0$ then V_o is held at zero volts and V_a in Fig. 5.28 is equal to V_{be14}. Thus, as V_i is taken positive, Q_{23}, Q_{18},

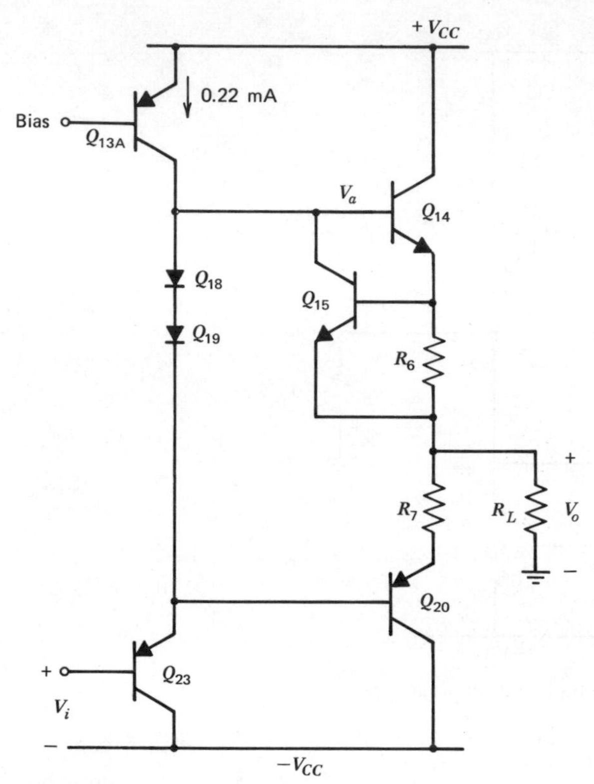

Fig. 5.28 Schematic of the 741 op amp showing partial short-circuit protection.

and Q_{19} will cut off and all of the current of Q_{13A} is fed to Q_{14}. If this is a high-β device then the output current can become destructively large.

$$I_{c14} = \beta_{14}|I_{C13A}| \tag{5.72}$$

If

$$\beta_{14} = 500$$

then

$$I_{c4} = 500 \times 0.22 = 110 \text{ mA}$$

If $V_{CC} = 15$ V, this current level gives a power dissipation in Q_{14} of

$$P_{c14} = V_{ce}I_c = 15 \times 110 \text{ mW} = 1.65 \text{ W}$$

This is sufficient to destroy the device. Thus the current under short-circuit conditions must be limited and this is achieved using R_6 and Q_{15} for positive V_o.

The short-circuit protection operates by sensing the output current with resistor R_6 of about 25 Ω. The voltage developed across R_6 is sensed by the base-emitter voltage of Q_{15}, which is normally off. When the current through R_6 reaches about 20 mA (the maximum safe level), Q_{15} begins to conduct appreciably

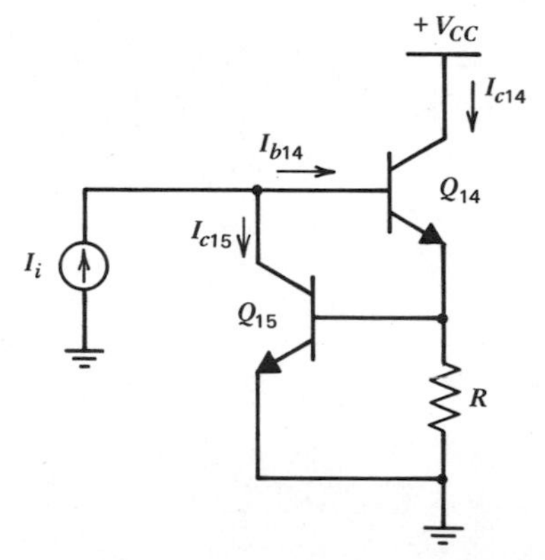

Fig. 5.29 Equivalent circuit for the calculation of the effect of Q_{15} on the transfer characteristic of Q_{14} in Fig. 5.28 when $R_L = 0$.

and diverts any further drive away from the base of Q_{14}. The drive current is thus harmlessly passed to the output instead of being multiplied by the β of Q_{14}.

The operation of this circuit can be seen by calculating the transfer characteristic of Q_{14} when driving a short-circuit load. This can be done using Fig. 5.29.

$$I_i = I_{b14} + I_{c15} \tag{5.73}$$

$$I_{c15} = I_{S15} \exp \frac{V_{be15}}{V_T} \tag{5.74}$$

Also

$$V_{be15} \simeq I_{c14} R \tag{5.75}$$

From (5.73)

$$I_{b14} = I_i - I_{c15}$$

But

$$I_{c14} = \beta_{14} I_{b14} = \beta_{14}(I_i - I_{c15}) \tag{5.76}$$

Substitution of (5.74) and (5.75) in (5.76) gives

$$I_{c14} + \beta_{14} I_{S15} \exp \frac{I_{c14} R}{V_T} = \beta_{14} I_i \tag{5.77}$$

The second term on the left side of (5.77) is due to Q_{15} and if this is negligible then $I_{c14} = \beta_{14} I_i$ as expected. The transfer characteristic of the stage can be plotted from (5.77) and this has been done in Fig. 5.30, using $\beta_{14} = 500$, $I_{S15} = 10^{-14}$ A and $R = 25\ \Omega$. For a maximum drive of $I_i = 0.22$ mA, the value of I_{c14} is effectively limited to 24 mA. For values of I_{c14} below 20 mA, Q_{15} has little effect on circuit operation.

Similar protection for negative output voltages in Fig. 5.28 is achieved by sensing the voltage across R_7 and diverting the base drive away from one of the preceding stages.

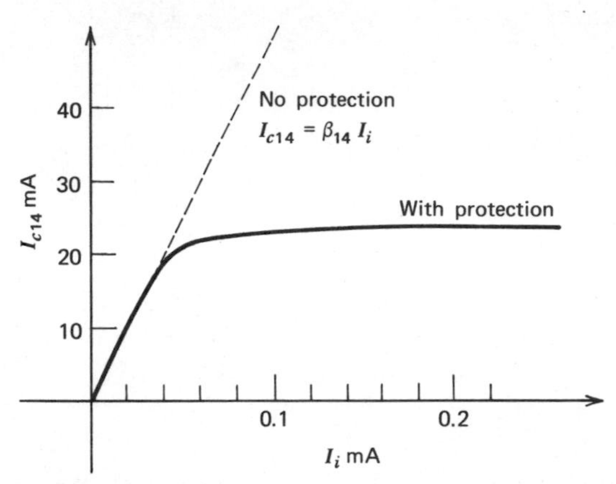

Fig. 5.30 Transfer characteristic of the circuit of Fig. 5.29 with and without protection transistor Q_{15} ($\beta_{14} = 500$).

PROBLEMS

5.1 A circuit as shown in Fig. 5.1 has $V_{CC} = 15$ V, $R_1 = R_2 = 0$, $R_3 = 5$ kΩ, $R_L = 2$ kΩ, $V_{CE(\text{sat})} = 0.2$ V, and $V_{BE(\text{on})} = 0.7$ V. All device areas are equal.
(a) Sketch the transfer characteristic from V_i to V_o.
(b) Repeat (a) if $R_L = 10$ kΩ.
(c) Sketch the waveform of V_o if a sinusoidal input voltage with an amplitude (zero to peak) of 10 V is applied at V_i in (a) and (b) above.

5.2 (a) For the circuit of Problem 5.1, sketch load lines in the I_c–V_{ce} plane for $R_L = 2$ kΩ and $R_L = 10$ kΩ.
(b) Calculate the maximum average sinusoidal output power that can be delivered to R_L (both values) before clipping occurs in (a) above. Sketch corresponding waveforms for I_{c1}, V_{ce1}, and P_{c1}.
(c) Calculate the circuit efficiency for each value of R_L in (b). (Neglect power dissipated in Q_3 and R_3.)
(d) Select R_L for maximum efficiency in this circuit and calculate the corresponding average output power with sinusoidal signals.

5.3 (a) Prove that any load line tangent to a power hyperbola makes contact with the hyperbola at the midpoint of the load line.
(b) Calculate the maximum possible instantaneous power dissipation in Q_1 for the circuit of Problem 5.1 with $R_L = 2$ kΩ and $R_L = 10$ kΩ.
(c) Calculate the *average* power dissipated in Q_1 for the circuit of Problem 5.1 with $R_L = 2$ kΩ and $R_L = 10$ kΩ. Assume that V_o is sinusoidal with an amplitude equal to the maximum possible before clipping occurs.

5.4 If $\beta_F = 100$ for Q_1 in Problem 5.1, calculate the average signal power delivered to Q_1 by its driver stage if V_o is sinusoidal with an amplitude equal to the maximum possible before clipping occurs. Repeat for $R_L = 10$ kΩ. Thus calculate the power gain of the circuit.

5.5 Calculate the incremental slope of the transfer characteristic of the circuit of Problem 5.1 at the quiescent point and at the extremes of the signal swing with a peak sinusoidal output of 1 V and $R_L = 2\ \text{k}\Omega$.

5.6 (a) For the circuit of Problem 5.1, draw load lines in the I_c–V_{ce} plane for $R_L = 0$ and $R_L = \infty$. Use an I_c scale from 0 to 30 mA. Also draw constant power hyperbolas for $P_c = 0.1$ W, 0.2 W, and 0.3 W. What is the maximum possible instantaneous power dissipation in Q_1 for the above values of R_L? Assume that the driver stage can supply a maximum base current to Q_1 of 0.3 mA and $\beta_F = 100$ for Q_1.

(b) If the maximum allowable instantaneous power dissipation in Q_1 is 0.2 W, calculate the minimum allowable value of R_L. (A graphical solution is the easiest.)

5.7 A common-emitter stage as shown in Fig. 5.8 has $I_Q = 1$ mA, $R_L = 3\ \text{k}\Omega$, $V_{CE(\text{sat})} = 0.1$ V, $V_{CC} = 5$ V, and $I_S = 10^{-15}$ A.

(a) Sketch the transfer characteristic from V_i to V_o. Repeat if $I_Q = 2$ mA.

(b) Calculate the maximum average output power that can be delivered to R_L before clipping occurs. Calculate the corresponding circuit efficiency and the *average* power dissipated in Q_1.

(c) Calculate the maximum *instantaneous* power dissipated in Q_1 with a large input signal applied.

5.8 Calculate the second and third harmonic distortion in the output voltage of the circuit of Problem 5.7 for a peak sinusoidal output voltage of 0.5 V and 1 V.

5.9 The circuit of Fig. 5.11 has $V_{CC} = 15$ V, $R_L = 2\ \text{k}\Omega$, $V_{BE(\text{on})} = 0.6$ V, and $V_{CE(\text{sat})} = 0.2$ V.

(a) Sketch the transfer characteristic from V_i to V_o assuming that the transistors turn on abruptly for $V_{be} = V_{BE(\text{on})}$.

(b) Sketch the output voltage waveform and the collector current waveform in each device for a sinusoidal input voltage of amplitude 1 V, 10 V, 20 V.

5.10 In the circuit of Fig. 5.14, $V_{CC} = 12$ V, $I_Q = 0.1$ mA, $R_L = 1\ \text{k}\Omega$, and for all devices $I_S = 10^{-15}$ A, $\beta = 150$. Calculate the value of V_i and the current in each device for $V_o = 0$, ± 5 V and ± 10 V. Then sketch the transfer characteristic from $V_o = 10$ V to $V_o = -10$ V.

5.11 For the circuit of Fig. 5.11 assume that $V_{CC} = 12$ V, $R_L = 1\ \text{k}\Omega$, and $V_{CE(\text{sat})} = 0.2$ V. Assume that there is sufficient sinusoidal input voltage available at V_i to drive V_o to its limits of clipping. Calculate the maximum average power that can be delivered to R_L before clipping occurs, the corresponding efficiency, and the maximum instantaneous device dissipation. Neglect crossover distortion.

5.12 For the circuit of Problem 5.11 calculate and sketch the waveforms of I_{c1}, V_{ce1}, and P_{c1} for device Q_1 over one cycle. Do this for output voltage amplitudes (zero to peak) of 11.5 V, 6 V, and 3 V. Neglect crossover distortion and assume sinusoidal signals.

5.13 For the output stage of Fig. 5.19 assume that $V_{CC} = 15$ V and for all devices $V_{CE(\text{sat})} = 0.2$ V, $V_{BE(\text{on})} = 0.7$ V, and $\beta_F = 50$.

(a) Calculate the maximum positive and negative limits of V_o for $R_L = 10\ \text{k}\Omega$ and $R_L = 2\ \text{k}\Omega$.

(b) Calculate the maximum average power that can be delivered to R_L before clipping occurs for $R_L = 10\ \text{k}\Omega$ and $R_L = 2\ \text{k}\Omega$. Calculate the corresponding circuit efficiency (for the output devices only) and the average power dissipated per output device. Neglect crossover distortion and assume sinusoidal signals.

5.14 For the output stage of Fig. 5.21*a* assume that $V_{CC} = 15$ V, $\beta_F(pnp) = 50$, $\beta_F(npn) = 200$, and for all devices $V_{BE(\text{on})} = 0.7$ V, $V_{CE(\text{sat})} = 0.2$ V, $I_S = 10^{-14}$ A. Assume that the collector current in Q_{13A} is 0.2 mA.
   (a) Calculate the maximum positive and negative limits of V_o for $R_L = 10$ kΩ, $R_L = 1$ kΩ, and $R_L = 200$ Ω.
   (b) Calculate the maximum average power that can be delivered to $R_L = 1$ kΩ before clipping occurs, and the corresponding circuit efficiency (for the output devices only). Also calculate the peak instantaneous power dissipation in each output device. Assume sinusoidal signals.

5.15 (a) For the circuit of Problem 5.14, calculate the maximum possible average output power that can be delivered to a load R_L if the instantaneous power dissipation per device must be less than 100 mW. Also specify the corresponding value of R_L and the circuit efficiency (for the output devices only). Assume sinusoidal signals.
   (b) Repeat (a) if the maximum instantaneous power dissipation per device is 200 mW.

5.16 For the circuit of Problem 5.14, calculate bias currents in Q_{23}, Q_{20}, Q_{19}, Q_{18}, and Q_{14} for $V_o = -10$ V with $R_L = 1$ kΩ. Use $I_S = 10^{-14}$ A for all devices.

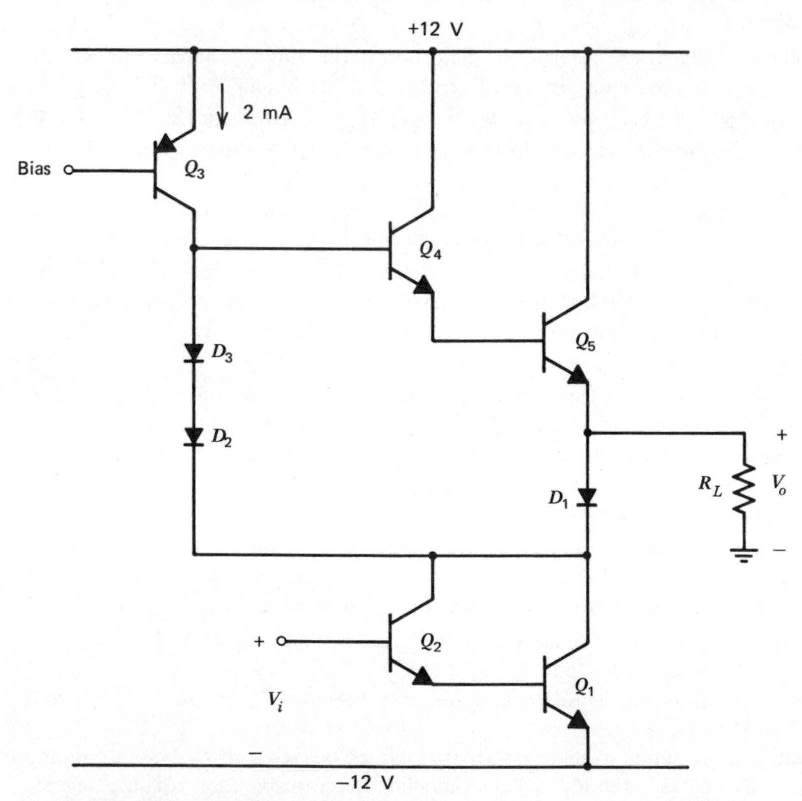

Fig. 5.31 All-*npn* Darlington output stage.

5.17 An all-*npn* Darlington output stage is shown in Fig. 5.31. For all devices $V_{BE(\text{on})} = 0.7$ V, $V_{CE(\text{sat})} = 0.2$ V, $\beta_F = 100$. The collector current in Q_3 is 2 mA.
(a) If $R_L = 8\ \Omega$, calculate the maximum positive and negative limits of V_o.
(b) Calculate the power dissipated in the circuit for $V_o = 0$ V.
(c) Calculate the maximum average power that can be delivered to $R_L = 8\ \Omega$ before clipping occurs and the corresponding efficiency of the *complete* circuit. Also calculate the maximum instantaneous power dissipated in each output transistor. Assume that feedback is used around the circuit so that V_o is approximately sinusoidal.

5.18 For the circuit of Fig. 5.26 assume that $V_{CC} = 15$ V, $\beta_F(pnp) = 30$, $\beta_F(npn) = 150$, $I_S(npn) = 10^{-14}$ A, $I_S(pnp) = 10^{-15}$ A, and for all devices $V_{BE(\text{on})} = 0.7$ V, $V_{CE(\text{sat})} = 0.2$ V. Assume that Q_5 and Q_6 are *npn* devices and the collector current in Q_7 is 0.15 mA.
(a) Calculate the maximum positive and negative limits of V_o for $R_L = 1\ \text{k}\Omega$.
(b) Calculate quiescent currents in Q_1–Q_7 for $V_o = 0$ V.
(c) Calculate the maximum average output power (sine wave) that can be delivered to R_L if the maximum instantaneous dissipation in any device is 100 mW. Calculate the corresponding value of R_L, and the peak currents in Q_3 and Q_4.

REFERENCES

1. E. M. Cherry and D. E. Hooper. *Amplifying Devices and Low-Pass Amplifier Design.* Wiley, New York, 1968, Chapter 9.
2. E. M. Cherry and D. E. Hooper. Op. cit., Ch. 16.
3. J. Millman and C. C. Halkias. *Integrated Electronics.* McGraw-Hill, New York, 1972, Chapter 18.
4. A. B. Grebene. *Analog Integrated Circuit Design.* Van Nostrand Reinhold, New York, 1972, Chapter 5–6.
5. T. M. Frederiksen and J. E. Solomon. "A High-Performance 3-Watt Monolithic Class-B Power Amplifier," *IEEE J. Solid-State Circuits,* Vol. SC-3, pp. 152–160, June 1968.
6. P. R. Gray. "A 15-W Monolithic Power Operational Amplifier," *IEEE J. Solid-State Circuits,* Vol. SC-7, pp. 474–480, December 1972.
7. E. L. Long and T. M. Frederiksen. "High-Gain 15-W Monolithic Power Amplifier with Internal Fault Protection," *IEEE J. Solid-State Circuits,* Vol. SC-6, pp. 35–44, February 1971.

CHAPTER 6

OPERATIONAL AMPLIFIERS

In the previous three chapters, the most important circuit building blocks utilized in analog integrated circuits have been studied. Most analog ICs consist primarily of these basic circuits connected in such a way as to perform the desired function. While the variety of standard and special-purpose custom integrated circuits that fall into the analog category is almost limitless, a few standard circuits stand out as perhaps having the widest application in systems of various kinds. These include operational amplifiers, voltage regulators, phase-locked loops, and A/D and D/A converters. In this chapter we will consider monolithic operational amplifiers, both as an example of the utilization of the previously discussed circuit building blocks and as an introduction to the design and application of this important class of analog circuit. Phase-locked loops are considered in Chapter 10, and voltage-regulator circuits are considered in Chapter 8. The design of A/D and D/A converters is not covered explicitly, but it involves application of the circuit techniques described throughout the book.

An ideal operational amplifier (op amp) is a differential input, single-ended output amplifier with infinite gain, infinite input resistance, and zero output resistance. A conceptual schematic diagram is shown in Fig. 6.1. While actual op amps do not have these ideal characteristics, their performance is usually sufficiently good that in most applications the circuit behavior closely approximates that of an ideal op amp.

Prior to the development of integrated-circuit technology, the op amp was a rather costly circuit whose principal application was analog computation. Monolithic-circuit technology has reduced the cost of such circuits by approximately two orders of magnitude since 1960, and as a result op amps are widely used in the design of systems of all types.

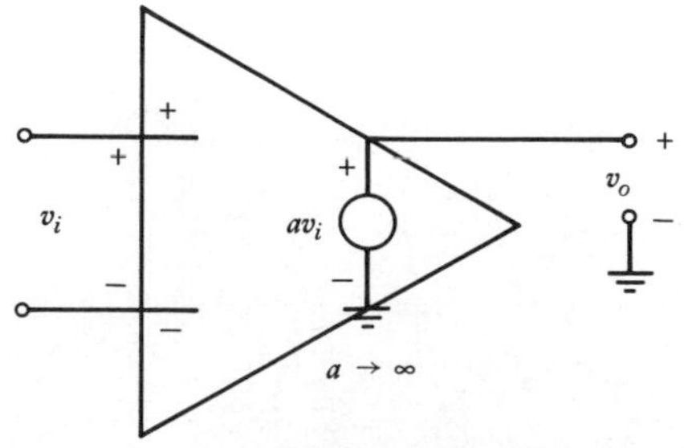

Fig. 6.1 Ideal operational amplifier.

In this chapter, we first explore several applications of operational amplifiers to illustrate their versatility in analog circuit and system design. A general-purpose monolithic operational amplifier, the 741, is then analyzed as an example, and the ways in which the performance of the circuit deviates from ideality are discussed. Finally, design considerations for improving the various aspects of monolithic op amp low-frequency performance are discussed. The high-frequency and transient response of operational amplifiers is discussed in Chapters 7 and 9.

6.1 APPLICATIONS OF OPERATIONAL AMPLIFIERS

Basic Feedback Concepts. Virtually all operational amplifier applications rely on the principles of *feedback*. The topic of feedback amplifiers is covered in detail in Chapter 8; we now consider a few basic concepts necessary for an understanding of operational-amplifier circuits. A generalized feedback amplifier is shown in Fig. 6.2. The block labeled a is called the forward or basic amplifier, and the block labeled f is called the feedback network. The gain of the basic amplifier when the feedback network is not present is called the *open-loop gain, a,* of the amplifier. The function of the feedback network is to sense the output signal, S_o, and develop a feedback signal S_{fb}, which is equal to fS_o, where f is usually less than unity. This feedback signal is subtracted from the input signal, S_i, and the difference, S_ϵ, is applied to the basic amplifier. The gain of the system when the feedback network is present is called the *closed-loop gain*. For the basic amplifier we have:

$$S_o = aS_\epsilon = a(S_i - S_{fb}) = a(S_i - fS_o) \tag{6.1}$$

and thus

$$\frac{S_o}{S_i} = \frac{a}{1 + af} \tag{6.2a}$$

Notice that as the product of a and f, called the *loop gain, T*, becomes large compared to unity, the closed-loop gain becomes very nearly equal to:

$$\lim_{af \to \infty} \frac{a}{1 + af} = \frac{1}{f} \tag{6.2b}$$

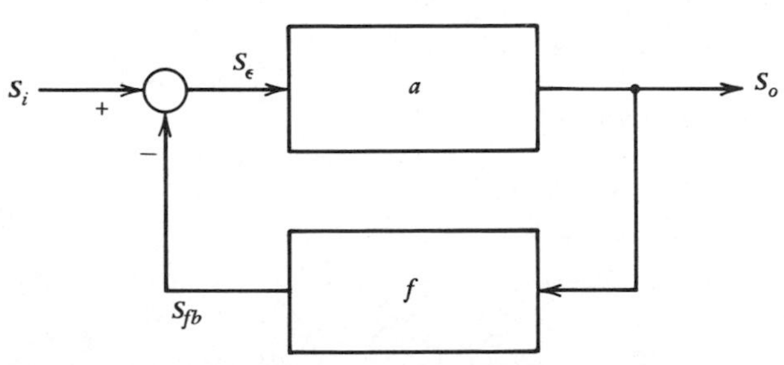

Fig. 6.2 A conceptual feedback amplifier.

Since the feedback network is composed of passive components, the value of f can be set to an arbitrary degree of accuracy, and will establish the gain at a value of $1/f$, independent of any variations in the open-loop gain a. This independence of closed-loop performance from the parameters of the active amplifier is the primary motivating factor in the wide use of operational amplifiers as active elements in analog circuits.

For the circuit shown in Fig. 6.2, the feedback signal tends to *reduce* the magnitude of S_ϵ below that of the open-loop case when a and f have the same sign. This is called *negative* feedback, and in the case of practical interest in this chapter.

With this brief introduction to feedback concepts, we proceed to a consideration of several useful operational amplifier configurations. Because of the simplicity of these circuits, it is most useful to analyze them directly using Kirchoff's laws rather than attempting to consider them as feedback amplifiers. In Chapters 8 and 9, more complicated feedback configurations are considered in which the use of feedback concepts as an analytical tool is more useful.

Inverting Amplifier. The inverting amplifier connection is shown in Fig. 6.3*a*.[1,2,3] We assume that the op amp input resistance is infinite, and that the output resistance is zero. Summing currents at node X,

$$\frac{V_s - V_i}{R_1} + \frac{V_o - V_i}{R_2} = 0 \tag{6.3a}$$

or

$$\frac{V_s}{R_1} + \frac{V_o}{R_2} = V_i\left(\frac{1}{R_1} + \frac{1}{R_2}\right) \tag{6.3b}$$

However, if the open-loop voltage gain is a,

$$V_i = \frac{-V_o}{a} \tag{6.4}$$

Combining (6.3b) and (6.4)

$$\frac{V_o}{V_s} = -\frac{R_2}{R_1}\left[\frac{1}{1 + \frac{1}{a}\left(1 + \frac{R_2}{R_1}\right)}\right] \tag{6.5}$$

If the gain of the op amp is large enough that

$$a\left(\frac{R_1}{R_1 + R_2}\right) \gg 1 \tag{6.6a}$$

then the closed-loop gain is

$$\frac{V_o}{V_s} \simeq -\frac{R_2}{R_1} \tag{6.6b}$$

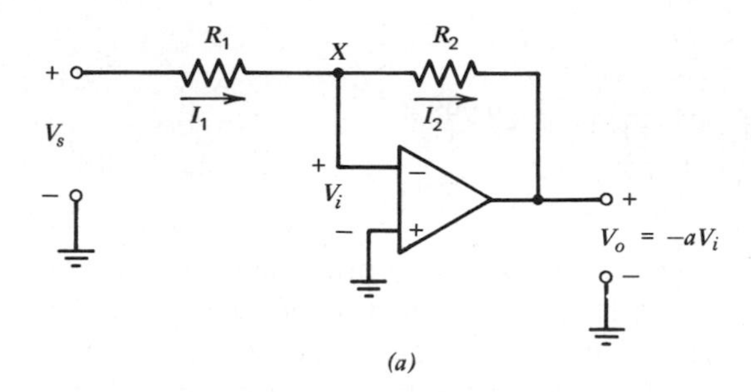

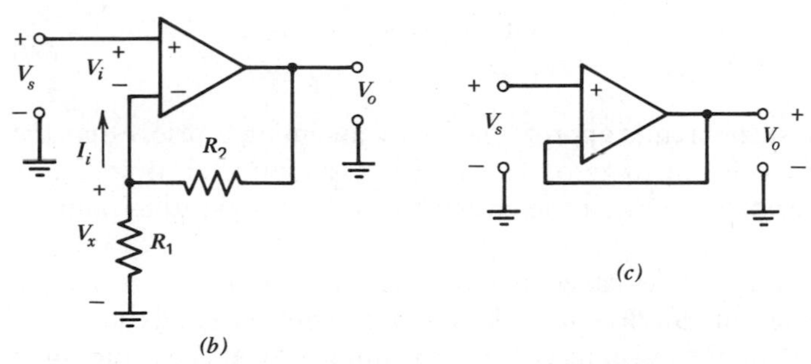

Fig. 6.3 (*a*) Inverting amplifier configuration. (*b*) Noninverting amplifier configuration. (*c*) Voltage-follower configuration.

Note that the closed-loop gain depends only on the external passive components, R_1 and R_2. Since these can be selected with arbitrary accuracy, a high degree of precision can be obtained in closed-loop performance independent of variations in the active device (op amp) parameters. For example, if the op amp gain were to change from 5×10^4 to 10^5, this 100 percent increase in gain would have almost no observable effect on closed-loop performance.

EXAMPLE

Calculate the gain of the circuit Fig. 6.3*a* for $a = 10^4$ and $a = 10^5$, and $R_1 = 1\ \text{k}\Omega$, $R_2 = 10\ \text{k}\Omega$

From (6.5)

$$\frac{V_o}{V_s} = A = -\frac{R_2}{R_1}\left[\frac{1}{1 + \dfrac{1}{a}\left(1 + \dfrac{R_2}{R_1}\right)}\right]$$

For $a = 10^4$,

$$A = -10\left(\frac{1}{1 + \dfrac{11}{10^4}}\right) = -9.9890$$

For $a = 10^5$

$$A = -10\left(\frac{1}{1 + \frac{11}{10^5}}\right) = -9.99890$$

The large gain of operational amplifiers allows the approximate analysis of circuits like that of Fig. 6.3*a* to be performed by the use of *summing point constraints.*[1] If the op amp is connected in a negative-feedback circuit, and if the gain of the op amp is very large, then for a finite value of output voltage the input voltage must approach zero since,

$$V_i = -\frac{V_o}{a} \tag{6.7}$$

Thus one can analyze such circuits approximately by assuming a priori that the op amp input voltage is driven to zero. An implicit assumption in doing so is that the feedback is negative, and that the circuit has a *stable* operating point at which (6.7) is valid.

The assumption that $V_i = 0$ is called a *summing point constraint.* A second constraint is that no current can flow into the op amp input terminals, since no voltage exists across the input resistance of the op amp if $V_i = 0$. This summing point approach allows a more intuitive understanding of the operation of the inverting amplifier configuration of Fig. 6.3*a*. Since the inverting input terminal is forced to ground potential, the resistor, R_1, serves to convert the voltage, V_s, to an input current of value V_s/R_1. But this current cannot enter the op amp, so it flows through R_2 producing voltage V_sR_2/R_1 across it. Notice that since the op amp input terminal is at ground potential, the input resistance of the overall circuit as seen by V_s is equal to R_1. Since the inverting input of the amplifier is forced to ground potential by the negative feedback, it is sometimes called a "virtual" ground.

Noninverting Amplifier. The noninverting circuit is shown in Fig. 6.3*b*.[1,2,3] Using the summing point constraints, $I_i = 0$, and

$$V_x = V_o\left(\frac{R_1}{R_1 + R_2}\right) \tag{6.8}$$

Also $V_i = 0$ and thus $V_s = V_x$. Rearranging (6.8) we obtain

$$V_o = V_s\left(1 + \frac{R_2}{R_1}\right) \tag{6.9}$$

In contrast to the inverting case, this circuit displays a very high input resistance because of the type of feedback used. (See Chapter 8.) Note also that unlike the inverting connection, the noninverting connection causes the common-mode input voltage of the operational amplifier to be equal to V_s. An important variation of this connection is the voltage follower, in which $R_2 = 0$, $R_1 = \infty$, to give a gain of 1. This circuit is shown in Fig. 6.3*c*.

Differential Amplifier The differential amplifier is used to amplify the *difference* between two voltages. The circuit is shown in Fig. 6.4.[1,2] For this circuit, $I_{i1} = 0$ and thus resistors R_1 and R_2 form a voltage divider. Voltage V_x is then given by:

$$V_x = V_1 \left(\frac{R_2}{R_1 + R_2} \right) \tag{6.10}$$

and since $V_i = 0$ for this circuit (a summing-point constraint) we have $V_x = V_y$. The current I_1 is thus:

$$I_1 = \left(\frac{V_2 - V_y}{R_1} \right) = I_2 \tag{6.11}$$

The output voltage is given by:

$$V_o = V_y - I_2 R_2 \tag{6.12}$$

Substituting (6.10) and (6.11) into (6.12), and using $V_y = V_x$ we find

$$V_o = V_1 \frac{R_2}{R_1 + R_2} - \left[\frac{V_2 - V_1 \left(\dfrac{R_2}{R_1 + R_2} \right)}{R_1} \right] R_2 \tag{6.13}$$

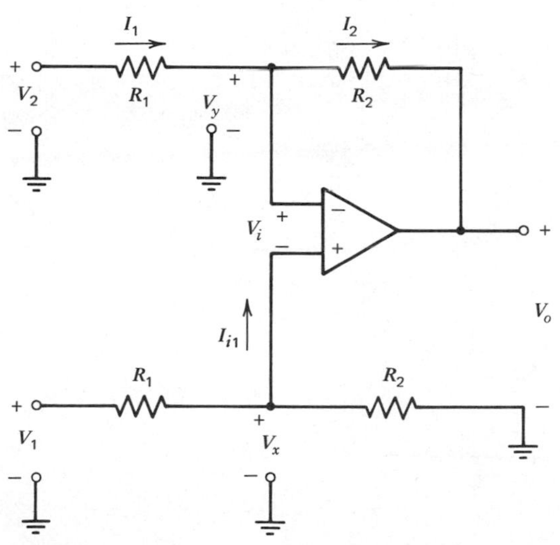

Fig. 6.4 Differential amplifier configuration.

Rearranging terms,

$$V_o = \frac{R_2}{R_1}[V_1 - V_2] \tag{6.14}$$

The circuit thus amplifies the *difference* voltage $(V_1 - V_2)$.

Differential amplifiers are often required to detect and amplify small differences between two sizable voltages, a typical application being measurement of the difference voltage between the two arms of a Wheatstone bridge. Note that as in the case of the noninverting amplifier, the op amp of Fig. 6.4 experiences a common-mode input that is almost equal to the common-mode voltage applied to the input terminals for $R_2 \gg R_1$.

Nonlinear Analog Operations. By including nonlinear elements in the feedback network, operational amplifiers can be used to perform nonlinear operations on one or more analog signals. The logarithmic amplifier, shown in Fig. 6.5, is an example of such an application. Log amplifiers find wide application in instrumentation systems where signals of very large dynamic range must be sensed and recorded. The operation of this circuit can again be understood by application of the summing point constraints. Because the input voltage of the op amp must be zero, the resistor, R, serves to convert the input voltage into a current. This same current must then flow into the collector of the transistor. Thus the circuit forces the collector current of the transistor to be proportional to the input voltage. Since the base-emitter voltage of a bipolar transistor in the forward active region is logarithmically related to the collector current, and since the output voltage is just the base-emitter voltage of the transistor, a logarithmic transfer characteristic is produced. In terms of equations:

$$I_1 = \frac{V_s}{R} = I_c = I_S\left(\exp\frac{V_{be}}{V_T} - 1\right) \tag{6.15}$$

and

$$V_o = -V_{be} \tag{6.16}$$

Thus

$$V_o = -V_T \ln \frac{V_s}{I_S R} \tag{6.17}$$

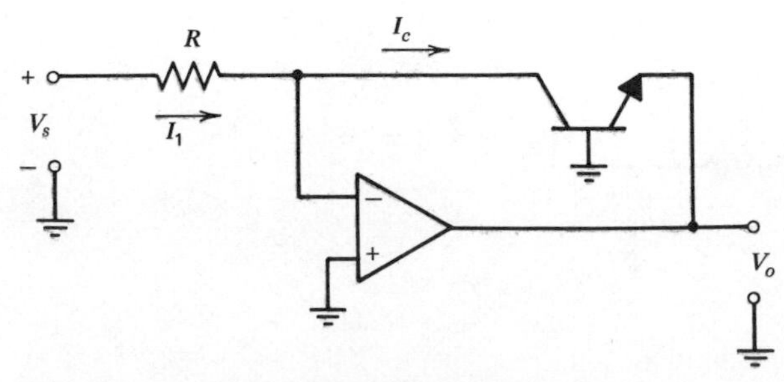

Fig. 6.5 Logarithmic amplifier configuration.

The log amplifier is only one example of a wide variety of op amp applications in which a nonlinear feedback element is used to develop a nonlinear transfer characteristic. For example, two log amplifiers can be used to develop the logarithm of two different signals, these voltages can be summed, and then the exponential function of the result can be developed using an inverting amplifier connection with R_1 replaced with a diode. The result is an analog multiplier, discussed in detail in Chapter 10. Other nonlinear operations such as limiting, rectification, peak detection, squaring, square rooting, raising to a power, and division can be performed in conceptually similar ways.

Integrator, Differentiator. The integrator and differentiator circuits, shown in Fig. 6.6, are examples of using op amps with reactive elements in the feedback network to realize a desired frequency response or time-domain response.[1,2] In the case of the integrator, the resistor, R, is used to develop a current I_1, which is proportional to the input voltage. This current flows into the capacitor, C, whose voltage is proportional to the integral of the current, I_1, with respect to time. Since the output voltage is equal to the negative of the capacitor voltage, the output is proportional to the integral of the input voltage with respect to time. In terms of equations:

$$I_1 = \frac{V_s}{R} = I_2 \tag{6.18}$$

and

$$V_o = -\frac{1}{C}\int_0^t I_2\, dt + V_o(0) \tag{6.19}$$

Combining (6.18) and (6.19),

$$V_o(t) = -\frac{1}{RC}\int_0^t V_s(t)\, dt + V_o(0) \tag{6.20}$$

The performance limitations of real op amps limit the range of V_o and the rates of change of V_o and V_s for which this relationship is maintained. In the case of the differentiator, the capacitor, C, is connected between the input and the virtual

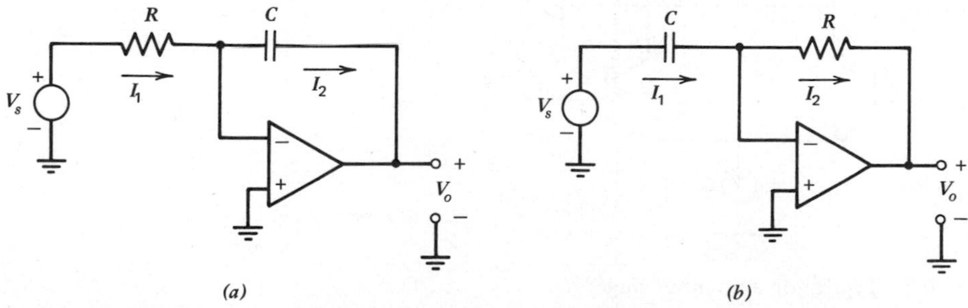

Fig. 6.6 (*a*) Integrator configuration. (*b*) Differentiator configuration.

ground point. The current through the capacitor is proportional to the time derivative of the voltage across it, which is equal to the input voltage. This current flows through the feedback resistor, R, producing a voltage at the output proportional to the capacitor current, which is proportional to the time rate of change of the input voltage. In terms of equations:

$$I_1 = C\frac{dV_s}{dt} = I_2 \tag{6.21}$$

$$V_o = -RI_2 = -RC\frac{dV_s}{dt} \tag{6.22}$$

6.2 DEVIATIONS FROM IDEALITY IN REAL OPERATIONAL AMPLIFIERS

Real operational amplifiers deviate from ideal behavior in significant ways. The principal effect of these deviations is to limit the frequency range of the signals that can be accurately amplified, to place a lower limit on the magnitude of dc signal that can be detected and to place an upper limit on the magnitudes of the impedance of the passive elements that can be used in the feedback network with the amplifier. The following is a summary of the important deviations from ideality and their effect in applications.

Input Bias Current. Most monolithic operational amplifiers contain a bipolar transistor input stage as shown in Fig. 6.7. Here Q_1 and Q_2 are the input transistors of the amplifier. The base currents of Q_1 and Q_2 flow into the amplifier input

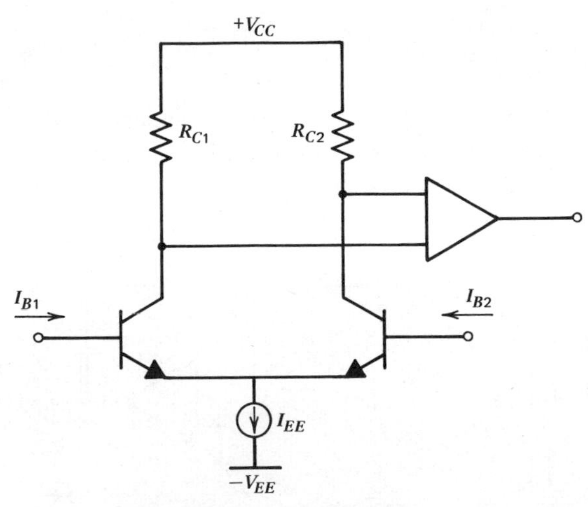

Fig. 6.7 Typical op amp input stage.

terminals and each is called an input bias current. The presence of this current violates the assumption made in summing-point analysis that the current into the input terminals is zero. Typical magnitudes for the bias current are 10 to 100 nA for bipolar input devices and 1 to 10 pA for JFET input devices. In dc inverting, noninverting, and differential amplifiers, this bias current can cause undesired voltage drops in the resistors forming the feedback network, with the result that a small residual dc voltage will appear at the output when the external input voltage is zero. In integrator circuits the input bias current is indistinguishable from the current being integrated and causes the output voltage to change at a constant rate even when the input is zero. However, if the input bias currents were equal in the two input leads, their effects could be canceled in most applications by including a balancing resistor in series with one of the input leads so that the same impedance is presented at both inputs. For example, the differential amplifier of Fig. 6.4 is insensitive to bias current as long as the currents flowing into the two inputs are equal. More serious, however, is the problem of random mismatches between the two input currents.

Input Offset Current. For the emitter-coupled pair shown in Fig. 6.7, the two input bias currents will be equal only if the two transistors have equal betas. Geometrically identical devices on the same IC die typically display beta mismatches that are described by a normal distribution with a standard deviation of around 5 to 10 percent of the mean value. This mismatch in the two currents is random from circuit to circuit and cannot be compensated by a fixed resistor. This aspect of circuit performance is characterized by the input offset current, defined as:

$$I_{OS} = I_{B1} - I_{B2} \tag{6.23}$$

The input bias current is then defined as the average of the two input currents.

$$I_{\text{bias}} = \frac{I_{B1} + I_{B2}}{2} \tag{6.24}$$

Input Offset Voltage. As discussed in Chapter 3, mismatches result in nonzero input offset voltage in the amplifier. The input offset voltage is the differential input voltage which must be applied to drive the output to zero. For monolithic op amps with bipolar transistor input devices, this offset is typically 2 to 5 mV, and can often be nulled with an external potentiometer. The input offset voltage is also a function of temperature, and this temperature sensitivity, or drift, does not necessarily go to zero when the input offset is nulled. In dc amplifier applications, the offset and drift place a lower limit on the magnitude of the dc voltage which can be accurately amplified with the amplifier.

Common-Mode Rejection Ratio (CMRR). The common-mode rejection ratio of an operational amplifier is the ratio of the differential-mode gain to the common-mode gain as defined in Chapter 3. A more meaningful characterization from an applications standpoint is to regard the CMRR as the change in input offset voltage that results from a unit change in common-mode input voltage. Assume, for example, that we apply zero common-mode input voltage to the amplifier and then apply just enough differential voltage to the input to drive the output voltage to zero. The dc voltage we have applied is just the input offset voltage, V_{OS}. If we keep the applied differential voltage constant and increase the common-mode input voltage by an amount ΔV_{ic}, the output voltage will change, by an amount:

$$\Delta V_o = A_{cm}\,\Delta V_{ic} \tag{6.25}$$

In order to drive the output voltage back to zero, we will have to change the differential input voltage by an amount:

$$\Delta V_{id} = \frac{\Delta V_o}{A_{dm}} = \frac{A_{cm}\,\Delta V_{ic}}{A_{dm}} \tag{6.26}$$

Thus we can regard the effect of noninfinite CMRR as causing a change in the input offset voltage whenever the common-mode input voltage is changed. Using (3.97) and (6.26),

$$\text{CMRR} = \left|\frac{A_{dm}}{A_{cm}}\right| = \left(\left.\frac{\Delta V_{id}}{\Delta V_{ic}}\right|_{V_o=0}\right)^{-1} = \left(\frac{\Delta V_{OS}}{\Delta V_{ic}}\right)^{-1} \approx \left(\left.\frac{\partial V_{OS}}{\partial V_{ic}}\right|_{V_o=0}\right)^{-1} \tag{6.27}$$

This expression was used in Chapter 4 to obtain (4.132). In circuits such as the differential amplifier of Fig. 6.7, an offset voltage is produced that is a function of the common-mode signal input, producing a voltage at the output that is indistinguishable from the desired signal. For a common-mode rejection ratio of 10^4 (or 80 dB), (6.27) shows that a 10-V common-mode signal produces a 1-mV change in the input offset voltage.

Input Resistance. In bipolar transistor input stages, the input resistance is typically in the 100 kΩ to 1 MΩ range. Usually, however, the voltage gain is large enough that this input resistance has little effect on circuit performance in closed-loop feedback configurations.

Output Resistance. Due to circuit limitations, the output resistance of general-purpose op amps is on the order of 40 to 100 Ω. Again, this resistance does not strongly affect closed-loop performance except as it affects stability under large capacitive loading, and in the case of power operational amplifiers that must drive a load resistance of small value.

Frequency Response. Because of parasitic junction capacitance and minority-carrier charge storage in devices making up the op amp circuit, the voltage gain decreases at high frequencies. This falloff must usually be enhanced by the addition of extra capacitance called compensation capacitance in the circuit to insure that the circuit does not oscillate. (See Chapter 9.) This aspect of op amp behavior is characterized by the unity-gain bandwidth, which is the frequency at which the open-loop voltage gain is equal to unity. For general-purpose amplifiers, this frequency is in the 1 to 20 MHz range. This topic is discussed in detail in Chapters 7 and 9.

A second aspect of op amp high-frequency behavior is a limitation of the rate at which the output voltage can change. This limitation results because of the limited current available within the circuit to charge the compensation capacitor. This maximum rate is called the slew rate, and is discussed more extensively in Chapter 9.

Operational Amplifier Equivalent Circuit. The effect of these deviations from ideality on the low-frequency performance of an operational amplifier in a particular application can be calculated using the equivalent circuit shown in Fig. 6.8. Here, the two current sources labeled I_{bias} represent the *average* value of dc current flowing into the input terminals. These would have the polarity shown for an *npn* transistor input stage, but would flow *out* of the amplifier terminals for a *pnp* transistor input stage. The current source labeled I_{OS} represents the *difference* between the currents flowing into the amplifier terminals. For example, if a particular circuit displayed a current of 1 μA flowing into the noninverting input terminal and a current of 1.5 μA flowing into the inverting input terminal, then the value of I_{bias} in Fig. 6.8 would be 1.25 μA, and the value of I_{OS} would be 0.5 μA.

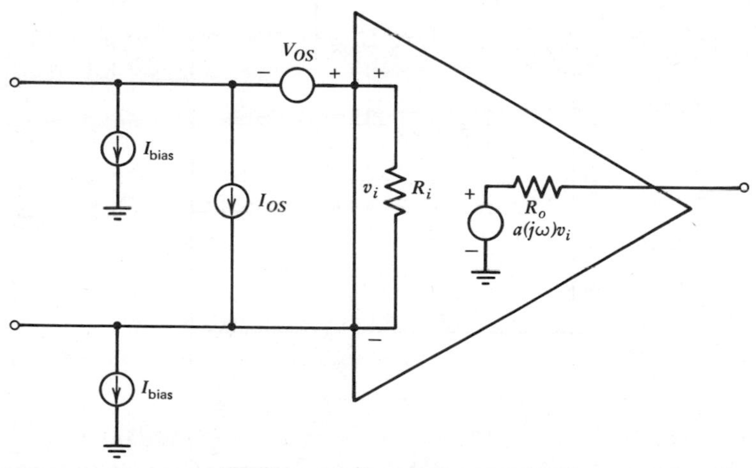

Fig. 6.8 Equivalent circuit for the operational amplifier including input offset voltage and current, input bias current, input and output resistance, and voltage gain.

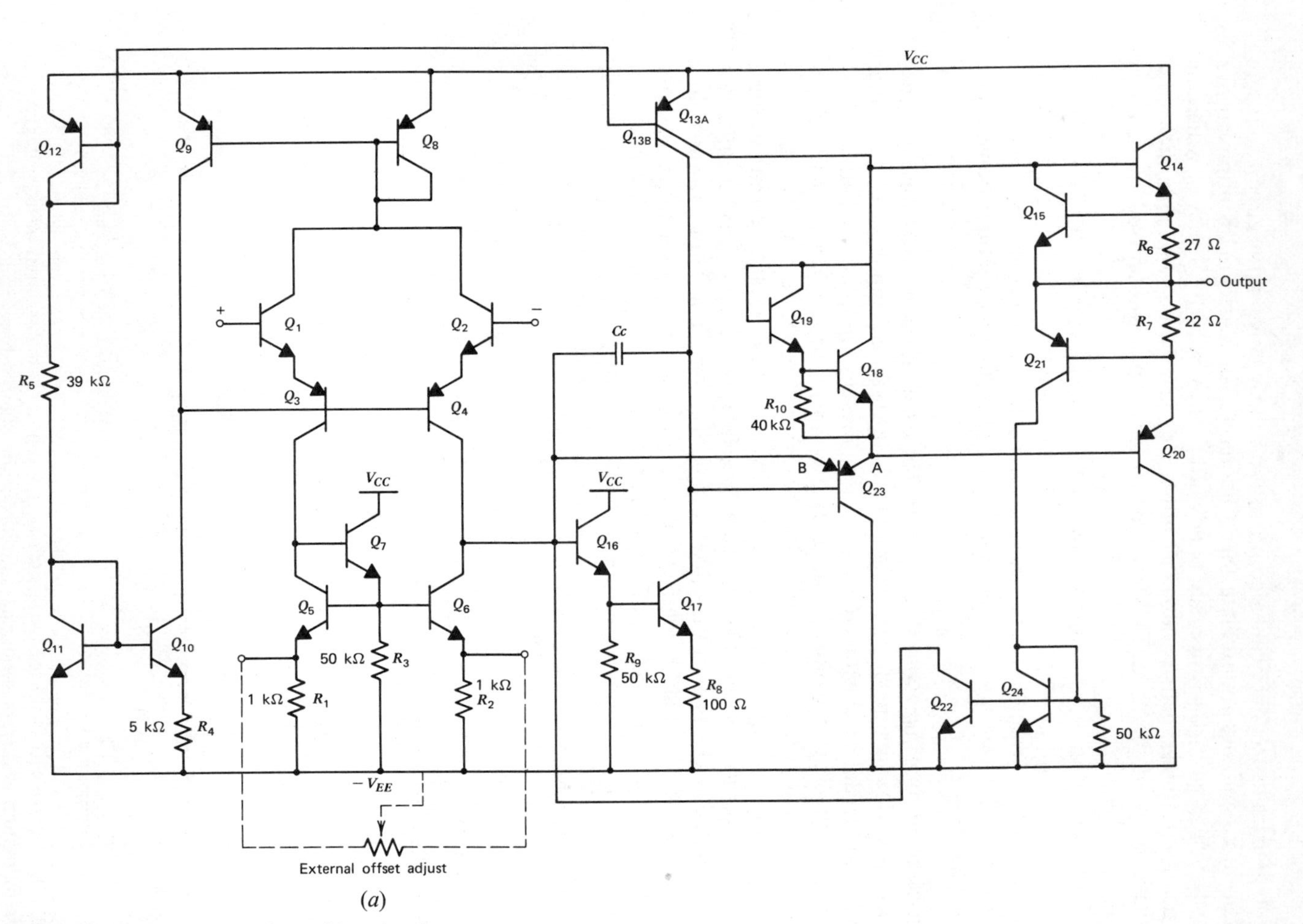

(*a*)

Fig. 6.9*a* 741 operational amplifier circuit.

6.3 ANALYSIS OF MONOLITHIC OPERATIONAL AMPLIFIERS

In Section 6.1 several applications were described in which op amps were utilized. In this section we will analyze perhaps the most widely used operational amplifier as of this writing—the 741. This circuit was first introduced in 1966 and is manufactured by virtually all of the integrated-circuit manufacturing firms in the United States. Its popularity arises from the fact that it is internally compensated (see Chapter 9) and is a relatively simple circuit that can be made to fit on a die less than 40 mils on a side. It has large voltage gain and good common-mode and differential-mode input voltage ranges.

The 741 circuit is shown in Fig. 6.9*a*[3]. The analysis is carried out in three parts. First the operation of the circuit is discussed qualitatively, and then a dc analysis is carried out to determine the quiescent currents and voltages in the circuit. Finally, a small-signal analysis is carried out to determine the voltage gain, input resistance, and output resistance.

Qualitative Description of Circuit Operation. A simplified conceptual circuit diagram is shown in Fig. 6.9*b*. The input transistors, Q_1 and Q_2, are emitter followers

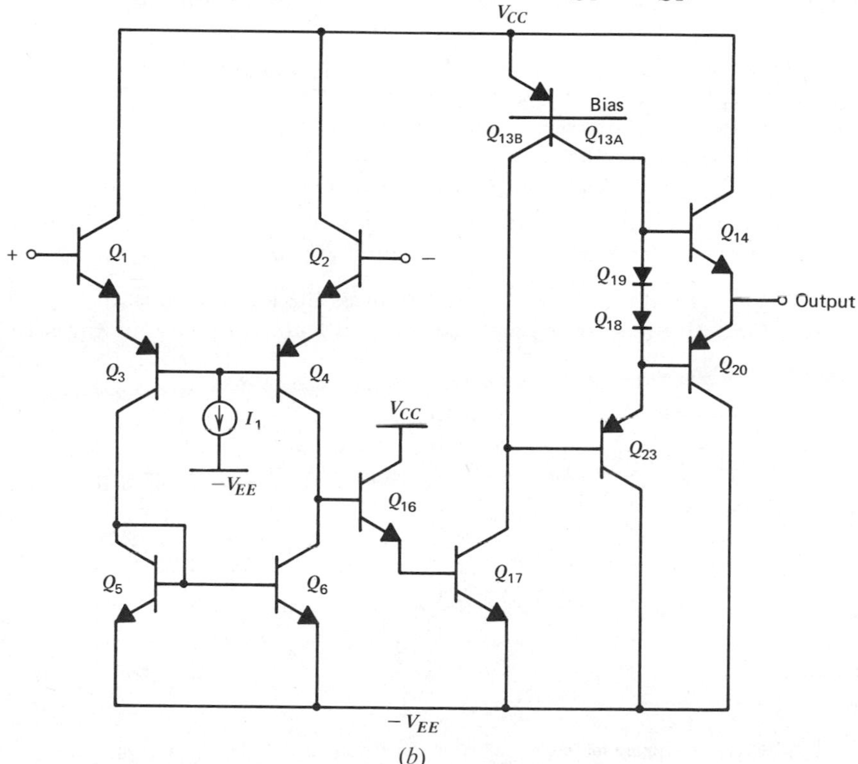

(*b*)

Fig. 6.9*b* Simplified, conceptual schematic diagram of the 741 amplifier.

that maintain high input resistance and low input current. These drive the emitters of the common-base differential pair of *pnp* devices Q_3 and Q_4. The transistors, Q_5 and Q_6, form an active load for Q_3 and Q_4. These six transistors taken together actually perform three separate functions that must be carried out in monolithic op amp realizations.

1. They provide a differential input that is relatively insensitive to common mode voltages, has high input resistance, and provides some voltage gain. The realization of some voltage gain in the input stage is desirable since the noise and offset voltage associated with the second and later stages are divided by this gain when referred to the input.
2. Level shifting. The *pnp* transistors produced by standard IC technology have poor frequency response as discussed in Chapter 2. Thus the most desirable approach to an op amp realization would be to use only *npn* transistors. Somewhere in the amplifier, however, the dc level of the signal path must be shifted in the negative direction. In general-purpose amplifiers like the 741, this is usually accomplished by inserting a lateral *pnp* transistor in the signal path. Note that in the 741, the collectors of the *pnp* devices Q_3 and Q_4 rest at a potential very near the negative supply.
3. Differential to single-ended conversion. Operational amplifiers have differential inputs and single-ended outputs, so within the circuit a conversion must be made to single-ended operation. One obvious approach would be to simply take *one* of the outputs of an emitter-coupled pair and feed that into a single-ended circuit. This approach tends to result in high sensitivity to common-mode input voltages as discussed in Chapter 4, so typically an active-load circuit is used as realized by transistors Q_5 and Q_6.

Transistor Q_{16} is an emitter follower that reduces the loading effect of Q_{17} on the output of the active-load stage. Transistor Q_{17} is a common-emitter amplifier that also has an active load formed by Q_{13B}. This amplifier stage provides large

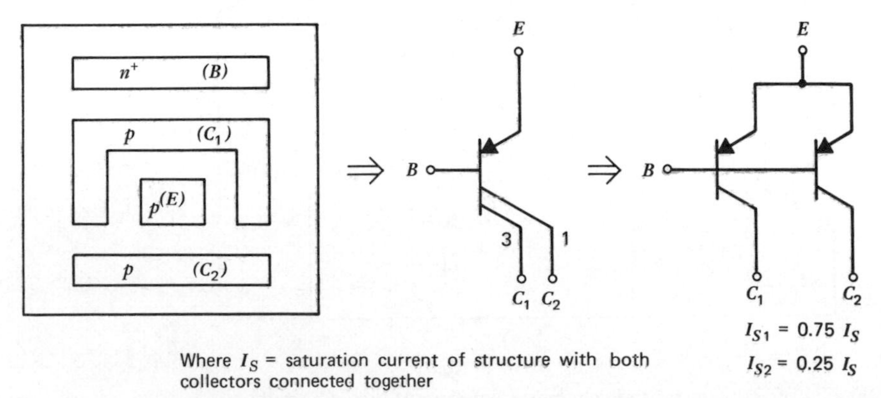

Fig. 6.10 Electrical equivalent for multicollector lateral *pnp*.

voltage gain. Transistor Q_{23} is another emitter follower that prevents the output stage from loading the output of the gain stage. Transistors Q_{14} and Q_{20} form the class AB output stage.

Transistor Q_{13} is a multicollector lateral *pnp*. The geometry of the device is shown in Fig. 6.10. Note that the collector ring has been split into two parts, one that faces on three-fourths of the emitter periphery and collects the holes injected from that periphery, and a second that faces on one-fourth of the emitter periphery and collects holes injected from that face. Thus the structure is analogous to two *pnp* transistors whose base-emitter junctions are connected in parallel and one of which has an I_S that is one-fourth that of a standard *pnp* transistor and the other an I_S that is three-fourths that of a standard *pnp* with fully enclosed emitter. This equivalence is depicted in Fig. 6.10.

6.3.1 dc Analysis of the 741 Operational Amplifier

The first step in evaluating the performance of the circuit is to determine the quiescent operating current and voltage of each of the transistors in the circuit. This dc analysis presents a special problem in operational amplifier circuits because of the very high gain involved. If we were to begin the dc analysis with the assumption that the two input terminals are grounded and then try to predict the output voltage, we would find that a small variation in the beta or output resistance of the devices in the circuit would cause large changes in the output voltage we predict, and in fact the calculation would usually show that the output stage was not in the active region but was saturated in one direction or the other. This problem exists in practice; for a voltage gain of 10^5, only 0.1 mV of input offset voltage is required to drive the output into saturation when the input voltage is zero. Thus the dc analysis must start with an assumption that the circuit is enclosed in a feedback loop which forces the *output* to some specified voltage, which is usually zero. We can then work backward and determine the operating points within the circuit.

A second assumption that simplifies the analysis is to assume that for the dc analysis the output resistance of the transistors does not greatly affect the dc currents flowing in the circuit. This is a reasonable assumption and results in 10 to 20 percent error in the calculated currents. Of course, the output resistance must be included in the small-signal analysis since it strongly affects the gain for the active-load amplifiers.

We first calculate the currents in the biasing current sources Q_{10} and Q_{13AB}. This subcircuit is shown in Fig. 6.11. Neglecting base currents and assuming that all transistors are in the forward-active region, we can calculate the reference current as

$$I_{\text{ref}} = \frac{V_{CC} - V_{EE} - 2V_{BE(\text{on})}}{39\ \text{k}\Omega} = 0.73\ \text{mA} \tag{6.28}$$

where we have assumed that $V_{BE(\text{on})} = 0.7$ V.

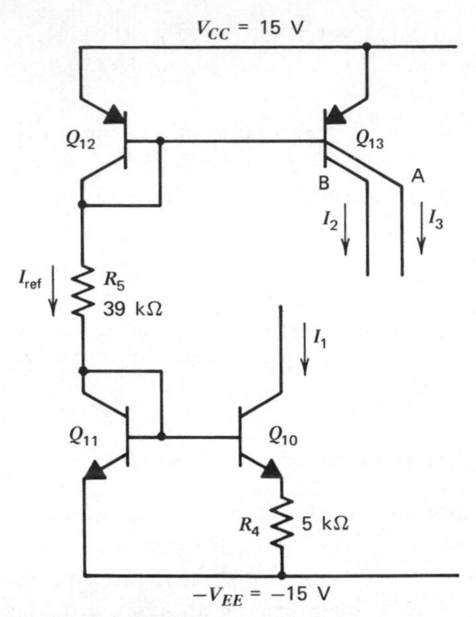

Fig. 6.11 Bias circuitry for the 741.

The combination of Q_{11} and Q_{10} forms a Widlar current source as discussed in Chapter 4. The output current, I_1, must be found by trial-and-error solution of the relation

$$V_T \ln \frac{I_{\text{ref}}}{I_1} = (5 \text{ k}\Omega) I_1 \tag{6.29}$$

the result of which is,

$$I_1 = 19 \ \mu\text{A} \tag{6.30}$$

The currents, I_2 and I_3, are three-fourths and one-fourth of the reference current, respectively,

$$I_2 = 0.55 \text{ mA} \qquad I_3 = 0.18 \text{ mA} \tag{6.31}$$

The circuit can now be simplified to the form shown in Fig. 6.12. We first determine the bias currents in the input stage. This subcircuit is shown in Fig. 6.13. For this analysis we will neglect the effects of the dc base currents flowing in the *npn* transistors since these typically have betas of several hundred. However, the betas of the lateral *pnp* transistors are typically much lower and we must consider the effects of base current in these. We designate the collector current of transistor Q_9 as I_{C9}. The collector current of Q_8 must be the same, and so the current I_A is the sum of that current plus the base currents of Q_8 and Q_9.

$$I_A = -I_{C9}\left(1 + \frac{2}{\beta_{pnp}}\right) \tag{6.32}$$

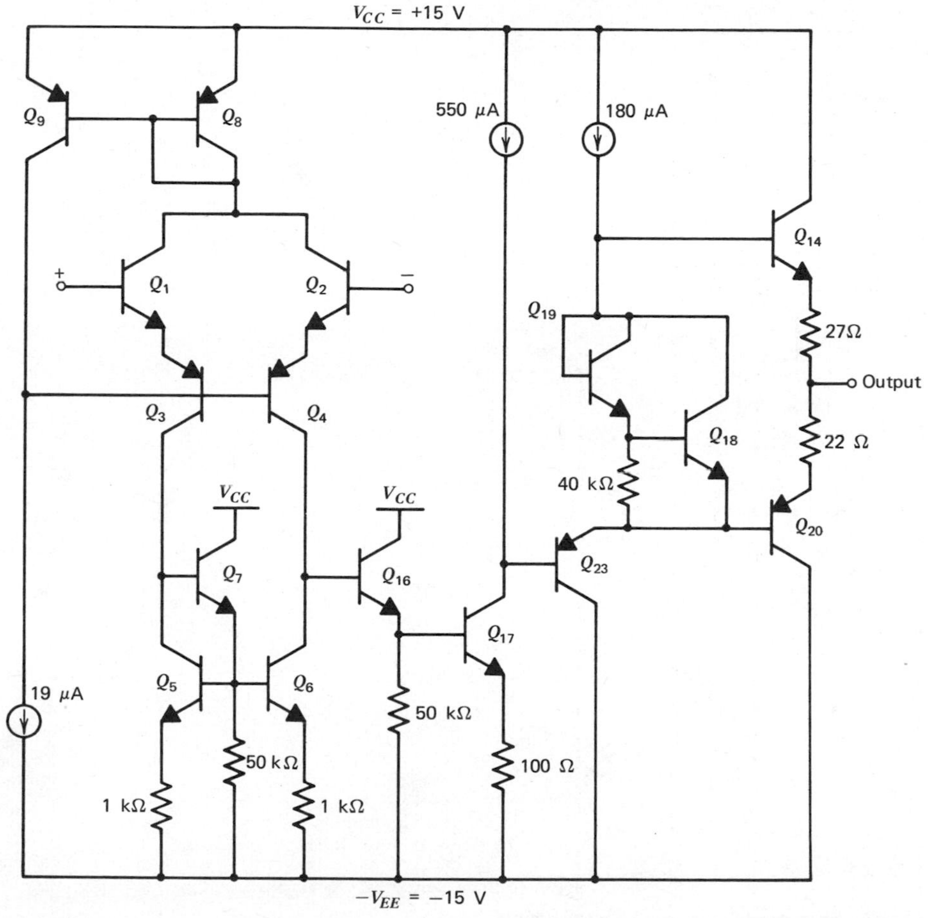

Fig. 6.12 Simplified schematic of the 741 with idealized biasing current sources.

If we neglect the base current of Q_1 and Q_2, then the emitter currents of Q_3 and Q_4 are,

$$I_{E3}, I_{E4} = \frac{I_A}{2} = \frac{-I_{C9}}{2}\left[1 + \frac{2}{\beta_{pnp}}\right] \tag{6.33}$$

The sum of the base currents of Q_3 and Q_4 and the collector current of Q_9 is equal to the biasing current of 19 μA. Thus,

$$19\ \mu\text{A} = -I_{C9} + \frac{I_{E3}}{1 + \beta_{pnp}} + \frac{I_{E4}}{1 + \beta_{pnp}} = -I_{C9}\left[1 + \left(\frac{1 + \dfrac{2}{\beta_{pnp}}}{1 + \beta_{pnp}}\right)\right] \tag{6.34}$$

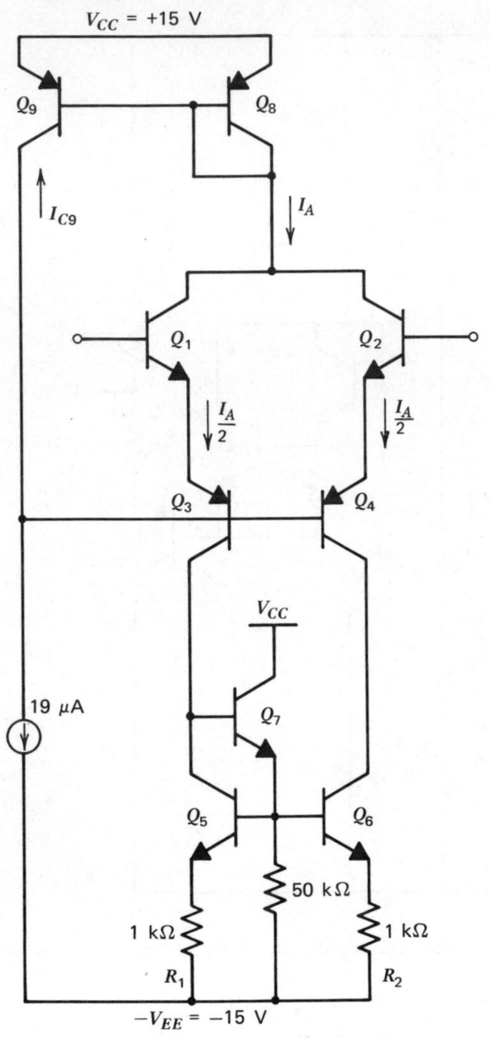

Fig. 6.13 Input stage of 741 with biasing source.

where (6.33) has used. Substitution of (6.34) in (6.32) gives

$$I_A = 19\ \mu\text{A}\left[\frac{1 + \dfrac{2}{\beta_{pnp}}}{1 + \left(\dfrac{1 + \dfrac{2}{\beta_{pnp}}}{1 + \beta_{pnp}}\right)}\right] \simeq 19\ \mu\text{A}\left(1 + \frac{1}{\beta_{pnp}}\right) \tag{6.35}$$

Thus if the *pnp* beta is reasonably large, input transistors Q_1 and Q_2 have a collector current of magnitude 9.5 μA each, and transistors Q_3, Q_4, Q_5, and Q_6 also have collector currents of a magnitude equal to approximately 9.5 μA. The biasing feedback loop formed by Q_8 and Q_9 thus stabilizes the bias current in each of the input devices at approximately one-half of the collector current of Q_{10}.

The magnitudes of the collector currents of transistors Q_5 and Q_6 are equal to those of transistors Q_3 and Q_4 if base currents are neglected. We must now calculate the dc collector current in transistor Q_7. The emitter current of Q_7 consists of the base current of Q_5 and Q_6, which we will neglect, and the current flowing in the 50-kΩ resistor. The voltage across this resistor is the sum of the base-emitter drop of Q_5 and Q_6 and the drop across the 1-kΩ resistors in series with the emitters of Q_5 and Q_6. The base-emitter drop, assuming an I_S of 10^{-14} A and a collector current of 9.5 μA, is 537 mV. The voltage drop across the 1-kΩ resistors is 9.5 mV, so that the current in the 50-kΩ resistor is 547 mV/50 kΩ or 11 μA. Thus the collector current of Q_7 is equal to 11 μA.

We next consider transistor Q_{16} in Fig. 6.12. The voltage at the base of Q_{17} is equal to the base-emitter voltage of Q_{17} plus the drop across the 100-Ω resistor. The collector current of Q_{17} must be equal to 550 μA, the current supplied by current source Q_{13B} if the output voltage of the amplifier is to be zero as we have assumed. The voltage at the base of Q_{17} with respect to V^- is:

$$V_{B17} = (550\ \mu\text{A})(100) + V_T \ln \frac{550 \times 10^{-6}}{10^{-14}} \tag{6.36}$$

$$= 55\text{ mV} + 642\text{ mV} \tag{6.37}$$

$$V_{B17} = 697\text{ mV} \tag{6.38}$$

The base current of Q_{17} is, assuming a beta of 250,

$$I_{B17} = \frac{550\ \mu\text{A}}{250} = 2.2\ \mu\text{A} \tag{6.39}$$

Thus the collector current of Q_{16} is the sum of the current in the 50-kΩ resistor in its emitter and the base current of Q_{17}, or

$$I_{C16} = \frac{697\text{ mV}}{50\text{ k}\Omega} + 2.2\ \mu\text{A} = 16\ \mu\text{A} \tag{6.40}$$

We now consider the output stage shown in Fig. 6.14, assuming for the time being that base currents are negligible. If this is true, all of the 180-μA current source current flows through transistor Q_{23}. We must now determine the dc bias currents in Q_{14}, Q_{18}, Q_{20}, and Q_{19}. We assume that the circuit is connected with feedback in such a way that the output is driven to zero volts, and the output current is zero. Thus I_{C14} and I_{C20} are approximately equal in magnitude. The

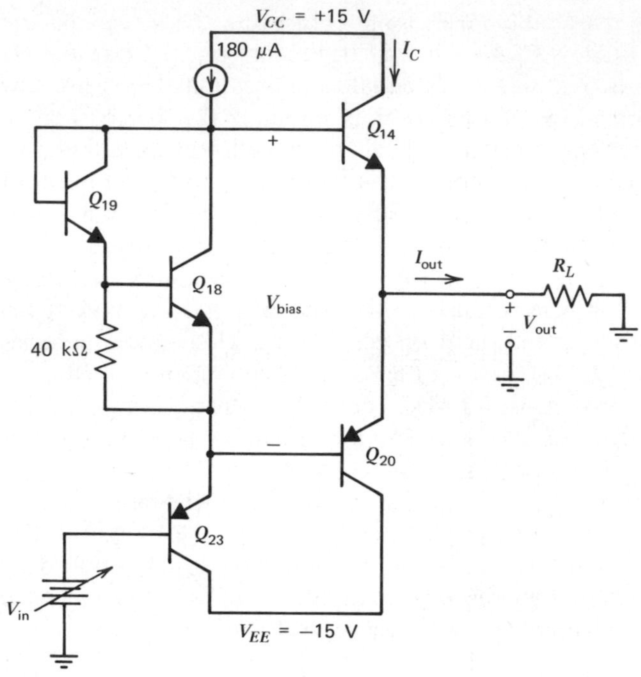

Fig. 6.14 Schematic of output stage of the 741. V_{out} is zero for the bias current calculation.

first task is to compute V_{bias}. The collector current of Q_{19} will be *approximately* 0.6 V/40 kΩ or 15 μA, if we neglect the base current of transistor Q_{18}. If this were true, the collector current of Q_{18} would be approximately (180 − 15) μA or 165 μA. Let us assume that this is true, and recalculate the collector current of Q_{19}. This current will be the sum of the base current of Q_{18} and the current in the 40-kΩ resistor.

$$I_{C19} = \frac{165\ \mu\text{A}}{\beta_F} + \frac{V_T \ln \dfrac{165\ \mu\text{A}}{10^{-14}\ \text{A}}}{40\ \text{k}\Omega} \tag{6.41}$$

$$I_{C19} = 15.6\ \mu\text{A} \tag{6.42}$$

This is very close to the original estimate. Now that we know the current in Q_{19} and Q_{18}, we can estimate the output transistor bias current using the relationships developed in Chapter 5 for this type of output stage. We found that

$$I_{C14} = -I_{C20} = \sqrt{I_{C19}I_{C18}}\sqrt{\frac{I_{S14}I_{S20}}{I_{S18}I_{S19}}} \tag{6.43}$$

Here we have neglected the voltage drop in the small resistors in series with the emitters of Q_{14} and Q_{20}. The actual quiescent current that flows in the output transistors depends on their I_S values, which depends on their physical geometry. Both of these transistors must carry large currents while maintaining good beta when a small value of load resistance is attached to the output, so they are made with larger geometry than the other transistors in the circuit. The specific geometry used varies with manufacturer but the I_S of these devices is typically about three times as large as a small geometry device. Thus,

$$I_{C14} = -I_{C20} = \sqrt{(15.6\ \mu\text{A})(165\ \mu\text{A})} \times 3 = 152\ \mu\text{A} \tag{6.44}$$

This completes the dc analysis of the circuit. Referring to the schematic diagram of Fig. 6.9*a*, we can see that the circuit contains several devices that are active only during overload conditions. Transistor Q_{15} turns on only when the voltage drop across the 27-Ω resistor in series with the output transistor exceeds about 550 mV, which occurs when the output sources a current of about 20 mA. When Q_{15} turns on it limits the current to the base of Q_{14} and the output current can increase no further. Thus the transistor performs short-circuit protection, preventing damage to the amplifier due to excess current flow and power dissipation should the output be shorted, for example, to the negative power supply. Transistors Q_{21}, Q_{22}, and Q_{24} perform a similar function for the case of the output sinking current. The extra emitter on the substrate *pnp* Q_{23} also avoids excess dissipation that can occur if Q_{17} is allowed to saturate. Assume that the inverting input terminal were overdriven so as to make it more positive than the non-inverting terminal by enough voltage to turn off Q_1. The current into the base of Q_{16} would be 19 μA. If no clamping were included, this current would be amplified by the beta of Q_{16}, which can be as high as 1000, giving a current of 19 mA. This current would flow into the base of Q_{17} (which would saturate) and finally to V^- through the 100-Ω resistor in the emitter of Q_{17}. This would result in a power dissipation of (19 mA) (30 V), or about 600 mW in Q_{16}. The extra emitter on Q_{23} prevents Q_{17} from saturating by diverting the base drive away from Q_{16} when V_{CB} of Q_{17} reaches zero volts. This eliminates the possibility of the high current condition that could damage Q_{16}.

6.3.2 Small-Signal Analysis of the 741 Operational Amplifier

Our next objective is to determine the small-signal properties of the amplifier. We will break the circuit up into its three stages—the input stage, gain stage, and output stage—and determine the input resistance, output resistance, and transconductance of each stage. Consider first the ac schematic of the input stage as shown in Fig. 6.15. Here we have assumed a pure differential-mode input to the circuit; we will consider common-mode inputs later. As a result, the bases of the *pnp* transistors are at ac ground. We first calculate the transconductance of the stage by shorting the output to ground and calculating the output current

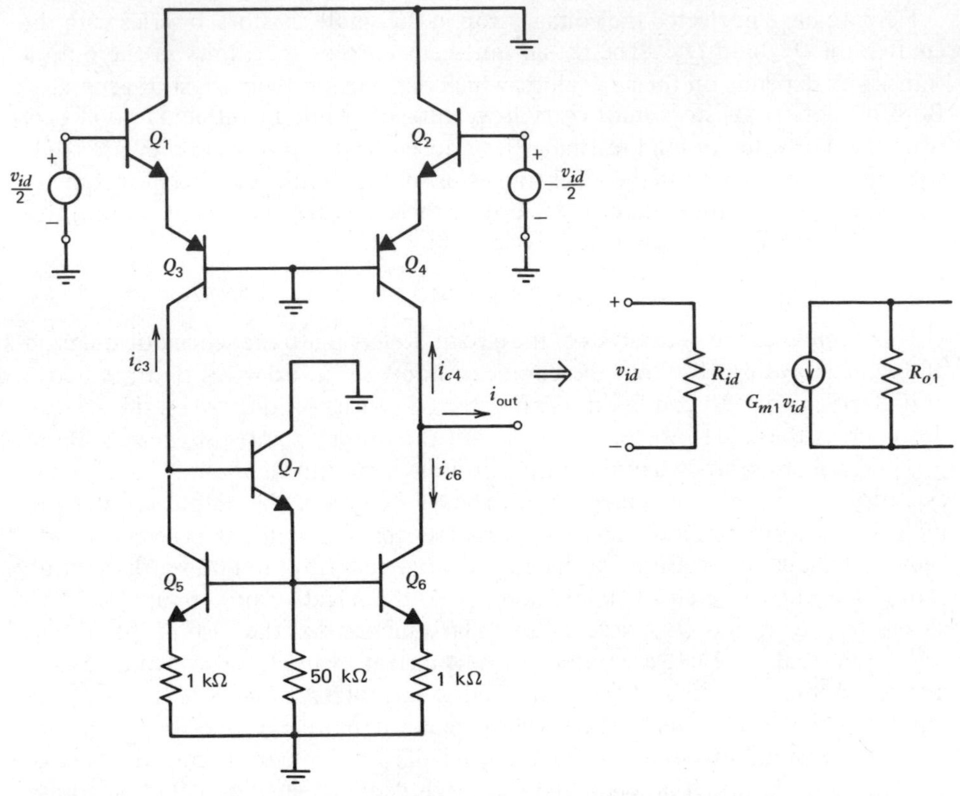

Fig. 6.15 Input stage ac schematic and small-signal, two-port equivalent circuit.

that results when a differential input is applied. Under the shorted output condition, no small-signal current flows in the output resistances of Q_4 and Q_6, so that the active-load circuit simply produces a current i_{c6} that is equal in magnitude and opposite in sign to the collector current i_{c3} of Q_3. Thus i_{out} is equal to $-(i_{c4} - i_{c3})$, and to calculate the transconductance, G_m, we need only consider the circuit shown in Fig. 6.16. Since the resistance presented to the collector of Q_3 is relatively small compared to r_o of Q_3, we need not consider the output resistances of the devices in this calculation. The small-signal equivalent half-circuit for Fig. 6.16*a* is shown in Fig. 6.16*b*. For the half-circuit,

$$\frac{v_{id}}{2} = v_1 + v_3 \tag{6.45}$$

and

$$v_1 g_{m1}\left(1 + \frac{1}{\beta_{01}}\right) = v_3 g_{m3}\left(1 + \frac{1}{\beta_{03}}\right) \tag{6.46}$$

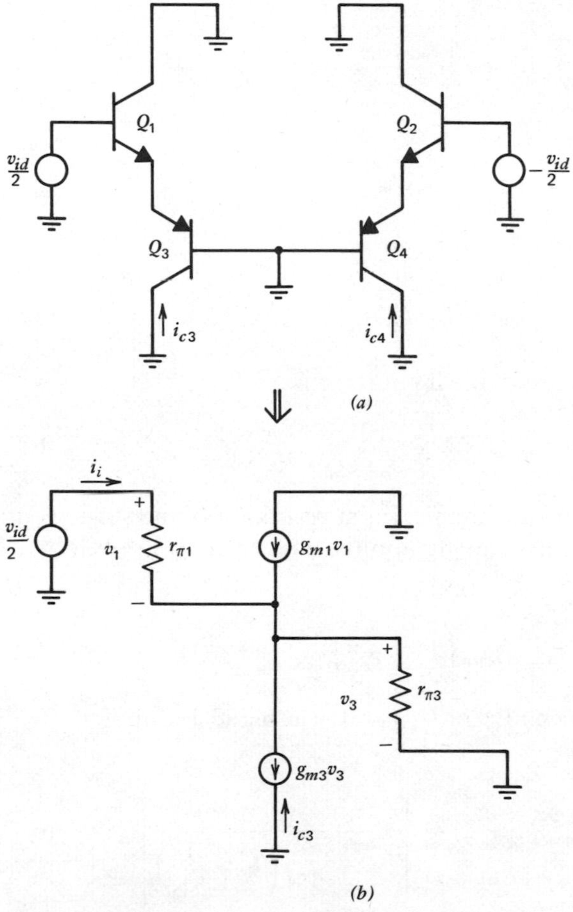

Fig. 6.16 (*a*) ac schematic and (*b*) small-signal equivalent half circuit for calculation of the transconductance of the input stage.

where β_{01} is the β_0 of Q_1 and β_{03} is the β_0 of Q_3. Combining these equations,

$$\frac{v_{id}}{2} = v_3 \left[\frac{g_{m3}\left(1 + \dfrac{1}{\beta_{03}}\right)}{g_{m1}\left(1 + \dfrac{1}{\beta_{01}}\right)} + 1 \right] \tag{6.47}$$

Since $|I_{C1}| = |I_{C3}|$, then $g_{m1} \simeq g_{m3}$. We also assume that $\beta_{01}, \beta_{03} \gg 1$. Thus

$$v_3 = \frac{v_{id}}{4} \tag{6.48}$$

and

$$i_{c3} = \frac{-g_{m3}v_{id}}{4} \tag{6.49}$$

From the symmetry of the circuit,

$$i_{c4} = +\frac{g_{m3}v_{id}}{4} \tag{6.50}$$

and, referring to Fig. 6.15,

$$i_{\text{out}} = -i_{c4} + i_{c3} = -\frac{g_{m3}v_{id}}{2} \tag{6.51}$$

Thus the overall transconductance of the input stage is

$$G_{m1} = -\frac{i_{\text{out}}}{v_{id}} = \frac{g_m}{2} = \frac{1}{5.4\ \text{k}\Omega} \tag{6.52}$$

This calculation also yields the differential input resistance of the stage. From Fig. 6.16*b*, the resistance, R_{eq}, seen looking into the emitter of Q_3 is given by

$$R_{\text{eq}} = \frac{1}{g_{m3}\left(1 + \dfrac{1}{\beta_{03}}\right)}$$

This resistance appears in the emitter of Q_1, so that using (3.27) we find

$$\frac{\left(\dfrac{v_{id}}{2}\right)}{i_i} = r_{\pi 1} + R_{\text{eq}}(\beta_{01} + 1) = \left[r_{\pi 1} + \frac{(\beta_{01} + 1)}{g_{m3}\left(1 + \dfrac{1}{\beta_{03}}\right)}\right] \tag{6.53}$$

Assuming again that $\beta_{03} \gg 1$, and that $g_{m1} = g_{m3}$,

$$\frac{v_{id}}{2} = \left(r_{\pi 1} + \frac{\beta_{01}}{g_{m3}}\right) i_i = (r_{\pi 1} + r_{\pi 1}) i_i \tag{6.54}$$

Solving for R_{id},

$$\frac{v_{id}}{i_i} = R_{id} = 4r_{\pi 1} = 2.7\ \text{M}\Omega \tag{6.55}$$

where $\beta_0 = 250$ is assumed.

The input resistance of the amplifier is thus four times the input resistance of one of the input transistors. In this calculation, we have neglected the fact that when v_{id} changes, the output voltage changes and produces feedback to the input through the output resistance of Q_4. This effect produces a difference between the input resistances seen at the two input terminals.

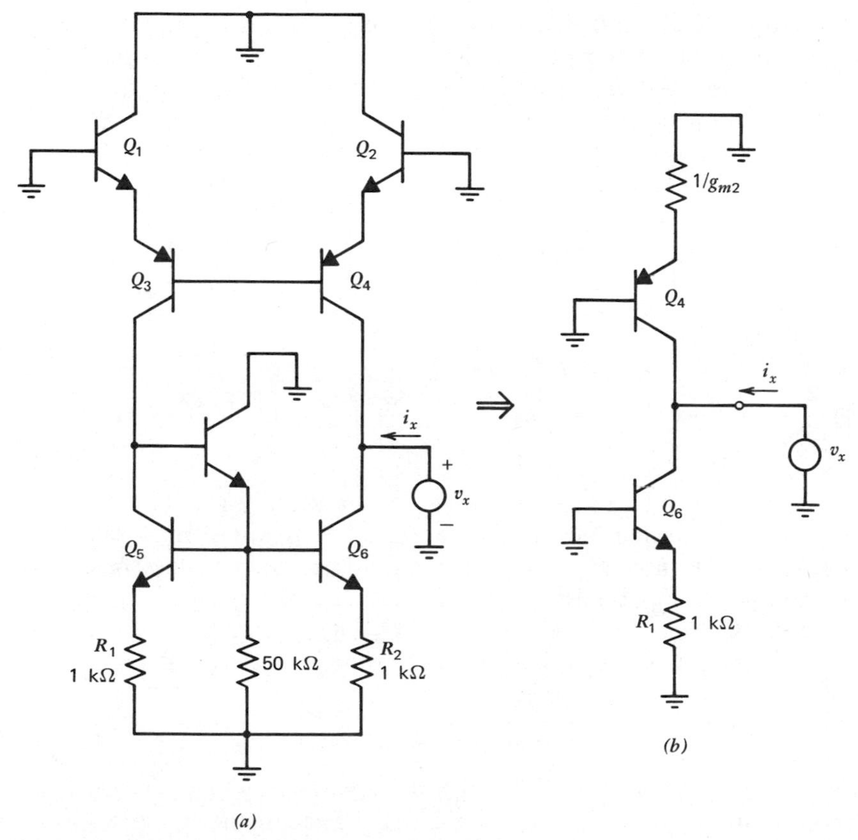

Fig. 6.17 (*a*) Test voltage source v_x applied to the output of the input stage for calculation of the output resistance. (*b*) Simplified circuit.

We now must calculate the output resistance R_{o1} of the input stage. The input voltage is set to zero as shown in Fig. 6.17*a*, and the resistance seen looking into the output of the stage is calculated. This calculation is greatly simplified by the following approximations. The Thévenin equivalent resistance, which is presented to the base of Q_6, is small compared to the r_π of Q_6, so little error is introduced by assuming that the base of Q_6 is grounded. Also, the only output resistances that are significant in determining R_{o1} are those of Q_4 and Q_6 since all the other collectors are returned to low impedance points. The circuit is not balanced, so that the node to which the bases of the *pnp* transistors are attached is not a virtual ground. A more detailed calculation shows, however, that this point may be considered a virtual ground without greatly changing the results. The calculation can now be carried out using the circuit of Fig. 6.17*b*, for which by inspection,

$$R_{o1} = (R_{\text{out}}|_{Q4} \| R_{\text{out}}|_{Q6}) \tag{6.56}$$

The incremental resistance in the emitter of Q_6 is equal to 1 kΩ, while the incremental resistance in the emitter of Q_4 is the r_e of Q_2. The output resistance of transistor current sources with a resistance in the emitter was considered in Chapter 4. Equation 4.20 can be used to give

$$R_{o1} = \left\{ r_{o4} \times \left[\frac{1 + g_{m4}\left(\dfrac{1}{g_{m2}}\right)}{1 + \dfrac{g_{m4}/g_{m2}}{\beta_{04}}} \right] \right\} \Bigg\| \left\{ r_{o6} \times \left[\frac{1 + (g_{m6})(1\ \text{k}\Omega)}{1 + \dfrac{(g_{m6})(1\ \text{k}\Omega)}{\beta_{06}}} \right] \right\} \tag{6.57}$$

Assuming $\beta_{04}, \beta_{06} \gg 1$,

$$R_{o1} = 2r_{o4} \| 1.36 r_{o6} \tag{6.58}$$

For $\eta_{npn} = 2 \times 10^{-4}$, $I_C = 9.5\ \mu\text{A}$, $\eta_{pnp} = 5 \times 10^{-4}$ we find

$$R_{o1} = 6.8\ \text{M}\Omega \tag{6.59}$$

Thus the equivalent circuit for the input stage is as shown in Fig. 6.18.

We now turn to the second stage, shown in Fig. 6.19*a*. Again we must calculate the input resistance, transconductance, and output resistance of the stage. We begin by calculating the input resistance. We first calculate the Thévenin equivalent resistance seen looking into the base of transistor Q_{17}, designated R_{eq1} in Fig. 6.19*a*. Utilizing the results from Chapter 3 for the input resistance of a common emitter amplifier with a resistance in the emitter,

$$R_{\text{eq1}} = [r_{\pi 17} + (\beta_{017} + 1)100\ \Omega] \tag{6.60a}$$

Here we have neglected the effects of the output resistance of Q_{17}. The positive feedback contributed by this resistance does actually decrease the input resistance slightly.

The circuit can now be reduced to the form shown in Fig. 6.19*b*. The input resistance of the stage is:

$$R_{i2} = r_{\pi 16} + (\beta_{016} + 1)(R_{\text{eq1}} \| 50\ \text{k}\Omega) \tag{6.60b}$$

Combining Equations 6.60a and 6.60b,

$$R_{i2} = r_{\pi 16} + (\beta_0 + 1)\{[r_{\pi 17} + (\beta_0 + 1)(100\ \Omega)] \| [50\ \text{k}\Omega]\} \tag{6.60c}$$

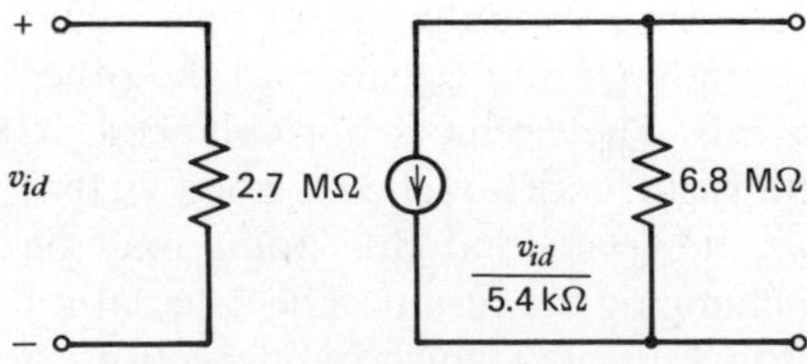

Fig. 6.18 Two-port equivalent circuit for the input stage.

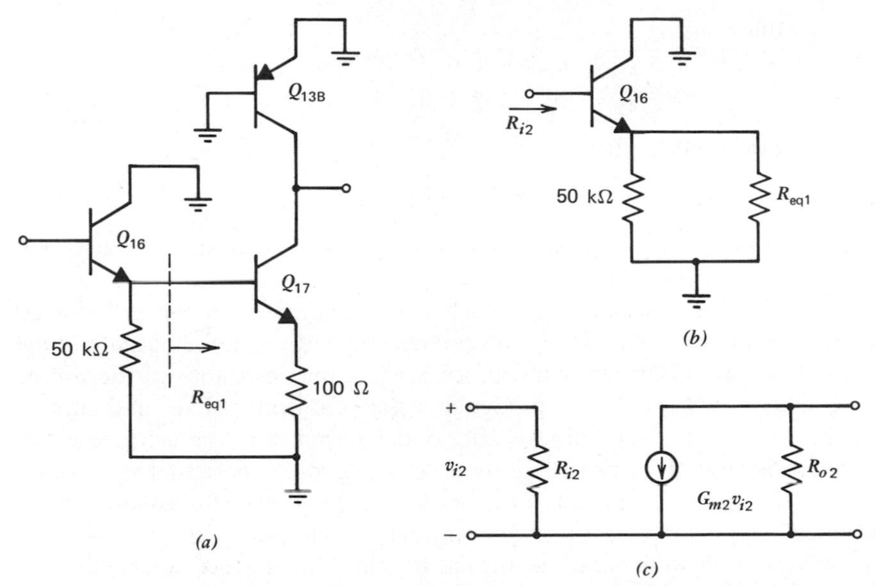

Fig. 6.19 (*a*) Small-signal equivalent circuit for second stage. (*b*) Circuit for calculation of input resistance. (*c*) Two-port equivalent circuit for second stage.

We assume that the transistors have a β_0 of 250. Transistor Q_{16} operates at a current of 16 μA while Q_{17} operates at 550 μA. Evaluating (6.60c) we find

$$R_{i2} = 460 \text{ k}\Omega + 251 \times (35 \text{ k}\Omega \| 50 \text{ k}\Omega) = 5.5 \text{ M}\Omega \tag{6.60d}$$

We next calculate the transconductance of the stage. If we assume that the voltage gain of the emitter follower Q_{16} is nearly unity, the transconductance G_{m2} of the stage is just that of transistor Q_{17} with the 100-Ω resistor in the emitter.

$$G_{m2} = \frac{g_{m17}}{1 + (g_{m17})R_E} = \left(\frac{1}{147\ \Omega}\right) \tag{6.61}$$

The output resistance of the stage (R_{o2}) is the output resistance of $Q_{13\text{B}}$ in parallel with that seen looking into the collector of Q_{17}. Again utilizing (4.20),

$$R_{o2} = r_{o13\text{B}} \| r_{o17} \left(\frac{1 + g_{m17}R_E}{1 + \dfrac{g_{m17}R_E}{\beta_0}} \right) \tag{6.62}$$

Assuming $\beta_0 \gg 1$,

$$R_{o2} = [r_{o13\text{B}}] \| \{ r_{o17}[1 + (g_{m17})(100\ \Omega)] \} \tag{6.63}$$

Assuming that

$$\eta_{pnp} = 5 \times 10^{-4} \quad (6.64)$$

$$\eta_{npn} = 2 \times 10^{-4} \quad (6.65)$$

(6.63) can be evaluated to give:

$$R_{o2} = 83\ \text{k}\Omega \quad (6.66)$$

Thus we have developed R_{i2}, R_{o2}, and G_{m2} for the second stage. A two-port equivalent circuit is shown in Fig. 6.19*c*.

We now turn to the output stage. A schematic diagram is shown in Fig. 6.20. The output is either sourcing or sinking current, depending on the output voltage and load. As a result, the input resistance and output resistance of the output stage is heavily dependent on the particular value of output voltage and current. As will become evident, the input resistance of the output stage is much larger than the output resistance of the preceding stage, and as a result the actual value of the input resistance of the output stage does not strongly influence the voltage gain of the circuit. Thus we simply assume, for example, the output current is 2 mA and that this current is flowing *out* of the output terminal. We further assume that the load resistance is 2 kΩ. Since the output current is flowing *out* of the output stage, transistor Q_{14} is in the active region and transistor Q_{20} is conducting only a small amount of current. The small-signal equivalent circuit for this case is shown in Fig. 6.20*b*. First, it is clear that the voltage gain is approximately unity since the circuit consists of two emitter followers in series. We now calculate the input resistance, R_{i3}, of the output stage.

We first calculate the resistance seen looking into the base of transistor Q_{14}, which is designated $R_{\text{eq}2}$ in Fig. 6.20*b*. Using the results from Chapter 3 on emitter followers,

$$R_{\text{eq}2} = r_{\pi 14} + (\beta_{014} + 1)(2\ \text{k}\Omega) \quad (6.67)$$

We next calculate the Thévenin equivalent resistance seen at the emitter of Q_{23}, looking toward diodes Q_{19} and Q_{18}. This resistance is designated $R_{\text{eq}3}$ in Fig. 6.20*b*, and is seen by inspection to be

$$R_{\text{eq}3} = r_{d18} + r_{d19} + r_{o13\text{A}} \| R_{\text{eq}2} \quad (6.68)$$

Finally, the input resistance of the stage is that of emitter follower Q_{23} with a resistance $R_{\text{eq}3}$ in its emitter, or:

$$R_{i3} = r_{\pi 23} + (\beta_{o23} + 1)R_{\text{eq}3} \quad (6.69)$$

Transistor Q_{23} and the two diodes operate at a current of 180 μA, while Q_{14} operates at 2 mA. Assuming an *npn* β_0 of 250 and a *pnp* β_0 of 50, (6.69) can be evaluated to yield:

$$R_{i3} = r_{\pi 23} + 51R_{\text{eq}3} \quad (6.70)$$

and thus

$$R_{i3} = 9.1\ \text{M}\Omega \quad (6.71)$$

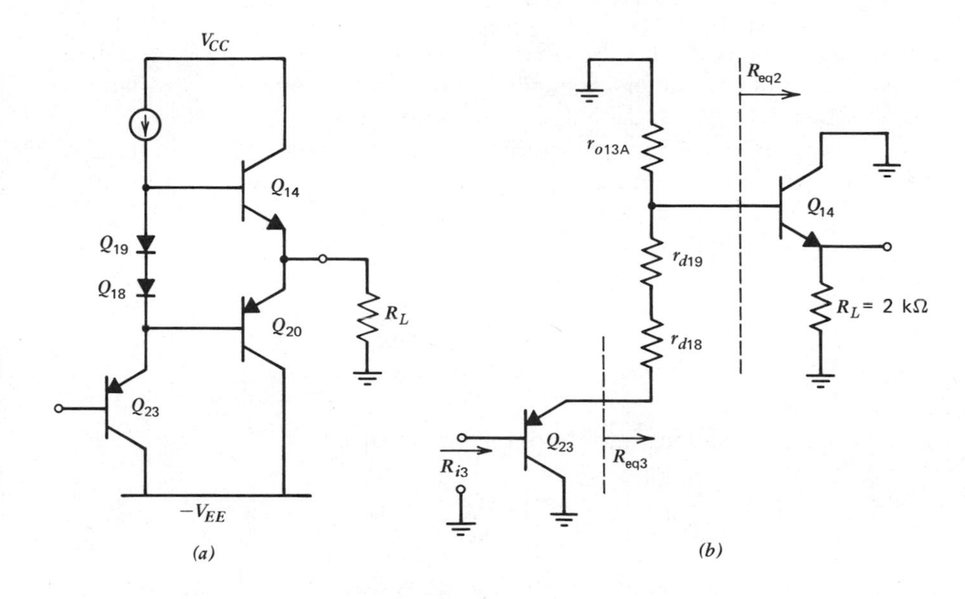

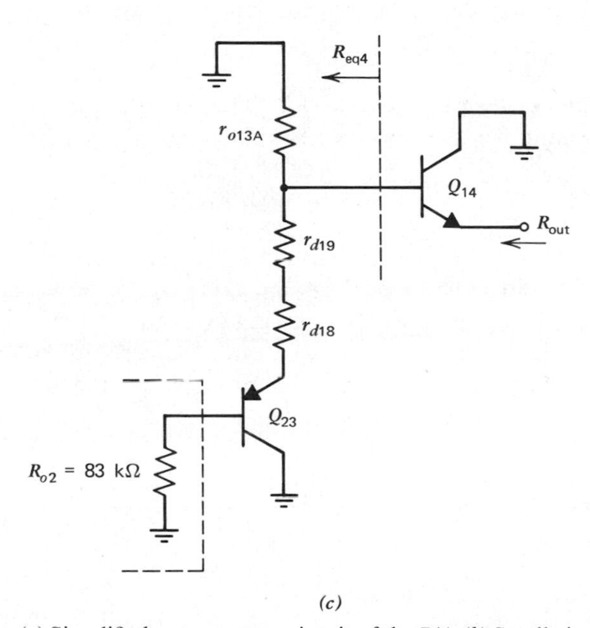

Fig. 6.20 (*a*) Simplified output-stage circuit of the 741. (*b*) Small-signal circuit when sourcing current. (*c*) Circuit for calculation of R_{out}.

This input resistance is much larger than the output resistance of the preceding stage, and as a result the gain of the amplifier is not strongly affected by variations in load resistance attached to the output of the amplifier.

We now calculate the output resistance of the output stage. We must include the output resistance of the preceding stage in this calculation; the small-signal equivalent circuit for performing the calculation is shown in Fig. 6.20*c*.

The resistance seen looking to the left from the base of Q_{14} is:

$$R_{eq4} = r_{o13A} \left\| \left[r_{d19} + r_{d18} + \frac{R_{o2} + r_{\pi 23}}{\beta_{o23} + 1} \right] \right. \tag{6.72}$$

$$= 2.06 \text{ k}\Omega \tag{6.73}$$

The resistance seen looking into the output terminal is then:

$$R_{out} = \frac{R_{eq4} + r_{\pi 14}}{\beta_{o14} + 1} = 21 \ \Omega \tag{6.74}$$

To this we must add the series current limiting resistance, R_6, which is 26 Ω. The actual value of R_{out} is thus 47 Ω. The output resistance will change with operating point and is heavily dependent on the current flowing in the output transistor.

Small-Signal Performance of the Complete Circuit. The small-signal equivalent circuit for the complete amplifier is shown in Fig. 6.21. The voltage gain is:

$$A_v = (G_{m1})(R_{o1} \| R_{i2})(G_{m2})(R_{o2} \| R_{i3}) \tag{6.75}$$

$$= 560 \times 564 = 315{,}000 \tag{6.76}$$

Note that both stages contribute about the same gain. The second stage loads the first, reducing its gain by about half. This loading gives the gain of the circuit a beta dependence that will cause the gain to vary with temperature and process variations. However, the output stage does not significantly load the output of the second stage and the voltage gain is almost independent of the load resistance

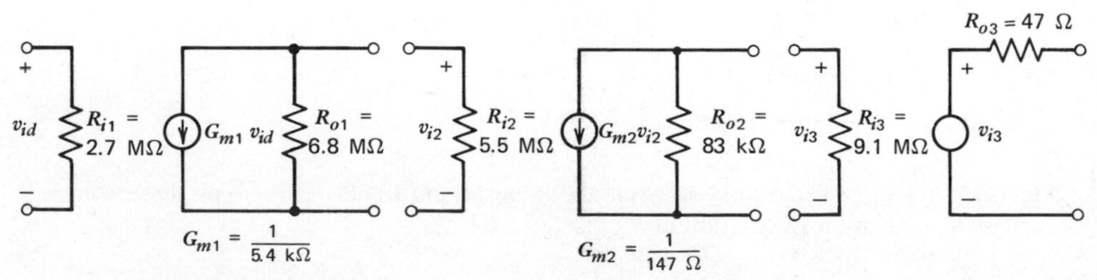

Fig. 6.21 Small-signal equivalent circuit for the complete 741.

attached to the output. The input and output resistances of the complete amplifier are:

$$R_{id} = 2.7\ \text{M}\Omega \tag{6.77}$$

$$R_o = 47\ \Omega \tag{6.78}$$

Inclusion of Second-Order Effects in the Analysis; Computer Analysis. The objective of the first-order approximate analysis procedure just carried out for the 741 was to gain an insight into the factors that are most important in affecting circuit performance. However, the simplifying assumptions made in the analysis limit the accuracy of the results both in terms of dc operating points and voltage gain. The principal violations of the assumptions made are as follows:

1. The output resistance of the transistors was neglected in the dc analysis. Actually, the collector current of a typical lateral *pnp* transistor increases about 30 percent when V_{CE} increases from zero to 15 V, assuming that the base-emitter voltage is held constant and that V_A is about 50 V. For the *npn* devices, the figure is 12 percent, assuming a V_A of 120 V. The net effect of this is to increase the bias currents in the circuit by from 10 to 30 percent over those calculated. This has an effect on the calculated small-signal impedance levels.
2. The variation of transistor beta with current has been neglected. As discussed in Chapter 1, the transistor current gain falls at low collector current levels due to recombination in the emitter-base, space-charge layer. Because of this, the input stage devices Q_1 to Q_6 and device Q_{16} have substantially lower small-signal current gain than was assumed in the analysis. The principal effect of this falloff is to reduce the voltage gain of the amplifier and increase the input bias current.

When a detailed quantitative prediction is required regarding the performance of such circuits, a computer analysis is generally less time consuming than attempting a hand analysis, taking into account the second-order effects. Such an analysis applied to the 741 yields the bias levels shown in Fig. 6.22.

6.3.3 Input Offset Voltage, Input Offset Current, and Common-Mode Rejection Ratio of the 741

Two important aspects of the performance of operational amplifiers are the input offset voltage and current. These deviations from ideality limit the ability of the circuit to amplify small dc signals accurately, since the offsets are indistinguishable from the signal. The calculation of these quantities is somewhat tedious and a calculation of the input offset voltage is contained in Appendix A6.1. In this section we discuss qualitatively the factors influencing the offset voltage and current of the 741 circuit.

The input offset voltage of differential amplifiers can be shown to be primarily dependent on the offset voltage of the first stage, provided that the voltage gain

Transistor	Collector Current, μA
Q_1	12.32
Q_2	12.38
Q_3	−12.30
Q_4	−12.36
Q_5	12.10
Q_6	12.11
Q_7	11.34
Q_8	−23.53
Q_9	−30.56
Q_{10}	31.05
Q_{11}	32.2
Q_{12}	−700.6
Q_{13A}	−221.6
Q_{13B}	−682.9
Q_{14}	172.7
Q_{15}	Off
Q_{16}	16.87
Q_{17}	687.3
Q_{18}	204.0
Q_{19}	16.38
Q_{20}	−173.4
Q_{21}	Off
Q_{22}	Off
Q_{23A}	−219.7
Q_{23B}	Off
Q_{24}	Off

Fig. 6.22 Computer-predicted dc bias levels of the 741 operational amplifier.

of the first stage is reasonably high.[1] The input stage of the 741 is somewhat complex, consisting of three pairs of transistors, Q_1–Q_2, Q_3–Q_4, and Q_5–Q_6. The stage provides level shifting and differential-to-single-ended conversion as well as differential amplification. As a result, mismatches in each of the three device pairs, as well as mismatches in the resistor pair R_1–R_2, contribute to the input offset voltage of the circuit. As shown in Appendix A6.1, a typical offset voltage of about 2.6 mV would be expected for the 741 assuming typical transistor mismatch and resistor mismatch data.

The input offset current of the circuit results primarily from mismatches in the beta of the two input transistors, Q_1 and Q_2. As calculated in Appendix A6.1, a typical input offset current of about 4 nA would be expected for the circuit. This figure is highly dependent on the actual matching of transistor beta achieved in the circuit for the low current level at which the input stage operates.

As discussed in Section 6.2, the common-mode rejection ratio can be regarded as the change in input offset voltage that results from a unit change in common-mode input voltage. This change in input offset voltage arises because of two separate effects. First, when the common-mode input voltage changes, current-source transistors Q_9 and Q_{10} experience a change in collector-emitter voltage. Because of the finite output resistance of these devices, their collector current changes, which results in a change in the bias current in the input stage. If resistors R_1 and R_2 were not present, this would not change the input offset voltage, but the presence of these resistors results in a change in the ratio of the collector currents of Q_5 and Q_6 if a mismatch exists between Q_5 and Q_6, or if a mismatch exists between R_1 and R_2. This change in the ratio of the collector currents of Q_5 and Q_6 results in a change in the ratio of the currents flowing in Q_1–Q_2 and in Q_3–Q_4, and a change in the input offset voltage.

The second source of input offset voltage change is caused by the fact that Q_1, Q_2, Q_3, and Q_4 all experience changes in their collector-emitter voltages when the input common-mode voltage is changed. As a result, if a base-width mismatch exists in these devices, the ratio of the saturation currents of Q_1 and Q_2 and the ratio of the saturation currents of Q_3 and Q_4 will vary also. These ratio variations result in changes in the input offset voltage.

Comparison of Calculated Performance Parameters and Experimentally Observed Values. The calculated and observed values of typical performance parameters of the 741 are shown below.

	Calculated	Observed (typical)
Open-loop gain	315,000	200,000
Input resistance	2.7 MΩ	2.0 MΩ
Input bias current	38 nA	80 nA
Input offset current	3.8 nA[a]	10 nA
Input offset voltage	2.6 mV[a]	2 mV
Output resistance	47 Ω	75 Ω

[a] See Appendix A6.1.

The differences between the calculated and observed parameters result from differences between typical device parameters obtained in production and those assumed in the analysis, and from the approximations made in the hand analysis. We now turn from the analysis problem to that of improving the performance parameters of the 741 by circuit modifications.

6.4 DESIGN CONSIDERATIONS IN MONOLITHIC OPERATIONAL AMPLIFIERS

Operational-amplifier design involves tradeoffs between the various dc, small-signal, and transient performance parameters as well as the die size and resultant cost. In the 741, the principal design objective was to obtain an internally compensated circuit with moderate dc and ac performance while maintaining a small inexpensive die. However, by the use of more complex circuitry, the dc, small-signal, and transient performance of the circuit can be greatly improved. Frequently, steps taken to improve the transient behavior degrade the dc parameters such as input offset voltage and input bias current. In this section we explore design approaches to the improvement of the input parameters: input offset voltage, input bias current, and input offset current of the 741. Since the frequency response and compensation of operational amplifiers is discussed extensively in Chapter 9, we defer the discussion of optimization of transient behavior and slew rate until then.

Applications of Precision Operational Amplifiers. The input characteristics of operational amplifiers, including the input offset voltage and drift, input bias current and drift, and input offset current and drift, determine the lower limit of the magnitude of the dc signal that can be accurately amplified. These performance parameters are most important in instrumentation applications where very small dc potential differences must often be measured. A typical example is the thermocouple amplifier shown in Fig. 6.23. A thermocouple is a junction of two dissimilar metals that produces a potential difference that varies with temperature. When used as temperature sensors, two junctions are used in series with one held at a reference temperature. The potential difference produced by the series combination

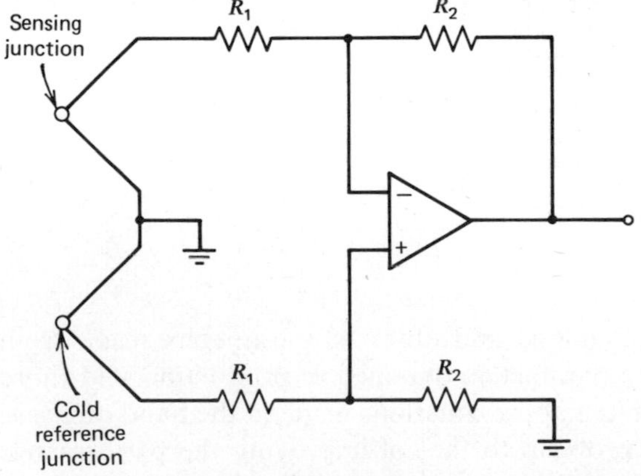

Fig. 6.23 Differential temperature-sensing amplifier using thermocouples.

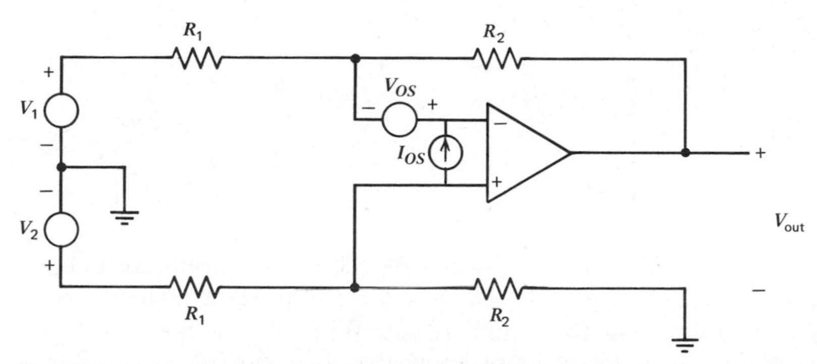

Fig. 6.24 Equivalent circuit of thermocouple amplifier including I_{OS} and V_{OS}.

of the junctions is then proportional to the temperature difference between them. These devices have a usable sensing temperature range that extends to several thousand degrees, and they are useful in furnace controllers and other such systems. For iron-constantan thermocouples the output voltage is about 50 μV/°C.

The offset voltage and current of the operational amplifier used will limit the resolution of this temperature-measuring system. The equivalent circuit with these imperfections included is shown in Fig. 6.24. By use of the summing point constraints the output can be calculated to be:

$$V_{\text{out}} = -\frac{R_2}{R_1}(\Delta V_i + V_{OS} + 2I_{OS}R_1) \tag{6.79}$$

where $\Delta V_i = (V_1 - V_2)$ and we have assumed for simplicity that $R_2 \gg R_1$. The output contains an error term that depends on the input characteristics of the amplifier. In this case the input offset current and voltage are the critical quantities, and not input bias current. If, for example, we use a 741 in this circuit with an iron-constantan thermocouple, the input offset voltage of 2.6 mV alone will give an error in the measured temperature of

$$T_\epsilon = \frac{2.6\text{ mV}}{50\ \mu\text{V/°C}} = 52°\text{C} \tag{6.80}$$

In such a critical application, an external potentiometer would be used to null the offset voltage of the 741 to zero to eliminate this large error. This is accomplished by inserting an external potentiometer between R_1 and R_2 in the 741 circuit. However, the important factor would then become the drift of the input offset voltage with temperature.

The offset voltage of the 741 is evaluated in Appendix A6.1. The drift behavior of the 741 can be evaluated by differentiating (6.101) from the appendix, which yields:

$$\frac{dV_{OS}}{dT} = \frac{V_{OS}}{T} + 2V_T \frac{d}{dT} \left[\frac{\dfrac{\Delta I_{S5-6}}{I_{S5-6}}}{\left(1 + \dfrac{I_{C5-6}R_{1-2}}{V_T}\right)} - \frac{\left(\dfrac{I_{C5-6}R_{1-2}}{V_T} \dfrac{\Delta R_{1-2}}{R_{1-2}}\right)}{\left(1 + \dfrac{I_{C5-6}R_{1-2}}{V_T}\right)} \right] \tag{6.81}$$

The terms involving I_S in (6.101) are assumed independent of temperature and the term involving β is usually negligible. The first term of (6.81) is forced to zero when the offset is nulled, but the second is not. While the saturation-current mismatch, $\Delta I_{S5-6}/I_{S5-6}$, is temperature insensitive, bias current I_{C5-6}, R_{1-2}, and V_T are all temperature dependent. Furthermore, the resistor mismatch term, $\Delta R_{1-2}/R_{1-2}$, will generally have nonzero temperature dependence because of the presence of the nulling potentiometer with different temperature coefficient than the diffused resistors themselves. The actually observed behavior is that nulling the 741 improves the drift somewhat over the unnulled state, and the residual offset voltage temperature coefficient has a typical value of about 3 to 5 μV/°C. This level of drift in the thermocouple temperature-sensor application would result in an error of 0.1°C in the sensed temperature for every degree of temperature change in the ambient temperature experienced by the operational amplifier. This sensitivity of sensed temperature to ambient temperature would be unacceptable in many precision process control systems. A need, therefore, exists for operational amplifiers with substantially better input offset voltage and input offset current drift than the 741. While in this particular example the offset voltage was most important, many applications involve high source impedances so that the input bias and offset current are also important.

6.4.1 Design of Low-Drift Operational Amplifiers

The offset voltage of multistage differential amplifiers depends primarily on the offset of the first stage, provided that stage has sufficiently high voltage gain. Thus the design of low-offset amplifiers is primarily a problem of designing an input stage in which as few component pairs as possible contribute to the stage offset voltage. In the 741, the input stage is relatively complicated since it is called on to provide gain, level shifting, and differential-to-single-ended conversion. A more optimum circuit from the standpoint of drift and offset is one which contains fewer devices to contribute to the offset voltage.[4] The simplest differential stage is the resistively loaded emitter-coupled pair shown in Fig. 6.25. The offset of this stage was considered in Chapter 3 and was found to be simply:

$$V_{OS} = V_T\left(-\frac{\Delta I_S}{I_S} - \frac{\Delta R_C}{R_C}\right) \tag{6.82}$$

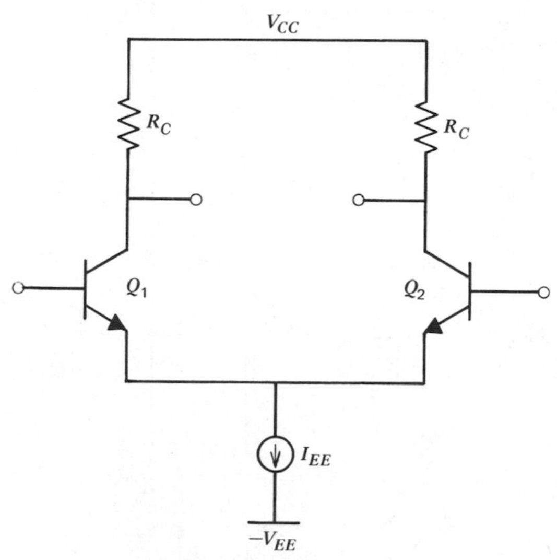

Fig. 6.25 Resistively loaded emitter-coupled pair.

The drift of this offset with temperature is given by:

$$\frac{dV_{OS}}{dt} = \frac{V_{OS}}{T} \tag{6.83}$$

For the same typical transistor and resistor mismatches assumed for the case of the 741, the standard deviation of the offset of the emitter-coupled pair alone is 1.56 mV and the drift unnulled is 5 μV/°C. This circuit is thus advantageous for use as the input stage of low-drift operational amplifiers.

One approach to the incorporation of this stage in the operational amplifier is to simply use an emitter-coupled pair followed by a 741 or equivalent. An example of such an approach is the 725 amplifier shown in Fig. 6.26. The gain of the input stage is chosen to be large enough that the second-stage drift does not contribute significantly to the offset, but yet the collector resistors must be kept small enough that the frequency response of the overall circuit is not excessively degraded by the response of the first stage.

If the collector resistors, R_C, could be adjusted to null the offset in the emitter-coupled pair input stage, then nulling of the offset would in principle simultaneously null the offset drift. Actually, the potentiometer used to adjust the collector resistor ratio generally displays different temperature dependence than the diffused resistors so that a new drift component is introduced because the R_C mismatch factor becomes temperature dependent. In practice, temperature drifts of offset on the order of 1 to 2 μV/°C are achievable, with external potentiometer nulling.

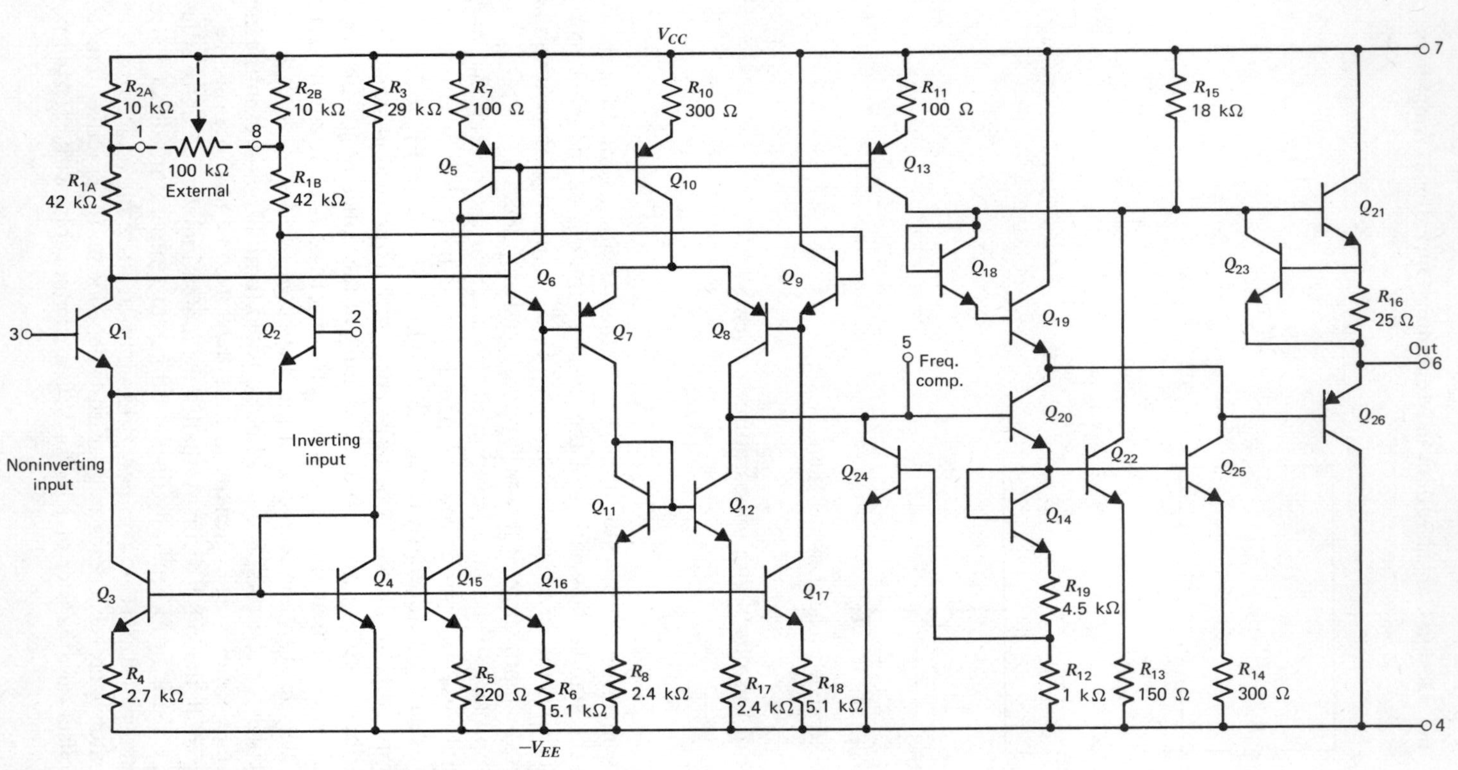

Fig. 6.26 Schematic diagram of the 725 low-drift operational amplifier.

Offset Trimming Techniques. An alternate approach to the realization of very low drift is the nulling of the offset voltage using on-chip resistances that are inserted or removed from the circuit under the control of an on-chip programmable read-only memory.[5] A schematic diagram of such a circuit is shown in Fig. 6.27. The collector resistors consist of a large portion in series with a smaller portion that is divided up into binary-weighted segments. The single-pole switches shown connected across the resistors are either fusible links of aluminum that can be opened by a current pulse, or in some cases are Zener diodes that can be caused to permanently change to a short-circuit state by passing a large current through

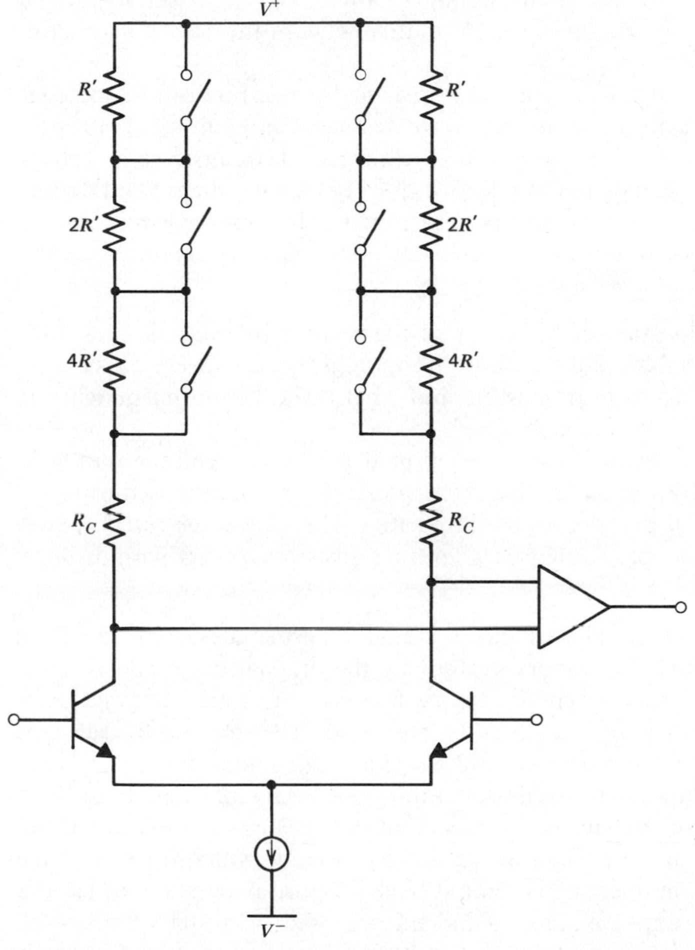

Fig. 6.27 Input-stage circuit for on-chip discrete trimming using fusible or shortable links. Resistance R' is much smaller than R_C.

them in the reverse direction. In either case, the offset can be nulled when the devices are tested in wafer form, and the nulling does not introduce additional drift components since the collector resistances are all diffused components of the same type. Since the resistors are binary weighted, the number of discrete steps into which the adjustment range is divided is proportional to 2^n where n is the number of fusible links in one collector. This is a powerful and low-cost technique for improving the input characteristics of monolithic precision operational amplifiers.

Layout and Device Matching Considerations. A basic objective in the design of precision circuits like the 725 is the minimization of the input offset voltage, which requires the minimization of the mismatch between the collector resistors and the input devices.

The accuracy with which two identical transistors or resistors can be matched has a first-order effect on the attainable performance in monolithic operational amplifiers as well as many other types of analog integrated circuits such as voltage regulators (Chapter 8), analog multipliers (Chapter 10), analog-digital and digital-analog converters, voltage comparators, and others. The observed random distribution of mismatches between two supposedly identical resistors or transistors is primarily the result of two factors:

1. Variations in the location of the edges of the resistor or transistor resulting from the limited resolution of the photolithographic process itself. This causes emitter area mismatches in transistors and length/width ratio mismatches in resistors.
2. Variations in sheet resistance and junction depth of the emitter and base diffusions across the wafer resulting from nonuniform conditions during the predeposition and/or diffusion of the impurities. This causes the resistor sheet resistance and the net base doping, Q_B, in the transistors to vary with distance across the die.

In addition to these random phenomena, systematic mismatches can occur from a number of causes, including errors in drawing the original mask and thermal gradients on the die. These systematic errors, however, are easily detected since they reoccur on each die, and can thus be corrected. The random fluctuations represent the basic factor limiting the attainable matching accuracy.

Since the photolithographic resolution limits the attainable matching, it is clear that the important parameter is the size of the resistor or transistor being defined in relation to the resolution. If the resistor is made wider, then the same amount of edge location uncertainty should have proportionately less effect and the matching should improve. This is indeed observed; a graph of observed resistor matching as a function of resistor width is shown in Fig. 6.28. Similarly, the matching of I_S observed in transistors improves markedly as the emitter size

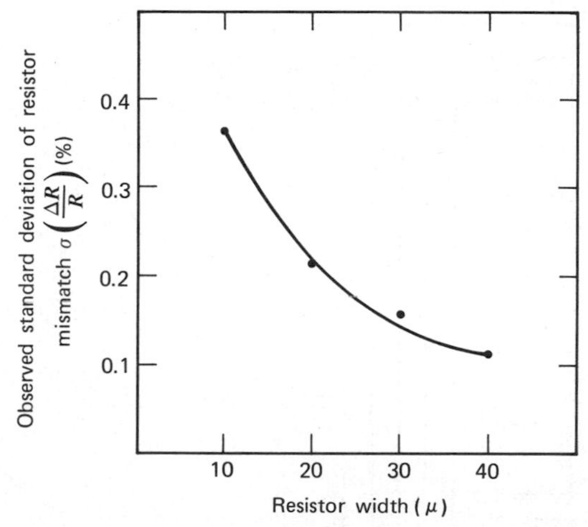

Fig. 6.28 Experimentally observed standard deviation in the mismatch distribution of ion-implanted resistors with sheet resistance 500 Ω/□ and length of 500 μ.

is made larger, as illustrated by the graph of observed matching data shown in Fig. 6.29. The increase in die size and resulting cost generally limits the size of the devices that can be used, but in the case of the precision operational amplifier, the input transistors and resistors are generally made as large as practicable.

The second factor limiting the matching, process-related gradients across the die, can be partially alleviated by use of appropriate geometries that cause the

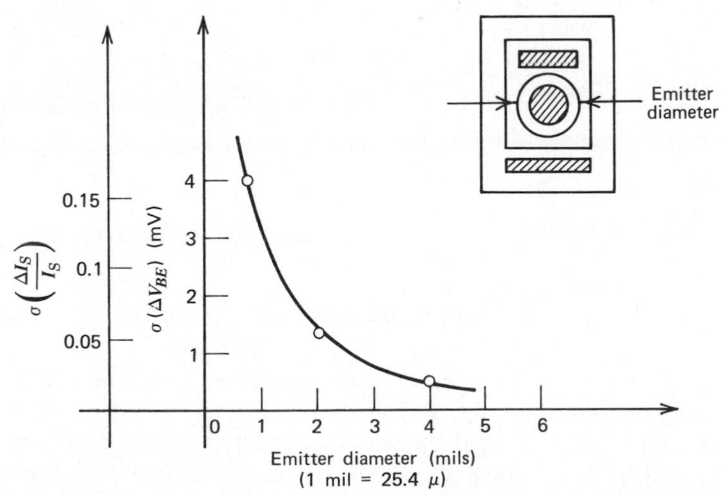

Fig. 6.29 Observed standard deviation in the mismatch distribution of V_{BE} and I_S as a function of emitter diameter.

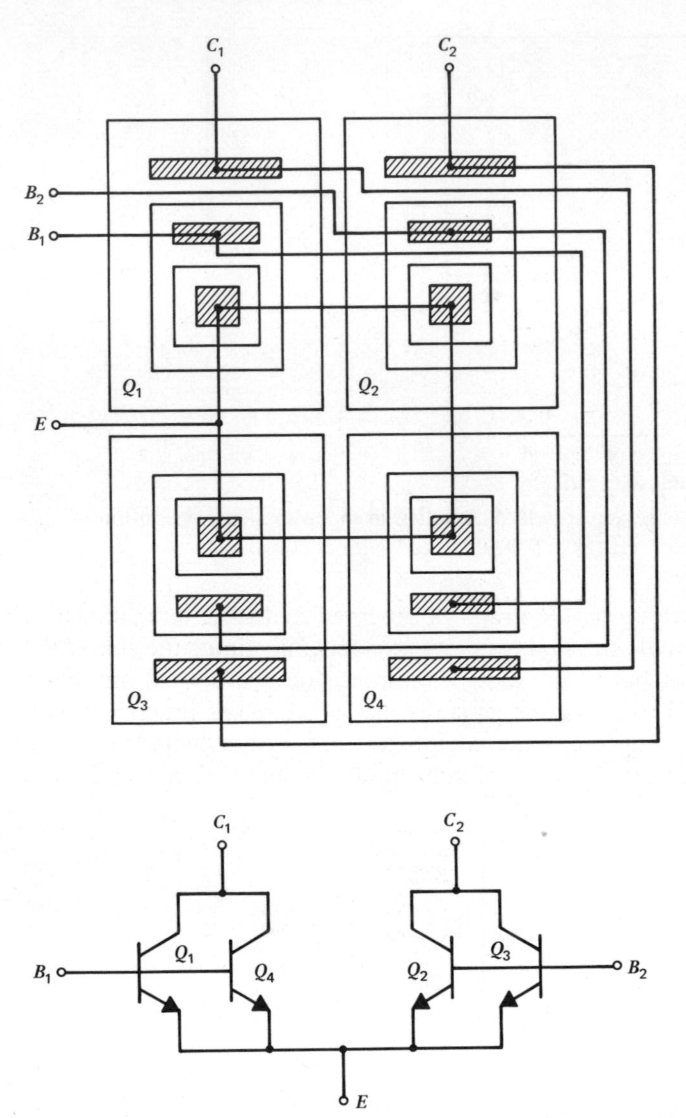

Fig. 6.30 Common-centroid geometry for the emitter-coupled pair.

device mismatch to be insensitive to certain types of process gradients. Such geometries are termed common centroid, and the simplest case is illustrated in Fig. 6.30. Here the transistors making up the pair are formed from cross-connected segments of a quad of transistors. In a geometric sense, the centroid of both composite devices lies at center of the structure. An example of an operational amplifier utilizing such a geometry is the 725 of which the circuit is shown in

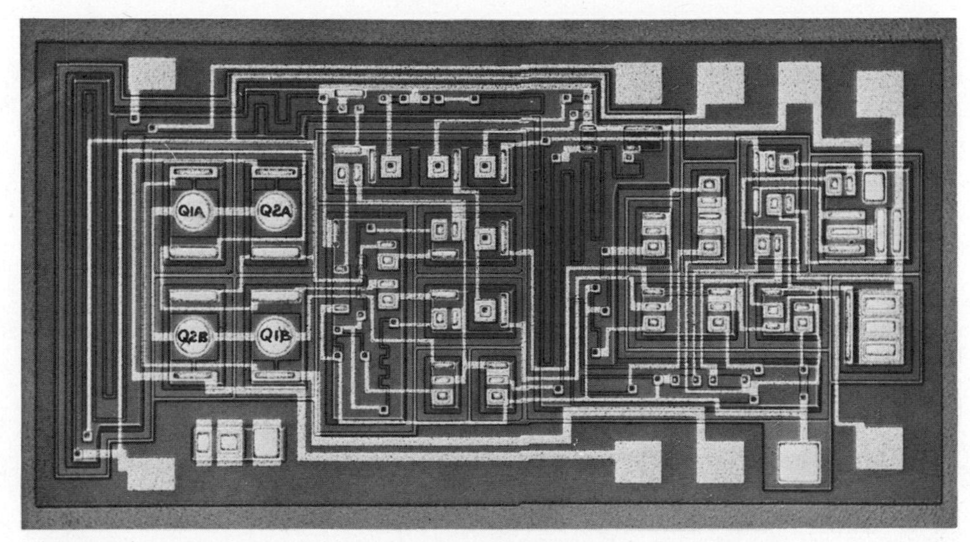

Fig. 6.31 725 die photograph. Note common-centroid input-device geometry.

Fig. 6.26. A die photograph is shown in Fig. 6.31. The input devices are split into a quad and cross connected to provide the common centroid geometry. This particular circuit achieves a typical input offset voltage of less than 1 mV.

6.4.2 Design of Low-Input-Current Operational Amplifiers

In instrumentation applications in which the signal source has a low internal impedance, the input offset voltage and its associated drift usually place a lower limit on the dc voltage that can be resolved. However, when the source impedance is high, the input bias current and input offset current of the operational amplifier flowing in the source resistance or the gain-setting resistors can be important in limiting the ability of the circuit to resolve small dc signals. Furthermore, many applications involve the direct sensing of currents; a good example is the photodiode amplifier shown in Fig. 6.32. Such photodetectors are used in a variety of applications where the level of ambient light must be sensed. Typical photodiode output currents vary from the picoampere level to the microampere level over the light flux range of interest, so that the input current of the operational amplifier in Figure 6.32 will limit the lower level of light flux that can be resolved. A need thus exists in this application as well as others for operational amplifiers with input currents much lower than that of the 741.

In the 741, the input current is the base current of the input devices. We might consider several alternatives in order to improve on this value, including reduction of the bias current in the input stage, insertion of a current source to cancel the input current, an increase in the beta of the input devices, or the use of a

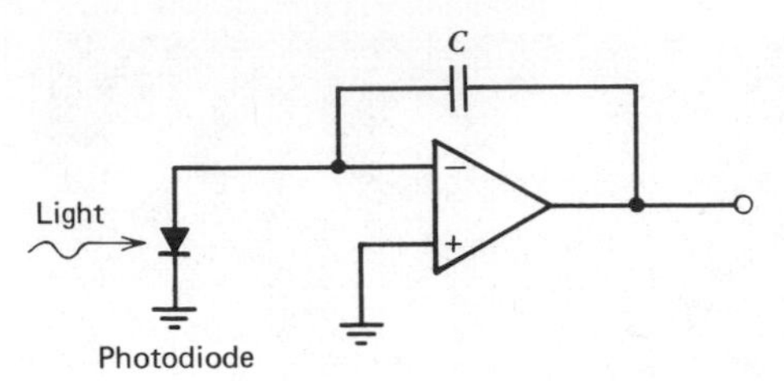

Fig. 6.32 Integrating photodiode amplifier.

different device such as a JFET or MOSFET as the input transistor. Reducing the bias current of the stage unfortunately degrades the frequency response of the circuit as will be discussed in Chapter 7, and furthermore the current gain of integrated-circuit *npn* transistors tends to fall rapidly at current levels below 1 μA unless special steps are added to the fabrication process. The techniques that have gained wide acceptance are the use of a canceling current source (bias-current cancellation), an increase in the beta of the input devices (superbeta transistors), and the use of JFETs or MOSFETs as the input transistors. We will discuss each of these approaches.

Bias-Current Cancellation. Since the input bias current of an operational amplifier with bipolar input transistors is equal to the collector current of the input transistors divided by the beta of the input transistors, a second current equal to the input current is derivable from the collector current of the input transistors by performing a division by beta. This current then can be fed back into the input terminal via a current source and in principle the input current could be completely canceled. A conceptual schematic of such an input stage is shown in Fig. 6.33. This technique indeed does result in a large reduction in the input bias current, the amount of the reduction being limited by the matching of the betas of the *npn* transistors in the circuit. It unfortunately does not improve the input offset current, however.

A practical input-current-cancellation circuit is shown in Figure 6.34. In this circuit, transistor Q_9, together with diodes Q_{10} and Q_{11}, serve to bias the emitters of Q_6 and Q_8 at a potential three diode drops more positive than the emitters of Q_1 and Q_2. Diodes Q_7 and Q_5 are forward biased since they carry the base currents of Q_3 and Q_4, and so the bases of Q_3 and Q_4 are two diode drops above the emitters of Q_1 and Q_2. The emitters of Q_3 and Q_4 are then one diode drop above the emitters of Q_1 and Q_2, and these input transistors operate at approximately zero collector-base bias. Assuming the beta on the *npn* devices is large, the collector current of Q_3 and Q_4 is approximately equal to that of input transistors Q_1 and Q_2. If the *npn* betas are all the same, then the base currents of Q_3 and Q_4 are the same as the base currents of Q_1 and Q_2. The *pnp* current sources, Q_5, Q_6, Q_7, and Q_8, then take this base current and supply an identical current back into the base of Q_1 and Q_2. If the *pnp* betas are large, the cancellation is precise.

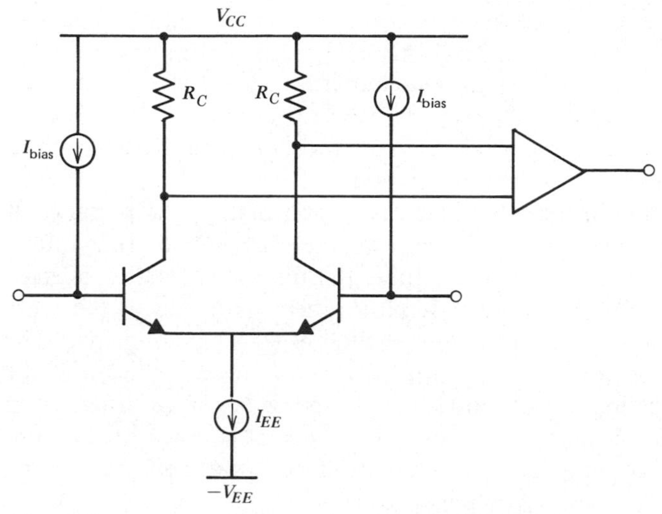

Fig. 6.33 Conceptual schematic of input-current cancellation scheme. The I_{bias} current sources are equal in value to the base current of the input devices.

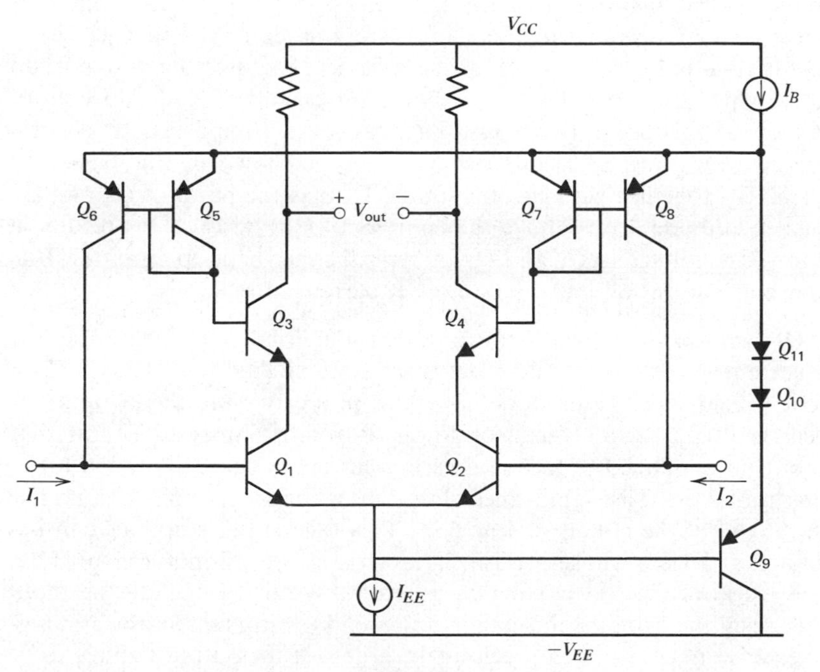

Fig. 6.34 Typical bias-current-cancellation circuit.

In practice, the *pnp* betas are rather low in this circuit because the *pnp* devices themselves operate at very low current level. Also, the betas of the *npn* transistors do not match precisely. As a result, the mismatch of from 5 to 20 percent typically exists between the base current of Q_1 and the collector current of Q_6, which is supposed to cancel it. The resulting residual input current is typically one-fifth to one-twentieth of the original uncompensated input current.

The input offset current, on the other hand, is made worse by the presence of the cancellation circuitry. The input offset current is the sum of the original mismatch in the base currents of Q_1 and Q_2, plus the mismatch in the collector currents of Q_6 and Q_8. The latter depends on the beta matching of the pairs Q_3–Q_4, Q_5–Q_6, and Q_7–Q_8, and on the I_S matching of Q_5–Q_6 and Q_7–Q_8. The offset current will be made worse by a factor of two to four compared with the uncompensated case. The input noise current is also made larger, as discussed in Chapter 11. A complete schematic of a typical bias current compensated amplifier, the 0P-07, is shown in Fig. 6.35. This circuit also incorporates offset trimming as described in Section 6.4.1 and shown in Fig. 6.27. The resulting circuit has a typical input offset voltage of 50 μV, a typical input bias current of 4 nA and a typical offset current of 4 nA.

Superbeta Transistors. A second approach to decreasing the input bias current is to increase the current gain of the input-stage transistors.[6] A practical method of achieving this result is use of superbeta transistors as described in Chapter 2. From a circuit standpoint, application of these devices requires the design of an input stage in which the input devices are never subjected to a collector-emitter voltage of more than one or two volts. This can be accomplished in a fairly straightforward way using a cascode configuration as shown in Fig. 6.36. The diodes D_1 and D_2, together with current source I_1, bias the bases of Q_3 and Q_4 at a potential two diode drops above the emitters of Q_1 and Q_2. This results in zero collector-base voltage in Q_1 and Q_2. A typical super-beta op amp (the 108) has an input bias current of 3 nA and an offset current of 0.5 nA.

FET-Input Operational Amplifiers. A third practical alternative for achieving low input bias current is the use of field-effect transistors as the input devices. As a class, junction field-effect transistors have the property that their input, or gate, current is in the 10-pA range or lower at room temperature, and that their ratio of transconductance to dc operating current is generally much lower than bipolar transistors. The significance of this latter consideration is best understood with the aid of the circuit of Fig. 6.37. This differential amplifier consists of a pair of active devices whose precise characteristics we will not define, and a pair of load resistors that do not match precisely. We will calculate the input offset voltage resulting from the mismatch in the load resistors under the assumption that the active devices match precisely. In order for the output voltage to be zero we must have:

$$I_1 R_1 = I_2 R_2 \tag{6.84}$$

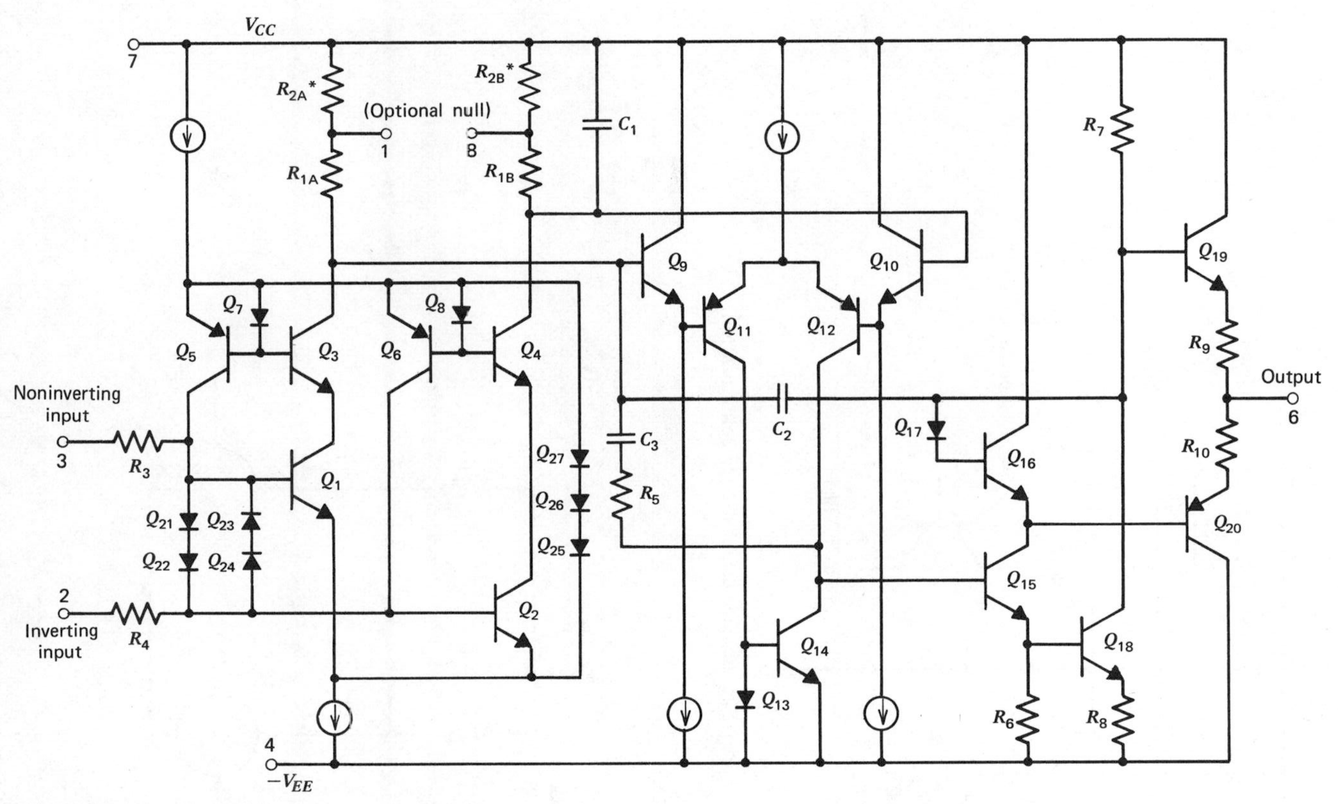

Fig. 6.35 Schematic diagram of the OP-07 input-current-compensated operational amplifier.

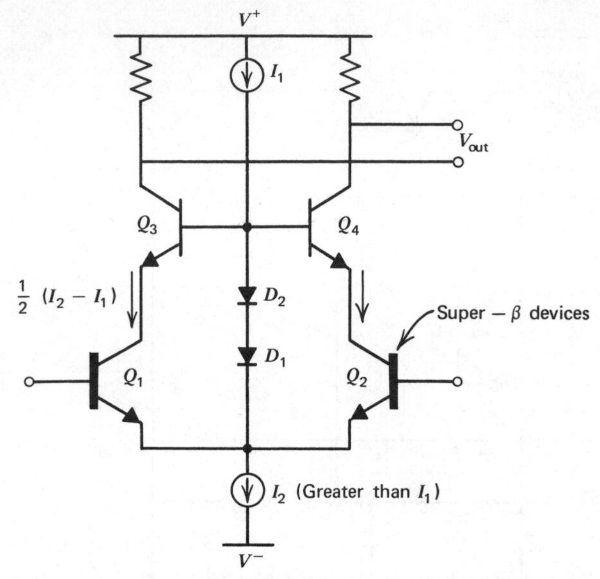

Fig. 6.36 Super-β input stage using cascode configuration.

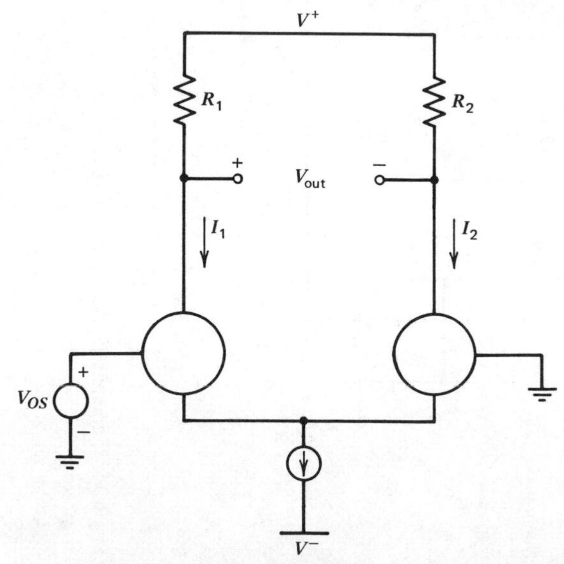

Fig. 6.37 Differential amplifier using arbitrary unilateral active devices.

Thus if a mismatch in R_1 and R_2 is present, we must have a corresponding mismatch in the currents in order for V_{out} to be zero.

$$-\frac{\Delta I}{I} = \frac{\Delta R}{R} \tag{6.85}$$

The input offset voltage V_{OS} required to bring about this current imbalance is given implicitly by:

$$-\Delta I = g_m V_{OS}$$

and substitution of this expression in (6.85) gives

$$V_{OS} = \frac{I}{g_m}\left(\frac{\Delta R}{R}\right) \tag{6.86}$$

Mismatches in the active devices will also be important, but for the sake of simplicity in this example we consider only the load element mismatch. Notice that for a given percentage resistor mismatch, the important parameter of the active devices that determines the input offset voltage the ratio of dc operating current to transconductance. For a bipolar transistor, this quantity is 26 mV, but for a uniform-channel JFET in the saturated region, we have

$$I_D = I_{DSS}\left(1 - \frac{V_{GS}}{V_P}\right)^2 \tag{6.87}$$

$$g_m = -2\frac{I_{DSS}}{V_P}\left(1 - \frac{V_{GS}}{V_P}\right) \tag{6.88}$$

and thus

$$\left|\frac{I_D}{g_m}\right| = \frac{V_P}{2}\sqrt{\frac{I_D}{I_{DSS}}} \tag{6.89}$$

For typical JFETs, the pinch-off voltage varies from 1 to 4 V. Thus for drain currents on the order of I_{DSS}, the I_D/g_m ratio is on the order of 1 V, or a factor of 40 larger than the bipolar transistor. Thus, to obtain low input offset voltage, load elements of the differential stage must be much more precisely matched than in the bipolar case, as must be the parameters of the JFETs themselves. The offset voltage of the source-coupled JFET pair was calculated in Chapter 3 in terms of the mismatches in channel width and length-width ratio.

Junction FETs can be fabricated on the same die with bipolar devices by a number of different techniques. As discussed in Chapter 2, a diffused JFET can be fabricated on the same die as bipolar transistors by exactly the same process used in superbeta processing. As in the case of the superbeta device, the structure can be formed by either performing two separate emitter diffusions into the same base diffusion or by performing the same emitter diffusion into two separate base diffusions. In both cases, the channel width of the resulting device is quite narrow,

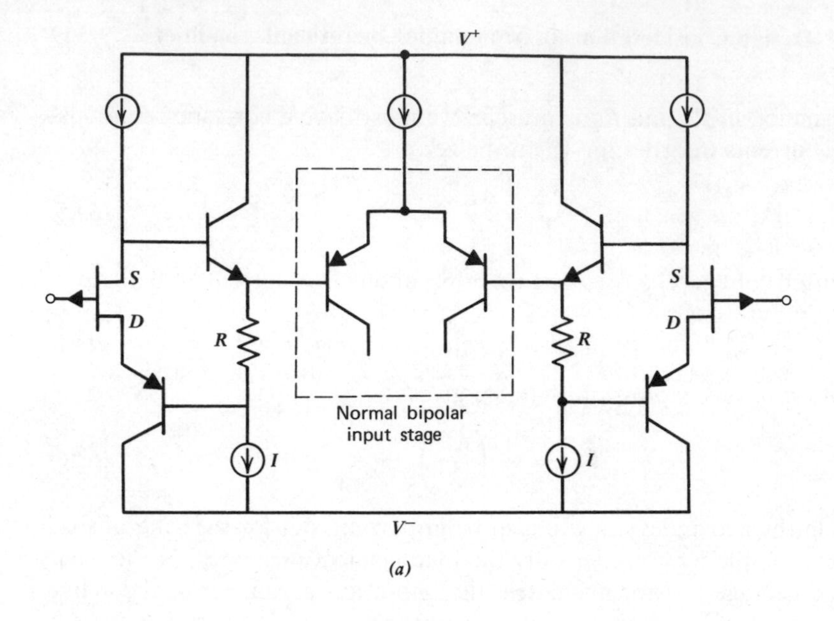

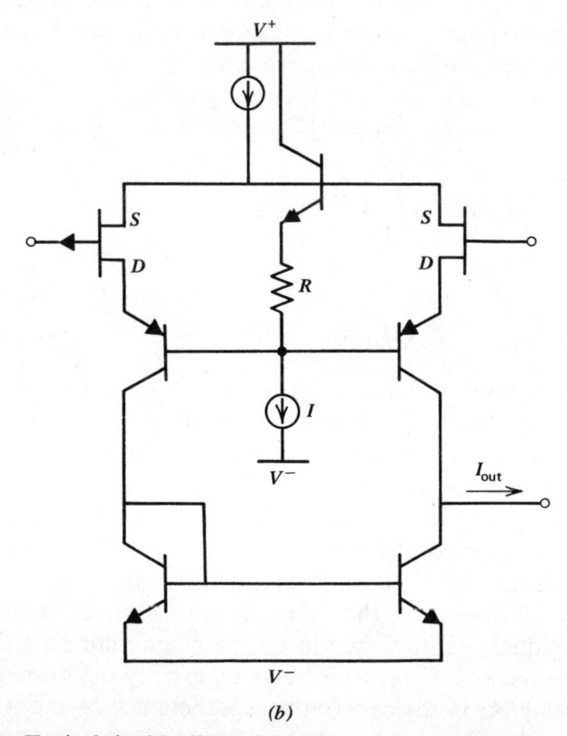

Fig. 6.38 Typical double-diffused JFET operational amplifier input stages. (*a*) Source follower with drain-voltage clamp. (*b*) Common source with drain-voltage clamp. In both cases, $IR < BV_{DSS} \approx 6$ V.

and a relatively small variation in conditions during the predeposition and diffusion will result in large variations in the channel width and the pinch-off voltage. Another drawback is that the breakdown voltage from drain to gate is limited to that of an emitter-base junction of a bipolar transistor, or about 7 V. Therefore, these devices, like superbeta transistors, must be used in an input stage that limits the source-drain voltage applied to the JFETs. Two typical double-diffused JFET input stages are shown in Fig. 6.38, and in these two circuits the current sources, I, produce voltage drops across resistors R of from 3 to 5 V. The top ends of the resistors are clamped to the JFET source potential by *npn* transistors, and the JFET drains are clamped to the potential on the lower end of the resistors by *pnp* transistors.

The drawbacks of the double-diffused JFET, then, are that the process required is somewhat difficult to control, the input-stage circuitry is complicated and consumes considerable die area, and that the device inherently results in high input offset voltage, typically on the order of 20 to 50 mV. These drawbacks can be overcome by fabricating the JFET with a sequence of ion-implantation steps of the type discussed in Chapter 2.[7] One *p*-type implantation is used to form the channel, and since the implantation process is capable of very uniform and precise implacement of impurities, the number of impurity atoms in the channel profile is precisely controlled. This allows precise pinch-off voltage control and matching compared to the double-diffused case. Also, since the peak doping density in the channel region is only approximately 10^{16} atoms/cm^3, the drain-gate breakdown voltage can be kept high. Op amps using such devices in the input stage have realized input offset voltages of 3 to 5 mV.

The use of ion-implanted JFETs permits the fabrication of monolithic JFET-input amplifiers with offset voltages comparable to those of bipolar circuits. Thus at room temperature these amplifiers provide a very desirable combination of low input offset voltage and input current, the latter typically on the order of 10 pA. However, the input current is that of a reverse-biased *pn* junction, so that at elevated temperatures the JFET input current becomes much larger, actually exceeding that of some superbeta operational amplifiers at temperatures above 100°C.

APPENDIX

A6.1 CALCULATION OF THE INPUT OFFSET VOLTAGE AND CURRENT OF THE 741

Two important aspects of the performance of the amplifier are its input offset current and voltage. As discussed earlier in this chapter, these parameters place a lower limit on the values of dc current and voltage than can be accurately detected and amplified by the circuit. Calculation of these performance parameters, however, is fundamentally different from calculation of other parameters such as voltage gain. Offset voltage and current are random parameters with mean values that are usually very near zero. These offsets arise from randomly occurring mismatches between the device pairs making up the input stage of the circuit, and are best described by a probability distribution with (ideally) zero mean and some standard deviation. The parameter of interest from the designer's standpoint is the standard deviation of the offset voltage and current distributions, which will dictate the limit he can place on the offset voltage and current of production units being tested while maintaining an acceptable yield. From the user's standpoint the details of the distribution are of little significance except as they affect the specified maximum offset voltage arrived at by the designer. The information available to the designer at the design stage is the distribution of mismatches in resistor values, transistor saturation current, and transistor beta. Given these distributions, his task is to design an input stage with the minimum offset while meeting the other circuit requirements. We will calculate the offset voltage and current of the 741 under the assumption that the various device mismatches are described by normal or Gaussian distributions.

A simplified input stage schematic diagram is shown in Fig. 6.39. In this analysis we will assume that the current gains of the *npn* transistors, Q_1, Q_2, Q_5, and Q_6, are large enough that their base currents can be neglected. The same assumption is made regarding Q_7 and Q_{16}.

Provided that the mismatches are small enough that they only slightly perturb the currents in the circuit, we can simplify the problem by considering each device pair mismatch independently and superposing the results. Referring to Fig. 6.39, we see that the problem is to determine the input voltage, V_{OS}, to drive I_{out} to zero in the presence of the various mismatches.

First consider a mismatch between Q_5 and Q_6 and between R_1 and R_2. The offset in the collector currents of such a configuration was investigated in Chapter 4, with the result that

$$\frac{\Delta I_{C5-6}}{I_{C5-6}} = \left(\frac{1}{1 + g_m R_{1-2}}\right)\frac{\Delta I_{S5-6}}{I_{S5-6}} - \left(\frac{g_m R_{1-2}}{1 + g_m R_{1-2}}\right)\frac{\Delta R_{1-2}}{R_{1-2}} \tag{6.90}$$

where $\Delta I_{C5-6}/I_{C5-6}$ is the fractional mismatch in the collector currents, I_{C5} and I_{C6}. Equation 6.90 can be written as

$$\frac{\Delta I_{C5-6}}{I_{C5-6}} = \left(\frac{1}{1 + \dfrac{I_{C5-6}R_{1-2}}{V_T}}\right)\left(\frac{\Delta I_{S5-6}}{I_{S5-6}}\right) - \left(\frac{\dfrac{I_{C5-6}R_{1-2}}{V_T}}{1 + \dfrac{I_{C5-6}R_{1-2}}{V_T}}\right)\left(\frac{\Delta R_{1-2}}{R_{1-2}}\right) \tag{6.91}$$

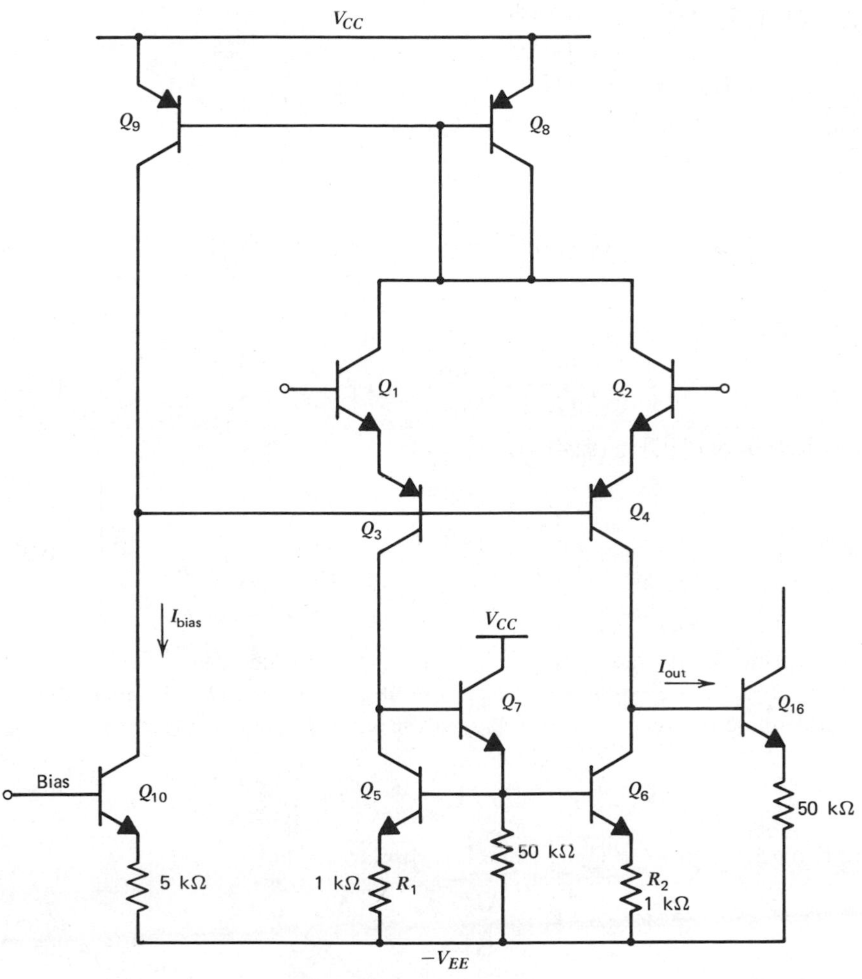

Fig. 6.39 Circuit for the calculation of V_{OS} of the 741.

In order to produce an equivalent difference current in Q_3 and Q_4 to drive I_{out} to zero, we must apply an input voltage of V_{OS1} where

$$\Delta I_{C3-4} = \frac{g_{m1-2}}{2} V_{OS1} = \Delta I_{C5-6} \tag{6.92}$$

Thus,

$$V_{OS1} = 2\frac{\Delta I_{C5-6}}{g_{m1-2}} \tag{6.93}$$

Using (6.91) in (6.93),

$$V_{OS1} = \frac{2}{g_{m1-2}}\left[(I_{C5-6})\left(\frac{1}{1+\dfrac{I_{C5-6}R_{1-2}}{V_T}}\right)\left(\frac{\Delta I_{S5-6}}{I_{S5-6}}\right) - (I_{C5-6})\left(\frac{\dfrac{I_{C5-6}R_{1-2}}{V_T}}{1+\dfrac{I_{C5-6}R_{1-2}}{V_T}}\right)\left(\frac{\Delta R_{1-2}}{R_{1-2}}\right)\right] \tag{6.94}$$

But

$$\frac{I_{C5-6}}{g_{m1-2}} = \frac{I_{C5-6}}{g_{m5-6}} = V_T \tag{6.95}$$

and substitution of (6.95) in (6.94) gives

$$V_{OS1} = 2V_T\left[\left(\frac{1}{1+\dfrac{I_{C5-6}R_{1-2}}{V_T}}\right)\left(\frac{\Delta I_{S5-6}}{I_{S5-6}}\right) - \left(\frac{\dfrac{I_{C5-6}R_{1-2}}{V_T}}{1+\dfrac{I_{C5-6}R_{1-2}}{V_T}}\right)\left(\frac{\Delta R_{1-2}}{R_{1-2}}\right)\right] \tag{6.96}$$

Now we consider mismatches in I_{S1-2} and I_{S3-4}. If the collector currents of Q_1, Q_2, Q_3, and Q_4 are equal, the offset contribution of the Q_1–Q_2 mismatch is just the difference between V_{BE1} and V_{BE2} at the same I_C. This is given by

$$V_{BE1} - V_{BE2} = V_T \ln\frac{I_{S2}}{I_{S1}} \approx -V_T\frac{\Delta I_{S1-2}}{I_{S1-2}} \tag{6.97}$$

The contribution of the Q_3–Q_4 mismatch is similar, so that:

$$V_{OS2} = -V_T\frac{\Delta I_{S1-2}}{I_{S1-2}} - V_T\frac{\Delta I_{S3-4}}{I_{S3-4}} \tag{6.98}$$

The final consideration is the beta mismatch in the *pnp* transistors. Assuming that the two collector currents, I_{C3} and I_{C4}, are equal, the emitter currents will be unequal by an amount equal to the difference in the base currents. Assuming $\Delta\beta_F \ll \beta_F$,

$$\frac{\Delta I_{E3-4}}{I_{E3-4}} = -\left(\frac{1}{\beta_{3-4}}\right)\left(\frac{\Delta\beta_{3-4}}{\beta_{3-4}}\right) \tag{6.99}$$

If we neglect base currents in the *npn* transistors, the two collector currents in Q_1 and Q_2 will differ by ΔI_{E3-4}. Then

$$V_{OS3} = V_{BE1} - V_{BE2} = V_T \ln\frac{I_{C1}}{I_{C2}} \cong V_T\frac{\Delta I_{C1-2}}{I_{C1-2}} = V_T\frac{\Delta I_{E3-4}}{I_{E3-4}}$$

and substitution of (6.99) yields

$$V_{OS3} = -V_T \frac{1}{\beta_{3-4}} \frac{\Delta\beta_{3-4}}{\beta_{3-4}} \tag{6.100}$$

The total input offset voltage is then the sum of the three components.

$$V_{OS} = V_{OS1} + V_{OS2} + V_{OS3}$$

$$V_{OS} = V_T \left[\left(\frac{2}{1 + \dfrac{I_{C5-6}R_{1-2}}{V_T}} \right) \left(\frac{\Delta I_{S5-6}}{I_{S5-6}} \right) - \left(\frac{2\dfrac{I_{C5-6}R_{1-2}}{V_T}}{1 + \dfrac{I_{C5-6}R_{1-2}}{V_T}} \right) \left(\frac{\Delta R_{1-2}}{R_{1-2}} \right) \right.$$

$$\left. - \frac{\Delta I_{S1-2}}{I_{S1-2}} - \frac{\Delta I_{S3-4}}{I_{S3-4}} - \left(\frac{1}{\beta_{3-4}} \right) \left(\frac{\Delta\beta_{3-4}}{\beta_{3-4}} \right) \right] \tag{6.101}$$

For a given set of mismatches this expression will give the input offset voltage. However, the information of interest to the designer is the distribution of the observed offset voltages over a large number of samples, and the information available is the distribution of the mismatch factors. If each of the quantities

$$\frac{\Delta I_{S5-6}}{I_{S5-6}} \qquad \frac{\Delta R_{1-2}}{R_{1-2}} \qquad \frac{\Delta I_{S1-2}}{I_{S1-2}} \qquad \frac{\Delta I_{S3-4}}{I_{S3-4}} \qquad \frac{\Delta\beta_{3-4}}{\beta_{3-4}}$$

are regarded as independent random variables with normal distributions, then, as discussed in Appendix A3.1, the standard deviation of the sum is given by

$$\sigma_{\text{sum}} = \sqrt{\sum_n \sigma_n^2} \tag{6.102}$$

Thus the standard deviation of the distribution for V_{OS} is calculated by taking the square root the sum of the squares of the individual contributions.

Assuming that the standard deviation of resistor matching is 1 percent, that of I_S matching is 5 percent, and that of beta matching is 10 percent, we obtain, with $R_{1-2} = 1\ \text{k}\Omega$, $I_{C5-6} = 9.5\ \mu\text{A}$, $\beta_{3-4} = 50$

$$\sigma_{V_{OS}} = V_T \sqrt{(0.073)^2 + (0.0052)^2 + (0.05)^2 + (0.05)^2 + \left(\frac{1}{50}\right)^2 (0.1)^2}$$

$$= V_T(0.103) = 2.6\ \text{mV} \tag{6.103}$$

The largest single offset contribution is thus the mismatch in the active load devices, Q_5 and Q_6. If, for example, the offset voltage has a standard deviation of 2.6 mV, then the fraction, Y, of all devices fabricated that will have an offset voltage of less than the 741 specification of 5 mV is given by

$$Y = \int_{-5}^{+5} \frac{1}{2.6\sqrt{2\pi}} \exp \frac{x^2}{2(2.6)^2} dx \tag{6.104}$$

This integral can be evaluated with the aid of Fig. 3.49, giving a value of .93. Thus a 7 percent yield loss will be suffered from offset voltage variations with the 5-mV offset specification.

The input offset *current* is determined by the beta mismatch between Q_1 and Q_2 and the mismatch in I_{C1} and I_{C2}. It is given approximately by

$$\begin{aligned} I_{OS} &= I_{B1} - I_{B2} \\ &= I_{B1-2}\left(-\frac{\Delta\beta_{1-2}}{\beta_{1-2}} - \frac{1}{\beta_{3-4}}\frac{\Delta\beta_{3-4}}{\beta_{3-4}} + .73\frac{\Delta I_{S5-6}}{I_{S5-6}} - .27\frac{\Delta R_{1-2}}{R_{1-2}}\right) \end{aligned} \tag{6.105}$$

Assuming a *pnp* beta of 50, and an *npn* beta of 250, $I_{B1-2} = 9.5\ \mu\text{A}/250 = 38$ nA and

$$\sigma_{I_{OS}} \cong (0.1)I_{B1-2} = 3.8 \text{ nA} \tag{6.106}$$

PROBLEMS

6.1 For the circuit of Fig. 6.40, determine the output current as a function of the input voltage. Assume that the transistor is in the forward-active region.

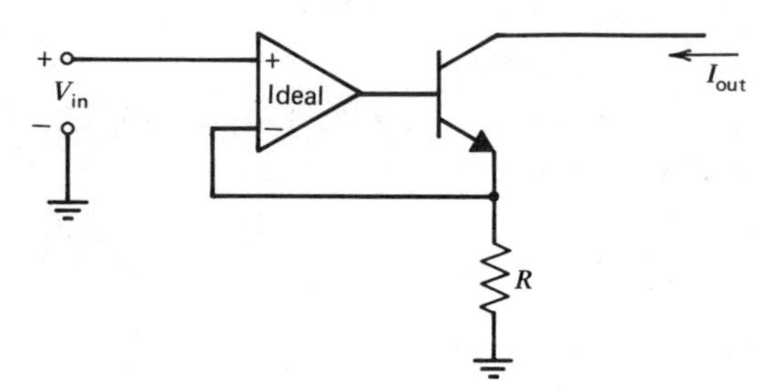

Fig. 6.40 Circuit for Problem 6.1

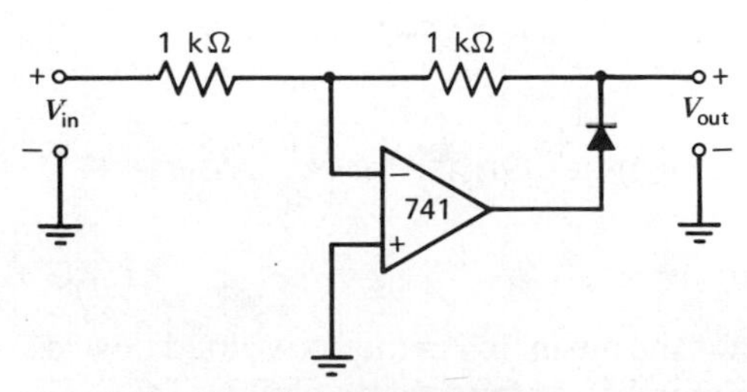

Fig. 6.41 Circuit for Problem 6.2

6.2 Determine the output voltage as a function of the input voltage for the circuit of Fig. 6.41.

6.3 Design a circuit that develops an output voltage proportional to the square root of the incoming analog voltage. Use ideal op amps and bipolar transistors.

6.4 In the circuit of Fig. 6.42, determine the correct value of R_x so that the output voltage is zero when the input voltage is zero. Assume a finite input bias current, but zero input offset current and input offset voltage.

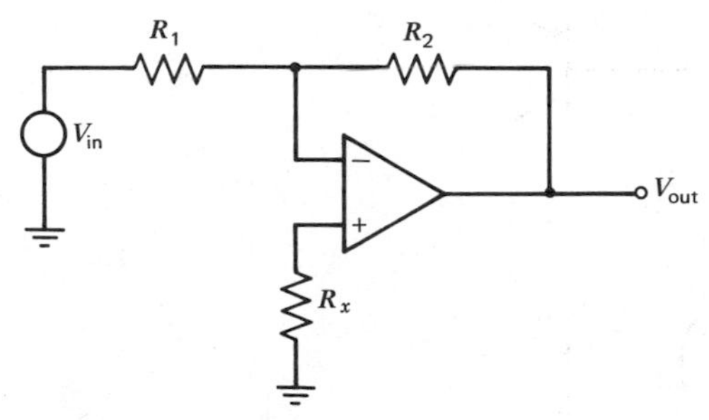

Fig. 6.42 Circuit for Problem 6.4

6.5 The differential instrumentation amplifier shown in Fig. 6.43 must have a voltage gain of 10^3 with an accuracy of 0.1 percent. Can the 741 meet the requirements of this application? How much open-loop voltage gain must the operational amplifier have? Assume the op amp open-loop gain has a tolerance of +100 percent, −50 percent. Neglect the effects of R_{in} and R_{out} in the operational amplifier.

6.6 The differential amplifier of Problem 6.5, once the offset is adjusted to zero, must have an input-referred offset voltage that is less than 1 mV in magnitude for common-mode input voltages between +10 and −10 Volts. What is the maximum CMRR allowable for the amplifier to achieve this? Does the 741 meet this requirement? (The specified CMRR for the 741 is 80 dB minimum.)

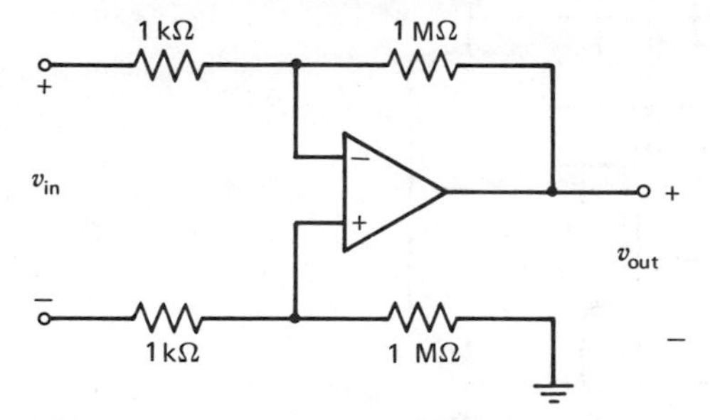

Fig. 6.43 Circuit for Problem 6.5

6.7 In the input stage of the 741, how much does the collector current of Q_1 and Q_2 change when the power-supply voltage is reduced from ±15 V to ±10 V. Neglect the effects of finite output resistance in the transistors.

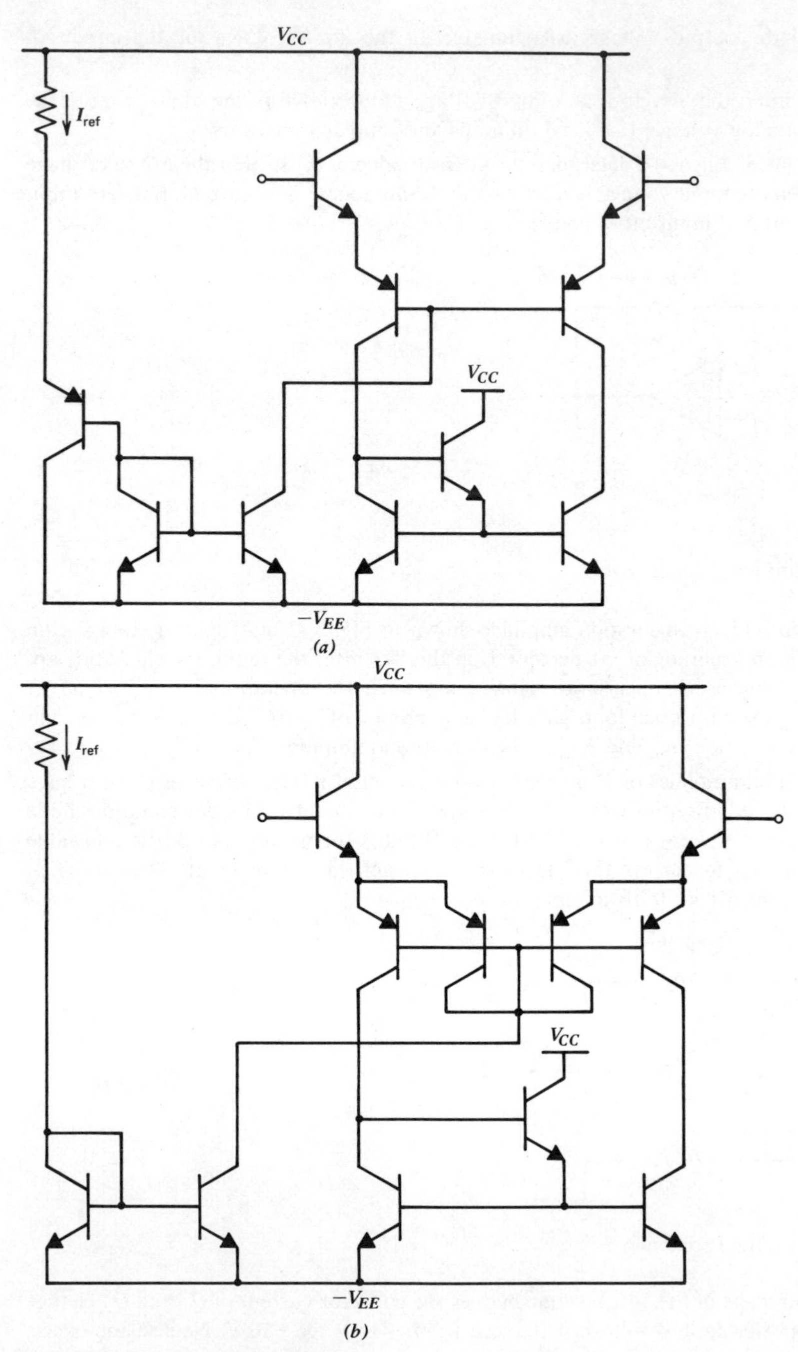

Fig. 6.44 Circuits for Problem 6.8

6.8 Shown in Fig. 6.44 are two alternate schemes for biasing the 741 input stage. In each case determine the required value of I_{ref} to give an input transistor collector current of 10 μA. Neglect *npn* base currents. Assume that the *pnp* beta is 50 and $V_A = \infty$ for all devices.

6.9 Shown in Fig. 6.45 is an alternate output-stage biasing scheme using a V_{BE} multiplier. Determine the bias current as a function of the resistor ratio, R_2/R_1. Determine the resistor ratio required to give a bias current of 50 μA. Does I_{bias} remain constant over temperature? Assume that the output transistors have a saturation current that is five times that of the small-geometry devices. Neglect the portion of the 200 μA flowing through R_2 and R_1, and neglect base currents. Assume $I_S = 10^{-15}$ A.

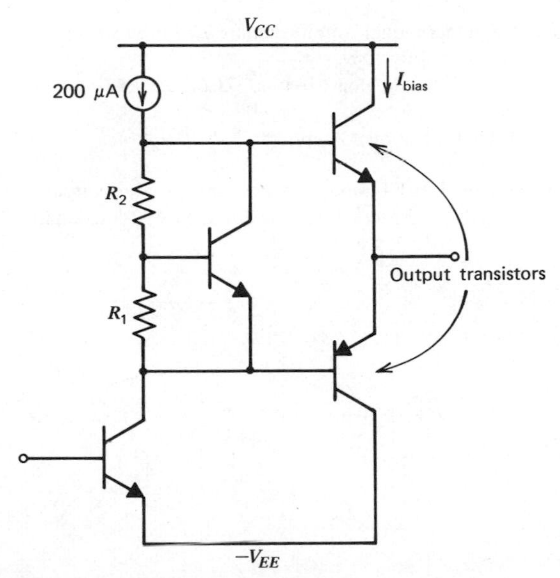

Fig. 6.45 Circuit for Problem 6.9

6.10 Determine the gain of the 741 if the bias current level in the input stage is doubled.

6.11 How is the gain of the 741 affected if the 100-Ω resistor in the emitter of Q_{17} is removed? Find the new value of voltage gain.

6.12 Determine the common-mode input voltage range of the 741. This is the range of dc common-mode input voltages over which the circuit operates normally with the offset voltage and input bias current within specifications.

6.13 The objective of this problem is the calculation of the offset voltage drift of the 741 using (6.81). First calculate the worst-case offset voltage assuming that $\Delta I_S/I_S = 0.05$, $\Delta R/R = 0.01$, $\Delta\beta/\beta = 0.1$ using (6.101). Next, use (6.81) to calculate the offset voltage drift. Use the results of Chapter 4 on the Widlar source to calculate the temperature coefficient of the biasing current source and thus show the second term in (6.81) is negligible if R_1 and R_2 are diffused resistors.

6.14 Determine the dc bias currents and small-signal voltage gain for the 725 operational amplifier shown in Fig. 6.26. Assume that $V_{CC} + 15\ V$, $-V_{EE} = -15$ V, and make the same assumptions made in the 741 analysis.

REFERENCES

1. G. E. Tobey, J. G. Graeme, and L. P. Huelsman. *Operational Amplifiers.* McGraw-Hill, New York, 1971.
2. J. V. Wait, L. P. Huelsman, and G. A. Korn. *Introduction to Operational Amplifier Theory and Applications.* McGraw-Hill, New York, 1975.
3. D. Fullagar. "A New High-Performance Monolithic Op Amp," *Fairchild Semiconductor Applications Brief,* May 1968.
4. G. Erdi. "A Low-Drift, Low-Noise Monolithic Operational Amplifier for Low-Level Signal Processing," *Fairchild Semiconductor Applications Brief,* July 1969.
5. G. Erdi. "A Precision Trim Technique for Monolithic Analog Circuits," *IEEE Journal of Solid-State Circuits,* Vol. SC-10, pp. 412–416, December 1975.
6. R. J. Widlar. "Design Techniques for Monolithic Operational Amplifiers," *IEEE Journal of Solid-State Circuits,* Vol. SC-4, pp. 184–191, August 1969.
7. R. W. Russell and D. D. Culmer. "Ion-Implanted JFET-Bipolar Monolithic Analog Circuits," *Digest of Technical Papers,* 1974 International Solid-State Circuits Conference, Philadelphia, pp. 140–141, February 1974.

CHAPTER 7

FREQUENCY RESPONSE OF INTEGRATED CIRCUITS

7.1 INTRODUCTION

The analysis of integrated-circuit behavior in previous chapters was concerned with low-frequency performance, and the effects of parasitic capacitance and charge storage in transistors were not considered. However, as the frequency of the signal being processed by a circuit increases, the capacitive elements in the circuit eventually dominate the behavior.

In this chapter the small-signal behavior of integrated circuits at high frequencies is considered. The frequency response of single-stage amplifiers is treated first, followed by an analysis of multistage amplifier response. Finally, the frequency response of the 741 operational amplifier is considered and those parts of the circuit that limit the frequency response are identified.

7.2 SINGLE-STAGE AMPLIFIER FREQUENCY RESPONSE

7.2.1 Single-Stage Differential Amplifier

A basic building block of analog integrated circuits is the differential stage shown in Fig. 7.1. For small-signal differential inputs at v_i, the common-emitter point, E, is a virtual ground and we can form the differential mode (DM) half circuit of Fig. 7.2*a*. The gain of this circuit is equal to the differential gain of the full circuit. The analysis is also applicable to a single transistor stage. The small-signal equivalent circuit of Fig. 7.2*a* is shown in Fig. 7.2*b* and, for compactness, the factor $\frac{1}{2}$ has been omitted from the input and output voltage. This does not alter the analysis in any way. Also, for simplicity, the collector-substrate capacitance of the transistor has been omitted, and the effect of this will be considered later.

An approximate analysis of the circuit of Fig. 7.2*b* can be made using the *Miller effect* approximation. This is done by considering the input impedance seen looking across the plane AA in Fig. 7.2*b*. To calculate this impedance we calculate the current, i_1, produced by the voltage, v_1.

$$i_1 = (v_1 - v_o)C_\mu s \tag{7.1}$$

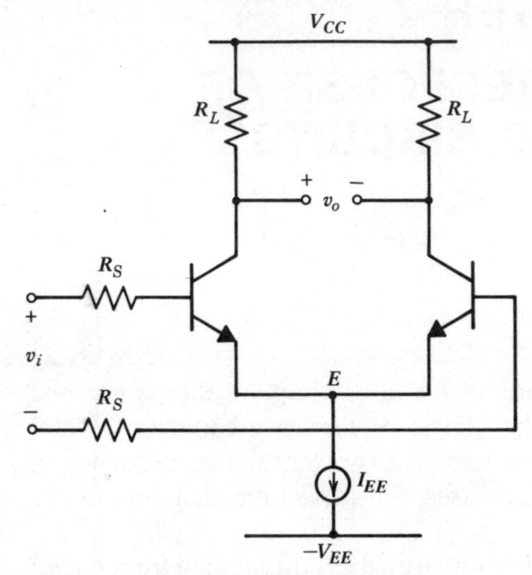

Fig. 7.1 Differential amplifier circuit.

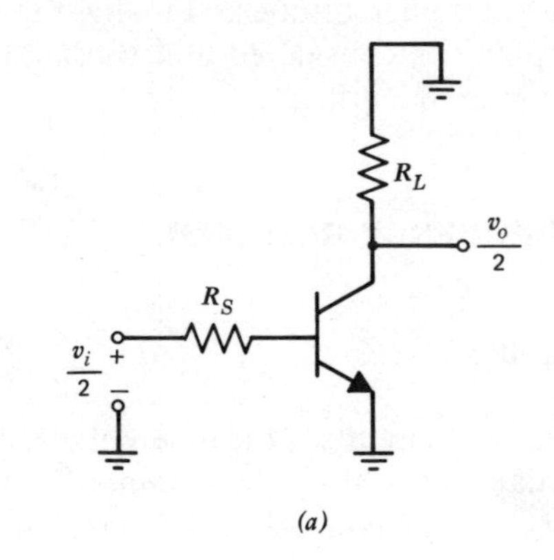

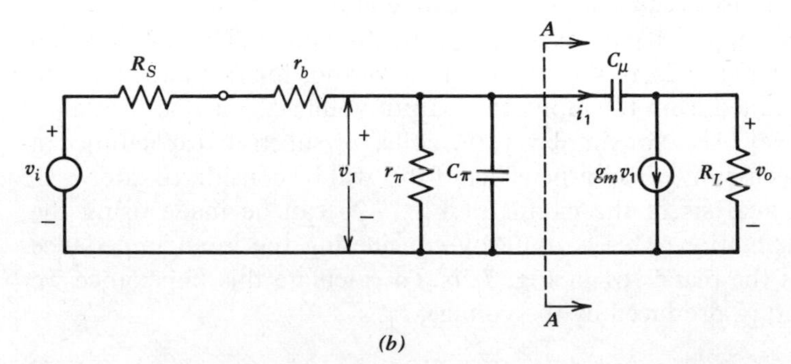

Fig. 7.2 (*a*) Differential-mode half circuit for Fig. 7.1. (*b*) Small-signal equivalent circuit for (*a*).

Summation of currents at the collector gives

$$g_m v_1 + \frac{v_o}{R_L} + (v_o - v_1)C_\mu s = 0 \tag{7.2}$$

The last term in (7.2) represents current fed forward to the output via C_μ, and this is usually negligible compared to the first two terms of (7.2). Using this approximation gives

$$v_o \simeq -g_m R_L v_1 \tag{7.3}$$

Substitution of (7.3) in (7.1) gives

$$i_1 = (1 + g_m R_L)C_\mu s v_1$$

and thus

$$\frac{i_1}{v_1} = (1 + g_m R_L)C_\mu s \tag{7.4}$$

Equation 7.4 indicates that the impedance seen looking across the plane AA is a capacitance of value

$$C_M = (1 + g_m R_L)C_\mu \tag{7.5}$$

This is called the *Miller* capacitance. Equation 7.5 can be written as

$$C_M = (1 + A_v)C_\mu \tag{7.6}$$

where A_v is the magnitude of *voltage gain* from the internal base to the collector. Since A_v is often $\gg 1$, the Miller capacitance is often $\gg C_\mu$. The physical origin of the Miller capacitance is found in the voltage gain of the circuit. A small input voltage v_1 produces a large output voltage $v_o = -A_v v_1$ *of opposite polarity*. Thus the voltage across C_μ is $(1 + A_v)v_1$ and a correspondingly large current i_1 flows in this capacitor.

We can now form a new equivalent circuit that is useful for calculating the *forward transmission* and input impedance of the circuit. This is shown in Fig. 7.3 using the Miller-effect approximation. Note that this equivalent circuit is *not* useful for calculating high-frequency reverse transmission or output impedance. From this circuit we can see that at high frequencies the common-emitter input impedance eventually approaches just r_b.

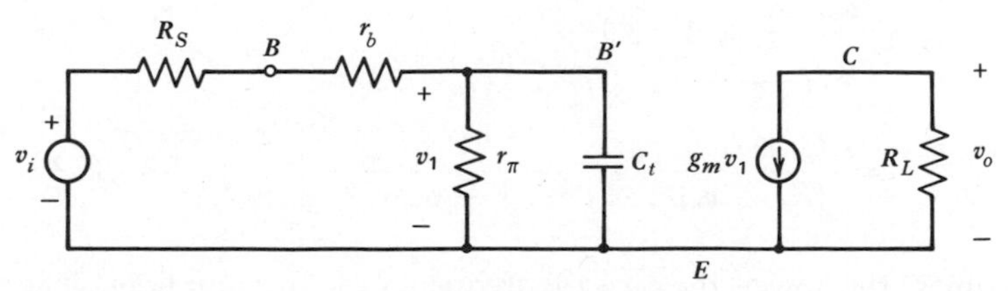

Fig. 7.3 Equivalent circuit for Fig. 7.2*b* using the Miller approximation.

In Fig. 7.3 the Miller capacitance adds directly to C_π of the transistor and thus *degrades* the frequency response of the amplifier, and this can be seen by calculating the gain of the amplifier as follows:

$$v_1 = \frac{\dfrac{r_\pi}{1 + r_\pi C_t s}}{\dfrac{r_\pi}{1 + r_\pi C_t s} + R_S + r_b} v_i \tag{7.7}$$

$$v_o = -g_m R_L v_1 \tag{7.8}$$

where

$$C_t = C_\pi + C_M \tag{7.9}$$

Substitution of (7.7) in (7.8) gives the differential-mode gain

$$A_{dm} = \frac{v_o}{v_i} = -g_m R_L \frac{r_\pi}{R_S + r_b + r_\pi} \frac{1}{1 + sC_t \dfrac{(R_S + r_b) r_\pi}{R_S + r_b + r_\pi}} \tag{7.10}$$

$$= \frac{K}{1 - \dfrac{s}{p_1}} \tag{7.11}$$

where K is the low-frequency voltage gain and p_1 is the pole of the circuit. Comparing (7.10) with (7.11) shows that

$$K = -g_m R_L \frac{r_\pi}{R_S + r_b + r_\pi} \tag{7.11a}$$

$$p_1 = -\frac{R_S + r_b + r_\pi}{(R_S + r_b) r_\pi} \frac{1}{C_t} \tag{7.11b}$$

This analysis indicates that the circuit has a single-pole response and substitution of $s = j\omega$ in (7.11) shows that the voltage gain is 3 dB below its low-frequency value at a frequency

$$\omega_{-3\,\mathrm{dB}} = |p_1|$$

$$= \frac{R_S + r_b + r_\pi}{(R_S + r_b) r_\pi} \frac{1}{C_t}$$

$$= \frac{R_S + r_b + r_\pi}{(R_S + r_b) r_\pi} \frac{1}{C_\pi + (1 + g_m R_L) C_\mu} \tag{7.12}$$

The larger C_t becomes, the lower the -3-dB frequency of the amplifier. The approximate value of $|p_1|$ can be estimated by assuming that R_S is greater than

r_π, and R_L is small. Then (7.12) gives

$$|p_1| \simeq \frac{1}{r_\pi C_\pi} = \frac{1}{\beta_0}\frac{g_m}{C_\pi} \simeq \frac{\omega_T}{\beta_0}$$

Larger values of R_L will give lower values of $|p_1|$. Smaller values of R_S will give larger values of $|p_1|$.

EXAMPLE

Using the Miller approximation, calculate the -3-dB frequency of a common-emitter transistor stage using the following parameters.

$$R_S = 1\ \text{k}\Omega \qquad r_b = 200\ \Omega \qquad I_C = 1\ \text{mA} \qquad \beta = 100$$
$$f_T = 400\ \text{MHz (at } I_C = 1\ \text{mA)} \qquad C_\mu = 0.5\ \text{pF} \qquad R_L = 5\ \text{k}\Omega$$

The transistor small-signal parameters are

$$r_\pi = \frac{\beta_0}{g_m} = 100 \times 26\ \Omega = 2.6\ \text{k}\Omega$$

$$\tau_T = \frac{1}{2\pi f_T} = 398\ \text{psec}$$

Using (1.129) from Chapter 1 gives

$$C_\pi + C_\mu = g_m \tau_T = \frac{1}{26} 398\ \text{pF} = 15.3\ \text{pF}$$

Thus

$$C_\pi - 14.8\ \text{pF}$$

Substitution of data in (7.5) gives, for the Miller capacitance,

$$C_M = (1 + g_m R_L)C_\mu = \left(1 + \frac{5000}{26}\right) 0.5\ \text{pF} = 96.7\ \text{pF}$$

This term is much greater than C_π and dominates the frequency response. Substitution of values in (7.12) gives

$$f_{-3\ \text{dB}} = \frac{1}{2\pi}\frac{1000 + 200 + 2600}{(1000 + 200)2600}\frac{10^{12}}{14.8 + 96.7}\ \text{Hz} = 1.74\ \text{MHz}$$

The low-frequency gain can be calculated from (7.10) as

$$A_{dm}|_{\omega = 0} = -g_m R_L \frac{r_\pi}{R_S + r_b + r_\pi} = \frac{-5000}{26}\frac{2.6}{5 + 0.2 + 2.6} = -64.1$$

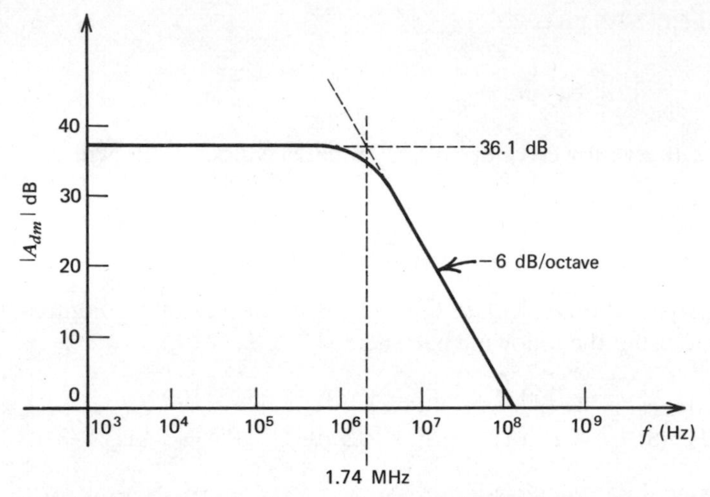

Fig. 7.4 Gain magnitude versus frequency for the circuit of Fig. 7.3 using typical device data in Equation 7.10.

The gain magnitude is thus 36.1 dB and the gain versus frequency on log scales is shown in Fig. 7.4 as calculated using (7.10) with $s = j\omega$.

The preceding calculations illustrate the usefulness of the Miller approximation, and it is instructive to return to the equivalent circuit of Fig. 7.2*b* to perform an exact analysis for comparison. For this purpose it is simpler to convert the input voltage source to its Norton equivalent as shown in Fig. 7.5 and define

$$R = (R_S + r_b)\|r_\pi \tag{7.13}$$

$$i_i = \frac{v_i}{R_S + r_b} \tag{7.14}$$

Summation of currents at B' gives

$$i_i = \frac{v_1}{R} + v_1 C_\pi s + (v_1 - v_o)C_\mu s \tag{7.15}$$

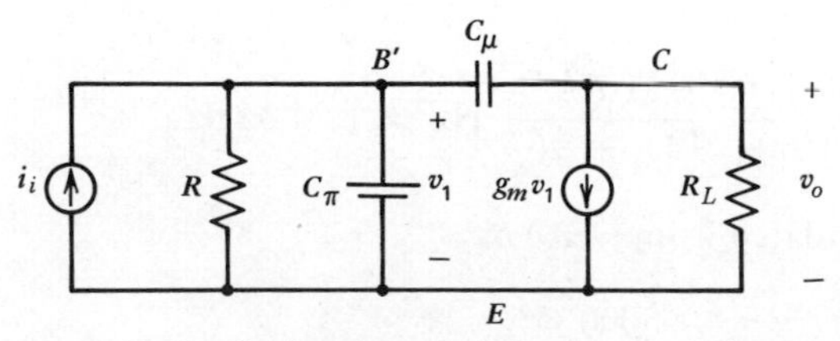

Fig. 7.5 Small-signal equivalent circuit of Fig. 7.2*a* using a Norton equivalent circuit at the input.

Summation of currents at C gives

$$g_m v_1 + \frac{v_o}{R_L} + (v_o - v_1)C_\mu s = 0 \tag{7.16}$$

Equation 7.16 can be written as

$$v_1(g_m - C_\mu s) = -v_o\left(\frac{1}{R_L} + C_\mu s\right)$$

and thus

$$v_1 = -v_o \frac{\frac{1}{R_L} + C_\mu s}{g_m - C_\mu s} \tag{7.17}$$

Substitution of (7.17) in (7.15) gives

$$i_i = -\left(\frac{1}{R} + C_\pi s + C_\mu s\right) \frac{\frac{1}{R_L} + C_\mu s}{g_m - C_\mu s} v_o - C_\mu s v_o$$

and the transfer function can be calculated as

$$\frac{v_o}{i_i} = -\frac{RR_L(g_m - C_\mu s)}{1 + s(C_\mu R_L + C_\mu R + C_\pi R + g_m R_L R C_\mu) + s^2 R_L R C_\mu C_\pi} \tag{7.18}$$

Substitution of i_i from (7.14) in (7.18) gives

$$\frac{v_o}{v_i} = -\frac{g_m R_L R}{R_S + r_b} \frac{1 - \frac{C_\mu}{g_m}s}{1 + s(C_\mu R_L + C_\mu R + C_\pi R + g_m R_L R C_\mu) + s^2 R_L R C_\mu C_\pi} \tag{7.19}$$

Substitution of R from (7.13) into (7.19) gives, for the low-frequency gain,

$$\left.\frac{v_o}{v_i}\right|_{\omega=0} = -g_m R_L \frac{r_\pi}{R_S + r_b + r_\pi} \tag{7.20}$$

as obtained in (7.10).

Equation 7.19 shows that the transfer function v_o/v_i has a positive real zero with magnitude g_m/C_μ. This is due to transmission of the signal directly through C_μ to the output. The effect of this zero is small except at very high frequencies and it will be neglected. The denominator of (7.19) shows the transfer function has two poles, and in practice these are usually widely separated in frequency. If the poles are at p_1 and p_2 we can write the denominator of (7.19) as

$$D(s) = \left(1 - \frac{s}{p_1}\right)\left(1 - \frac{s}{p_2}\right) \tag{7.21}$$

and thus

$$D(s) = 1 - s\left(\frac{1}{p_1} + \frac{1}{p_2}\right) + \frac{s^2}{p_1 p_2} \tag{7.22}$$

We now assume that the poles are widely separated, and we let the lower-frequency pole be p_1 (the dominant pole) and the higher-frequency pole be p_2.

Then $|p_2| \gg |p_1|$ and (7.22) becomes

$$D(s) \simeq 1 - \frac{s}{p_1} + \frac{s^2}{p_1 p_2} \tag{7.23}$$

If the coefficient of s in (7.23) is compared with that in (7.19) we can identify

$$\begin{aligned} p_1 &= -\frac{1}{RC_\pi + C_\mu(R_L + R + g_m R_L R)} \\ &= -\frac{1}{R}\frac{1}{C_\pi + C_\mu\left[(1 + g_m R_L) + \dfrac{R_L}{R}\right]} \end{aligned} \tag{7.24}$$

If the value of R from (7.13) is substituted in (7.24) then the dominant pole is

$$p_1 = -\frac{R_S + r_b + r_\pi}{(R_S + r_b)r_\pi}\frac{1}{C_\pi + C_\mu\left[(1 + g_m R_L) + \dfrac{R_L}{R}\right]} \tag{7.25}$$

The value of $|p_1|$ is almost identical to $\omega_{-3\text{dB}}$ given in (7.12) by the Miller approximation. The last term in the denominator of (7.25) is the only difference and this is usually small. This result shows that the Miller-effect calculation is equivalent to calculating the dominant pole of the amplifier and neglecting higher-frequency poles. It will be shown later that this gives a good estimate of $\omega_{-3\text{dB}}$ in most circuits.

Let us now calculate the nondominant pole by comparing the coefficient of s^2 in (7.23) with that in (7.19) giving

$$p_2 = \frac{1}{p_1}\frac{1}{R_L R C_\mu C_\pi} \tag{7.26}$$

Substitution of p_1 from (7.24) in (7.26) gives

$$p_2 = -\left(\frac{1}{R_L C_\mu} + \frac{1}{RC_\pi} + \frac{1}{R_L C_\pi} + \frac{g_m}{C_\pi}\right) \tag{7.27}$$

The last term of (7.27) is $g_m/C_\pi = \omega_T$ and thus $|p_2| > \omega_T$. Consequently $|p_2|$ is a very high frequency and it is almost always true that $|p_1| \gg |p_2|$. In the s plane, the poles of the amplifier are thus widely separated as shown in Fig. 7.6.

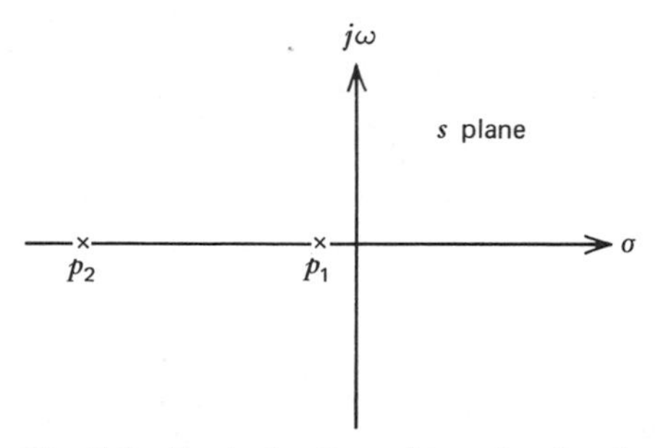

Fig. 7.6 Typical pole positions for the circuit of Fig. 7.5.

EXAMPLE

Calculate the nondominant pole frequency using the data given in the previous example. From (7.27),

$$p_2 = -\left(\frac{1}{R_L C_\mu} + \frac{1}{RC_\pi} + \frac{1}{R_L C_\pi} + \omega_T\right) \tag{7.28}$$

Now

$$R = (R_S + r_b)\|r_\pi = 1200\|2600\ \Omega = 821\ \Omega$$

Use of (7.28) gives

$$\begin{aligned} p_2 &= -\left(\frac{10^{12}}{5000 \times 0.5} + \frac{10^{12}}{821 \times 14.8} + \frac{10^{12}}{5000 \times 14.8} + 2\pi \times 400 \times 10^6\right) \text{rad/sec} \\ &= -(4 \times 10^8 + 0.8 \times 10^8 + 0.1 \times 10^8 + 25 \times 10^8)\ \text{rad/sec} \\ &= -30 \times 10^8\ \text{rad/sec} \\ &= -476\ \text{MHz} \end{aligned}$$

This is far beyond the −3-dB frequency of the amplifier and only has a significant effect on the transfer function of the circuit at frequencies near the unity gain frequency of the circuit. If the curve of Fig. 7.4 were extended, a second break frequency would occur at 476 MHz. Although negligible for most purposes, such effects can however be important as will be illustrated in the considerations of feedback amplifier compensation in Chapter 9.

In Chapter 3 the importance of the common-mode (CM) gain of a differential amplifier was described. It was shown that low values of CM gain are desirable so that the circuit can reject undesired signals that are applied equally to both inputs. Such undesired CM signals may have high-frequency components and thus the frequency response of the CM gain can be important. The CM frequency response of the differential circuit of Fig. 7.1 can be calculated from the CM half circuit shown in Fig. 7.7. In Fig. 7.7*a*, R_E and C_E are the equivalent output resistance

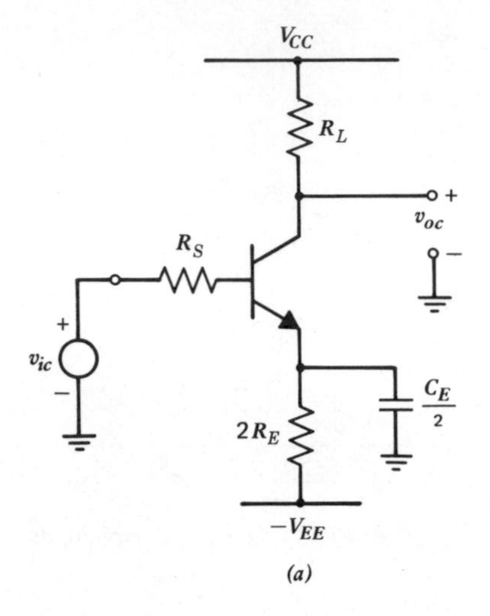

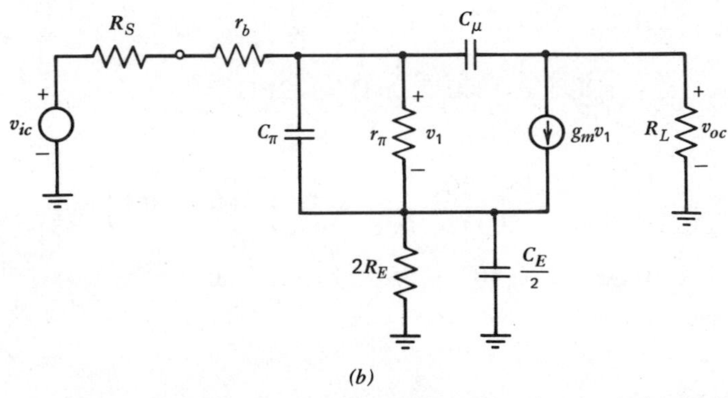

(b)

Fig. 7.7 (*a*) Common-mode half circuit for the circuit of Fig. 7.1 including parasitic emitter capacitance. (*b*) Small-signal equivalent circuit of (*a*).

and capacitance of the current source, I_{EE}. Since impedances common to the two devices are doubled in the CM half circuit, R_E and C_E become $2R_E$ and $C_E/2$. The small-signal equivalent circuit of Fig. 7.7*a* is shown in Fig. 7.7*b*.

The complete analysis of Fig. 7.7*b* is quite complex. However the important aspects of the frequency response can be calculating by making some approximations. Consider the time constant $R_E C_E$. The resistance, R_E, is the output resistance of the current source and is usually equal to the r_o of a transistor. At low bias current levels this may be of the order of 5 MΩ. The capacitor, C_E, is the C_{cs} of the current-source transistor and is about 2 pF. Thus the time constant $R_E C_E$ is 10 μsec using this data, and the break frequency corresponding to this time constant is $1/2\pi R_E C_E = 16$ kHz. Below this frequency the emitter impedance is dominated by R_E, and above this frequency C_E dominates. Thus as the frequency of operation is increased, the emitter impedance will exhibit frequency variation well before the rest of the circuit. We now calculate the frequency response assuming that this is so and also that C_E is the only significant capacitance. Since the emitter impedance is so high we can write, for the CM gain,

$$A_{cm} = \frac{v_{oc}}{v_{ic}} \simeq -\frac{R_L}{Z_E} \tag{7.29}$$

where

$$Z_E = \frac{2R_E}{1 + sC_E R_E} \tag{7.30}$$

and Z_E is the impedance from emitter to ground in Fig. 7.7*b*. Substitution of (7.30) in (7.29) gives

$$A_{cm}(s) = \frac{v_{oc}}{v_{ic}}(s) \simeq -\frac{R_L}{2R_E}(1 + sC_E R_E) \tag{7.31}$$

In the frequency domain

$$A_{cm}(j\omega) \simeq -\frac{R_L}{2R_E}(1 + j\omega C_E R_E) \tag{7.32}$$

Equation 7.31 shows that the CM gain expression contains a *zero*, which causes CM gain to rise at 6 dB/octave above a frequency $\omega = 1/R_E C_E$. This is undesirable because the CM gain should ideally be as small as possible. The increase in CM gain cannot continue indefinitely, however, as the other capacitors in the circuit of Fig. 7.7*b* eventually become important. This causes the CM gain to fall again at very high frequencies and this behavior is shown in the plot of CM gain versus frequency in Fig. 7.8*a*.

The differential-mode (DM) gain A_{dm} of the circuit of Fig. 7.1 is plotted versus frequency in Fig. 7.8*b* using (7.10). As discussed earlier, A_{dm} begins to fall off at a frequency given by $f = 1/2\pi RC_t$ where $R = (R_S + r_b)\|r_\pi$. An important differential amplifier parameter is the common-mode rejection ratio CMRR defined as

$$\text{CMRR} = \left|\frac{A_{dm}}{A_{cm}}\right| \tag{7.33}$$

and this is plotted as a function of frequency in Fig. 7.8*c* by simply taking the ratio of DM and CM gains. This quantity begins to decrease at frequency $f = 1/2\pi R_E C_E$ when $|A_{cm}|$ begins to increase. The rate of decrease of CMRR further increases when $|A_{dm}|$ begins to fall with increasing frequency. Thus differential amplifiers are far less able to reject CM signals as the frequency of those signals increases.

7.2.2 Emitter-Follower Frequency Response

The emitter follower is widely used in integrated-circuit design as a buffer stage, level-shift stage, and output stage and its frequency response is thus of considerable interest. Consider the emitter-follower ac schematic of Fig. 7.9*a* and the small-signal

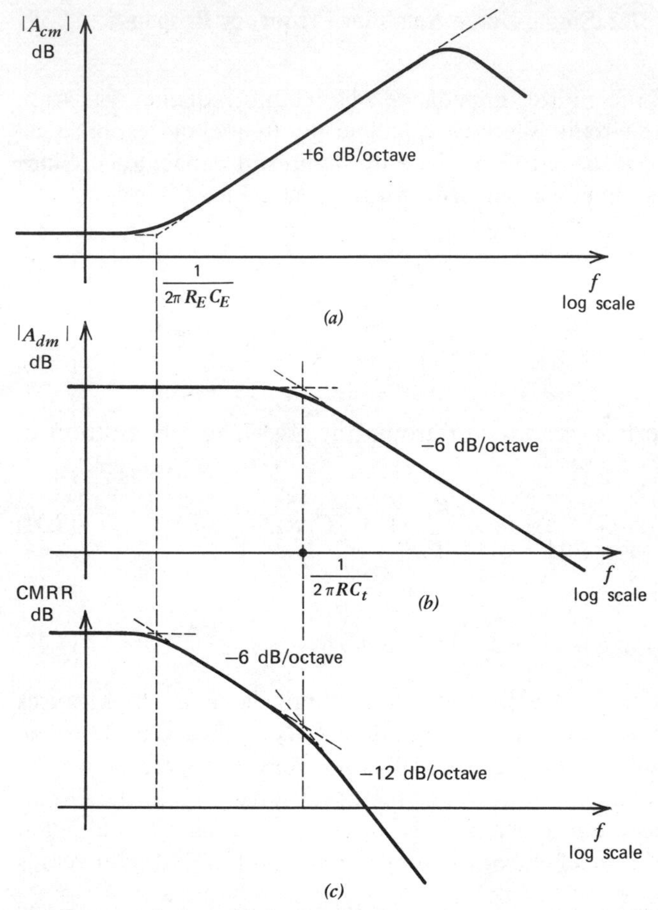

Fig. 7.8 Variation with frequency of the gain parameters of the circuit of Fig. 7.1. (*a*) Common-mode gain. (*b*) Differential-mode gain. (*c*) Common-mode rejection ratio.

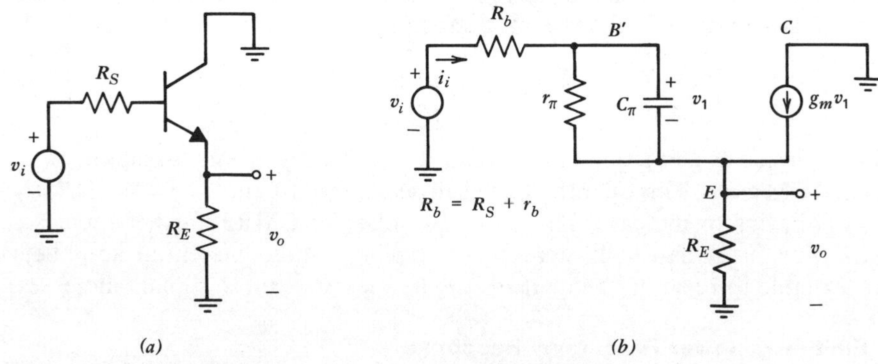

Fig. 7.9 Emitter-follower circuit (*a*) *ac* schematic. (*b*) Small-signal equivalent circuit.

equivalent circuit of Fig. 7.9*b*. The emitter follower is assumed driven from a voltage source v_i with source resistance R_S. The base resistance of the transistor and the source resistance are lumped together as $R_b = R_S + r_b$. For simplicity the effect of C_μ is initially neglected, a reasonable approximation if the source resistance and r_b are small. The effect of C_μ is to form a low-pass circuit with R_b and to cause the gain to decrease at very high frequencies. From Fig. 7.9*b*

$$v_i = i_i R_b + v_1 + v_o \tag{7.34}$$

$$i_i = \frac{v_1}{z_\pi} \tag{7.35}$$

$$z_\pi = \frac{r_\pi}{1 + sC_\pi r_\pi} \tag{7.36}$$

$$i_i + g_m v_1 = \frac{v_o}{R_E} \tag{7.37}$$

Using (7.35) and (7.36) in (7.37) gives

$$\frac{v_1}{r_\pi}(1 + sC_\pi r_\pi) + g_m v_1 = \frac{v_o}{R_E}$$

and thus

$$v_1 = \frac{v_o}{R_E} \frac{1}{g_m + \dfrac{1}{r_\pi}(1 + sC_\pi r_\pi)} \tag{7.38}$$

Using (7.38) and (7.35) in (7.34) gives

$$v_i = \left(\frac{R_b}{z_\pi} + 1\right) \frac{v_o}{R_E} \frac{1}{g_m + \dfrac{1}{r_\pi}(1 + sC_\pi r_\pi)} + v_o$$

and collecting terms in this equation we find

$$\frac{v_o}{v_i} = \frac{g_m R_E + \dfrac{R_E}{r_\pi}}{1 + g_m R_E + \dfrac{R_b + R_E}{r_\pi}} \frac{1 - \dfrac{s}{z_1}}{1 - \dfrac{s}{p_1}} \tag{7.39}$$

where

$$z_1 \simeq -\frac{g_m}{C_\pi} = -\omega_T \tag{7.40}$$

$$p_1 = -\frac{1}{C_\pi R_1} \tag{7.41}$$

$$R_1 = r_\pi \left\| \frac{R_b + R_E}{1 + g_m R_E} \right. \tag{7.42}$$

Equation 7.39 shows that as expected, the low-frequency voltage gain is approximately unity if $g_m R_E \gg 1$. The high-frequency gain is controlled by the presence of a pole at p_1 and a zero at z_1. For typical parameter values the zero is at a slightly higher frequency than the pole and both are approximately equal to the ω_T of the device. In particular, if $g_m R_E \gg 1$ and $R_b \ll R_E$ in (7.42) then $R_1 \simeq 1/g_m$ and in (7.41) $p_1 \simeq -g_m/C_\pi \simeq -\omega_T$. However, if R_S is large then R_b in (7.42) may become large compared to R_E, and the pole frequency will be significantly less than ω_T.

EXAMPLE

Calculate the transfer function for an emitter follower with $C_\pi = 10$ pF, $C_\mu = 0$, $R_E = 2$ kΩ, $R_S = 50\ \Omega$, $r_b = 150\ \Omega$, $\beta = 100$, and $I_C = 1$ mA.

From the data, $g_m = 38$ mA/V, $r_\pi = 2.6$ kΩ, and $R_b = R_S + r_b = 200\ \Omega$. Since $C_\mu = 0$, the ω_T of the device is

$$\omega_T = \frac{g_m}{C_\pi} = \frac{1}{26}\frac{10^{12}}{10} = 3.85 \times 10^9 \text{ rad/sec}$$

and thus

$$f_T = 612 \text{ MHz}$$

From (7.40) the zero of the transfer function is

$$z_1 = -3.85 \times 10^9 \text{ rad/sec}$$

From (7.42)

$$R_1 = 2.6 \text{ k}\Omega \left\| \frac{200 + 2000}{1 + \dfrac{2000}{26}} \Omega \simeq 28\ \Omega\right.$$

The pole frequency from (7.41) is

$$p_1 = -\frac{10^{12}}{10}\frac{1}{28} \text{ rad/sec} = -3.57 \times 10^9 \text{ rad/sec}$$

The pole and zero are thus quite closely spaced as shown in the s-plane plot of Fig. 7.10*a*.

The low-frequency gain of the circuit from (7.39) is

$$\frac{v_o}{v_i} = \frac{g_m R_E + \dfrac{R_E}{r_\pi}}{1 + g_m R_E + \dfrac{R_b + R_E}{r_\pi}} = \frac{\dfrac{2000}{26} + \dfrac{2000}{2600}}{1 + \dfrac{2000}{26} + \dfrac{2200}{2600}} = 0.986$$

The parameters derived above can be used in (7.39) to plot the circuit gain versus frequency and this has been done in Fig. 7.10*b*. The gain is flat with frequency until near f_T at about 600 MHz where a decrease of 0.7 dB occurs. The analysis

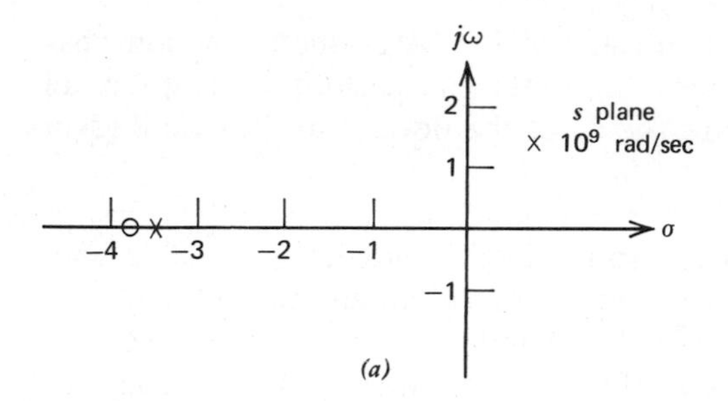

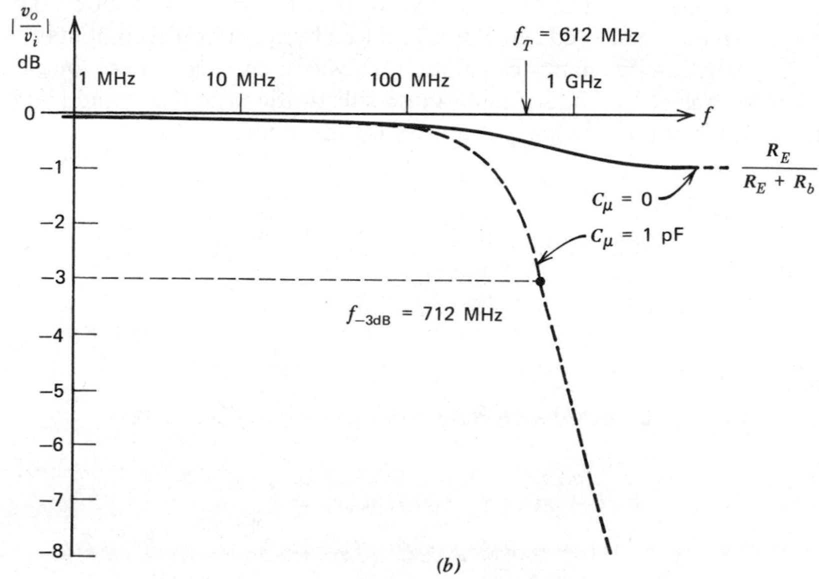

Fig. 7.10 (*a*) Pole-zero plot for an emitter follower with $C_\mu = 0$, $C_\pi = 10$ pF, $R_E = 2$ kΩ, $R_S = 50$ Ω, $r_b = 150$ Ω, $\beta = 100$, and $I_C = 1$ mA. (*b*) Voltage gain versus frequency for the emitter follower in (*a*) with $C_\mu = 0$ and with $C_\mu = 1$ pF.

predicts the gain is then flat as frequency is increased further. This is because the input signal is simply fed forward to R_E via C_π at very high frequencies.

By inspecting Fig. 7.9*b* we can see that the high-frequency gain is asymptotic to $R_E/(R_E + R_b)$ since C_π becomes a short circuit. This forces $v_1 = 0$ and thus the controlled generator $g_m v_1$ is also zero. If a value of $C_\mu = 1$ pF is included in the equivalent circuit the more realistic dotted frequency response of Fig. 7.10*b* is obtained. Since the collector is grounded, C_μ is connected from B' to ground and thus high-frequency signals are attenuated by voltage division between R_b and C_μ.

As a result the circuit has a -3-dB frequency of 712 MHz due to the low-pass action of R_b and C_μ. However, the bandwidth of the emitter follower is still quite large and band-widths of the order of the f_T of the device can be obtained in practice.

The preceding considerations have shown the large bandwidths available from the emitter-follower circuit. One of the primary uses of an emitter follower is as a buffer circuit due to its high input impedance and low output impedance. The behavior of these terminal impedances as a function of frequency is thus significant and will now be examined.

In Chapter 3 the terminal impedances of the emitter follower were calculated using a circuit similar to that of Fig. 7.9*b* except that C_π was not included. The results obtained there can be used here if r_π is replaced by z_π, a parallel combination of r_π and C_π. In the low-frequency calculation, β_0 was used as a symbol for $g_m r_\pi$ and thus is now replaced by $g_m z_\pi$. Using these substitutions in (3.23) and (3.26), and including r_b, we obtain at high frequencies for the emitter follower

$$z_i = r_b + z_\pi + (g_m z_\pi + 1)R_E \tag{7.43}$$

$$z_o = \frac{z_\pi + R_S + r_b}{1 + g_m z_\pi} \tag{7.44}$$

where

$$z_\pi = \frac{r_\pi}{1 + sC_\pi r_\pi} \tag{7.45}$$

Consider first the input impedance. Substituting (7.45) in (7.43) gives

$$\begin{aligned} z_i &= r_b + \frac{r_\pi}{1 + sC_\pi r_\pi} + \left(\frac{g_m r_\pi}{1 + sC_\pi r_\pi} + 1\right) R_E \\ &= r_b + \frac{(1 + g_m R_E) r_\pi}{1 + sC_\pi r_\pi} + R_E \\ &= r_b + \frac{(1 + g_m R_E) r_\pi}{1 + s\dfrac{C_\pi}{1 + g_m R_E}(1 + g_m R_E) r_\pi} + R_E \\ &= r_b + \frac{R}{1 + sCR} + R_E \end{aligned} \tag{7.46}$$

where

$$R = (1 + g_m R_E) r_\pi \tag{7.46a}$$

and

$$C = \frac{C_\pi}{1 + g_m R_E} \tag{7.46b}$$

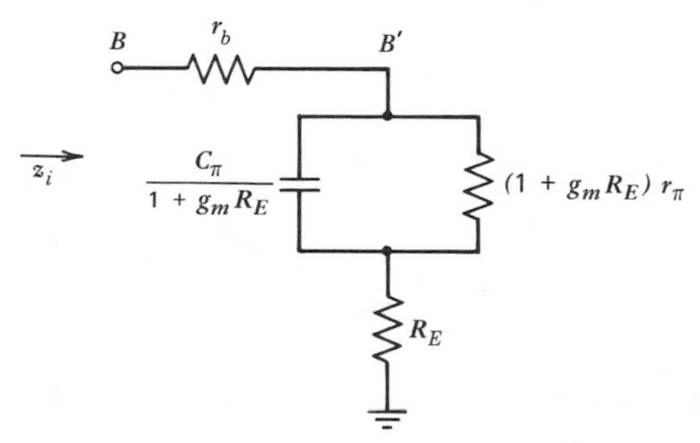

Fig. 7.11 Equivalent circuit for the input impedance of an emitter follower with $C_\mu = 0$.

Thus z_i can be represented as a parallel R-C circuit in series with r_b and R_E as shown in Fig. 7.11. The effective input capacitance is $C_\pi/(1 + g_m R_E)$ and is much less than C_π for typical values of $g_m R_E$. The collector-base capacitance C_μ may dominate the input capacitance and can be added to this circuit from B' to ground. Thus, at high frequencies, the input impedance of the emitter follower becomes capacitive and its magnitude decreases.

The emitter-follower high-frequency output impedance can be calculated by substituting (7.45) in (7.44). Before doing this it is instructive to examine (7.45) to determine the high and low frequency limits on $|z_o|$. At low frequencies, $z_\pi = r_\pi$ and

$$z_o\big|_{\omega=0} \simeq \frac{1}{g_m} + \frac{R_S + r_b}{\beta_0} \tag{7.47}$$

At high frequencies, $z_\pi \to 0$ because C_π becomes a short circuit and thus

$$z_o\big|_{\omega=\infty} = R_S + r_b \tag{7.48}$$

Thus z_o is resistive at very low and very high frequencies and its behavior in between depends on parameter values. At very low collector currents, $1/g_m$ is large and if $1/g_m > (R_S + r_b)$ then a comparison of (7.47) and (7.48) shows that $|z_o|$ *decreases* as frequency increases and the output impedance appears capacitive. However, at collector currents of more than several hundred microamperes we usually find $1/g_m < (R_S + r_b)$ and $|z_o|$ *increases* with frequency. This represents *inductive* behavior that can have a major influence on the circuit behavior, particularly when it drives capacitive loads.

Assuming that the collector bias current is such that z_o is inductive, we can postulate an equivalent circuit for z_o as shown in Fig. 7.12. At low frequencies the inductor is a short circuit and

$$z_o\big|_{\omega=0} = R_1 \| R_2 \tag{7.49}$$

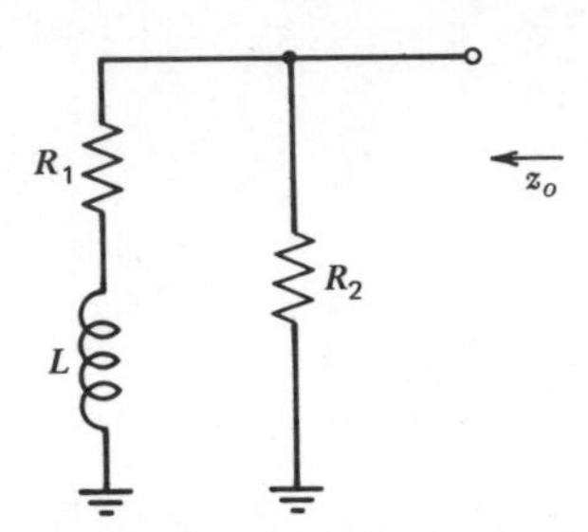

Fig. 7.12 Equivalent circuit for the output impedance of an emitter follower at moderate current levels.

At high frequencies the inductor is an open circuit and

$$z_o\big|_{\omega=\infty} = R_2 \tag{7.50}$$

If we assume that $z_o\big|_{\omega=0} \ll z_o\big|_{\omega=\infty}$ then $R_1 \ll R_2$ and we can simplify (7.49) to

$$z_o\big|_{\omega=0} \simeq R_1 \tag{7.51}$$

The impedance of the circuit of Fig. 7.12 can be expressed as

$$z = \frac{(R_1 + sL)R_2}{R_1 + R_2 + sL} \simeq \frac{(R_1 + sL)R_2}{R_2 + sL} \tag{7.52}$$

assuming that $R_1 \ll R_2$.

The actual emitter-follower output impedance can be calculated by substituting (7.45) in (7.44) and using

$$R_b = r_b + R_S \tag{7.53}$$

This gives

$$\begin{aligned} z_o &= \frac{\dfrac{r_\pi}{1 + sC_\pi r_\pi} + R_b}{1 + \dfrac{g_m r_\pi}{1 + sC_\pi r_\pi}} \\ &= \frac{r_\pi + R_b + sC_\pi r_\pi R_b}{\beta_0 + 1 + sC_\pi r_\pi} \\ &\simeq \frac{\left(\dfrac{1}{g_m} + \dfrac{R_b}{\beta_0} + sC_\pi r_\pi \dfrac{R_b}{\beta_0}\right) R_b}{R_b + sC_\pi r_\pi \dfrac{R_b}{\beta_0}} \end{aligned} \tag{7.54}$$

where $\beta_0 \gg 1$ is assumed.

Comparing (7.54) with (7.52) shows that with the assumptions made in this analysis the emitter-follower output impedance can be represented by the circuit of Fig. 7.12 with

$$R_1 = \frac{1}{g_m} + \frac{R_b}{\beta_0} \tag{7.55}$$

$$R_2 = R_b \tag{7.56}$$

$$L = C_\pi r_\pi \frac{R_b}{\beta_0} \tag{7.57}$$

Note that the effect of C_μ was neglected in this calculation, and this is a reasonable approximation for low to moderate values of R_b.

The preceding calculations have shown that the input and output impedances of the emitter follower are frequency dependent. Thus, even though the bandwidth of the circuit transfer function may be large, the variation of the terminal impedances with frequency may impose restrictions on the useful bandwidth of the circuit.

EXAMPLE

Calculate the elements in the equivalent circuits for input and output impedance of the emitter follower in the previous example.

In Fig. 7.11 the input capacitance is

$$\frac{C_\pi}{1 + g_m R_E} = \frac{10}{1 + \dfrac{2000}{26}} \text{ pF} = 0.13 \text{ pF}$$

The resistance in shunt with this capacitance is

$$(1 + g_m R_E) r_\pi = \left(1 + \frac{2000}{26}\right) 2.6 \text{ k}\Omega = 202 \text{ k}\Omega$$

In addition, $r_b = 150\ \Omega$ and $R_E = 2\ \text{k}\Omega$. The elements in the output equivalent circuit of Fig. 7.12 can be calculated from (7.55), (7.56), and (7.57) as

$$R_1 = \left(26 + \frac{200}{100}\right) \Omega = 28\ \Omega$$

$$R_2 = 200\ \Omega$$

$$L = 10^{-11} \times 2600 \times \frac{200}{100} \text{ H} = 52 \text{ nH}$$

Note that the assumption $R_1 \ll R_2$ is valid in this case.

7.2.3 Common-Base Amplifier Frequency Response

The common-base (CB) amplifier configuration is shown in ac schematic form in Fig. 7.13*a*. A common-base stage has a low input impedance, high output impedance, approximately unity current gain, and wide bandwidth. These stages find use in wideband applications and also in applications requiring low input impedance. As described in Chapter 1 the transistor breakdown voltage is maximum in this configuration. When this property is combined with the wideband property, it is apparent why these stages are often used as high-voltage wideband output stages driving oscilloscope deflection plates. Common-base stages are also useful for level shifting as in the 741 op amp.

Comparing Fig. 7.13*a* with Fig. 7.9*a* shows that the input impedance of the common-base stage is the same as the output impedance of the emitter follower with $R_S = 0$. Thus the common-base stage input impedance is low at low frequencies, and for collector bias currents of several hundred microamperes or more becomes inductive at high frequencies because of the presence of r_b. As shown in Chapter 3, the output resistance of the common-base stage at low frequencies is approximately $\beta_0 r_o$ and is extremely large. At high frequencies the output impedance is dominated by C_μ (and C_{cs} for *npn* transistors) and is capacitive.

A small-signal equivalent circuit of the CB stage is shown in Fig. 7.13*b* where C_{cs} has been omitted. (C_{cs} should be included for *npn* transistors and is readily handled as it simply shunts the load resistance.) The input voltage source is represented by its Norton equivalent, and, since the CB stage input impedance is quite low, R_S can often be neglected. This will be done in this analysis. Another good approximation if r_b is not too large is to assume that C_μ simply shunts R_L. In this analysis C_μ will be neglected and the output signal will be taken as the output current, i_o. The output voltage, v_o, is obtained by considering i_o to flow in the parallel combination of R_L, C_μ, and C_{cs}. Note that in this circuit there is no feedback capacitance from collector to emitter to cause the Miller effect in the way that C_μ does in the common-emitter stage. As a consequence, the frequency

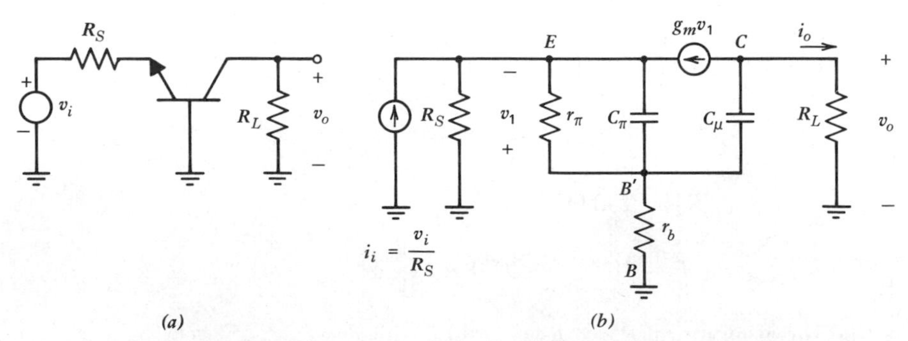

Fig. 7.13 Common-base circuit. (*a*) ac schematic. (*b*) Small-signal equivalent circuit.

response of the common-base stage is much less affected by large values of R_L than is the case for the common-emitter stage.

The analysis of the circuit of Fig. 7.13*b* proceeds by summation of currents at *E*.

$$i_i + \frac{v_1}{z_\pi} + g_m v_1 = 0 \tag{7.58}$$

where

$$z_\pi = \frac{r_\pi}{1 + sC_\pi r_\pi} \tag{7.59}$$

From (7.58) and (7.59)

$$i_i = -v_1\left(g_m + \frac{1}{r_\pi} + sC_\pi\right) \tag{7.60}$$

Now

$$i_o = -g_m v_1 \tag{7.61}$$

Substituting (7.60) in (7.61) gives

$$\frac{i_o}{i_i} \simeq \frac{\alpha_0}{1 + s\dfrac{C_\pi}{g_m}} \tag{7.62}$$

where

$$\alpha_0 = \frac{\beta_0}{1 + \beta_0} \tag{7.63}$$

This analysis shows that the CB stage current gain has a low-frequency value $\alpha_0 \simeq 1$ and a pole at $p_1 = -(g_m/C_\pi) \simeq -\omega_T$. The CB stage is thus a wideband unity-current-gain amplifier with low input impedance and high output impedance. It can be seen from the polarities in Fig. 7.13 that there is zero phase shift between v_i and v_o in the CB stage at low frequencies. This can be compared with the case of the common-emitter stage of Fig. 7.2, which has 180° phase shift between v_i and v_o at low frequencies.

7.3 MULTISTAGE AMPLIFIER FREQUENCY RESPONSE

The above analysis of the frequency behavior of single-stage circuits indicates the complexity that can arise even with simple circuits. The complete analysis of the frequency response of multistage circuits with many capacitive elements rapidly becomes very difficult and the answers so complicated that little use can be made of the results. For this reason approximate methods of analysis have been developed to aid in the circuit design phase, and computer simulation is used to verify the final design. One such method of analysis is the *zero-value time constant* analysis that will now be described. First some ideas regarding dominant poles are developed.

7.3.1 Dominant-Pole Approximation

For any electronic circuit we can derive a transfer function $A(s)$ by small-signal analysis to give

$$A(s) = \frac{N(s)}{D(s)} = \frac{a_0 + a_1 s + a_2 s^2 + \cdots + a_m s^m}{1 + b_1 s + b_2 s^2 + \cdots + b_n s^n} \tag{7.64}$$

where $a_0, a_1 \ldots a_m$, and $b_1, b_2 \ldots b_n$ are constants. Very often the transfer function contains poles only (or the zeros are unimportant). In this case we can factor the denominator of (7.64) to give

$$A(s) = \frac{K}{\left(1 - \dfrac{s}{p_1}\right)\left(1 - \dfrac{s}{p_2}\right) \cdots \left(1 - \dfrac{s}{p_n}\right)} \tag{7.65}$$

where K is a constant and $p_1, p_2, \ldots p_n$ are the poles of the transfer function.

It is apparent from (7.65) that

$$b_1 = \sum_{i=1}^{n} \left(-\frac{1}{p_i}\right) \tag{7.66}$$

An important practical case occurs when one pole is dominant. That is, when

$$|p_1| \ll |p_2|, |p_3|, \ldots \quad \text{so that} \quad \left|\frac{1}{p_1}\right| \gg \left|\sum_{i=2}^{n} \left(-\frac{1}{p_i}\right)\right|$$

This situation is shown in the s plane in Fig. 7.14 and in this case it follows from (7.66) that

$$b_1 \simeq \left|\frac{1}{p_1}\right| \tag{7.67}$$

If we return now to (7.65) and calculate the gain magnitude in the frequency domain we obtain

$$|A(j\omega)| = \frac{K}{\sqrt{\left[1 + \left(\dfrac{\omega}{p_1}\right)^2\right]\left[1 + \left(\dfrac{\omega}{p_2}\right)^2\right] \cdots \left[1 + \left(\dfrac{\omega}{p_n}\right)^2\right]}} \tag{7.68}$$

If a dominant pole exists then (7.68) can be approximated by

$$|A(j\omega)| \simeq \frac{1}{\sqrt{1 + \left(\dfrac{\omega}{p_1}\right)^2}} \tag{7.69}$$

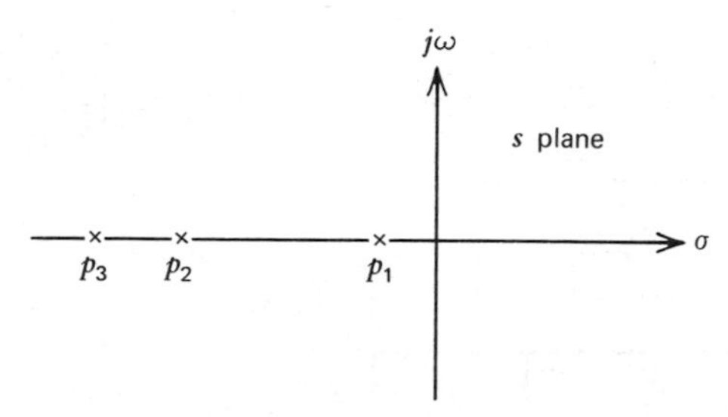

Fig. 7.14 Pole plot for a circuit with a dominant pole.

This approximation will be quite accurate at least until $\omega \simeq |p_1|$, and thus (7.69) will accurately predict the -3-dB frequency and we can write

$$\omega_{-3\text{ dB}} \simeq |p_1| \tag{7.70}$$

Use of (7.67) in (7.70) gives

$$\omega_{-3\text{ dB}} \simeq \frac{1}{b_1} \tag{7.71}$$

for a dominant-pole situation.

7.3.2 Zero-Value Time Constant Analysis

This is an approximate method of analysis that allows an estimate to be made of the dominant-pole frequency (and thus the -3-dB frequency) of complex circuits. Considerable saving in computational effort is achieved because a full analysis of the circuit is not required. The method will be developed by considering a practical example.

Consider the equivalent circuit shown in Fig. 7.15. This is a single-stage transistor amplifier with resistive source and load impedances. The feedback capacitance is split into two parts (C_x and C_μ) as shown. This is a slightly better approximation to the actual situation than the single collector-base capacitor we have been using, but is rarely used in hand calculations because of the analysis complexity. For purposes of analysis, the capacitor voltages, v_1, v_2, and v_3 are chosen as variables. The external input, v_i, is removed and the circuit excited with three independent current sources i_1, i_2, and i_3 across the capacitors as shown in Fig. 7.15. We can show that with this choice of variables the circuit equations are of the form

$$i_1 = (g_{11} + sC_\pi)v_1 + g_{12}v_2 + g_{13}v_3 \tag{7.72}$$

$$i_2 = g_{21}v_1 + (g_{22} + sC_\mu)v_2 + g_{23}v_3 \tag{7.73}$$

$$i_3 = g_{31}v_1 + g_{32}v_2 + (g_{33} + sC_x)v_3 \tag{7.74}$$

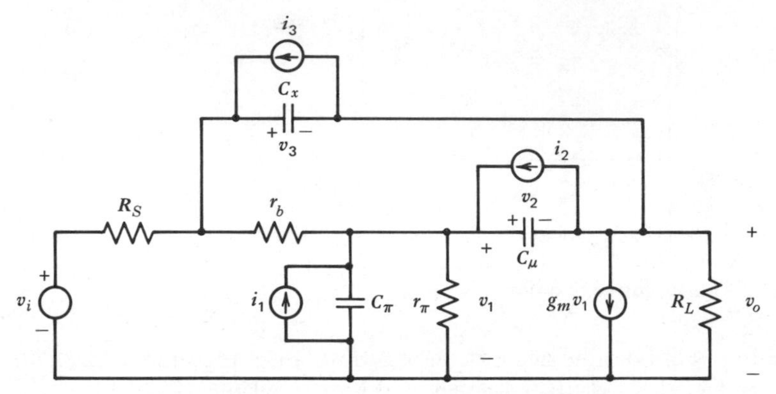

Fig. 7.15 Small-signal equivalent circuit of a common-emitter stage with internal feedback capacitors C_μ and C_x.

where the g terms are conductances. Note that the terms involving s contributed by the capacitors are associated only with their respective capacitor voltage variables and only appear on the diagonal of the system determinant.

The poles of the circuit transfer function are the zeros of the determinant Δ of the circuit equations, which can be written in the form

$$\Delta(s) = K_3 s^3 + K_2 s^2 + K_1 s + K_0 \tag{7.75}$$

where the coefficients K are composed of terms from the above equations. For example, K_3 is the sum of the coefficients of all terms involving s^3 in the expansion of the determinant. Equation 7.75 can be expressed as

$$\Delta(s) = K_0(1 + b_1 s + b_2 s^2 + b_3 s^3) \tag{7.76}$$

where this form corresponds to (7.64). Note that this is a third-order determinant because there are three capacitors in the circuit. The term K_0 in (7.75) is the value of $\Delta(s)$ if all capacitors are zero ($C_x = C_\mu = C_\pi = 0$). This can be seen from equations 7.72, 7.73, and 7.74. Thus

$$K_0 = \Delta|_{C_\pi = C_\mu = C_x = 0}$$

and it is useful to define

$$K_0 \triangleq \Lambda_0 \tag{7.77}$$

Consider now the term $K_1 s$ in (7.75). This is the sum of all the terms involving s that are obtained when the system determinant is evaluated. However from (7.72) to (7.74) it is apparent that s only occurs when associated with a capacitance. Thus the term $K_1 s$ can be written as

$$K_1 s = h_1 s C_\pi + h_2 s C_\mu + h_3 s C_x \tag{7.78}$$

where the h terms are constants. The term h_1 can be evaluated by expanding the determinant of Equations 7.72 to 7.74 about the first row.

$$\Delta(s) = (g_{11} + sC_\pi)\,\Delta_{11} + g_{12}\,\Delta_{12} + g_{13}\,\Delta_{13} \tag{7.79}$$

where Δ_{11}, Δ_{12}, and Δ_{13} are cofactors of the determinant. Inspection of (7.72), (7.73), and (7.74) shows that C_π occurs only in the first term of (7.79). Thus the coefficient of $C_\pi s$ in (7.79) is found by evaluating Δ_{11} with $C_\mu = C_x = 0$, which will eliminate the other capacitive terms in Δ_{11}. But this coefficient of $C_\pi s$ is just h_1 in (7.78), and so

$$h_1 = \Delta_{11}|_{C_\mu = C_x = 0} \tag{7.80}$$

Now consider expansion of the determinant about the second row. This must give the same value for the determinant, and thus

$$\Delta(s) = g_{21}\,\Delta_{21} + (g_{22} + sC_\mu)\,\Delta_{22} + g_{23}\,\Delta_{23} \tag{7.81}$$

In this case C_μ occurs only in the second term of (7.81). Thus the coefficient of $C_\mu s$ in this equation is found by evaluating Δ_{22} with $C_\pi = C_x = 0$, which will eliminate the other capacitive terms. This coefficient of $C_\mu s$ is just h_2 in (7.78), and thus

$$h_2 = \Delta_{22}|_{C_\pi = C_x = 0} \tag{7.82}$$

Similarly by expanding about the third row it follows that

$$h_3 = \Delta_{33}|_{C_\pi = C_\mu = 0} \tag{7.83}$$

Combining (7.78) with (7.80), (7.82), and (7.83) gives

$$K_1 = (\Delta_{11}|_{C_\mu = C_x = 0} \times C_\pi) + (\Delta_{22}|_{C_\pi = C_x = 0} \times C_\mu) + (\Delta_{33}|_{C_\mu = C_\pi = 0} \times C_x) \tag{7.84}$$

and

$$b_1 = \frac{K_1}{K_0} = \frac{\Delta_{11}|_{C_\mu = C_x = 0}}{\Delta_0} \times C_\pi + \frac{\Delta_{22}|_{C_\pi = C_x = 0}}{\Delta_0} \times C_\mu + \frac{\Delta_{33}|_{C_\mu = C_\pi = 0}}{\Delta_0} \times C_x \tag{7.85}$$

where the boundary conditions on the determinants are the same as in (7.84). Now consider putting $i_2 = i_3 = 0$ in Fig. 7.15. Solving Equations 7.72 to 7.74 for v_1 gives

$$v_1 = \frac{\Delta_{11} i_1}{\Delta(s)}$$

and thus

$$\frac{v_1}{i_1} = \frac{\Delta_{11}}{\Delta(s)} \tag{7.86}$$

Equation 7.86 is an expression for the driving-point impedance at the C_π node pair. Thus

$$\frac{\Delta_{11}|_{C_\mu = C_x = 0}}{\Delta_0}$$

is the driving-point *resistance* at the C_π node pair with *all* capacitors equal to zero because

$$\frac{\Delta_{11}|_{C_\mu = C_x = 0}}{\Delta_0} = \left.\frac{\Delta_{11}}{\Delta}\right|_{C_\mu = C_x = C_\pi = 0} \tag{7.87}$$

We now define

$$R_{\pi 0} = \left.\frac{\Delta_{11}}{\Delta_0}\right|_{C_\mu = C_x = 0} \tag{7.88}$$

Similarly,

$$\frac{\Delta_{22}|_{C_\pi = C_x = 0}}{\Delta_0}$$

is the driving-point *resistance* at the C_μ node pair with all capacitors put equal to zero and is represented by $R_{\mu 0}$. Thus we can write from (7.85)

$$b_1 = R_{\pi 0} C_\pi + R_{\mu 0} C_\mu + R_{x 0} C_x \tag{7.89}$$

The time constants in (7.89) are called "zero-value time constants" because all capacitors are put equal to zero to perform the calculation.

We showed previously that if there are no dominant zeros in the circuit transfer function, and if there *is* a dominant pole, then

$$\omega_{-3\text{ dB}} \simeq \frac{1}{b_1}$$

Thus

$$\omega_{-3\text{ dB}} \simeq \frac{1}{\sum T_0} \tag{7.90}$$

where $\sum T_0$ is the sum of the zero-value time constants. Although derived in terms of a specific example, this result is true in *any* circuit for which the various assumptions made in this analysis are valid.

Consider the circuit of Fig. 7.15. By inspection,

$$R_{\pi 0} = r_\pi \| (R_S + r_b) \tag{7.91}$$

In order to calculate $R_{\mu 0}$ it is necessary to write some simple circuit equations. We apply a test current i at the C_μ terminals as shown in Fig. 7.16 and calculate the resulting v.

$$v_1 = R_{\pi 0} i \tag{7.92}$$

$$v_o = -(i + g_m v_1) R_L \tag{7.93}$$

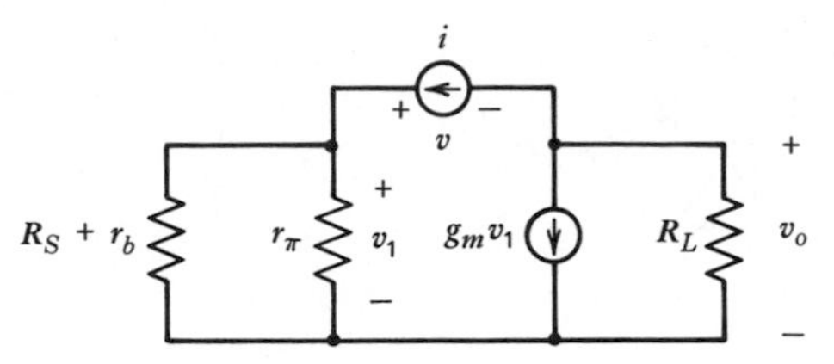

Fig. 7.16 Equivalent circuit for the calculation of $R_{\mu 0}$ for Fig. 7.15.

Substituting (7.92) in (7.93) gives

$$v_o = -(i + g_m R_{\pi 0} i) R_L \tag{7.94}$$

Now

$$R_{\mu 0} = \frac{v}{i}$$

and

$$R_{\mu 0} = \frac{v_1 - v_o}{i} \tag{7.95}$$

Substitution of (7.92) and (7.94) in (7.95) gives

$$R_{\mu 0} = R_{\pi 0} + R_L + g_m R_L R_{\pi 0} \tag{7.96}$$

R_{x0} can be calculated in a similar fashion, and it is apparent that $R_{x0} \simeq R_{\mu 0}$ if $r_b \ll r_\pi$. This justifies the common practice of lumping C_x in with C_μ if r_b is small. Assuming that this is done, (7.90) gives, for the -3-dB frequency,

$$\omega_{-3\text{ dB}} = \frac{1}{R_{\pi 0} C_\pi + R_{\mu 0} C_\mu} \tag{7.97}$$

Using (7.96) in (7.97) gives

$$\omega_{-3\text{ dB}} = \frac{1}{R_{\pi 0}\left\{C_\pi + C_\mu\left[(1 + g_m R_L) + \dfrac{R_L}{R_{\pi 0}}\right]\right\}} \tag{7.98}$$

Equation 7.98 is identical with the result obtained in (7.25) by exact analysis. [Recall that R in (7.25) is the same quantity as $R_{\pi 0}$ in (7.98).] However the zero-value time constant analysis gives the result with *much less* effort. It does *not* give any information on the nondominant pole.

As a further illustration of the uses and limitations of the zero-value time constant approach, consider the emitter-follower circuit of Fig. 7.9 where only the capacitance, C_π, has been included. The value of $R_{\pi 0}$ can be calculated by

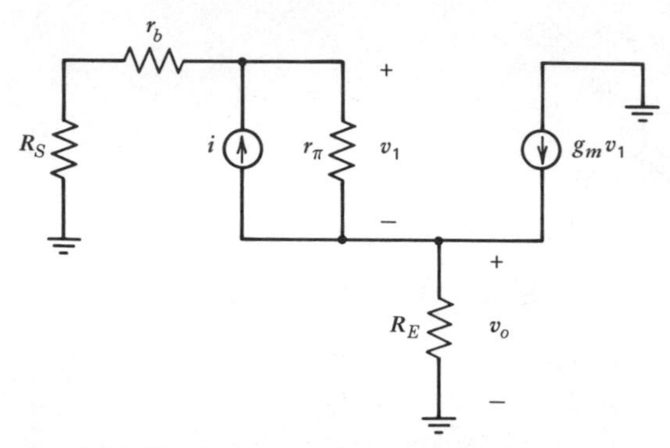

Fig. 7.17 Equivalent circuit for the calculation of $R_{\pi 0}$ for the emitter follower.

inserting a current source i as shown in Fig. 7.17 and calculating the resulting voltage, v_1.

$$i = \frac{v_1}{r_\pi} + \frac{v_1 + v_o}{R_S + r_b} \tag{7.99}$$

$$\frac{v_1}{r_\pi} - i + g_m v_1 = \frac{v_o}{R_E} \tag{7.100}$$

Substituting (7.100) in (7.99) gives

$$i = \frac{v_1}{r_\pi} + \frac{v_1}{R_S + r_b} + \frac{R_E}{R_S + r_b}\left(\frac{v_1}{r_\pi} + g_m v_1 - i\right)$$

and this equation can be expressed as

$$i = \frac{v_1}{r_\pi} + v_1 \frac{1 + g_m R_E}{R_S + r_b + R_E}$$

Finally $R_{\pi 0}$ can be calculated as

$$R_{\pi 0} = \frac{v_1}{i} = r_\pi \left\| \frac{R_S + r_b + R_E}{1 + g_m R_E} \right. \tag{7.101}$$

Thus the dominant pole of the emitter follower is at

$$\omega = \frac{1}{R_{\pi 0} C_\pi} \tag{7.102}$$

This is in agreement with the result obtained in (7.41) by exact analysis and requires less effort. However the zero-value time constant approach tells us nothing of the zero that exact analysis showed. Because of the dominant zero, the dominant-pole

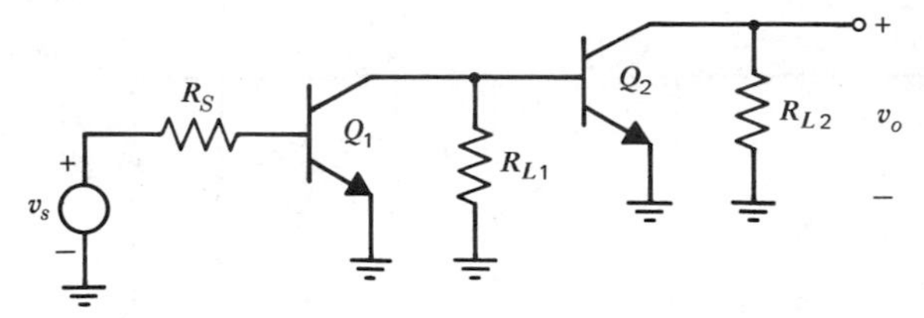

Fig. 7.18 Two-stage, common-emitter cascade amplifier.

frequency is *not* the −3-dB frequency in this case. This shows that care must be exercised in interpreting the results of zero-value time constant analysis. However it *is* a useful technique and with experience the designer can recognize circuits that are likely to contain dominant zeros. *Such circuits usually have a capacitive path directly coupling input and output as* C_π *does in the emitter follower.*

7.3.3 Common-Emitter Cascade Frequency Response

The real advantages of the zero-value time constant approach appear when circuits containing more than one device are analyzed. For example, consider the two-stage common-emitter amplifier shown in Fig. 7.18. This could be a single-ended circuit or a differential half circuit. The conventional analysis of this circuit to find the dominant pole and −3 dB frequency is extremely arduous, but the zero-value time constant analysis is quite straightforward as shown below. In order to show typical numerical calculations, specific parameter values are assumed. In this example, as in others in this chapter, parasitic capacitance associated with resistors is neglected. This is usually a reasonable approximation for diffused resistors of several thousand ohms or less.

EXAMPLE

Calculate the −3-dB frequency of the circuit of Fig. 7.18 assuming the following parameter values.

$$R_S = 10\text{ k}\Omega \qquad r_{b1} = r_{b2} = 400\ \Omega \qquad r_{\pi 1} = 20\text{ k}\Omega$$
$$C_{\pi 1} = 5\text{ pF} \qquad C_{\pi 2} = 10\text{ pF} \qquad C_{\mu 1} = C_{\mu 2} = 1\text{ pF} \qquad R_{L1} = 10\text{ k}\Omega$$
$$C_{cs1} = C_{cs2} = 2\text{ pF} \qquad r_{\pi 2} = 10\text{ k}\Omega \qquad R_{L2} = 5\text{ k}\Omega$$
$$g_{m1} = 3\text{ mA/V} \qquad g_{m2} = 6\text{ mA/V}$$

The small-signal equivalent circuit of Fig. 7.18 is shown in Fig. 7.19. The zero-value time constants for this circuit are determined by calculating the resistance seen by each capacitor across its own terminals. However, significant effort can be

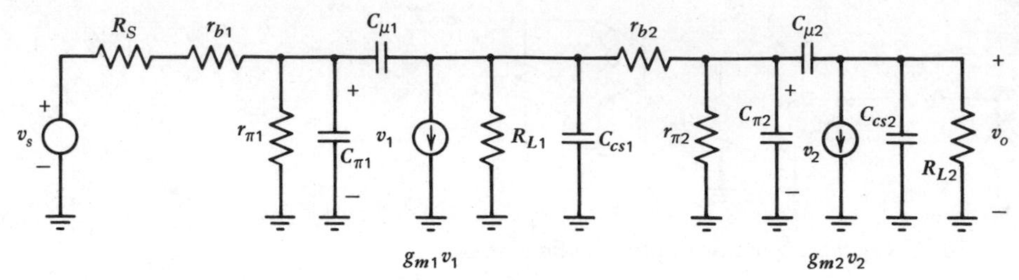

Fig. 7.19 Small-signal equivalent circuit of Fig. 7.18.

saved by recognizing that certain capacitors in the circuit are in similar configurations and the same formula can be applied to each. For example, consider $C_{\mu1}$ and $C_{\mu2}$. Equation 7.96 can be applied to both of these.

$$C_{\mu1}R_{\mu o1} = C_{\mu1}R_{\pi o1}\left(1 + g_{m1}R_{L1\text{eff}} + \frac{R_{L1\text{eff}}}{R_{\pi o1}}\right) \tag{7.103}$$

$$C_{\mu2}R_{\mu o2} = C_{\mu2}R_{\pi o2}\left(1 + g_{m2}R_{L2\text{eff}} + \frac{R_{L2\text{eff}}}{R_{\pi o2}}\right) \tag{7.104}$$

where $R_{L\text{eff}}$ is the effective load resistance seen by each device and

$$R_{L1\text{eff}} = R_{L1}\|(r_{b2} + r_{\pi2}) = 5.1\text{ k}\Omega$$
$$R_{L2\text{eff}} = R_{L2} = 5\text{ k}\Omega$$

The value of $R_{\pi o}$ for each device can be calculated using (7.91)

$$R_{\pi o1} = r_{\pi1}\|(R_{S1} + r_{b1}) = 20\text{ k}\Omega\|(10.4\text{ k}\Omega) = 6.84\text{ k}\Omega$$
$$R_{\pi o2} = r_{\pi2}\|(R_{S2} + r_{b2})$$

where

$$R_{S2} = R_{L1} = 10\text{ k}\Omega$$

Substituting data gives

$$R_{\pi o2} = 10\text{ k}\Omega\|(10.4\text{ k}\Omega) = 5.1\text{ k}\Omega$$

Thus

$$C_{\pi1}R_{\pi o1} = 5 \times 6.84\text{ nsec} = \underline{34.2\text{ nsec}}$$
$$C_{\pi2}R_{\pi o2} = 10 \times 5.1\text{ nsec} = \underline{51\text{ nsec}}$$

Substituting values in (7.103) and (7.104) gives

$$C_{\mu1}R_{\mu o1} = 1 \times 6.84\left(1 + 3 \times 5.1 + \frac{5.1}{6.84}\right)\text{nsec}$$
$$= 6.84(1 + 15.3 + 0.75)\text{ nsec}$$
$$= \underline{116.6}\text{ nsec}$$

$$C_{\mu 2}R_{\mu 02} = 1 \times 5.1\left(1 + 6 \times 5 + \frac{5}{5.1}\right) \text{nsec}$$

$$= 5.1(1 + 30 + 1) \text{ nsec}$$

$$= \underline{163.2 \text{ nsec}}$$

Finally the zero-value time constants for the collector-substrate capacitors are

$$C_{cs1}R_{cs01} = C_{cs1}R_{L1\text{eff}} = 2 \times 5.1 \text{ nsec} = \underline{10.2} \text{ nsec}$$

$$C_{cs2}R_{cs02} = C_{cs2}R_{L2\text{eff}} = 2 \times 5 \text{ nsec} = \underline{10} \text{ nsec}$$

If we assume that the circuit transfer function has a dominant pole, the -3-dB frequency can be estimated as

$$\omega_{-3\text{dB}} = \frac{1}{\sum T_0}$$

$$= \frac{10^9}{34.2 + 51 + 116.6 + 163.2 + 10.2 + 10} \text{ rad/sec}$$

$$= \frac{10^9}{385.2} \text{ rad/sec}$$

$$= 2.6 \times 10^6 \text{ rad/sec}$$

and therefore

$$f_{-3\text{ dB}} = 413 \text{ kHz}$$

A computer simulation of this circuit using the program SLIC gave a -3-dB frequency of 456 kHz, which is close to the calculated value. The computer gave four negative real poles with magnitudes 463 kHz, 4.37 MHz, 41.06 MHz, and 212 MHz. There were two negative real zeros with magnitudes 478 MHz and 955 MHz. From the computer, the sum of the reciprocals of the pole magnitudes was 385.2 nsec, which exactly equals the sum of the zero-value time constants as calculated by hand.

The above analytical result was obtained with relatively small effort and the calculation has focused on the contributions to the -3-dB frequency from the various capacitors in the circuit. In this example, as is usually the case in a common-emitter cascade of this kind, the collector-base capacitance, C_μ, is the major contributor to the dominant pole of the circuit. *One of the major benefits of the zero-value time constant analysis is the information it gives on the circuit elements that most affect the -3-dB frequency of the circuit.*

In the preceding calculation we assumed that the circuit of Fig. 7.19 had a dominant pole. The significance of this assumption will be examined in more detail. Assume for purposes of illustration that capacitors $C_{\mu 1}$, $C_{\mu 2}$, C_{cs1}, and C_{cs2}

in Fig. 7.19 are zero and, furthermore, that $r_{\pi 1} = r_{\pi 2}$ and $C_{\pi 1} = C_{\pi 2}$. Then the circuit has two identical stages and each will contribute a pole at the same frequency. That is, a dominant-pole situation does *not* exist because the circuit has two identical poles. It can be shown, however, that the inclusion of the capacitors, $C_{\mu 1}$ and $C_{\mu 2}$, tends to cause these poles to split apart, and produces a dominant-pole situation (see Section 9.3.2). For this reason, most practical circuits of this kind do have a dominant pole and the zero-value time constant analysis gives a good estimate of $\omega_{-3\text{ dB}}$. However, even if the circuit has two identical poles, the zero-value time constant analysis is still useful. Equations 7.66 and 7.89 are valid in general and thus it is *always* true that

$$\sum T_0 = \sum_{i=1}^{n} \left(-\frac{1}{p_i} \right) \tag{7.105}$$

That is, the sum of the zero-value time constants equals the sum of the reciprocals of all the poles whether or not a dominant pole exists. Consider a circuit with two identical negative real poles with magnitudes ω_x. Then the circuit gain magnitude is

$$|G(j\omega)| = \frac{G_0}{1 + \left(\dfrac{\omega}{\omega_x} \right)^2} \tag{7.106}$$

The -3-dB frequency of this circuit is the frequency where $|G(j\omega)| = G_0/\sqrt{2}$ and this can be shown to be

$$\omega_{-3\text{ dB}} = \sqrt{\sqrt{2} - 1}\, \omega_x = 0.64\omega_x \tag{7.107}$$

The zero-value time constant approach predicts

$$\sum T_0 = \frac{2}{\omega_x}$$

and thus

$$\omega_{-3\text{ dB}} = \frac{1}{\sum T_0} = 0.5\omega_x \tag{7.108}$$

Even in this extreme case the prediction is only 22 percent in error and gives a pessimistic estimate.

7.3.4 Cascode Frequency Response

The cascode connection is a multiple-device configuration that is useful in high-frequency applications. An ac schematic is shown in Fig. 7.20 and consists of a common-emitter stage driving a common-base stage. The cascode derives its advantage at high frequencies from the fact that the collector load for the common-emitter stage is the very low input impedance of the common-base stage. This

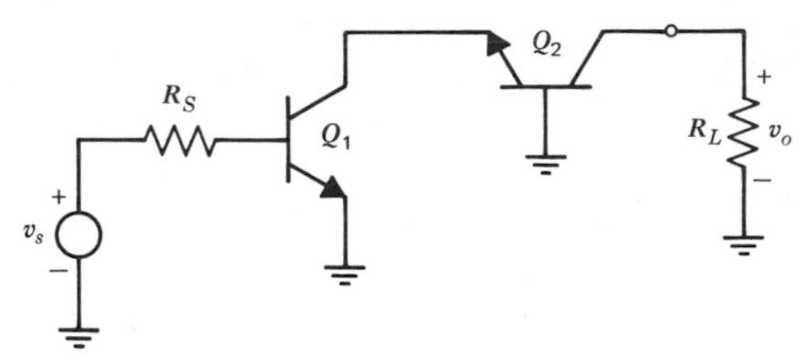

Fig. 7.20 Cascode circuit connection.

was shown in previous sections to be

$$\frac{1}{g_m} + \frac{r_b}{\beta_0 + 1}$$

at low frequencies, and this can usually be approximated as $1/g_m$. If Q_1 and Q_2 have equal collector bias currents, the transconductance of Q_1 is g_m, and, since its load resistance is $1/g_m$, the voltage gain of Q_1 is unity. Thus the influence of the Miller effect on Q_1 is minimal, even for large values of R_L. Since the common-base stage, Q_2, has a wide bandwidth (see Section 7.2.3) the cascode circuit overall has good high-frequency performance when compared to the simple common-emitter stage, especially for large R_L (see Problem 7.17).

A useful characteristic of the cascode is the small amount of reverse transmission that occurs in the circuit. The common-base stage provides this good isolation that is required in high-frequency tuned-amplifier applications.

Another characteristic of the cascode that finds use in circuit design is its high output impedance. This is used to advantage in current-source design as described in Chapter 4 where it was shown that the output resistance of a cascode is approximately β_0 times that of a common-emitter stage.

As an example of the calculation of the -3-dB frequency of a wideband integrated circuit, consider the cascode amplifier of Fig. 7.21. This can be used as one stage of a more complex wideband amplifier, or with the addition of an emitter-follower output stage it forms a useful general-purpose wideband integrated circuit as it stands. In this circuit the -3-dB bandwidth is of prime importance and the factors determining the frequency response of this circuit will now be examined.

In the circuit of Fig. 7.21, the input differential pair is biased using a resistor R_3. If common-mode rejection is an important consideration, R_3 can be replaced with an active current source. Emitter resistors R_E are used in the input stage to stabilize the gain, increase the input impedance, and broad-band the circuit. This amounts to local series feedback as described in Chapter 8. The resistive divider composed of R_1 and R_2 sets the bias voltage at the bases of Q_3 and Q_4,

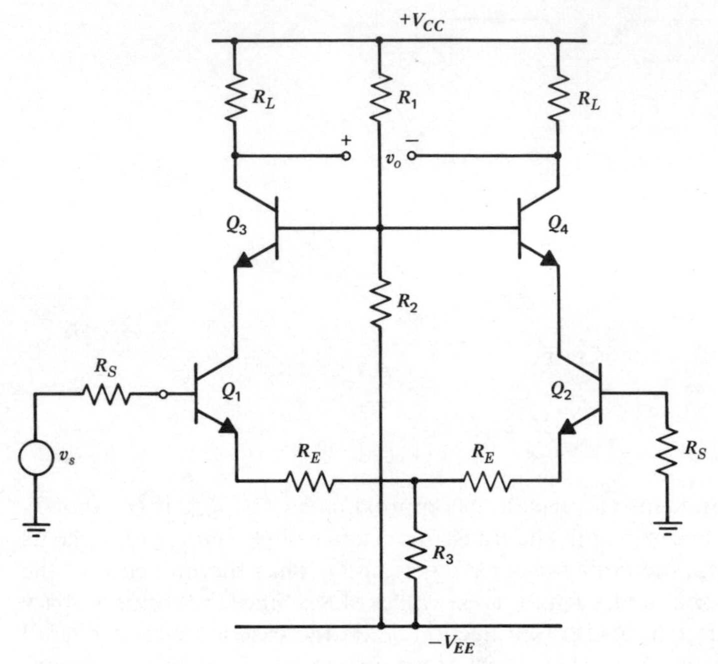

Fig. 7.21 Cascode differential amplifier.

and this voltage is chosen to give adequate collector-emitter bias voltage for each device.

For purposes of analysis, the circuit is assumed driven with source resistance R_S from each base to ground. This has the advantage of minimizing dc offset due to base current in Q_1 and Q_2, but if the base of Q_2 is grounded the frequency response of the circuit is not greatly affected if R_S is small. The circuit of Fig. 7.21 can be analyzed using the ac differential half circuit of Fig. 7.22*a*. Note that in forming the differential half circuit, the common base point of Q_3 and Q_4 is assumed to be a virtual ground for differential signals. The frequency response $(v_o/v_s)(j\omega)$ of the circuit of Fig. 7.22*a* will be the same as that of Fig. 7.21 if R_3 in Fig. 7.21 is large enough to give a reasonable value of common-mode rejection. The small-signal equivalent circuit of Fig. 7.22*a* is shown in Fig. 7.22*b*.

EXAMPLE

Calculate the low-frequency, small-signal gain and -3 dB frequency of the circuit of Fig. 7.21 using the following data. $R_S = 1\ \text{k}\Omega$, $R_E = 75\ \Omega$, $R_3 = 4\ \text{k}\Omega$, $R_L = 1\ \text{k}\Omega$, $R_1 = 4\ \text{k}\Omega$, $R_2 = 10\ \text{k}\Omega$, and $V_{CC} = V_{EE} = 10$ V. Device data is $\beta = 200$, $V_{BE(\text{on})} = 0.7$ V, $\tau_F = 0.25$ nsec, $r_b = 200\ \Omega$, r_c (active region) $= 150\ \Omega$,

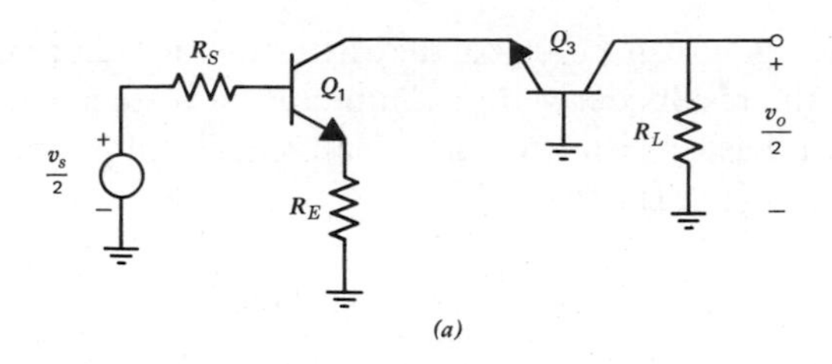

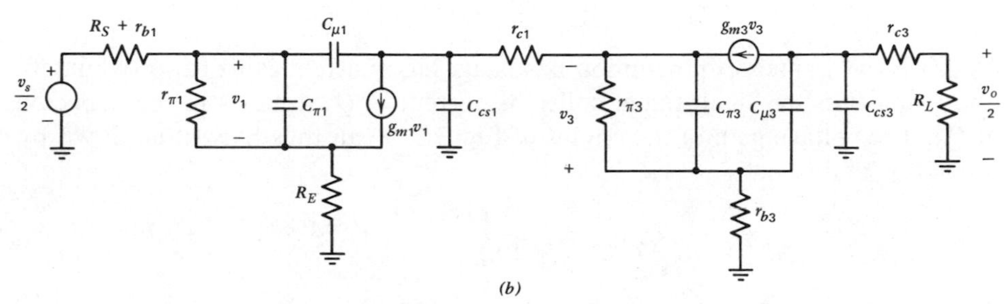

Fig. 7.22 (*a*) ac differential half circuit of Fig. 7.21. (*b*) Small-signal equivalent of the circuit in (*a*).

$C_{je0} = 1.3$ pF, $C_{\mu 0} = 0.6$ pF, $\psi_0 = 0.6$ V (collector-base), $C_{cs0} = 2$ pF, and $\psi_0 = 0.58$ V (collector-substrate).

The dc bias conditions are first calculated neglecting transistor base current. The voltage at the base of Q_3 and Q_4 is

$$V_{B3} = V_{CC} - \frac{R_1}{R_1 + R_2}(V_{CC} + V_{EE}) = 10 - \frac{4}{14} \times 20 = 4.3 \text{ V}$$

The voltage at the collectors of Q_1 and Q_2 is

$$V_{C1} = V_{B3} - V_{BE3(\text{on})} = 3.6 \text{ V}$$

Assuming the bases of Q_1 and Q_2 are grounded, we can calculate the collector currents of Q_1 and Q_2 as

$$I_{C1} = \frac{V_{EE} - V_{BE(\text{on})}}{2R_3 + R_E} = \frac{10 - 0.7}{8.075} \text{ mA} = 1.15 \text{ mA}$$

Since we assume that $\alpha \simeq 1$, we have

$$I_{C1} = I_{C2} = I_{C3} = I_{C4} = 1.15 \text{ mA}$$

The dc analysis is completed by noting that the voltage at the collectors of Q_3 and Q_4 is

$$V_{C3} = V_{CC} - I_{C3}R_L = 10 - 1.15 = 8.85 \text{ } V$$

The low-frequency gain can be calculated from the ac differential half circuit of Fig. 7.22*a*, using the results derived in Chapter 3 for a stage with emitter resistance. If we neglect base resistance, the small-signal transconductance of Q_1 including R_E is given by (3.31) as

$$G_{m1} = \frac{g_{m1}}{1 + g_{m1}R_E} = 10.24 \text{ mA/V}$$

The small-signal input resistance of Q_1 including R_E is given by (3.28) as

$$R_{i1} = r_{\pi 1}(1 + g_{m1}R_E) = 19.5 \text{ k}\Omega$$

As shown in Chapter 3, the common-base stage has a current gain of approximately unity, and thus the small-signal collector current of Q_1 appears in the collector of Q_3. The voltage gain of the circuit of Fig. 7.22*a* can thus be written down by inspection and is

$$\frac{v_o}{v_s} = -\frac{R_{i1}}{R_{i1} + R_S} G_m R_L = -\frac{19.5}{19.5 + 1} \times 10.24 \times 1 = -9.74$$

In order to calculate the -3-dB frequency of the circuit, the parameters in the small-signal equivalent circuit of Fig. 7.22*b* must be determined. The resistive parameters are $g_{m1} = g_{m3} = qI_{C1}/kT = 44.2$ mA/V, $r_{\pi 1} = r_{\pi 3} = \beta/g_{m1} = 4525\ \Omega$, $r_{c1} = r_{c3} = 150\ \Omega$, $r_{b1} = r_{b3} = 200\ \Omega$, and $R_S + r_b = 1.2$ kΩ. Because of the low resistances in the circuit, transistor output resistances are neglected.

The capacitive elements in Fig. 7.22*b* are calculated as described in Chapter 1. First consider base-emitter, depletion-layer capacitance, C_{je}. As described in Chapter 1, the value of C_{je} in the forward-active region is difficult to estimate and a reasonable approximation is to double C_{je0}. This gives $C_{je} = 2.6$ pF. From (1.104) the base-charging capacitance for Q_1 is

$$C_{b1} = \tau_F g_{m1} = 0.25 \times 10^{-9} \times 44.2 \times 10^{-3} \text{ F} = 11.1 \text{ pF}$$

Use of (1.118) gives

$$C_{\pi 1} = C_{b1} + C_{je1} = 13.7 \text{ pF}$$

Since the collector currents of Q_1 and Q_3 are equal, $C_{\pi 3} = C_{\pi 1} = 13.7$ pF.

The collector-base capacitance $C_{\mu 1}$ of Q_1 can be calculated using (1.21) and noting that the collector-base bias voltage of Q_1 is $V_{CB1} = 3.6$ V. Thus

$$C_{\mu 1} = \frac{C_{\mu 0}}{\sqrt{1 + \dfrac{V_{CB}}{\psi_0}}} = \frac{0.6}{\sqrt{1 + \dfrac{3.6}{0.6}}} \text{ pF} = 0.23 \text{ pF}$$

The collector-substrate capacitance of Q_1 can also be calculated using (1.21) with a collector-substrate voltage of $V_{CS} = V_{C1} + V_{EE} = 13.6$ V. (The substrate is assumed connected to the negative supply voltage.) Thus we have

$$C_{cs1} = \frac{C_{cs0}}{\sqrt{1 + \dfrac{V_{CS}}{\psi_0}}} = \frac{2}{\sqrt{1 + \dfrac{13.6}{0.58}}} \text{ pF} = 0.40 \text{ pF}$$

Similar calculations show the parameters of Q_3 are $C_{\mu 3} = 0.20$ pF and $C_{cs3} = 0.35$ pF.

The -3-dB frequency of the circuit can now be estimated by calculating the zero-value time constants for the circuit. First consider $C_{\pi 1}$. The resistance seen across its terminals is given by (7.101), which was derived for the emitter follower. The presence of resistance in series with the collector of Q_1 makes no difference to the calculation because of the infinite impedance of the current generator $g_{m1}v_1$. Thus from (7.101)

$$\begin{aligned} R_{\pi 01} &= r_{\pi 1} \left\| \frac{R_S + r_{b1} + R_E}{1 + g_{m1}R_E} \right. \\ &= 4525 \left\| \frac{1000 + 200 + 75}{1 + 44.2 \times 0.075} \right. \Omega \\ &= 4525 \| 295 \ \Omega \\ &= 277 \ \Omega \end{aligned}$$

Note that the effect of R_E is to *reduce* $R_{\pi 01}$ and this *increases* the bandwidth of the circuit by reducing the zero-value time constant associated with $C_{\pi 1}$. This has a value

$$C_{\pi 1}R_{\pi 01} = 13.7 \times 0.277 \text{ nsec} = \underline{3.79 \text{ nsec}}$$

The collector-substrate capacitance of Q_1 sees a resistance equal to r_{c1} plus the common-base stage input resistance, which is

$$R_{i3} = \frac{1}{g_{m3}} + \frac{r_{b3}}{\beta + 1} = 23.6 \ \Omega$$

and thus C_{cs1} sees a resistance

$$R_{cs01} = R_{i3} + r_{c1} = 174 \ \Omega$$

The zero-value time constant is

$$C_{cs1}R_{cs01} = 0.4 \times 0.174 \text{ nsec} = \underline{0.07 \text{ ns}}$$

The zero-value time constant associated with $C_{\mu 1}$ of Q_1 can be determined by calculating the resistance, $R_{\mu 01}$, seen across the terminals of $C_{\mu 1}$ using the equivalent circuit of Fig. 7.23*a*. This is conveniently transformed to the circuit

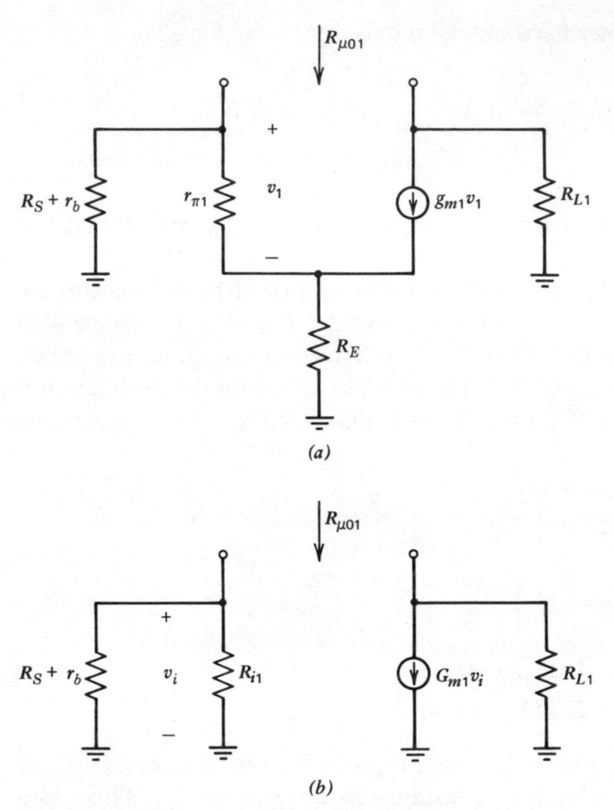

Fig. 7.23 (*a*) Circuit for the calculation of $R_{\mu o1}$ for Q_1. (*b*) Equivalent circuit for the circuit in (*a*).

of Fig. 7.23*b* where the transistor with emitter generation is represented by parameters R_{i1} and G_{m1}, which were defined previously. The circuit of Fig. 7.23*b* is in the form of a common-emitter stage as shown in Fig. 7.16 and the formula derived for that case can be used now. Thus, from (7.96),

$$R_{\mu o1} = R_1 + R_{L1} + G_{m1}R_{L1}R_1 \tag{7.109}$$

where

$$R_1 = R_{i1}\|(R_S + r_b) = 19.5\|1.2 \text{ k}\Omega = 1.13 \text{ k}\Omega$$

The load resistance, R_{L1}, is just r_{c1} plus the input resistance of Q_3, and using the previously calculated values we obtain

$$R_{L1} = 174\ \Omega$$

Thus, in (7.109),

$$R_{\mu o1} = 1.13 + 0.17 + 10.24 \times 1.13 \times 0.17 = 3.27 \text{ k}\Omega$$

The zero-value time constant associated with $C_{\mu 1}$ is thus

$$C_{\mu 1}R_{\mu 01} = 0.23 \times 3.27 \text{ nsec} = \underline{0.75 \text{ nsec}}$$

Note that because of the low input impedance of the common-base stage, R_{L1} is small and the contribution of $C_{\mu 1}$ to the sum of the zero value time constants is much smaller than that due to $C_{\pi 1}$.

The time constant associated with $C_{\pi 3}$ of Q_3 can be calculated by recognizing that Equation 7.101 derived for the emitter follower also applies here. The effective source resistance, R_S, is zero as the base is grounded, and the effective emitter resistance, R_E, is infinite because the collector of Q_1 is connected to the emitter of Q_3. Thus, (7.101) gives

$$R_{\pi 03} = r_{\pi 3} \left\| \frac{1}{g_{m3}} \right. = 22.6\ \Omega$$

The zero-value time constant associated with $C_{\pi 3}$ is thus

$$C_{\pi 3}R_{\pi 03} = 13.7 \times 0.023 \text{ nsec} = \underline{0.32 \text{ nsec}}$$

The time constant associated with collector-base capacitance $C_{\mu 3}$ of Q_3 can be calculated using (7.109) with G_{m1} equal to zero since the effective value of R_E is infinite in this case. In (7.109) the effective value of R_1 is just r_b and thus

$$R_{\mu 03} = r_b + R_{L3}$$

where

$$R_{L3} = r_{c3} + R_L$$

and R_{L3} is the load resistance seen by Q_3. Thus

$$R_{\mu 03} = 200 + 150 + 1000 = 1.35\ \text{k}\Omega$$

and the time constant is

$$C_{\mu 3}R_{\mu 03} = 0.2 \times 1.35 \text{ nsec} = \underline{0.27 \text{ nsec}}$$

Finally the collector-substrate capacitance of Q_3 sees a resistance

$$R_{cs03} = r_{c3} + R_L = 1.15\ \text{k}\Omega$$

and

$$C_{cs3}R_{cs03} = 0.35 \times 1.15 \text{ ns} = \underline{0.4 \text{ nsec}}$$

The sum of the zero-value time constants is thus

$$\sum T_0 = (3.79 + 0.07 + 0.75 + 0.32 + 0.27 + 0.4) = 5.60 \text{ nsec}$$

The -3-dB frequency is estimated as

$$f_{-3\text{ dB}} = \frac{1}{2\pi \sum T_0} = 28.4 \text{ MHz}$$

Computer simulation of this circuit using the program SLIC gave a -3-dB frequency of 34.7 MHz. The computer showed six poles, of which the first two were negative real poles with magnitudes 35.8 MHz and 253 MHz. The zero-value time constant analysis has thus given a reasonable estimate of the -3-dB frequency and has also shown that the major limitation on the circuit frequency response comes from $C_{\pi 1}$ of Q_1. The circuit can thus be broadbanded even further by increasing resistance R_E in the emitter of Q_1, since the calculation of $R_{\pi 01}$ showed that increasing R_E will reduce the value of $R_{\pi 01}$. Note that this will reduce the gain of the circuit.

Further useful information regarding the circuit frequency response can be obtained from the previous calculations by recognizing that Q_3 in Fig. 7.22 effectively isolates $C_{\mu 3}$ and C_{cs3} from the rest of the circuit. In fact, if r_{b3} is zero then these two capacitors are connected in parallel across the output and will contribute a separate pole to the transfer function. This pole can be estimated by summing zero-value time constants for $C_{\mu 3}$ and C_{cs3} alone to give $\tau = 0.67$ nsec. This corresponds to a pole at $f = 1/2\pi\tau = 237$ MHz, which is very close to the second pole calculated by the computer. The dominant pole would then be estimated by summing the rest of the time constants to give 4.93 nsec, which corresponds to a pole at 32.3 MHz. This is also close to the computer-calculated value. This technique can be used anytime there is a *high degree of isolation* between various portions of a circuit. The zero-value time constants may be summed for each separate section and the dominant pole of that section may thus be estimated.

7.4 ANALYSIS OF THE FREQUENCY RESPONSE OF THE 741 op amp

Up to this point the analysis of the frequency response of integrated circuits has been limited to fairly simply configurations. The reason for this is apparent in previous sections where the large amount of calculation required to estimate the dominant pole of some simple circuits was illustrated. A complete frequency analysis by hand of a large integrated circuit is thus out of the question. However, a circuit designer often needs insight into the frequency response of large circuits such as the 741 op amp, and, by making some sensible approximations, the methods of analysis described above can be used to provide such information. We will now illustrate this by analyzing the frequency response of the 741.

7.4.1 High-Frequency Equivalent Circuit of the 741

A schematic of the 741 is shown in Fig. 6.9*a*. Its frequency response is dominated by the 30-pF integrated capacitor, C_c, which is a compensation capacitor designed to prevent the circuit from oscillating when connected in a feedback loop. The choice of C_c and its function is described in Chapter 9.

Since the 741 contains over 20 interconnected transistors, a complete analysis is not attempted even using zero-value time constant techniques. In order to obtain an estimate of the frequency response of this circuit, the circuit designer must be able to recognize those parts of the circuit that have little or no influence on the frequency response and to discard these from the analysis. As a general rule, elements involved in the bias circuit can often be eliminated, and those portions of the circuit that are differential can be replaced by a half circuit. This approach leads to the ac schematic of Fig. 7.24, which is adequate for an approximate calculation of the high-frequency behavior of the 741. All bias elements have been eliminated except where they contribute parasitic elements to the gain path, and these are represented by R_{p1}, C_{p1}, R_{p2}, and C_{p2}. In the output stage, either Q_{14} or Q_{20} will be conducting, depending on whether the output voltage is positive or negative, and the frequency response of the circuit will be slightly different in these two cases. Transistor Q_{14} is assumed conducting in the schematic of Fig. 7.24.

The major approximation in Fig. 7.24 is neglect of the frequency limitations of the input-stage active load that has been eliminated from the gain path. However, the output resistance and capacitance of Q_6 are included as part of R_{p1} and C_{p1}. More detailed calculation and computer simulation show that these are reasonable approximations.

As mentioned above, the frequency response of the 741 is dominated by C_c, and the -3-dB frequency can be estimated by considering the effect of this capacitance alone. However, as described in Chapter 9, the presence of poles other than the dominant one has a crucial effect on the behavior of the circuit when feedback is applied. The magnitude of the nondominant poles is thus of considerable interest and methods of estimating their magnitudes are described. First we calculate the -3-dB frequency of the circuit.

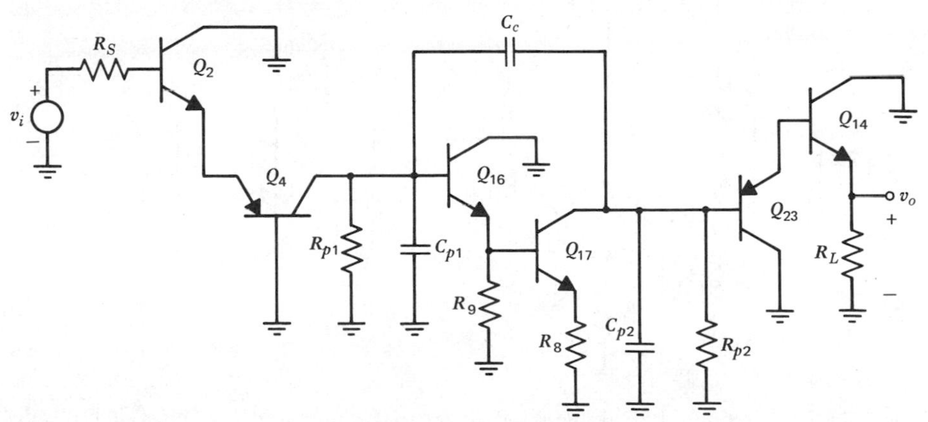

Fig. 7.24 ac schematic of the high-frequency gain path of the 741.

7.4.2 Calculation of the −3-dB Frequency of the 741

The −3-dB frequency of the circuit of Fig. 7.24 can be estimated by calculating the zero-value resistance, R_{c0}, seen by C_c. This can be calculated from the ac circuit of Fig. 7.25*a* where the input resistance of Q_{23} is assumed very high and is neglected. The calculation of R_{c0} can be greatly simplified by representing the circuit of Fig. 7.25*a* as shown in Fig. 7.25*b*. Quantities R_{ic}, R_{oc}, and G_{mc} are the input resistance, output resistance, and transconductance, respectively, of the circuit of Fig. 7.25*a*. These quantities were calculated in Chapter 6 using nominal device data but a more accurate calculation allowing for β variation with bias current yields the following values.

$$R_{ic} = R_{o4}\|R_{p1}\|R_{i16} = 1.95\ \text{M}\Omega$$
$$R_{oc} = R_{o17}\|R_{p2} = 86.3\ \text{k}\Omega$$
$$G_{mc} = 6.39\ \text{mA/V}$$

Note that R_{p1} is the effective output resistance of Q_6 and R_{p2} is the effective output resistance of Q_{13B} in the schematic of Fig. 6.8*a*.

Since the circuit of Fig. 7.25*b* is topologically the same as that of Fig. 7.16, Equation 7.96 can now be used to estimate R_{c0}.

$$\begin{aligned} R_{c0} &= R_{ic} + R_{oc} + G_{mc}R_{ic}R_{oc} \qquad (7.110) \\ &= (1.95 + 0.086 + 6.39 \times 10^{-3} \times 86.3 \times 10^3 \times 1.95)\ \text{M}\Omega \\ &= 1.08 \times 10^9\ \Omega \end{aligned}$$

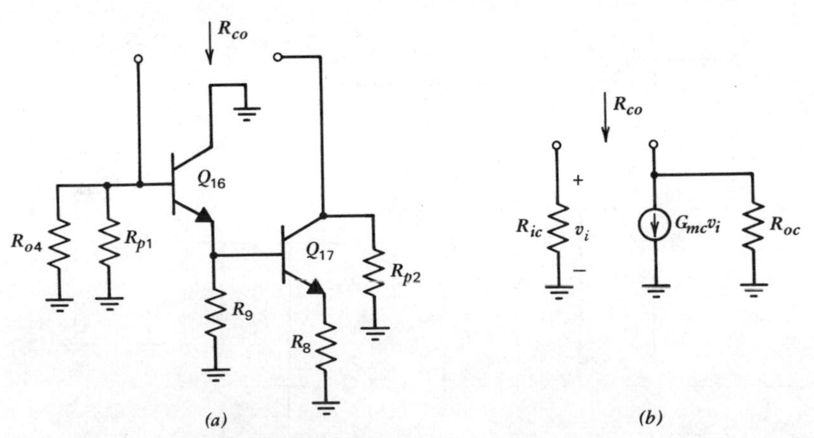

Fig. 7.25 (*a*) Circuit for the calculation of the zero-value time constant for C_c. (*b*) Equivalent circuit for the circuit in (*a*).

This extremely large resistance when combined with $C_c = 30$ pF gives a time constant

$$C_c R_{c0} = 30 \times 10^{-12} \times 1.08 \times 10^9 \text{ sec} = 32.4 \times 10^{-3} \text{ sec}$$

This totally dominates the sum of the zero-value time constants and gives a -3-dB frequency when compensated of

$$f_{-3\text{dB}} = \frac{1}{2\pi C_c R_{c0}} = 4.9 \text{ Hz}$$

A computer simulation of the complete 741 gave $f_{-3\text{dB}} = 5.0$ Hz.

An alternative means of calculating the effect of the frequency compensation is using the Miller effect as described in Section 7.2.1. The compensation capacitor is connected from the base of Q_{16} to the collector of Q_{17} and the magnitude of the voltage gain between these two points can be calculated from the equivalent circuit of Fig. 7.25*b*.

$$A_v = G_{mc} R_{oc} \tag{7.111}$$

From (7.6) the Miller capacitance seen at the base of Q_{16} is

$$C_M = (1 + A_v) C_c \tag{7.112}$$

and substitution of (7.111) in (7.112) gives

$$\begin{aligned} C_M &= (1 + G_{mc} R_{oc}) C_c \qquad (7.113) \\ &= (1 + 6.39 \times 10^{-3} \times 86.3 \times 10^3) \times 30 \text{ pF} \\ &= 16{,}540 \text{ pF} \end{aligned}$$

This extremely large effective capacitance at the base of Q_{16} swamps all other capacitances and, when combined with resistance $R_{ic} = 1.95$ MΩ from the base of Q_{16} to ground, gives a -3-dB frequency for the circuit of

$$\begin{aligned} f_{-3\text{ dB}} &= \frac{1}{2\pi C_M R_{ic}} \\ &= \frac{1}{2\pi \times 16{,}540 \times 10^{-12} \times 1.95 \times 10^6} \text{ Hz} \\ &= 4.9 \text{ Hz} \end{aligned}$$

This is the same value as predicted by the zero-value time constant approach.

Note that an additional capacitor is introduced into the circuit along with C_c. This is the capacitance described in Chapter 2 from the underside of C_c to the substrate. In this case, capacitor C_c is connected so that the parasitic capacitance exists from the base of Q_{16} to ground and is thus swamped by the Miller effect due to C_c. The parasitic capacitance has a relatively large value (about 14 pF) because of the large area of the 30-pF capacitor. This capacitor consumes an area 16 mil square in a chip that is 56 mil square, and thus occupies an area about 13 times that of a typical transistor in the circuit.

It is interesting to note that if the compensation capacitor is removed and the zero-value time constants of the circuit are calculated, then the -3-dB frequency of the circuit is found to be 18.9 kHz. This is dominated by the capacitance, C_{p1}, which is about 3.4 pF and is composed largely of collector-substrate capacitance from Q_6 and Q_{22}. The resistance seen by C_{p1} is $R_{ic} = 1.95\ \text{M}\Omega$ as calculated above, giving a time constant of 6.6 μsec.

7.4.3 Nondominant Poles of the 741

The calculations above have shown that the 30-pF compensation capacitor produces a dominant pole in the 741 with a magnitude of 4.9 Hz. From the complexity of the circuit it is evident that there will be a large number of higher-frequency poles that we now consider.

Transistors Q_{16} and Q_{17} form the gain stage around which the compensation capacitor is connected. After C_c is connected, the transfer function of this pair contains a pole with a magnitude of 4.9 Hz plus higher-frequency poles. As shown in Chapter 9, these higher-frequency poles are much less significant after C_c is connected, but they still contribute phase shift at the unity-gain frequency of the amplifier (which is about 1.25 MHz). The exact calculation of these higher-frequency poles is quite difficult, however.

Other sources of higher-frequency poles are the active load, which has been omitted from Fig. 7.24, and also the lateral *pnp* emitter follower Q_{23}. Computer simulation shows that both of these parts of the circuit contribute phase shift at the unity-gain frequency of 1.25 MHz. As in the previous case, hand calculation of the frequency response of these portions of the circuit is difficult and requires consideration of all parasitic elements.

There is, however, one portion of the circuit of Fig. 7.24 that contributes a nondominant pole that can be calculated. This is the lateral *pnp* common-base stage, Q_4. This stage is driven by the *npn* emitter follower, Q_2, which may be assumed to have a frequency response that is much broader than that of Q_4 because the f_T of an *npn* device is much higher than that of a lateral *pnp*. Neglecting frequency effects in Q_2, we may assume that Q_4 is fed by v_i in series with $1/g_{m2}$, which is the resistance seen looking into the emitter of Q_2 if R_S is small. In addition, the collector of Q_4 may be assumed to be feeding an ac short circuit, since the large Miller capacitance produced by C_c results in a very low impedance at the collector of Q_4. The small-signal equivalent circuit of Q_4 can thus be drawn as in Fig. 7.26 and this is effectively isolated from the rest of the circuit. As a result, this stage contributes a separate pole that can now be estimated using zero-value time constant techniques. Note that r_{b4} and $C_{\mu 4}$ are neglected.

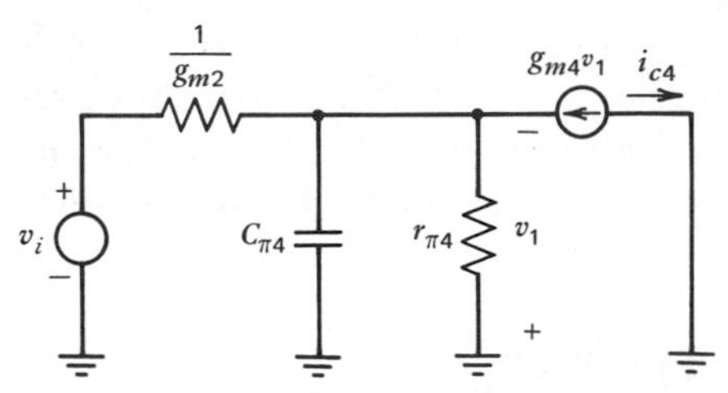

Fig. 7.26 Small-signal equivalent circuit of Q_4 in Fig. 7.24.

We first calculate the small-signal parameters of the circuit of Fig. 7.26. The bias levels were calculated in Chapter 6 as $I_{C2} = -I_{C4} = 12\ \mu\text{A}$ and thus $g_{m2} = g_{m4} = 0.46$ mA/V. A typical value of C_{je} for a lateral *pnp* in forward bias is 0.6 pF. Assuming $\tau_F = 25$ nsec for the lateral *pnp*, we can calculate the base charging capacitance for Q_4 using (1.104) in Chapter 1.

$$C_{b4} = \tau_{F4} g_{m4} = 25 \times 10^{-9} \times \frac{12 \times 10^{-6}}{26 \times 10^{-3}}\ \text{F} = 11.6\ \text{pF}$$

Using (1.118) gives

$$C_{\pi 4} = C_{b4} + C_{je4} = 12.2\ \text{pF}$$

Even at this low bias current, the C_π of the lateral *pnp* is still dominated by C_b because the transit time τ_F is so large.

The frequency of the pole contributed by Q_4 can now be calculated by determining the resistance, $R_{\pi 04}$, seen by $C_{\pi 4}$. This is simply $1/g_{m2}$ in parallel with the input resistance of Q_4, which is approximately $1/g_{m4}$. Thus

$$R_{\pi 04} = \tfrac{1}{2}\frac{1}{g_{m2}} = 1087\ \Omega$$

and, consequently,

$$C_{\pi 4} R_{\pi 04} = 12.2 \times 10^{-12} \times 1087\ \text{sec} = 0.0126\ \mu\text{sec}$$

The magnitude of the pole contributed by Q_4 is thus

$$\frac{1}{2\pi C_{\pi 4} R_{\pi 04}} = 12.6\ \text{MHz}$$

A negative real pole at -15 MHz appears in the poles and zeros of the 741 as calculated by computer and described in Chapter 9. This is the pole contributed by Q_4, and the slight difference in magnitude from the value calculated above is due to the more accurate modeling in the computer. Note that a pole with magnitude 12.6 MHz contributes 6° of phase shift at the unity-gain frequency of 1.25 MHz and this is a significant amount. However nondominant poles contributed by Q_{16}, Q_{17}, Q_{23}, and the active load also produce significant phase shift at the unity-gain frequency, and an accurate estimate of the total circuit phase shift can only be made by computer simulation or direct measurement. However, the calculations of this section allow the designer to isolate those parts of the circuit contributing excess phase shift and to make design changes where necessary to improve this aspect of circuit performance.

7.5 RELATION BETWEEN FREQUENCY RESPONSE AND TIME RESPONSE

In this chapter the effect of increasing signal frequency on circuit performance has been illustrated by considering the circuit response to a sinusoidal input signal. In practice, however, an amplifier may be required to amplify nonsinusoidal signals such as pulse trains or square waves. In addition, such signals are often used in testing circuit frequency response. The response of a circuit to such input signals is thus of some interest and will now be calculated.

Initially we consider a circuit whose small-signal transfer function can be approximated by a single-pole expression

$$\frac{v_o}{v_i}(s) = \frac{K}{1 - \dfrac{s}{p_1}} \tag{7.114}$$

where K is the low-frequency gain and p_1 is the pole of the transfer function. As described earlier, the -3-dB frequency of this circuit for sinusoidal signals is $\omega_{-3\text{ dB}} = -p_1$. Now consider a small input voltage step of amplitude v_a applied to the circuit. If we assume that the circuit responds linearly, we can use (7.114) to calculate the circuit response using $v_i(s) = v_a/s$. Thus

$$v_o(s) = \frac{Kv_a}{s}\frac{1}{1 - \dfrac{s}{p_1}} = Kv_a\left(\frac{1}{s} + \frac{1}{s - p_1}\right)$$

and the circuit response to a step input is

$$v_o(t) = Kv_a(1 - e^{p_1 t}) \tag{7.115}$$

The output voltage thus approaches Kv_a and the time constant of the exponential in (7.115) is $1/p_1$. Equation 7.115 is sketched in Fig. 7.27*a* together with v_i. The rise time of the output is usually specified by the time taken to go from 10 percent to 90 percent of the final value. From (7.115) we have

$$0.1Kv_a = Kv_a(1 - e^{p_1 t_1}) \tag{7.116}$$

$$0.9Kv_a = Kv_a(1 - e^{p_1 t_2}) \tag{7.117}$$

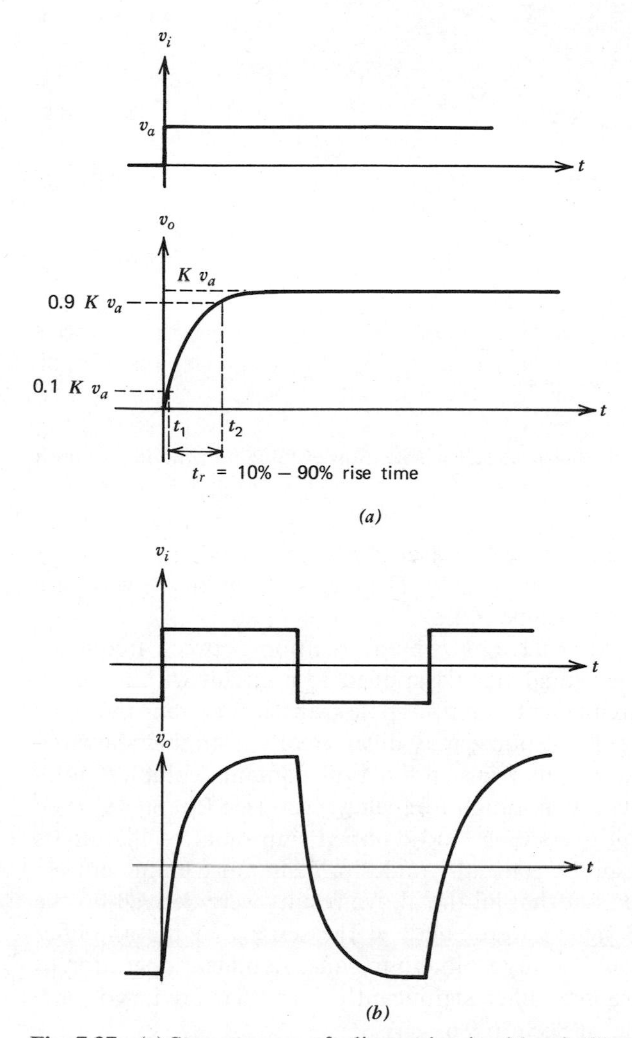

Fig. 7.27 (*a*) Step response of a linear circuit with gain K and a single-pole transfer function. (*b*) Response of a linear circuit with a single-pole transfer function when a square-wave input is applied.

From (7.116) and (7.117) we obtain, for the 10 percent to 90 percent rise time,

$$t_r = t_2 - t_1 = -\frac{1}{p_1} \ln 9 = \frac{2.2}{\omega_{-3\text{ dB}}} = \frac{0.35}{f_{-3\text{ dB}}} \tag{7.118}$$

This equation shows that the pulse rise time is directly related to the -3-dB frequency of the circuit. For example, if $f_{-3\text{ dB}} = 10$ MHz then (7.118) predicts

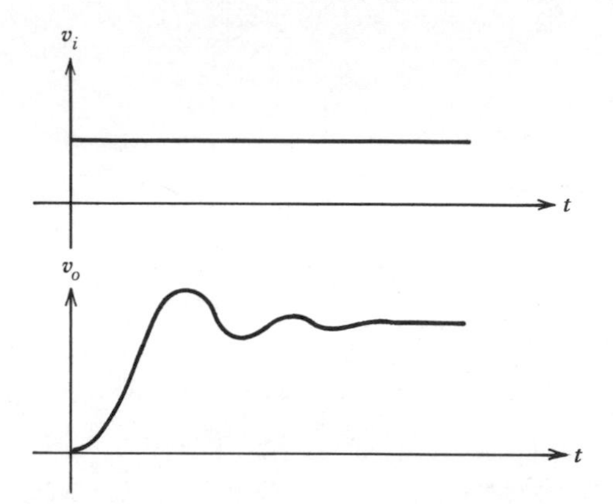

Fig. 7.28 Typical step response of a linear circuit whose transfer function contains complex poles.

$t_r = 35$ nsec. If a square wave is applied to a circuit with a single-pole transfer function the response is as shown in Fig. 7.27*b*. The edges of the square wave are rounded as described above for a single pulse.

The calculations in this section have shown the relation between frequency response and time response for small signals applied to a circuit with a single-pole transfer function. For circuits with multiple-pole transfer functions the same general trends apply but the pulse response may differ greatly from that shown in Fig. 7.27. In particular if the circuit transfer function contains complex poles leading to a frequency response with a high-frequency peak (see Chapter 9) then the pulse response will exhibit overshoot[4] and damped sinusoidal oscillation as shown in Fig. 7.28. Such a response is usually undesirable in pulse amplifiers.

Finally it should be pointed out that all the above results were derived on the assumption that the applied signals were small in the sense that the amplifier acted linearly. If the applied pulse is large enough to cause nonlinear operation of the circuit, the pulse response may differ significantly from that predicted here. This point is discussed further in Section 9.6.

PROBLEMS

7.1 (a) Use the Miller approximation to calculate the −3-dB frequency of the small-signal voltage gain of a common-emitter transistor stage as shown in Fig. 7.2*a* using these parameters:

$R_S = 5\text{ k}\Omega \quad r_b = 300\ \Omega \quad I_C = 0.5\text{ mA} \quad \beta = 200$
$f_T = 500\text{ MHz (at } I_C = 0.5\text{ mA)} \quad C_\mu = 0.3\text{ pF}$
$R_L = 3\text{ k}\Omega \quad C_{cs} = 0 \quad V_A = \infty$

(b) Calculate the nondominant pole frequency for the circuit in (a).

7.2 Calculate an expression for the output impedance of the circuit of Problem 7.1 as seen by R_L and form an equivalent circuit. Plot the magnitude of this impedance on log scales from $f = 1$ kHz to $f = 100$ MHz.

7.3 Repeat Problem 7.2 for $R_S = 0$ and $R_S = \infty$.

7.4 A differential amplifier as shown in Fig. 7.1 has $I_{EE} = 1$ mA and data as given in Problem 7.1. If the current source has resistance $R_E = 300$ kΩ and capacitance $C_E = 2$ pF as defined in Fig. 7.7*a*, calculate the CM and DM gain and CMRR as a function of frequency. Sketch the magnitude of these quantities in decibels from $f = 10$ kHz to $f = 20$ MHz, using a log frequency scale.

7.5 A lateral *pnp* emitter follower has $R_S = 250\ \Omega$, $r_b = 200\ \Omega$, $\beta = 50$, $I_C = -300\ \mu$A, $f_T = 4$ MHz, $R_E = 4$ kΩ, $C_\mu = 0$, and $r_o = \infty$. Calculate the small-signal voltage gain as a function of frequency. Sketch the magnitude of the voltage gain in decibels from $f = 10$ kHz to $f = 20$ MHz, using a log frequency scale.

7.6 Calculate the value of the elements in the small-signal equivalent circuit of the input and output impedances of the emitter follower of Problem 7.5. Sketch the magnitudes of the above impedances as a function of frequency from $f = 10$ kHz to $f = 20$ MHz, using log scales.

7.7 A common-base stage has the following parameters: $I_C = 0.5$ mA, $C_\pi = 10$ pF, $C_\mu = 0.3$ pF, $r_b = 200\ \Omega$, $\beta = 100$, $r_o = \infty$, $R_L = 0$, and $R_S = \infty$.
(a) Calculate an expression for the small-signal current gain of the stage as a function of frequency and thus determine the frequency where the current gain is 3 dB below its low-frequency value.
(b) Calculate the value of the elements in the small-signal equivalent circuit of the input and output impedances of the stage and sketch the magnitudes of these impedances from $f = 100$ kHz to $f = 100$ MHz, using log scales.

7.8 The ac schematic of a common-emitter stage is shown in Fig. 7.29 where $R_S = 10$ kΩ and $R_L = 5$ kΩ. Calculate the low-frequency small-signal voltage gain v_o/v_i and use the zero-value time constant method to estimate the -3-dB frequency.
Data. $\beta = 200$, $f_T = 600$ MHz at $I_C = 1$ mA, $C_\mu = 0.2$ pF, $C_{je} = 2$ pF, $C_{cs} = 1$ pF, $r_b = 0$, $r_o = \infty$, and $I_C = 1$ mA.

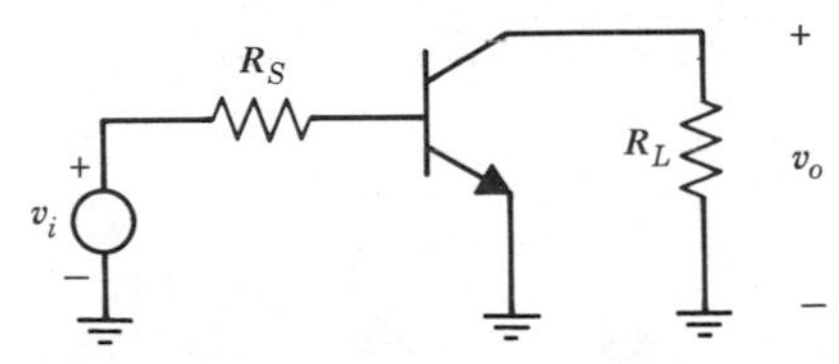

Fig. 7.29 ac schematic of a common-emitter stage.

7.9 Repeat Problem 7.8 if an emitter resistor of value 300 Ω is included in the circuit.

7.10 Repeat Problem 7.8 if a resistor of value 30 kΩ is connected between collector and base of the transistor.

7.11 A Darlington stage and a common-collector-common-emitter cascade are shown schematically in Fig. 7.30 where $R_S = 100$ kΩ and $R_L = 3$ kΩ.

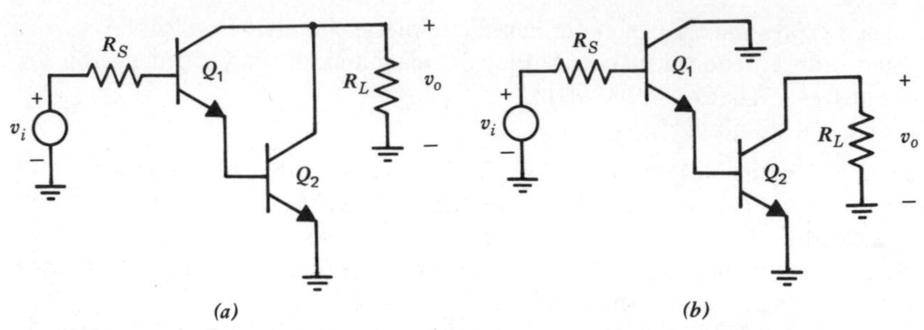

Fig. 7.30 ac schematics of (*a*) Darlington stage and (*b*) common-collector, common-emitter stage.

(a) Calculate the low-frequency small-signal voltage gain v_o/v_i for each circuit.

(b) Use the zero-value time constant method to calculate the −3-dB frequency of the gain of each circuit.

Data. $\beta = 100$, $f_T = 500$ MHz at $I_C = 1$ mA, $C_\mu = 0.4$ pF, $C_{je} = 2$ pF, $C_{cs} = 1$ pF, $r_b = 0$, $r_o = \infty$, $I_{C1} = 10\ \mu$A, and $I_{C2} = 1$ mA. (Values of C_μ, C_{cs}, and C_{je} are at the bias point.)

7.12 Repeat Problem 7.11 if a bleed resistor of 15 kΩ is added from the emitter of Q_1 to ground, which increases the collector bias current in Q_1 to 50 μA.

7.13 Repeat Problem 7.11 if the signal input is a current source of value i_i applied at the base of Q_1. The transfer function is then a transresistance v_o/i_i.

7.14 An amplifier stage is shown in Fig. 7.31 where bias current I_B is adjusted so that $V_o = 0$ V dc.

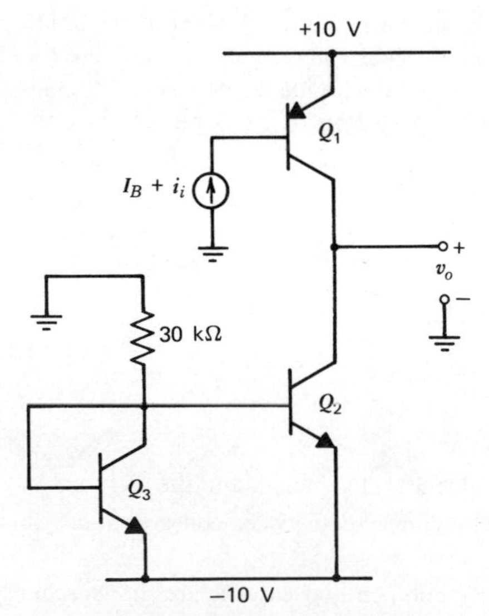

Fig. 7.31 Amplifier stage.

(a) Calculate the low-frequency, small-signal transresistance v_o/i_i and use the zero-value time constant method to estimate the -3-dB frequency.
Data. npn: $\beta = 200$, $f_T = 500$ MHz at $I_C = 1$ mA, $C_{\mu 0} = 0.7$ pF, $C_{je} = 3$ pF (at the bias point), $C_{cs0} = 2$ pF, $r_b = 0$, $V_A = 120$ V, and $\psi_0 = 0.55$ V for all junctions.
pnp: $\beta = 50$, $f_T = 4$ MHz at $I_C = -0.5$ mA, $C_{\mu 0} = 1.0$ pF, $C_{je} = 3$ pF (at the bias point), $C_{bs0} = 2$ pF, $r_b = 0$, $V_A = 50$ V, and $\psi_0 = 0.55$ V for all junctions.
(b) Repeat (a) if a 20-pF capacitor is connected from collector to base of Q_1.

7.15 A differential circuit employing active loads is shown in Fig. 7.32. Bias current I_B is adjusted so that the collectors of Q_1 and Q_2 are at $+5$ V dc. Calculate the low-frequency, small-signal voltage gain v_o/v_i and use the zero-value time constant method to estimate the -3-dB frequency. Use device data as in Problem 7.14.

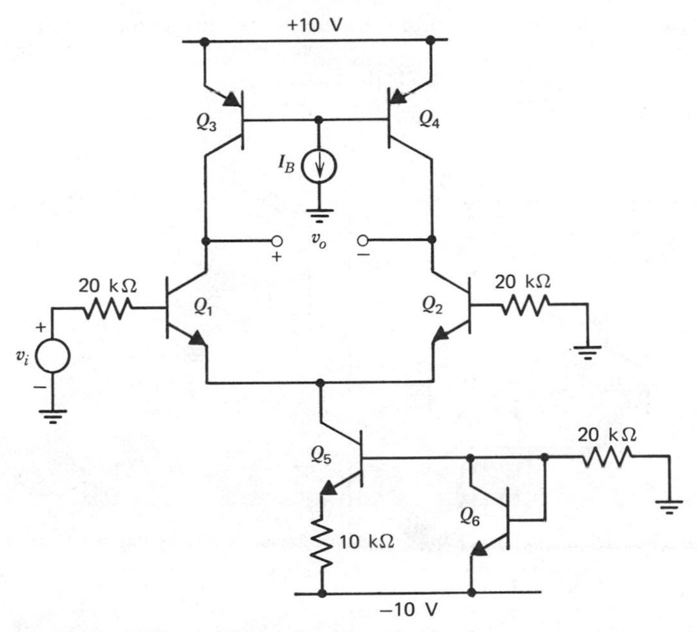

Fig. 7.32 Differential circuit with active loads.

7.16 The ac schematics of a common-emitter stage and a cascode stage are shown in Fig. 7.33 where $R_S = 5$ kΩ and $R_L = 3$ kΩ.
(a) Calculate the low-frequency, small-signal voltage gain v_o/v_i for each circuit.
(b) Use the zero-value time constant method to calculate and compare the -3-dB frequencies of the gain of the two circuits.
(c) Estimate the 10 to 90 percent rise time for each circuit for a small step input and sketch the output voltage waveform over 0 to 300 nsec for a 1-mV step input.
Data. $I_C = 1$ mA, $\beta = 100$, $r_b = 0$, $C_{cs} = 1$ pF, $C_\mu = 0.4$ pF, $f_T = 500$ MHz at $I_C = 1$ mA, and $r_o = \infty$.

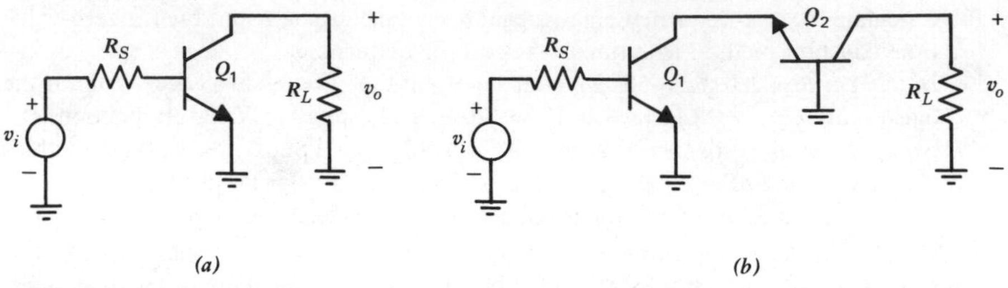

Fig. 7.33 ac schematic of (*a*) common-emitter stage and (*b*) cascode stage.

7.17 An amplifier stage is shown in Fig. 7.34.

(a) Calculate the low-frequency, small-signal voltage gain v_o/v_i.

(b) Use the zero-value time constant method to calculate the -3-dB frequency of the circuit gain.

Data. $C_{cs0} = 2$ pF, $C_{\mu 0} = 0.5$ pF, $C_{je} = 4$ pF (in forward bias), $f_T = 500$ MHz at $I_C = 2$ mA, $\beta = 200$, $r_b = 0$, $r_o = \infty$, and $\psi_0 = 0.55$ V for all junctions.

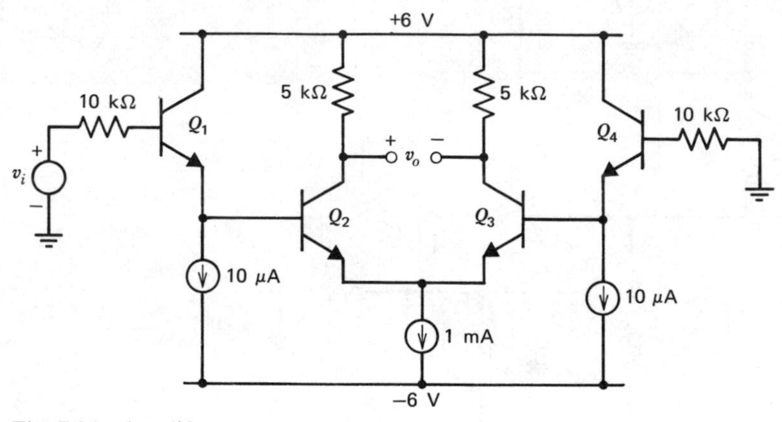

Fig. 7.34 Amplifier stage.

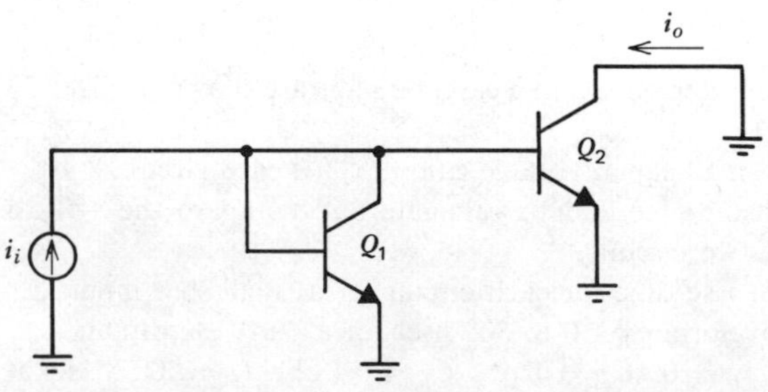

Fig. 7.35 ac schematic of a wideband monolithic current amplifier.

7.18 The ac schematic of a wideband monolithic current amplifier is shown in Fig. 7.35. The emitter area of Q_2 is four times that of Q_1 and corresponding bias currents are $I_{C1} = 1$ mA and $I_{C2} = 4$ mA. Calculate the low-frequency, small-signal current gain i_o/i_i and use the zero-value time constant method to estimate the -3-dB frequency. Calculate the 10 to 90 percent rise time for a small step input.
Data at the operating point.
Q_1: $\beta = 200$, $\tau_F = 0.2$ nsec, $C_\mu = 0.2$ pF, $C_{je} = 1$ pF, $C_{cs} = 1$ pF, $r_b = 0$, and $r_o = \infty$.
Q_2: $\beta = 200$, $\tau_F = 0.2$ nsec, $C_\mu = 0.8$ pF, $C_{je} = 4$ pF, $C_{cs} = 4$ pF, $r_b = 0$, and $r_o = \infty$.

7.19 A two-stage amplifier is shown in Fig. 7.36. Calculate the low-frequency, small-signal gain and use the zero-value time constant method to estimate the -3-dB frequency. Calculate the 10 to 90 percent rise time for a small step input.
Data. $\beta = 200$, $f_T = 600$ MHz at $I_C = 1$ mA, $C_\mu = 0.2$ pF, $C_{je} = 2$ pF, $C_{cs} = 1$ pF, $r_b = 0$, $V_{BE(\text{on})} = 0.6$ V, and $r_o = \infty$. (Values of C_μ, C_{cs}, and C_{je} are at the bias point.)

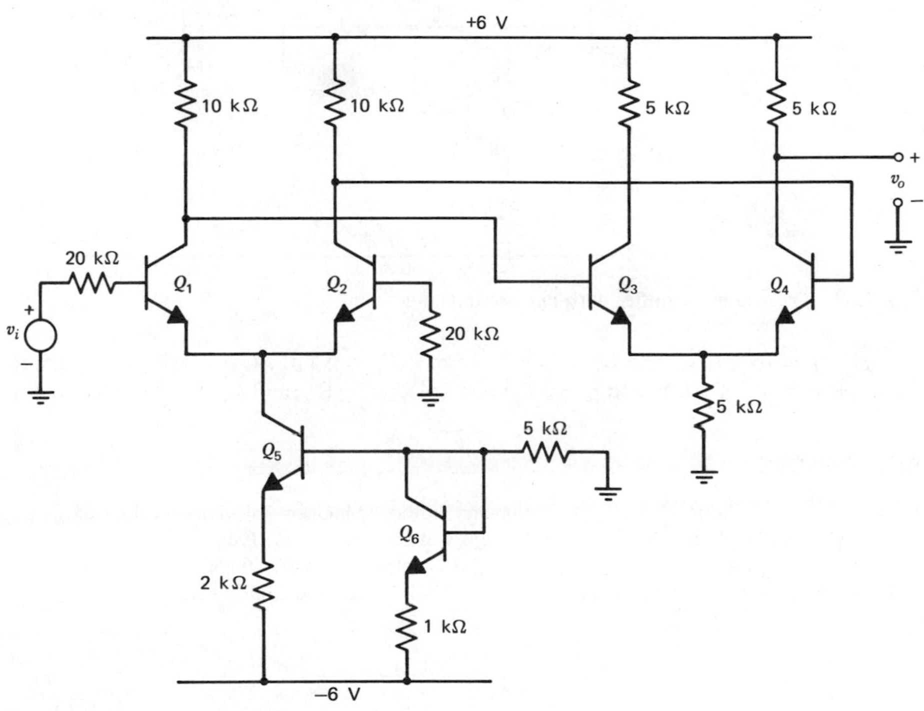

Fig. 7.36 Two-stage amplifier.

7.20 A two-stage amplifier is shown in Fig. 7.37. Calculate the low-frequency, small-signal voltage gain v_o/v_i and use the zero-value time constant method to estimate the -3-dB frequency. Use a half circuit for the differential pair.
Data. npn: $\beta = 200$, $f_T = 400$ MHz at $I_C = 1$ mA, $C_\mu = 0.3$ pF, $C_{je} = 3$ pF, $C_{cs} = 1.5$ pF, $r_b = 0$, $V_{BE(\text{on})} = 0.6$ V, and $r_o = \infty$.

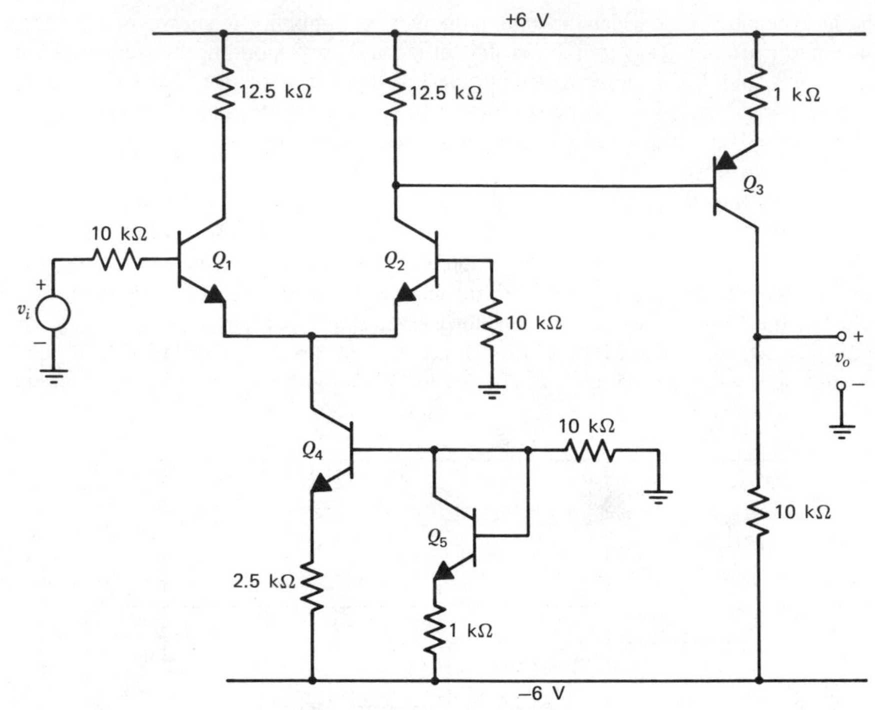

Fig. 7.37 Two-stage amplifier with *pnp* second stage.

pnp: $\beta = 100$, $f_T = 6$ MHz at $I_C = -0.5$ mA, $C_\mu = 0.3$ pF, $C_{je} = 3$ pF, $C_{bs} = 1.5$ pF, $r_b = 0$, $V_{BE(\text{on})} = 0.6$ V, and $r_o = \infty$. (Values of C_μ, C_{cs}, C_{bs}, and C_{je} are at the bias point.)

REFERENCES

1. D. O. Pederson. *Electronic Circuits.* Preliminary Edition, McGraw-Hill, New York, 1965.
2. P. E. Gray and C. L. Searle. *Electronic Principles.* Wiley, New York, 1969.
3. R. D. Thorton et al. *Multistage Transistor Circuits.* Wiley, New York, 1965.
4. K. Ogata. *Modern Control Engineering.* Prentice-Hall, Englewood Cliffs, N. J., 1970.

CHAPTER 8

FEEDBACK

Negative feedback is widely used in amplifier design because it produces several important benefits. One of the most significant is that negative feedback stabilizes the gain of the amplifier against parameter changes in the active devices due to supply voltage variation, temperature changes, or device aging. A second benefit is that negative feedback allows the designer to modify the input and output impedances of the circuit in any desired fashion. Another significant benefit of negative feedback is the reduction in signal waveform distortion that it produces, and for this reason almost all high-quality audio amplifiers employ negative feedback around the power output stage. Finally, negative feedback can produce an increase in the bandwidth of circuits and is widely used in broadband amplifiers.

However, the benefits of negative feedback listed above are accompanied by two disadvantages. First, the gain of the circuit is reduced in almost direct proportion to the other benefits achieved. Thus, it is often necessary to make up the decrease in gain by adding extra amplifier stages with a consequent increase in hardware cost. The second potential problem associated with the use of feedback is the tendency for oscillation to occur in the circuit, and careful attention by the designer is often required to overcome this problem.

In this chapter, the various benefits of negative feedback are considered, together with a systematic classification of feedback configurations. The problem of feedback-induced oscillation and its solution is considered in Chapter 9.

8.1 IDEAL FEEDBACK EQUATION

Consider the idealized feedback configuration of Fig. 8.1. In this figure S_i and S_o are input and output signals that may be voltages or currents. The feedback network (which is usually linear and passive) has a transfer function f and feeds back a signal S_{fb} to the input. At the input, signal S_{fb} is subtracted from input signal S_i at the input differencing node. Error signal S_ϵ is the difference between S_i and S_{fb}, and S_ϵ is fed to the basic amplifier with transfer function a. Note that another common convention is to assume that S_i and S_{fb} are added together in an input *summing* node, and this leads to some sign changes in the analysis. It should be pointed out that negative-feedback amplifiers in practice have an input differencing node and thus the convention assumed here is more convenient for amplifier analysis.

From Fig. 8.1,

$$S_o = aS_\epsilon \tag{8.1}$$

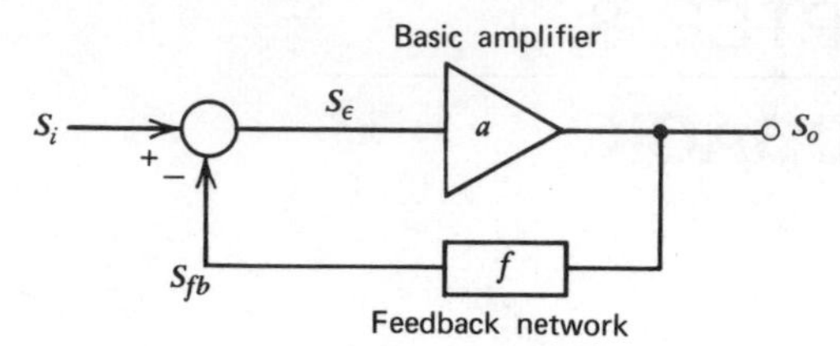

Fig. 8.1 Ideal feedback configuration.

assuming that the feedback network does not load the basic amplifier. Also

$$S_{fb} = fS_o \tag{8.2}$$

$$S_\epsilon = S_i - S_{fb} \tag{8.3}$$

Substituting (8.2) in (8.3) gives

$$S_\epsilon = S_i - fS_o \tag{8.4}$$

Substituting (8.4) in (8.1) gives

$$S_o = aS_i - afS_o$$

and thus

$$\frac{S_o}{S_i} = A = \frac{a}{1 + af} \tag{8.5}$$

Equation 8.5 is the fundamental equation for negative feedback circuits where A is the overall gain with feedback applied. (A is often called the *closed-loop gain.*)

It is useful to define a quantity T called the *loop gain* such that

$$T = af \tag{8.6}$$

and

$$\frac{S_o}{S_i} = A = \frac{a}{1 + T} \tag{8.7}$$

T is the total gain around the feedback loop. If $T \gg 1$, then, from (8.5), gain A is given by

$$A \simeq \frac{1}{f} \tag{8.8}$$

That is, for large values of loop gain T, the overall amplifier gain is determined by the feedback transfer function, f. Since the feedback network is usually formed from stable, passive elements the value of f is well defined and so is the overall amplifier gain.

The feedback loop operates by forcing S_{fb} to be nearly equal to S_i. This is achieved by amplifying the difference $S_\epsilon = S_i - S_{fb}$ and the feedback loop then effectively minimizes error signal S_ϵ. This can be seen by substituting (8.5) in (8.4) to obtain

$$S_\epsilon = S_i - f\frac{aS_i}{1 + af}$$

and this leads to

$$\frac{S_\epsilon}{S_i} = \frac{1}{1 + af} = \frac{1}{1 + T} \tag{8.9}$$

As T becomes much greater than 1, S_ϵ becomes much less than S_i. In addition, substituting (8.5) in (8.2) gives

$$S_{fb} = fS_i\frac{a}{1 + af}$$

and thus

$$\frac{S_{fb}}{S_i} = \frac{T}{1 + T} \tag{8.10}$$

If $T \gg 1$, then S_{fb} is approximately equal to S_i. That is, feedback signal S_{fb} is a replica of the input signal. Since S_{fb} and S_o are directly related by (8.2), it follows that if $|f| < 1$ then S_o is an amplified replica of S_i. This is the aim of a feedback amplifier.

8.2 GAIN SENSITIVITY

In most practical situations, gain a of the basic amplifier is not well defined. It is dependent on temperature, active-device operating conditions, and transistor parameters such as β. As mentioned previously the negative-feedback loop reduces variations in overall amplifier gain due to variations in a. This effect may be examined by differentiating (8.5) to obtain

$$\frac{dA}{da} = \frac{(1 + af) - af}{(1 + af)^2}$$

and this reduces to

$$\frac{dA}{da} = \frac{1}{(1 + af)^2} \tag{8.11}$$

If a changes by δa, then A changes by δA where

$$\delta A = \frac{\delta a}{(1 + af)^2}$$

The fractional change in A is

$$\frac{\delta A}{A} = \frac{1 + af}{a}\frac{\delta a}{(1 + af)^2}$$

This can be expressed as

$$\frac{\delta A}{A} = \frac{\dfrac{\delta a}{a}}{1 + af} = \frac{\dfrac{\delta a}{a}}{1 + T} \tag{8.12}$$

Equation 8.12 shows that the fractional change in A is reduced by $(1 + T)$ compared to the fractional change in a. For example, if $T = 100$ and a changes by 10 percent due to temperature change, then the overall gain, A, changes by only 0.1 percent using (8.12).

8.3 EFFECT OF NEGATIVE FEEDBACK ON DISTORTION

The results above show that even if the basic-amplifier gain, a, changes, the negative feedback keeps overall gain A approximately constant. This suggests that feedback should be effective in reducing distortion because distortion is caused by changes in the slope of the basic-amplifier transfer characteristic. The feedback should tend to reduce the effect of these slope changes since A is relatively independent of a. This is illustrated below.

Suppose the basic amplifier has a transfer characteristic with a nonlinearity as shown in Fig. 8.2. It is assumed that two regions exist, each with constant but different slopes a_1 and a_2. When feedback is applied, the overall gain will still be given by (8.5) but the appropriate value of a must be used, depending on which region of Fig. 8.2 is being traversed. Thus the *overall* transfer characteristic with feedback applied will also have two regions of different slope as shown in Fig. 8.3. However slopes A_1 and A_2 are almost equal because of the effect of the negative feedback. This can be seen by substituting in (8.5) to give

$$A_1 = \frac{a_1}{1 + a_1 f} \simeq \frac{1}{f} \tag{8.13}$$

$$A_2 = \frac{a_2}{1 + a_2 f} \simeq \frac{1}{f} \tag{8.14}$$

Thus the transfer characteristic of the feedback amplifier of Fig. 8.3 shows much less nonlinearity than the original basic-amplifier characteristic of Fig. 8.2.

Note that the horizontal scale in Fig. 8.3 has been *compressed* as compared to Fig. 8.2 in order to allow easy comparison of the two graphs. This scale change is necessary because the negative feedback *reduces the gain.* The reduction in gain by the factor $(1 + T)$, which accompanies the use of negative feedback, presents few serious problems, since the gain can easily be made up by placing a preamplifier in front of the feedback amplifier. Since the preamplifier handles much smaller signals than does the output amplifier, distortion is usually not a problem in that amplifier.

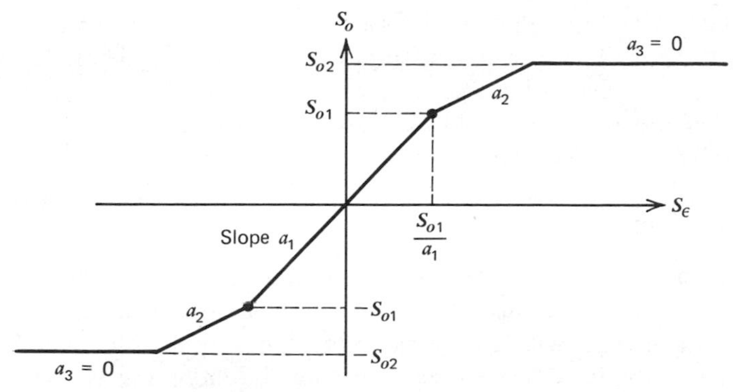

Fig. 8.2 Basic-amplifier transfer characteristic.

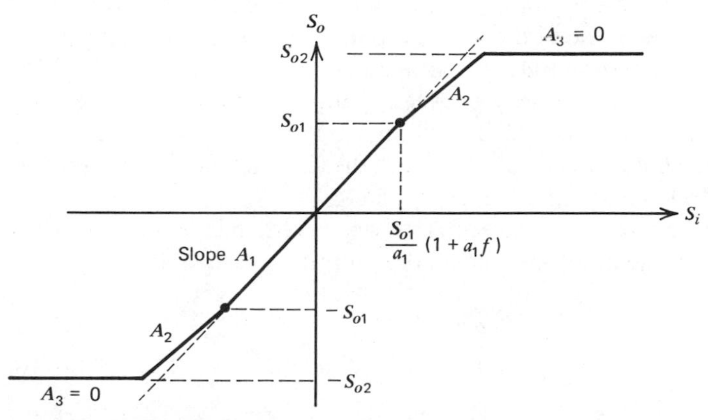

Fig. 8.3 Feedback-amplifier transfer characteristic corresponding to the basic-amplifier characteristic of Fig. 8.2.

One further point that should be made about Figs. 8.2 and 8.3 is that both show hard saturation of the output amplifier (i.e., the output becomes independent of the input) at an output signal level of S_{o2}. Since the incremental slope $a_3 = 0$ in that region, negative feedback cannot improve the situation as $A_3 = 0$ also, using (8.5).

8.4 FEEDBACK CONFIGURATIONS

The treatment in the above sections was based on the idealized configuration shown in Fig. 8.1. Practical feedback amplifiers are composed of circuits that have current or voltage signals as inputs and produce output currents or voltages. In order to pursue feedback amplifier design at a practical level, it is necessary to specify the details of the feedback sampling process and the circuits used to realize this operation. There are four basic feedback amplifier connections. These are specified according to whether the output signal S_o, which is sampled, is a current or a voltage and whether the feedback signal, S_{fb}, is a current or a voltage. It is apparent that four combinations exist and these are now considered.

8.4.1 Series-Shunt Feedback

Suppose it is required to design a feedback amplifier that stabilizes a voltage transfer function. That is, a given input voltage should produce a well-defined proportional output voltage. This will require sampling the output voltage and feeding back a proportional voltage for comparison with the incoming voltage. This situation is shown schematically in Fig. 8.4. The basic amplifier has gain a, and the feedback network is a two-port with transfer function f that *shunts* the output of the basic amplifier to sample v_o. *Ideally* the impedance $z_{22f} = \infty$, and the feedback network does not load the basic amplifier. The feedback voltage, v_{fb}, is connected in *series* with the input to allow comparison with v_i and, ideally, $z_{11f} = 0$. The signal, v_ϵ, is the *difference* between v_i and v_{fb} and is fed to the basic amplifier. The basic amplifier and feedback circuits are assumed *unilateral* in that the basic amplifier transmits only from v_ϵ to v_o and the feedback network transmits only from v_o to v_{fb}. This point will be taken up later.

This feedback is called *series-shunt* feedback because the feedback network is connected in *series* with the input and *shunts* the output.

From Fig. 8.4,

$$v_o = av_\epsilon \tag{8.15}$$

$$v_{fb} = fv_o \tag{8.16}$$

$$v_\epsilon = v_i - v_{fb} \tag{8.17}$$

From (8.15), (8.16), and (8.17),

$$\frac{v_o}{v_i} = \frac{a}{1 + af} \tag{8.18}$$

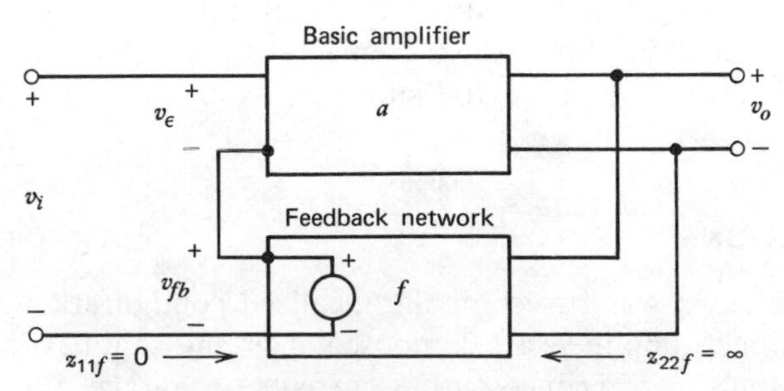

Fig. 8.4 Series-shunt feedback configuration.

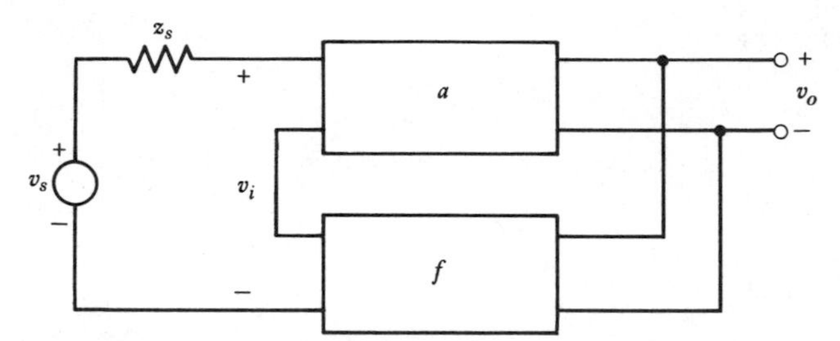

Fig. 8.5 Series-shunt configuration fed from a finite source impedance.

Thus the ideal feedback equation applies. Equation 8.18 indicates that the transfer function that is stabilized is v_o/v_i, as desired. If the circuit is fed from a high source impedance as shown in Fig. 8.5 the ratio v_o/v_i is still stabilized [and given by (8.18)] *but* now v_i is given by

$$v_i = \frac{Z_i}{Z_i + z_s} v_s \tag{8.19}$$

where Z_i is the input impedance seen by v_i. If $z_s \approx Z_i$, then v_i depends on Z_i, which is *not* usually well defined since it often depends on active-device parameters. Thus the overall gain, v_o/v_i, will not be stabilized. Consequently, the full benefits of gain stabilization are achieved for a series-shunt feedback amplifier when the source impedance is low compared to the input impedance of the closed-loop amplifier. The ideal driving source is a voltage source.

Consider now the effect of series-shunt feedback on the terminal impedances of the amplifier. Assume the basic amplifier has input and output impedances z_i and z_o as shown in Fig. 8.6. Again assume the feedback network is ideal and feeds back a voltage fv_o as shown. Both networks are unilateral. The applied voltage, v_i, produces input current i_i and output voltage v_o. From Fig. 8.6

$$v_o = av_\epsilon \tag{8.20}$$

$$v_i = v_\epsilon + fv_o \tag{8.21}$$

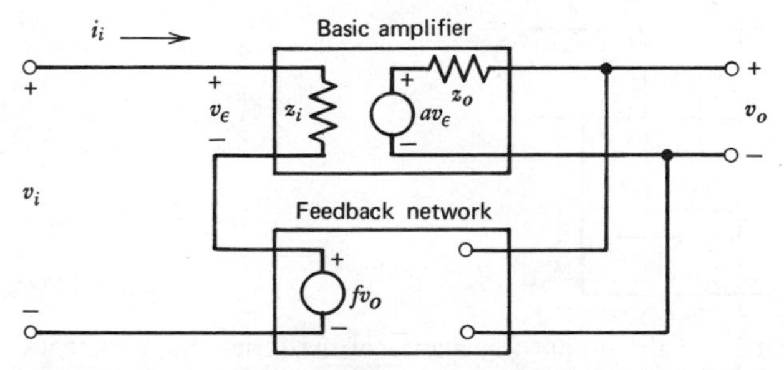

Fig. 8.6 Series-shunt configuration with finite impedances in the basic amplifier.

Substituting (8.20) in (8.21) gives

$$v_i = v_\epsilon + afv_\epsilon = v_\epsilon(1 + af) \tag{8.22}$$

Also

$$i_i = \frac{v_\epsilon}{z_i} \tag{8.23}$$

Substituting (8.22) in (8.23) gives

$$i_i = \frac{v_i}{z_i}\frac{1}{1 + af} \tag{8.24}$$

Thus, from (8.24), input impedance Z_i with feedback applied is

$$Z_i = \frac{v_i}{i_i} = (1 + T)\, z_i \tag{8.25}$$

Series feedback at the input *always* raises the input impedance by $(1 + T)$.

The effect of series-shunt feedback on the output impedance can be calculated using the circuit of Fig. 8.7. The input voltage is removed (the input is shorted) and a voltage v applied at the output. From Fig. 8.7,

$$v_\epsilon + fv = 0 \tag{8.26}$$

$$i = \frac{v - av_\epsilon}{z_o} \tag{8.27}$$

Substituting (8.26) in (8.27) gives

$$i = \frac{v + afv}{z_o} \tag{8.28}$$

From (8.28) output impedance Z_o with feedback applied is

$$Z_o = \frac{v}{i} = \frac{z_o}{1 + T} \tag{8.29}$$

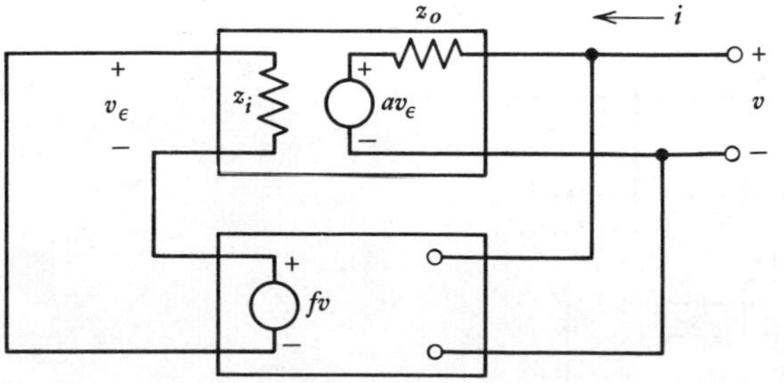

Fig. 8.7 Circuit for the calculation of the output impedance of the series-shunt feedback configuration.

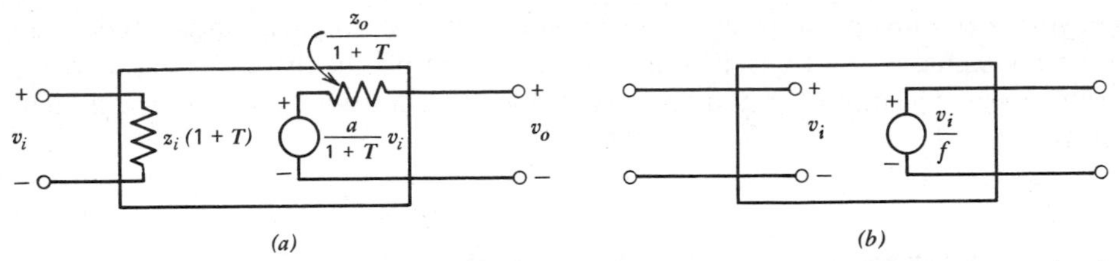

Fig. 8.8 (*a*) Equivalent circuit of a series-shunt feedback amplifier. (*b*) Equivalent circuit of a series-shunt feedback amplifier for $a \to \infty$.

Shunt feedback at the output *always* lowers the output impedance by $(1 + T)$. This makes the output a better voltage source so that series-shunt feedback produces a *good voltage amplifier*. It stabilizes v_o/v_i, raises Z_i, and lowers Z_o.

The original series-shunt feedback amplifier of Fig. 8.6 can now be represented as shown in Fig. 8.8*a* using (8.18), (8.25), and (8.29). As the forward gain, a, approaches infinity, the equivalent circuit approaches that of Fig. 8.8*b*, which is an ideal voltage amplifier.

8.4.2 Shunt-Shunt Feedback

This configuration is shown in Fig. 8.9. The feedback network again shunts the output of the basic amplifier and samples v_o and, ideally, $z_{22f} = \infty$ as before. However, the feedback network now *shunts* the input of the main amplifier as well, and feeds back a proportional current fv_o. Ideally, $z_{11f} = \infty$ so that the feedback

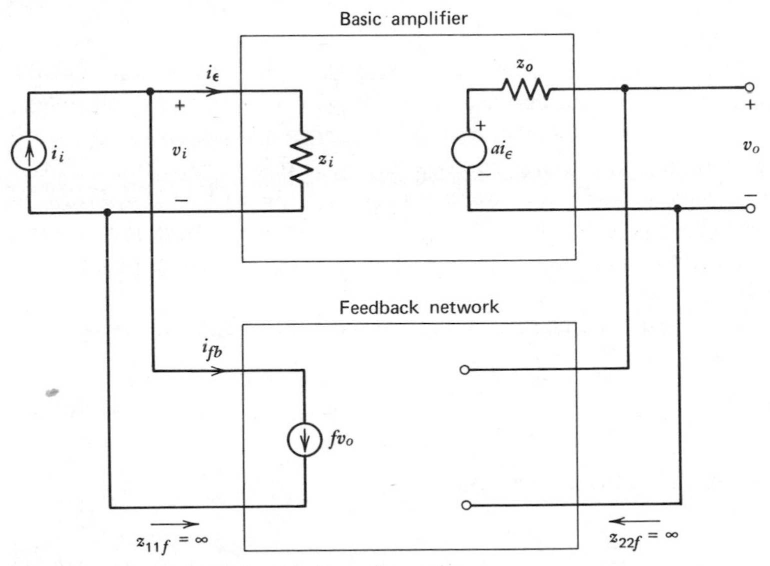

Fig. 8.9 Shunt-shunt feedback configuration.

network does not produce any shunt loading on the amplifier input. Since the feedback signal is a current, it is more convenient to deal with an error *current* i_ϵ at the input. The input signal in this case is ideally a current i_i and this is assumed. From Fig. 8.9,

$$a = \frac{v_o}{i_\epsilon} \tag{8.30}$$

where a is a *transresistance*

$$f = \frac{i_{fb}}{v_o} \tag{8.31}$$

where f is a *transconductance.*

$$v_o = ai_\epsilon \tag{8.32}$$

$$i_\epsilon = i_i - i_{fb} \tag{8.33}$$

Substitution of i_{fb} from (8.31) in (8.33) gives

$$i_\epsilon = i_i - fv_o \tag{8.34}$$

Substitution of (8.32) in (8.34) gives

$$\frac{v_o}{a} = i_i - fv_o$$

Rearranging terms we find

$$\frac{v_o}{i_i} = \frac{a}{1 + af} = A \tag{8.35}$$

Again the ideal feedback equation applies. Note that although a and f have dimensions of resistance and conductance, the loop gain, $T = af$, is *dimensionless.* This is always true.

In this configuration, if the source impedance, z_s, is finite a division of input current i_i occurs between z_s and the amplifier input and the ratio v_o/i_i will not be as well defined as (8.35) suggests. The full benefits of negative feedback for a shunt-shunt feedback amplifier are thus obtained for $z_s \gg Z_i$, which approaches a current-source drive.

The input impedance of the circuit of Fig. 8.9 can be calculated using (8.32) and (8.35) to give

$$i_\epsilon = \frac{i_i}{1 + af} \tag{8.36}$$

The input impedance, Z_i, with feedback is

$$Z_i = \frac{v_i}{i_i} \tag{8.37}$$

Substituting (8.36) in (8.37) gives

$$Z_i = \frac{v_i}{i_\epsilon} \frac{1}{1 + af} = \frac{z_i}{1 + T} \tag{8.38}$$

Thus shunt feedback at the input *reduces* the amplifier input impedance by $(1 + T)$. This is always true.

It is easily shown that the output impedance in this case is

$$Z_o = \frac{z_o}{1 + T} \tag{8.39}$$

as before, for shunt feedback at the output.

Shunt-shunt feedback has made this amplifier a good *transresistance* amplifier. The transfer function, v_o/i_i, has been stabilized and both Z_i and Z_o are lowered.

The original shunt-shunt feedback amplifier of Fig. 8.9 can now be represented as shown in Fig. 8.10*a* using (8.35), (8.38), and (8.39). As forward gain a approaches

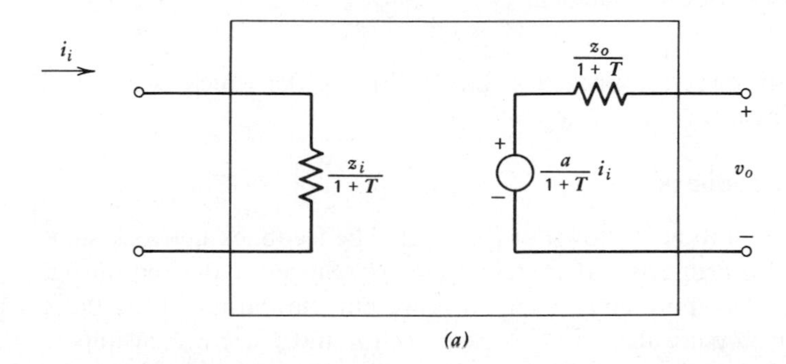

(*a*)

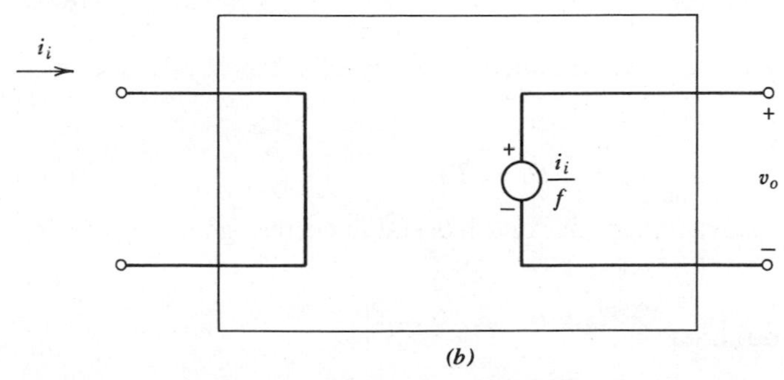

(*b*)

Fig. 8.10 (*a*) Equivalent circuit of a shunt-shunt feedback amplifier. (*b*) Equivalent circuit of a shunt-shunt feedback amplifier for $a \to \infty$.

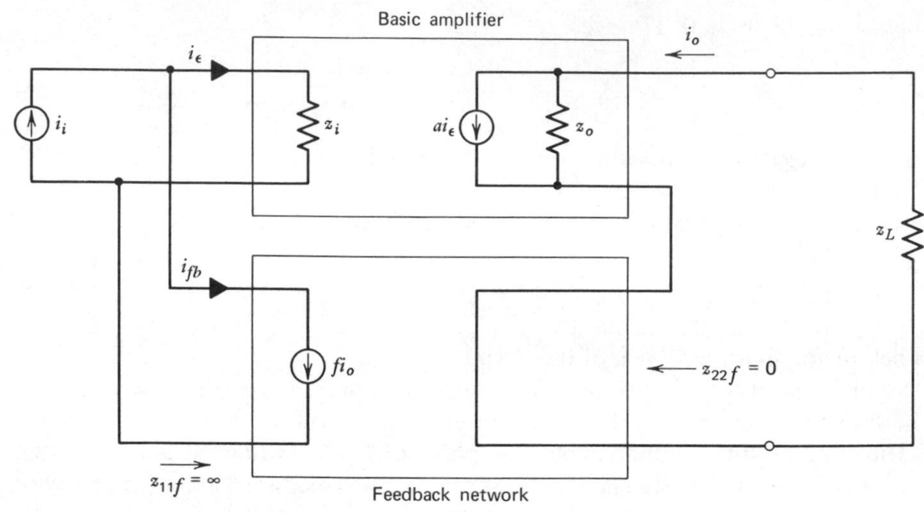

Fig. 8.11 Shunt-series feedback configuration.

infinity, the equivalent circuit approaches that of Fig. 8.10*b*, which is an ideal transresistance amplifier.

8.4.3 Shunt-Series Feedback

The shunt-series configuration is shown in Fig. 8.11. The feedback network samples i_o and feeds back a proportional current $i_{fb} = fi_o$. Since the desired output signal is a current i_o, it is more convenient to represent the output of the basic amplifier with a Norton equivalent. In this case both a and f are dimensionless current ratios, and the ideal source is a current source i_i. If $z_L \ll z_o$ it can be shown that

$$\frac{i_o}{i_i} = \frac{a}{1 + af} \tag{8.40}$$

$$Z_i = \frac{z_i}{1 + T} \tag{8.41}$$

$$Z_o = z_o(1 + T) \tag{8.42}$$

This amplifier is a good *current* amplifier and has stable current gain i_o/i_i, low Z_i, and high Z_o.

8.4.4 Series-Series Feedback

The series-series configuration is shown in Fig. 8.12. The feedback network samples i_o and feeds back a proportional voltage v_{fb} in *series* with the input. The

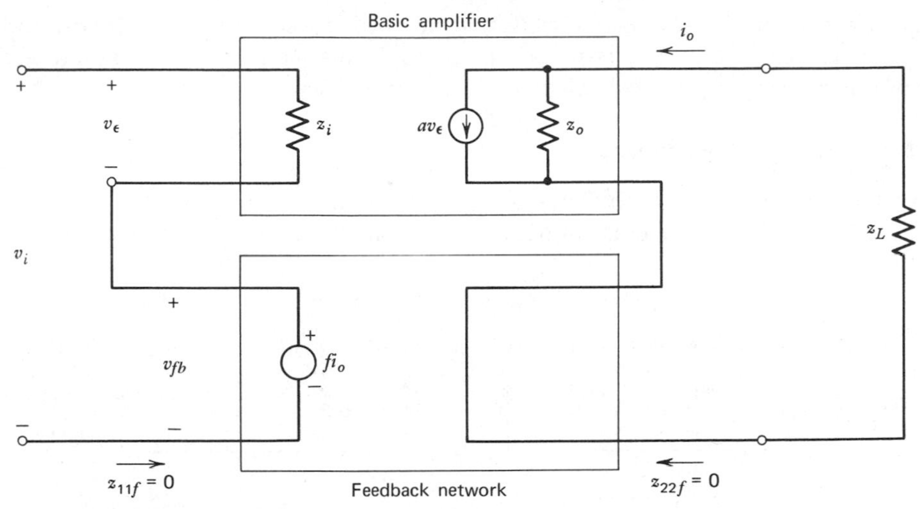

Fig. 8.12 Series-series feedback configuration.

forward gain, a, is a transconductance and f is a transresistance, and the ideal driving source is a voltage source v_i. If $z_L \ll z_o$ it can be shown that

$$\frac{i_o}{v_i} = \frac{a}{1 + af} \tag{8.43}$$

$$Z_i = z_i(1 + T) \tag{8.44}$$

$$Z_o = z_o(1 + T) \tag{8.45}$$

This amplifier is a good transconductance amplifier and has a stabilized gain i_o/v_i, and has high Z_i and Z_o.

8.5 PRACTICAL CONFIGURATIONS AND THE EFFECT OF LOADING

In practical feedback amplifiers, the feedback network causes loading at the input and output of the basic amplifier, and the division into basic amplifier and feedback network is not as obvious as the above treatment implies. In such cases, the circuit can always be analyzed by writing circuit equations for the whole amplifier and solving for the transfer function and terminal impedances. However this procedure becomes very tedious and difficult in most practical cases, and the equations so complex that one loses sight of the important aspects of circuit performance. Thus it is profitable to identify a basic amplifier and feedback network in such cases and then to use the ideal feedback equations derived above. In general it will be necessary to include the loading effect of the feedback network on the basic amplifier, and methods of including this loading in the calculations are now considered. The *method* will be developed through the use of two-port representations of the circuits involved, although this method of representation is not necessary for practical calculations, as we will see.

8.5.1 Shunt-Shunt Feedback

Consider the shunt-shunt feedback amplifier of Fig. 8.9. The effect of nonideal networks may be included as shown in Fig. 8.13a where finite input and output

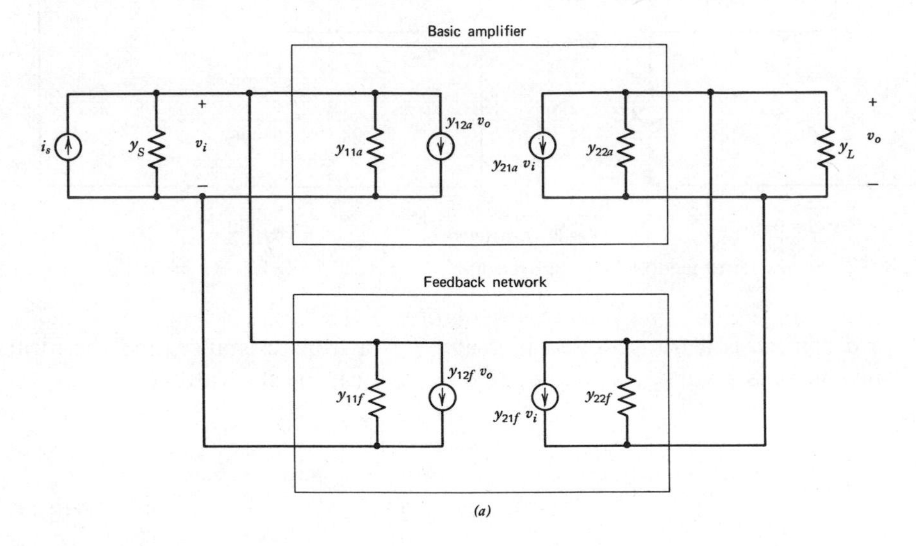

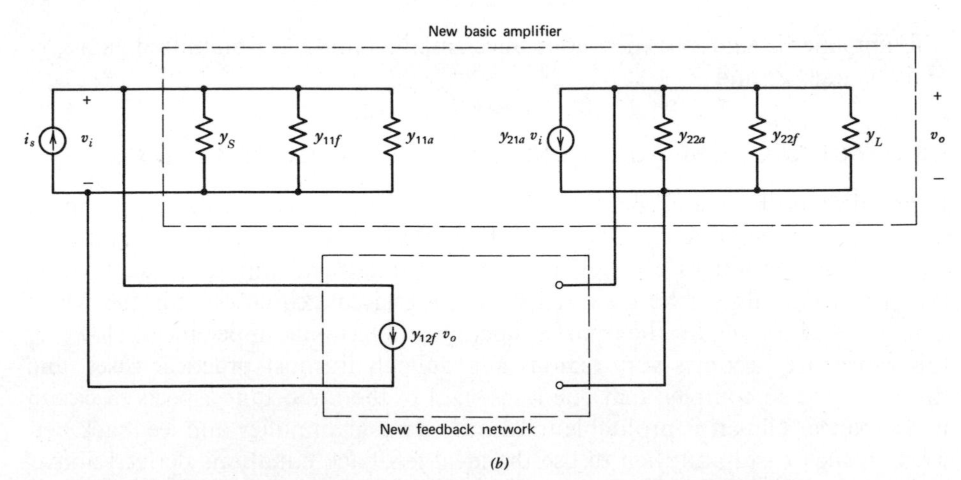

Fig. 8.13 (a) Shunt-shunt feedback configuration using the y-parameter representation. (b) Circuit of (a) redrawn with generators $y_{21f}v_i$ and $y_{12a}v_o$ omitted.

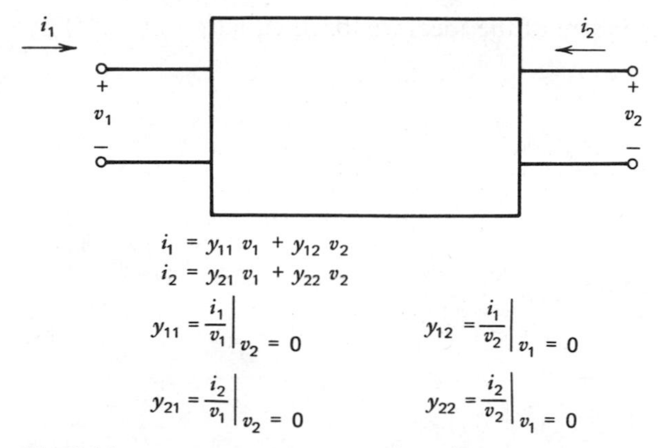

Fig. 8.14 y-parameter representation of a two-port.

admittance is assumed in both forward and feedback paths, as well as reverse transmission in each. Finite source and load admittances y_S and y_L are assumed. The most convenient two-port representation in this case is use of the short-circuit admittance parameters or y parameters,[1] as used in Fig. 8.13*a*. The reason for this is that the basic amplifier and the feedback network are connected in parallel at input and output, and thus have identical *voltages* at their terminals. The y parameters specify the response of a network by expressing the terminal currents in terms of the terminal voltages, and this results in very simple calculations when two networks have identical terminal voltages. This will be evident in the circuit calculations to follow. The y-parameter representation is illustrated in Fig. 8.14.

From Fig. 8.13*a*, at the input

$$i_s = (y_S + y_{11a} + y_{11f})v_i + (y_{12a} + y_{12f})v_o \tag{8.46}$$

Summation of currents at the output gives

$$0 = (y_{21a} + y_{21f})v_i + (y_L + y_{22a} + y_{22f})v_o \tag{8.47}$$

It is useful to define

$$y_i = y_S + y_{11a} + y_{11f} \tag{8.48}$$

$$y_o = y_L + y_{22a} + y_{22f} \tag{8.49}$$

Solving (8.46) and (8.47) by using (8.48) and (8.49) gives

$$\frac{v_o}{i_s} = \frac{-(y_{21a} + y_{21f})}{y_i y_o - (y_{21a} + y_{21f})(y_{12a} + y_{12f})} \tag{8.50}$$

This equation can be put in the form of the ideal feedback equation of (8.35) by dividing by $y_i y_o$ to give

$$\frac{v_o}{i_s} = \frac{\dfrac{-(y_{21a} + y_{21f})}{y_i y_o}}{1 + \dfrac{-(y_{21a} + y_{21f})}{y_i y_o}(y_{12a} + y_{12f})} \tag{8.51}$$

Comparing (8.51) with (8.35) gives

$$a = -\frac{y_{21a} + y_{21f}}{y_i y_o} \tag{8.52}$$

$$f = y_{12a} + y_{12f} \tag{8.53}$$

At this point, a number of approximations can be made that greatly simplify the calculations. First, we assume that the signal transmitted by the basic amplifier is much greater than the signal fed forward by the feedback network. Since the former has gain (usually large) while the latter has loss, this is almost invariably a valid assumption. This means that

$$|y_{21a}| \gg |y_{21f}| \tag{8.54}$$

Second, we assume that the signal fed back by the feedback network is much greater than the signal fed back through the basic amplifier. Since most active devices have very small reverse transmission, the basic amplifier has a similar characteristic and this assumption is almost invariably quite accurate. This assumption means that

$$|y_{12a}| \ll |y_{12f}| \tag{8.55}$$

Using (8.54) and (8.55) in (8.51) gives

$$\frac{v_o}{i_s} = A \simeq \frac{\dfrac{-y_{21a}}{y_i y_o}}{1 + \left(\dfrac{-y_{21a}}{y_i y_o}\right) y_{12f}} \tag{8.56}$$

Comparing (8.56) with (8.35) gives

$$a = -\frac{y_{21a}}{y_i y_o} \tag{8.57}$$

$$f = y_{12f} \tag{8.58}$$

A circuit representation of (8.57) and (8.58) can be found as follows. Equations 8.54 and 8.55 mean that in Fig. 8.13*a* the feedback generator of the basic amplifier and the forward-transmission generator of the feedback network may be neglected. If this is done the circuit may be redrawn, as in Fig. 8.13*b* where the terminal admittances y_{11f} and y_{22f} of the feedback network have been absorbed into the basic amplifier, together with source and load impedances y_S and y_L. The new basic amplifier thus *includes the loading effect* of the original feedback network, and the new feedback network is an ideal one as used in Fig. 8.9. If the transfer function of the basic amplifier of Fig. 8.13*b* is calculated (by first removing the feedback network) the result given in (8.57) is obtained. Similarly, the transfer function of the feedback network of Fig. 8.13*b* is given by (8.58). Thus Fig. 8.13*b* is a *circuit representation* of (8.57) and (8.58).

Since Fig. 8.13*b* has a direct correspondence with Fig. 8.9, all the results derived in Section 8.4.2 for Fig. 8.9 can now be used. The loading effect of the feedback network on the basic amplifier is now included by simply shunting input and output with y_{11f} and y_{22f} respectively. As shown in Fig. 8.14, these terminal admittances of the feedback network are calculated with the other port of the network short circuited. In practice, loading term y_{11f} is simply obtained by shorting the output node of the amplifier and calculating the feedback circuit input admittance. Similarly term y_{22f} is calculated by shorting the input node of the amplifier and calculating the feedback circuit output admittance. The feedback transfer function f given by (8.58) is the short-circuit reverse transfer admittance of the feedback network and is defined in Fig. 8.14. This is readily calculated in practice, and is often obtained by inspection. Note that the use of y parameters in further calculations is *not* necessary. Once the circuit of Fig. 8.13*b* is established, any convenient network analysis method may be used to calculate gain a of the basic amplifier. We have simply used the two-port representation as a general means of illustrating how loading effects may be included in the calculations.

For example, consider the common shunt-shunt feedback circuit using an op amp as shown in Fig. 8.15*a*. The equivalent circuit is shown in Fig. 8.15*b* and is redrawn in Fig. 8.15*c* to allow for loading of the feedback network on the basic amplifier. The y parameters of the feedback network can be found from Fig. 8.15*d*.

$$y_{11f} = \left.\frac{i_1}{v_1}\right|_{v_2=0} = \frac{1}{R_F} \tag{8.59}$$

$$y_{22f} = \left.\frac{i_2}{v_2}\right|_{v_1=0} = \frac{1}{R_F} \tag{8.60}$$

$$y_{12f} = \left.\frac{i_1}{v_2}\right|_{v_1=0} = -\frac{1}{R_F} = f \tag{8.61}$$

Using (8.54), we neglect y_{21f}.
The basic-amplifier gain a can be calculated from Fig. 8.15*c* by putting $i_{fb} = 0$ to give

$$v_1 = \frac{z_i R_F}{z_i + R_F} i_i \tag{8.62}$$

$$v_o = -\frac{R}{R + z_o} a_v v_1 \tag{8.63}$$

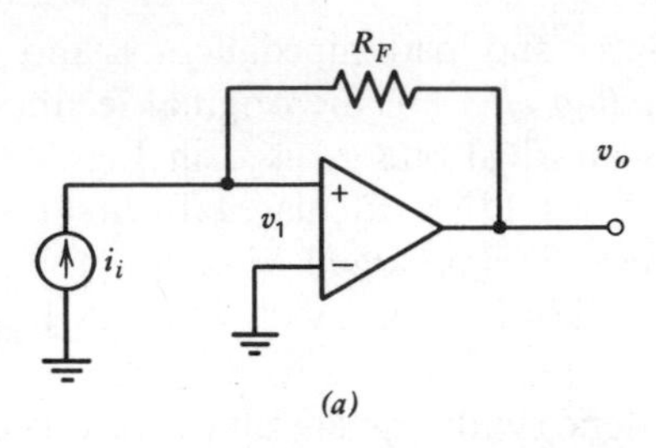

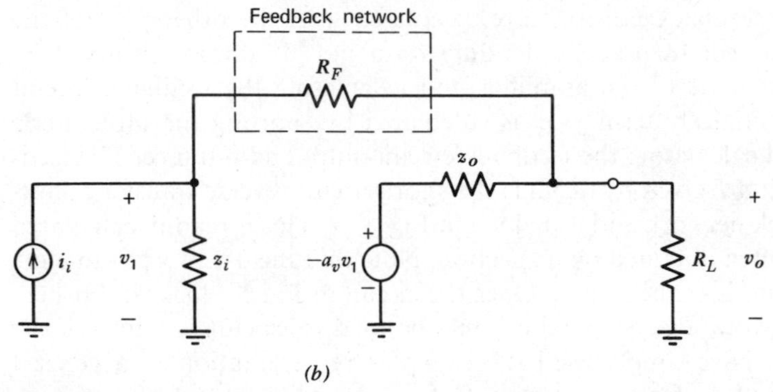

Fig. 8.15 (*a*) Shunt-shunt feedback circuit using an op amp as the gain element. (*b*) Equivalent circuit of (*a*). (*c*) Division of the circuit in (*b*) into forward and feedback paths. (*d*) Circuit for the calculation of the *y* parameters of the feedback network of the circuit in (*b*).

where

$$R = R_F \| R_L \tag{8.64}$$

Substituting (8.62) in (8.63) gives

$$\frac{v_o}{i_i} = a = -\frac{R}{R + z_o} a_v \frac{z_i R_F}{z_i + R_F} \tag{8.65}$$

Using the formulas derived in Section 8.4.2 we can now calculate all parameters of the feedback circuit. The input and output impedances of the basic amplifier now *include* the effect of feedback loading and it is *these impedances* that are divided by $(1 + T)$ as described in Section 8.4.2. Thus the input impedance of the basic amplifier of Fig. 8.15*c* is

$$z_{ia} = R_F \| z_i = \frac{R_F z_i}{R_F + z_i} \tag{8.66}$$

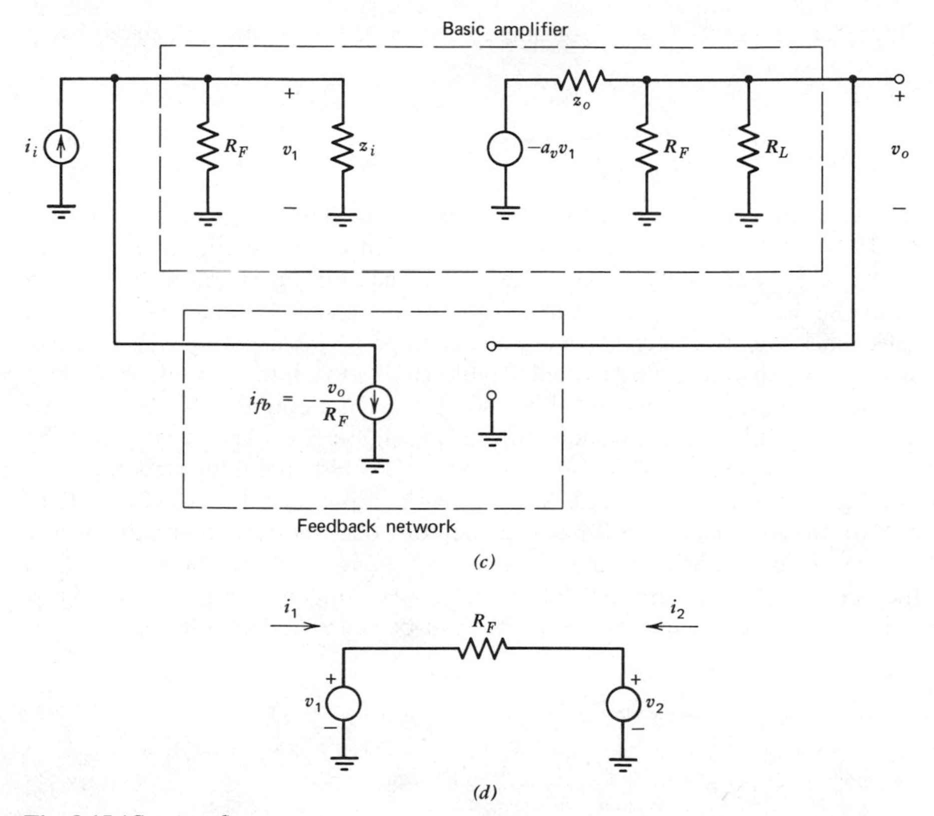

Fig. 8.15 (*Continued*)

When feedback is applied the input impedance is

$$Z_i = \frac{z_{ia}}{1 + T} \tag{8.67}$$

Similarly for the output impedance of the basic amplifier

$$z_{oa} = z_o \| R_F \| R_L \tag{8.68}$$

When feedback is applied this becomes

$$Z_o = \frac{z_o \| R_F \| R_L}{1 + T} \tag{8.69}$$

Note that these calculations can be made using the circuit of Fig. 8.15*c* *without* further need of two-port *y* parameters.

Since the loop gain, T, is of considerable interest this is now calculated using (8.61) and (8.65),

$$T = af = \frac{R_F R_L}{R_F R_L + z_o R_F + z_o R_L} a_v \frac{z_i}{z_i + R_F} \tag{8.70}$$

It often occurs during certain phases of a design that we wish to calculate T only. This can be done directly from the equivalent circuit of Fig. 8.15*b* without need to make the transformation to Fig. 8.15*c*. The loop gain is just the total gain around the feedback loop and can be calculated from the circuit of Fig. 8.16. The feedback loop is broken at some convenient point and a test signal is inserted. The signal which is then transmitted around the feedback loop is calculated and the ratio of the transmitted signal to the input signal equals $-T$. This can be seen from Fig. 8.1. If the feedback loop in that figure is broken at any point (at the input of the basic amplifier, for example) and a test signal inserted, the ratio of the signal transmitted around the loop to the input signal is $-af = -T$. In Fig. 8.16, the loop has been broken at the controlled voltage generator (where there are no loading effects) and a test voltage v_x is inserted. The signal which is then produced at the terminals of the controlled voltage generator is the signal that is transmitted around the loop. This can be calculated as follows:

$$v_1 = \frac{z_i}{z_i + R_F} v_2 \tag{8.71}$$

$$v_2 = v_x \frac{R_L \| (R_F + z_i)}{R_L \| (R_F + z_i) + z_o} \tag{8.72}$$

Substituting (8.72) in (8.71) gives

$$\frac{v_1}{v_x} = \frac{z_i}{z_i + R_F} \frac{R_L(R_F + z_i)}{R_L(R_F + z_i) + z_o(R_L + R_F + z_i)} \tag{8.73}$$

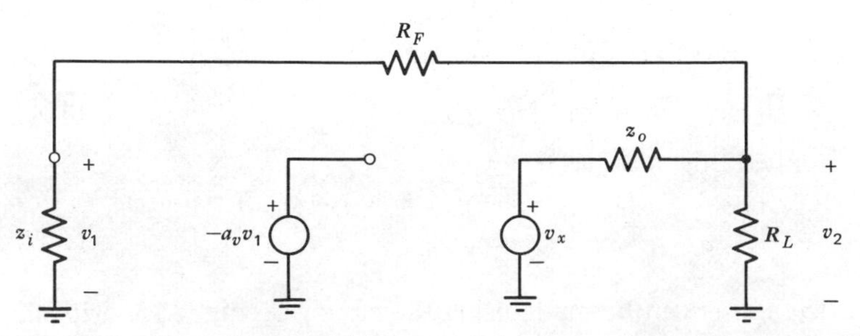

Fig. 8.16 Circuit for the calculation of the loop gain of the circuit of Fig. 8.15*b*.

The signal transmitted around the loop in Fig. 8.16 is $-a_v v_1$, and the ratio of this signal to input signal v_x is $-T$. Thus

$$T = \frac{a_v v_1}{v_x} \tag{8.74}$$

Substituting (8.73) in (8.74) gives

$$T = \frac{R_F R_L}{R_F R_L + z_o R_F + z_o R_L \dfrac{R_F}{R_F + z_i}} a_v \frac{z_i}{z_i + R_F} \tag{8.75}$$

This is very close to the value calculated from Fig. 8.15*c* and given by (8.70). The small error introduced by the previous approximations are apparent in comparing (8.70) and (8.75). Equation 8.75 is exact. Either method can be used to calculate T but the method described above is often easier if T is all that is required.

EXAMPLE

Assuming that the circuit of Fig. 8.15*a* is realized using a 741 op amp with $R_F = 1\ \text{M}\Omega$ and $R_L = 10\ \text{k}\Omega$, calculate the terminal impedances, loop gain, and overall gain of the feedback amplifier at low frequencies. Typical 741 data is $z_i = 2\ \text{M}\Omega$, $z_o = 75\ \Omega$, $a_v = 200{,}000$.

From (8.66) the low-frequency input impedance of the basic amplifier including loading is

$$z_{ia} = \frac{10^6 \times 2 \times 10^6}{10^6 + 2 \times 10^6}\ \Omega = 666.7\ \text{k}\Omega \tag{8.76}$$

From (8.68), the low-frequency output impedance of the basic amplifier is

$$z_{oa} = 75\ \Omega \| 1\ \text{M}\Omega \| 10\ \text{k}\Omega \simeq 75\ \Omega \tag{8.77}$$

The low-frequency loop gain can be calculated from (8.70) as

$$\begin{aligned} T &= \frac{10^6 \times 10^4}{10^6 \times 10^4 + 75 \times 10^6 + 75 \times 10^4} \times 200{,}000 \times \frac{2 \times 10^6}{2 \times 10^6 + 10^6} \\ &= 133{,}333 \end{aligned} \tag{8.78}$$

The loop gain in this case is quite large. Note that a finite source resistance at the input would reduce this significantly.

The input impedance with feedback applied is found by substituting (8.76) and (8.78) in (8.67) to give

$$Z_i = \frac{666.7 \times 10^3}{133{,}333}\ \Omega = 5\ \Omega$$

The output impedance with feedback applied is found by substituting (8.77) and (8.78) in (8.69) to give

$$Z_o = \frac{75}{133{,}333}\,\Omega = 0.000563\ \Omega$$

In practice, second-order effects in the circuit may result in a larger value of Z_o.

The overall transfer function with feedback can be found approximately from (8.8) as

$$\frac{v_o}{i_i} = A \simeq \frac{1}{f} \tag{8.79}$$

Using (8.61) in (8.79) gives

$$\frac{v_o}{i_i} = A \simeq -R_F$$

Substituting for R_F we obtain

$$\frac{v_o}{i_i} = A \simeq -1\ \text{M}\Omega \tag{8.80}$$

A more exact value for A can be calculated from (8.5). Since the loop gain is very large in this case it is useful to transform (8.5) as follows

$$A = \frac{1}{f}\frac{1}{1 + \dfrac{1}{af}} \tag{8.81}$$

$$= \frac{1}{f}\frac{1}{1 + \dfrac{1}{T}} \tag{8.82}$$

Since T is so high in this example, A differs very little from $1/f$. Substituting $T = 133{,}333$ and $1/f = -1\ \text{M}\Omega$ in (8.82) we obtain

$$A = -999{,}992\ \Omega \tag{8.83}$$

For most practical purposes, (8.80) is sufficiently accurate.

8.5.2 Series-Series Feedback

Consider the series-series feedback connection of Fig. 8.12. The effect of nonideal networks can be calculated using the representation of Fig. 8.17*a*. In this case the most convenient two-port representation is the use of the open-circuit impedance parameters or z parameters because the basic amplifier and the feedback

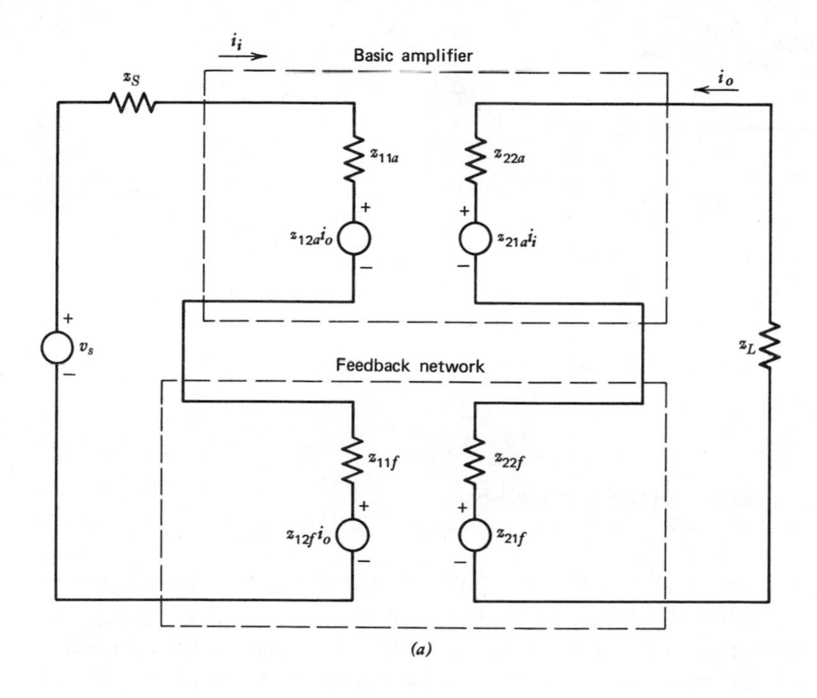

(a)

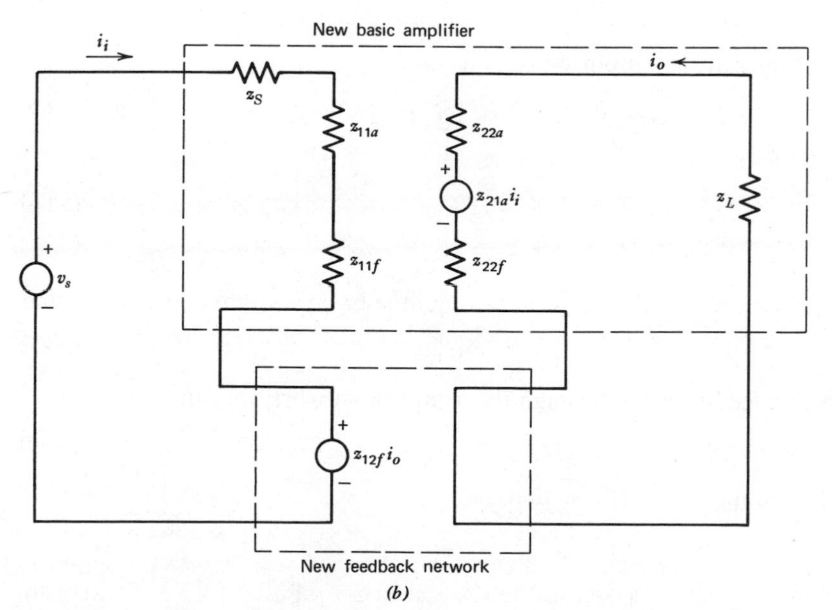

(b)

Fig. 8.17 (a) Series-series feedback configuration using the z-parameter representation. (b) Circuit of (a) redrawn with generators $z_{21f}i_i$ and $z_{12a}i_o$ omitted.

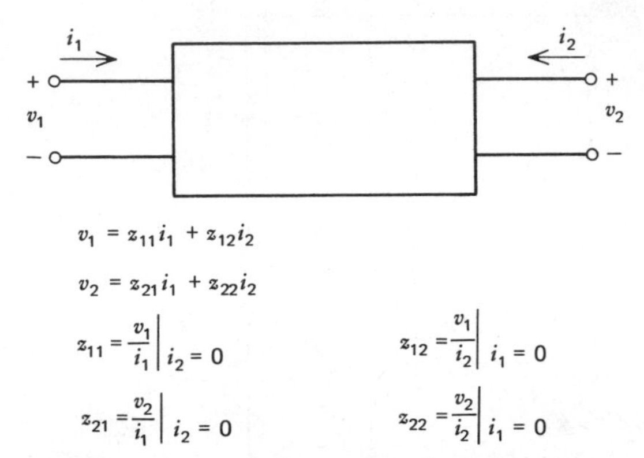

Fig. 8.18 z-parameter representation of a two-port.

network are now connected in series at input and output and thus have identical *currents* at their terminals. As shown in Fig. 8.18, the z parameters specify the network by expressing terminal voltages in terms of terminal currents and this results in simple calculations when the two networks have common terminal currents. The calculation in this case proceeds as the exact dual of that in Section 8.5.1. From Fig. 8.17, summation of voltages at the input gives

$$v_s = (z_S + z_{11a} + z_{11f})i_i + (z_{12a} + z_{12f})i_o \tag{8.84}$$

Summing voltages at the output we obtain

$$0 = (z_{21a} + z_{21f})i_i + (z_L + z_{22a} + z_{22f})i_o \tag{8.85}$$

It is useful to define

$$z_i = z_S + z_{11a} + z_{11f} \tag{8.86}$$

$$z_o = z_L + z_{22a} + z_{22f} \tag{8.87}$$

Again neglecting reverse transmission through the basic amplifier we assume that

$$|z_{12a}| \ll |z_{12f}| \tag{8.88}$$

Also neglecting feed-forward through the feedback network we can write

$$|z_{21a}| \gg |z_{21f}| \tag{8.89}$$

With these assumptions it follows that

$$\frac{i_o}{v_s} = A \simeq \frac{\dfrac{-z_{21a}}{z_i z_o}}{1 + \left(\dfrac{-z_{21a}}{z_i z_o}\right) z_{12f}} = \frac{a}{1 + af} \tag{8.90}$$

where

$$a = -\frac{z_{21a}}{z_i z_o} \tag{8.91}$$

$$f = z_{12f} \tag{8.92}$$

A circuit representation of a in (8.91) and f in (8.92) can be found by removing generators $z_{21f}i_i$ and $z_{12a}i_o$ from Fig. 8.17*a* in accord with (8.88) and (8.89). This gives the approximate representation of Fig. 8.17*b* where the new basic amplifier includes the loading effect of the original feedback network. The new feedback network is an ideal one as used in Fig. 8.12. The transfer function of the basic amplifier of Fig. 8.17*b* is the same as in (8.91), and the transfer function of the feedback network of Fig. 8.17*b* is given by (8.92). Thus Fig. 8.17*b* is a circuit representation of (8.91) and (8.92).

Since Fig. 8.17*b* has a direct correspondence with Fig. 8.12, all the results of Section 8.4.4 can now be used. The loading effect of the feedback network on the basic amplifier is included by connecting the feedback-network terminal impedances, z_{11f} and z_{22f}, in series at input and output of the basic amplifier. Terms z_{11f} and z_{22f} are defined in Fig. 8.18 and are obtained by calculating the terminal impedances of the feedback network with the other port *open circuited.* Feedback function f given by (8.92) is the *reverse transfer impedance* of the feedback network.

Consider, for example, the series-series feedback triple of Fig. 8.19*a*, which is useful as a wideband feedback amplifier. R_{E2} is usually a small resistor that samples the output current, i_o, and the resulting voltage across R_{E2} is sampled by the divider, R_F and R_{E1}, to produce a feedback voltage across R_{E1}. Usually $R_F \gg R_{E1}$ and R_{E2}.

The two-port theory derived above cannot be applied directly in this case because the basic amplifier cannot be represented by a two-port. However the techniques developed previously using two-port theory can be used with minor modification by first noting that the feedback network can be represented by a two-port as shown in Fig. 8.19*b*. One problem with this circuit is that the feedback generator, $z_{12f}i_{e3}$, is in the emitter of Q_1 and not in the input lead where it can be compared directly with v_s. This problem can be overcome by considering the small-signal equivalent of the input portion of this circuit as shown in Fig. 8.20. For this circuit,

$$v_s = i_i z_S + v_{be} + i_{e1} z_{11f} + z_{12f} i_{e3} \tag{8.93}$$

Using

$$i_{e3} = \frac{i_o}{\alpha_3} \tag{8.94}$$

in (8.93) gives

$$v_s - z_{12f}\frac{i_o}{\alpha_3} = i_i z_S + v_{be} + i_{e1} z_{11f} \tag{8.95}$$

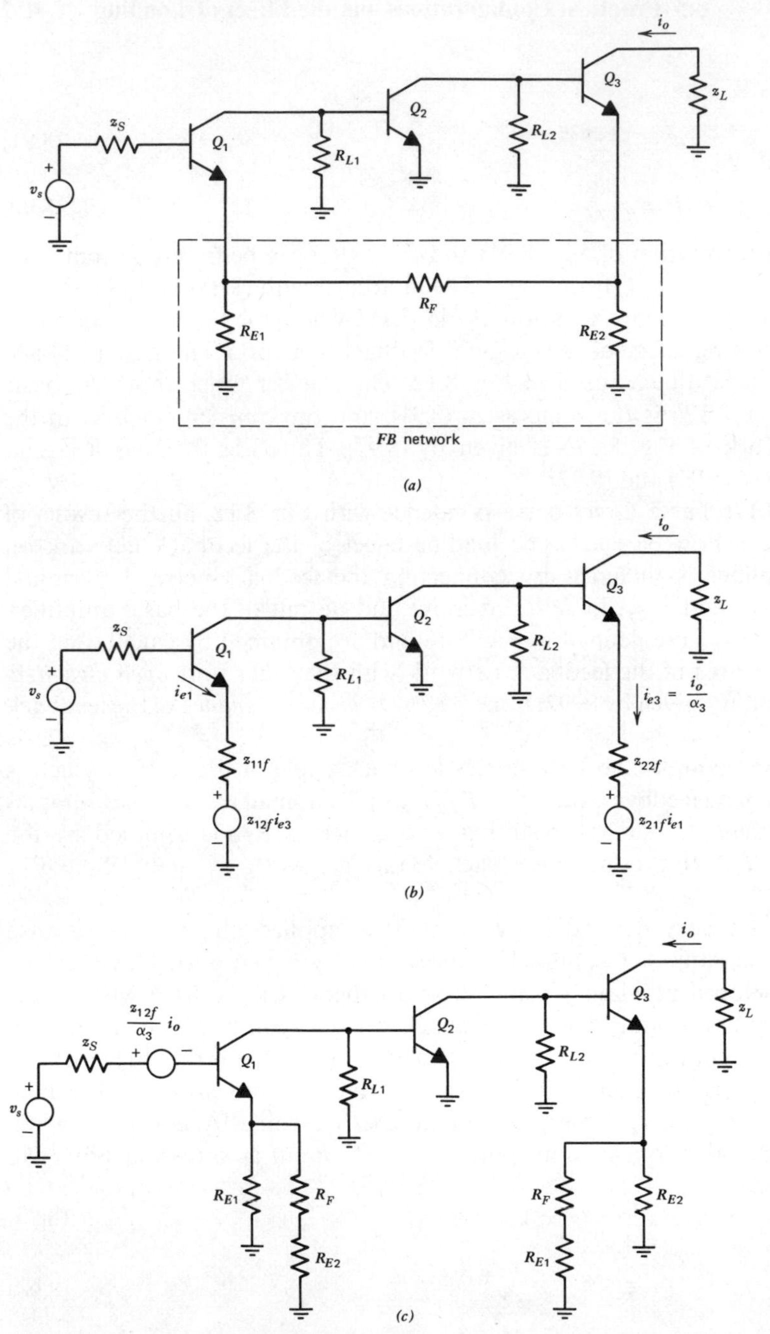

Fig. 8.19 (*a*) Series-series feedback triple. (*b*) Circuit of (*a*) redrawn using a two-port *z*-parameter representation of the feedback network. (*c*) Approximate representation of the circuit in (*b*).

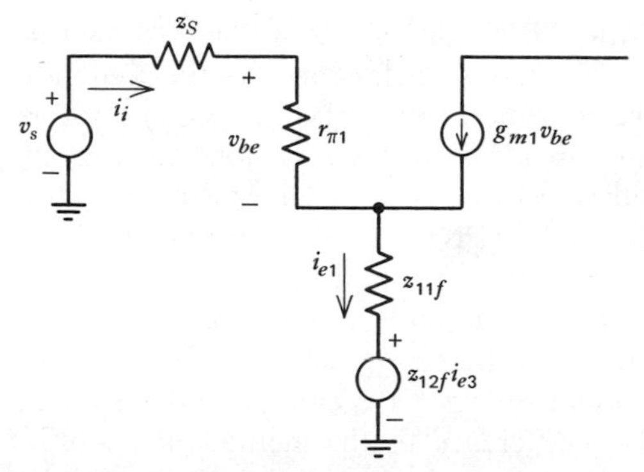

Fig. 8.20 Small-signal equivalent circuit of the input stage of the circuit in Fig. 8.19*b*.

where the quantities in these equations are *small-signal* quantities. Equation 8.95 shows that the feedback voltage generator, $z_{12f}(i_o/\alpha_3)$ can be *moved back* in series with the input lead; if this was done, exactly the same equation would result. (See Fig. 8.19*c*.) Note that the common-base current gain α_3 of Q_3 appears in this feedback expression because the output current is sampled by R_{E2} in the *emitter* of Q_3 in order to feed back a correcting signal to the input. This problem is common to most circuits employing series feedback at the output, and the α_3 of Q_3 is *outside* the feedback loop. There are many applications where this is not a problem, since $\alpha \simeq 1$. However if high gain precision is required, variations in α_3 can cause difficulties.

The z parameters of the feedback network can be determined from Fig. 8.21 as

$$z_{12f} = \left.\frac{v_1}{i_2}\right|_{i_1=0} = \frac{R_{E1}R_{E2}}{R_{E1} + R_{E2} + R_F} \tag{8.96}$$

$$z_{22f} = \left.\frac{v_2}{i_2}\right|_{i_1=0} = R_{E2}\|(R_{E1} + R_F) \tag{8.97}$$

$$z_{11f} = \left.\frac{v_1}{i_1}\right|_{i_2=0} = R_{E1}\|(R_F + R_{E2}) \tag{8.98}$$

Using (8.89), we neglect z_{21f}.

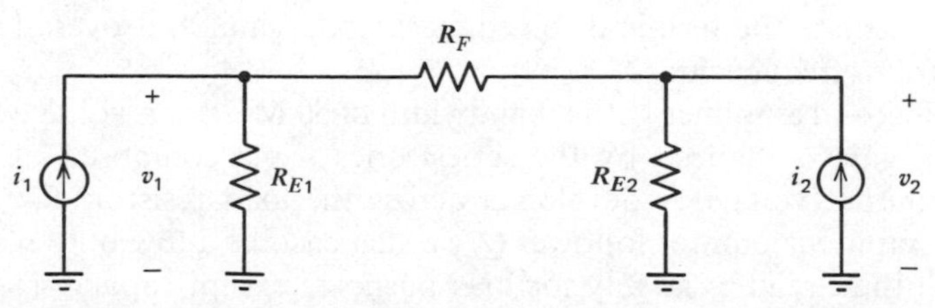

Fig. 8.21 Circuit for the calculation of the z parameters of the feedback network of the circuit in Fig. 8.19*a*.

From the above results we can redraw the circuit of Fig. 8.19*b* as shown in Fig. 8.19*c*. As in previous calculations, the signal fed forward via the feedback network (in this case $z_{21f}i_{e1}$) is neglected. The feedback voltage generator is placed in series with the input lead and an ideal differencing node then exists at the input. The effect of feedback loading on the basic amplifier is represented by the impedances in the emitters of Q_1 and Q_3. Note that this case does differ somewhat from the example of Fig. 8.17*b* in that the impedances z_{11f} and z_{22f} of the feedback network appear in *series* with the input lead in Fig. 8.17*b*, whereas in Fig. 8.19*c* these impedances appear in the emitters of Q_1 and Q_3. This is due to the fact that the basic amplifier of the circuit of Fig. 8.19*a* cannot be represented by two-port z parameters but makes no difference to the method of analysis. Since the feedback voltage generator in Fig. 8.19*c* is directly in series with the input and is proportional to i_o, a direct correspondence with Fig. 8.17*b* can be established and the results of Section 8.4.4 can be applied. There is no further need of the z parameters and by inspection we can write

$$\frac{i_o}{v_s} = A = \frac{a}{1 + af} \tag{8.99}$$

where

$$f = \frac{z_{12f}}{\alpha_3} = \frac{1}{\alpha_3} \frac{R_{E1}R_{E2}}{R_{E1} + R_{E2} + R_F} \tag{8.100}$$

and a is the transconductance of circuit of Fig. 8.19*c* with the feedback generator $[(z_{12f}/\alpha_3)i_o]$ removed.

The *input impedance* seen by v_s with feedback applied is $(1 + af) \times$ (input impedance of the basic amplifier of Fig. 8.19*c* *including* feedback loading).

The *output impedance* with feedback applied is $(1 + af)$ times the output impedance of the basic amplifier *including* feedback loading.

If the loop gain, $T = af$, is large then the gain with feedback applied is

$$A = \frac{i_o}{v_s} \simeq \frac{1}{f} = \alpha_3 \frac{R_{E1} + R_{E2} + R_F}{R_{E1}R_{E2}} \tag{8.101}$$

EXAMPLE

A commercial integrated circuit[2] based on the series-series triple is the MC 1553 shown in Fig. 8.22*a*. Calculate the terminal impedances, loop gain, and overall gain of this amplifier at low frequencies.

The MC 1553 is a wideband amplifier with a bandwidth of 50 MHz at a voltage gain of 50. The circuit gain is realized by the series-series triple composed of Q_1, Q_2, and Q_3. The output voltage is developed across the load resistor, R_C, and is then fed to the output via emitter follower Q_4, which ensures a low output impedance. The rest of the circuit is largely for bias purposes except capacitors C_P, C_F, and C_B. Capacitors C_P and C_F are small capacitors of several picofarads

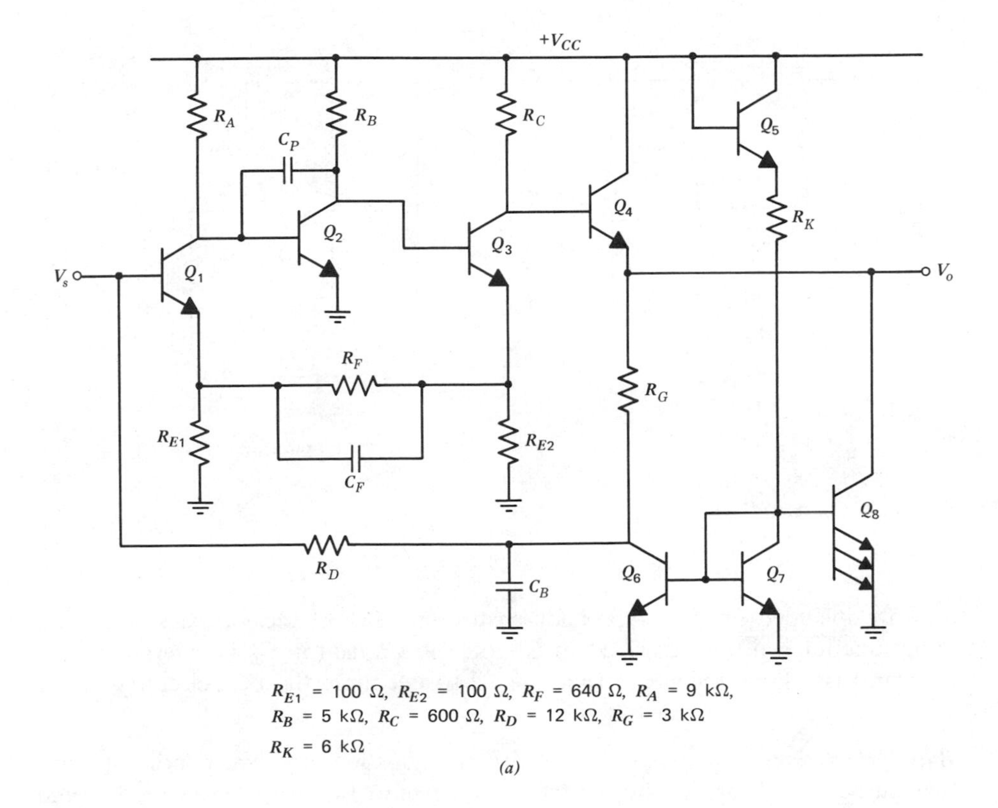

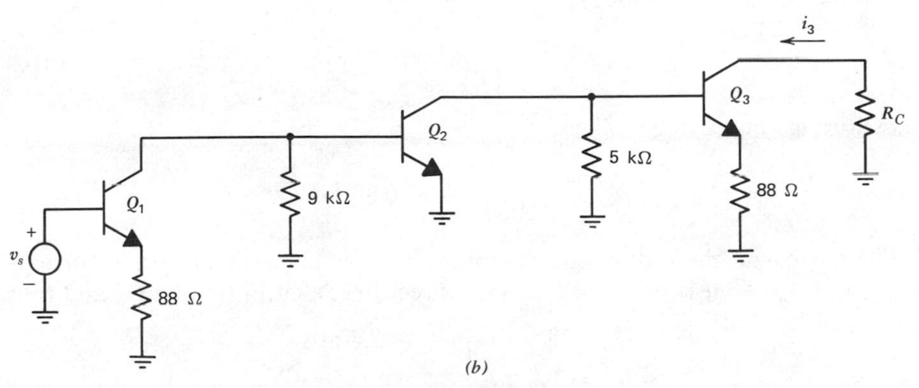

Fig. 8.22 (*a*) Circuit of the MC 1553 wideband integrated circuit. (*b*) Basic amplifier of the series-series triple in (*a*). (*c*) Small-signal equivalent circuit of the basic amplifier in (*b*).

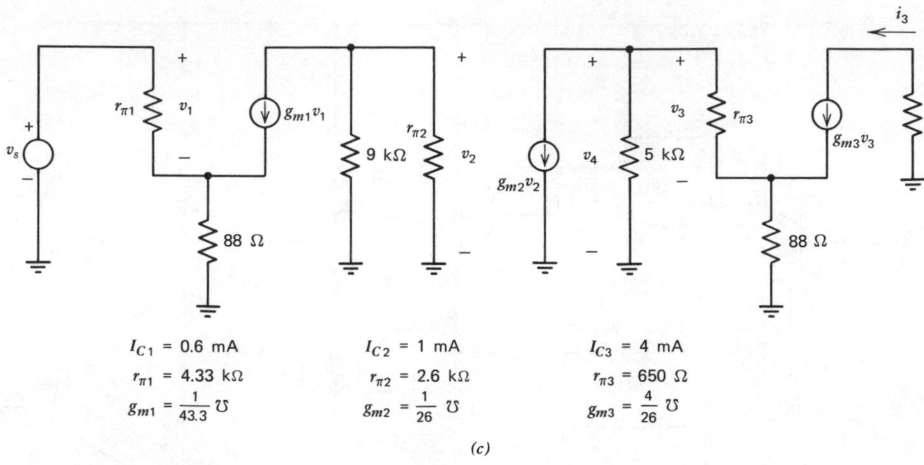

Fig. 8.22 (*Continued*)

and are included on the chip. They ensure stability of the feedback loop and their function will be described in Chapter 9. Capacitor C_B is external to the chip and is a large bypass capacitor used to decouple the bias circuitry at the signal frequencies of interest.

Bias Calculation. The analysis of the circuit begins with the bias conditions. The bias current levels are set by the reference current, I_{RK}, in the resistor, R_K, and assuming $V_{BE(\text{on})} = 0.6$ V and $V_{CC} = 6$ V we obtain

$$I_{RK} = \frac{V_{CC} - 2V_{BE(\text{on})}}{R_K} \tag{8.102}$$

Substituting data in (8.102) gives

$$I_{RK} = \frac{6 - 1.2}{6000} \text{ A} = 0.80 \text{ mA}$$

The current in the output emitter follower Q_4 is determined by the currents in Q_6 and Q_8. Transistor Q_8 has an area three times that of Q_7 and Q_6 and thus

$$I_{C8} = 3 \times 0.8 \text{ mA} = 2.4 \text{ mA}$$

$$I_{C6} = 0.8 \text{ mA}$$

where β is assumed large in these bias calculations. If the base current of Q_1 is small, all of I_{C6} and I_{C8} flow through Q_4 and

$$I_{C4} \simeq I_{C6} + I_{C8} \tag{8.103}$$

Thus

$$I_{C4} \simeq 3.2 \text{ mA}$$

Transistor Q_8 supplies most of the bias current to Q_4 and this device functions as a class A emitter-follower output stage of the type described in Section 5.2. The function of Q_6 is to allow formation of a negative-feedback bias loop for stabilization of the dc operating point, and resistor R_G is chosen to cause sufficient dc voltage drop to allow connection of R_D back to the base of Q_1. Transistors Q_1, Q_2, Q_3, and Q_4 are then connected in a negative-feedback bias loop and the dc conditions can be ascertained approximately as follows.

If we assume that Q_2 is on and conducting, the voltage at the collector of Q_1 is about 0.6 V and the voltage across R_A is 5.4 V. Thus the current through R_A is

$$I_{RA} = \frac{5.4}{R_A}$$

$$= \frac{5.4}{9000} = 0.6 \text{ mA} \tag{8.104}$$

If β is assumed high, it follows that

$$I_{C1} \simeq I_{RA} = 0.6 \text{ mA} \tag{8.105}$$

Since the voltage across R_{E1} is small, the voltage at the base of Q_1 is approximately 0.6 V and if the base current of Q_1 is small, this is also the voltage at the collector of Q_6 since any voltage across R_D will be small. The dc output voltage can be written

$$V_O = V_{C6} + I_{C6}R_G \tag{8.106}$$

Substitution of data gives

$$V_O = (0.6 + 0.8 \times 3) \text{ V} = 3 \text{ V}$$

The voltage at the base of Q_4 (collector of Q_3) is obviously V_{BE} above V_O and is thus 3.6 V. The collector current of Q_3 is

$$I_{C3} \simeq \frac{V_{CC} - V_{C3}}{R_C} \tag{8.107}$$

Substitution of parameter values gives

$$I_{C3} \simeq \frac{6 - 3.6}{600} \text{ A} = 4 \text{ mA}$$

The voltage at the base of Q_3 (collector of Q_2) is

$$V_{B3} \simeq -I_{E3}R_{E2} + V_{BE(\text{on})} \tag{8.108}$$

Thus

$$V_{B3} = V_{C2} \simeq (4 \times 0.1 + 0.6) \text{ V} = 1 \text{ V}$$

I_{C2} may be calculated from

$$I_{C2} \simeq \frac{V_{CC} - V_{C2}}{R_B} \tag{8.109}$$

and substitution of parameter values gives

$$I_{C2} \simeq \frac{6-1}{5000} \text{ A} = 1 \text{ mA}$$

Ac Calculation. The ac analysis can now proceed using the methods previously developed in this chapter. For purposes of ac analysis, the feedback triple composed of Q_1, Q_2, and Q_3 in Fig. 8.22*a* is identical to the circuit of Fig. 8.19*a*, and the results derived previously for the latter circuit are directly applicable to the triple in Fig. 8.22*a*. To obtain the voltage gain of the circuit of Fig. 8.22*a*, we simply multiply the transconductance of the triple by the load resistor, R_C, since the gain of the emitter follower Q_4 is almost exactly unity. Note that resistor R_D is assumed grounded for ac signals by the large capacitor, C_B, and thus has no influence on the ac circuit operation, except for a shunting effect at the input that will be discussed later. From (8.100) the feedback factor, f, of the series-series triple of Fig. 8.22*a* is

$$f = \frac{1}{0.99} \frac{100 \times 100}{100 + 100 + 640} \Omega = 12.0 \, \Omega \tag{8.110}$$

where $\beta = 100$ has been assumed.

If the loop gain is large, the transconductance of the triple of Fig. 8.22*a* can be calculated from (8.101) as

$$\frac{i_{o3}}{v_s} \simeq \frac{1}{f} = \frac{1}{12} \text{ A/V} \tag{8.111}$$

where i_{o3} is the small-signal collector current in Q_3 in Fig. 8.22*a*. If the input impedance of the emitter follower, Q_4, is large, the load resistance seen by Q_3 is $R_C = 600 \, \Omega$ and the voltage gain of the circuit is

$$\frac{v_o}{v_s} = -\frac{i_{o3}}{v_s} \times R_C \tag{8.112}$$

Substituting (8.111) in (8.112) gives

$$\frac{v_o}{v_s} = -50.0 \tag{8.113}$$

Consider now the loop gain of the circuit of Fig. 8.22*a*. This can be calculated by using the basic-amplifier representation of Fig. 8.19*c* to calculate the forward gain, *a*. Fig. 8.19*c* is redrawn in Fig. 8.22*b* using data from this example, assuming that $z_S = 0$ and omitting the feedback generator. The small-signal, low-frequency equivalent circuit is shown in Fig. 8.22*c* assuming $\beta = 100$ and it is a straightforward calculation to show that the gain of the basic amplifier is

$$a = \frac{i_3}{v_s} = 20.3 \text{ A/V} \tag{8.114}$$

Combination of (8.110) and (8.114) gives

$$T = af = 12 \times 20.3 = 243.6 \tag{8.115}$$

The transconductance of the triple can now be calculated more accurately from (8.99) as

$$\frac{i_{o3}}{v_s} = \frac{a}{1 + T} = \frac{20.3}{244.6}\ \text{A/V} = 0.083\ \text{A/V} \tag{8.116}$$

Substitution of (8.116) in (8.112) gives for the overall voltage gain

$$\frac{v_o}{v_s} = -\frac{i_{o3}}{v_s} R_C = 0.083 \times 600 = 49.8 \tag{8.117}$$

This is close to the approximate value given by (8.113).

The input resistance of the basic amplifier is readily determined from Fig. 8.22*c* to be

$$r_{ia} = 13.2\ \text{k}\Omega \tag{8.118}$$

The input resistance when feedback is applied is

$$R_i = r_{ia}(1 + T) \tag{8.119}$$

Substituting (8.118) and (8.115) in (8.119) gives

$$R_i = 13.2 \times 244.6\ \text{k}\Omega = 3.23\ \text{M}\Omega \tag{8.120}$$

As expected, series feedback at the input results in a high input resistance. In this example, however, the bias resistor, R_D, directly shunts the input for ac signals and is *outside* the feedback loop. Since $R_D = 12\ \text{k}\Omega$ and is much less than R_i, the resistor, R_D, determines the input resistance for this circuit.

Finally, the output resistance of the circuit is of some interest. The output resistance of the triple can be calculated from Fig. 8.22*c* by including output resistance r_o in the model for Q_3. The resistance obtained is then multiplied by $(1 + T)$ and the resulting value is much greater than the collector load resistor of Q_3, which is $R_C = 600\ \Omega$. The output resistance of the full circuit is thus essentially the output resistance of emitter follower Q_4 fed from a 600-Ω source resistance, and this is

$$R_o = \frac{1}{g_{m4}} + \frac{R_C}{\beta_4} = \left(\frac{26}{3.2} + \frac{600}{100}\right)\Omega = 14\ \Omega \tag{8.121}$$

8.5.3 Series-Shunt Feedback

Series-shunt feedback is shown schematically in Fig. 8.4 and the basic amplifier and the feedback network have the same input current and the same output voltage. A two-port representation that uses input current and output voltage as

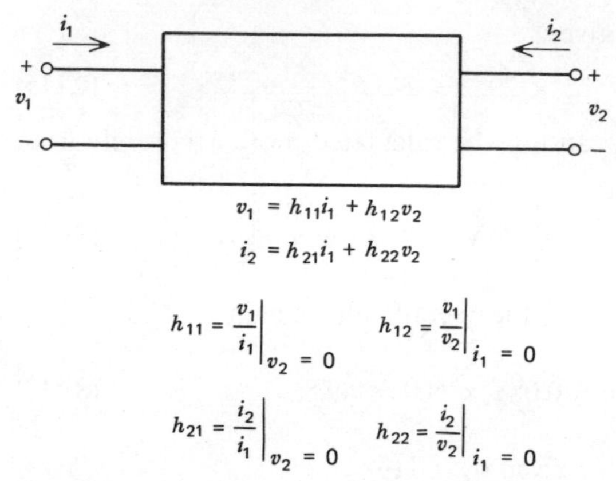

Fig. 8.23 *h*-parameter representation of a two-port.

the independent variables is the hybrid *h*-parameter representation shown in Fig. 8.23. The *h* parameters can be used to represent nonideal circuits in a series-shunt feedback as shown in Fig. 8.24*a*. Summation of voltages at the input of this figure gives

$$v_s = (z_S + h_{11a} + h_{11f})i_i + (h_{12a} + h_{12f})v_o \tag{8.122}$$

Summing currents at the output

$$0 = (h_{21a} + h_{21f})i_i + (y_L + h_{22a} + h_{22f})v_o \tag{8.123}$$

We now define

$$z_i = z_S + h_{11a} + h_{11f} \tag{8.124}$$

$$y_o = y_L + h_{22a} + h_{22f} \tag{8.125}$$

and make the same assumptions as in previous examples

$$|h_{12a}| \ll |h_{12f}| \tag{8.126}$$

$$|h_{21a}| \gg |h_{21f}| \tag{8.127}$$

It can then be shown that

$$\frac{v_o}{v_s} = A \simeq \frac{-\dfrac{h_{21a}}{z_i y_o}}{1 + \left(-\dfrac{h_{21a}}{z_i y_o}\right) h_{12f}} = \frac{a}{1 + af} \tag{8.128}$$

where

$$a = -\frac{h_{21a}}{z_i y_o} \tag{8.129}$$

$$f = h_{12f} \tag{8.130}$$

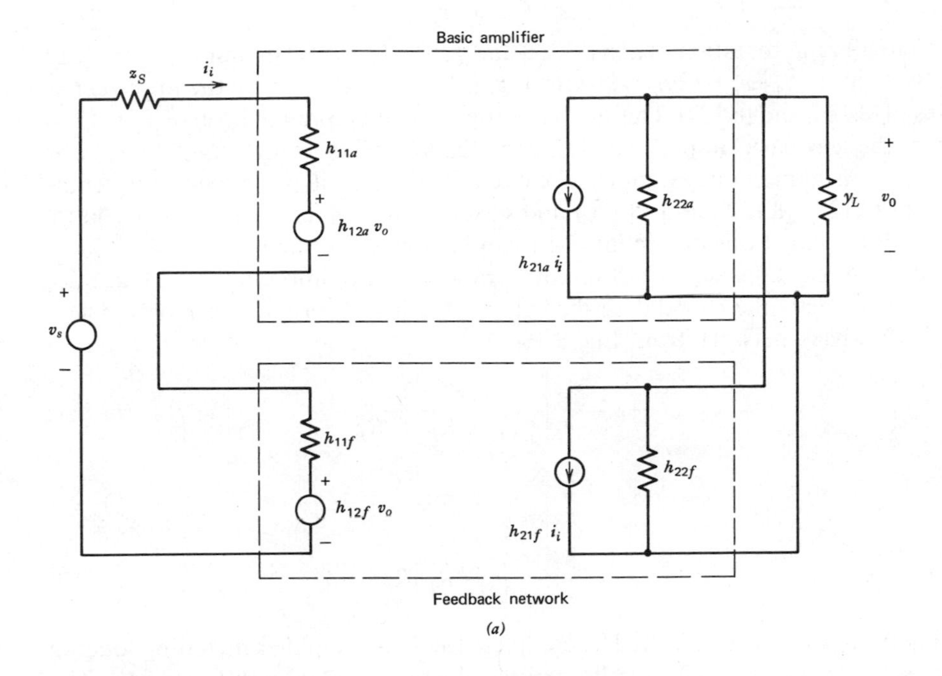

(a)

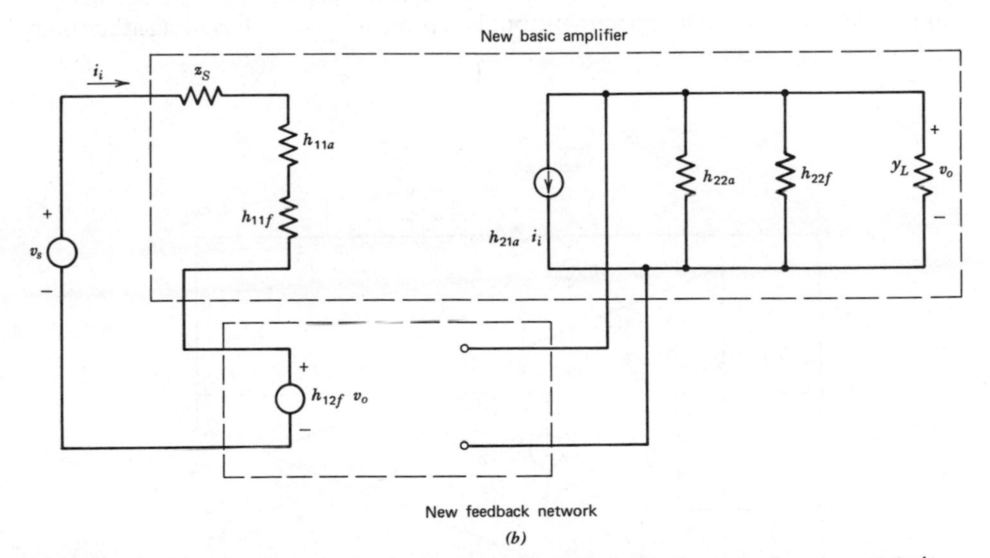

(b)

Fig. 8.24 (*a*) Series-shunt feedback configuration using the *h*-parameter representation. (*b*) Circuit of (*a*) redrawn with generators $h_{21f}i_i$ and $h_{12a}v_o$ omitted.

A circuit representation of a in (8.129) and f in (8.130) can be found by removing the generators $h_{12a}v_o$ and $h_{21f}i_i$ from Fig. 8.24a as suggested by the approximations of (8.126) and (8.127). This gives the approximate representation of Fig. 8.24b where the new basic amplifier includes the loading effect of the original feedback network. As in previous examples, the circuit of Fig. 8.24b is a circuit representation of (8.128), (8.129), and (8.130) and has the form of an ideal feedback loop. Thus all the equations of Section 8.4.1 can be applied to the circuit.

For example, consider the common series-shunt op amp circuit of Fig. 8.25, which fits exactly the model described above. We first determine the h parameters of the feedback network from Fig. 8.26,

$$h_{22f} = \left.\frac{i_2}{v_2}\right|_{i_1=0} = \frac{1}{R_F + R_E} \tag{8.131}$$

$$h_{12f} = \left.\frac{v_1}{v_2}\right|_{i_1=0} = \frac{R_E}{R_E + R_F} \tag{8.132}$$

$$h_{11f} = \left.\frac{v_1}{i_1}\right|_{v_2=0} = R_E \| R_F \tag{8.133}$$

Using (8.127), we neglect h_{21f}. The complete feedback amplifier including loading effects is shown in Fig. 8.27 and has a direct correspondence with Fig. 8.24b. (The only difference is that the op amp output is represented by a Thévenin rather than a Norton equivalent.)

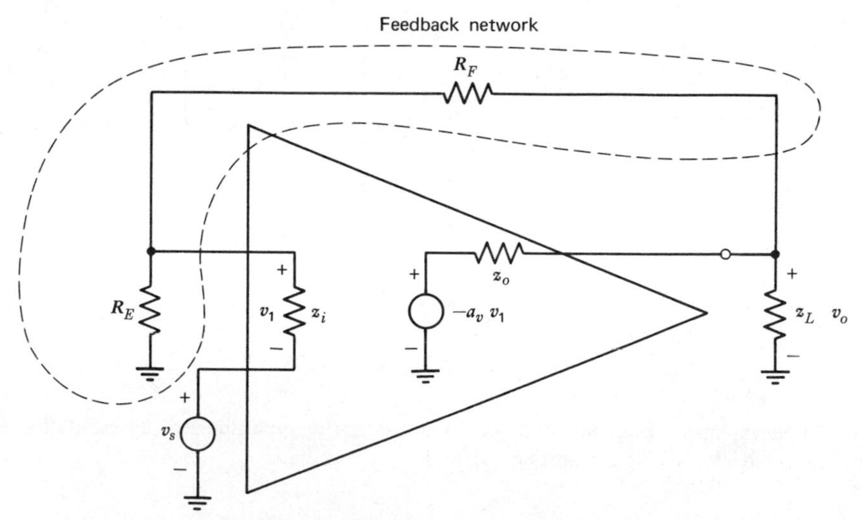

Fig. 8.25 Series-shunt feedback circuit using an op amp as the gain element.

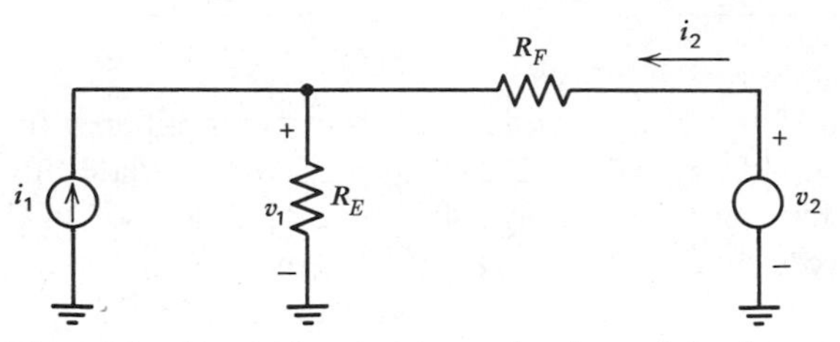

Fig. 8.26 Circuit for the determination of the h parameters of the feedback network in Fig. 8.25.

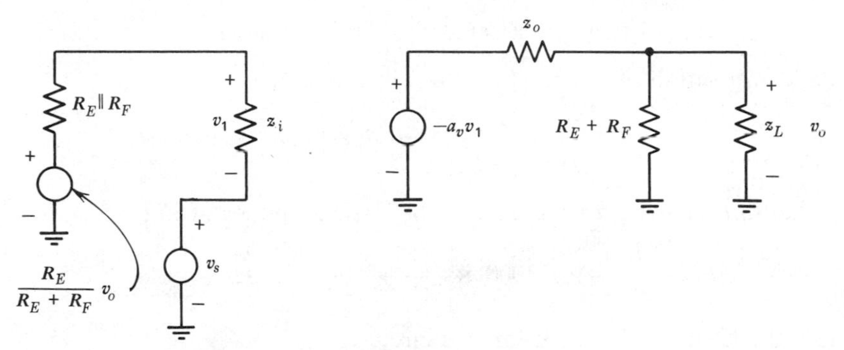

Fig. 8.27 Equivalent circuit for Fig. 8.25.

The gain, a, of the basic amplifier can be calculated from Fig. 8.27 by initially disregarding the feedback generator to give

$$a = \frac{z_i}{z_i + R} a_v \frac{z_{LX}}{z_{LX} + z_o} \tag{8.134}$$

where

$$R = R_E \| R_F \tag{8.135}$$

$$z_{LX} = z_L \| (R_E + R_F) \tag{8.136}$$

Also

$$f = \frac{R_E}{R_E + R_F} \tag{8.137}$$

Thus the overall gain of the feedback circuit is

$$A = \frac{v_o}{v_s} = \frac{a}{1 + af} \tag{8.138}$$

and A can be evaluated using (8.134) and (8.137).

EXAMPLE

Assume that the circuit of Fig. 8.25 is realized, using a differential amplifier with low-frequency parameters $z_i = 100\ \text{k}\Omega$, $z_o = 10\ \text{k}\Omega$, and $a_v = 3000$. Calculate the input impedance of the feedback amplifier at low frequencies if $R_E = 5\ \text{k}\Omega$, $R_F = 20\ \text{k}\Omega$, and $z_L = 10\ \text{k}\Omega$. Note that z_o in this case is not small as is usually the case for an op amp, and this situation can arise in some applications.

This problem is best approached by first calculating the input impedance of the basic amplifier and then multiplying by $(1 + T)$ as indicated by (8.25) to calculate the input impedance of the feedback amplifier. By inspection from Fig. 8.27 the input impedance of the basic amplifier is

$$z_{ia} = z_i + R_E \| R_F = (100 + 5\|20)\ \text{k}\Omega = 104\ \text{k}\Omega$$

The parallel combination of z_L and $(R_F + R_E)$ in Fig. 8.27 is

$$z_{LX} = \frac{10 \times 25}{35} = 7.14\ \text{k}\Omega$$

Substitution ın (8.134) gives, for the gain of the basic amplifier of Fig. 8.27,

$$a = \frac{100}{100 + 4} \times 3000 \times \frac{7.14}{7.14 + 10} = 1202$$

From (8.132) the feedback factor, f, for this circuit is

$$f = \frac{5}{5 + 20} = 0.2$$

and thus the loop gain is

$$T = af = 1202 \times 0.2 = 240$$

The input impedance of the feedback amplifier is thus

$$Z_i = z_{ia}(1 + T) = 104 \times 241\ \text{k}\Omega = 25\ \text{M}\Omega$$

In this example the loading effect produced by $(R_F + R_E)$ on the output has a significant effect on the gain, a, of the basic amplifier and thus on the input impedance of the feedback amplifier.

As another example of a series-shunt feedback circuit, consider the series-series triple of Fig. 8.19*a* but assume the output signal is taken as the voltage at the emitter of Q_3, as shown in Fig. 8.28. This is an example of how the same circuit can realize two different feedback functions if the output is taken from different nodes. As in the case of Fig. 8.19*a*, the basic amplifier of Fig. 8.28 cannot be represented as a two-port. However, the feedback network *can* be represented as a two-port and the appropriate parameters are the h parameters as shown in Fig. 8.29. The h parameters for this feedback network are given by (8.131), (8.132), and (8.133) and h_{21f} is neglected. The analysis of Fig. 8.29 then proceeds in the

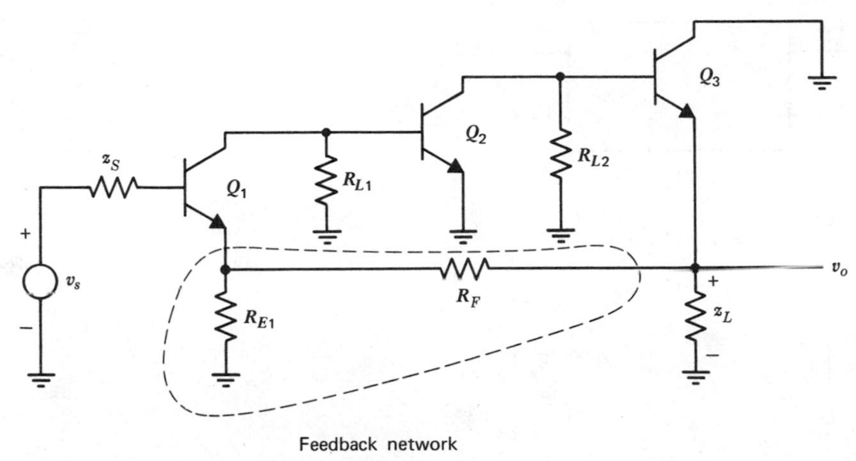

Fig. 8.28 Series-shunt feedback circuit.

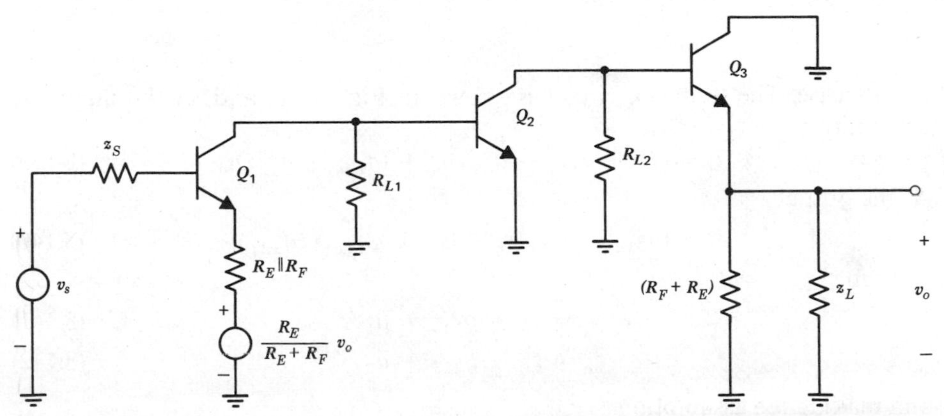

Fig. 8.29 Circuit equivalent to that of Fig. 8.28 using a two-port representation of the feedback network.

usual manner, but one last point is worth noting in this circuit. If the circuits of Fig. 8.29 and Fig. 8.19*c* are compared (assuming that R_{E2} in Fig. 8.19*c* equals z_L in Fig. 8.29 to make them identical), a minor difference is apparent. The resistance in the emitter of Q_1 is $R_{E1} \| (R_F + R_{E2})$ in Fig. 8.19*c* but is $R_{E1} \| R_F$ in Fig. 8.29. This difference is insignificant and is a measure of the level of approximation involved in this analysis.

8.5.4 Shunt-Series Feedback

Shunt-series feedback is shown schematically in Fig. 8.11. In this case the basic amplifier and the feedback network have common input voltages and output currents, and hybrid g parameters as defined in Fig. 8.30 are best suited for use

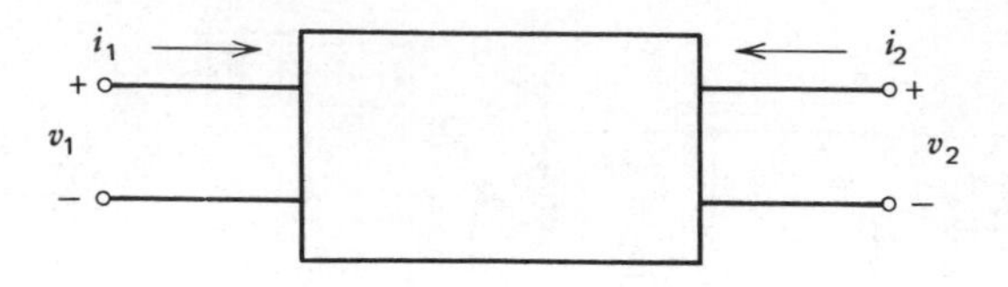

$$i_1 = g_{11}v_1 + g_{12}i_2$$

$$v_2 = g_{21}v_1 + g_{22}i_2$$

$$g_{11} = \left.\frac{i_1}{v_1}\right|_{i_2 = 0} \qquad g_{12} = \left.\frac{i_1}{i_2}\right|_{v_1 = 0}$$

$$g_{21} = \left.\frac{v_2}{v_1}\right|_{i_2 = 0} \qquad g_{22} = \left.\frac{v_2}{i_2}\right|_{v_1 = 0}$$

Fig. 8.30 *g*-parameter representation of a two-port.

in this case. The feedback circuit is shown in Fig. 8.31*a*, and, at the input, we find that

$$i_s = (y_S + g_{11a} + g_{11f})v_i + (g_{12a} + g_{12f})i_o \tag{8.139}$$

At the output

$$0 = (g_{21a} + g_{21f})v_i + (z_L + g_{22a} + g_{22f})i_o \tag{8.140}$$

Defining

$$y_i = y_S + g_{11a} + g_{11f} \tag{8.141}$$

$$z_o = z_L + g_{22a} + g_{22f} \tag{8.142}$$

and making the assumptions

$$|g_{12a}| \ll |g_{12f}| \tag{8.143}$$

$$|g_{21a}| \gg |g_{21f}| \tag{8.144}$$

we find

$$\frac{i_o}{i_s} = A \simeq \frac{-\dfrac{g_{21a}}{y_i z_o}}{1 + \left(-\dfrac{g_{21a}}{y_i z_o}\right) g_{12f}} = \frac{a}{1 + af} \tag{8.145}$$

where

$$a = -\frac{g_{21a}}{y_i z_o} \tag{8.146}$$

$$f = g_{12f} \tag{8.147}$$

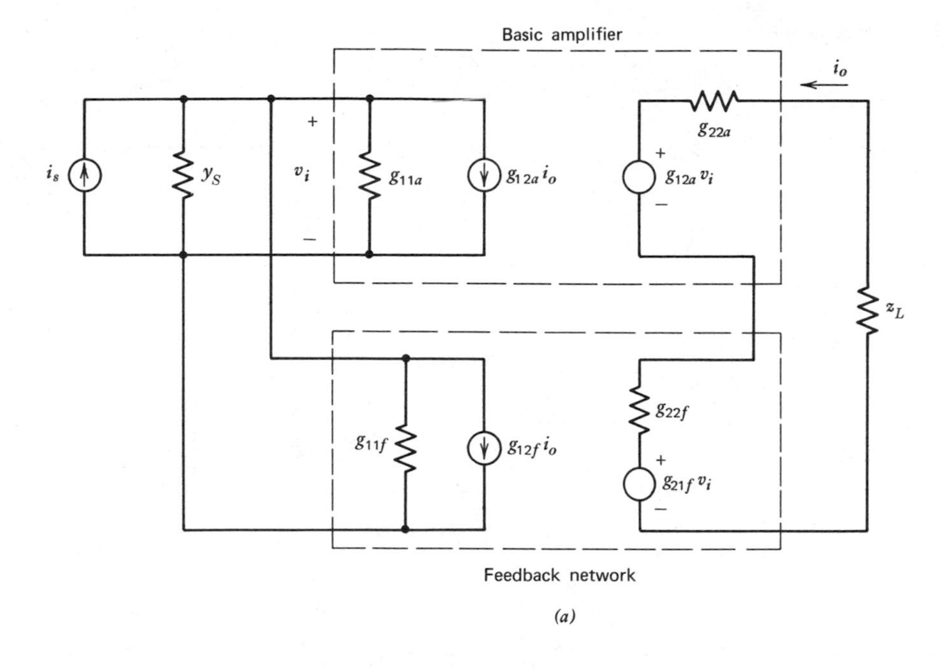

(a)

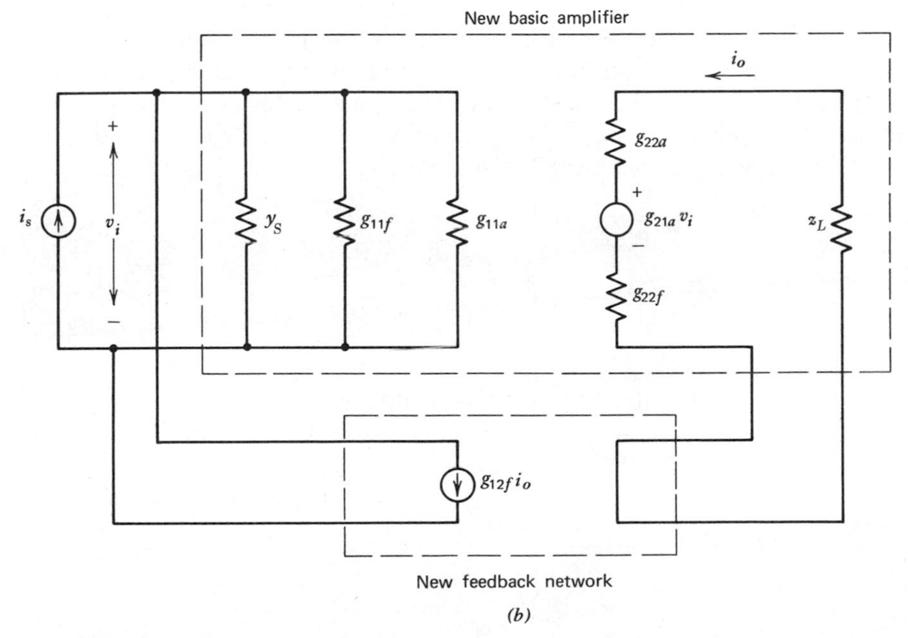

(b)

Fig. 8.31 (a) Shunt-series feedback configuration using the g-parameter representation. (b) Circuit of (a) redrawn with generators $g_{21f}v_i$ and $g_{12a}i_o$ omitted.

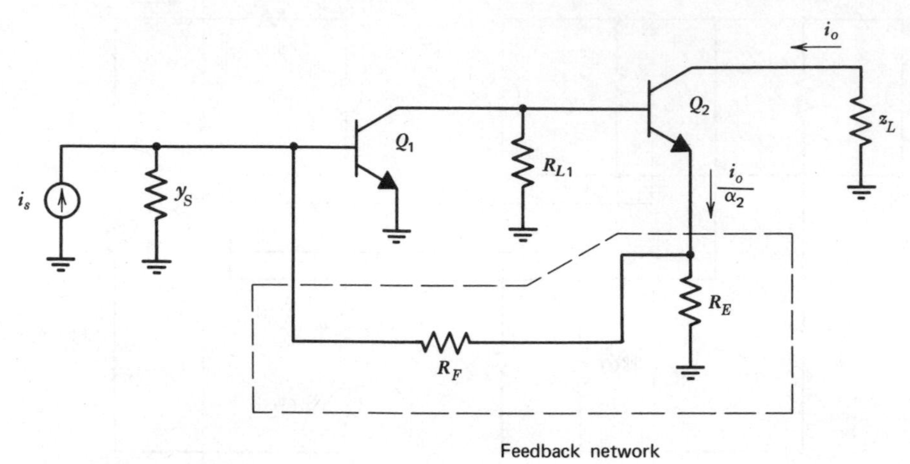

Fig. 8.32 Current feedback pair.

Following the procedure for the previous examples, we can find a circuit representation for this case by eliminating the generators, $g_{21f}v_i$ and $g_{12a}i_o$, to obtain the approximate representation of Fig. 8.31*b*. Since this has the form of an ideal feedback circuit, all the results of Section 8.4.3 may now be used.

A common shunt-series feedback amplifier is the current-feedback pair of Fig. 8.32. Since the basic amplifier of Fig. 8.32 cannot be represented by a two-port, the representation of Fig. 8.31*b* cannot be used directly. However, as in previous examples, the feedback network *can* be represented by a two-port and the g parameters can be calculated using Fig. 8.33 to give

$$g_{11f} = \left.\frac{i_1}{v_1}\right|_{i_2=0} = \frac{1}{R_F + R_E} \tag{8.148}$$

$$g_{12f} = \left.\frac{i_1}{i_2}\right|_{v_1=0} = -\frac{R_E}{R_E + R_F} \tag{8.149}$$

$$g_{22f} = \left.\frac{v_2}{i_2}\right|_{v_1=0} = R_E \| R_F \tag{8.150}$$

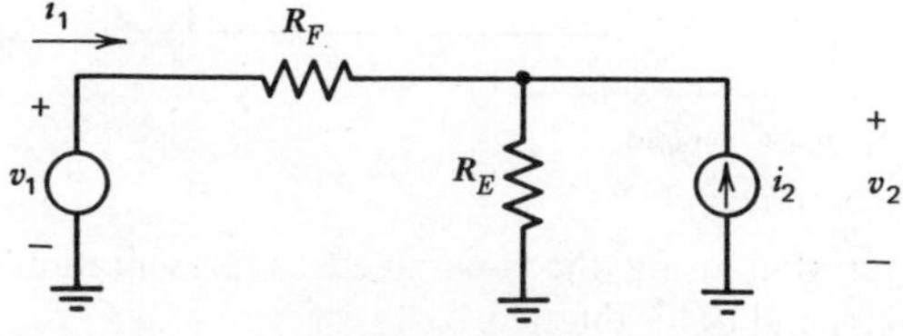

Fig. 8.33 Circuit for the calculation of the g parameters of the feedback network in Fig. 8.32.

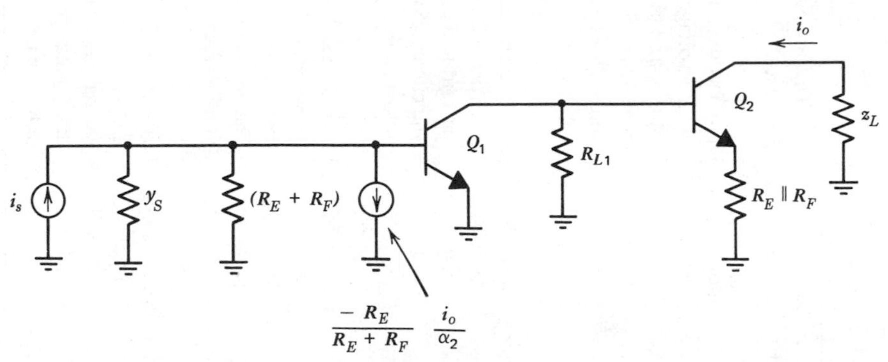

Fig. 8.34 Circuit equivalent to that of Fig. 8.32 using a two-port representation of the feedback network.

Using (8.144), we neglect g_{21f}. Assuming that $g_{21f}v_i$ is negligible, we can redraw the circuit of Fig. 8.32 as shown in Fig. 8.34. This circuit has an ideal input differencing node and the feedback function can be identified as

$$f = -\frac{R_E}{R_E + R_F}\frac{1}{\alpha_2} \tag{8.151}$$

The gain, a, of the basic amplifier is determined by calculating the current gain of the circuit of Fig. 8.34 with the feedback generator removed. The overall current gain with feedback applied can then be calculated from (8.40).

8.5.5 Summary

The results derived above regarding practical feedback circuits and the effect of feedback loading can be summarized as follows.

First, input and output variables must be identified and the feedback identified as shunt or series at input and output.

The feedback function, f, is found by the following procedure. If the feedback is *shunt* at the *input*, short the input feedback node to ground and calculate the feedback *current*. If the feedback is *series* at the *input*, *open circuit* the input feedback node and calculate the feedback *voltage*. In both these cases, if the feedback is *shunt* at the *output*, drive the feedback network with a *voltage* source. If the feedback is *series* at the *output*, drive with a current source.

The effect of feedback loading on the basic amplifier is found as follows. If the feedback is *shunt* at the *input*, *short* the input feedback node to ground to find the feedback loading on the *output*. If the feedback is *series* at the input, *open* circuit the input feedback node to calculate *output* feedback loading. Similarly, if the feedback is *shunt* or *series* at the output, then *short* or *open* the output feedback node to calculate feedback loading on the input.

These results along with other information are summarized in Table 8.1.

Table 8.1

Feedback Connection	Two-Port Parameter Representation	Output Variable	Input Variable	Transfer Function Stabilized	Z_i	Z_o	To Calculate Feedback Loading: At Input	To Calculate Feedback Loading: At Output	To Calculate Feedback Function f
Shunt-shunt	y	v_o	i_s	$\frac{v_o}{i_s}$ Transresistance	Low	Low	Short output feedback node	Short input feedback node	Drive feedback network with a voltage at its output and calculate the current flowing into a short at its input
Shunt-series	g	i_o	i_s	$\frac{i_o}{i_s}$ Current gain	Low	High	Open circuit output feedback node	Short input feedback node	Drive the feedback network with a current and calculate the current flowing into a short
Series-shunt	h	v_o	v_s	$\frac{v_o}{v_s}$ Voltage gain	High	Low	Short output feedback node	Open circuit input feedback node	Drive the feedback network with a voltage and calculate the voltage produced into an open circuit
Series-series	z	i_o	v_s	$\frac{i_o}{v_s}$ Transconductance	High	High	Open circuit output feedback node	Open circuit input feedback node	Drive the feedback network with a current and calculate the voltage produced into an open circuit

8.6 SINGLE-STAGE FEEDBACK

The considerations of feedback circuits in this chapter have been mainly directed toward the general case of circuits with multiple stages in the basic amplifier. However, in dealing with some of these circuits (such as the series-series triple of Fig. 8.19*a*), equivalent circuits were derived in which one or more stages contained an emitter resistor. (See Fig. 8.19*c*.) Such a stage represents *in itself* a feedback circuit as will be shown below. Thus the circuit of Fig. 8.19*c* contains feedback loops within a feedback loop, and this has a direct effect on the amplifier performance. For example, the emitter resistor of Q_3 in Fig. 8.19*c* has a linearizing effect on Q_3, and so does the overall feedback when the loop is closed. It is thus important to calculate the effects of both *local* and *overall* feedback loops. The term *local* feedback is often used instead of *single-stage* feedback. Local feedback is often used in isolated stages as well as being found inside overall feedback loops. In this section, the low-frequency characteristics of the two basic single-stage feedback circuits will be analyzed.

8.6.1 Local Series Feedback or Emitter Degeneration

A local series-feedback stage is shown in Fig. 8.35*a*. This can be recognized as a degenerate series-series feedback configuration as described in Sections 8.4.4 and 8.5.2. Instead of attempting to use the generalized forms of those sections, we can perform a more straightforward calculation in this simple case by working directly from the low-frequency, small-signal equivalent circuit shown in Fig. 8.35*b*. For simplicity, source impedance is assumed zero but can be lumped in with r_b if desired. If a Thévenin equivalent across the plane *AA* is calculated, Fig. 8.35*b* can be redrawn as in Fig. 8.35*c*. The existence of an input differencing node is apparent and quantity $i_o R_E$ is identified as the feedback voltage, v_{fb}. Writing equations for Fig. 8.35*c* we find

$$v_1 = \frac{r_\pi}{r_\pi + r_b + R_E}(v_s - v_{fb}) \tag{8.152}$$

$$i_o = g_m v_1 \tag{8.153}$$

$$v_{fb} = i_o R_E \tag{8.154}$$

These equations are of the form of the ideal feedback equations where v_1 is the error voltage, i_o is the output signal, and v_{fb} is the feedback signal. From (8.152) and (8.153)

$$i_o = \frac{g_m r_\pi}{r_\pi + r_b + R_E}(v_s - v_{fb}) \tag{8.155}$$

and thus we can identify

$$a = \frac{g_m r_\pi}{r_\pi + r_b + R_E} = \frac{g_m}{1 + \dfrac{r_b + R_E}{r_\pi}} \tag{8.156}$$

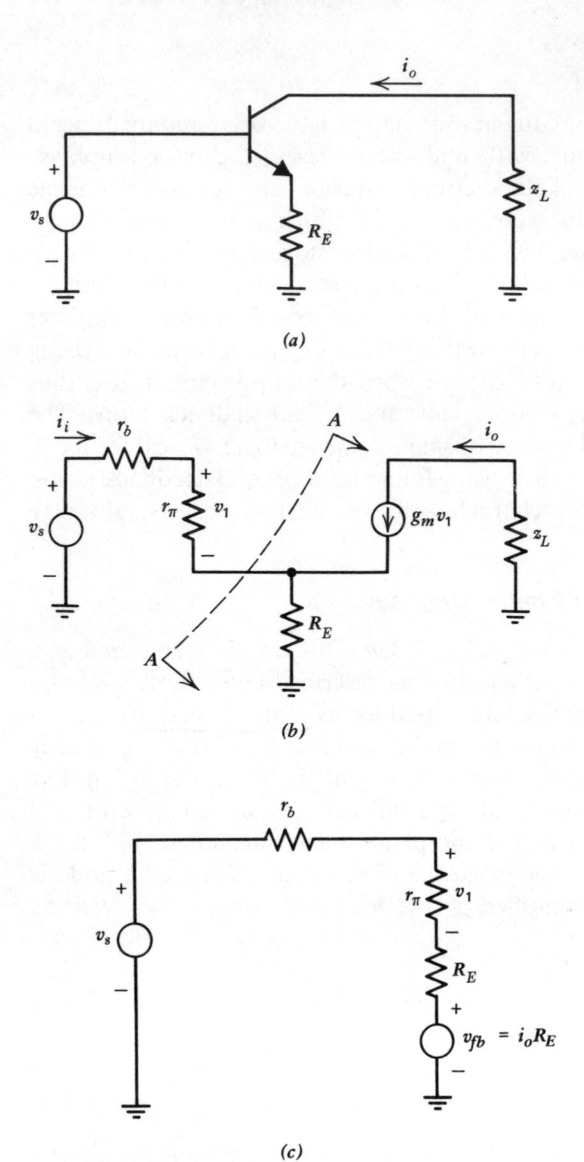

Fig. 8.35 (*a*) Single-stage series feedback circuit. (*b*) Low-frequency equivalent circuit of (*a*). (*c*) Circuit equivalent to that in (*b*).

From (8.154)

$$f = R_E \tag{8.157}$$

Thus for the complete circuit

$$\frac{i_o}{v_s} = A = \frac{a}{1 + af} = \frac{1}{R_E} \frac{1}{1 + \dfrac{1}{R_E}\left(\dfrac{1}{g_m} + \dfrac{r_b + R_E}{\beta_0}\right)} \tag{8.158}$$

and $A \simeq 1/R_E$ for large loop gain.

The loop gain is given by $T = af$ and thus

$$T = \frac{g_m R_E}{1 + \dfrac{r_b + R_E}{r_\pi}} \tag{8.159}$$

If $(r_b + R_E) \ll r_\pi$ we find

$$T \simeq g_m R_E \tag{8.160}$$

The input resistance of the circuit is given by

$$\begin{aligned} \text{Input resistance} &= (1 + T) \times (\text{input resistance with } v_{fb} = 0) \\ &= (1 + T)(r_b + r_\pi + R_E) \end{aligned} \tag{8.161}$$

Using (8.159) in (8.161) gives

$$\text{Input resistance} = r_b + R_E + r_\pi(1 + g_m R_E) \tag{8.162}$$

$$= r_b + r_\pi + (\beta_0 + 1)R_E \tag{8.163}$$

We can also show that if the output resistance, r_o, of the transistor is included, the output resistance of the circuit is given by

$$\text{Output resistance} \simeq r_o \left(1 + \frac{g_m R_E}{1 + \dfrac{r_b + R_E}{r_\pi}}\right) \tag{8.164}$$

Both input and output resistance are increased by the application of emitter feedback, as expected.

EXAMPLE

Calculate the low-frequency parameters of the series-feedback stage represented by Q_3 in Fig. 8.22*b*. The relevant parameters are as follows.

$$R_E = 88\ \Omega \qquad r_\pi = 650\ \Omega \qquad g_m = \frac{4}{26}\text{ A/V} \qquad \beta_0 = 100 \qquad r_b = 0$$

The loading produced by Q_3 at the collector of Q_2 is given by the input resistance expression of (8.163) and is

$$R_{i3} = (650 + 101 \times 88)\ \Omega = 9.54\ \text{k}\Omega \tag{8.165}$$

The output resistance seen at the collector of Q_3 can be calculated from (8.164) using $r_b = 5$ kΩ to allow for the finite source resistance in Fig. 8.22*b*. If we assume that $r_o = 25$ kΩ at $I_C = 4$ mA for Q_3, then, from (8.164),

$$R_{o3} = 25\left(1 + \frac{\dfrac{4}{26}88}{1 + \dfrac{5088}{650}}\right)\text{k}\Omega = 63\ \text{k}\Omega \tag{8.166}$$

In the example of Fig. 8.22*b*, the above output resistance would be multiplied by the loop gain of the series-series triple.

Finally, when ac voltage v_4 at the collector of Q_2 in Fig. 8.22*c* is determined, output current i_3 in Q_3 can be calculated using (8.158).

$$\frac{i_3}{v_4} = \frac{1}{88}\,\frac{1}{1 + \dfrac{1}{88}\left(\dfrac{26}{4} + \dfrac{88}{100}\right)}\ \text{A/V} = \frac{1}{95.4}\ \text{A/V} \tag{8.167}$$

Note that since the voltage v_4 exists at the base of Q_3, the effective source resistance in the above calculation is zero.

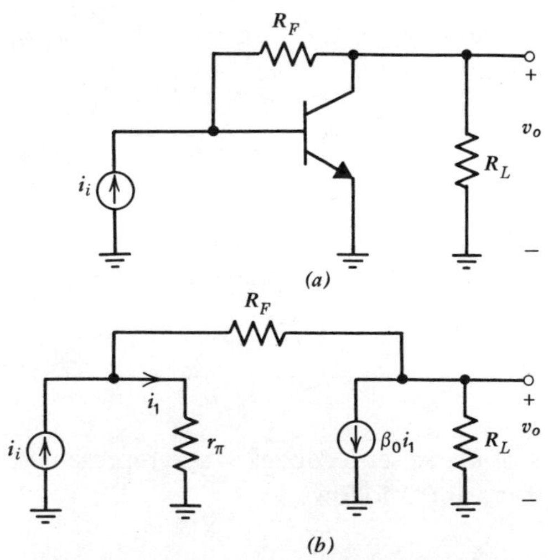

Fig. 8.36 (*a*) Single-stage shunt feedback circuit. (*b*) Low-frequency equivalent circuit of (*a*).

8.6.2 Local Shunt Feedback

A local shunt-feedback stage is shown in Fig. 8.36*a*, and a low-frequency, small-signal circuit is shown in Fig. 8.36*b* where base resistance and source resistance are omitted for simplicity. As in the case of local series feedback, a straightforward calculation can be done directly from the equivalent circuit. The loop gain of Fig. 8.36*b* is easily calculated by inserting a test current source in place of the controlled generator and calculating the loop transmission. The result is

$$T = \beta_0 \frac{R_L}{R_L + R_F + r_\pi} \tag{8.168}$$

Analysis of Fig. 8.36*b* shows that the transfer function with feedback applied is

$$A = \frac{v_o}{i_i}$$

$$\simeq \frac{-R_F}{1 + \dfrac{R_L + R_F + r_\pi + R_L r_\pi / R_F}{\beta_0 R_L}} \tag{8.169}$$

$$\simeq \frac{-R_F}{1 + \dfrac{1}{T}} \tag{8.169a}$$

For large loop gain (8.169a) becomes

$$\frac{v_o}{i_i} \simeq -R_F \tag{8.170}$$

The output resistance is best calculated directly from Fig. 8.36*b* and is

$$R_o = \frac{R_F + r_\pi}{\beta_0 + 1} \simeq \frac{1}{g_m} + \frac{R_F}{\beta_0 + 1} \tag{8.171}$$

Thus R_o is low as expected, and has a lower limit of $1/g_m$ as R_F is reduced or β_0 increased.

Finally, the input resistance can be calculated from Fig. 8.36*b* as

$$R_i \simeq r_\pi \left\| \frac{R_F + R_L}{1 + g_m R_L} \right. \tag{8.172}$$

The input resistance is low, as expected for shunt feedback at the input. Note that (8.172) gives a result similar to the Miller effect using feedback resistance instead of capacitance.

EXAMPLE

Calculate the low-frequency parameters of Fig. 8.36*a* if $R_F = 10\ \text{k}\Omega$, $R_L = 5\ \text{k}\Omega$, $\beta_0 = 100$, $I_C = 1$ mA, and $r_\pi = 2.6\ \text{k}\Omega$.

From (8.168) the shunt-feedback loop gain is

$$T = 100 \frac{5}{5 + 10 + 2.6} = 28.4 \tag{8.173}$$

From (8.169a) the transfer function is

$$\frac{v_o}{i_i} = -\frac{10\ \text{k}\Omega}{1 + \dfrac{1}{28.4}} = -9.66\ \text{k}\Omega \tag{8.174}$$

The output resistance from (8.171) is

$$R_o = \left(26 + \frac{10{,}000}{101}\right)\Omega = 125\ \Omega \tag{8.175}$$

The input resistance from (8.172) is

$$R_i = 2600 \left\| \frac{10{,}000 + 5000}{1 + \dfrac{5000}{26}} \right. \Omega = 75.4\ \Omega \tag{8.176}$$

8.7 THE VOLTAGE REGULATOR AS A FEEDBACK CIRCUIT

As a final example of a practical feedback circuit, the operation of a voltage regulator will be examined. This section is introduced for the dual purpose of illustrating the use of feedback in practice and for describing the elements of voltage regulator design.

Voltage regulators are widely used components that accept a poorly specified (possibly fluctuating) dc input voltage and produce from it a constant, well-specified output voltage that can then be used as a supply voltage for other circuits.[3] In this way, fluctuations in the supply voltage are essentially eliminated and this usually results in improved performance for circuits powered from such a supply.

The most common type of voltage regulator is the "series" regulator shown schematically in Fig. 8.37. The name "series" comes from the fact that the output voltage is controlled by a power transistor in series with the output. This is the last stage of a high-gain voltage amplifier as shown in Fig. 8.37.

Many of the techniques discussed in previous chapters are utilized in the design of circuits of this kind. A stable reference voltage V_R is generated using either a Zener diode or the bandgap reference, as described in Chapter 4. This is then fed to the noninverting input of the high-gain amplifier where it is compared with a

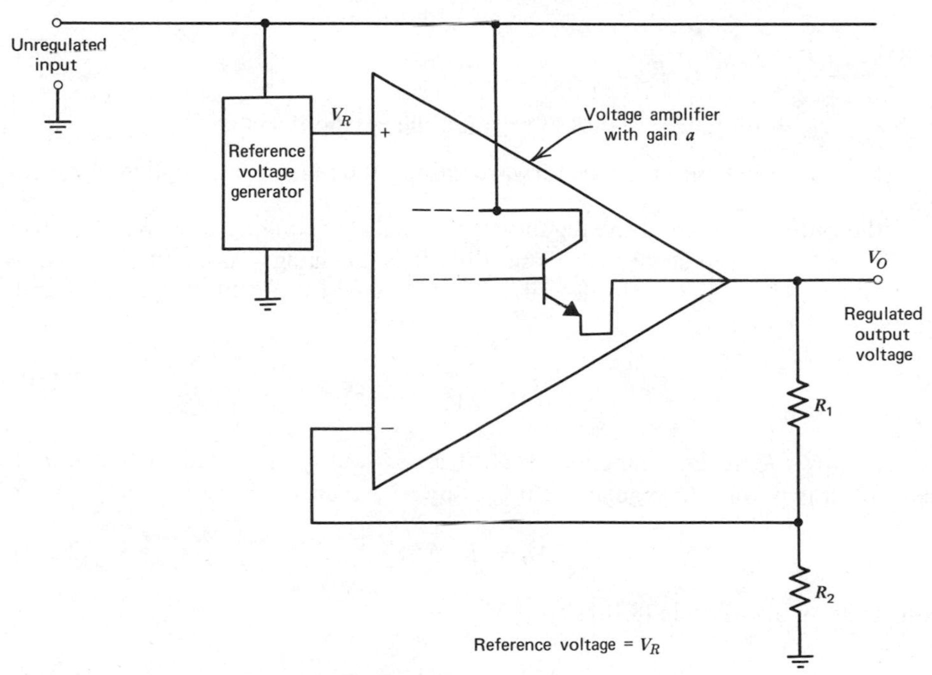

Fig. 8.37 Schematic of a series voltage regulator.

sample of the output taken by resistors R_1 and R_2. This is recognizable as a series-shunt feedback arrangement, and, using (8.137), we find that for large loop gain

$$V_O = V_R \frac{R_1 + R_2}{R_2} \tag{8.177}$$

The output voltage can be varied by changing ratio R_1/R_2.

The characteristics required in the amplifier of Fig. 8.37 are those of a good op amp as described in Chapter 6. In particular, low drift and offset are essential so that the output voltage V_O is as stable as possible. Note that the series-shunt feedback circuit will present a high input impedance to the reference generator, which is desirable to minimize loading effects. In addition, a very low output impedance will be produced at V_O, which is exactly the requirement for a good voltage source. If the effects of feedback loading are neglected (usually a good assumption in such circuits), the low-frequency output resistance of the regulator is given by (8.29) as

$$R_o = \frac{r_{oa}}{1 + T} \tag{8.178}$$

where $$T = a\frac{R_2}{R_1 + R_2} \tag{8.179}$$

r_{oa} = output resistance of the amplifier without feedback

a = magnitude of the forward gain of the regulator amplifier

If the output voltage of the regulator is varied by changing ratio R_1/R_2, then (8.178) and (8.179) indicate that T and thus R_o also change. Assuming that V_R is constant and $T \gg 1$, we can describe this behavior by substituting (8.177) and (8.179) in (8.178) to give

$$R_o = \frac{r_{oa}}{aV_R} V_O \tag{8.180}$$

which shows R_o to be a function of V_O if a, V_R, and r_{oa} are fixed. If the output current drawn from the regulator changes by ΔI_O, then V_O changes by ΔV_O where

$$\Delta V_O = R_o \Delta I_O \tag{8.181}$$

Substitution of (8.180) in (8.181) gives

$$\frac{\Delta V_O}{V_O} = \frac{r_{oa}}{aV_R} \Delta I_O \tag{8.182}$$

This equation allows calculation of the *load regulation* of the regulator. This is a widely used specification, which gives the percentage change in V_O for a specified change in I_O and should be as small as possible.

Another common regulator specification is the *line regulation*, which is the percentage change in output voltage for a specified change in input voltage. Since V_O is directly proportional to V_R, the line regulation is determined by the change in reference voltage V_R with changes in input voltage and depends on the particular reference circuit used.

As an example of a practical regulator, consider the circuit diagram of the 723 monolithic voltage regulator shown in Fig. 8.38. The correspondence to Fig. 8.37 can be recognized, with the portion of Fig. 8.38 to the left of the dotted line being the reference voltage generator. The divider resistors, R_1 and R_2 in Fig. 8.37, are labeled R_A and R_B in Fig. 8.38 and are external to the chip. The output power transistor, Q_{15}, is on the chip and is Darlington connected with Q_{14} for high gain. Differential pair Q_{11} and Q_{12}, together with active load Q_8, contribute most of the gain of the amplifier. Resistor R_C couples the reference voltage to the comparator amplifier and C_2 is an external capacitor, which is needed to prevent oscillation in the high-gain feedback loop. Its function is discussed in Chapter 9.

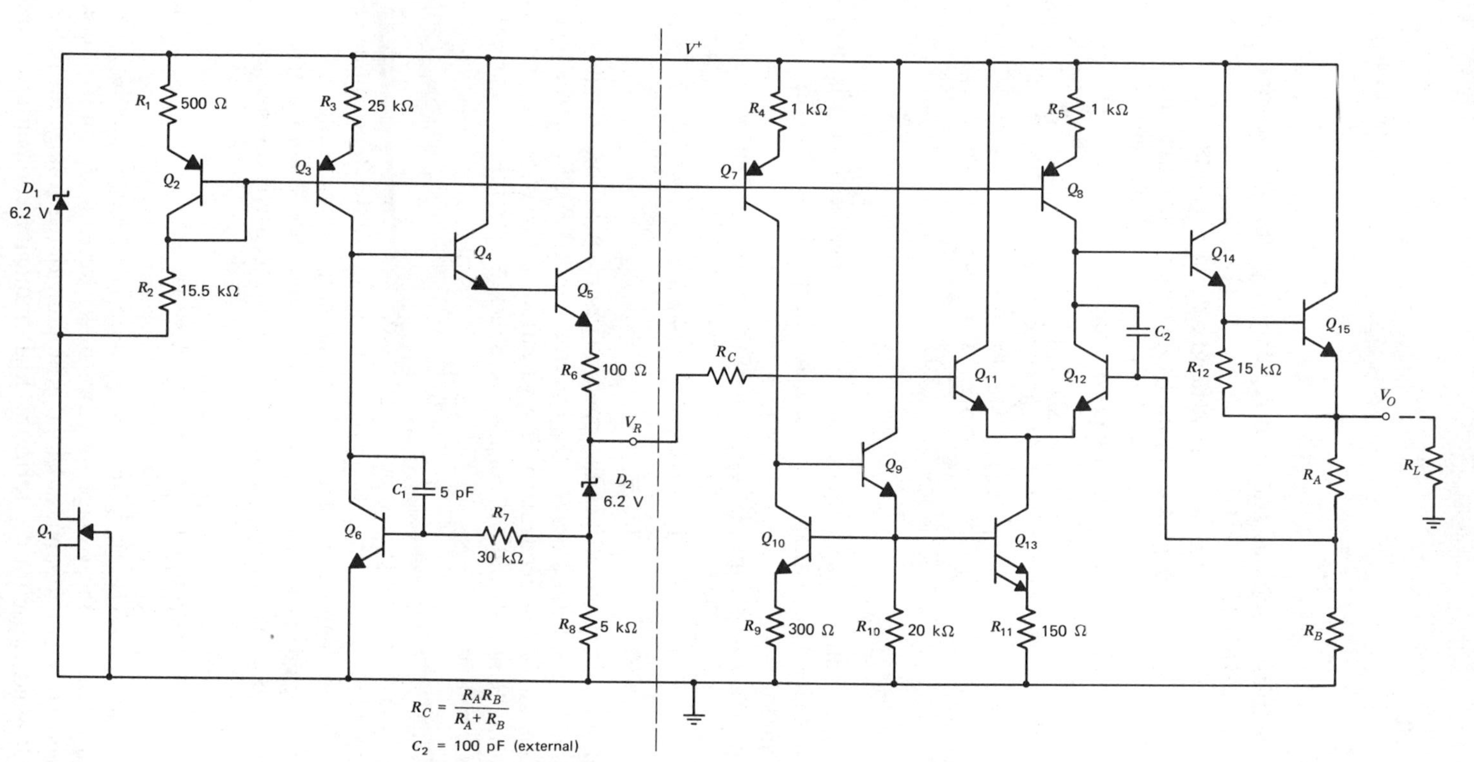

$R_C = \frac{R_A R_B}{R_A + R_B}$

C_2 = 100 pF (external)

Fig. 8.38 Circuit diagram of the 723 monolithic voltage regulator.

EXAMPLE

Calculate the bias conditions and load regulation of the 723. Assume that the total supply voltage is 15 V.

The bias calculation begins at the left hand side of Fig. 8.38. The monolithic *n*-channel JFET Q_1 has its gate connected to its source and thus functions as a constant-current source that biases the Zener diode, D_1. The Zener D_1 produces a voltage drop of about 6.2 V, which sets up a reference current in Q_2:

$$I_{C2} = -\frac{6.2 - V_{BE2}}{R_1 + R_2}$$

$$= -\frac{6.2 - 0.6}{16{,}000} \text{ A}$$

$$= -348\ \mu\text{A} \tag{8.183}$$

Note that I_{C2} is almost independent of supply voltage because it is dependent only on the Zener diode voltage, and the Zener itself is biased by Q_1 to an almost constant current.

The voltage across R_1 and Q_2 establishes the currents in current sources Q_3, Q_7, and Q_8.

$$I_{C7} = I_{C8} = -174\ \mu\text{A} \tag{8.184}$$

$$I_{C3} = -10.5\ \mu\text{A} \tag{8.185}$$

Current source Q_3 establishes the operating current in the voltage reference circuit composed of transistors Q_4, Q_5, Q_6, resistors R_6, R_7, R_8, and Zener diode D_2. This circuit can be recognized as a variation of the Wilson current source described in Chapter 4, and the negative feedback loop forces the current in Q_6 to equal I_{C3} so that

$$I_{C6} = 10.5\ \mu\text{A} \tag{8.186}$$

where the base current of Q_4 has been neglected.

The output reference voltage is composed of the sum of the Zener diode voltage, D_2, plus the base-emitter voltage of Q_6. The temperature coefficients of these tend to cancel, giving a reference voltage of about 6.8 V with a temperature coefficient of about 0.2 mV/°C. The current in the Zener is established by V_{BE6} and R_8 giving

$$I_{D2} = \frac{V_{BE6}}{R_8} = \frac{600}{5}\ \mu\text{A} = 120\ \mu\text{A} \tag{8.187}$$

The Darlington pair, Q_4, Q_5, gives a large loop gain that results in a very low output impedance at the voltage reference node. Resistor R_6 limits the current and protects Q_5 in case of an accidential grounding of the voltage reference node. Resistor R_7 and capacitor C_1 form the high-frequency compensation required to prevent oscillation in the feedback loop. Note that the feedback is shunt at the output node. Any changes in reference-node voltage (due to loading for example) are detected at the base of Q_6, amplified, and fed to the base of Q_4 and thus back to the output where the original change is opposed.

The biasing of the output amplifier is achieved via current sources Q_7 and Q_8. The current in Q_7 also appears in Q_{10} (neglecting the base current of Q_9). Transistor Q_{13} has an area twice that of Q_{10} and one half the emitter resistance. Thus

$$I_{C13} = 2I_{C10} = 2I_{C7} = 348\ \mu\text{A} \tag{8.188}$$

Transistor Q_9 provides current gain to minimize the effect of base current in Q_{10} and Q_{13}. This type of current source was discussed in Chapter 4.

The bias current in each half of the differential pair Q_{11}, Q_{12} is thus

$$I_{C11} = I_{C12} = \tfrac{1}{2}I_{C13} = 174\ \mu\text{A} \tag{8.189}$$

Since Q_8 and R_5 are identical to Q_7 and R_4 the current in Q_8 is given by

$$I_{C8} = -174\ \mu\text{A} \tag{8.190}$$

Transistor Q_8 functions as an active load for Q_{12}, and, since the collector currents in these two devices are nominally equal, the input offset voltage for the differential pair is nominally zero.

The current in output power transistor Q_{15} depends on the load resistance but can go as high as 150 mA before a current-limit circuit (not shown) prevents further increase. Resistor R_{12} provides bleed current so that Q_{14} always has at least 0.04 mA of bias current, even when the current in Q_{15} is low and/or its current gain is large.

In order to calculate the load regulation of the 723, Equation 8.182 indicates that it is necessary to calculate the open-loop gain and output resistance of the regulator amplifier. For this purpose, a differential ac equivalent circuit of this amplifier is shown in Fig. 8.39. Load resistance R_{L12} is the output resistance presented by Q_8, which is

$$R_{L12} \simeq r_{o8}(1 + g_{m8}R_5) \tag{8.191}$$

Assuming that the Early voltage of Q_8 is 100 V and $I_{C8} = -174\ \mu\text{A}$ we can calculate the value of R_{L12} as

$$R_{L12} = \frac{100}{0.174}\left(1 + \frac{0.174}{26}1000\right)\text{k}\Omega = 4.42\ \text{M}\Omega \tag{8.192}$$

Since $g_{m11} = g_{m12}$, the impedance in the emitter of Q_{12} halves the transconductance and gives an effective output resistance of

$$R_{o12} = \left(1 + g_{m12}\frac{1}{g_{m11}}\right)r_{o12} \tag{8.193}$$

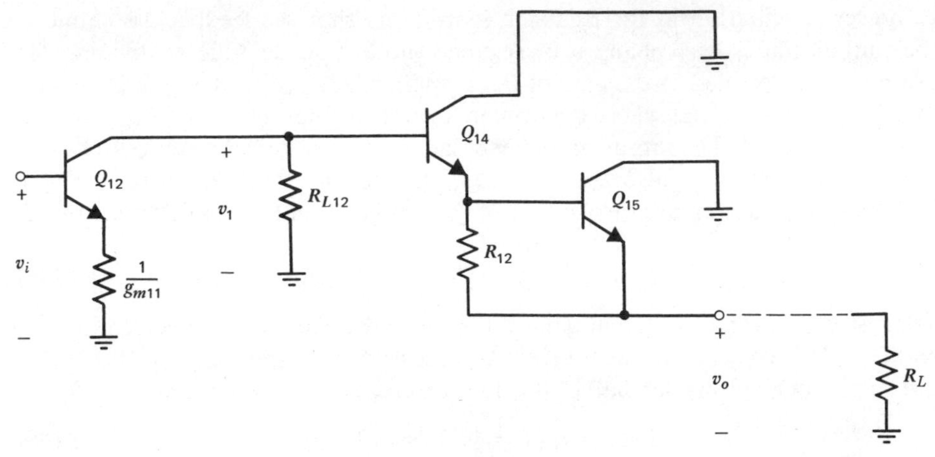

Fig. 8.39 ac equivalent circuit of the regulator amplifier of the 723 voltage regulator.

where r_{o12} is the output resistance of Q_{12} alone and is 575 kΩ if the Early voltage is 100 V.

Thus

$$R_{o12} = 1.15\ \text{M}\Omega \tag{8.194}$$

The external load resistance, R_L, determines the load current and thus the bias currents in Q_{14} and Q_{15}. However, R_L is not included in the small-signal calculation of output resistance because this quantity is the resistance seen by R_L looking back into the circuit. Thus, for purposes of calculating the ac output resistance, R_L may be assumed infinite and the output Darlington pair then produces no loading at the collector of Q_{12}. The voltage gain of the circuit may then be calculated as

$$a = \left|\frac{v_o}{v_i}\right| = \left|\frac{v_1}{v_i}\right| = \frac{g_{m12}}{2}(R_{o12}\|R_{L12})$$

$$= \frac{0.174}{26}\,\tfrac{1}{2}(1.15\|4.42) \times 10^6$$

$$= 3054 \tag{8.195}$$

The output resistance, r_{oa}, of the circuit of Fig. 8.39 is the output resistance of a Darlington emitter follower. If R_{12} is assumed large compared with $r_{\pi 15}$ then

$$r_{oa} = \frac{1}{g_{m15}} + \frac{1}{\beta_{15}}\left(\frac{1}{g_{m14}} + \frac{R_s}{\beta_{14}}\right) \tag{8.196}$$

where

$$R_s = R_{o12}\|R_{L12} = 913\ \text{k}\Omega \tag{8.197}$$

If we assume collector bias currents of 20 mA in Q_{15} and 0.5 mA in Q_{14} together with $\beta_{15} = \beta_{14} = 100$, then substitution in (8.196) gives

$$r_{oa} = \left[1.3 + \frac{1}{100}\left(52 + \frac{913{,}000}{100}\right)\right]\Omega$$

$$= (1.3 + 92)\ \Omega$$

$$= 93\ \Omega \tag{8.198}$$

Substituting for r_{oa} and a in (8.182) and using $V_R = 6.8$ V, we obtain, for the load regulation,

$$\frac{\Delta V_O}{V_O} = \frac{93}{3054 \times 6.8}\Delta I_O = 4.5 \times 10^{-3}\,\Delta I_O \tag{8.199}$$

where ΔI_O is in amps.

If ΔI_O is 50 mA, then (8.199) gives

$$\frac{\Delta V_O}{V_O} = 2 \times 10^{-4} = 0.02 \text{ percent} \tag{8.200}$$

This answer is close to the value of 0.03 percent given on the specification sheet. Note the extremely small percentage change in output voltage for a 50 mA-change in load current.

PROBLEMS

8.1 (a) In a feedback amplifier, forward gain $a = 100{,}000$ and feedback factor $f = 10^{-3}$. Calculate overall gain A and the percentage change in A if a changes by 10 percent.
(b) Repeat (a) if $f = 0.1$.

8.2 For the characteristic of Fig. 8.2 the following data apply.

$$S_{o2} = 15\text{ V} \qquad S_{o1} = 7\text{ V} \qquad a_1 = 50{,}000 \qquad a_2 = 20{,}000$$

(a) Calculate and sketch the overall transfer characteristic of Fig. 8.3 for the above amplifier when placed in a feedback loop with $f = 10^{-4}$.
(b) Repeat (a) with $f = 0.1$.

8.3 (a) For the conditions in Problem 8.2b, sketch the output voltage waveform, S_o, and the error voltage waveform, S_ε, if a sinusoidal input voltage S_i with amplitude 1.5 V is applied.
(b) Repeat (a) with an input amplitude of 2 V.

8.4 Verify equations (8.40), (8.41), and (8.42) for a shunt-series feedback amplifier.

8.5 Verify equations (8.43), (8.44), and (8.45) for a series-series feedback amplifier.

8.6 Calculate input impedance, output impedance, loop gain, and overall gain for the shunt-shunt feedback amplifier of Fig. 8.15a with data $R_F = 100$ kΩ and $R_L = 15$ kΩ. For the op amp, assume that $R_i = 500$ kΩ, $R_o = 200$ Ω, and $a_v = 75{,}000$.

8.7 The ac schematic of a shunt-shunt feedback amplifier is shown in Fig. 8.40. All collector currents are 1 mA and $\beta = 200$, $V_A = 50$ V, and $r_b = 0$.
(a) Calculate the overall gain v_o/i_i, the loop gain, the input impedance, and the output impedance at low frequencies.
(b) If the circuit is fed from a source resistance of 1 kΩ, what is the new output resistance of the circuit?

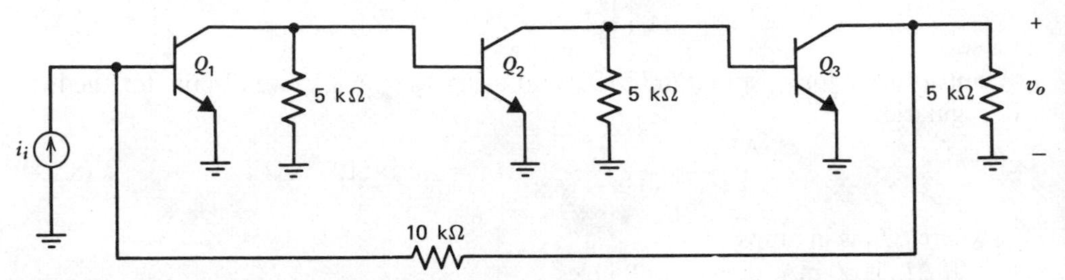

Fig. 8.40 ac schematic of a shunt-shunt feedback amplifier.

8.8 The half circuit of a balanced monolithic series-series triple is shown in Fig. 8.19*a*. Calculate the input impedance, output impedance, loop gain, and overall gain of the half circuit at low frequencies using the following data.

$$R_{E1} = R_{E2} = 290\ \Omega \qquad R_F = 1.9\ \text{k}\Omega \qquad R_{L1} = 10.6\ \text{k}\Omega \qquad R_{L2} = 6\ \text{k}\Omega$$

For the transistors, $I_{C1} = 0.5$ mA, $I_{C2} = 0.77$ mA, $I_{C3} = 0.73$ mA, $\beta = 120$, $r_b = 0$, and $V_A = 40$ V.

8.9 Repeat Problem 8.8 if the output signal is taken as the voltage at the emitter of Q_3.

8.10 A feedback amplifier is shown in Fig. 8.41. Device data are as follows: $\beta_{npn} = 200$, $\beta_{pnp} = 100$, $V_{BE(\text{on})} = 0.7$ V, $r_b = 0$, and $V_A = \infty$. If the dc input voltage is zero, calculate the overall gain v_o/v_i, the loop gain, and the input and output impedance at low frequencies.

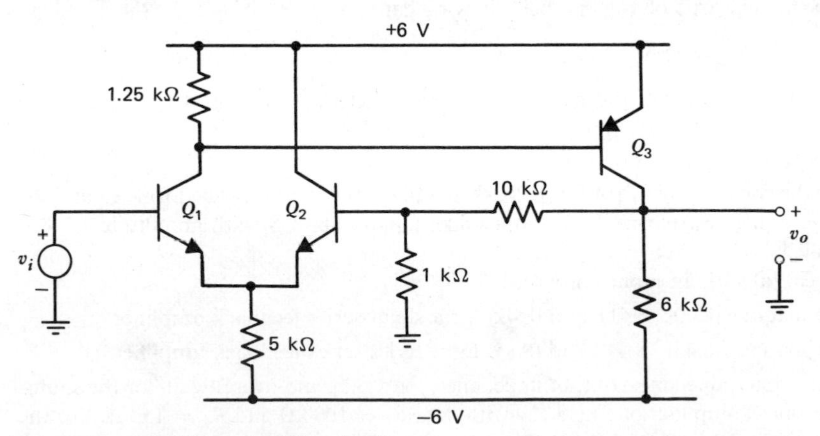

Fig. 8.41 Feedback amplifier circuit.

8.11 A balanced monolithic series-shunt feedback amplifier is shown in Fig. 8.42.
(a) If the common-mode input voltage is zero, calculate the bias current in each device. Assume that β is high.
(b) Calculate the voltage gain, input impedance, output impedance, and loop gain of the circuit at low frequencies using the following data.

$$\beta = 100 \qquad r_b = 50\ \Omega \qquad V_A = \infty \qquad V_{BE(\text{on})} = 0.7\text{ V}$$

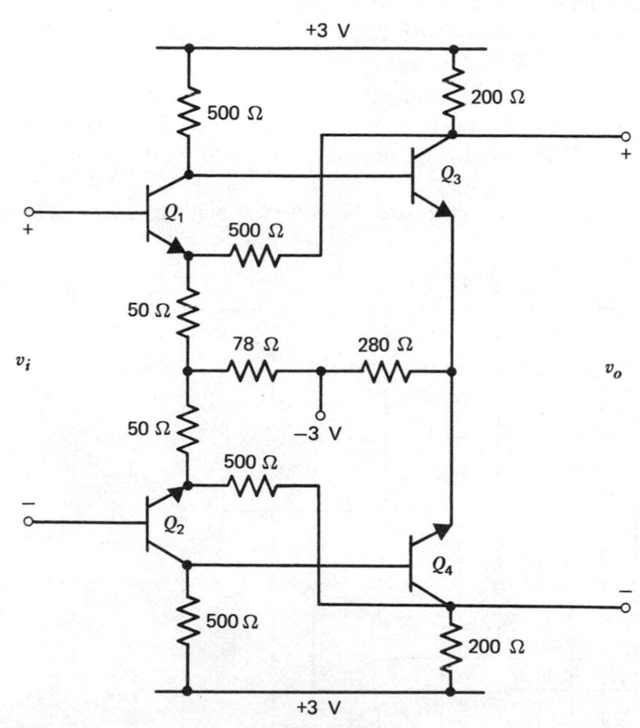

Fig. 8.42 Balanced series-shunt feedback amplifier.

8.12 How does the loop gain of the circuit of Fig. 8.42 change as the following circuit elements change? Discuss qualitatively.
(a) 50 Ω emitter resistor of the input stage.
(b) 500 Ω feedback resistor.
(c) 200 Ω load resistor on the output.

8.13 The ac schematic of a shunt-series feedback amplifier is shown in Fig. 8.32. Element values are $R_F = 1\text{ k}\Omega$, $R_E = 100\ \Omega$, $R_{L1} = 4\text{ k}\Omega$, $R_S = 1/y_S = 1\text{ k}\Omega$, and $z_L = 0$. *Device data.* $\beta = 200$, $r_b = 0$, $I_{C1} = I_{C2} = 1\text{ mA}$, $V_A = 100\text{ V}$.
(a) Calculate the overall gain i_o/i_i, the loop gain, and the input and output impedances at low frequencies.
(b) If the value of R_{L1} changes by +10 percent, what is the approximate change in overall gain and input impedance?

8.14 (a) Repeat Problem 8.13(a) with $R_F = 5$ kΩ, $R_E = 200$ Ω, $R_{L1} = 10$ kΩ, and $y_S = 0$.
(b) If the collector current of Q_1 increases by 20 percent, what will be the approximate change in overall gain and output resistance?

8.15 Calculate the transconductance, input impedance, output impedance, and loop gain at low frequencies of a local series-feedback stage with parameters: $R_E = 200$ Ω, $\beta = 150$, $I_C = 1$ mA, $r_b = 200$ Ω, $V_A = 80$ V.

8.16 Calculate the transresistance, input impedance, output impedance, and loop gain at low frequencies of a local shunt-feedback stage with parameters: $R_F = 2$ kΩ, $R_L = 2$ kΩ, $\beta = 200$, $I_C = 1$ mA, $r_b = 0$, $V_A = 100$ V.

8.17 A commercial wideband monolithic feedback amplifier (the 733) is shown in Fig. 8.43. This consists of a local series-feedback stage feeding a two-stage shunt-shunt feedback amplifier. The current output of the input stage acts as a current drive to the shunt-shunt output stage.
(a) Assuming all device areas are equal, calculate the collector bias current in each device.

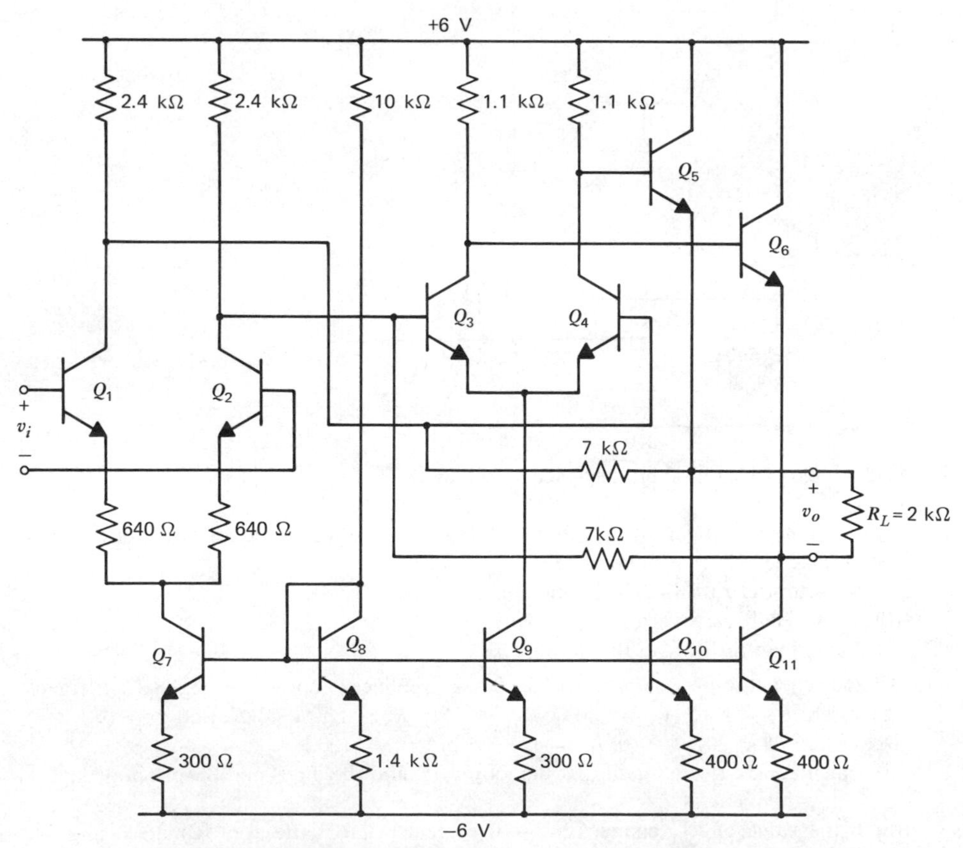

Fig. 8.43 Circuit diagram of the 733 wideband monolithic amplifier.

(b) Calculate input impedance, output impedance, and overall gain v_o/v_i for this circuit at low frequencies with $R_L = 2\ \text{k}\Omega$. Also calculate the loop gain of the output stage. *Data.* $\beta = 100$, $r_b = 0$, $r_o = \infty$.

8.18 If the 723 voltage regulator is used to realize an output voltage $V_o = 10$ V with a 1-kΩ load, calculate the output resistance and the loop gain of the regulator. If a 500-Ω load is connected to the regulator in place of the 1-kΩ load, calculate the new value of V_o.

REFERENCES

1. C. A. Desoer and E. S. Kuh. *Basic Circuit Theory.* McGraw-Hill, New York, 1969.
2. J. E. Solomon and G. R. Wilson. "A Highly Desensitized, Wideband Monolithic Amplifier," *IEEE J. Solid-State Circuits*, Vol. SC-1, pp. 19–28, September 1966.
3. A. B. Grebene. *Analog Integrated Circuit Design.* Van Nostrand Reinhold, New York, 1972, Chapter 6.

General Reference

P. E. Gray and C. L. Searle. *Electronic Principles.* Wiley, New York, 1969.

CHAPTER 9

FREQUENCY RESPONSE AND STABILITY OF FEEDBACK AMPLIFIERS

9.1 INTRODUCTION

In Chapter 8, we considered the effects of negative feedback on circuit parameters such as gain and terminal impedance. We saw that application of negative feedback resulted in a number of performance improvements, such as reduced sensitivity of gain to active-device parameter changes and reduction of distortion due to circuit nonlinearities.

In this chapter, we see the effect of negative feedback on the frequency response of a circuit. The possibility of *oscillation* in feedback circuits is illustrated, and methods of overcoming these problems by *compensation* of the circuit are described. Finally, the effect of compensation on the large-signal high-frequency performance of feedback amplifiers is investigated.

9.2 RELATION BETWEEN GAIN AND BANDWIDTH IN FEEDBACK AMPLIFIERS

Chapter 8 showed that the performance improvements produced by negative feedback were obtained at the expense of a reduction in gain by a factor $(1 + T)$, where T is the loop gain. The performance specifications that were improved were also changed by the factor $(1 + T)$.

In addition to the above effects, negative feedback also tends to *broadband* the amplifier. Consider first a feedback circuit as shown in Fig. 9.1 with a simple basic amplifier whose gain function contains a single pole

$$a(s) = \frac{a_0}{1 - \dfrac{s}{p_1}} \tag{9.1}$$

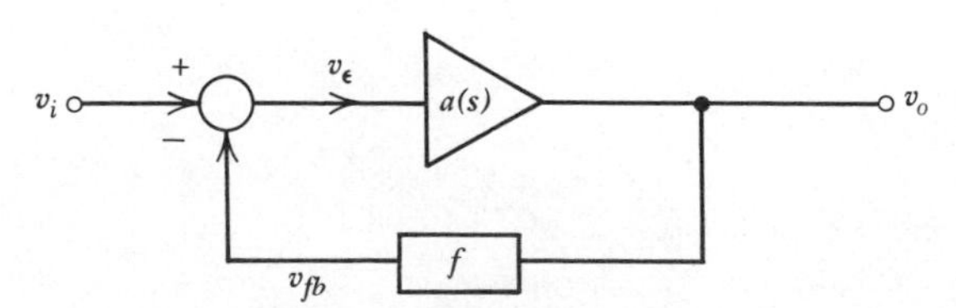

Fig. 9.1 Feedback circuit configuration.

where a_0 is the low-frequency gain of the basic amplifier and p_1 is the basic-amplifier pole in radians per second. Assume that the feedback path is purely resistive and thus feedback function f is a constant. Since Fig. 9.1 is an ideal feedback arrangement the *overall* gain is

$$A(s) = \frac{v_o}{v_i} = \frac{a(s)}{1 + a(s)f} \tag{9.2}$$

where the loop gain is $T(s) = a(s)f$

Substitution of (9.1) in (9.2) gives

$$A(s) = \frac{\dfrac{a_0}{1 - \dfrac{s}{p_1}}}{1 + \dfrac{a_0 f}{1 - \dfrac{s}{p_1}}} = \frac{a_0}{1 - \dfrac{s}{p_1} + a_0 f} = \frac{a_0}{1 + a_0 f} \frac{1}{1 - \dfrac{s}{p_1}\dfrac{1}{1 + a_0 f}} \tag{9.3}$$

From (9.3) the low-frequency gain, A_0, is

$$A_0 = \frac{a_0}{1 + T_0} \tag{9.4}$$

where
$$T_0 = a_0 f = \text{low-frequency loop gain} \tag{9.5}$$

The -3-dB bandwidth of the feedback circuit (i.e., the new pole frequency) is $(1 + a_0 f) \cdot |p_1|$ from (9.3). Thus the feedback has reduced the low-frequency gain by a factor $(1 + T_0)$, which is consistent with the results of Chapter 8, but it is now apparent that the -3-dB frequency of the circuit has been *increased* by the same quantity $(1 + T_0)$. Note that the gain-bandwidth product is constant. These results are illustrated in the Bode plots of Fig. 9.2 where the magnitudes of $a(j\omega)$ and $A(j\omega)$ are plotted versus frequency on log scales. It is apparent that the gain curves for any value of T_0 are contained in an envelope bounded by the curve of $|a(j\omega)|$.

Because the use of negative feedback allows the designer to trade gain for bandwidth, negative feedback is widely used as a method for designing broadband amplifiers. The gain reduction that occurs is made up by using additional gain stages, which in general are also feedback amplifiers.

Let us now examine the effect of the feedback on the pole of the overall transfer function, $A(s)$. It is apparent from (9.3) that as the low-frequency loop gain, T_0, is increased, the magnitude of the pole of $A(s)$ increases. This is illustrated in Fig. 9.3, which shows the *locus* of the pole of $A(s)$ in the s plane as T_0 varies. The pole

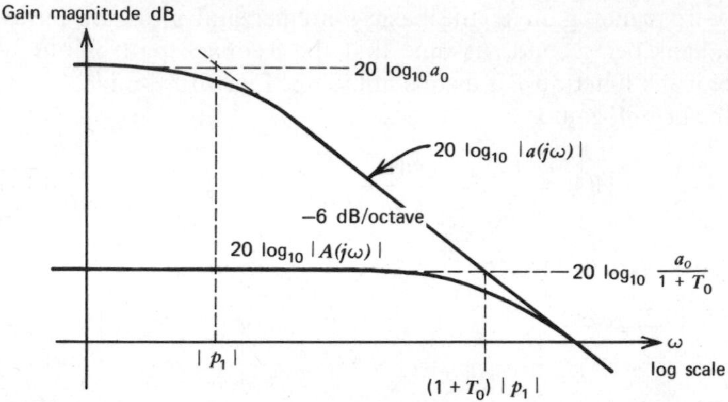

Fig. 9.2 Gain magnitude versus frequency for the basic amplifier and the feedback amplifier.

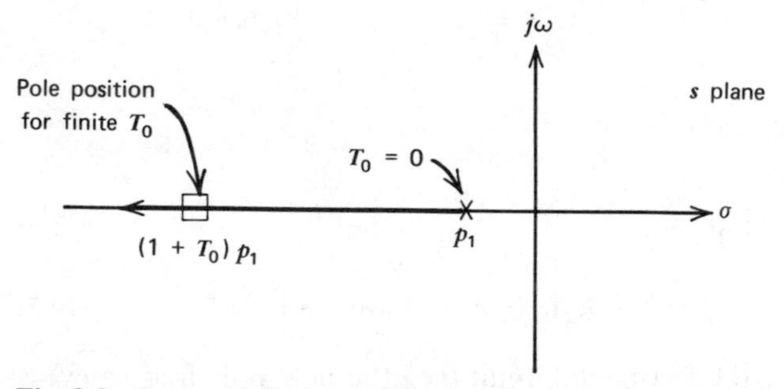

Fig. 9.3 Locus of the pole of the circuit of Fig. 9.1 as loop gain T_0 varies.

starts at p_1 for $T_0 = 0$ and moves out along the negative real axis as T_0 is made positive. Fig. 9.3 is a simple *root-locus* diagram and will be discussed further in Section 9.5.

9.3 INSTABILITY AND THE NYQUIST CRITERION[1,2,3]

In the above simple example the basic amplifier was assumed to have a single-pole transfer function and this situation is closely approximated in practice by internally compensated op amps such as the 741. However, many amplifiers (including the 741 before compensation is added) have multipole transfer functions that cause deviations from the above results. The process of compensation is designed to overcome these problems as will be seen later.

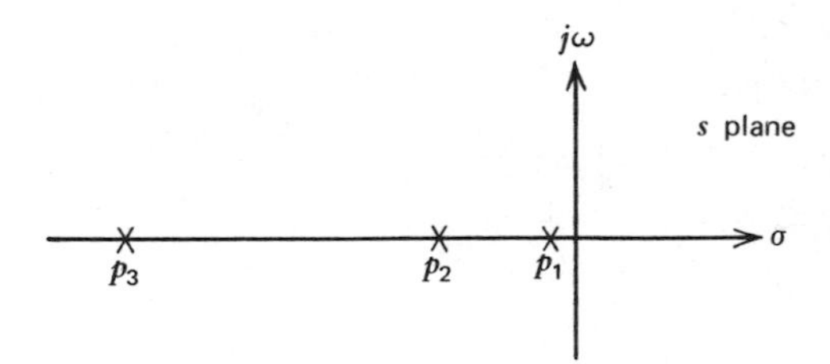

Fig. 9.4 Poles of an amplifier in the s plane.

Consider an amplifier with a three-pole transfer function

$$a(s) = \frac{a_0}{\left(1 - \dfrac{s}{p_1}\right)\left(1 - \dfrac{s}{p_2}\right)\left(1 - \dfrac{s}{p_3}\right)} \tag{9.6}$$

where $|p_1|$, $|p_2|$, and $|p_3|$ rad/sec are the pole frequencies. The poles are shown in the s plane in Fig. 9.4 and gain magnitude $|a(j\omega)|$ and phase ph $a(j\omega)$ are plotted versus frequency in Fig. 9.5 assuming about a factor of 10 separation between the poles. Only asymptotes are shown for the magnitude plot. At frequencies above the first pole frequency $|p_1|$ the plot of $|a(j\omega)|$ falls at 6 dB/octave and ph $a(j\omega)$ approaches $-90°$. Above $|p_2|$ these become 12 dB/octave and $-180°$ and above $|p_3|$ they become 18 dB/octave and $-270°$. The frequency where ph $a(j\omega) = -180°$ has special significance and is marked ω_{180}, and the value of $|a(j\omega)|$ at this frequency is a_{180}. If the three poles are fairly widely separated (factor of 10 or more) the phase shifts at frequencies $|p_1|$, $|p_2|$, and $|p_3|$ are approximately $-45°$, $-135°$, and $-225°$, respectively. This will now be assumed for simplicity. In addition, the gain magnitude will be assumed to follow the asymptotic curve and the effect of these assumptions in practical cases will be considered later.

Now consider this amplifier connected in a feedback loop as in Fig. 9.1 with f constant. Since f is constant, the loop gain, $T(j\omega) = a(j\omega)f$, will have the same variation with frequency as $a(j\omega)$. A plot of $af(j\omega) = T(j\omega)$ in magnitude and phase on a polar plot (with ω as a parameter) can thus be drawn using the data of Fig. 9.5 and the magnitude of f. Such a plot for this example is shown in Fig. 9.6 (not to scale) and is called a *Nyquist diagram.* The variable on the curve is frequency and varies from $\omega = -\infty$ to $\omega = \infty$. For $\omega = 0, |T(j\omega)| = T_0$ and ph $T(j\omega) = 0$ and the curve meets the real axis with an intercept T_0. As ω increases, Fig. 9.5 shows that $|a(j\omega)|$ decreases and ph $a(j\omega)$ becomes negative and thus the plot is in the fourth quadrant. As $\omega \to \infty$, ph $a(j\omega) \to -270°$ and $|a(j\omega)| \to 0$. Consequently the plot is asymptotic to the origin and is tangent to the imaginary axis. At the frequency ω_{180} the phase is $-180°$ and the curve crosses the negative real axis. If $|a(j\omega_{180})f| > 1$ at this point the Nyquist diagram will encircle the point $(-1,0)$

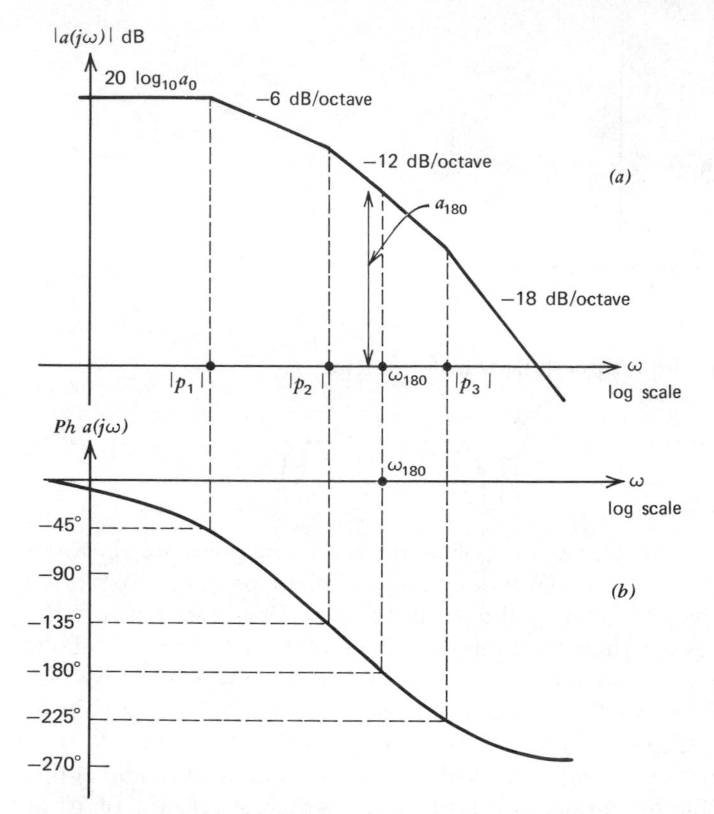

Fig. 9.5 Gain and phase versus frequency for a circuit with a three-pole transfer function.

as shown, and this has particular significance, as will now become apparent. For the purposes of this treatment, the *Nyquist criterion* for stability of the amplifier can be stated as follows:

"If the Nyquist plot encircles the point $(-1,0)$ the amplifier is unstable."

This criterion simply amounts to a mathematical test for poles of transfer function $A(s)$ in the right half plane. If the Nyquist plot encircles the point $(-1,0)$ the amplifier has poles in the right half plane and the circuit will *oscillate*. In fact the number of encirclements of the point $(-1,0)$ gives the number of right-half-plane poles and in this example there are two. The significance of poles in the right half plane can be seen by assuming that a circuit has a pair of complex poles at $(\sigma_1 \pm j\omega_1)$ where σ_1 is positive. The transient response of the circuit then contains a term $K_1 \exp \sigma_1 t \sin \omega_1 t$, which represents a *growing* sinusoid if σ_1 is positive. (K_1 is a constant representing initial conditions.) This term is then present even if no further input is applied, and a circuit behaving in this way is said to be *unstable* or *oscillatory*.

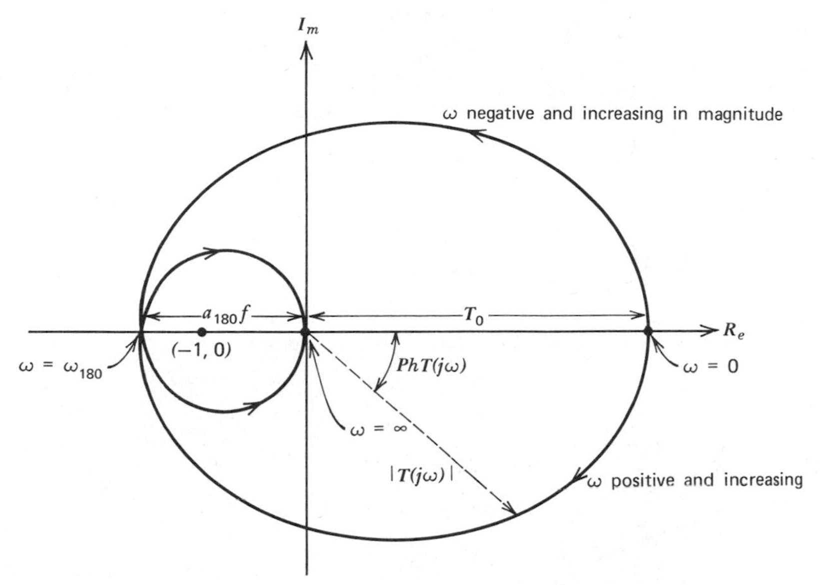

Fig. 9.6 Nyquist diagram [polar plot of $T(j\omega)$ in magnitude and phase] corresponding to the characteristic of Fig. 9.5 (not to scale).

The significance of the point $(-1,0)$ can be appreciated if the Nyquist diagram is assumed to pass through this point. Then at the frequency ω_{180}, $T(j\omega) = a(j\omega)f = -1$ and $A(j\omega) = \infty$ using (9.2) in the frequency domain. The feedback amplifier is thus calculated to have a forward gain of infinity and this indicates the onset of instability and oscillation. This situation corresponds to poles of $A(s)$ on the $j\omega$ axis in the s plane. If T_0 is then increased by increasing a_0 or f, the Nyquist diagram expands *linearly* and then encircles $(-1,0)$. This corresponds to poles of $A(s)$ in the right half plane as shown in Fig. 9.7.

From the above criterion for stability, a *simpler* test can be derived that is useful in most common cases.

"If $|T(j\omega)| > 1$ at the frequency where ph $T(j\omega) = -180°$, then the amplifier is unstable." The validity of this criterion for the example considered here is apparent from inspection of Fig. 9.6 and application of the Nyquist criterion.

In order to examine the effect of feedback on the stability of an amplifier, consider the three-pole amplifier with gain function given by (9.6) to be placed in a negative-feedback loop with f constant. The gain (in decibels) and phase of the amplifier are shown again in Fig. 9.8, and also plotted is the quantity $20 \log_{10} 1/f$. The value of $20 \log_{10} 1/f$ is approximately equal to the low-frequency gain in decibels with feedback applied since

$$A_0 = \frac{a_0}{1 + a_0 f} \tag{9.7}$$

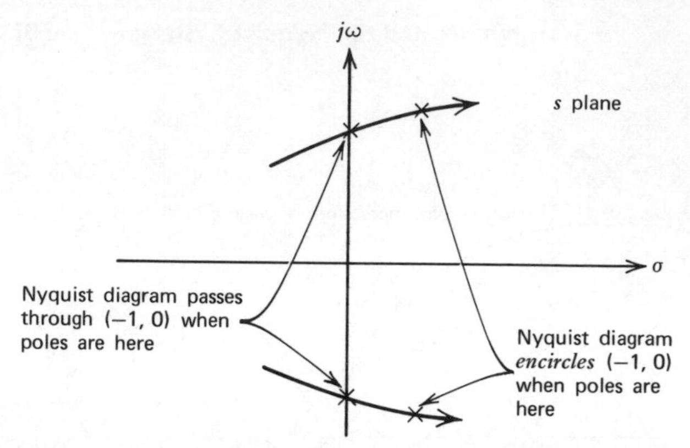

Fig. 9.7 Pole positions corresponding to different Nyquist diagrams.

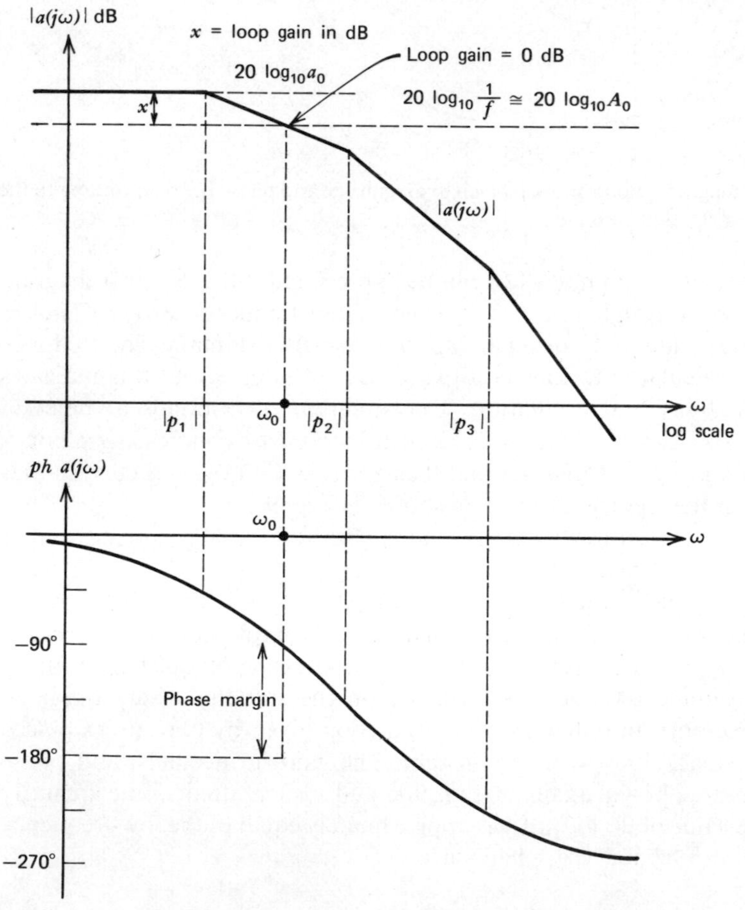

Fig. 9.8 Amplifier gain and phase versus frequency showing the phase margin.

and thus

$$\frac{1}{f} \simeq A_0 \tag{9.8}$$

if

$$T_0 = a_0 f \gg 1$$

Consider the vertical distance between the curve of $20 \log_{10} |a(j\omega)|$ and the line $20 \log_{10} 1/f$ in Fig. 9.8. Since the vertical scale is in decibels this quantity is

$$x = 20 \log_{10} |a(j\omega)| - 20 \log_{10} 1/f \tag{9.9}$$

$$= 20 \log_{10} |a(j\omega) f|$$

$$= 20 \log_{10} |T(j\omega)| \tag{9.10}$$

Thus the distance, x, is a *direct measure in decibels* of the loop-gain magnitude, $|T(j\omega)|$. The point where the curve of $20 \log_{10} |a(j\omega)|$ intersects line $20 \log_{10} 1/f$ is the point where the loop-gain magnitude $|T(j\omega)|$ is 0-dB or *unity*, and the curve of $|a(j\omega)|$ in decibels in Fig. 9.8 can thus be considered a curve of $|T(j\omega)|$ in decibels *if the dotted line at 20* $\log_{10} 1/f$ is taken as the new zero axis.

The simple example of Section 9.1 showed that the gain curve versus frequency with feedback applied ($20 \log_{10} |A(j\omega)|$) follows the $20 \log_{10} A_0$ line until it intersects the gain curve $20 \log_{10} |a(j\omega)|$. At higher frequencies the curve $20 \log_{10} |A(j\omega)|$ simply follows the curve of $20 \log_{10} |a(j\omega)|$ for the basic amplifier. The reason for this is now apparent in that at the higher frequencies the loop gain $|T(j\omega)| \to 0$ and the feedback then has *no influence* on the gain of the amplifier.

Figure 9.8 shows that the loop-gain magnitude $|T(j\omega)|$ is unity at frequency ω_0. At this frequency the phase of $T(j\omega)$ has not reached $-180°$ for the case shown, and using the modified Nyquist criterion stated above we conclude that *this feedback loop is stable.* Obviously $|T(j\omega)| < 1$ at the frequency where ph $T(j\omega) = -180°$. If the polar Nyquist diagram is sketched for this example, it docs *not* encircle point $(-1,0)$.

As $|T(j\omega)|$ is made closer to unity at the frequency where ph $T(j\omega) = -180°$, the amplifier has a *smaller margin* of stability and this can be specified in two ways. The most common is the *phase margin*, which is defined as follows:

Phase margin $= 180° +$ (ph $T(j\omega)$ at frequency where $|T(j\omega)| = 1$). The phase margin is indicated in Fig. 9.8 and must be greater than 0° for stability.

Another measure of stability is the *gain margin.* This is defined to be $|T(j\omega)|$ in decibels at the frequency where ph $T(j\omega) = -180°$, and this must be less than 0 dB for stability.

The significance of the phase-margin magnitude is now explored. For the feedback amplifier considered in Section 9.1 where the basic amplifier had a single-pole response, the phase margin is obviously 90° if the low-frequency loop gain is reasonably large. This is illustrated in Fig. 9.9 and results in a very stable amplifier. A typical lower allowable limit for the phase margin in practice is 45°, with a value of 60° being more common.

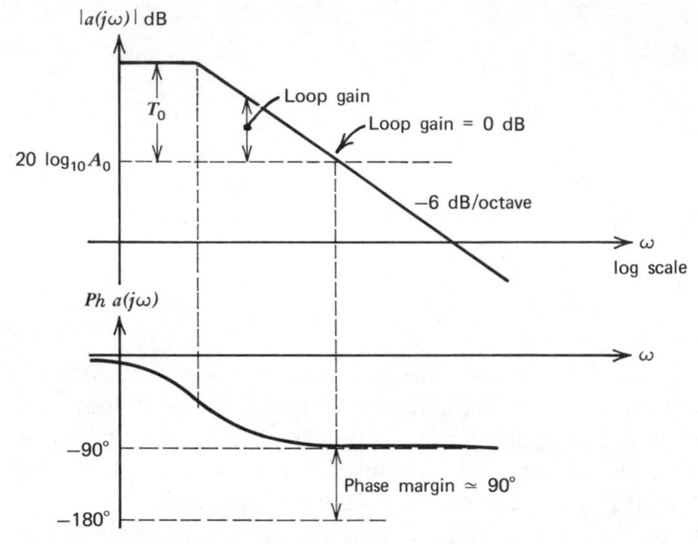

Fig. 9.9 Gain and phase versus frequency for a single-pole basic amplifier showing the phase margin for a low-frequency loop gain T_0.

Consider a feedback amplifier with a phase margin of 45° and a feedback function f that is real (and thus constant). Then

$$\text{ph } T(j\omega_0) = -135^\circ \tag{9.11}$$

where ω_0 is the frequency defined by

$$|T(j\omega_0)| = 1 \tag{9.12}$$

Now $|T(j\omega_0)| = |a(j\omega_0)f| = 1$ implies that

$$|a(j\omega_0)| = \frac{1}{f} \tag{9.13}$$

assuming that f is real.

The overall gain is

$$A(j\omega) = \frac{a(j\omega)}{1 + T(j\omega)} \tag{9.14}$$

Substitution of (9.11) and (9.12) in (9.14) gives

$$A(j\omega_0) = \frac{a(j\omega_0)}{1 + e^{-j135^\circ}} = \frac{a(j\omega_0)}{1 - 0.7 - 0.7j} = \frac{a(j\omega_0)}{0.3 - 0.7j}$$

and thus

$$|A(j\omega_0)| = \frac{|a(j\omega_0)|}{0.76} = \frac{1.3}{f} \tag{9.15}$$

using (9.13).

The frequency, ω_0, where $|T(j\omega_0)| = 1$ is the nominal -3-dB point for a single-pole basic amplifier, but in this case there is 2.4 dB($1.3\times$) of *peaking* above the low-frequency gain of $1/f$.

Consider a phase margin of 60°. At the frequency, ω_0, in this case

$$\text{ph } T(j\omega_0) = -120^\circ \tag{9.16}$$

and

$$|T(j\omega_0)| = 1 \tag{9.17}$$

Following a similar analysis we obtain

$$|A(j\omega_0)| = \frac{1}{f}$$

In this case there is no peaking at $\omega = \omega_0$, but there has also been no gain reduction at this frequency.

Finally the case where the phase margin is 90° can be similarly calculated. In this case

$$\text{ph } T(j\omega_0) = -90^\circ \tag{9.18}$$

and

$$|T(j\omega_0)| = 1 \tag{9.19}$$

A similar analysis gives

$$|A(j\omega_0)| = \frac{0.7}{f} \tag{9.20}$$

As expected in this case, the gain at frequency ω_0 is 3 dB below the midband value.

These results are illustrated in Fig. 9.10 where the normalized overall gain versus frequency is shown for various phase margins. The plots are drawn assuming the response is dominated by the first two poles of the transfer function, except for the case of the 90° phase margin, which has one pole only. As the phase margin diminishes, the gain peak becomes larger until the gain approaches infinity and oscillation occurs for phase margin = 0°. The gain peak usually occurs close to the frequency where $|T(j\omega)| = 1$, but for a phase margin of 60° there is 0.2 dB of peaking just below this frequency. Note that after the peak the gain curves approach an asymptote of -12 dB/octave for phase margins other than 90°. This is because the open-loop gain falls at -12 dB/octave due to the presence of two poles in the transfer function.

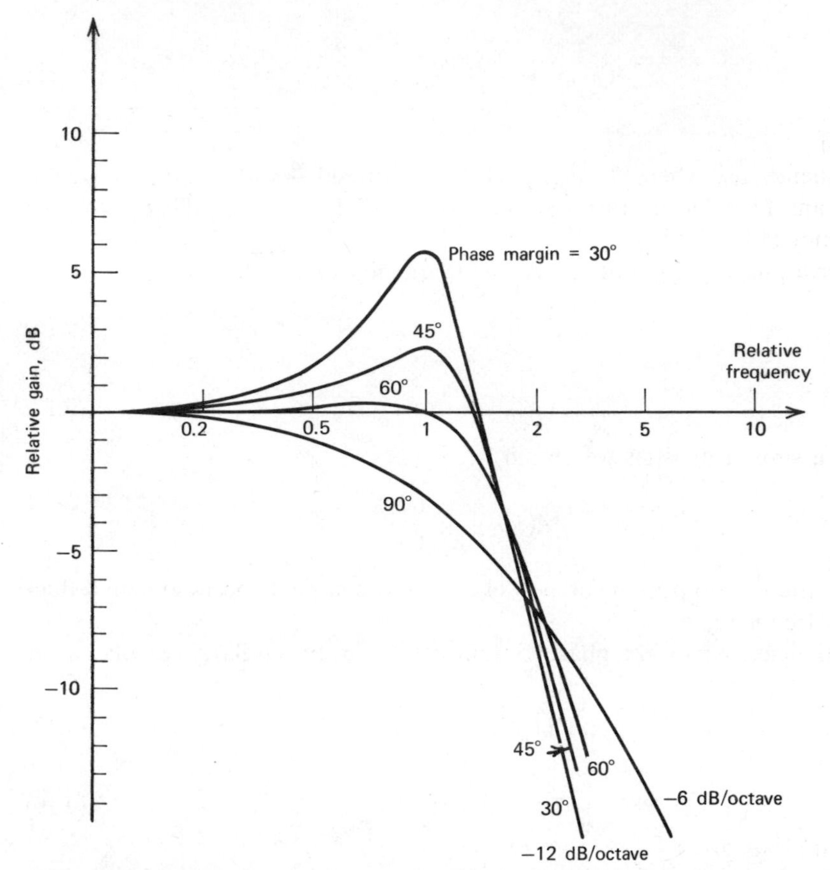

Fig. 9.10 Normalized overall gain for feedback amplifiers versus normalized frequency for various phase margins. Frequency is normalized to the frequency where the loop gain is unity.

9.4 COMPENSATION

9.4.1 Theory of Compensation

Consider again the amplifier whose gain and phase is shown in Fig. 9.8. For the feedback circuit in which this was assumed to be connected, the forward gain was A_0 as shown in Fig. 9.8 and the phase margin was positive. Thus the circuit was stable. It is apparent, however, that if the amount of feedback is increased by making f larger (and thus A_0 smaller), oscillation will eventually occur. This is shown in Fig. 9.11 where f_1 is chosen to give a zero phase margin and the corresponding overall gain is $A_1 \simeq 1/f_1$. If the feedback is increased to f_2 (and $A_2 \simeq 1/f_2$ is the overall gain) the phase margin is negative and the circuit will oscillate. Thus if this amplifier is to be used in a feedback loop with loop gain

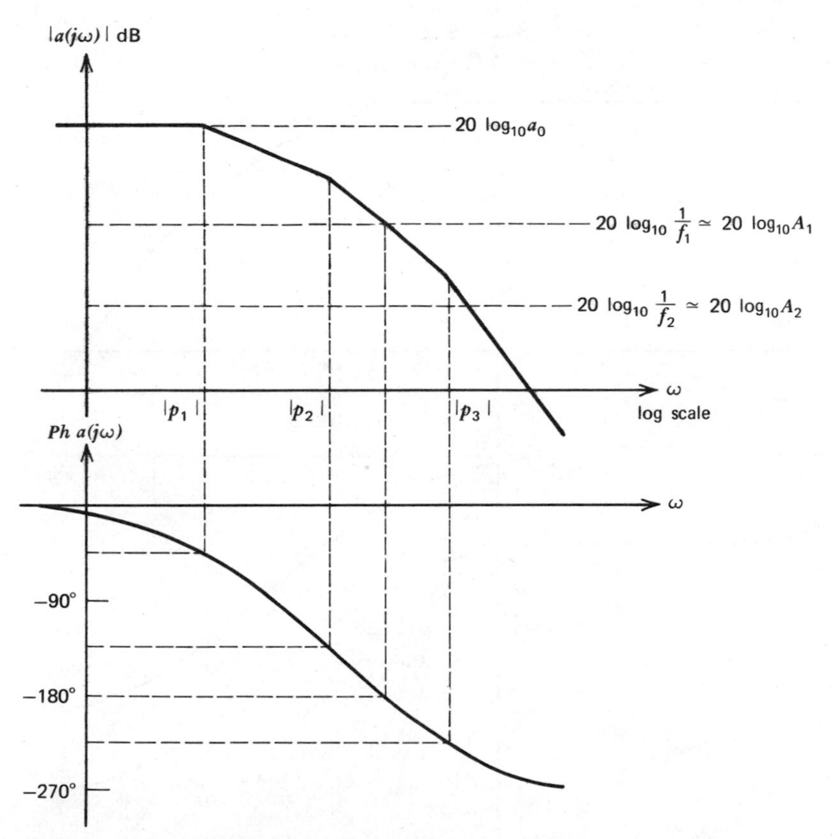

Fig. 9.11 Gain and phase versus frequency for a three-pole basic amplifier. Feedback factor f_1 gives a zero phase margin and factor f_2 gives a negative phase margin.

larger than $a_0 f_1$, efforts must be made to increase the phase margin. This process is known as *compensation.* Note that without compensation, the forward gain of the feedback amplifier cannot be made less than $A_1 \simeq 1/f_1$ because of the oscillation problem.

The simplest and most common method of compensation is to reduce the bandwidth of the amplifier (often called *narrowbanding*). That is, a dominant pole is deliberately introduced into the amplifier to *force* the phase shift to be less than $-180°$ when the loop gain is unity. This involves a direct sacrifice of the frequency capability of the amplifier.

The most difficult case to compensate is for $f = 1$, which is a unity-gain feedback configuration. In this case the loop-gain curve is identical to the gain curve of the basic amplifier. Consider this situation and assume that the basic amplifier has the same characteristic as in Fig. 9.11. To compensate the amplifier, we introduce a new dominant pole at a frequency $|p_D|$ as shown in Fig. 9.12 and

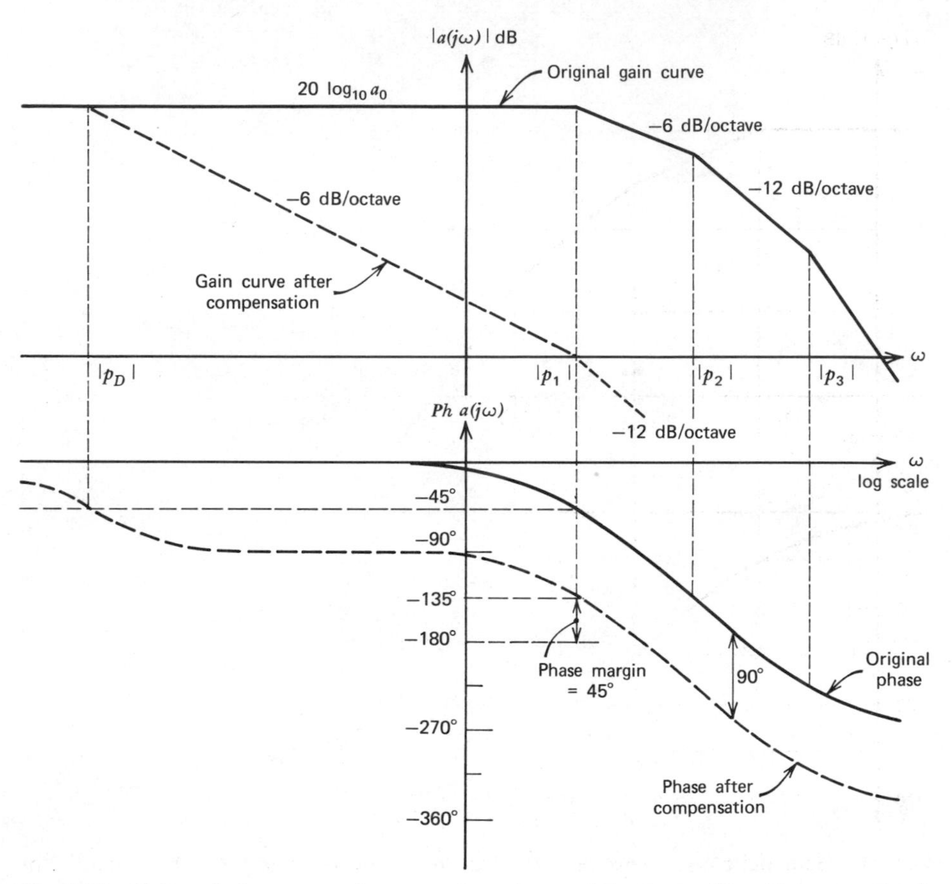

Fig. 9.12 Gain and phase versus frequency for a three-pole basic amplifier. Compensation for unity-gain feedback operation ($f = 1$) is achieved by introduction of a negative real pole with magnitude $|p_D|$.

assume that this does not affect the original amplifier poles at frequencies $|p_1|$, $|p_2|$, and $|p_3|$. This is often not the case, but is true in some circuits such as the 702 op amp.

The introduction of the dominant pole at frequency $|p_D|$ into the amplifier gain function causes the gain to decrease at 6 dB/octave until frequency $|p_1|$ is reached, and over this region the amplifier phase shift asymptotes to $-90°$. If frequency $|p_D|$ is chosen so that the gain $|a(j\omega)|$ is unity at frequency $|p_1|$ as shown, then the loop gain is also unity at frequency $|p_1|$ for the assumed case of unity feedback with $f = 1$. The phase margin in this case is then 45°, which means that the amplifier is stable. The original amplifier would have been *unstable* in such a feedback connection.

The price that has been paid for achieving stability in this case is that with the feedback removed, the basic amplifier has a unity-gain bandwidth of only $|p_1|$, which is much less than before. Also, with feedback applied, the loop gain now begins to decrease at a frequency $|p_D|$, and all the benefits of feedback diminish as the loop gain decreases. For example, in Chapter 8 it was shown that shunt feedback at the input or output of an amplifier *reduces* the basic terminal impedance by $[1 + T(j\omega)]$. Since $T(j\omega)$ is frequency dependent, the terminal impedance of a shunt-feedback amplifier will begin to *rise* when $|T(j\omega)|$ begins to decrease. Thus the high-frequency terminal impedance will appear *inductive*, as in the case of z_o for an emitter follower, which was calculated in Chapter 7. (See Problem 9.8.)

EXAMPLE

Calculate the dominant-pole frequency required to give unity-gain compensation of the 702 op amp with a phase margin of 45°. The low-frequency gain is $a_0 = 3600$ and the circuit has poles at $-(p_1/2\pi) = 1$ MHz, $-(p_2/2\pi) = 4$ MHz and $-(p_3/2\pi) = 40$ MHz.

In this example, the second pole, p_2, is sufficiently close to p_1 to produce significant phase shift at the amplifier -3-dB frequency. The approach to this problem will be to use the approximate results developed above to obtain an initial estimate of the required dominant-pole frequency and then to empirically adjust this frequency to obtain the required results.

The results of Fig. 9.12 indicate that a dominant pole with magnitude $|p_D|$ should be introduced so that gain $a_0 = 3600$ is reduced to unity at $|p_1/2\pi| = 1$ MHz with a 6-dB/octave decrease as a function of frequency. Since 6 dB/octave indicates direct proportionality, this gives

$$\left|\frac{p_D}{2\pi}\right| = \frac{1}{a_0}\left|\frac{p_1}{2\pi}\right| = \frac{10^6}{3600}\text{ Hz} = 278\text{ Hz}$$

This would give a transfer function

$$a(j\omega) = \frac{3600}{\left(1 + \dfrac{j\omega}{|p_D|}\right)\left(1 + \dfrac{j\omega}{|p_1|}\right)\left(1 + \dfrac{j\omega}{|p_2|}\right)\left(1 + \dfrac{j\omega}{|p_3|}\right)} \tag{9.21}$$

where the pole frequencies are in radians per second. Equation 9.21 gives a unity-gain frequency [where $|a(j\omega)| = 1$] of 780 kHz. This is slightly below the design value of 1 MHz because the actual gain curve is 3 dB below the asymptote at the break frequency $|p_1|$. At 780 kHz, the phase shift obtained from (9.21) is $-139°$ instead of the desired $-135°$ and this includes a contribution of $-11°$ from pole p_2. Although this result is close enough for most purposes, a phase margin of precisely 45° can be achieved by empirically reducing $|p_D|$ until (9.21) gives a phase shift of $-135°$ at the unity gain frequency. This occurs for $|p_D/2\pi| = 260$ Hz, which gives a unity-gain frequency of 730 kHz.

Consider now the performance of the amplifier whose characteristic is shown in Fig. 9.12 (with dominant pole at frequency $|p_D|$) when used in a feedback loop with $f < 1$ (i.e., overall gain $A_0 > 1$). This is shown in Fig. 9.13. The loop gain now falls to unity at frequency ω_x and the phase margin of the circuit is now approximately 90°. The -3-dB bandwidth of the feedback circuit is ω_x. The circuit now has more compensation than is needed, and, in fact, bandwidth is being

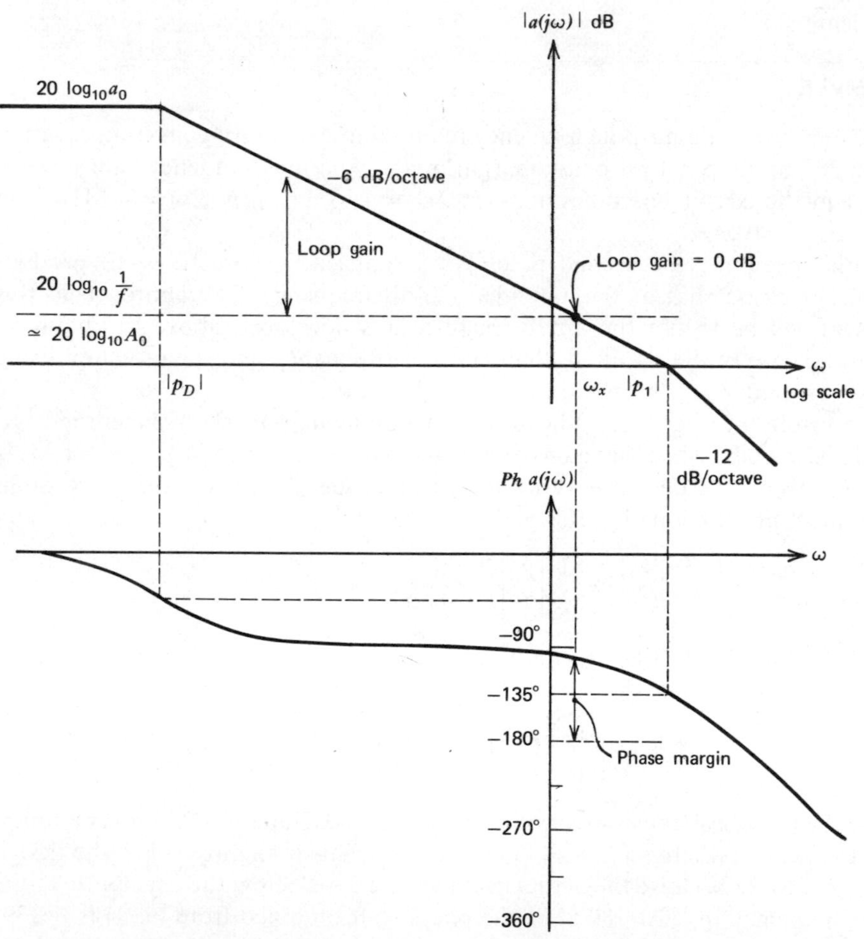

Fig. 9.13 Gain and phase versus frequency for an amplifier compensated for use in a feedback loop with $f = 1$ and a phase margin of 45°. The phase margin is shown for operation in a feedback loop with $f < 1$.

wasted. Thus, although it is convenient to compensate an amplifier for unity gain and then use it unchanged for other applications (as is done in many op amps) this procedure is quite wasteful of bandwidth. Fixed-gain amplifiers that are designed for applications where maximum bandwidth is required are usually compensated for a specified phase margin (typically 45 to 60°) at the required gain value. However, op amps are general purpose circuits that have to allow for operation with differing feedback networks with f values ranging from 0 to 1. Optimum bandwidth is achieved in such circuits if the compensation is added by the user who tailors it to the gain value required, and this gives much higher bandwidths for high gain values, as seen in Fig. 9.14. This shows compensation

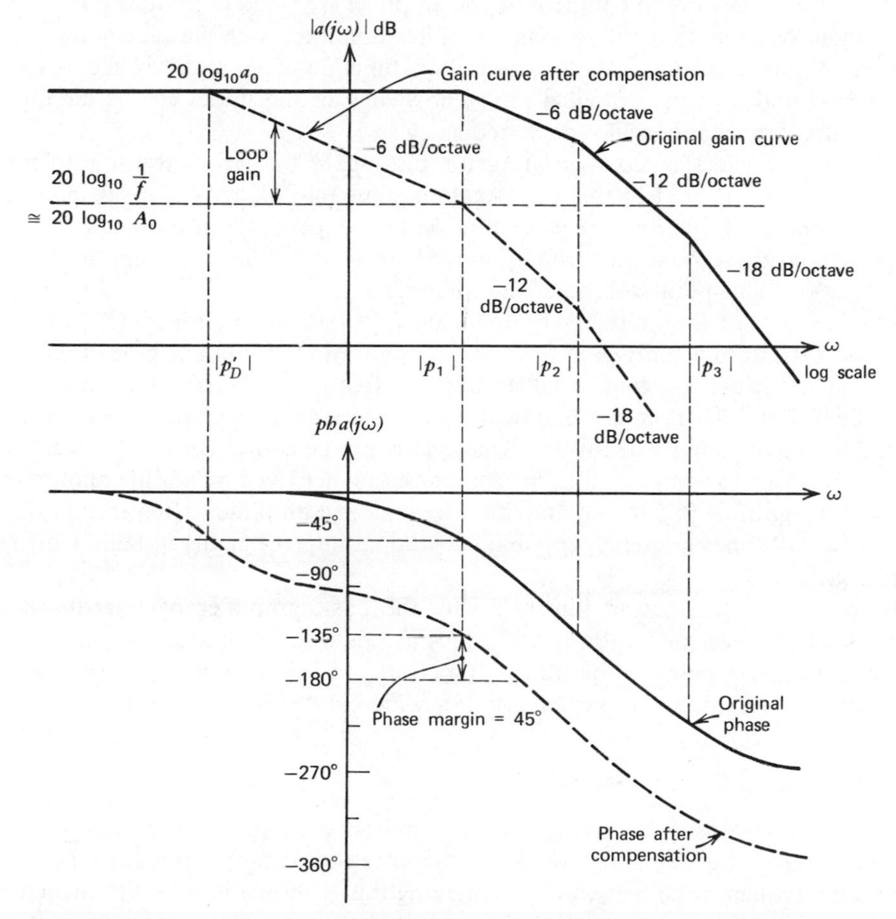

Fig. 9.14 Gain and phase versus frequency for an amplifier compensated for use in a feedback loop with $f < 1$ and a phase margin of 45°. Compensation is achieved by adding a new pole p'_D to the amplifier.

of the amplifier characteristic of Fig. 9.11 for operation in a feedback circuit with forward gain A_0. A dominant pole is added at frequency $|p'_D|$ to give a phase margin of 45°. Frequency $|p'_D|$ is obviously $\gg |p_D|$, and the -3-dB bandwidth of the feedback amplifier is nominally $|p_1|$, at which frequency the loop gain is 0 dB (disregarding peaking). The -3-dB frequency from Fig. 9.13 would be only $\omega_x = |p_1|/A_0$ if unity-gain comepnsation had been used. Obviously, since A_0 can be large, the improvement in bandwidth is significant.

In the compensation schemes discussed above, an additional dominant pole was assumed to be added to the amplifier, and the original amplifier poles were assumed to be unaffected by this procedure. In terms of circuit bandwidth, a much more efficient way to compensate the amplifier is to add capacitance to the circuit in such a way that the original amplifier dominant pole frequency $|p_1|$ is reduced so that it performs the compensation function. This requires access to the internal nodes of the amplifier, and knowledge of the nodes in the circuit where added capacitance will reduce frequency $|p_1|$.

Consider the effect of compensating for unity-gain operation the amplifier characteristic of Fig. 9.11 in this way. Again assume that higher-frequency poles p_2 and p_3 are unaffected by this procedure. In fact, depending on the method of compensation, these poles are usually moved up or down in frequency by the compensation. This point will be taken up later.

Compensation of the amplifier by reducing $|p_1|$ is shown in Fig. 9.15. For a 45°-phase margin in a unity-gain feedback configuration, dominant pole magnitude $|p'_1|$ must cause the gain to fall to unity at frequency $|p_2|$ (the second pole magnitude). Thus the nominal bandwidth in a unity-gain configuration is $|p_2|$, and the loop gain is unity at this frequency. This can be contrasted with a bandwidth of $|p_1|$, as shown in Fig. 9.12 for compensation achieved by adding another pole with magnitude $|p_D|$ to the amplifier. In practical amplifiers, frequency $|p_2|$ is often 5 or 10 times frequency $|p_1|$ and substantial improvements in bandwidth are thus achieved.

The results of this section illustrate why the basic amplifier of a feedback circuit is usually designed with as few stages as possible. Each stage of gain inevitably adds more poles to the transfer function and this makes the compensation problem more difficult, particularly if a wide bandwidth is required.

9.4.2 Methods of Compensation

In order to compensate a circuit by the common method of narrowbanding described above, it is necessary to add capacitance to create a dominant pole at the desired frequency. One method of achieving this is shown in Fig. 9.16, which is a schematic of the first two stages of the MC1530 op amp. A large capacitor C is connected between the collectors of the input stage. The output stage, which is relatively broadband, is not shown. A differential half-circuit of Fig. 9.16 is

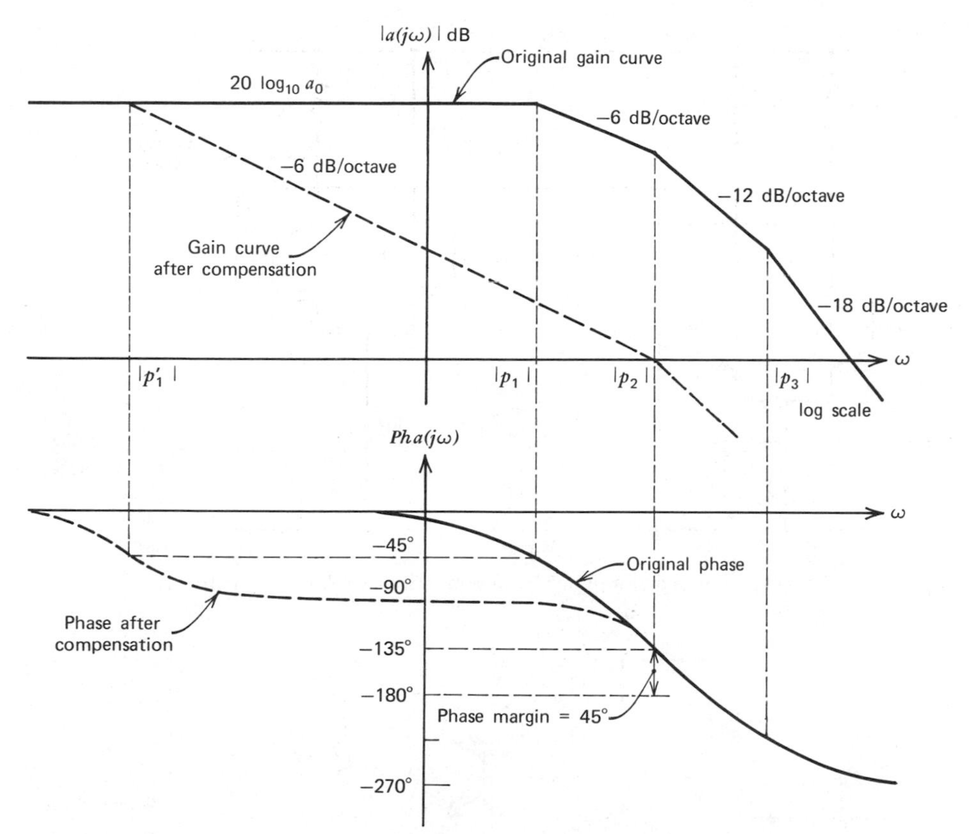

Fig. 9.15 Gain and phase versus frequency for an amplifier compensated for use in a feedback loop with $f = 1$ and a phase margin of 45°. Compensation is achieved by reducing the frequency $|p_1|$ of the dominant pole of the original amplifier.

shown in Fig. 9.17 and it should be noted that the compensation capacitor is doubled in the half circuit. The major contributions to the dominant pole of a circuit of this type (if R_S is not large) come from the input capacitance of Q_4 and Miller capacitance associated with Q_4. Thus the compensation as shown will reduce the frequency of the dominant pole of the original amplifier so that it performs the required compensation function. Almost certainly, however, the higher-frequency poles of the amplifier will also be changed by the addition of C. In practice, the best method of approaching the compensation design is to use computer simulation or measurement to determine the original pole positions. A first estimate of C is made on the assumption that the higher-frequency poles do not change in frequency and a new computer simulation or measurement is made with C included to check this assumption. Another estimate of C is then

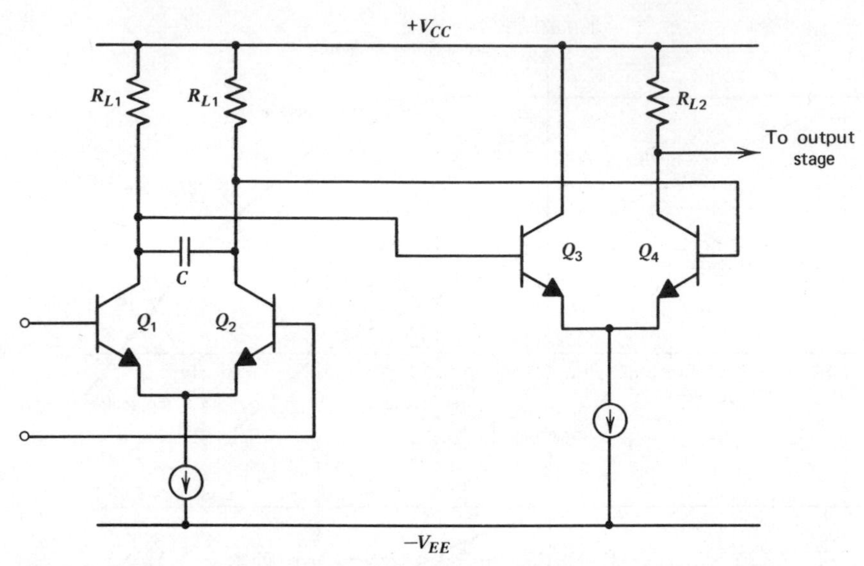

Fig. 9.16 Compensation of an amplifier by introduction of a large capacitor C.

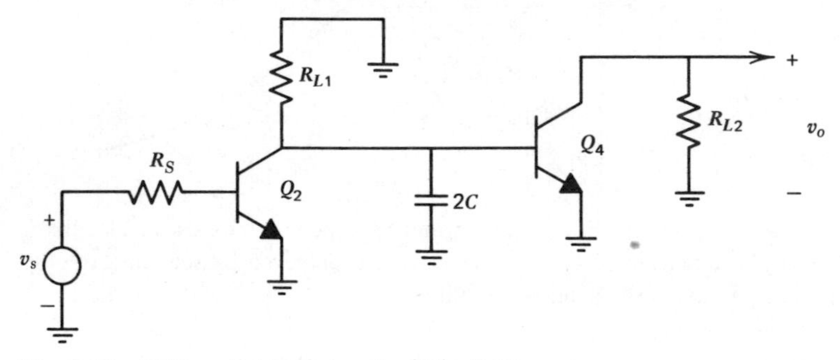

Fig. 9.17 Differential half circuit of Fig. 9.16.

made on the basis of the new data and this process usually converges after several iterations.

The magnitude of the dominant pole of Fig. 9.17 can be estimated using zero-value time constant analysis. However, if the value of C required is very large, this capacitor will dominate and a good estimate of the dominant pole can be made by considering C only and ignoring other circuit capacitance. In that case the dominant-pole magnitude is

$$|p_D| = \frac{1}{2CR} \tag{9.22}$$

where
$$R = R_{L1} \| R_{i4} \tag{9.23}$$
and
$$R_{i4} = r_{b4} + r_{\pi 4} \tag{9.24}$$
One disadvantage of the above method of compensation is that the value of C required is quite large (typically >1000 pF) and cannot be realized on a monolithic chip.

Many modern op amps have unity-gain compensation included on the monolithic chip and require no further compensation from the user. (The sacrifice in bandwidth caused by this technique when using gain other than unity was described above.) In order to realize an internally compensated monolithic op amp, compensation must be achieved using capacitance less than about 50 pF. This can be achieved using *Miller multiplication* of the capacitance as shown in Fig. 9.18 for the 741 op amp. The 30-pF compensation capacitor is connected around the Darlington pair Q_{16}–Q_{17} and produces a pole with magnitude 4.9 Hz, as shown in Chapter 7 using zero-value time constant analysis. The resulting gain and phase curves for the 741 are shown in Fig. 9.19. These were generated using the program SLIC and typical 741 device data. Use of this data shows that the unity-gain frequency is 1.25 MHz, the phase margin is 80° and the low-frequency gain is 108 dB. It should be pointed out that somewhat different performance is obtained from some commercial 741 circuits because of different integrated circuit processes giving different device parameters.

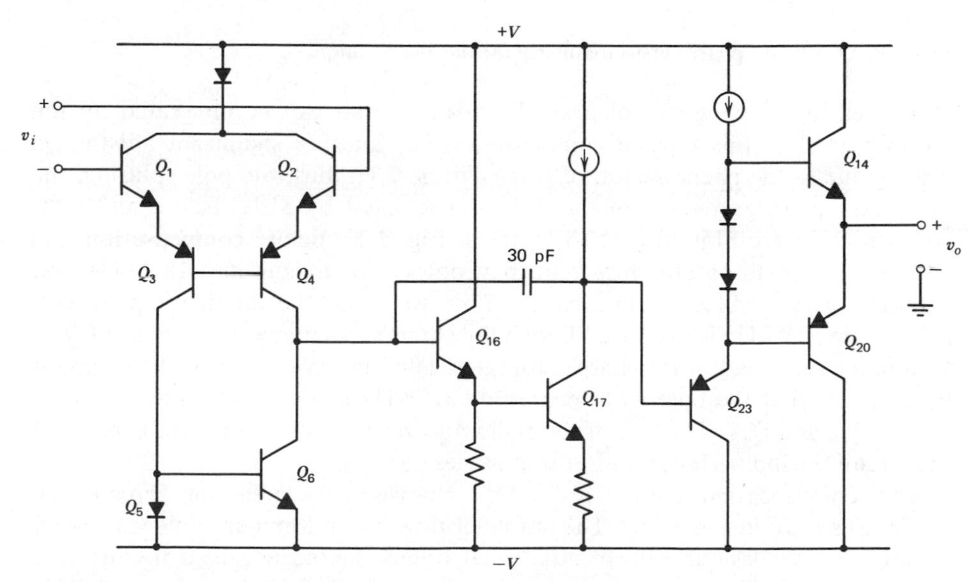

Fig. 9.18 Simplified schematic of the 741 op amp showing compensation by Miller effect using a 30-pF capacitor connected around the Darlington pair Q_{16}–Q_{17}.

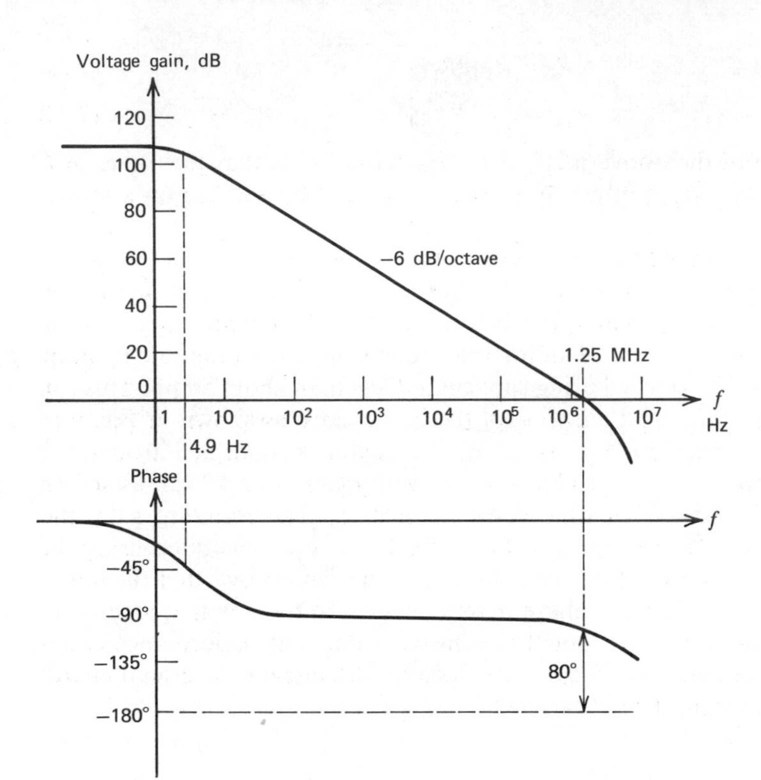

Fig. 9.19 Gain and phase versus frequency for the 741 op amp.

As well as allowing use of a small capacitor that can be integrated on the monolithic chip, this type of compensation has another significant advantage. This is due to the phenomenon of *pole splitting*[4]. To illustrate pole splitting, the important poles and zeros of the 741 as calculated by SLIC before and after compensation are plotted (not to scale) in Fig. 9.20. Before compensation, the circuit has two important low-frequency poles with magnitudes 18.9 kHz and 328 kHz. The calculations in Chapter 7 showed that the dominant pole with magnitude 18.9 kHz is produced largely by shunt capacitance at the base of Q_{16}. Computer runs made with charge storage deleted in various parts of the circuit have shown that the pole with magnitude 328 kHz is also contributed by transistors Q_{16} and Q_{17}. The rest of the poles and zeros come from various parts of the circuit, including input and output stages.

After compensation, there is a dramatic change in the poles and zeros of the circuit as shown in Fig. 9.20*b*. The amplifier now has a dominant pole with magnitude 5 Hz as desired, but an additional bonus has been gained because the original second most dominant pole with magnitude 328 kHz has been effectively removed. The second and higher most dominant poles now form a cluster with

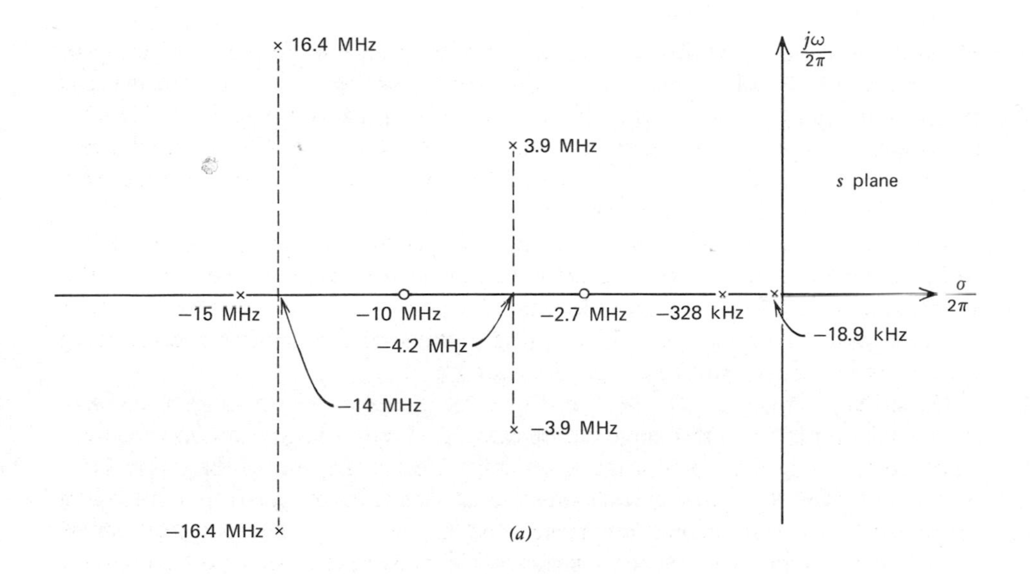

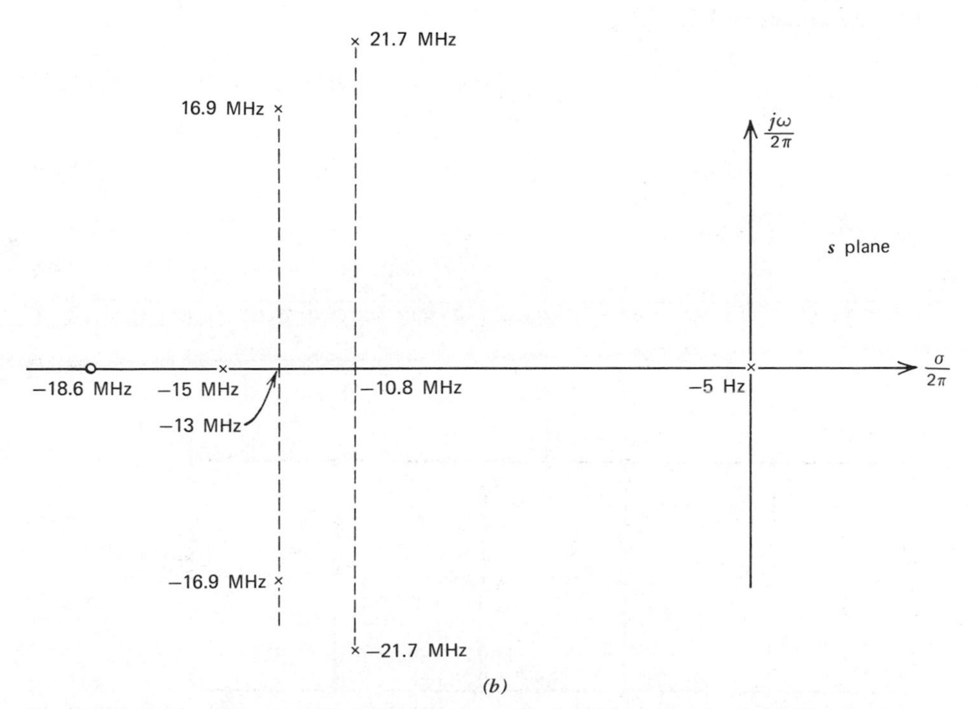

Fig. 9.20 Poles and zeros of the 741 op amp as calculated by computer. (*a*) Before compensation. (*b*) After compensation. (Not to scale.)

magnitudes 10 to 15 MHz and include complex poles. Since the amplifier gain must be made to fall to unity at a frequency below the second most dominant pole (for adequate phase margin), the removal of the pole with magnitude 328 kHz has greatly increased the realizable bandwidth of the circuit. If this pole did not move, the compensation capacitor would have to be adjusted to cause the gain of the 741 to fall to unity at a frequency *less than* 328 kHz.

The splitting of the two low-frequency poles of the 741 described above is a rather complex process involving other higher-frequency poles and zeros in the Darlington pair. However the process involved can be illustrated by assuming the Darlington pair is replaced by a single transistor. A similar pole-splitting process then occurs as will now be illustrated.

If the Darlington pair, Q_{16}–Q_{17} of Fig. 9.18, is replaced by a single transistor, the transfer function of this stage can be calculated from the approximate equivalent circuit of Fig. 9.21. The stage is fed from a current i_s out of the active load. Resistors R_1 and R_2 represent total shunt resistance at input and output, including transistor input and output resistance, and C_1 and C_2 represent total shunt capacitance at input and output. Capacitor C represents transistor collector-base capacitance plus compensation capacitance.

For the circuit of Fig. 9.21,

$$i_s = \frac{v_1}{R_1} + v_1 C_1 s + (v_1 - v_o)Cs \tag{9.25}$$

$$g_m v_1 + \frac{v_o}{R_2} + v_o C_2 s + (v_o - v_1)Cs = 0 \tag{9.26}$$

From (9.25) and (9.26)

$$\frac{v_o}{i_s} = -\frac{(g_m - Cs)R_2R_1}{1 + s[(C_2 + C)R_2 + (C_1 + C)R_1 + g_m R_2 R_1 C] + s^2 R_2 R_1 (C_2 C_1 + CC_2 + CC_1)} \tag{9.27}$$

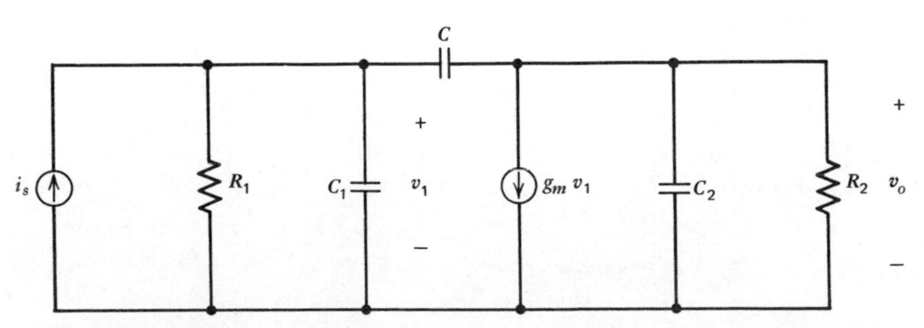

Fig. 9.21 Small-signal equivalent circuit of a single transistor stage. Feedback capacitor C includes compensation capacitance.

The circuit has a two-pole transfer function. If p_1 and p_2 are the poles of the circuit then the denominator of (9.27) can be written

$$D(s) = \left(1 - \frac{s}{p_1}\right)\left(1 - \frac{s}{p_2}\right) \tag{9.28}$$

$$= 1 - s\left(\frac{1}{p_1} + \frac{1}{p_2}\right) + \frac{s^2}{p_1 p_2} \tag{9.29}$$

and thus

$$D(s) \simeq 1 - \frac{s}{p_1} + \frac{s^2}{p_1 p_2} \tag{9.30}$$

if the poles are widely separated, which is usually true. Note that p_1 is assumed to be the dominant pole.

If the coefficients in (9.27) and (9.30) are equated then

$$p_1 = -\frac{1}{(C_2 + C)R_2 + (C_1 + C)R_1 + g_m R_2 R_1 C} \tag{9.31}$$

and this can be approximated by

$$p_1 \simeq -\frac{1}{g_m R_2 R_1 C} \tag{9.32}$$

since the Miller effect due to C will be dominant if C is large. Equation 9.31 is the same result for the dominant pole as is obtained using zero-value time constant analysis.

The nondominant pole, p_2, can now be estimated by equating coefficients of s^2 in (9.27) and (9.30) and using (9.32):

$$p_2 \simeq -\frac{g_m C}{C_2 C_1 + C(C_2 + C_1)} \tag{9.33}$$

Equation 9.32 indicates that the dominant-pole magnitude, $|p_1|$, *decreases* as C *increases* whereas (9.33) shows that $|p_2|$ *increases* as C *increases*. Thus, increasing C causes the poles to *split apart*. As C becomes very much greater than C_1 and C_2, the value of $|p_2|$ approaches $g_m/(C_1 + C_2)$, which is usually a relatively high frequency (tens of MHz for typical parameter values).

It is interesting to note that the poles of the circuit of Fig. 9.21 for $C = 0$ are at

$$p_1 = -\frac{1}{R_1 C_1} \tag{9.34}$$

$$p_2 = -\frac{1}{R_2 C_2} \tag{9.35}$$

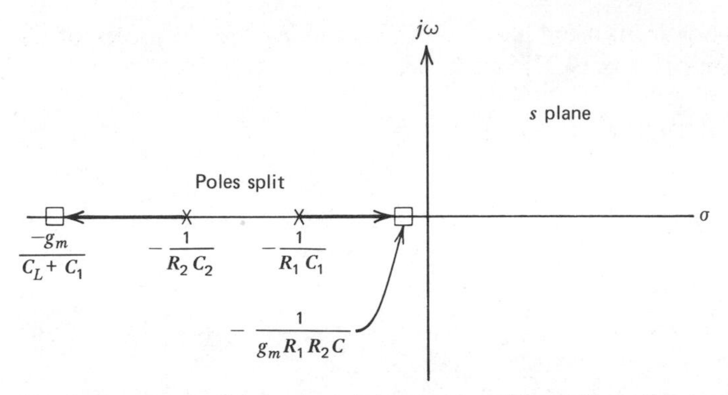

Fig. 9.22 Locus of the poles of the circuit of Fig. 9.21 as C is increased from zero.

Thus as C increases from zero, the locus of the poles of the circuit of Fig. 9.21 is as shown in Fig. 9.22.

The calculations above have shown how compensation of an amplifier by addition of a large Miller capacitance to a single transistor stage causes the non-dominant pole to move to a much higher frequency. A similar process occurs in the 741 and is the reason for the elimination of the pole with magnitude 328 kHz after compensation, as shown in Fig. 9.20. It is interesting to consider the performance attainable if the 741 is compensated by simply adding shunt capacitance to ground at the base of Q_{16} in Fig. 9.18. In order to achieve the same phase margin of 80° using the same device parameters, it was found by computer simulation that a compensation capacitance of 0.3 μF was required. This gave a dominant pole at 0.27 Hz and a unity-gain frequency of only 63 kHz. The second most dominant pole had a magnitude 294 kHz. Thus very little movement of the second most dominant pole had occurred, and the change that did occur was a *decrease* in the pole magnitude. As a consequence, the realizable bandwidth of the circuit when compensated in this way is only 1/20 of that obtained with Miller effect compensation. The other obvious disadvantage is that a capacitance of 0.3 μF cannot be included on the monolithic chip.

The dominant pole in this case can be calculated using data developed in Chapter 7. There it was shown that the total shunt resistance from the base of Q_{16} to ground was 1.95 MΩ, and combination of this resistance with 0.3 μF of capacitance gives a pole with magnitude

$$\frac{1}{2\pi RC} = \frac{1}{2\pi \times 1.95 \times 10^{6} \times 0.3 \times 10^{-6}} = 0.27 \text{ Hz}$$

This agrees exactly with the computer result.

The effect of different methods of compensation described above can be seen in the circuit of Fig. 9.21. If C_1 is made large in that circuit to produce a dominant pole, then the pole can be calculated from (9.31) as $p_1 \simeq -1/R_1C_1$. The non-dominant pole can be estimated by equating coefficients of s^2 in (9.27) and (9.30) and using the above value of p_1. This gives $p_2 \simeq -1/R_2(C_2 + C)$. This value of p_2 is approximately the same as that given by (9.35), which is for $C = 0$ and is *before* pole splitting occurs. Thus, creation of a dominant pole in the circuit of Fig. 9.21 by making C_1 large will result in a second pole magnitude $|p_2|$, which is much smaller than that obtained if the dominant pole is created by increasing C. The same general trend is true in the more complex situation that exists in the 741.

The results derived in this section are useful in further illuminating the considerations of Section 7.3.3. In that section, it was stated that in a common-emitter cascade, the existence of collector-base capacitance C_μ tends to cause pole splitting and to produce a dominant-pole situation. If the equivalent circuit of Fig. 9.21 is taken as a representative section of a cascade of common-emitter stages (C_2 is the input capacitance of the following stage and r_b is neglected) and capacitor C is taken as C_μ, the calculations of this section show that the presence of C_μ does, in fact, tend to produce a dominant-pole situation because of the pole splitting that occurs. Thus, the zero-value time constant approach gives a good estimate of $\omega_{-3\text{ dB}}$ in such circuits.

9.5 ROOT-LOCUS TECHNIQUES[5,6]

To this point the considerations of this chapter have been mainly concerned with calculations of feedback amplifier stability and compensation using frequency-domain techniques. Such techniques are widely used because they allow the design of feedback amplifier compensation without requiring excessive design effort. The *root-locus* technique involves calculation of the actual poles and zeros of the amplifier, and their movement in the s plane as the low-frequency, loop-gain magnitude T_0 is changed. This method thus gives more information about the amplifier performance than is given by frequency-domain techniques, but also requires more computational effort. In practice, some problems can be solved equally well using either method, whereas others yield more easily to one or the other. The circuit designer needs skill in applying both methods. The root-locus technique will be first illustrated with a simple example.

9.5.1 Root Locus for a Three-Pole Transfer Function

Consider an amplifier whose transfer function has three identical poles. The transfer function can be written as

$$a(s) = \frac{a_0}{\left(1 - \dfrac{s}{p_1}\right)^3} \tag{9.36}$$

where a_0 is the low-frequency gain and $|p_1|$ is the pole frequency. Consider this amplifier placed in a negative-feedback loop as in Fig. 9.1, where the feedback network has a transfer function f, which is a constant. If we assume that the effects of feedback loading are small, the overall gain with feedback is

$$A(s) = \frac{a(s)}{1 + a(s)f} \tag{9.37}$$

Using (9.36) in (9.37) gives

$$A(s) = \frac{\dfrac{a_0}{\left(1 - \dfrac{s}{p_1}\right)^3}}{1 + \dfrac{a_0 f}{\left(1 - \dfrac{s}{p_1}\right)^3}} = \frac{a_0}{\left(1 - \dfrac{s}{p_1}\right)^3 + T_0} \tag{9.38}$$

where $T_0 = a_0 f$ is the low-frequency loop gain.

The *poles* of $A(s)$ are the *roots* of the equation

$$\left(1 - \frac{s}{p_1}\right)^3 + T_0 = 0 \tag{9.39}$$

That is

$$\left(1 - \frac{s}{p_1}\right)^3 = -T_0$$

and thus

$$1 - \frac{s}{p_1} = \sqrt[3]{-T_0} = -\sqrt[3]{T_0} \quad \text{or} \quad \sqrt[3]{T_0}\, e^{j60} \quad \text{or} \quad \sqrt[3]{T_0}\, e^{-j60}$$

Thus the three roots of (9.39) are

$$\begin{aligned} s_1 &= p_1(1 + \sqrt[3]{T_0}) \\ s_2 &= p_1(1 - \sqrt[3]{T_0}\, e^{j60^\circ}) \\ s_3 &= p_1(1 - \sqrt[3]{T_0}\, e^{-j60^\circ}) \end{aligned} \tag{9.40}$$

These three roots are the poles of $A(s)$ and (9.38) can be written as

$$A(s) = \frac{a_0}{1 + T_0} \frac{1}{\left(1 - \dfrac{s}{s_1}\right)\left(1 - \dfrac{s}{s_2}\right)\left(1 - \dfrac{s}{s_3}\right)} \tag{9.41}$$

Equations 9.40 allow calculation of the poles of $A(s)$ for any value of low-frequency loop gain T_0. For $T_0 = 0$, all three poles are at p_1 as expected. As T_0 is made finite, one pole moves out along the negative real axis while the other two leave the axis at an angle of 60° and move toward the right half plane. The *locus of the roots* (or the *root locus*) is shown in Fig. 9.23, and each point of this root locus can be identified with the corresponding value of T_0. One point of significance on the root locus is the value of T_0 at which the two complex poles cross into the right half plane, as this is the value of loop gain causing *oscillation.* From the equation for s_2 in (9.40), this is where Re $(s_2) = 0$ from which we obtain

$$1 - \text{Re}\,(\sqrt[3]{T_0}\, e^{j60^\circ}) = 0$$

That is,

$$\sqrt[3]{T_0} \cos 60^\circ = 1$$

and

$$T_0 = 8$$

Thus, *any* amplifier with three identical poles becomes unstable for low-frequency loop gain T_0 greater than 8. This is quite a restrictive condition and emphasizes the need for compensation if larger values of T_0 are required. Note that not only does the root-locus technique give the value of T_0 causing instability, it also allows calculation of the amplifier poles for values of $T_0 < 8$, and thus allows calculation of both sinusoidal *and* transient response of the amplifier.

The frequency of oscillation can be found from Fig. 9.23 by calculating the distance

$$\omega_0 = |p_1| \tan 60^\circ = 1.732\, |p_1| \tag{9.42}$$

Thus, when the poles just enter the right half plane their imaginary part has a magnitude $1.732\, |p_1|$ and this will be the frequency of the increasing sinusoidal response. That is, if the complex poles are at $(\sigma \pm j\omega_0)$ where σ is small and positive, the transient response of the circuit contains a term $Ke^{\sigma t} \sin \omega_0 t$, which represents a *growing* sinusoid. (K is an initial condition.)

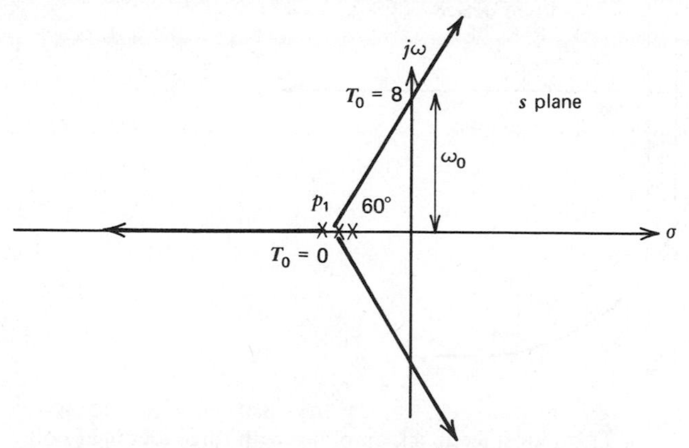

Fig. 9.23 Root locus for a feedback amplifier with three identical poles in $T(s)$.

It is useful to calculate the value of T_0 causing instability in this case by using the frequency-domain approach and the Nyquist criterion. From (9.36) the loop gain is

$$T(j\omega) = \frac{a_0 f}{\left(1 + \dfrac{j\omega}{|p_1|}\right)^3} = \frac{T_0}{\left(1 + j\dfrac{\omega}{|p_1|}\right)^3} \tag{9.43}$$

The magnitude and phase of $T(j\omega)$ as a function of ω are sketched in Fig. 9.24. The frequency ω_{180} where the phase shift of $T(j\omega)$ is $-180°$ can be calculated from (9.43) as

$$180 = 3 \arctan \frac{\omega_{180}}{|p_1|}$$

and this gives

$$\omega_{180} = 1.732\,|p_1| \tag{9.44}$$

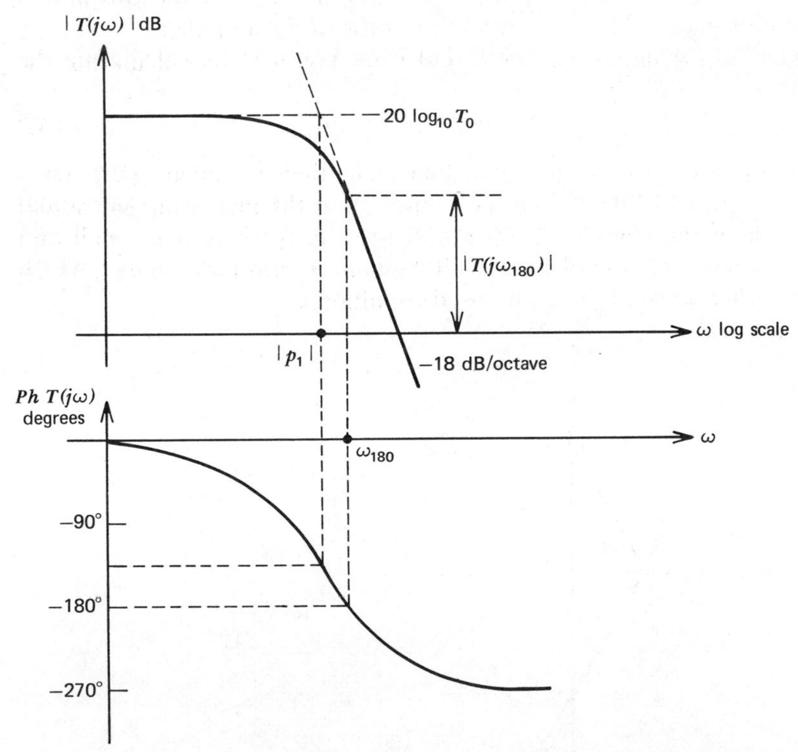

Fig. 9.24 Magnitude and phase of $T(j\omega)$ for a feedback amplifier with three identical poles in $T(s)$.

Comparing (9.42) with (9.44) shows that

$$\omega_{180} = \omega_0 \tag{9.45}$$

The loop-gain magnitude at ω_{180} can be calculated from (9.43) as

$$|T(j\omega_{180})| = \frac{T_0}{\left|1 + j\frac{\omega_{180}}{p_1}\right|^3} = \frac{T_0}{8} \tag{9.46}$$

using (9.44). The Nyquist criterion for stability indicates that it is necessary that $|T(j\omega_{180})| < 1$. This requires that $T_0 < 8$, the same result as obtained using root-locus techniques.

9.5.2 Rules for Root-Locus Construction

In the above simple example, it was possible to calculate exact expressions for the amplifier poles as a function of T_0, and thus to plot the root loci exactly. In most practical cases this is quite difficult since the solution of third- or higher-order polynomial equations is required. Consequently rules have been developed that allow the root loci to be sketched without requiring exact calculation of the pole positions, and much of the useful information is thus obtained without extensive calculation.

In general, the basic-amplifier transfer function and the feedback function may be expressed as a ratio of polynomials in s.

$$a(s) = a_0 \frac{1 + a_1 s + a_2 s^2 + \cdots}{1 + b_1 s + b_2 s^2 + \cdots} \tag{9.47}$$

This can be written as

$$a(s) = a_0 \frac{N_a(s)}{D_a(s)} \tag{9.48}$$

Also assume that

$$f(s) = f_0 \frac{1 + c_1 s + c_2 s^2 + \cdots}{1 + d_1 s + d_2 s^2 + \cdots} \tag{9.49}$$

This can be written as

$$f(s) = f_0 \frac{N_f(s)}{D_f(s)} \tag{9.50}$$

Loading produced by the feedback network on the basic amplifier is assumed to be included in (9.47). It is further assumed that the low-frequency loop gain, $a_0 f_0$, can be changed without changing the frequency response of $a(s)$ or $f(s)$.

The overall gain when feedback is applied is

$$A(s) = \frac{a(s)}{1 + a(s)f(s)} \tag{9.51}$$

Using (9.48) and (9.50) in (9.51) gives

$$A(s) = \frac{a_0 N_a(s) D_f(s)}{D_f(s) D_a(s) + T_0 N_a(s) N_f(s)} \tag{9.52}$$

where

$$T_0 = a_0 f_0 \tag{9.53}$$

is the low-frequency loop gain.

Equation 9.52 shows that the *zeros* of $A(s)$ are the *zeros* of $a(s)$ *and* the *poles* of $f(s)$. From (9.52) it is apparent that the *poles* of $A(s)$ are the roots of

$$D_f(s) D_a(s) + T_0 N_a(s) N_f(s) = 0 \tag{9.54}$$

Consider the two extreme cases.

(a) Assume that there is no feedback and that $T_0 = 0$. Then, from (9.54), the *poles* of $A(s)$ are the *poles* of $a(s)$ and $f(s)$. However the *poles* of $f(s)$ are also *zeros* of $A(s)$ and these cancel, leaving the *poles* of $A(s)$ composed of the *poles* of $a(s)$ as expected. The *zeros* of $A(s)$ are the *zeros* of $a(s)$ in this case.

(b) Let $T_0 \to \infty$. Then, (9.54) becomes

$$N_a(s) N_f(s) = 0 \tag{9.55}$$

This equation shows that the *poles* of $A(s)$ are now the zeros of $a(s)$ and the *zeros* of $f(s)$. However the *zeros* of $a(s)$ are also *zeros* of $A(s)$ and these cancel, leaving the *poles* of $A(s)$ composed of the *zeros* of $f(s)$. The *zeros* of $A(s)$ are the *poles* of $f(s)$ in this case.

Rule 1. The branches of the root locus start at the poles of $T(s) = a(s)f(s)$ where $T_0 = 0$, and terminate on the zeros of $T(s)$ where $T_0 = \infty$. If $T(s)$ has more poles than zeros, some of the branches of the root locus will terminate at infinity.

Examples of loci terminating at infinity are shown in Figs. 9.3 and 9.23.

More rules for the construction of root loci can be derived by returning to (9.54) and dividing it by $D_f(s)D_a(s)$. Poles of $A(s)$ are roots of

$$1 + T_0 \frac{N_a(s)}{D_a(s)} \frac{N_f(s)}{D_f(s)} = 0$$

That is

$$T_0 \frac{N_a(s)}{D_a(s)} \frac{N_f(s)}{D_f(s)} = -1$$

The complete expression including poles and zeros is

$$T_0 \frac{\left(1 - \frac{s}{z_{a1}}\right)\left(1 - \frac{s}{z_{a2}}\right)\cdots\left(1 - \frac{s}{z_{f1}}\right)\left(1 - \frac{s}{z_{f2}}\right)\cdots}{\left(1 - \frac{s}{p_{a1}}\right)\left(1 - \frac{s}{p_{a2}}\right)\cdots\left(1 - \frac{s}{p_{f1}}\right)\left(1 - \frac{s}{p_{f2}}\right)\cdots} = -1 \tag{9.56}$$

where

$z_{a1}, z_{a2} \cdots$ are zeros of $a(s)$
$z_{f1}, z_{f2} \cdots$ are zeros of $f(s)$
$p_{a1}, p_{a2} \cdots$ are poles of $a(s)$
$p_{f1}, p_{f2} \cdots$ are poles of $f(s)$

Equation 9.56 can be written as

$$T_0 \frac{(-p_{a1})(-p_{a2})\cdots(-p_{f1})(-p_{f2})\cdots}{(-z_{a1})(-z_{a2})\cdots(-z_{f1})(-z_{f2})\cdots} \times \frac{(s - z_{a1})(s - z_{a2})\cdots(s - z_{f1})(s - z_{f2})\cdots}{(s - p_{a1})(s - p_{a2})\cdots(s - p_{f1})(s - p_{f2})\cdots} = -1 \tag{9.57}$$

If the poles and zeros of $a(s)$ and $f(s)$ are restricted to the left half plane [this does *not* restrict the poles of $A(s)$] then $-p_{a1}$, $-p_{a2}$, and so on are *positive* numbers and (9.57) can be written

$$T_0 \frac{|p_{a1}|\cdot|p_{a2}|\cdots|p_{f1}|\cdot|p_{f2}|\cdots}{|z_{a1}|\cdot|z_{a2}|\cdots|z_{f1}|\cdot|z_{f2}|\cdots} \frac{(s - z_{a1})(s - z_{a2})\cdots(s - z_{f1})(s - z_{f2})\cdots}{(s - p_{a1})(s - p_{a2})\cdots(s - p_{f1})(s - p_{f2})\cdots} = -1 \tag{9.58}$$

Values of complex variable s satisfying (9.58) are *poles* of closed-loop function $A(s)$. Equation 9.58 requires the fulfilment of two conditions simultaneously, and these conditions are used to determine points on the root locus.

The *phase condition* for values of s satisfying (9.58) is

$$\underline{/s - z_{a1}} + \underline{/s - z_{a2}}\cdots + \underline{/s - z_{f1}} + \underline{/s - z_{f2}} + \cdots - (\underline{/s - p_{a1}} + \underline{/s - p_{a2}}\cdots + \underline{/s - p_{f1}} + \underline{/s - p_{f2}}) = (2n - 1)\pi \tag{9.59}$$

The magnitude condition for values of s satisfying (9.58) is

$$T_0 \frac{|p_{a1}|\cdot|p_{a2}|\cdots|p_{f1}|\cdot|p_{f2}|\cdots}{|z_{a1}|\cdot|z_{a2}|\cdots|z_{f1}|\cdot|z_{f2}|\cdots} \frac{|s - z_{a1}|\cdot|s - z_{a2}|\cdots|s - z_{f1}|\cdot|s - z_{f2}|\cdots}{|s - p_{a1}|\cdot|s - p_{a2}|\cdots|s - p_{f1}|\cdot|s - p_{f2}|\cdots} = 1 \tag{9.60}$$

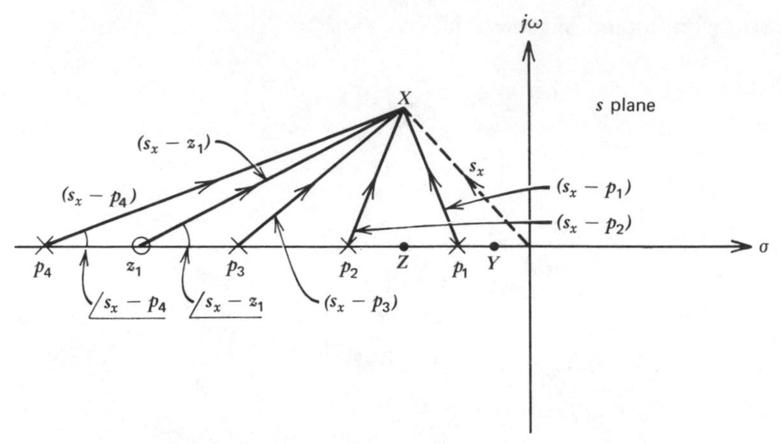

Fig. 9.25 Poles and zeros of loop gain $T(s)$ of a feedback amplifier. Vectors are drawn to the point X to determine if this is on the root locus.

Consider an amplifier with poles and zeros of $T(s)$ as shown in Fig. 9.25. In order to determine if some arbitrary point X is on the root locus, the phase condition of (9.59) is used. Note that the vectors of (9.59) are formed by drawing lines *from* the various poles and zeros of $T(s)$ to the point X and the angles of these vectors are then substituted in (9.59) to check the phase condition. This is readily done for points Y and Z on the axis.

At Y

$$\underline{/s_Y - z_1} = 0^\circ$$
$$\underline{/s_Y - p_1} = 0^\circ$$

and so on.

All angles are zero for point Y and thus the phase condition is not satisfied. This is obviously the case for all points to the right of p_1.

At Z,

$$\underline{/s_Z - z_1} = 0^\circ$$
$$\underline{/s_Z - p_1} = 180^\circ$$
$$\underline{/s_Z - p_2} = 0^\circ$$
$$\underline{/s_Z - p_3} = 0^\circ$$
$$\underline{/s_Z - p_4} = 0^\circ$$

In this case the phase condition of (9.59) *is* satisfied and points on the axis between p_1 and p_2 *are* on the locus. By similar application of the phase condition it is readily shown that the locus exists on the real axis between p_3 and z_1 and to the left of p_4. The following rule is true in general.

Rule 2. The locus is situated along the real axis whenever there is an odd number of poles and zeros of $T(s)$ to the right.

Consider again the situation in Fig. 9.25. Rule 1 indicates that branches of the locus must start at p_1, p_2, p_3, and p_4. Rule 2 indicates that the locus exists between p_3 and z_1, and thus the branch beginning at p_3 ends at z_1. Rule 2 also indicates that the locus exists to the left of p_4, and thus the branch beginning at p_4 moves out to infinity. The branches beginning at p_1 and p_2 must also terminate at infinity, and this is only possible if these branches *break away* from the axis as shown in Fig. 9.26. This can be stated as follows.

Rule 3. All segments of loci that lie on the real axis between pairs of poles (or pairs of zeros) of $T(s)$ must, at some internal break point, branch out from the real axis.

The following rules can be derived.

Rule 4. The locus is symmetrical with respect to the real axis (because complex roots occur only in conjugate pairs).

Rule 5 Branches of the locus that leave the real axis do so at *right angles*, as illustrated in Fig. 9.26.

Rule 6. If branches of the locus break away from the real axis, they do so at a point where the vector sum of reciprocals of distances to the poles of $T(s)$ equals the vector sum of reciprocals of distances to the zeros of $T(s)$.

Rule 7. Branches of the locus that terminate at infinity do so asymptotically to straight lines with angles to the real axis of $[(2n - 1)\pi]/(N_p - N_z)$ where N_p is the number of poles of $T(s)$ and N_z is the number of zeros of $T(s)$.

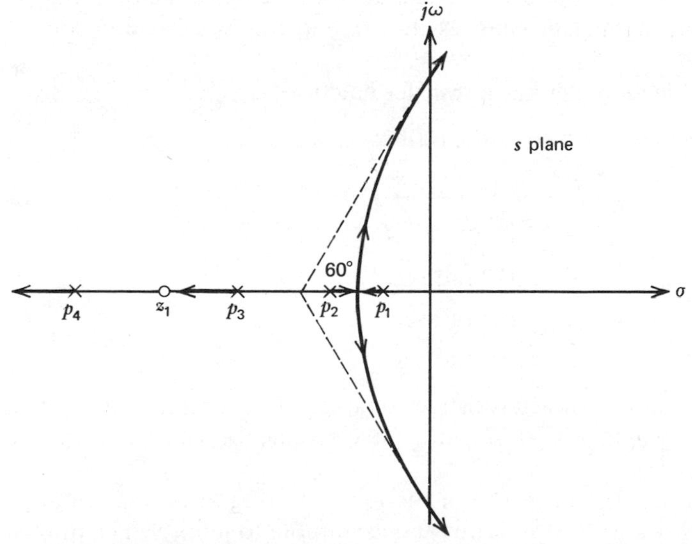

Fig. 9.26 Root-locus construction for the poles and zeros of Fig. 9.25.

Rule 8. The asymptotes of branches that terminate at infinity all intersect on the real axis at a point given by

$$\sigma_a = \frac{\sum[\text{poles of } T(s)] - \sum[\text{zeros of } T(s)]}{N_p - N_z} \tag{9.61}$$

A number of other rules have been developed for sketching root loci, but those described above are adequate for most requirements in amplifier design. The rules are used to obtain a rapid idea of the shape of the root locus in any situation, and to calculate amplifier performance in simple cases. More detailed calculation on circuits exhibiting complicated pole-zero patterns generally require computer calculation of the root locus.

Note that the above rules are all based on the *phase condition* of (9.59). Once the locus has been sketched, it can then be *calibrated* with values of low-frequency loop gain T_0 calculated at any desired point using the *magnitude condition* of (9.60).

The procedures described above will now be illustrated with examples.

EXAMPLE

In Section 9.5.1 the root locus was calculated for an amplifier with three identical poles. This example was chosen because it was analytically tractable. Now consider a more practical case where the amplifier has three nonidentical poles and resistive feedback is applied. It is required to plot the root locus for this amplifier as feedback factor f is varied (thus varying T_0) and it is assumed that variations in f do not cause significant changes in the basic-amplifier transfer function, $a(s)$. Referring to Chapter 8 will show that, for most practical circuits, feedback factor f can be varied without causing significant changes in $a(s)$, and this is especially true of op amp circuits.

Assume that the basic amplifier has a transfer function

$$a(s) = \frac{100}{\left(1 - \dfrac{s}{p_1}\right)\left(1 - \dfrac{s}{p_2}\right)\left(1 - \dfrac{s}{p_3}\right)} \tag{9.62}$$

where

$$p_1 = -1 \times 10^6 \text{ rad/sec}$$
$$p_2 = -2 \times 10^6 \text{ rad/sec}$$
$$p_3 = -4 \times 10^6 \text{ rad/sec}$$

Since the feedback circuit is assumed resistive, loop gain $T(s)$ contains three poles. The root locus is shown in Fig. 9.27, and, for convenience, the numbers are normalized to 10^6 rad/sec.

Rules 1 and 2 indicate that branches of the locus starting at poles p_1 and p_2 move toward each other and then split out and asymptote to infinity. The branch starting at pole p_3 moves out along the negative real axis to infinity.

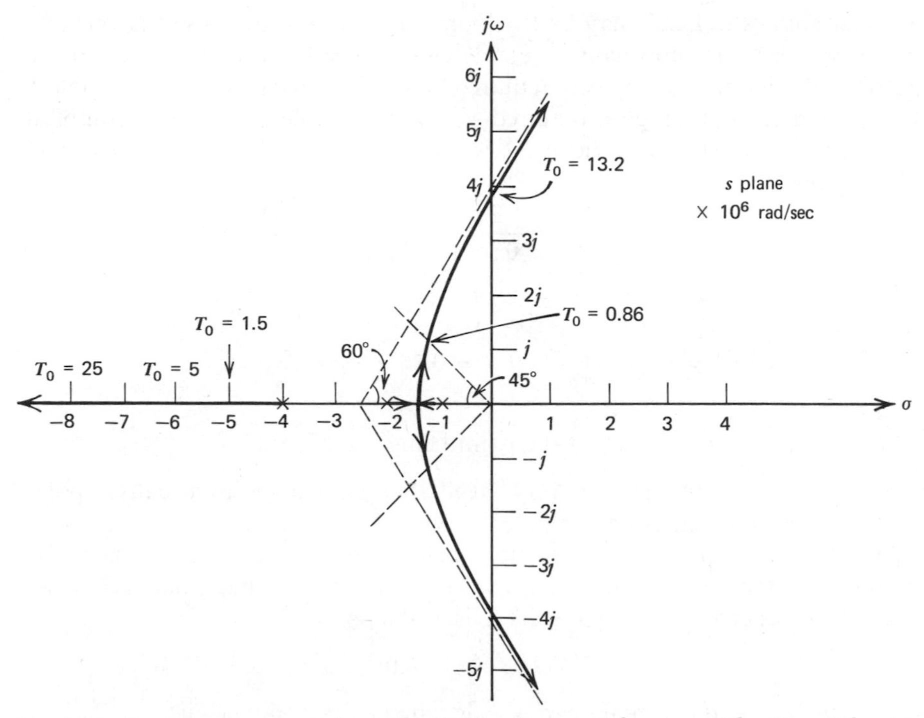

Fig. 9.27 Root-locus example for poles of $T(s)$ at -1×10^6, -2×10^6, -4×10^6 rad/sec.

The breakaway point for the locus between p_1 and p_2 can be calculated using rule 6. If σ_i is the coordinate of the breakaway point then

$$\frac{1}{\sigma_i + 1} + \frac{1}{\sigma_i + 2} + \frac{1}{\sigma_i + 4} = 0 \tag{9.63}$$

Solving this quadratic equation for σ_i gives $\sigma_i = -3.22$ or -1.45. The value -1.45 is the only possible solution because the breakaway point lies between -1 and -2 on the real axis.

The angles of the asymptotes to the real axis can be found using rule 7 and are $\pm 60^\circ$ and 180°. The asymptotes meet the real axis at a point whose coordinate is σ_a given by (9.61), and using (9.61) gives

$$\sigma_a = \frac{(-1 - 2 - 4) - 0}{3} = -2.33$$

When these asymptotes are drawn, the locus can be sketched as in Fig. 9.27 noting, from rule 5, that the locus leaves the real axis at right angles. The locus can now be calibrated for loop gain by using the magnitude condition of (9.60). Aspects of interest about the locus may be the loop gain required to cause the poles to become complex, the loop gain required for poles with an angle of 45° to the negative real axis and the loop gain required for oscillation (right half-plane poles).

Consider first the loop gain required to cause the poles to become complex. This is a point on the locus on the real axis at $\sigma_i = -1.45$. Substituting $s = -1.45$ in (9.60) gives

$$T_0 \frac{1 \times 2 \times 4}{0.45 \times 0.55 \times 2.55} = 1 \tag{9.64}$$

where

$$|p_1| = 1 \qquad |p_2| = 2 \qquad |p_3| = 4$$
$$|s - p_1| = 0.45 \qquad |s - p_2| = 0.55 \qquad |s - p_3| = 2.55$$

and

$$s = -1.45 \text{ at the point being considered}$$

From (9.64), $T_0 = 0.08$. Thus a very small loop-gain magnitude causes poles p_1 and p_2 to come together and split.

The loop gain required to cause right half-plane poles can be estimated by assuming that the locus coincides with the asymptote at that point. Thus we assume the locus crosses the imaginary axis at the point

$$j2.33 \tan 60° = 4.0j$$

Then the loop gain at this point can be calculated using (9.60) to give

$$T_0 \frac{1 \times 2 \times 4}{4.1 \times 4.5 \times 5.7} = 1 \tag{9.65}$$

where

$$|s - p_1| = 4.1 \qquad |s - p_2| = 4.5 \qquad |s - p_3| = 5.7$$

and

$$s = 4j \text{ at this point on the locus}$$

From (9.65) $T_0 = 13.2$. Since $a_0 = 100$ for this amplifier [from (9.62)] the overall gain of the feedback amplifier for $T_0 = 13.2$ is

$$A_0 = \frac{a_0}{1 + T_0} = 7.04$$

and

$$f = \frac{T_0}{a_0} = 0.132$$

The loop gain when the complex poles make an angle of 45° with the negative real axis can be calculated by making the assumption that this point has the same real-axis coordinate as the breakaway point. Then, using (9.60) with

$s = (-1.45 + 1.45j)$, we obtain

$$T_0 \frac{1 \times 2 \times 4}{1.52 \times 1.55 \times 2.93} = 1$$

and thus

$$T_0 = 0.86$$

Finally, the loop gain required to move the locus out from pole p_3 is of interest. When the real-axis pole is at -5 the loop gain can be calculated using (9.60) with $s = -5$ to give

$$T_0 \frac{1 \times 2 \times 4}{1 \times 3 \times 4} = 1$$

That is,

$$T_0 = 1.5$$

When this pole is at -6 the loop gain is

$$T_0 \frac{1 \times 2 \times 4}{2 \times 4 \times 5} = 1$$

and thus

$$T_0 = 5$$

These values are marked on the root locus of Fig. 9.27.

In this example, it is useful to compare the prediction of instability at $T_0 = 13.2$ with the results using the Nyquist criterion. The loop gain in the frequency domain is

$$T(j\omega) = \frac{T_0}{\left(1 + \dfrac{j\omega}{10^6}\right)\left(1 + \dfrac{j\omega}{2 \times 10^6}\right)\left(1 + \dfrac{j\omega}{4 \times 10^6}\right)} \tag{9.66}$$

A series of trial substitutions shows that $\angle T(j\omega) = -180^\circ$ for $\omega = 3.8 \times 10^6$ rad/sec. Note that this is close to the value of 4×10^6 rad/sec where the root locus was assumed to cross the $j\omega$ axis. Substitution of $\omega = 3.8 \times 10^6$ in (9.66) gives, for the loop gain at that frequency,

$$|T(j\omega)| = \frac{T_0}{11.6} \tag{9.67}$$

Thus, for stability, the Nyquist criterion requires that $T_0 < 11.6$ and this is close to the answer obtained from the root locus. If the point on the $j\omega$ axis where the root locus crossed had been determined more accurately, it would have been found to be at 3.8×10^6 rad/sec and both methods would predict instability for $T_0 > 11.6$.

It should be pointed out that the root locus of Fig. 9.27 shows the movement of the *poles* of the feedback amplifier as T_0 changes. The theory developed in

Section 9.5.2 showed that the *zeros* of the feedback amplifier are the *zeros* of the basic amplifier and the *poles* of the feedback network. In this case there are no zeros in the feedback amplifier, but this is not always the case. It should be kept in mind that if the basic amplifier has zeros in its transfer function, these may be an important part of the overall transfer function.

9.5.3 Root Locus for Dominant-Pole Compensation

Consider an op amp that has been compensated by creation of a dominant pole at p_1. If we assume the second most dominant pole is at p_2 and neglect the effect of higher-order poles, the root locus when resistive feedback is applied is as shown in Fig. 9.28. Using rules 1 and 2 indicates that the root locus exists on the axis between p_1 and p_2, and the breakaway point is readily shown to be

$$\sigma_i = \frac{p_1 + p_2}{2} \tag{9.68}$$

using rule 6. Using rules 7 and 8 shows that the asymptotes are at 90° to the real axis and meet the axis at σ_i.

As T_0 is increased, the branches of the locus come together and then split out to become complex. As T_0 becomes large, the imaginary part of the poles becomes large, and the circuit will then have a high-frequency *peak* in its overall gain function $A(j\omega)$. This is consistent with the previous viewpoint of gain peaking that occurred with diminishing phase margin.

Assume that maximum bandwidth in this amplifier is required, but that little or no peaking is allowed. This means that with maximum loop gain applied, the

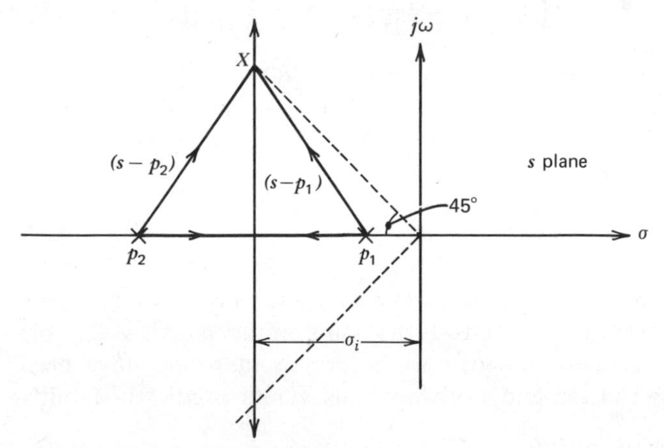

Fig. 9.28 Root locus for an op amp with two poles in its transfer function. The feedback is assumed resistive.

poles should not go beyond the points marked X on the locus where an angle of 45° is made between the negative real axis and a line drawn from X to the origin. At X, the loop gain can be calculated using (9.60):

$$T_0 \frac{|p_1| \cdot |p_2|}{|s - p_1| \cdot |s - p_2|} = 1 \tag{9.69}$$

If p_1 is a dominant pole, we can assume that $|p_1| \ll |p_2|$ and $\sigma_i = p_2/2$. For poles at 45°, $|s - p_1| = |s - p_2| \simeq \sqrt{2}\,|p_2|/2$. Thus, (9.69) becomes

$$T_0 = \frac{1}{|p_1| \cdot |p_2|}\left(\sqrt{2}\frac{|p_2|}{2}\right)^2$$

This gives

$$T_0 = \frac{1}{2}\frac{|p_2|}{|p_1|} \tag{9.70}$$

for the value of T_0 required to produce poles at X in Fig. 9.28. The effect of narrow-banding the amplifier is now apparent. As $|p_1|$ is made smaller, it requires a larger value of T_0 to move the poles out to 45°. From (9.70), the dominant-pole magnitude, $|p_1|$, required to ensure adequate performance with a given T_0 and $|p_2|$, can be calculated.

EXAMPLE

Estimate the loop gain required to produce poles at 45° to the negative real axis for the 741 op amp before and after compensation with 0.3 μF connected from the base of Q_{16} to ground.

Before compensation, thc two most important poles of the 741 have magnitudes 18.9 kHz and 328 kHz. If we assume that these have a dominant effect on the root locus near the origin, the loop gain required to produce poles at 45° can be calculated from (9.70) as

$$T_0 = \frac{1}{2}\frac{328}{18.9} = 8.7 \text{ (18.8 dB)}$$

Thus, without compensation, the amount of loop gain that can be used is severely limited.

After compensation with 0.3 μF, previous calculations showed a dominant pole with magnitude 0.27 kHz with the second most dominant pole with magnitude 294 kHz. If we assume that these two poles dominate the behavior, (9.70) gives for the loop gain required to produce poles at 45° to the real axis as

$$T_0 = \frac{1}{2}\frac{294{,}000}{0.27} = 544{,}000 = 115 \text{ dB}$$

Since the open-loop gain of the amplifier is 108 dB, even unity-gain feedback ($f = 1$) will not produce peaking in this case, for T_0 would then be only 108 dB.

9.5.4 Root Locus for Feedback-Zero Compensation

The techniques of compensation described earlier in this chapter involved modification of the basic amplifier only. This is the universal method used with op amps that must be compensated for use with a wide variety of feedback networks chosen by the user. However, this method is quite wasteful of bandwidth as was apparent in the calculations.

In this section, a different method of compensation will be described that involves modification of the feedback path and is generally limited to fixed-gain amplifiers. This method finds application in the compensation of wideband feedback amplifiers where bandwidth is of prime importance. An example is the shunt-series feedback amplifier of Fig. 8.32, which is known as a *current feedback pair.* The method is generally useful in amplifiers of this type where the feedback is over two stages, and in circuits such as the series-series triple of Fig. 8.19*a*.

A shunt-series feedback amplifier including a feedback capacitor C_F is shown in Fig. 9.29. The basic amplifier including feedback loading for this circuit is shown in Fig. 9.30. Capacitors C_F at input and output have only a minor effect on the circuit transfer function. The feedback circuit for this case is shown in Fig. 9.31 and feedback function f is given by

$$f = \frac{i_1}{i_2} = -\frac{R_E}{R_F + R_E} \frac{1 + R_F C_F s}{1 + \dfrac{R_E R_F}{R_E + R_F} C_F s} \tag{9.71}$$

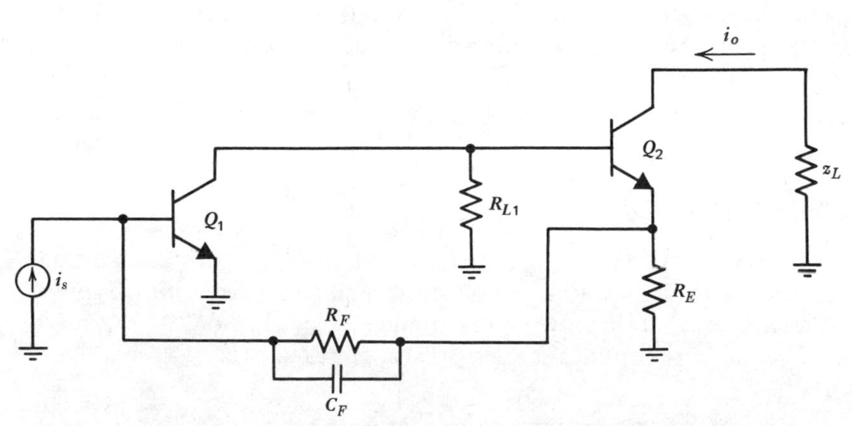

Fig. 9.29 Shunt-series feedback amplifier including a feedback capacitor C_F.

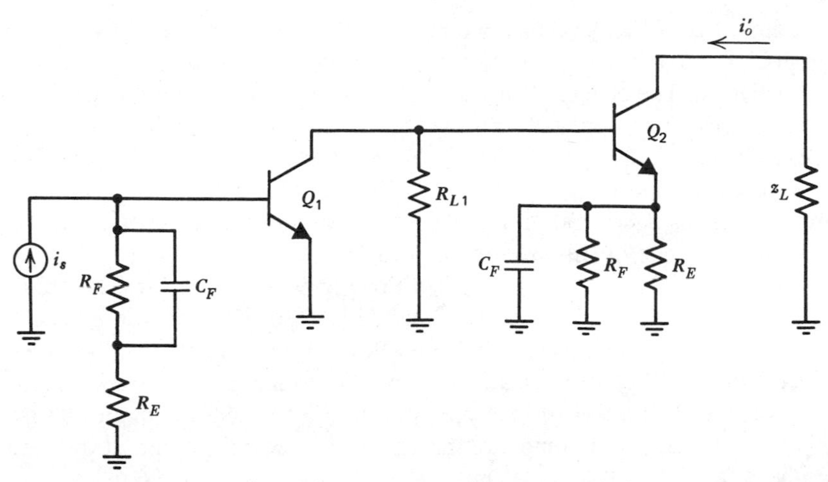

Fig. 9.30 Basic amplifier including feedback loading for the circuit of Fig. 9.29.

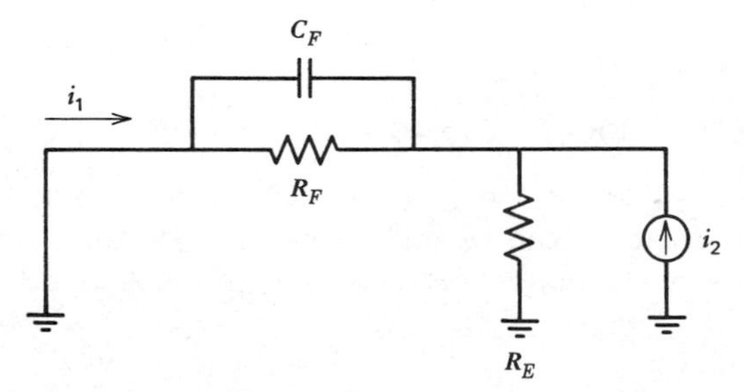

Fig. 9.31 Circuit for the calculation of feedback function f for the amplifier of Fig. 9.29.

Feedback function f thus contains a zero at a frequency

$$\omega_z = \frac{1}{R_F C_F} \tag{9.72}$$

and a pole at a frequency

$$\omega_p = \frac{R_E + R_F}{R_E} \frac{1}{R_F C_F} \tag{9.73}$$

Quantity $(R_E + R_F)/R_E$ is approximately the low-frequency gain of the overall circuit with feedback applied, and, since it is usually true that $(R_E + R_F)/R_E \gg 1$, the pole frequency given by (9.73) is usually much larger than the zero frequency. This will be assumed and the pole will be neglected, but if $(R_E + R_F)/R_E$ becomes comparable to unity the pole will be important and must be included.

The basic amplifier of Fig. 9.30 has two important poles contributed by Q_1 and Q_2. Although higher frequency poles exist, these do not have a dominant influence and will be neglected. The effects of this assumption will be investigated later. The loop gain of the circuit of Fig. 9.29 thus contains two forward-path poles and a feedback zero, giving rise to the root locus of Fig. 9.32. For purposes of illustration, the two poles are assumed to be $p_1 = -10 \times 10^6$ rad/sec and $p_2 = -20 \times 10^6$ rad/sec and the zero is $z = -50 \times 10^6$ rad/sec. For convenience in the calculations, the numbers will be normalized to 10^6 rad/sec.

Assume now that the loop gain of the circuit of Fig. 9.29 can be varied without changing the parameters of the basic amplifier of Fig. 9.30. Then a root locus can be plotted as the loop gain changes, and using rules 1 and 2 indicates that the root locus exists on the axis between p_1 and p_2, and to the left of z. The root locus must thus break away from the axis between p_1 and p_2 at σ_1 as shown, and return again at σ_2. One branch then extends to the right along the axis to end at the zero while the other branch heads towards infinity on the left. Using rule 6 gives

$$\frac{1}{\sigma_1 + 10} + \frac{1}{\sigma_1 + 20} = \frac{1}{\sigma_1 + 50} \tag{9.74}$$

Solution of (9.74) for σ_1 gives

$$\sigma_1 = -84.6 \quad \text{or} \quad -15.4$$

Obviously $\sigma_1 = -15.4$ and the other value is $\sigma_2 = -84.6$. Note that these points are equidistant from the zero, and, in fact, it can be shown that in this example the portion of the locus that is off the real axis is a *circle* centered on the zero. An

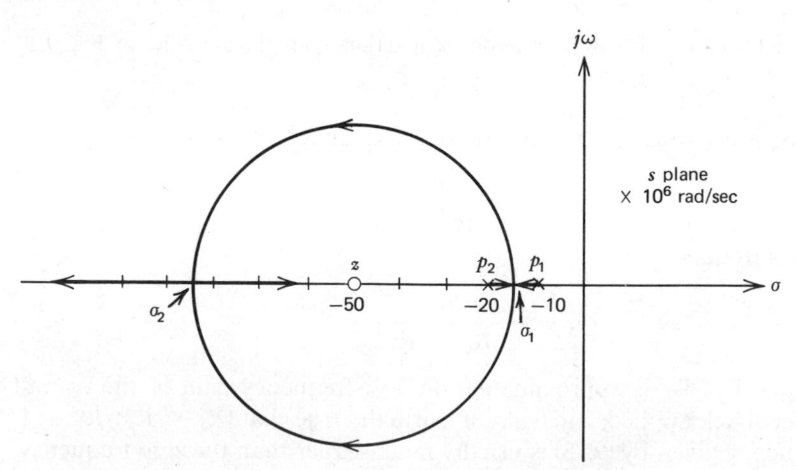

Fig. 9.32 Root locus for the circuit of Fig. 9.29 assuming the basic amplifier contributes two poles to $T(s)$ and the feedback circuit contributes one zero.

aspect of the root-locus diagrams that is a useful aid in sketching the loci is apparent from Fig. 9.27 and Fig. 9.32. The locus tends to *bend toward* zeros as if attracted and tends to *bend away* from poles as if repelled.

The effectiveness of the feedback zero in compensating the amplifier is apparent from Fig. 9.32. If we assume that the amplifier has poles p_1 and p_2 and there is no feedback zero, then when feedback is applied the amplifier poles will split out and move parallel to the $j\omega$ axis. For practical values of loop gain T_0, this would result in "high Q" poles near the $j\omega$ axis, which would give rise to an excessively *peaked* response. In practice, oscillation can occur because higher-frequency poles do exist and these would tend to give a locus of the kind of Fig. 9.27 where the remote poles cause the locus to bend and enter the right half plane. (Note that this behavior is consistent with the alternative approach of considering a diminished phase margin to be causing a peaked response and eventual instability.) The inclusion of the feedback zero, however, *bends* the locus away from the $j\omega$ axis and allows the designer to position the poles in any desired region.

An important point that should be stressed is that the root locus of Fig. 9.32 gives the *poles* of the feedback amplifier. The zero in that figure is a zero of loop gain $T(s)$ and thus must be included in the root locus. However, the zero is contributed by the *feedback network* and is *not* a zero of the overall feedback amplifier. As pointed out in Section 9.5.2 the *zeros* of the overall feedback amplifier are the *zeros* of basic amplifier $a(s)$ and the *poles* of feedback network $f(s)$. Thus the transfer function of the overall feedback amplifier in this case has two poles and no zeros as shown in Fig. 9.33, and the poles are assumed placed at 45° to the axis by appropriate choice of z. Since the feedback zero affects the root locus but does *not* appear as a zero of the overall amplifier, it has been called a *phantom zero*.

On the other hand, if the zero, z, were contributed by the basic amplifier the situation would be different. For the same zero, the root locus would be identical but the transfer function of the overall feedback amplifier would then *include* the

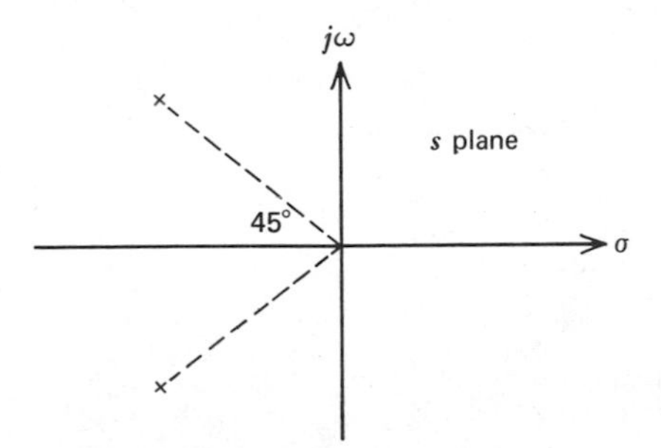

Fig. 9.33 Poles of the transfer function of the feedback amplifier of Fig. 9.29. The transfer function contains no zeros.

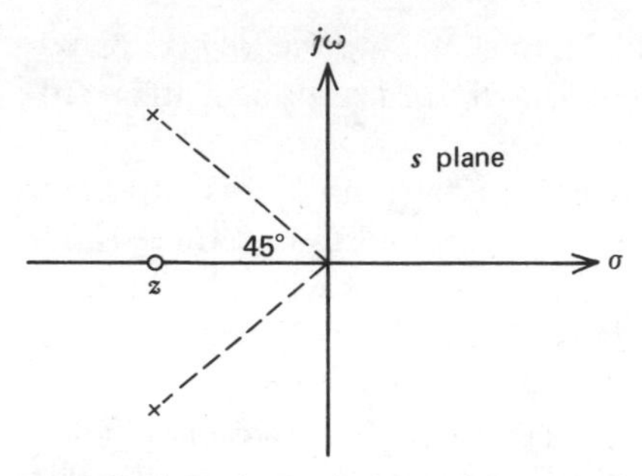

Fig. 9.34 Poles and zeros of the transfer function of the feedback amplifier of Fig. 9.29 if the zero is assumed contributed by the basic amplifier.

zero as shown in Fig. 9.34. This zero would then have a significant effect on the amplifier characteristics. This point is made simply to illustrate the difference between forward path and feedback-path zeros. There is no practical way to introduce a useful forward-path zero in this situation.

Before leaving this subject, we mention the effect of higher-frequency poles on the root locus of Fig. 9.32, and this is illustrated in Fig. 9.35. A remote pole p_3 will cause the locus to deviate from the original as shown and produce poles with a larger imaginery part than expected. The third pole, which is on the real axis, may also be significant in the final amplifier. Acceptable performance can usually be obtained by modifying the value of z from that calculated above.

Finally, the results derived in this chapter explain the function of capacitors C_P and C_F in the circuit of the MC 1553 series-series triple of Fig. 8.22a, which was

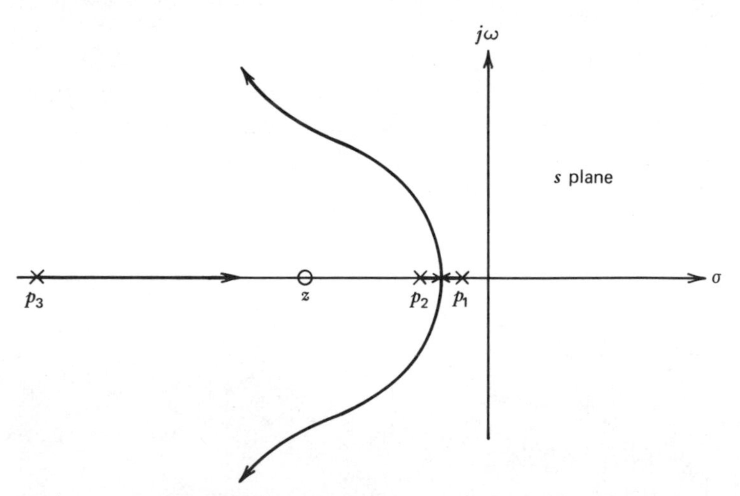

Fig. 9.35 Root locus of the circuit of Fig. 9.29 when an additional pole of the basic amplifier is included. (Not to scale.)

described in Chapter 8. Capacitor C_P causes pole splitting to occur in stage Q_2 and produces a dominant pole in the basic amplifier, which aids in the compensation. However, as described above, a large value of C_P will cause significant loss of bandwidth in the amplifier, and so a feedback zero is introduced via C_F, which further aids in the compensation by moving the root locus away from the axis. The final design is a combination of two methods of compensation in an attempt to find an optimum solution.

9.6 SLEW RATE[4]

The previous sections of this chapter have been concerned with the small-signal behavior of feedback amplifiers at high frequencies. However, the behavior of feedback circuits with large input signals (either step inputs or sinusoidal signals) is also of interest, and the effect of frequency compensation on the large-signal, high-frequency performance of feedback amplifiers is now considered.

9.6.1 Origin of Slew-Rate Limitations

A common test of the high-frequency, large-signal performance of an amplifier is to apply a step input voltage from 0 to +5 V as shown in Fig. 9.36. This figure shows an op amp in a unity-gain feedback configuration and will be used for purposes of illustration in this development. Suppose initially that the circuit operates linearly when this input is applied, and further that the circuit has a single-pole transfer function given by

$$\frac{V_o}{V_i}(s) = \frac{A}{1 + s\tau} \tag{9.75}$$

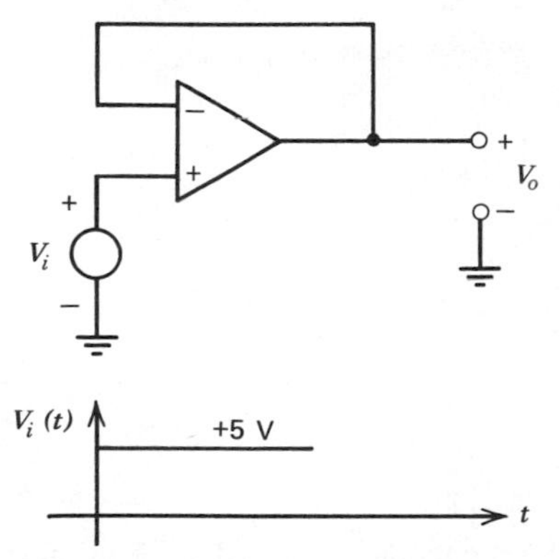

Fig. 9.36 Circuit for testing slew-rate performance.

where

$$\tau = \frac{1}{2\pi f_o} \tag{9.76}$$

and f_o is the -3-dB frequency. Since the circuit is connected as a voltage follower, the low-frequency gain, A, will be close to unity. If we assume that this is so, the response of the circuit to this step input $[V_i(s) = 5/s]$ is given by

$$V_o(s) = \frac{1}{1 + s\tau}\frac{5}{s} \tag{9.77}$$

using (9.75). Equation 9.77 can be factored to the form

$$V_o(s) = \frac{5}{s} - \frac{5}{s + \frac{1}{\tau}} \tag{9.78}$$

From (9.78)

$$V_o(t) = 5(1 - e^{-t/\tau}) \tag{9.79}$$

The predicted response from (9.79) is shown in Fig. 9.37*a* using data for the 741 op amp with $f_o \simeq 1.3$ MHz. This shows an exponential rise of $V_o(t)$ to 5 V and the output reaches 90 percent of its final value in about 0.3 μsec.

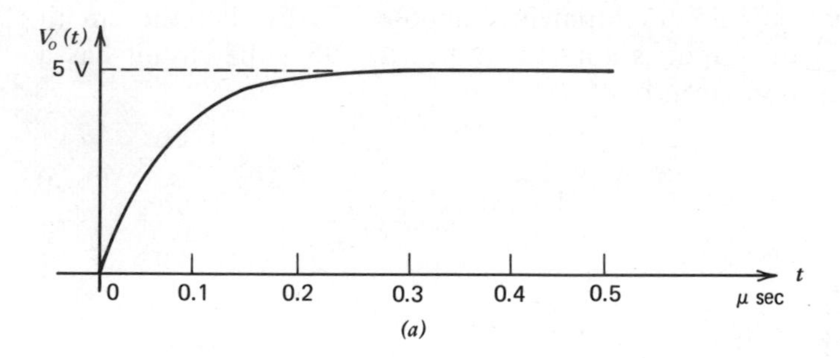

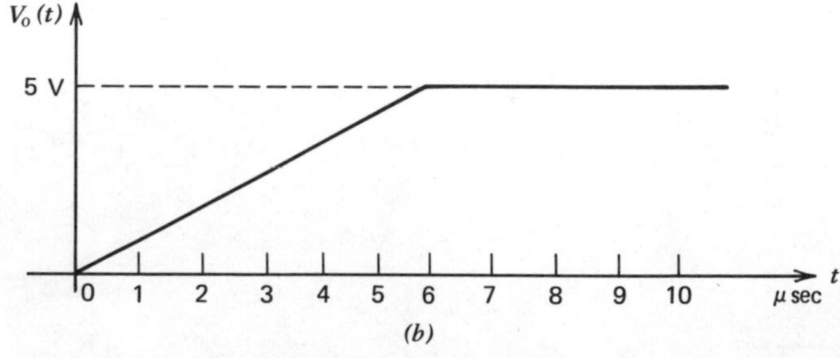

Fig. 9.37 Response of the circuit of Fig. 9.36 when a 5-V step input is applied. (*a*) Response predicted by (9.79) for the 741 op amp. (*b*) Measured response for the 741.

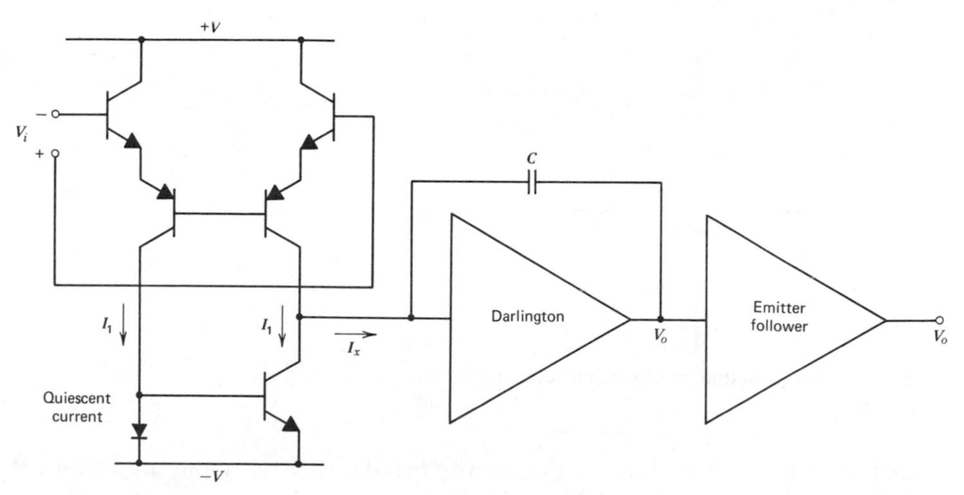

Fig. 9.38 Simplified schematic of the 741 op amp for slew-rate calculations.

A typical measured output for a 741 op amp in such a test is shown in Fig. 9.37*b* and exhibits a completely different response. The output voltage is a slow ramp of almost constant slope and takes about 5 μsec to reach 90 percent of its final value. Obviously the small-signal linear analysis is inadequate for predicting the circuit behavior under these conditions. The response shown in Fig. 9.37*b* is typical of op amp performance with a large input step voltage applied. The rate of change of output voltage dV_o/dt in the region of constant slope is called the *slew rate* and is usually specified in V/μsec.

The reason for the discrepancy between predicted and observed behavior noted above can be appreciated by examining the circuit of Fig. 9.36 and considering the responses of Fig. 9.37. At $t = 0$, the input voltage steps to +5 V but the output voltage cannot respond instantaneously and is initially zero. Thus the op amp input has a 5-V differential signal applied to it that is sufficient to drive the input stage completely out of its linear range of operation. This can be seen by taking the 741 as an example and a schematic for use in this analysis is shown in Fig. 9.38. The compensation capacitor, C, connected around the Darlington pair causes this stage to act as an integrator, and the current from the input stage, which charges the compensation point, is I_x. The large-signal transfer characteristic from input voltage V_i to I_x is that of a differential pair with four junctions in series and is shown in Fig. 9.39. This characteristic is similar to that derived for a simple differential pair in Section 3.4.1. From Fig. 9.39 it can be seen that the maximum current available to charge C is $2I_1$ where I_1 is the quiescent collector current per device in the input stage. Note that I_x is within 10 percent of its final value for $V_i = 120$ mV, and thus when differential signals of 5 V are applied as described above the input stage "limits" and $|I_x| \simeq 2I_1$. The circuit thus operates *nonlinearly*

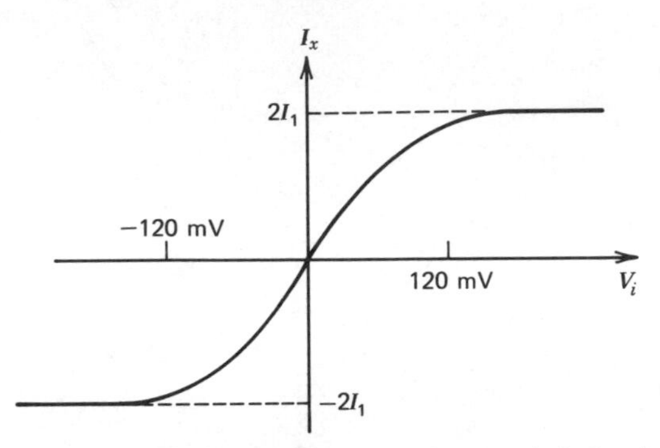

Fig. 9.39 Large-signal transfer characteristic for the 741 input stage.

and the linear analysis fails to predict the behavior. If the input stage did act linearly, the input voltage of 5 V would produce a very large current I_x to charge the compensation capacitor. The fact that this is limited to the fairly small value of $2I_1$ is the reason for the slew rate being much less than a linear analysis would predict.

Consider a large input voltage applied to the circuit of Fig. 9.38 so that $I_x = 2I_1$. Then the Darlington stage acts as an integrator with an input current $2I_1$ and the output voltage V_o can be written as

$$V_o = \frac{1}{C}\int 2I_1\, dt \tag{9.80}$$

and thus

$$\frac{dV_o}{dt} = \frac{2I_1}{C} \tag{9.81}$$

Equation 9.81 predicts a constant rate of change of V_o during the slewing period, which is in agreement with the experimental observation. For the 741, $I_1 = 12\ \mu\text{A}$ and $C = 30\text{ pF}$ giving $dV_o/dt = 0.8\text{ V}/\mu\text{sec}$, which is close to the measured value. It is apparent that the slew-rate limitation occurs because of the limited current available to charge the compensation capacitor during a transient. Thus, the worst-case slew rate is found when the circuit is compensated for unity-gain operation, and the circuit compensation should be specified when comparing slew-rate performance. The above calculation of slew rate was performed on the circuit of Fig. 9.38, which has no overall feedback. Since the input stage is completely cut off during the slewing period, the presence of a feedback connection to the input does not affect the circuit operation during this time. Thus, the slew rate of the amplifier is the same whether feedback is applied or not.

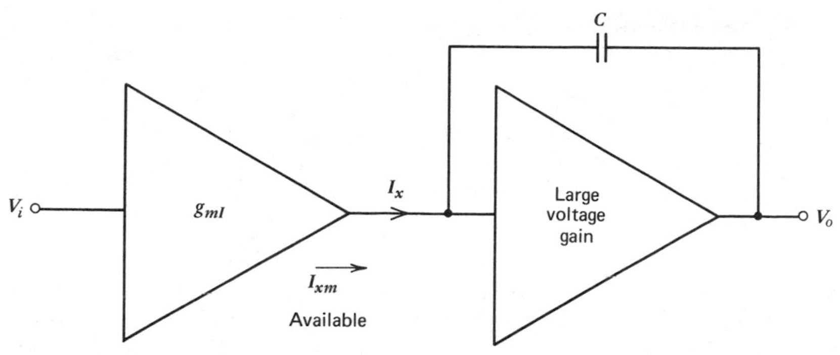

Fig. 9.40 Generalized representation of an op amp for slew-rate calculations.

9.6.2 Methods of Improving Slew Rate

In order to examine methods of slew-rate improvement, a more general analysis is required. This can be performed using the circuit of Fig. 9.40, which is a general representation of an op amp circuit. The input stage has a small-signal transconductance g_{mI} and, with a large input voltage, can deliver a maximum current I_{xm} to the compensation point. The compensation is shown as the Miller effect using the capacitor, C, since this representation describes most modern integrated-circuit op amps.

From Fig. 9.40 and using (9.81) we can calculate the slew rate for a large input voltage as

$$\frac{dV_o}{dt} = \frac{I_{xm}}{C} \tag{9.82}$$

Consider now *small-signal* operation. For the input stage, the small-signal transconductance is

$$\frac{\Delta I_x}{\Delta V_i} = g_{mI} \tag{9.83}$$

For the compensation stage (which acts as an integrator) the transfer function at high frequencies is

$$\frac{\Delta V_o}{\Delta I_x} = \frac{1}{sC}$$

and in the frequency domain

$$\frac{\Delta V_o}{\Delta I_x}(j\omega) = \frac{1}{j\omega C} \tag{9.84}$$

Combining (9.83) and (9.84) gives

$$\frac{\Delta V_o}{\Delta V_i}(j\omega) = \frac{g_{mI}}{j\omega C} \tag{9.85}$$

In our previous consideration of compensation, it was shown that the small-signal, open-loop voltage gain $(\Delta V_o/\Delta V_i)(j\omega)$ must fall to unity at or before the frequency of the second most dominant pole, (ω_2). If we assume, for ease of calculation, that the circuit is compensated for unity-gain operation with 45° phase margin as shown in Fig. 9.15, the gain $(\Delta V_o/\Delta V_i)(j\omega)$ as given by (9.85) must fall to unity at frequency ω_2. (Compensation capacitor C must be chosen to ensure this occurs.) Thus, from (9.85),

$$1 = \frac{g_{mI}}{\omega_2 C}$$

and thus

$$\frac{1}{C} = \frac{\omega_2}{g_{mI}} \tag{9.86}$$

Note that (9.86) was derived on the basis of a *small-signal* argument. This can now be substituted in the *large-signal* equation (9.82) to give

$$\text{Slew rate} = \frac{dV_o}{dt} = \frac{I_{xm}}{g_{mI}}\omega_2 \tag{9.87}$$

Equation 9.87 allows consideration of the effect of circuit parameters on slew rate and it is apparent that, for a given ω_2, ratio I_{xm}/g_{mI} must be increased if slew rate is to be increased. In the case of the 741 op amp, we have $I_{xm} = 2I_1$, $g_{mI} = qI_1/2kT$ and substitution in (9.87) gives

$$\text{Slew rate} = 4\frac{kT}{q}\omega_2 \tag{9.88}$$

Since both I_{xm} and g_{mI} are proportional to bias current I_1, the influence of I_1 cancels in the equation and slew rate is *independent* of I_1 for a given ω_2. Obviously, however, increasing ω_2 will increase the slew rate and this course is followed in most high-slew-rate circuits. The limit here is set by the frequency characteristics of the transistors in the IC process, and further improvements depend on circuit modifications as described below.

The above calculation has shown that varying the input-stage bias current of a 741-type op amp does not change the circuit slew rate. However, (9.87) indicates that for a given I_{xm}, slew rate can be increased by *reducing* the input-stage transconductance. One way this can be achieved is by including emitter-degeneration resistors to reduce g_{mI} as shown in Fig. 9.41. The small-signal transconductance

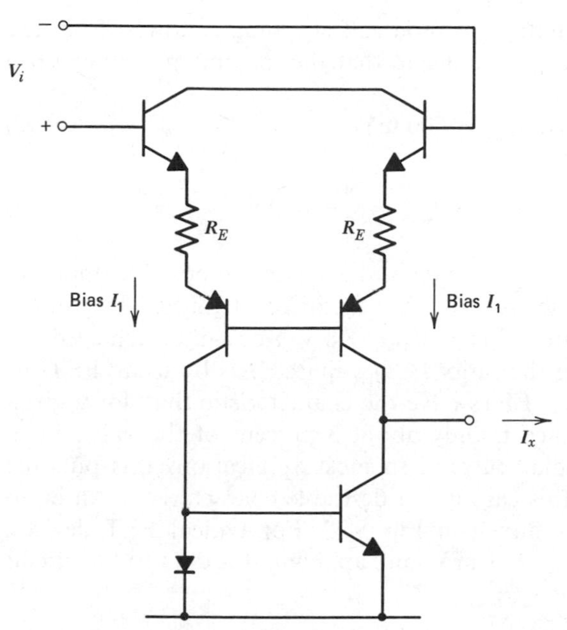

Fig. 9.41 Inclusion of emitter resistors in the input stage of the 741 op amp in order to improve the slew rate.

of this input stage can be shown to be

$$g_{mI} = \frac{\Delta I_x}{\Delta V_i} = \frac{g_{m1}}{2} \frac{1}{1 + \dfrac{g_{m1} R_E}{2}} \tag{9.89}$$

where

$$g_{m1} = \frac{q I_1}{kT} \tag{9.90}$$

The value of I_{xm} is still $2I_1$. Substituting (9.89) in (9.87) gives

$$\text{Slew rate} = \frac{4kT}{q} \omega_2 \left(1 + \frac{g_{m1} R_E}{2}\right) \tag{9.91}$$

Thus the slew rate is increased by the factor $(1 + g_{m1} R_E/2)$ over the value given by (9.88). The fundamental reason for this is that, for a given bias current I_1, reducing g_{mI} reduces the compensation capacitor C required, as shown by (9.86).

The practical limit to this technique is due to the fact that the emitter resistors of Fig. 9.41 have a dc voltage across them, and mismatches in the resistor values give rise to an input dc offset voltage. The use of large-area resistors (at least 1 mil wide) can give resistors whose values match to within 0.2 percent (1 part in 500). If the maximum contribution to input offset voltage allowed from the resistors is 1 mV, then these numbers indicate that the maximum voltage drop allowed is

$$I_1 R_E|_{\max} = 500 \text{ mV} \tag{9.92}$$

Thus

$$g_{m1} R_E|_{\max} = \frac{q}{kT} I_1 R_E|_{\max} = \frac{500}{26} = 20 \tag{9.93}$$

Using (9.93) in (9.91) shows that given this data, the maximum possible improvement in slew rate by use of emitter resistors is a factor of 11 times. Thus, in the case of 741-type circuits, slew rates of $11 \times 0.8 = 8.8$ V/μsec can be achieved.

Another method of improving the ratio, I_{xm}/g_{mI} in (9.87), is by using FETs as the input devices of the amplifier. FETs have the characteristic that, for a given bias current, the transconductance is only about 3 percent of the value for a bipolar transistor at the same bias current. In most applications this puts the FET at a disadvantage, but in this case it is a desirable characteristic. An input stage using *p*-channel JFETs is shown in Fig. 9.42. For typical FET devices, $g_m = 1$ mA/V for a bias current $I_1 = 1$ mA, and, applying this data to the circuit in Fig. 9.42, we obtain

$$g_{mI} = \frac{\Delta I_x}{\Delta V_i} = 1 \text{ mA/V} \tag{9.94}$$

$$I_{xm} = 2I_1 = 2 \text{ mA} \tag{9.95}$$

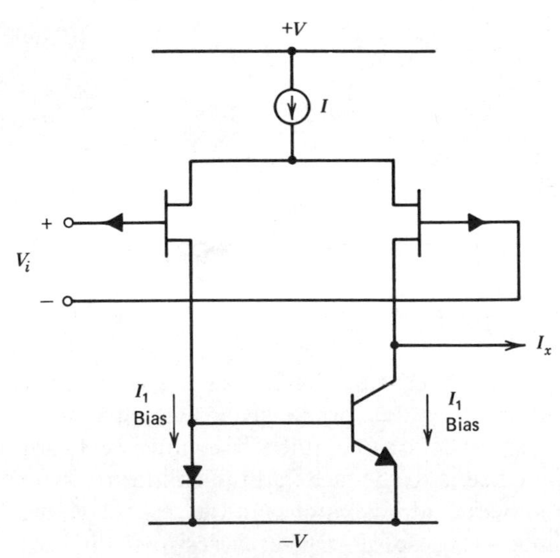

Fig. 9.42 Op amp input stage using *p*-channel JFETs.

Using this data in (9.87) with $\omega_2 = 8 \times 10^6$ rad/sec (1.3 MHz) gives

$$\text{Slew rate}\Big|_{\text{FET}} = \frac{2 \times 10^{-3}}{10^{-3}} \times 8 \times 10^6 \text{ V/sec} = 16 \text{ V/}\mu\text{sec}$$

This can be compared with the result given in (9.88) for the 741 op amp using bipolar input devices. Substituting $\omega_2 = 8 \times 10^6$ rad/sec in (9.88) gives

$$\text{Slew rate}\Big|_{\text{bipolar}} = 0.83 \text{ V/}\mu\text{sec}$$

As in the case of emitter-degeneration resistors, the limitation on the use of FET input devices is an increase in the input offset voltage of the circuit. As described in Chapter 3, FETs tend to have higher input offset voltages than do bipolar transistors by at least a factor of 3 and often much more.

Finally, in this description of methods of slew-rate improvement, we mention the nonlinear or "Class B" input stage described by Hearn[7]. In this technique, the small-signal transconductance of the input stage is left essentially unchanged, but the limit, I_{xm}, on the maximum current available for charging the compensation capacitor is greatly increased. This is done by providing alternative paths in the input stage that become operative for large inputs and deliver large charging currents to the compensation point. This has resulted in slew rates of the order of 30 V/μsec, and, as in the previous cases, the limitation is an increase in input offset voltage.

9.6.3 Effect of Slew-Rate Limitations on Large-Signal Sinusoidal Performance

The slew-rate limitations described above can also affect the performance of the circuit when handling large sinusoidal signals at higher frequencies. Consider the circuit of Fig. 9.36 with a large sinusoidal signal applied as shown in Fig. 9.43*a*. Since the circuit is connected as a voltage follower, the output voltage, V_o, will be forced to follow the V_i waveform. The maximum value of dV_i/dt occurs as the waveform crosses the axis, and if V_i is given by

$$V_i = \hat{V}_i \sin \omega t \tag{9.96}$$

then

$$\frac{dV_i}{dt} = \omega \hat{V}_i \cos \omega t$$

and

$$\left.\frac{dV_i}{dt}\right|_{\text{max}} = \omega \hat{V}_i \tag{9.97}$$

As long as the value of $dV_i/dt|_{\text{max}}$ given by (9.97) is less than the slew-rate limit, the output voltage will closely follow the input. However if the product, $\omega \hat{V}_i$, is greater than the slew-rate limit, the output voltage will be unable to follow the

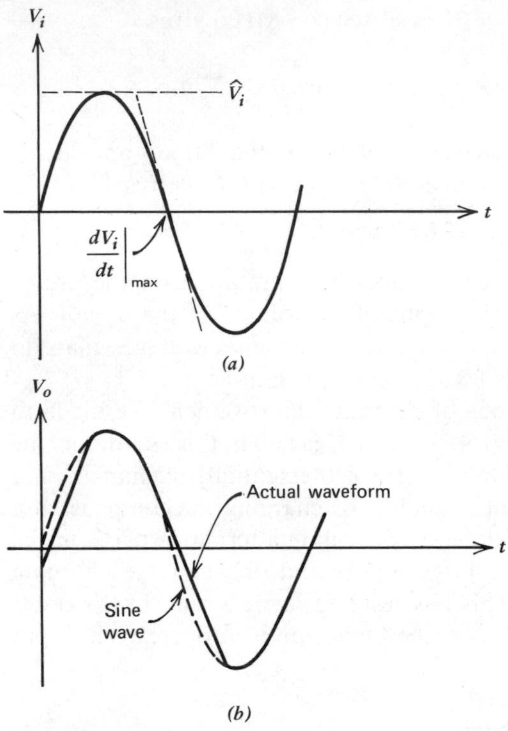

Fig. 9.43 (*a*) Large sinusoidal input voltage applied to the circuit of Fig. 9.36. (*b*) Output voltage resulting from input (*a*) showing slew limiting.

input, and waveform distortion of the kind shown in Fig. 9.43*b* will result. If a sine wave with $\hat{V}_i$ equal to the supply voltage is applied to the amplifier, slew limiting will eventually occur as the sine-wave frequency is increased. The frequency at which this occurs is called the "full-power" bandwidth of the circuit. (In practice, a value of $\hat{V}_i$ slightly less than the supply voltage is used to avoid clipping distortion of the type described in Chapter 5.)

EXAMPLE

Calculate the "full-power" bandwidth of the 741. Use $\hat{V}_i$ = supply voltage = 15 V. From (9.97) put

$$\omega \hat{V}_i = \text{slew rate}$$

This gives

$$\omega = \frac{0.8 \text{ V}/\mu\text{sec}}{15 \text{ V}} = 53.3 \times 10^3 \text{ rad/sec}$$

Thus

$$f = 8.5 \text{ kHz}$$

This means that a 741 op amp with a sinusoidal output of 15 V amplitude will begin to show slew-limiting distortion if the frequency exceeds 8.5 kHz.

PROBLEMS

9.1 An amplifier has a low-frequency forward gain of 200 and its transfer function has three negative real poles with magnitudes 1 MHz, 2 MHz, and 4 MHz. Calculate and sketch the Nyquist diagram for this amplifier if it is placed in a negative feedback loop with $f = 0.05$. Is the amplifier stable? Explain.

9.2 For the amplifier in Problem 9.1, calculate and sketch plots of gain (in decibels) and phase versus frequency (log scale) with no feedback applied. Thus, determine the value of f that just causes instability and the value of f giving a 60° phase margin.

9.3 If an amplifier has a phase margin of 20°, how much does the closed-loop gain peak (above the low-frequency value) at the frequency where the loop-gain magnitude is unity?

9.4 An amplifier has a low-frequency forward gain of 40,000 and its transfer function has three negative real poles with magnitudes 2 kHz, 200 kHz, and 4 MHz.

(a) If this amplifier is connected in a feedback loop with f constant and with low-frequency gain $A_0 = 400$, estimate the phase margin.

(b) Repeat (a) if A_0 is 200 and then 100.

9.5 An amplifier has a low-frequency forward gain of 5000 and its transfer function has three negative real poles with magnitudes 300 kHz, 2 MHz, and 25 MHz.

(a) Calculate the dominant-pole frequency required to give unity-gain compensation of this amplifier with a 45° phase margin if the original amplifier poles remain fixed. What is the resulting bandwidth of the circuit with the feedback applied?

(b) Repeat (a) for compensation in a feedback loop with a forward gain of 20 dB and 45° phase margin.

9.6 The amplifier of Problem 9.5 is to be compensated by reducing the frequency of the lowest frequency pole.

(a) Calculate the dominant-pole frequency required for unity-gain compensation with 45° phase margin, and the corresponding bandwidth of the circuit with the feedback applied. Assume that the remaining poles do not move.

(b) Repeat (a) for compensation in a feedback loop with a forward gain of 20 dB and 45° phase margin.

9.7 Repeat Problem 9.6 for the amplifier of Problem 9.4.

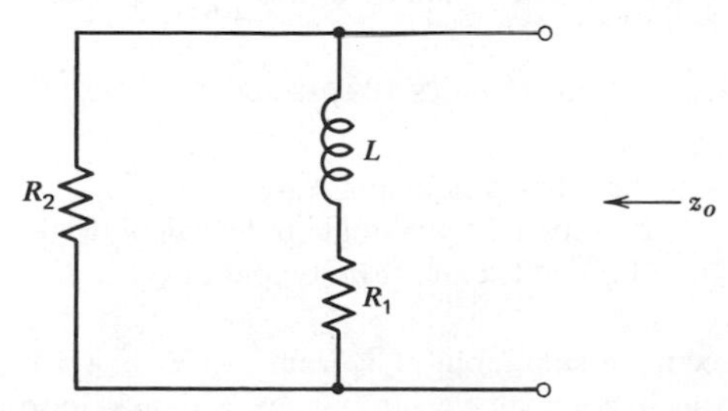

Fig. 9.44 Circuit representation of the output impedance of a series-shunt feedback circuit.

9.8 An op amp has a low-frequency voltage gain of 100,000 and a frequency response with a single negative-real pole with magnitude 5 Hz. This amplifier is to be connected in a series-shunt feedback loop with $f = 0.01$ giving a low-frequency forward voltage gain $A_0 \simeq 100$. If the output impedance without feedback is resistive with a value of 100 Ω, show that the output impedance of the feedback circuit can be represented as shown in Fig. 9.44 and calculate the values of these elements. Sketch the magnitude of the output impedance of the feedback circuit on log scales from 1 Hz to 100 kHz.

9.9 An op amp is fabricated using the 741 circuit but with different processing so that the transfer function before compensation has three negative real poles with magnitudes 30 kHz, 500 kHz, and 10 MHz. The circuit is compensated in the same manner as the 741, and pole-splitting causes the second most dominant pole to become negligible. Calculate the value of capacitance required to achieve a 60° phase margin in a unity-gain feedback connection and calculate the frequency where the resulting open-loop gain is 0 dB. Use low-frequency data as given in Chapter 7 for the 741 and assume that the pole with magnitude 10 MHz is unaffected by the compensation. The low-frequency gain of the 741 is 108 dB.

9.10 Repeat Problem 9.9 if the circuit is compensated by using shunt capacitance to ground at the base of Q_{16}. Assume that this affects only the lowest frequency pole.

9.11 An amplifier has gain $a_0 = 200$ and its transfer function has three negative real poles with magnitudes 1 MHz, 3 MHz, and 4 MHz. Calculate and sketch the root locus when feedback is applied as f varies from 0 to 1. Estimate the value of f causing instability.

9.12 Calculate and sketch the root locus for the amplifier of Problem 9.4 as f varies from 0 to 1. Estimate the value of f causing instability and check using the Nyquist criterion.

9.13 For the circuit of Fig. 9.29 parameter values are $R_F = 5$ kΩ, $R_E = 50$ Ω, and $C_F = 1.5$ pF. The basic amplifier of the circuit is shown in Fig. 9.30 and has two negative real poles with magnitudes 3 MHz and 6 MHz. The low-frequency *current gain* of the basic amplifier is 4000. Assuming that the loop gain of the circuit of Fig. 9.29 can be varied without changing the parameters of the basic amplifier, sketch root loci for this circuit as f varies from 0 to 0.01 both with and without C_F. Thus, estimate the pole positions of the current-gain transfer function of the feedback amplifier of Fig. 9.29 with the values of R_F and R_E specified above, and both with and without C_F. Sketch graphs in each case of gain magnitude versus frequency on log scales from $f = 10$ kHz to $f = 100$ MHz.

9.14 An op amp has two negative real open-loop poles with magnitudes 100 Hz and 120 kHz and a negative real zero with magnitude 100 kHz. The low-frequency open-loop voltage gain of the op amp is 100 dB. If this amplifier is placed in a negative feedback loop, sketch the root locus as f varies from 0 to 1. Calculate the poles and zeros of the feedback amplifier for $f = 10^{-3}$ and $f = 1$.

9.15 Repeat Problem 9.14 if the circuit has poles with magnitudes 100 Hz and 100 kHz and a zero with magnitude 120 kHz.

9.16 The input stages of an op amp are shown in the schematic of Fig. 9.45.

(a) Assuming that the frequency response is dominated by a single pole, calculate the frequency where small-signal voltage gain $|(\Delta V_o/\Delta V_i)(j\omega)|$ is unity and also the slew rate of the amplifier at V_o.

(b) Sketch response $V_o(t)$ from 0 to 20 μsec for a step input at V_i from −5 V to +5 V. Assume that the circuit is connected in a noninverting unity-gain feedback loop.

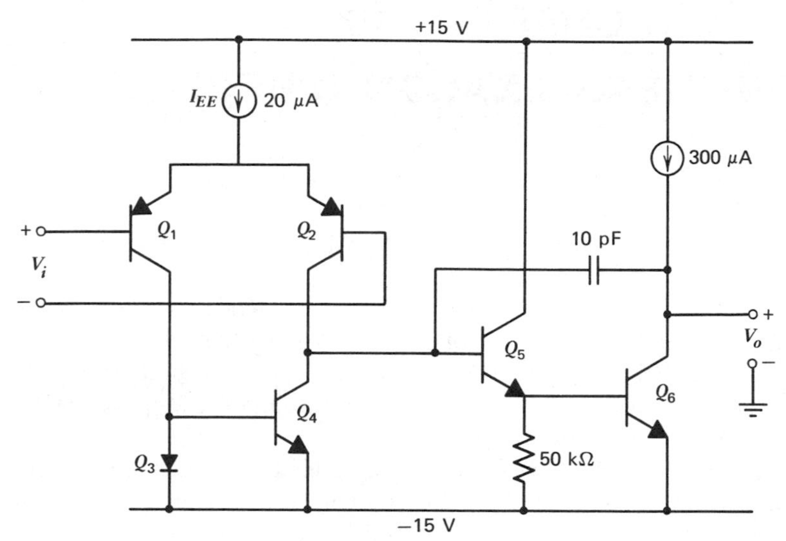

Fig. 9.45 Input stages of an op amp.

9.17 Repeat Problem 9.16 if the circuit of Fig. 9.45 is compensated by a capacitor of 0.05 μF connected from the base of Q_5 to ground. Assume that the voltage gain from the base of Q_5 to V_o is 500.

9.18 The slew rate of the circuit of Fig. 9.45 is to be increased by using 10 kΩ resistors in the emitters of Q_1 and Q_2. If the same unity-gain frequency is to be achieved, calculate the new value of compensation capacitor required and the improvement in slew rate.

9.19 Repeat Problem 9.18 if JFETs replace Q_1 and Q_2. Assume that the JFETs are biased to 300 μA each ($I_{EE} = 600$ μA) at which bias value the JFETs have $g_m = 300$ μA/V.

9.20 (a) Calculate the full-power bandwidth of the circuit of Fig. 9.45.
(b) If this circuit is connected in a noninverting unity-gain feedback loop, sketch output waveform V_o if V_i is a sinusoid of 10 V amplitude and frequency 45 kHz.

REFERENCES

1. K. Ogata. *Modern Control Engineering.* Prentice-Hall, Englewood Cliffs, N. J., 1970, Chapter 8.
2. P. E. Gray and C. L. Searle. *Electronic Principles.* Wiley, New York, 1969, Chapter 20.
3. R. D. Thornton et al. *Multistage Transistor Circuits.* Wiley, New York, 1965.
4. J. E. Solomon. "The Monolithic Op Amp: A Tutorial Study," *IEEE J. Solid-State Circuits,* Vol. SC-9, pp. 314–332, December 1974.
5. K. Ogata. Op cit.
6. P. E. Gray and C. L. Searle. Op cit. Chapter 19.
7. W. E. Hearn. "Fast Slewing Monolithic Operational Amplifier," *IEEE J. Solid-State Circuits,* Vol. SC-6, pp. 20–24, February 1971.

CHAPTER 10

NONLINEAR ANALOG CIRCUITS

10.1 INTRODUCTION

Chapters 1 through 9 have dealt almost entirely with analog circuits whose primary function is linear amplification of signals. While some of the circuits discussed (such as class AB output stages) were actually nonlinear in their operation, the operations performed on the signal passing through the amplifier were well approximated by linear relations.

Nonlinear operations on continuous-valued analog signals are often required in instrumentation, communication, and control-system design. These operations include rectification, modulation, demodulation, frequency translation, multiplication, and division. In this chapter we analyze the most commonly used techniques for performing these operations within a monolithic integrated circuit. We first discuss the use of diodes together with active elements to perform precision rectification. We then discuss the use of the bipolar transistor to synthesize nonlinear analog circuits and analyze the Gilbert multiplier cell, which is the basis for a wide variety of such circuits. Next we consider the application of this building block as a small-signal analog multiplier, as a modulator, as a phase comparator, and as a large-signal, four-quadrant multiplier.

Following the multiplier discussion, we introduce a highly useful circuit technique for performing demodulation of FM and AM signals and, at the same time, performing bandpass filtering. This circuit, the phase-locked loop (PLL), is particularly well-suited to monolithic construction. After exploring the basic concepts involved, the behavior of the PLL in the locked condition is analyzed. The capture transient is then considered, and, finally, an actual phase-locked-loop integrated circuit is analyzed.

10.2 PRECISION RECTIFICATION

Perhaps the most basic nonlinear operation performed on time-varying signals is rectification. An ideal half-wave rectifier is a circuit that passes signal currents or voltages of only one polarity while blocking signal voltages or currents of the other polarity. The transfer characteristic of an ideal half-wave rectifier is shown in Fig. 10.1. Also shown in Fig. 10.1 is the transfer characteristic of a second useful rectifier, the full-wave type. Practical rectifiers can be divided into two categories. The first class are termed power rectifiers, and these are used to con-

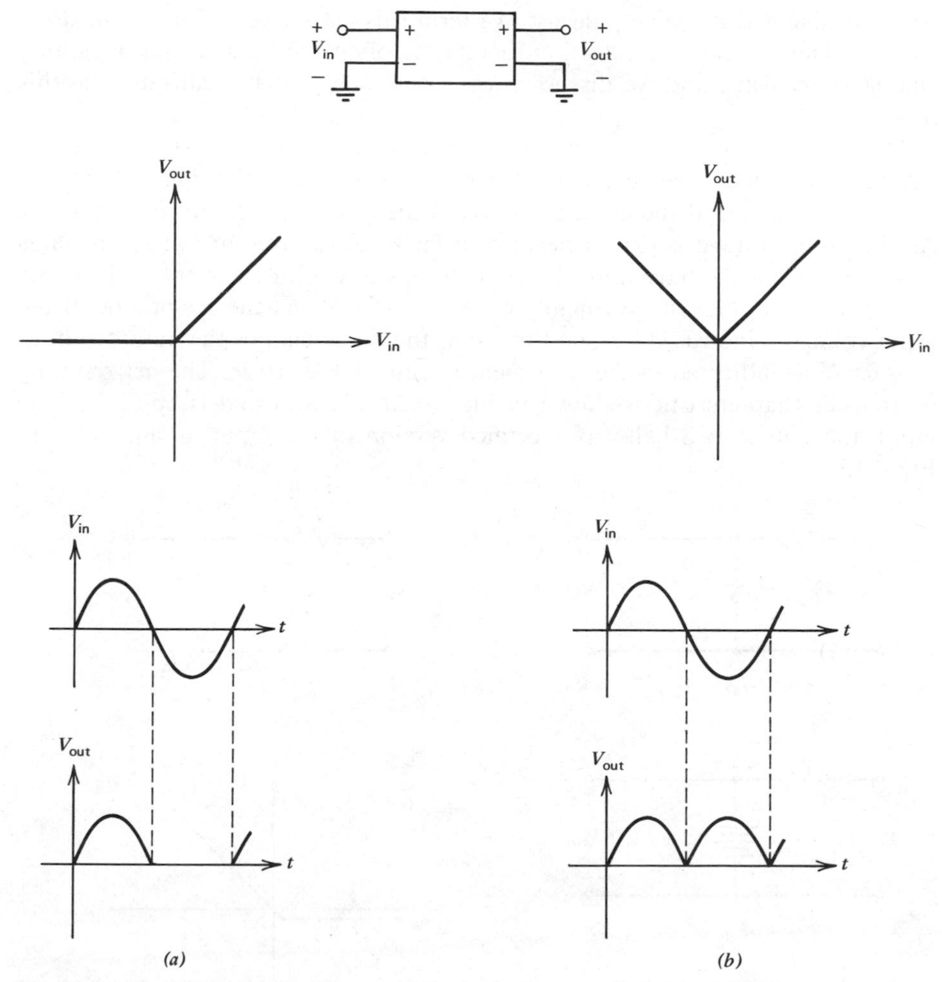

Fig. 10.1 Rectifier transfer characteristic and response for sinusoidal input waveform. (*a*) Half-wave rectifier. (*b*) Full-wave rectifier.

vert ac power to dc form. These circuits almost always use silicon diodes to perform the rectification, and the performance objectives are high efficiency and low cost. We will not consider this class of rectifier explicitly since most realizations of this type of high-power circuit utilize discrete components.

The second class of rectifiers has, as its objective, not the conversion of power but the extraction of information from a signal. Full-wave rectifiers of this type are used, for example, in the determination of the rms value of a signal, in certain types of demodulators, and in instrumentation systems that must sense signals of both positive and negative polarity. We term this class of rectifier the *precision rectifier*. The precision rectification function is often required within integrated analog subsystems, and we discuss circuit approaches to its realization in this section.

The simplest form of diode half-wave rectifier is shown in Fig. 10.2*a*. When the input voltage is positive, the diode is reverse biased, and the signal at the output is equal to that at the input as shown in the equivalent circuit of Fig. 10.2*b*. As the input voltage is driven negative from zero, the output voltage remains equal to the input voltage until the diode begins to conduct current. This occurs at a value of input voltage roughly equal to -0.6 V. As the magnitude of the input voltage is increased beyond this value, the diode clamps the output voltage at -0.6 V as indicated in the equivalent circuit of Fig. 10.2*c*. The net resulting dc transfer characteristic is shown in Fig. 10.2*d*. If a sinusoid is applied to the input, the output is a half-wave rectified version of the input as illustrated in Fig. 10.2*e*.

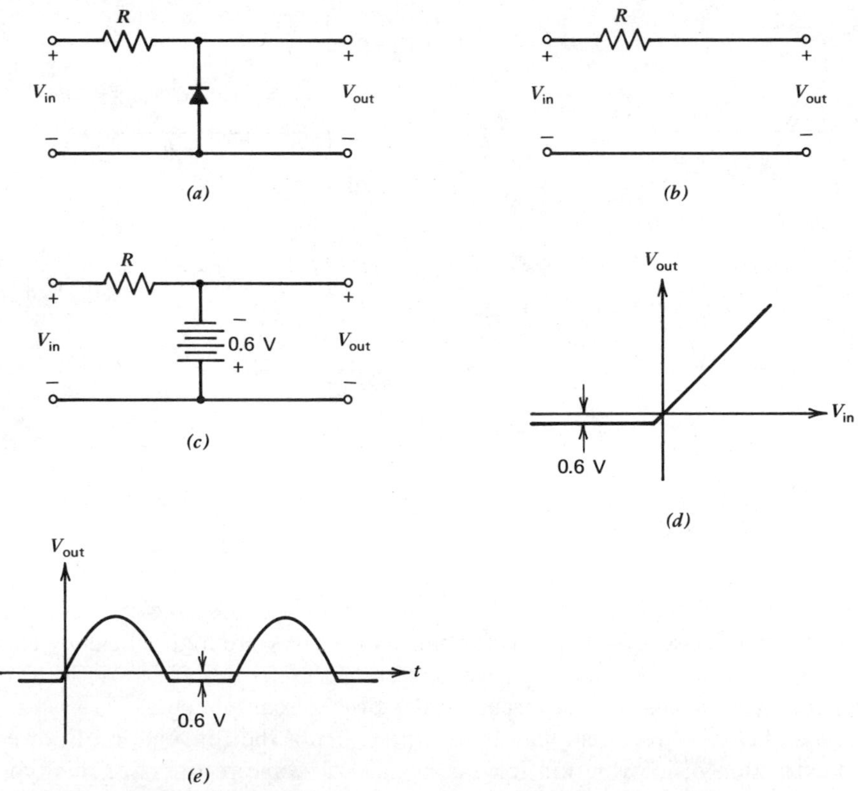

Fig. 10.2 Simple diode half-wave rectifier. (*a*) Rectifier circuit. (*b*) Equivalent circuit for $V_{in} > -0.6$ V. (*c*) Equivalent circuit for $V_{in} < -0.6$ V. (*d*) dc transfer characteristic. (*e*) Response to sinusoidal input.

An important drawback of this rectifier circuit is the fact that the rectification is not precise; the forward drop in the diode causes the circuit to pass the signal when it has a value between zero and -0.6 V. If the signal amplitude were, for example, only 2 or 3 V, the output waveform would not be a precise replica of the rectified version of the input waveform as shown in Fig. 10.1. In many signal-processing applications this error would be too large to be acceptable.

The performance of the rectifier can be greatly enhanced by the addition of active elements. Consider the active rectifier of Fig. 10.3. The diode has been replaced by a subcircuit consisting of a diode together with an operational amplifier with a gain a. We first consider the I-V characteristic of the diode op amp combination. If we assume that the op amp is ideal, then the current into its input terminals is zero and all of the current (I) flows through the diode and into the output terminal of the op amp.

The forward drop in the diode is equal to the difference between the op amp input voltage, V_i, and the op amp output voltage, V_o, so that the diode current I is:

$$I = -I_S\left[\exp\left(\frac{V_o - V_i}{V_T}\right) - 1\right] \tag{10.1}$$

The op amp has a gain a, so that:

$$V_o = -aV_i \tag{10.2}$$

Note that the output voltage of the circuit, V_{out}, is not the output voltage of the op amp, V_o. The output of the op amp serves only to drive the diode. The output voltage of the circuit is equal to the *input* voltage of the op amp. For the case in which the diode is forward biased, this voltage can be found by combining (10.1) and (10.2).

$$V_i = -\frac{V_T}{a+1}\ln\left(\frac{-I}{I_S} + 1\right) = V_{\text{out}} \tag{10.3}$$

Notice that the forward voltage drop is reduced by a factor $(a + 1)$ compared with the diode alone. Since a voltage gain a of many thousands is readily obtained in an op amp, the factor that determines the forward drop in this circuit will actually be the input offset voltage of the op amp itself.

When the input voltage is made positive, the op amp drives the diode into the reverse-biased state, and the behavior of the circuit is not well described by (10.3). Because of the presence of the diode, no current larger than the reverse leakage current of the diode can flow *into* the composite device. Thus, when the input voltage of Fig. 10.3 becomes positive, no current flows in the resistor and op amp input voltage V_i becomes equal to V_{in}. This causes the output voltage of the op amp to be driven in the negative direction until the output stage saturates. Thus, for positive input voltages, input voltage V_{in} is applied directly across the op amp input terminals and the amplifier saturates.

The dc transfer characteristic of the rectifier of Fig. 10.3a is shown in Fig. 10.3b. Note that the curve closely approaches that of an ideal rectifier.

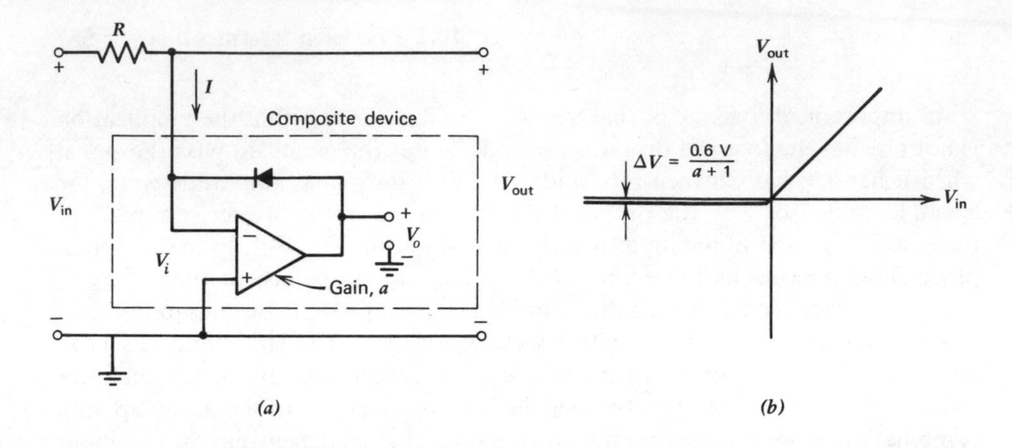

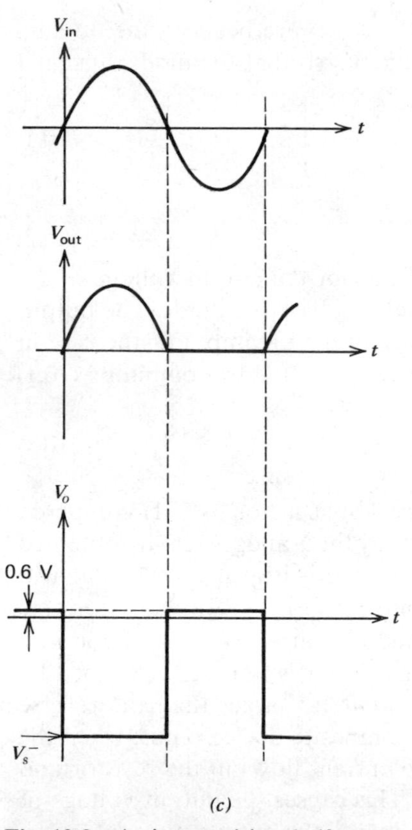

Fig. 10.3 Active precision half-wave rectifier. (*a*) Rectifier circuit. (*b*) dc transfer characteristic. The output voltage for $V_{in} < 0$ is in the microvolt range if the op amp has no offset. (*c*) Input and output waveforms for the precision half-wave rectifier. V_o is the output voltage of the op amp. During the positive excursions of the input V_o takes on a value of V_s^-, which is the output voltage at which the op amp output stage saturates.

Improved Precision Half-Wave Rectifier. The rectifier circuit of Fig. 10.3 has the property that for positive inputs the operational amplifier output saturates in the negative direction. As illustrated in Fig. 10.3*c*, the op amp output voltage is required to change instantaneously from this saturated voltage (V_s^-) to +0.6 V when the input waveform passes through zero. Because of the limited slew rate of real operational amplifiers (see Chapter 9), this cannot occur, and the output waveform will not be a precisely rectified version of the input waveform as the frequency of the input sinusoid is increased. An alternate circuit that greatly alleviates this problem is shown in Fig. 10.4. This circuit is similar to the original one except that one additional diode D_2 and one additional resistor are added.

For input voltages less than zero, operation of the circuit is exactly the same as the circuit of Fig. 10.3. The equivalent circuit for this condition is shown in Fig. 10.4*b*. Diode D_1 is forward biased and the op amp is in the active region. The inverting input of the op amp is clamped at ground by the feedback through D_1, and, since no current flows in R_2, the output voltage is also at ground. When the input voltage is made positive, no current can flow in the reverse direction through D_1 so the output voltage of the op amp, V_o, is driven in the negative direction. This reverse biases D_1 and forward biases D_2. The resulting equivalent

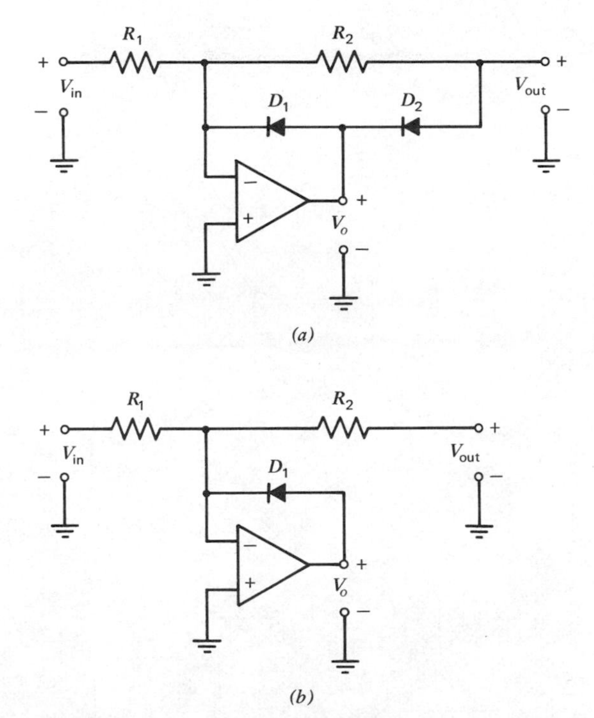

Fig. 10.4 Improved precision rectifier. (*a*) Rectifier circuit. (*b*) Equivalent circuit for $V_{in} < 0$. (*c*) Equivalent circuit for $V_{in} > 0$.

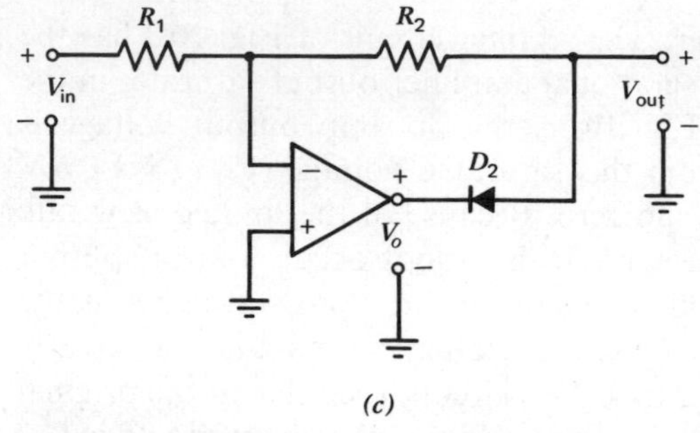

Fig. 10.4 (*Continued*)

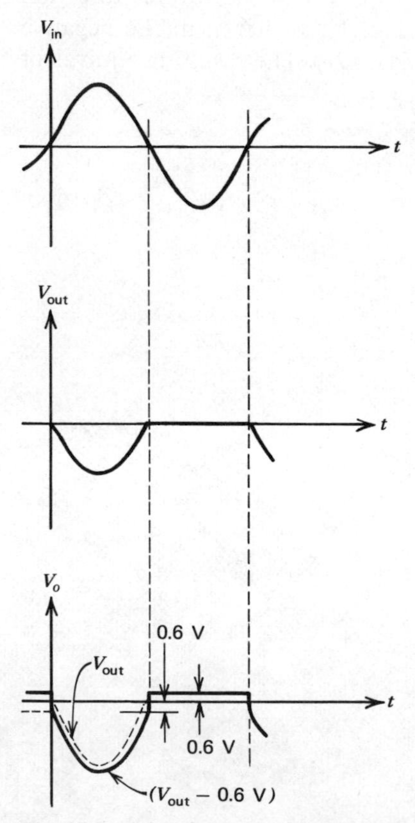

Fig. 10.5 Waveforms within the improved precision rectifier for a sinusoidal input.

circuit is shown in Fig. 10.4*c* and is simply an inverting amplifier with a forward-biased diode in series with the output lead of the op amp. Because of the large gain of the op amp, this diode has no effect on its behavior as long as it is forward biased, and so the circuit behaves as an inverting amplifier giving an output voltage of

$$V_{\text{out}} = -\frac{R_2}{R_1} V_{\text{in}} \tag{10.4}$$

As shown in Fig. 10.5, the output voltage of the operational amplifier need only change in value by approximately two diode drops when the input signal changes from positive to negative. Actually, this is but one of a wide variety of active-rectifier circuits, some of which are capable of even faster operation.[1]

10.3 ANALOG MULTIPLIERS EMPLOYING THE BIPOLAR TRANSISTOR

In analog-signal processing the need often arises for a circuit which takes two analog inputs and produces an output proportional to their product. Such circuits are termed *analog multipliers.* In the following sections we examine several analog multipliers which depend on the exponential transfer function of bipolar transistors.

10.3.1 The Emitter-Coupled Pair as a Simple Multiplier

The emitter-coupled pair, shown in Fig. 10.6, was shown in Chapter 3 to produce output currents that were related to the differential input voltage by

$$I_{c1} = \frac{I_{EE}}{1 + \exp\left(-\dfrac{V_{id}}{V_T}\right)} \tag{10.5}$$

$$I_{c2} = \frac{I_{EE}}{1 + \exp\left(\dfrac{V_{id}}{V_T}\right)} \tag{10.6}$$

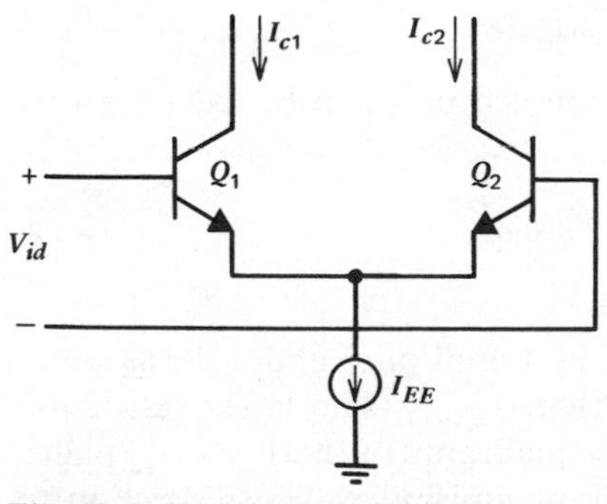

Fig. 10.6 Emitter-coupled pair.

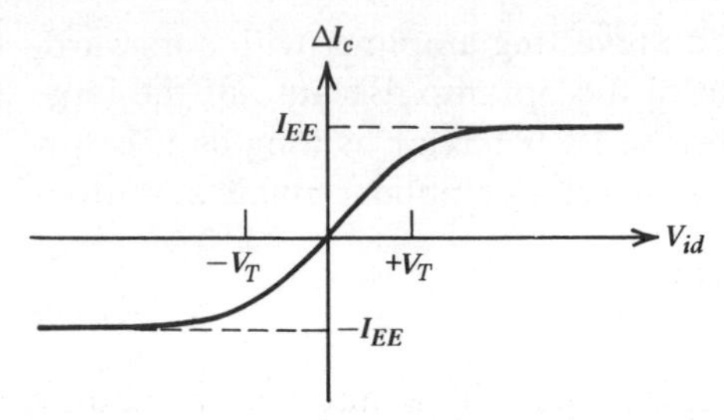

Fig. 10.7 dc transfer characteristic of emitter-coupled pair.

where base current has been neglected. Equations 10.5 and 10.6 can be combined to give the *difference* between the two output currents:

$$\Delta I_c = I_{c1} - I_{c2} = I_{EE} \tanh\left(\frac{V_{id}}{2V_T}\right) \tag{10.7}$$

This relationship is plotted in Fig. 10.7 and shows that the emitter-coupled pair by itself can be used as a primitive multiplier. We first assume that the differential input voltage V_{id} is much less than V_T. If this is true, we can utilize the approximation:

$$\tanh \frac{V_{id}}{2V_T} \approx \frac{V_{id}}{2V_T} \qquad \frac{V_{id}}{2V_T} \ll 1 \tag{10.8}$$

And (10.7) becomes:

$$\Delta I_c = I_{EE}\left(\frac{V_{id}}{2V_T}\right) \tag{10.9}$$

The current, I_{EE}, is actually the bias current for the emitter-coupled pair. With the addition of more circuitry, we can make I_{EE} proportional to a second input signal V_{i2} as shown in Fig. 10.8. Thus we have

$$I_{EE} \cong K_o(V_{i2} - V_{BE(\text{on})}) \tag{10.10}$$

The differential output current of the emitter-coupled pair can be calculated by substituting (10.10) in (10.9) to give

$$\Delta I_c = \frac{K_o V_{id}[V_{i2} - V_{BE(\text{on})}]}{2V_T} \tag{10.11}$$

Thus we have produced a circuit that functions as a multiplier under the assumption that V_{id} is small, and that V_{i2} is greater than $V_{BE(\text{on})}$. The latter restriction means that the multiplier only functions in two quadrants of the $V_{id} - V_{i2}$ plane, and this type of circuit is termed a two-quadrant multiplier. The restriction to

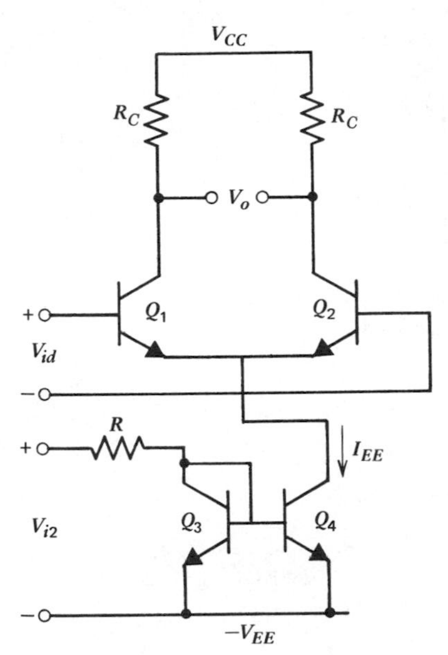

Fig. 10.8 Two-quadrant analog multiplier.

two quadrants of operation is a severe one for many communications applications and most practical multipliers allow four-quadrant operation. The Gilbert multiplier cell,[2] shown in Fig. 10.9, is a modification of the emitter-coupled cell, which allows four-quadrant multiplication. It is the basis for most integrated-circuit balanced multiplier systems. The series connection of an emitter-coupled pair with two cross-coupled, emitter-coupled pairs produces a particularly useful transfer characteristic as shown in the next section.

10.3.2 dc Analysis of the Gilbert Multiplier Cell

In the following analysis, we assume that the transistors are identical, that the output resistance of the transistors and that of the biasing current source can be neglected, and that the base currents can be neglected. For the Gilbert cell shown in Fig. 10.9, the collector currents of Q_3 and Q_4 are, using (10.5) and (10.6),

$$I_{c3} = \frac{I_{c1}}{1 + \exp\left(\dfrac{-V_1}{V_T}\right)} \tag{10.12}$$

$$I_{c4} = \frac{I_{c1}}{1 + \exp\left(\dfrac{V_1}{V_T}\right)} \tag{10.13}$$

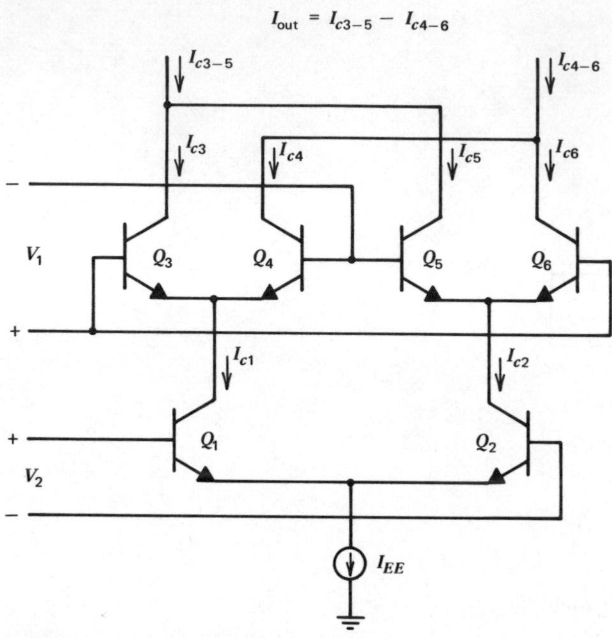

Fig. 10.9 Gilbert multiplier circuit.

Similarly the collector currents of Q_5 and Q_6 are given by

$$I_{c5} = \frac{I_{c2}}{1 + \exp\left(\frac{V_1}{V_T}\right)} \tag{10.14}$$

$$I_{c6} = \frac{I_{c2}}{1 + \exp\left(-\frac{V_1}{V_T}\right)} \tag{10.15}$$

The two currents, I_{c1} and I_{c2}, can be related to V_2 by again using (10.5) and (10.6):

$$I_{c1} = \frac{I_{EE}}{1 + \exp\left(-\frac{V_2}{V_T}\right)} \tag{10.16}$$

$$I_{c2} = \frac{I_{EE}}{1 + \exp\left(\frac{V_2}{V_T}\right)} \tag{10.17}$$

Combining (10.12) through (10.17), we obtain expressions for collector currents I_{c3}, I_{c4}, I_{c5}, and I_{c6} in terms of input voltages V_1 and V_2.

$$I_{c3} = \frac{I_{EE}}{\left[1 + \exp\left(-\frac{V_1}{V_T}\right)\right]\left[1 + \exp\left(-\frac{V_2}{V_T}\right)\right]} \tag{10.18}$$

$$I_{c4} = \frac{I_{EE}}{\left[1 + \exp\left(-\frac{V_2}{V_T}\right)\right]\left[1 + \exp\left(\frac{V_1}{V_T}\right)\right]} \tag{10.19}$$

$$I_{c5} = \frac{I_{EE}}{\left[1 + \exp\left(\frac{V_1}{V_T}\right)\right]\left[1 + \exp\left(\frac{V_2}{V_T}\right)\right]} \tag{10.20}$$

$$I_{c6} = \frac{I_{EE}}{\left[1 + \exp\left(\frac{V_2}{V_T}\right)\right]\left[1 + \exp\left(-\frac{V_1}{V_T}\right)\right]} \tag{10.21}$$

The differential output current is then given by:

$$\begin{aligned} \Delta I &= I_{c3-5} - I_{c4-6} = I_{c3} + I_{c5} - (I_{c6} + I_{c4}) \\ &= (I_{c3} - I_{c6}) - (I_{c4} - I_{c5}) \qquad (10.22) \\ &= I_{EE}\left[\tanh\left(\frac{V_1}{2V_T}\right)\right]\left[\tanh\left(\frac{V_2}{2V_T}\right)\right] \qquad (10.23) \end{aligned}$$

The dc transfer characteristic, then, is the product of the hyperbolic tangent of the two input voltages.

Practical applications of the multiplier cell can be divided into three categories according to the magnitude relative to V_T of applied signals V_1 and V_2. If the magnitude of V_1 and V_2 are kept small with respect to V_T, the hyperbolic tangent function can be approximated as linear and the circuit behaves as a multiplier, developing the product of V_1 and V_2. However, by including nonlinearity to compensate for the hyperbolic tangent function in series with each input, the range of input voltages over which linearity is maintained can be greatly extended. This technique is used in so-called four-quadrant analog multipliers.

The second class of applications is distinguished by the application to one of the inputs of a signal that is large compared to V_T, causing the transistors to which that signal is applied to behave like switches rather than near-linear devices. This effectively multiplies the applied small signal by a square wave, and in this mode of operation the circuit acts as a modulator.

In the third class of applications, the signals applied to both inputs are large compared to V_T, and all six transistors in the circuit behave as nonsaturating switches. This mode of operation is useful for the detection of phase differences between two amplitude-limited signals, as is required in phase-locked loops, and is sometimes called the phase-detector mode.

We first consider the application of the circuit as an analog multiplier of two continuous signals.

10.3.3 The Gilbert Cell as an Analog Multiplier

As mentioned earlier, the hyperbolic-tangent function may be represented by the infinite series:

$$\tanh x = x - \frac{x^3}{3} \cdots . \qquad (10.24)$$

Assuming that x is much less than one, the hyperbolic tangent can then be approximated by

$$\tanh x \approx x \qquad (10.25)$$

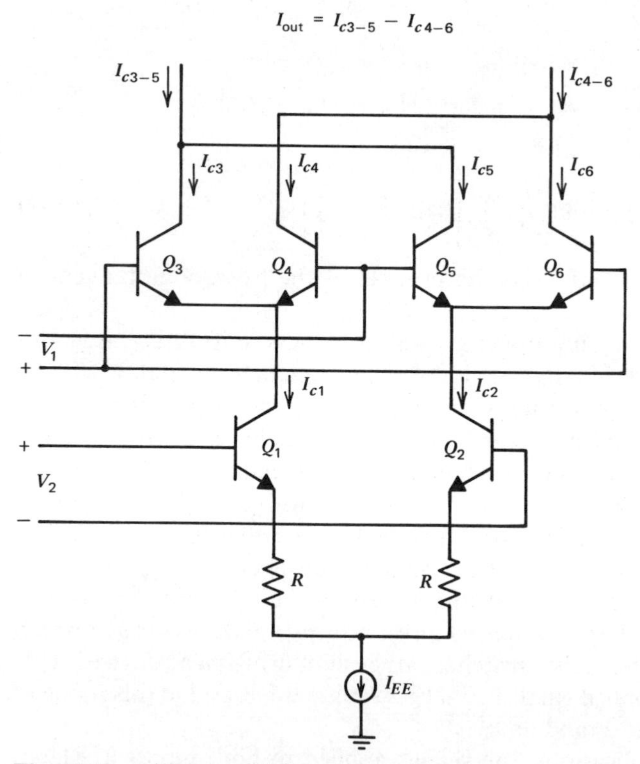

Fig. 10.10 Gilbert multiplier with emitter degeneration applied to improve input voltage range on V_2 input.

Applying this relation to (10.23),

$$\Delta I \approx I_{EE}\left(\frac{V_1}{2V_T}\right)\left(\frac{V_2}{2V_T}\right) \qquad V_1, V_2 \ll V_T \tag{10.26}$$

Thus for small-amplitude signals, the circuit performs an analog multiplication. Unfortunately, the amplitudes of the input signals are often much larger than V_T, but larger signals can be accommodated in this mode in a number of ways. In the event that only one of the signals is large compared to V_T, emitter degeneration can be utilized in the lower emitter-coupled pair, increasing the linear input range for V_2 as shown in Fig. 10.10. Unfortunately, this cannot be done with the cross-coupled pairs Q_3—Q_6 because the degeneration resistors destroy the required nonlinear relation between I_c and V_{be} in those devices.

An alternate approach is to introduce a nonlinearity that predistorts the input signals to compensate for the hyperbolic tangent transfer characteristic of the basic cell. The required nonlinearity is an inverse hyperbolic tangent characteristic, and a hypothetical example of such a system is shown in Fig. 10.11. Fortunately, this particular nonlinearity is straightforward to generate.

Referring to Fig. 10.12, we assume for the time being that the circuitry within the box develops a differential output current that is linearly related to the input

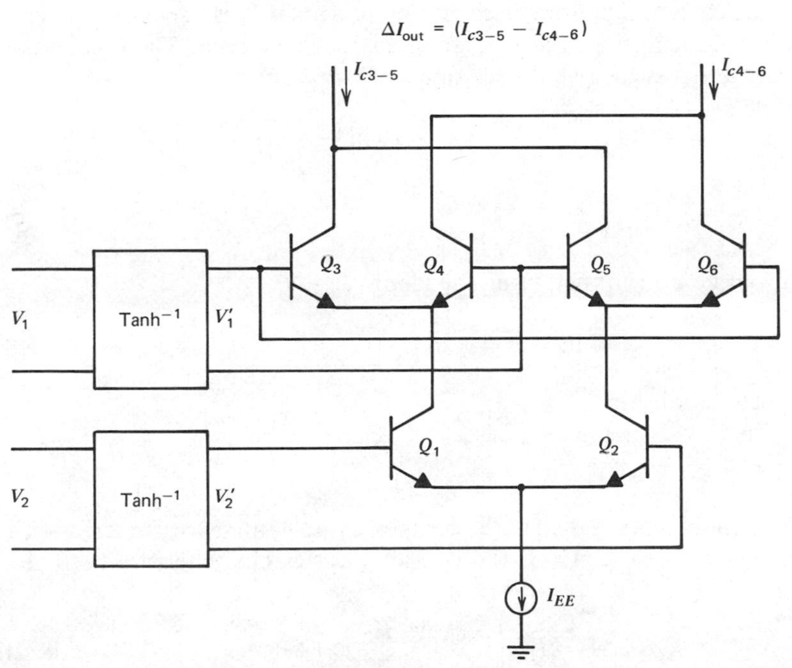

Fig. 10.11 Gilbert multiplier with predistortion circuits.

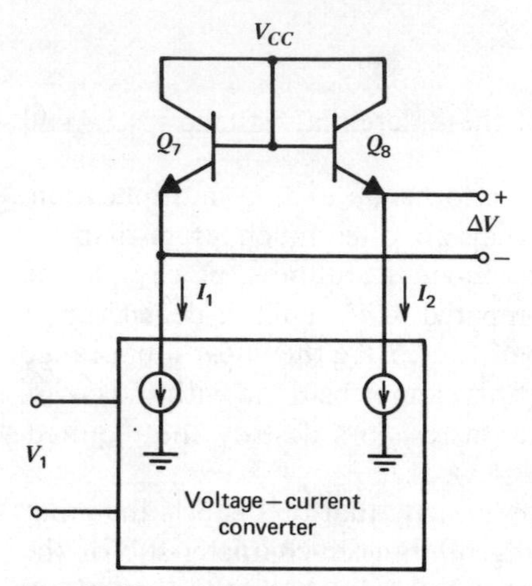

Fig. 10.12 Inverse hyperbolic tangent circuit.

voltage, V_1. Thus:

$$I_1 = I_{o1} + K_1 V_1 \tag{10.27}$$

$$I_2 = I_{o1} - K_1 V_1 \tag{10.28}$$

Here I_{o1} is the dc current that flows in each output leads if V_1 is equal to zero, and K_1 is the transconductance of the voltage-to-current converter. The differential voltage developed across the two diode-connected transistors is

$$\begin{aligned}\Delta V &= V_T \ln\left(\frac{I_{o1} + K_1 V_1}{I_S}\right) - V_T \ln\left(\frac{I_{o1} - K_1 V_1}{I_S}\right) \\ &= V_T \ln\left(\frac{I_{o1} + K_1 V_1}{I_{o1} - K_1 V_1}\right)\end{aligned} \tag{10.29}$$

This function can be transformed using the identity:

$$\tanh^{-1} x = \tfrac{1}{2} \ln\left(\frac{1 + x}{1 - x}\right) \tag{10.30}$$

into the desired relationship.

$$\Delta V = 2V_T \tanh^{-1}\left(\frac{K_1 V_1}{I_{o1}}\right) \tag{10.31}$$

Thus if this functional block is used as the compensating nonlinearity in series with each input as shown in Fig. 10.11, the overall transfer characteristic becomes, using (10.23),

$$\Delta I = I_{EE}\left(\frac{K_1 V_1}{I_{o1}}\right)\left(\frac{K_2 V_2}{I_{o2}}\right) \tag{10.32}$$

where I_{o2} and K_2 are the parameters of the functional block following V_2.

Equation 10.32 shows that the differential output current is *directly proportional* to the product $V_1 V_2$, and in principle, this relationship holds for all values of V_1 and V_2 for which the two output currents of the differential voltage-to-current converters are positive. For this to be true, I_1 and I_2 must always be positive, and, from (10.27) and (10.28), we have

$$-\frac{I_{o1}}{K_1} < V_1 < \frac{I_{o1}}{K_1} \tag{10.33}$$

$$-\frac{I_{o2}}{K_2} < V_2 < \frac{I_{o2}}{K_2} \tag{10.34}$$

Note that the inclusion of a compensating nonlinearity on the V_2 input simply makes the collector currents of Q_1 and Q_2 directly proportional to input voltage V_2 rather than to its hyperbolic tangent. Thus the combination of the pair Q_1–Q_2 and the compensating nonlinearity on the V_2 input is redundant, and the output currents of the voltage-to-current converter on the V_2 input can be fed directly into the emitters of the Q_3–Q_4 and Q_5–Q_6 pairs with exactly the same results. The multiplier then takes on the form shown in Fig. 10.13.

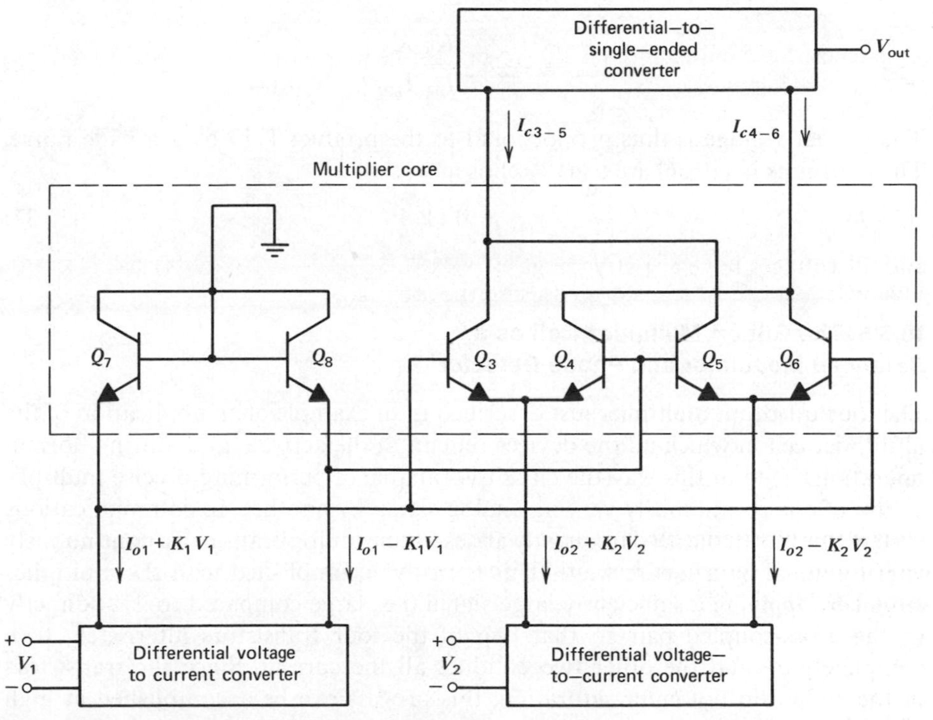

Fig. 10.13 Complete four-quadrant multiplier.

10.3.4 A Complete Analog Multiplier[3]

In order to be useful in a wide variety of applications, the multiplier circuit must develop an output voltage that is referenced to ground and can take on both positive and negative values. The transistors, Q_3, Q_4, Q_5, Q_6, Q_7, and Q_8, shown in Fig. 10.13, are referred to as the multiplier core and produce a differential current output that then must be amplified, converted to a single-ended signal, and referenced to ground. An output amplifier is thus required, and the complete multiplier consists of two voltage-current converters, the "core" transistors, and an output current-to-voltage amplifier. While the core configuration of Fig. 10.13 is common to most four-quadrant transconductance multipliers, the rest of the circuitry can be realized in a variety of ways.

The most common configurations used for the voltage-current converters are emitter-coupled pairs with emitter degeneration as shown in Fig. 10.10. The differential-to-single-ended converter of Fig. 10.13 is often realized with an op amp circuit of the type shown in Fig. 6.4. If this circuit has a transresistance given by

$$\frac{V_{\text{out}}}{\Delta I} = K_3 \tag{10.35}$$

then substitution in (10.32) gives for the overall multiplier characteristic

$$V_{\text{out}} = I_{EE} K_3 \frac{K_1}{I_{o1}} \frac{K_2}{I_{o2}} V_1 V_2 \tag{10.36}$$

The output voltage is thus proportional to the product $V_1 V_2$ over a wide range. The constants in (10.36) are usually chosen so that

$$V_{\text{out}} = 0.1 V_1 V_2 \tag{10.37}$$

and all voltages have a ± 10V range.

10.3.5 The Gilbert Multiplier Cell as a Balanced Modulator and Phase Detector

The four-quadrant multiplier just described is an example of an application of the multiplier cell in which all the devices remain in the active region during normal operation. Used in this way the circuit is capable of performing precise multiplication of one continuously varying analog signal by another. In communications systems, however, the need frequently arises for the multiplication of a continuously varying signal by a square wave. This is easily accomplished with the multiplier circuit by applying a sufficiently large signal (i.e., large compared to $2V_T$) directly to the cross-coupled pair so that two of the four transistors alternately turn completely off and the other two conduct all the current. Since the transistors in the circuit do not enter saturation, this process can be accomplished at high speed. A set of typical waveforms that might result when a sinusoid is applied to the small-signal input and a square wave to the large-signal input is shown in Fig. 10.14. Note that since the devices in the multiplier are being switched on and off by the incoming square wave, the amplitude of the output waveform is independent of the amplitude of the square wave as long as it is large enough to cause the devices in the multiplier circuit to be fully on or fully off. Thus the circuit in this mode does not perform a linear multiplication of two waveforms, but actually causes the output voltage of the circuit produced by the small-signal input to be alternately multiplied by $+1$ and -1.

The spectrum of the output may be developed directly from the Fourier series of the two inputs. For the low-frequency modulating sinusoidal input,

$$V_m(t) = V_m \cos \omega_m t \tag{10.38}$$

Small – signal input

$V_m(t)$

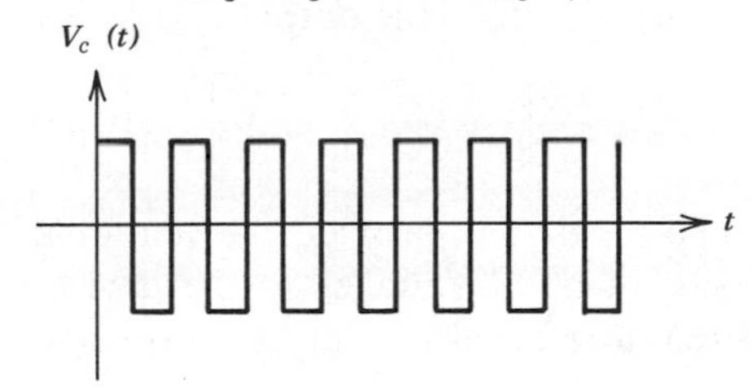

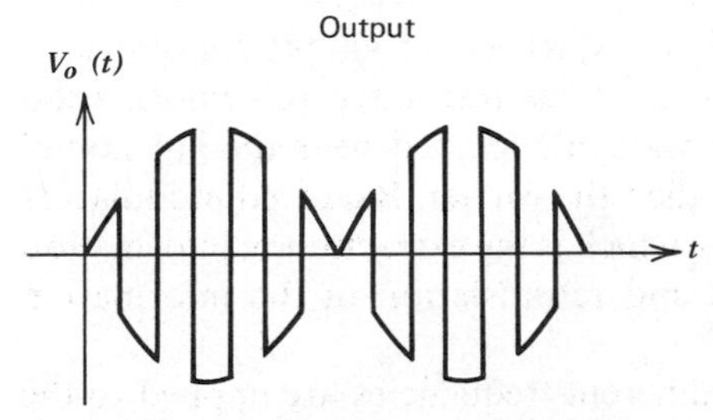

Fig. 10.14 Input and output waveforms for a phase detector with large input signals.

and for the high-frequency square wave input, which we assume has an amplitude of ± 1 as discussed above,

$$V_c(t) = \sum_{n=1}^{\infty} A_n \cos n\omega_c t, \; A_n = \frac{\sin n\pi/2}{n\pi/4} \tag{10.39}$$

Thus the output signal is

$$V_o(t) = K[V_c(t)V_m(t)] = K \sum_{n=1}^{\infty} A_n V_m \cos \omega_m t \cos n\omega_c t \tag{10.40}$$

$$= K \sum_{n=1}^{\infty} \frac{A_n V_m}{2} [\cos (n\omega_c + \omega_m)t + \cos (n\omega_c - \omega_m)t] \tag{10.41}$$

where K is the magnitude of the gain of the multiplier from the small-signal input to the output.

The spectrum has components located at frequencies ω_m above and below each of the harmonics of ω_c, but no component at the carrier frequency ω_c or its harmonics. The spectrum of the input signals and the resulting output signal is shown in Fig. 10.15. The lack of an output component at the carrier frequency is a very useful property of balanced modulators. The signal is usually filtered following the modulation process so that only the components near ω_c are retained.

If a dc component is added to the modulating input, the result is a signal component in the output at the carrier frequency and its harmonics. If the modulating signal is given by:

$$V_m(t) = V_m(1 + M \cos \omega_m t) \tag{10.42}$$

where the parameter M is called the modulation index, then the output is given by

$$V_o(t) = K \sum_{n=1}^{\infty} A_n V_m \left[\cos (n\omega_c t) + \frac{M}{2} \cos (n\omega_c + \omega_m)t + \frac{M}{2} \cos (n\omega_c - \omega_m)t \right] \tag{10.43}$$

This dc component can be introduced intentionally to provide conventional amplitude modulation or it can be the result of offset voltages in the devices within the modulator, which results in undesired carrier feedthrough in suppressed-carrier modulators.

Note that the balanced modulator actually performs a frequency translation. Information contained in the modulating signal, $V_m(t)$, was originally concentrated at the modulating frequency ω_m. The modulator has translated this information so that it is now contained in spectral components located near the harmonics of the high-frequency signal $V_c(t)$, usually called the carrier. Balanced modulators are also useful for performing demodulation, which is the extraction of information from the frequency band near the carrier and retranslation of the information back down to low frequencies.

In frequency translation, signals at two different frequencies are applied to the two inputs, and the sum or difference frequency component is taken from the

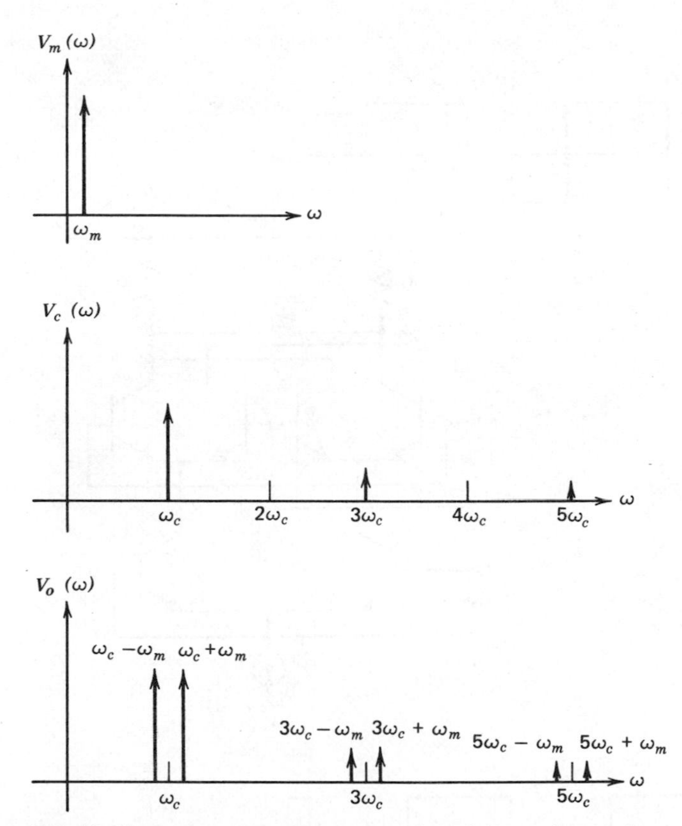

Fig. 10.15 Input and output spectra for a balanced modulator.

output. If unmodulated signals of identical frequency ω_0 are applied to the two inputs, the circuit behaves as a *phase detector* and produces an output whose dc component is proportional to the phase difference between the two inputs. For example, consider the two input waveforms in Fig. 10.16, which are applied to the Gilbert multiplier shown in the same figure. We assume first for simplicity that both inputs are large in magnitude so that all the transistors in the circuit are behaving as switches. The output waveform that results is shown in Fig. 10.16*c*, and consists of a dc component and a component at twice the incoming frequency. The dc component of this waveform is given by:

$$V_{\text{average}} = \frac{1}{2\pi}\int_0^{2\pi} V_o(t)d(\omega_0 t) \tag{10.44}$$

$$= \frac{-1}{\pi}(A_1 - A_2) \tag{10.45}$$

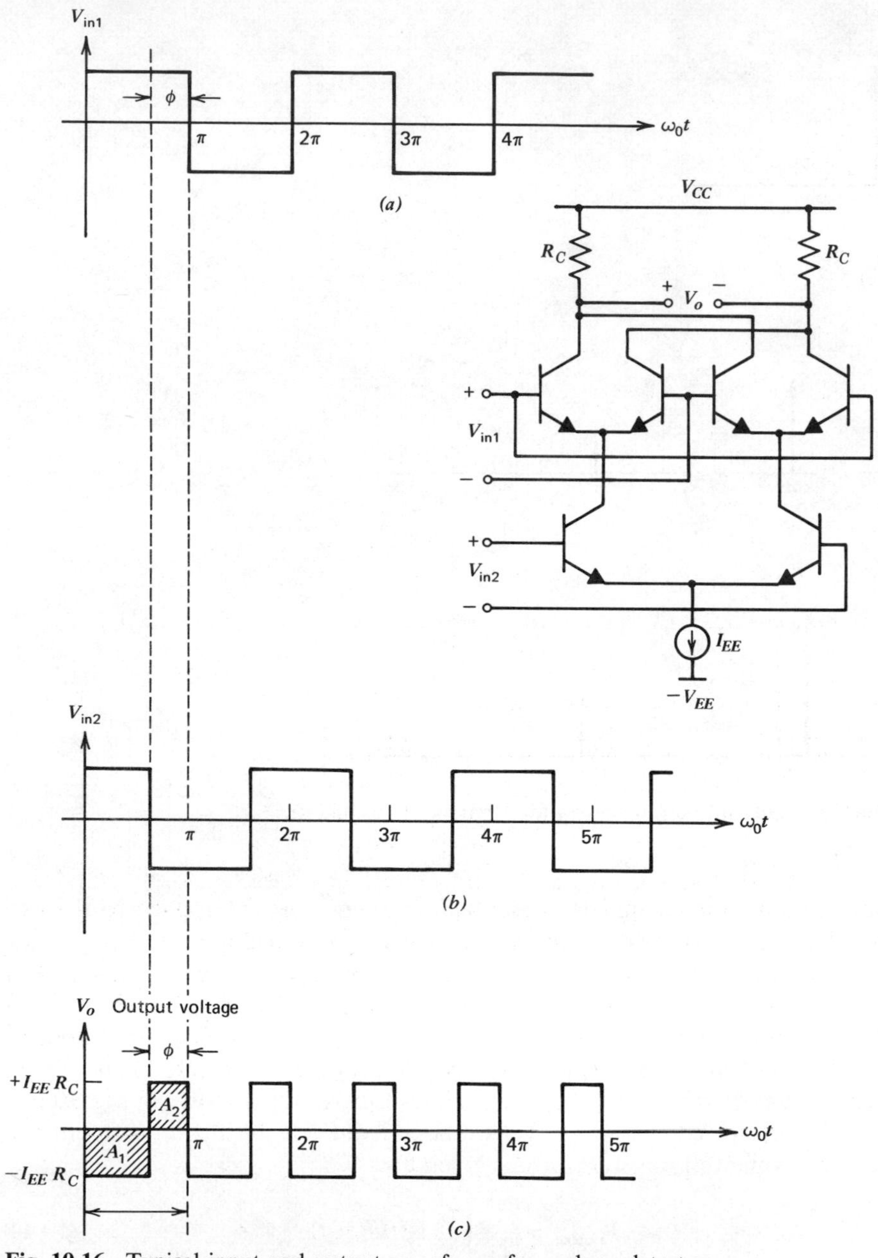

Fig. 10.16 Typical input and output waveforms for a phase detector.

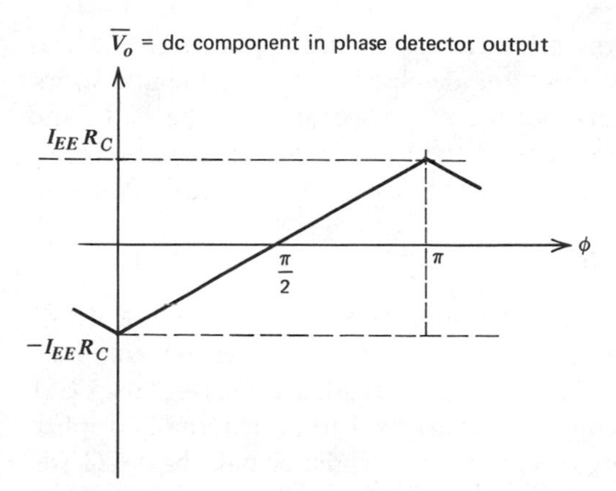

Fig. 10.17 Phase detector output versus phase difference.

where areas A_1 and A_2 are as indicated in Fig. 10.16*c*. Thus,

$$V_{\text{average}} = -\left[I_{EE}R_C\frac{(\pi - \phi)}{\pi} - \frac{I_{EE}R_C\phi}{\pi}\right] \tag{10.46}$$

$$= I_{EE}R_C\left(\frac{2\phi}{\pi} - 1\right) \tag{10.47}$$

This phase relationship is plotted in Fig. 10.17. This phase demodulation technique is widely used in phase-locked loops.

We assumed above that the input waveforms were large in amplitude and were square waves. If the input signal amplitude is large, the actual waveform shape is unimportant since the multiplier simply switches from one state to the other at the zero crossings of the waveform. For the case in which the amplitude of one or both of the input signals has an amplitude comparable to or smaller than V_T, the circuit still acts as a phase detector. However, the output voltage then depends both on the phase difference *and* on the *amplitude* of the two input waveforms. The operation of the circuit in this mode is considered further in Section 10.4.3.

10.4 PHASE-LOCKED LOOPS (PLL)

The phase-locked loop concept was first developed in the 1930s.[4] It has since been used in communications systems of many types, particularly in satellite communications systems. Until recently, however, phase-locked systems have been too complex and costly for use in most consumer and industrial systems where performance requirements are more modest and other approaches are more economical. The PLL is particularly amenable to monolithic construction, however, and integrated-circuit, phase-locked loops can now be fabricated at very low cost.[5] Their use has become attractive for many applications such as FM demodulators, stereo demodulators, tone detectors, frequency synthesizers, and others. In this section we first explore the basic operation of the PLL, and then consider analytically the performance of the loop in the locked condition. We then discuss some applications, and finally the design of monolithic PLLs.

10.4.1 Phase-Locked Loop Concepts

A block diagram of the basic phase-locked loop system is shown in Fig. 10.18. The elements of the system are a phase comparator, a loop filter, an amplifier, and a voltage-controlled oscillator. The voltage-controlled oscillator, or VCO, is simply an oscillator whose frequency is proportional to an externally applied voltage. When the loop is locked on an incoming periodic signal, the VCO frequency is exactly equal to that of the incoming signal. The phase detector produces a dc or low-frequency signal proportional to the phase difference between the incoming signal and the VCO output signal. This phase-sensitive signal is then passed through the loop filter and amplifier, and is applied to the control input of the VCO. If, for example, the frequency of the incoming signal shifts slightly, the phase difference between the VCO signal and the incoming signal will begin to increase with time. This will change the control voltage on the VCO in such a way as to bring the VCO frequency back to the same value as the incoming signal. Thus the loop can maintain lock when the input signal frequency changes, and the VCO input voltage is proportional to the frequency of the incoming signal. This behavior makes PLLs particularly useful for the demodulation of FM signals, where the frequency of the incoming signal varies in time and contains the desired information. The range of input signal frequencies over which the loop can maintain lock is called the "*lock range*."

An important aspect of PLL performance is the capture process, by which the loop goes from the unlocked, free-running condition to that of being locked on a signal. In the unlocked condition, the VCO runs at the frequency corresponding to zero applied dc voltage at its control input. This frequency is called the center frequency, or free-running frequency. When a periodic signal is applied that has a frequency near the free-running frequency, the loop may or may not lock on it depending on a number of factors. The capture process is inherently nonlinear in nature and we will describe the transient only in a qualitative way.

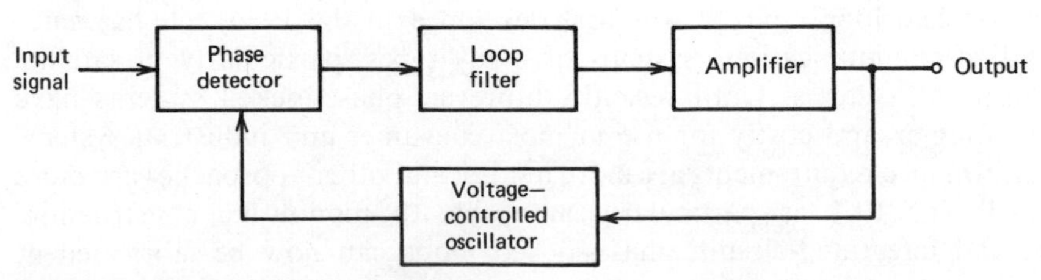

Fig. 10.18 Phase-locked-loop system.

First assume that the loop is opened between the loop filter and the VCO control input, and that a signal whose frequency is near, but not equal to, the free-running frequency is applied to the input of the PLL. The phase detector is usually of the type discussed in the last section, but for this qualitative discussion we assume that the phase detector is simply an analog multiplier which multiplies the two sinusoids together. Thus the output of the multiplier-phase detector contains the sum and difference frequency components, and we assume that the sum frequency component is sufficiently high in frequency that it is filtered out by the low-pass filter. The output of the low-pass filter, then, is a sinusoid with a frequency equal to the difference between the VCO free-running frequency and the incoming signal frequency.

Now assume that the loop is suddenly closed, and the difference frequency sinusoid is now applied to the VCO input. This will cause the VCO frequency itself to become a sinusoidal function of time. Let us assume that the incoming frequency was lower than the free-running frequency. Since the VCO frequency is varying as a function of time, it will alternately move *closer* to the incoming signal frequency and *further away* from the incoming signal frequency. The output of the phase detector is a near-sinusoid whose frequency is the *difference* between the VCO frequency and the input frequency. When the VCO frequency moves away from the incoming frequency, this sinusoid moves to a higher frequency. When the VCO frequency moves closer to the incoming frequency, the sinusoid moves to a lower frequency. If we examine the effect of this on the phase detector output, we see that the *frequency* of this sinusoidal difference-frequency waveform is reduced when its incremental amplitude is negative, and increased when its amplitude is positive. This causes the phase detector output to have an asymmetrical waveform during capture, as shown in Fig. 10.19. This assymmetry in the waveform introduces a dc component in the phase detector output that shifts

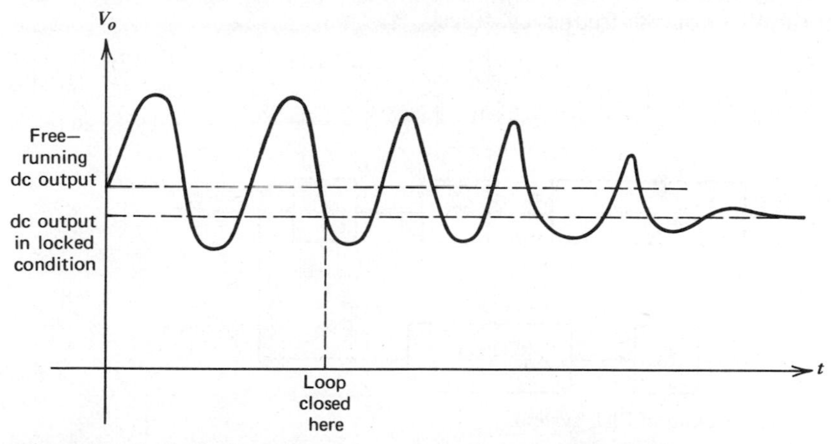

Fig. 10.19 Typical phase detector output during capture transient.

the average VCO frequency toward the incoming signal frequency, so that the difference frequency gradually decreases. Once the system becomes locked, of course, the difference frequency becomes zero and only a dc voltage remains at the loop-filter output.

The capture range of the loop is that range of input frequencies around the center frequency onto which the loop will become locked from an unlocked condition. The pull-in time is the time required for the loop to capture the signal. Both these parameters depend on the amount of gain in the loop itself, and the bandwidth of the loop filter. The objective of the loop filter is to filter out difference components resulting from interfering signals far away from the center frequency. It also provides a memory for the loop in case lock is momentarily lost due to a large interfering transient. Reducing the loop filter bandwidth thus improves the rejection of out-of-band signals, but, at the same time, the capture range is decreased, the pull-in time becomes longer, and the loop phase margin becomes poorer.

10.4.2 The Phase-Locked Loop in the Locked Condition

Under locked conditions, a linear relationship exists between the output voltage of the phase detector and the phase difference between the VCO and the incoming signal. This fact allows the loop to be analyzed using standard linear feedback concepts when in the locked condition. A block diagram representation of the system in this mode is shown in Fig. 10.20. The gain of the phase comparator is K_D V/rad of phase difference, the loop-filter transfer function is $F(s)$, and any gain in the forward loop is represented by A. The VCO "gain" is K_O rad/sec per volt.

If a constant input voltage is applied to the VCO control input, the output frequency of the VCO remains constant. However, the phase comparator is sensitive to the difference between the *phase* of the VCO output and the *phase* of the incoming signal. The phase of the VCO output is actually equal to the time integral of the VCO output frequency, since

$$\omega_{\text{osc}}(t) = \frac{d\phi_{\text{osc}}(t)}{dt} \tag{10.48}$$

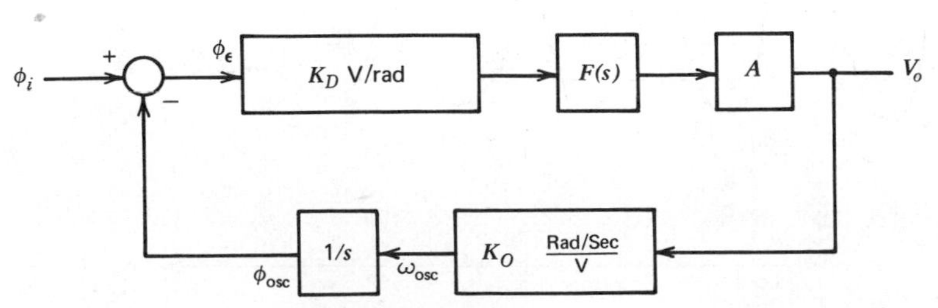

Fig. 10.20 Block diagram of PLL system.

and thus

$$\phi_{osc}(t) = \phi_{osc}\big|_{t=0} + \int_0^t \omega_{osc}(t)\, dt \tag{10.49}$$

Thus an integration inherently takes place within the phase-locked loop. This integration is represented by the $1/s$ block in Fig. 10.20.

For practical reasons, the VCO is actually designed so that when the VCO input voltage (i.e., V_o) is zero, the VCO frequency is not zero. The relation between the VCO output frequency ω_{osc}, and V_o is actually:

$$\omega_{osc} = \omega_o + K_O V_o$$

where ω_o is the free-running frequency that results when $V_o = 0$.

The system can be seen from Fig. 10.20 to be a classical linear feedback control system.[6] The closed-loop transfer function is given by:

$$\frac{V_o}{\phi_i} = \frac{K_D F(s) A}{1 + K_D F(s) A \dfrac{K_O}{s}} \tag{10.50}$$

$$= \frac{s K_D F(s) A}{s + K_D K_O A F(s)} \tag{10.51}$$

Usually we are interested in the response of this loop to *frequency* variations at the input, so that the input variable is frequency rather than phase. Since

$$\omega_i = \frac{d\phi_i}{dt} \tag{10.52}$$

then

$$\omega_i(s) = s\phi_i(s) \tag{10.53}$$

and

$$\frac{V_o}{\omega_i} = \frac{1}{s}\frac{V_o}{\phi_i} = \frac{K_D F(s) A}{s + K_D K_O A F(s)} \tag{10.54}$$

We first consider the case in which the loop filter is removed entirely, and $F(s)$ is unity. This is called a first-order loop and we have

$$\frac{V_o}{\omega_i} = \left(\frac{K_v}{s + K_v}\right)\left(\frac{1}{K_O}\right) \tag{10.55}$$

where

$$K_v = K_O K_D A \tag{10.56}$$

Thus the loop inherently produces a first-order, low-pass transfer characteristic. Remember that we regard the input variable as the frequency ω_i of the incoming signal. The response calculated above, then, is really the response from the frequency modulation on the incoming carrier to the loop voltage output.

The constant above (K_v) is termed the loop bandwidth. If the loop is locked on a carrier signal, and the frequency of that carrier is made to vary sinusoidally in time with a frequency ω_m, then a sinusoid of frequency ω_m will be observed at the loop output. When ω_m is increased above K_v, the magnitude of the sinusoid at the output falls. The loop bandwidth, K_v, then, is the effective bandwidth for the *modulating* signal that is being demodulated by the PLL. In terms of the loop parameters, K_v is simply the product of the phase detector gain, VCO gain, and any other electrical gain in the loop. The root locus of this single pole as a function of loop gain K_v is shown in Fig. 10.21*a*. The frequency response is also shown in this figure. The response of the loop to variations in input frequency is illustrated in Fig. 10.21*b* and by the following example.

EXAMPLE

A PLL has a K_O of 2π(1 kHz/V), a K_v of 500 $\sec^{-1}$, and a free-running frequency of 500 Hz.

(a) For a constant input signal frequency of 250 Hz and 1 kKz, find V_o.

$$V_o = \frac{\omega_i - \omega_o}{K_O} \quad \text{where } \omega_o = \text{oscillator free-running frequency}$$

$$\text{At 250 Hz, } V_o = \frac{2\pi(250) - 2\pi(500)}{2\pi(1\ \text{kHz/V})} = -0.25\ \text{V}$$

$$\text{At 1 kHz, } V_o = \frac{2\pi(1\ \text{kHz}) - 2\pi(500)}{2\pi(1\ \text{kHz/V})} = +0.5\ \text{V}$$

(b) Now the input signal is frequency modulated, so that

$$\omega_i(t) = (2\pi)\ 500\ \text{Hz}\ [1 + 0.1 \sin(2\pi \times 10^2)t]$$

Find the output signal $V_o(t)$. From (10.55) we have

$$\frac{V_o(j\omega)}{\omega_i(j\omega)} = \frac{1}{K_O}\left(\frac{K_v}{K_v + j\omega}\right) = \frac{1}{K_O}\left[\frac{K_v}{K_v + j(2\pi \times 10^2)}\right]$$

$$= \frac{1}{2\pi(1\ \text{kHz/V})}\left(\frac{500}{500 + j628}\right)$$

$$= \frac{1}{2\pi(1\ \text{kHz/V})}(0.39 - j0.48)$$

The *magnitude* of $\omega_i(j\omega)$ is

$$|\omega_i(j\omega)| = (0.1)(500\ \text{Hz})(2\pi) = (50)(2\pi)$$

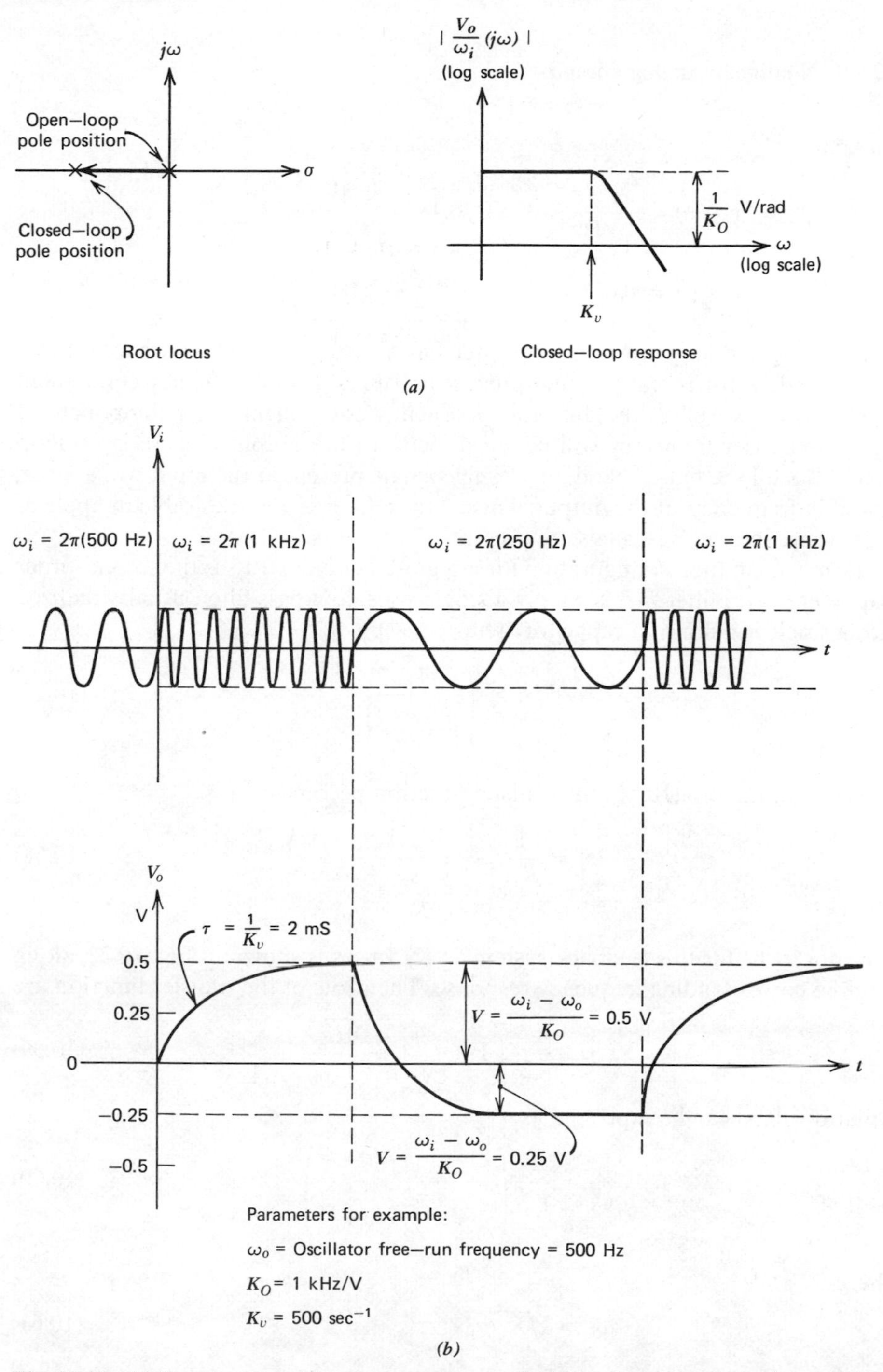

Fig. 10.21 (*a*) Root locus and frequency response of a first-order, phase-locked loop. (*b*) Response of loop output voltage to step changes in input frequency, example first-order loop.

Therefore,

$$V_o(j\omega) = \frac{50 \text{ Hz}}{1 \text{ kHz}}(0.39 - j0.48) = \frac{50}{1000}(0.62 \angle -51^\circ)$$

and

$$V_o(t) = 0.031 \sin(2\pi \times 10^2 t - 51^\circ)$$

Operating the loop with no loop filter has several practical drawbacks. Since the phase detector is really a multiplier, it produces a sum frequency component at its output as well as the difference frequency component. This component at twice the carrier frequency will be fed directly to the output if there is no loop filter. Also, all the out-of-band interfering signals present at the input will appear, shifted in frequency, at the output. Thus, a loop filter is very desirable in applications where interfering signals are present.

The most common configuration for integrated circuit PLLs is the second-order loop. Here, loop filter $F(s)$ is simply a single-pole, low-pass filter, usually realized with a single resistor and capacitor. Thus

$$F(s) = \left(\frac{1}{1 + \dfrac{s}{\omega_1}} \right) \tag{10.57}$$

By substituting into (10.54), the transfer function becomes

$$\frac{V_o}{\omega_i}(s) = \frac{1}{K_O} \left(\frac{1}{1 + \dfrac{s}{K_v} + \dfrac{s^2}{\omega_1 K_v}} \right) \tag{10.58}$$

The root locus for this feedback system as K_v varies is shown in Fig. 10.22, along with the corresponding frequency response. The roots of the transfer function are

$$s = -\frac{\omega_1}{2}\left(1 \pm \sqrt{1 - \frac{4K_v}{\omega_1}}\right) \tag{10.59}$$

Equation 10.58 can be expressed as

$$\frac{V_o}{\omega_i} = \frac{1}{K_O} \left(\frac{1}{\dfrac{s^2}{\omega_n{}^2} + \dfrac{2\zeta}{\omega_n} s + 1} \right) \tag{10.60}$$

where

$$\omega_n = \sqrt{K_v \omega_1} \tag{10.61}$$

$$\zeta = \tfrac{1}{2}\sqrt{\frac{\omega_1}{K_v}} \tag{10.62}$$

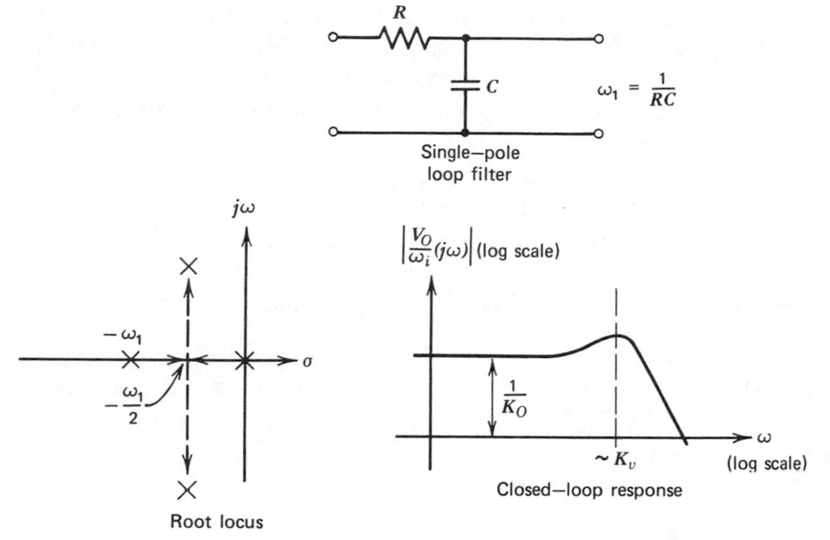

Fig. 10.22 Root locus and frequency response of a second-order, phase-locked loop.

The basic factor setting the loop bandwidth is K_v as in the first-order case. The frequency ω_1 of the additional pole is then made as low as possible without causing an unacceptable amount of peaking in the frequency response. This peaking is of concern both because it distorts the demodulated FM output and because it causes the loop to ring, or experience a poorly-damped oscillatory response, when a transient disturbs the loop. A good compromise is using a maximally flat low-pass pole configuration in which the poles are placed on radials angled 45° from the negative real axis. For this response, the damping factor ζ should be equal to $1/\sqrt{2}$. Thus:

$$\frac{1}{\sqrt{2}} = \frac{1}{2}\sqrt{\frac{\omega_1}{K_v}} \tag{10.63}$$

and

$$\omega_1 = 2K_v \tag{10.64}$$

The -3-dB frequency of the transfer function $(V_O/\omega_i)(j\omega)$ is then:

$$\omega_{-3\text{ dB}} = \omega_n = \sqrt{K_v\omega_1} = \sqrt{2}\,K_v \tag{10.65}$$

A disadvantage of the second-order loop as discussed thus far is that the -3-dB bandwidth of the loop is basically dictated by loop gain K_v as shown by (10.65). As we will show, the loop gain also sets the lock range, so that with the simple filter used above these two parameters are constrained to be comparable. Situations do arise in phase-locked communications in which a wide lock range is desired for tracking large signal-frequency variations, yet a narrow loop bandwidth is desired for rejecting out-of-band signals. Using a very small ω_1 would accomplish this were it not for the fact that this produces underdamped loop response. By adding a zero to the loop filter, the loop-filter pole can be made small while still maintaining good loop damping.

The effect of the addition of a zero on the loop response is best seen by examining the open-loop response of the circuit. Shown in Fig. 10.23*a* is the open-loop response of the circuit with no loop filter. Because of the integration inherent in the loop, the response has a -20-dB/decade slope throughout the frequency range and crosses unity gain at K_v. In Fig. 10.23*b*, a loop filter in which ω_1 is much

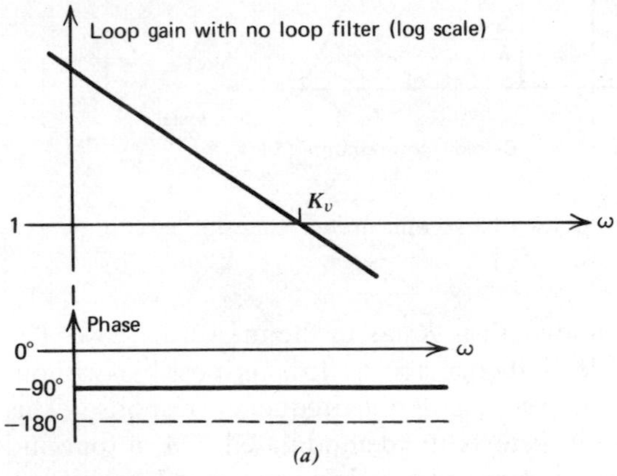

Fig. 10.23*a* PLL open-loop response with no loop filter.

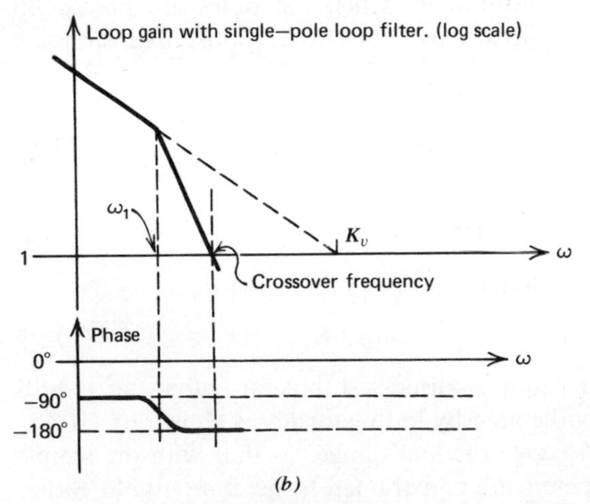

Fig. 10.23*b* PLL open-loop response with a single-pole filter and $\omega_1 \ll K_v$.

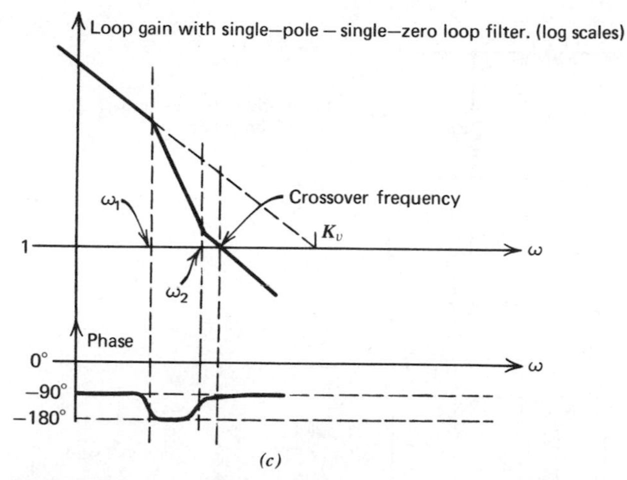

Fig. 10.23c PLL open-loop response with zero added in loop filter at $s = -\omega_2$.

less than K_v has been added, and, as a result, the loop phase shift is very nearly 180° at the crossover frequency. The result is a sharp peak in the closed-loop frequency response at the crossover frequency. By adding a zero in the loop filter at ω_2, as shown in Fig. 10.23*c*, the loop phase margin can be greatly improved. Note that for this case the loop bandwidth, which is equal to the crossover frequency, is much lower than K_v. This ability to set loop bandwidth and K_v independently is an advantage of this type of loop filter. An R-C circuit that provides the necessary pole and zero in the filter response is shown in Fig. 10.23*d*. The root locus for this loop filter and the resulting closed-loop response are also shown.

Loop Lock Range. The loop lock range is the range of input frequencies about the center frequency for which the loop maintains lock. In most cases, it is limited by the fact that the phase comparator has a limited phase comparison range; once the phase difference between the input signal and the VCO output reaches more than 90°, the phase comparator ceases to behave linearly. The transfer characteristic of a typical phase comparator is shown in Fig. 10.17. It is clear from this figure that in order to maintain lock, the phase difference between the VCO output and the incoming signal must be kept between zero and π. If the phase difference is equal to either zero or π, then the magnitude of the dc voltage at the output of the phase comparator is:

$$V_{o\,\max} = \pm K_D\left(\frac{\pi}{2}\right) \tag{10.66}$$

This dc voltage is amplified by the electrical gain, A, and the result is applied to the VCO input, producting a frequency shift away from the free-running center

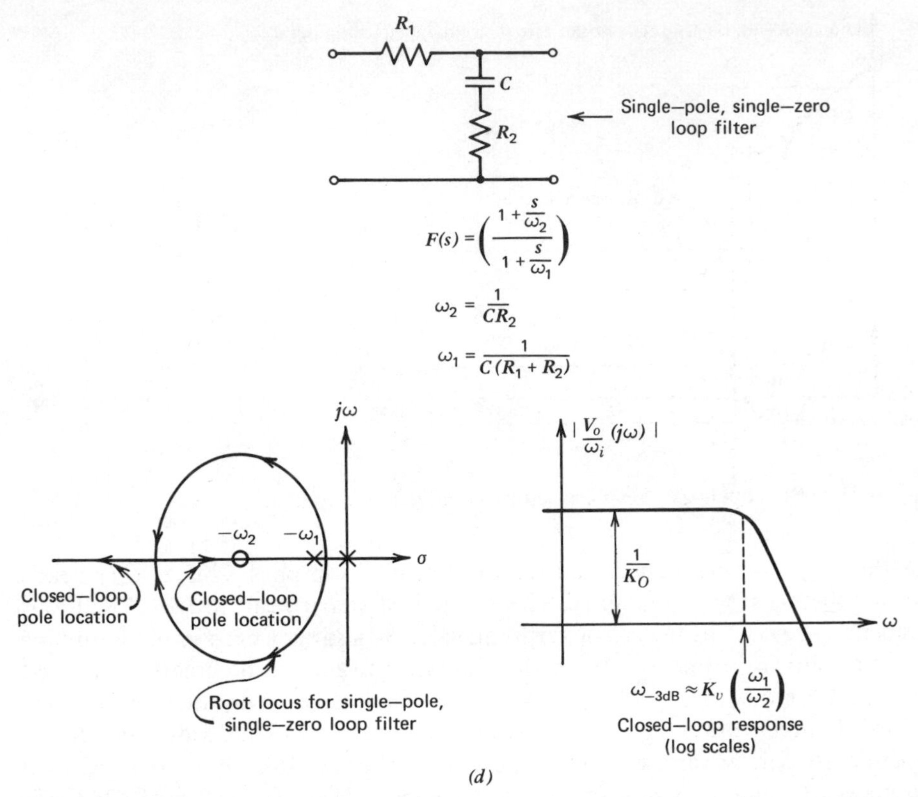

(*d*)

Fig. 10.23*d* Root locus and frequency response of a second-order PLL with a zero. Frequency response shown is for large loop gain such that poles are located as shown in the root locus.

frequency of

$$\Delta\omega_{osc} = K_D A K_O \left(\frac{\pi}{2}\right) = (K_v \pi/2) \tag{10.67}$$

If the input frequency is now shifted away from the free-running frequency, more voltage will have to be applied to the VCO in order for the VCO frequency to shift accordingly. However, the phase detector can produce no more dc output voltage to shift the VCO frequency further, so the loop will lose lock. The lock range, ω_L, is then given by

$$\omega_L = K_v \frac{\pi}{2} \tag{10.68}$$

This is the frequency range on either side of the free-running frequency for which the loop will track input frequency variations. It is a parameter that depends only on the dc gain in the loop and is independent of the properties of the loop filter.

The capture range is the range of input frequencies for which the initially unlocked loop will lock on an input signal when initially in an unlocked condition and is always less than the lock range. When the input frequency is swept through a range around the center frequency, the output voltage as a function of input frequency displays a hysterisis effect as shown in Fig. 10.24. As discussed earlier, the capture range is difficult to predict analytically. As a very rough rule of thumb, the approximate capture range can be estimated using the following procedure: refer to Fig. 10.18 and assume that the loop is opened at the loop-amplifier output and that a signal with a frequency not equal to the free-running VCO frequency is applied at the input of the PLL. The sinusoidal *difference* frequency component that appears at the output of the phase detector has the value

$$V_p(t) = \frac{\pi}{2} K_D \cos(\omega_i - \omega_{osc})t \qquad (10.69)$$

where ω_i is the input signal frequency and ω_{osc} is the VCO free-running frequency.

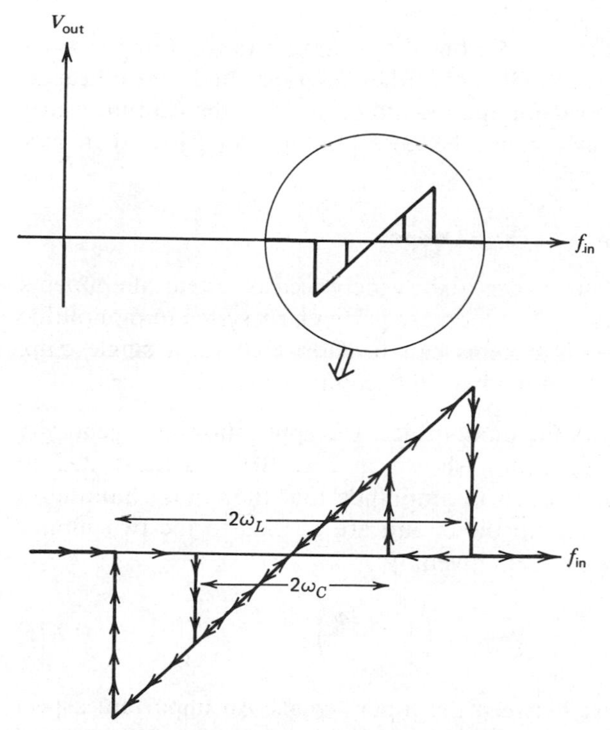

Fig. 10.24 PLL output versus frequency of input.

This component is passed through the loop filter and the output from the loop amplifier resulting from this component is

$$V_o(t) = \frac{\pi}{2} K_D A |F[j(\omega_i - \omega_{osc})]| \cos [(\omega_i - \omega_{osc})t + \phi.] \tag{10.70}$$

where $\phi = \angle F[j(\omega_i - \omega_{osc})]$

The output from the loop amplifier thus consists of a sinusoid at the difference frequency whose *amplitude* is reduced by the loop filter. In order for capture to occur, the magnitude of the voltage that must be applied to the VCO input is:

$$|V_{osc}| = \frac{\omega_i - \omega_{osc}}{K_O} \tag{10.71}$$

The capture process itself is rather complex, but the capture range can be estimated by setting the magnitudes of (10.70) and (10.71) equal. The result is that capture is likely to occur if the following inequality is satisfied:

$$|(\omega_i - \omega_{osc})| < \frac{\pi}{2} K_D K_O A |F[j(\omega_i - \omega_{osc})]| \tag{10.72}$$

This equation implicitly gives an estimation of the capture range. For the first-order loop, where $F(s)$ is unity, it predicts that the lock range and capture range are approximately equal, and that for the second-order loop the capture range is significantly less than the lock range because $|F[j(\omega_i - \omega_{osc})]|$ is then less than unity.

10.4.3 Integrated-Circuit Phase-Locked Loops

The principal reason that PLLs have come to be widely used as system components is that the elements of the phase-locked loop are particularly suited to monolithic construction, and complete PLL systems can be fabricated on a single chip. We now discuss the design of the individual PLL components.

Phase Detector. Phase detectors for monolithic PLL applications are generally of the Gilbert multiplier configuration shown in Fig. 10.9. As illustrated in Fig. 10.16, if two signals large enough in amplitude that they cause limiting in the emitter-coupled pairs making up the circuit are applied to the two inputs, the output will contain a dc component given by

$$V_{average} = -I_{EE}R_C\left(1 - \frac{2\phi}{\pi}\right) \tag{10.73}$$

where ϕ is the phase difference between the input signals. An important aspect of the performance of this phase detector is that if the amplitude of the applied signal at V_{in2} is small compared to the thermal voltage V_T, the circuit behaves as a balanced modulator, and the dc component of the output depends on the amplitude of the low-level input. The output waveform is then a sinusoid multiplied by a synchronous square wave, as shown in Fig. 10.25. In the limiting case when the small input is small compared to V_T, the dc component in the output becomes, referring to Fig. 10.25,

$$V_{\text{average}} = \frac{1}{\pi} g_m R_C V_i \left[\int_0^{\phi} (\sin \omega t) d(\omega t) - \int_{\phi}^{\pi} (\sin \omega t) d(\omega t) \right] \tag{10.74}$$

$$= -\frac{2 g_m R_C V_i \cos \phi}{\pi} \tag{10.75}$$

where R_C is the collector resistor in the Gilbert multiplier and g_m is the transconductance of the transistors. The phase detector output voltage then becomes

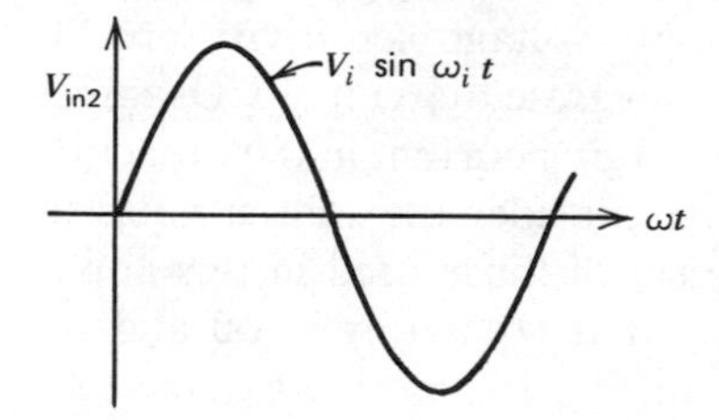

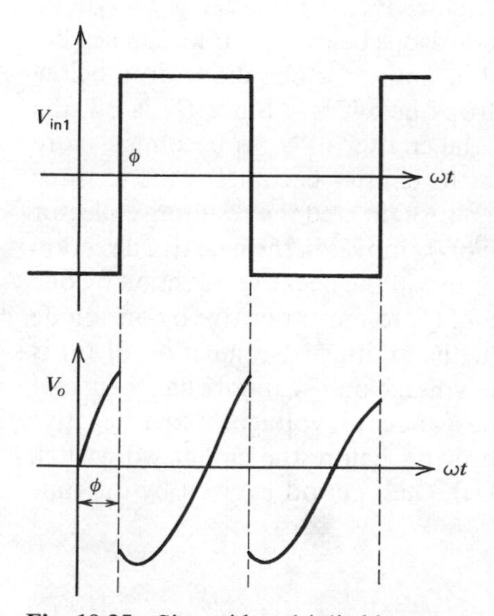

Fig. 10.25 Sinusoid multiplied by a synchronous square wave.

proportional to the amplitude V_i of the incoming signal, and if the signal amplitude varies then the loop gain of the phase-locked-loop changes. Thus when the signal amplitude varies, it is often necessary to precede the phase detector with an amplifier/limiter to avoid this problem. In FM demodulators, for example, any amplitude modulation appearing on the incoming frequency-modulated signal will be demodulated, producing an erroneous output.

In PLL applications the frequency response of the phase detector is usually not the limiting factor in the usable operating frequency range of the loop itself. At high operating frequencies, the parasitic capacitances of the devices result in a feedthrough of the carrier frequency, giving an erroneous component in the output at the center frequency. This component is removed by the loop filter, however, and does not greatly affect loop performance. The VCO is usually the limiting factor in the operating frequency range.

Voltage-controlled Oscillator. The operating frequency range, FM distortion, center-frequency drift, and center-frequency, supply-voltage sensitivity are all determined by the performance of the VCO. Integrated-circuit VCOs most often are simply R-C multivibrators in which the charging current in the capacitor is varied in response to the control input. We first consider the emitter-coupled multivibrator shown in Fig. 10.26*a*, which is typical of those used in this application. We calculate the period by first assuming that Q_1 is turned off and Q_2 is turned on. The circuit then appears as shown in Fig. 10.26*b*. We assume that current I is large so that the voltage drop IR is large enough to turn on diode Q_6. Thus the base of Q_4 is one diode drop below V_{CC}, the emitter is two diode drops below V_{CC}, and the base of Q_1 is two diode drops below V_{CC}. If we can neglect the base current of Q_3, its base is at V_{CC} and its emitter is one diode drop below V_{CC}. Thus the emitter of Q_2 is two diode drops below V_{CC}. Since Q_1 is off, the current I_1 is charging the capacitor so that the emitter of Q_1 is becoming more negative. Q_1 will turn on when the voltage at its emitter becomes equal to three diode drops below V_{CC}. Transistor Q_1 will then turn on, and the resulting collector current in Q_1 turns on Q_5. As a result, the base of Q_3 moves in the negative direction by one diode drop, causing the base of Q_2 to move in the negative direction by one diode drop. Q_2 will turn off, causing the base of Q_1 to move positive by one diode drop because Q_6 also turns off. As a result, the emitter-base junction of Q_2 is reverse biased by one diode drop because the voltage on C cannot change instantaneously. Current I_1 must now charge the capacitor voltage in the negative direction by an amount equal to two diode drops before the circuit will switch back again. Since the circuit is symmetrical, the half period is given by the time required to charge the capacitor and is

$$\frac{T}{2} = \frac{Q}{I_1} \tag{10.76}$$

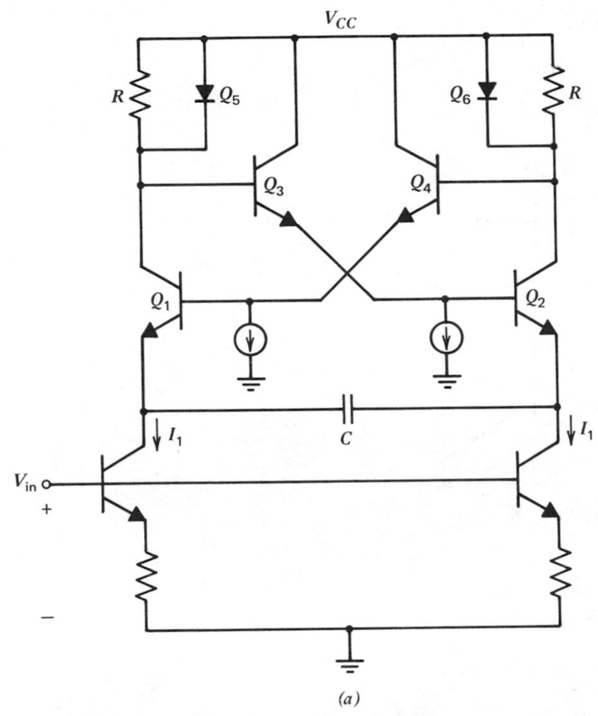

Fig. 10.26*a* Voltage-controlled, emitter-coupled multivibrator.

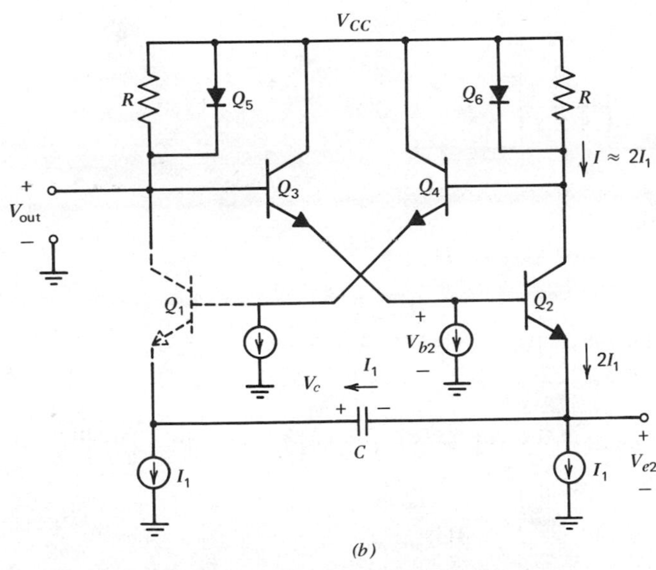

Fig. 10.26*b* Equivalent circuit during one half-cycle.

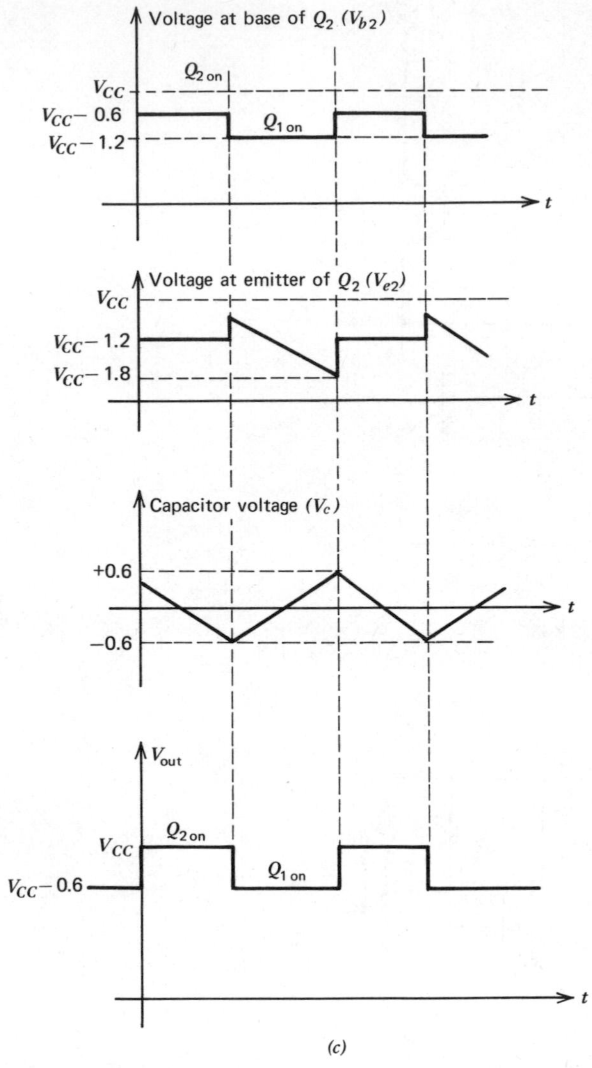

Fig. 10.26*c* Waveforms within the emitter-coupled multivibrator.

where $Q = C\,\Delta V = 2CV_{BE(\text{on})}$ is the charge on the capacitor. The frequency of the oscillator is thus

$$f = \frac{1}{T} = \frac{I_1}{4CV_{BE(\text{on})}} \tag{10.77}$$

The various waveforms in the circuit are shown in Fig. 10.26*c*. This emitter-coupled configuration is nonsaturating and contains only *npn* transistors. Furthermore, the voltage swings within the circuit are small. As a result, the circuit is capable of operating up to approximately 100 MHz for typical integrated-circuit transistors. However, the usable frequency range is limited to a value lower than this because the center frequency drift with temperature variations becomes large at the higher frequencies. This drift occurs because the switching transients themselves become a large percentage of the period of the oscillation, and the duration of the switching transients depends on circuit parasitics, circuit resistances, transistor transconductance, and transistor input resistance, which are all temperature sensitive.

While the emitter-coupled configuration is capable of high operating speed, it displays considerable sensitivity of center frequency to temperature even at low frequencies, since the period is dependent on $V_{BE(\text{on})}$. Utilizing (10.77), we can calculate the temperature coefficient of the period as:

$$\frac{1}{\omega_{\text{osc}}}\frac{d\omega_{\text{osc}}}{dT} = -\frac{1}{V_{BE(\text{on})}}\frac{dV_{BE(\text{on})}}{dT} = \frac{+2\text{ mV/°C}}{600\text{ mV}} = +3300\text{ ppm/°C} \quad (10.78)$$

This temperature sensitivity of center frequency can be compensated by causing current I_1 to be temperature sensitive in such a way that its effect is equal and opposite to the effect of the variation of $V_{BE(\text{on})}$.

10.4.4 Analysis of the 560B Monolithic Phase-Locked Loop

The first practical monolithic PLL was the 560/561/562 series circuits that became commercially available from several manufacturers in 1970 and 1971. These circuits contain a phase detector of the Gilbert type, an emitter-coupled, temperature-compensated VCO, and provision for use of an external R-C circuit to perform the loop filter function. A schematic diagram of the 560B is shown in Fig. 10.27*a*. Transistors Q_1 to Q_{14} form the bias reference circuit that provides a 13-V supply-independent, Zener-referenced voltage for the phase detector and the VCO. It also provides a dc reference potential for the differential inputs, through R_3 and R_4, so that the input can be ac coupled. Transistors Q_{13} and Q_{14} provide a reference voltage for the current sources in the VCO, and transistor Q_{11} provides the bias current for the phase detector.

The VCO is composed of transistors Q_{23} and Q_{24} and emitter followers Q_{27} and Q_{28}. The phase detector is composed of transistors Q_{15} to Q_{20} and the differential output of the phase detector is fed to the VCO through emitter followers Q_{21} and Q_{22}, and emitter-coupled pair Q_{35} to Q_{38}.

We will now analyze the circuit with the objective of determining the K_D and K_O parameters of the loop. We begin by analyzing the bias reference circuit shown separately in Fig. 10.27*b*. The bias circuit generates a regulated 13V subsupply for the VCO and the phase comparator so that the bias voltages in these circuits remain constant when the supply voltage changes over the allowed 16- to 26-V range. We will assume the supply voltage is 16 V; the only current level in the circuit affected by supply voltage is that in the chain through R_{18} and Q_3 to Q_6.

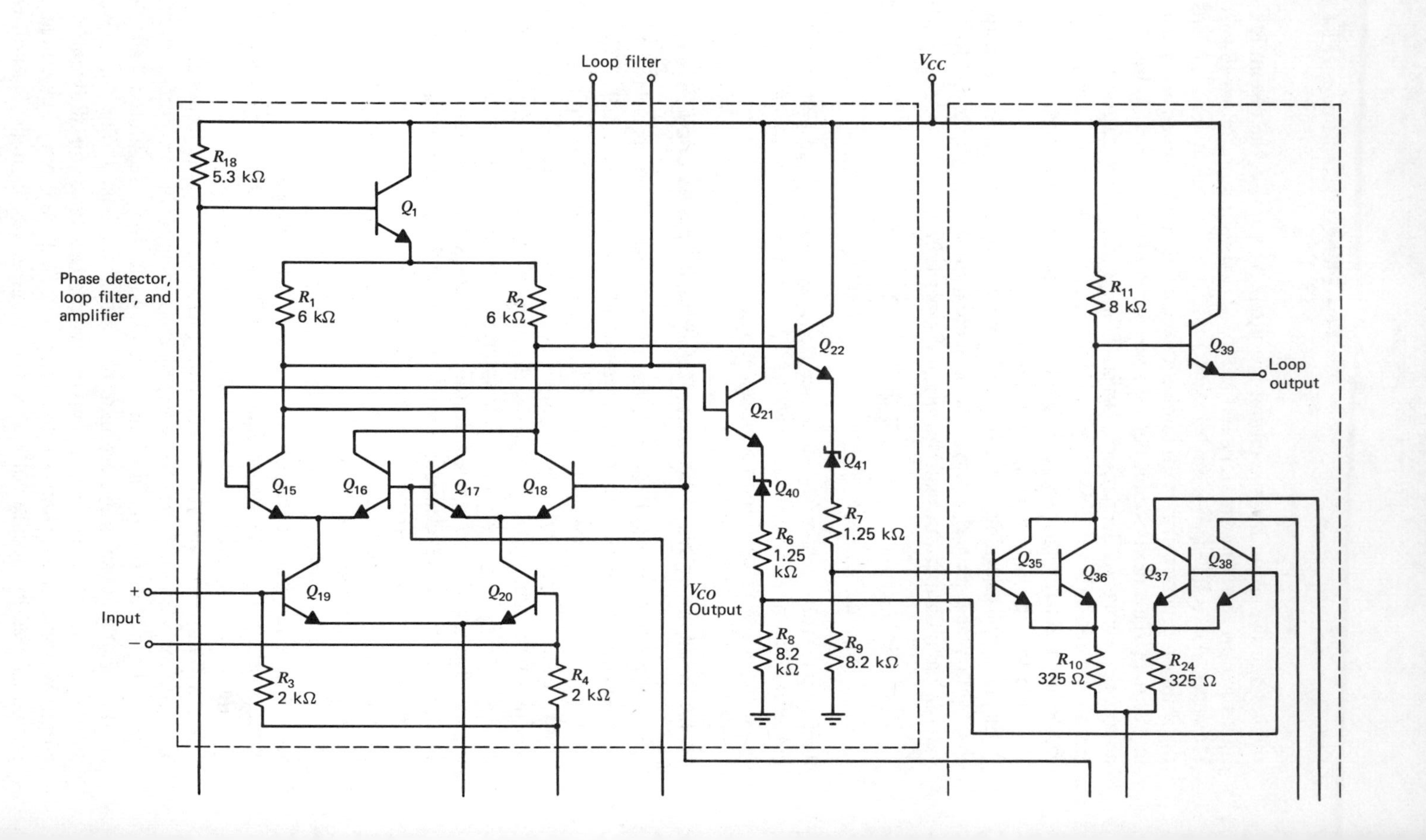

Loop filter
V_{CC}
R_{18} 5.3 kΩ
Q_1
Phase detector, loop filter, and amplifier
R_1 6 kΩ
R_2 6 kΩ
Q_{22}
Q_{21}
R_{11} 8 kΩ
Q_{39}
Loop output
Q_{15}
Q_{16}
Q_{17}
Q_{18}
Q_{40}
Q_{41}
R_7 1.25 kΩ
R_6 1.25 kΩ
Q_{35}
Q_{36}
Q_{37}
Q_{38}
+
Input
−
Q_{19}
Q_{20}
V_{CO} Output
R_8 8.2 kΩ
R_9 8.2 kΩ
R_{10} 325 Ω
R_{24} 325 Ω
R_3 2 kΩ
R_4 2 kΩ

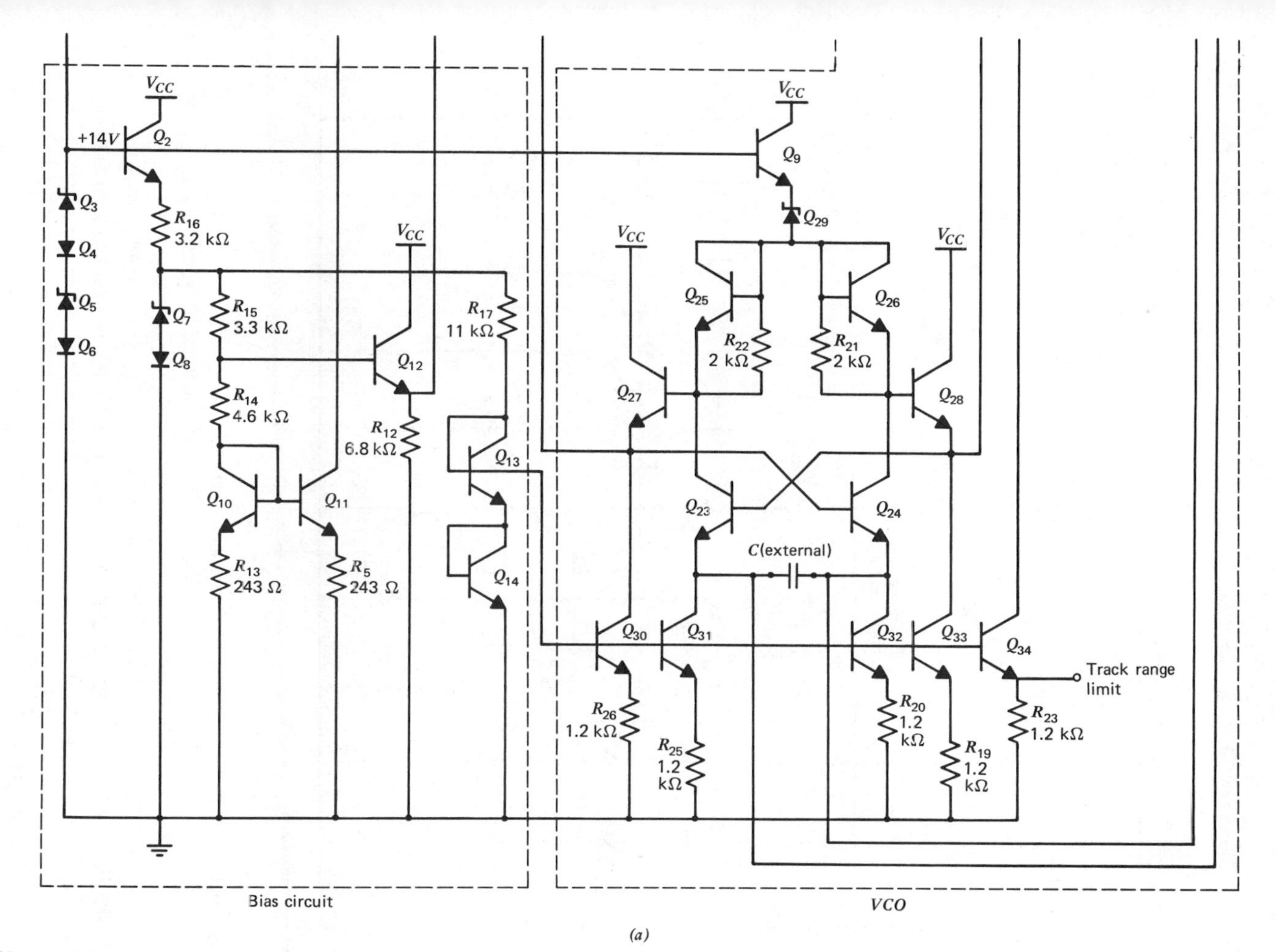

Fig. 10.27*a* 560B phase-locked loop circuit schematic.

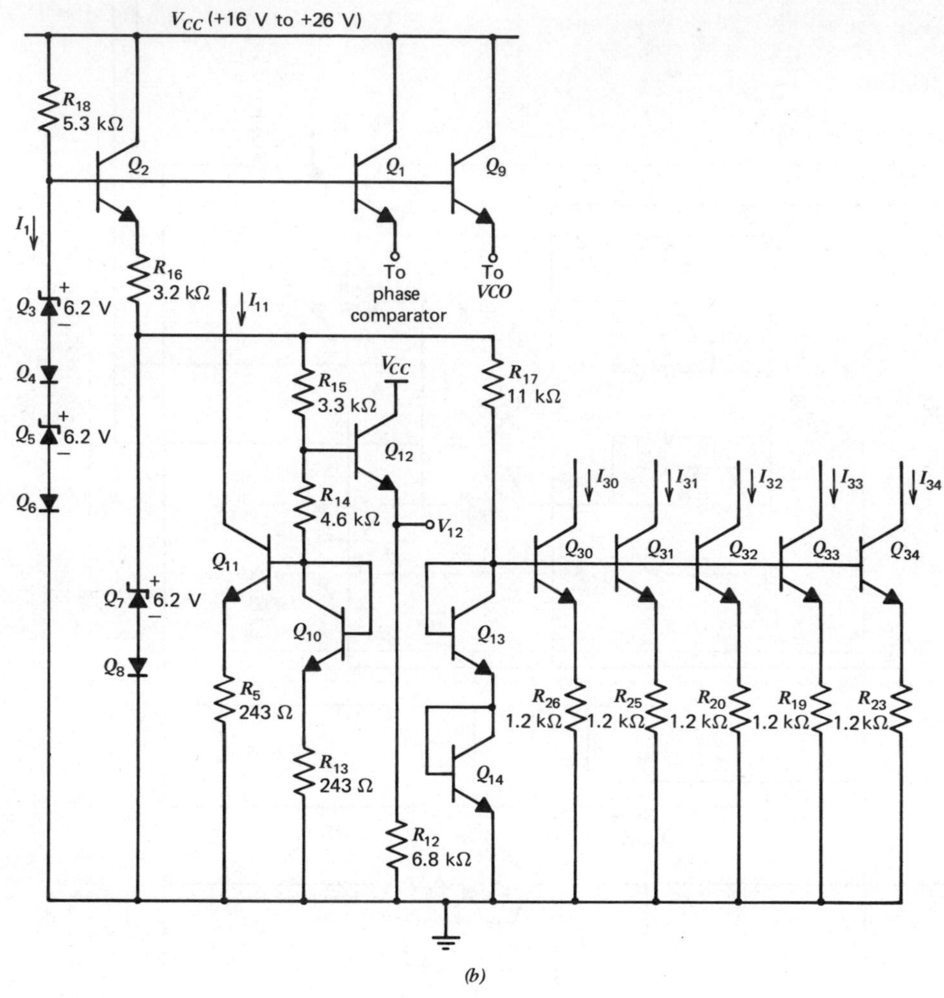

Fig. 10.27b Bias circuit for the 560B phase-locked loop.

Zener diodes Q_3 and Q_5 are reverse-biased, emitter-base junctions with breakdown voltages near 6.2 V so that

$$I_1 = \frac{V_{CC} - V_{Z3} - V_{BE4} - V_{Z5} - V_{BE6}}{R_{18}} \tag{10.79}$$

$$I_1 = \frac{(V_{CC} - 6.2 - 0.6 - 6.2 - 0.6)}{5.3\text{ k}\Omega} = \frac{16 - 13.6}{5.3\text{ k}\Omega} = 0.45\text{ mA} \tag{10.80}$$

Thus the voltage on the emitters of Q_1, Q_2, and Q_9 is 13.0 V. Assuming that Q_7 is conducting current, we can see that the voltage at the top of Q_7 is 6.8 V. Thus the current in R_{17} is

$$I_{R17} = \frac{V_{Z7} + V_{BE8} - V_{BE13} - V_{BE14}}{R_{17}} \tag{10.81}$$

$$I_{R17} = \frac{6.2 + 0.6 - 0.6\text{ V} - 0.6\text{ V}}{11\text{ k}\Omega} = 0.51\text{ mA} \tag{10.82}$$

The current in Q_{11}, which is the bias current for the phase comparator, is equal to that in Q_{10}, which is

$$I_{C10} = I_{C11} = \frac{V_{Z7} + V_{BE8} - V_{BE10}}{R_{13} + R_{14} + R_{15}} \tag{10.83}$$

$$= \frac{6.8\text{ V} - 0.6\text{ V}}{8.14\text{ k}\Omega} = 0.77\text{ mA} \tag{10.84}$$

The bias voltage, V_{12}, which supplies the dc reference for the input terminals, is then

$$V_{12} = V_{BE10} + (V_{BE8} + V_{Z7} - V_{BE10})\left(\frac{R_{14} + R_{13}}{R_{15} + R_{14} + R_{13}}\right) - V_{BE12} \tag{10.85}$$

and thus

$$V_{12} = 0.6\text{ V} + (6.8 - 0.6)\left(\frac{4.84}{3.3 + 4.84}\right)\text{V} - 0.6\text{ V} = 3.69\text{ V} \tag{10.86}$$

The voltage applied to the bases of current-source transistors Q_{30} to Q_{34} is equal to two diode drops above ground. The voltage across the 1.2-kΩ resistors R_{26} through R_{25}, R_{20}, R_{19}, and R_{23} is thus one diode drop, giving a current of approximately 0.5 mA. It is important to note that these currents have a large negative temperature coefficient, given by:

$$\frac{dI}{dT} = \frac{d}{dT}\left[\frac{V_{BE(\text{on})}}{R}\right] = \frac{V_{BE(\text{on})}}{R}\left[\frac{1}{V_{BE(\text{on})}}\frac{dV_{BE(\text{on})}}{dT} - \frac{1}{R}\frac{dR}{dT}\right] \tag{10.87}$$

$$= \frac{V_{BE(\text{on})}}{R}\left(\frac{-2\text{ mV/°C}}{600\text{ mV}} - 1500 \times 10^{-6}\right) \tag{10.88}$$

and thus

$$\frac{1}{I}\frac{dI}{dT} = -4800\text{ ppm/°C} \tag{10.89}$$

This temperature sensitivity partially temperature compensates the center frequency drift of the VCO, which results from the temperature variation in $V_{BE(\text{on})}$ within the VCO.

We now analyze the phase detector shown in Fig. 10.28. If the base currents of Q_{21} and Q_{22} are negligible, and a large, limiting signal is applied to the input, then the voltages appearing at the bases of Q_{21} and Q_{22} take on one of two values:

$$V_{\text{hi}} = 13\text{ V} \tag{10.90}$$

$$V_{\text{lo}} = 13\text{ V} - (I_{C11})(6\text{ k}\Omega) = 8.38\text{ V} \tag{10.91}$$

The amplitude of the differential square wave output is thus 4.62 V. Since I_{C11} has a temperature coefficient that is approximately the negative of that of a diffused resistor $(R_{13} + R_{14} + R_{15})$, and the output voltage swing is the product of this current with R_1 and R_2, the net temperature coefficient of the output amplitude is small. The purpose of the Q_{21}–Q_{40}–R_6–R_8 and Q_{22}–Q_{41}–R_7–R_9 bias

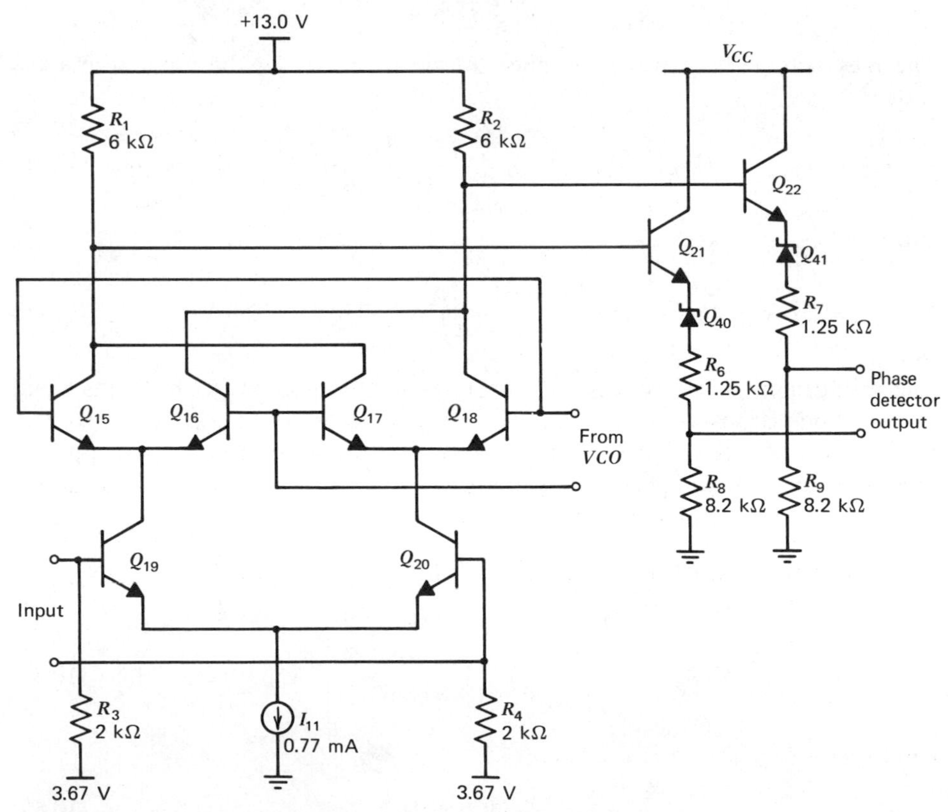

Fig. 10.28 Phase comparator of the 560B phase-locked loop.

strings is to level shift the signal prior to driving the VCO input. The voltages at the bases of Q_{35} and Q_{38} take on two values:

$$V_{hi} = (13\text{ V} - V_{BE21} - V_{Z40})\left(\frac{R_8}{R_6 + R_8}\right) \tag{10.92}$$

$$= (13\text{ V} - 0.6\text{ V} - 6.2\text{ V})\left(\frac{8.2}{8.2 + 1.25}\right) = 5.38\text{ V} \tag{10.93}$$

$$V_{lo} = (8.38\text{ V} - V_{BE21} - V_{Z40})\left(\frac{R_8}{R_6 + R_8}\right) = 1.37\text{ V} \tag{10.94}$$

The peak amplitude of the phase detector differential output is thus 4.0 V. Using (10.47) we find that the dc component of the differential output is

$$V_{\text{average}} = -(4.0\text{ V})\left(1 - \frac{2\phi}{\pi}\right) \qquad 0 < \phi < \pi \tag{10.95}$$

$$= K_D\left(\phi - \frac{\pi}{2}\right) \qquad 0 < \phi < \pi \tag{10.96}$$

Thus

$$K_D = \frac{4.0\text{ V}}{\pi/2} = 2.55\text{ V/rad} \tag{10.97}$$

If the signal applied to the input is small so that limiting does not occur, then the phase detector gain becomes, utilizing (10.75),

$$K_D = \frac{V_i}{\pi}(2g_{m19}R_1)\frac{R_8}{R_8 + R_6} \tag{10.98}$$

$$= \frac{V_i}{\pi}\left[(2)\left(\frac{0.38\text{ mA}}{26\text{ mV}}\right)(6\text{ k}\Omega)\right]\left(\frac{8.2}{9.45}\right) \tag{10.99}$$

$$= 48.4V_i\,(\text{rad})^{-1} \tag{10.100}$$

where V_i is the peak amplitude of the sinusoidal input. In terms of the rms value,

$$K_D = 68.4V_i|_{\text{rms}}\,(\text{rad})^{-1} \tag{10.101}$$

Thus for a 1-mV rms input, for example, the value of K_D is equal to 0.068 V/rad.

We now analyze the voltage-controlled oscillator shown in Fig. 10.29. We initially assume that the input differential voltage is zero, so the four emitter-coupled transistors (Q_{35} to Q_{38}) all conduct the same current. The collector currents of Q_{37} and Q_{38} are thus 125 μA, and the total current flowing from the emitters of Q_{23} and Q_{24} is 625 μA. This emitter-coupled multivibrator circuit

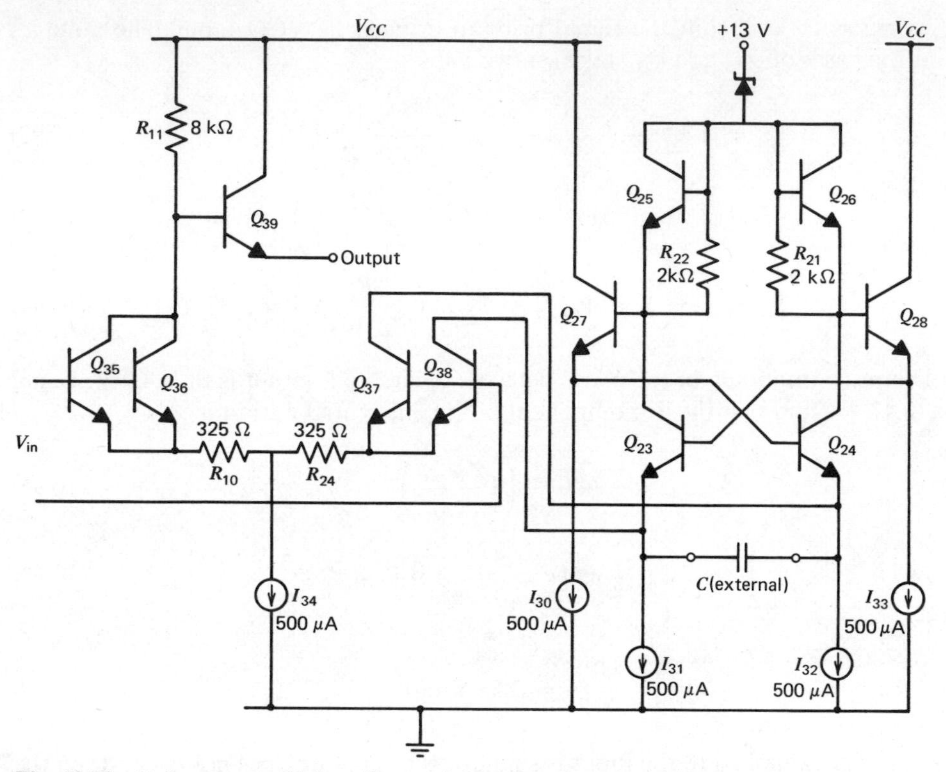

Fig. 10.29 Voltage-controlled oscillator of the 560B phase-locked loop.

was analyzed earlier, and the free-running frequency is given by (10.77) as

$$f_o = \frac{I}{4CV_{BE(\text{on})}} \tag{10.102}$$

Assuming a $V_{BE(\text{on})}$ of 0.6 V, the VCO center frequency is thus

$$f_o = \frac{625 \times 10^{-6}}{4C \times 0.6} = \frac{0.26 \times 10^{-3}}{C} \tag{10.103}$$

The required value of C to give a frequency f_o is

$$C = \frac{260}{f_o}\ \mu\text{F} \tag{10.104}$$

where f_o is in hertz and C is in microfarads. An important aspect of PLL performance is the center frequency sensitivity to temperature. Again, utilizing (10.77), we obtain

$$\frac{1}{f_o}\frac{df_o}{dT} = \frac{1}{I}\frac{dI}{dT} - \frac{1}{V_{BE}}\frac{dV_{BE}}{dT} - \frac{1}{C}\frac{dC}{dT} \tag{10.105}$$

The bias circuit that produces I_{31}, I_{32}, and I_{34} was analyzed previously, where it was determined that

$$\frac{1}{I}\frac{dI}{dT} = \frac{1}{V_{BE}}\frac{V_{BE}}{dT} - \frac{1}{R}\frac{dR}{dT} \tag{10.106}$$

Consequently, for this circuit, substituting (10.106) in (10.105) gives

$$\frac{1}{f_o}\frac{df_o}{dT} = -\frac{1}{R}\frac{dR}{dT} - \frac{1}{C}\frac{dC}{dT} \tag{10.107}$$

Assuming that an external capacitor with a low temperature coefficient is used, the center frequency temperature variation is related to the temperature variation of the diffused resistors making up the circuit. This temperature coefficient is usually on the order of 1000 to 1500 ppm/°C. This rather high drift can be reduced by redesigning the bias circuit to temperature compensate the resistor variation.

Since the collector currents of Q_{37} and Q_{38} can vary between zero and 250 μA, the total current charging the capacitor can vary from 500 to 750 μA, and the VCO frequency can vary ± 20 percent about its center frequency. Regarding Q_{37} and Q_{38} together as one device, we can calculate the ratio of the small-signal input voltage of the emitter-coupled pair to the total small-signal output current flowing in Q_{37}–Q_{38}.

$$\frac{i_o}{v_i} = \frac{1}{2}\left(\frac{g_{m37-38}}{1 + g_{m37-38}R_{24}}\right) = \frac{1}{2}\frac{\left(\dfrac{1}{104\ \Omega}\right)}{1 + \left(\dfrac{325}{104}\right)} = \frac{1}{2}\left(\frac{1}{430\ \Omega}\right) \tag{10.108}$$

Half of the output current of Q_{37}–Q_{38} charges the capacitor at any given time, so that

$$\frac{dI}{dV_{\text{in}}} = \frac{1}{2}\frac{i_o}{v_i} = \frac{1}{2}\frac{1}{860\ \Omega} \tag{10.109}$$

where V_{in} is the input to the pair Q_{35}–Q_{38}, and I is the current that sets the frequency of the VCO. From (10.77)

$$\frac{df}{dI} = \frac{1}{4CV_{BE(\text{on})}} \tag{10.110}$$

Equations (10.109) and (10.110) can be combined to give

$$K_O = \frac{df}{dV_{\text{in}}}$$

$$= \frac{df}{dI}\frac{dI}{dV_{\text{in}}}$$

$$= \frac{1}{8}\frac{1}{860CV_{BE(\text{on})}} \tag{10.111}$$

Substituting of (10.77) in (10.111) gives

$$K_O = \frac{f_o}{1720I_o} \tag{10.112}$$

where $I_o = 625\ \mu\text{A}$ is the value of I when $V_{\text{in}} = 0$, and f_o is the corresponding value of the VCO frequency. This is called the VCO free-running frequency. Thus

$$K_O = \frac{f_o}{1.08}\ \text{Hz/V}$$

$$= 0.15\ \omega_o\ \text{rad/V sec} \tag{10.113}$$

where $f_o = \omega_o/2\pi$. The value of K_O thus depends on the free-running frequency of the VCO through C in (10.111).

The analysis of the 560B is now complete. By combining the calculated values for K_O and K_D, the inherent loop bandwidth K_v for large inputs is found to be

$$K_v = K_D K_O$$

$$= 2.55 \times 0.15\ \omega_o\ \text{rad/sec}$$

$$= 0.38\ \omega_o\ \text{rad/sec} \tag{10.114}$$

Note that $A = 1$ in this case, since all the gain in the loop has been absorbed into K_D and K_O. Equation 10.114 shows that if no other limiting factor is present, the loop will be able to track carrier signals whose frequency deviates from the nominal value by an amount up to 0.38 times the free-running value of the VCO. Another consequence of the large inherent loop bandwidth is that for first-order loops and second-order loops with no zero in the loop filter, the bandwidth of the loop when used as a demodulator is quite broad.

In the 560B circuit, the lock range of the PLL is actually limited intentionally by limiting the operating frequency range of the VCO to within about 20 percent of the center frequency. This is accomplished by controlling the bias current in the emitter-coupled pair, Q_{35}–Q_{38}, which is the current that produces the variation in VCO frequency. This current can be either increased or decreased by the user in the actual circuit with external components so that the lock range is controllable externally. The advantage of this method of limiting the lock range is that it is independently controllable from the loop bandwidth, K_v.

The actual loop bandwidth, K_v, can be controlled by the user of the PLL in several ways. First, external resistors can be inserted in parallel with R_1 and R_2 in Fig. 10.28 to reduce their effective value. This directly reduces K_D and thus K_v. Second, a loop filter that contains a zero can be used. This allows placing the pole of the loop filter at a frequency far below the inherent loop bandwidth K_v without excessive peaking in the loop frequency response.

PROBLEMS

10.1 Determine the dc transfer characteristic of the circuit shown in Fig. 10.30. The Zener diodes have a reverse breakdown voltage of 6.2 V and zero incremental resistance in the breakdown region. In the forward direction $V_{BE(\text{on})} = 0.6$ V.

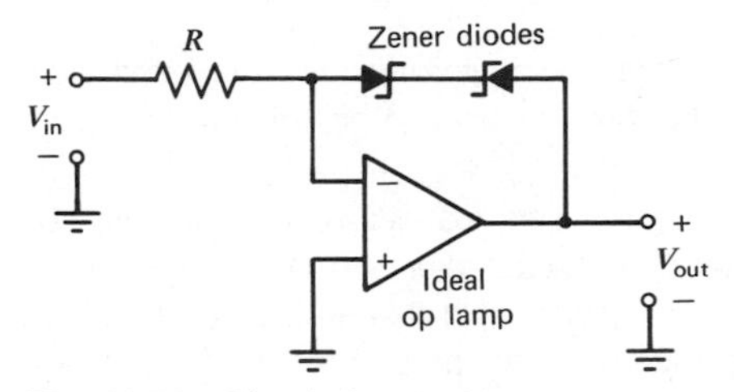

Fig. 10.30 Circuit for Problem 10.1.

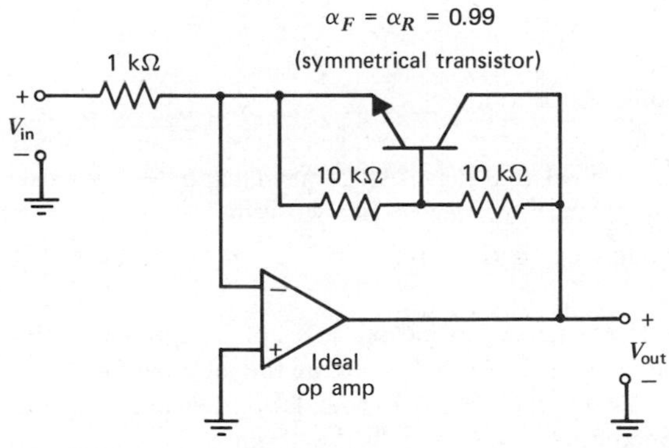

Fig. 10.31 Circuit for Problem 10.2.

10.2 Determine and sketch the dc transfer characteristic of the circuit shown in Fig. 10.31. Assume that $V_{BE(\text{on})} = 0.6$ V.

10.3 Determine and sketch the dc transfer characteristic of the circuit of Fig. 10.32.

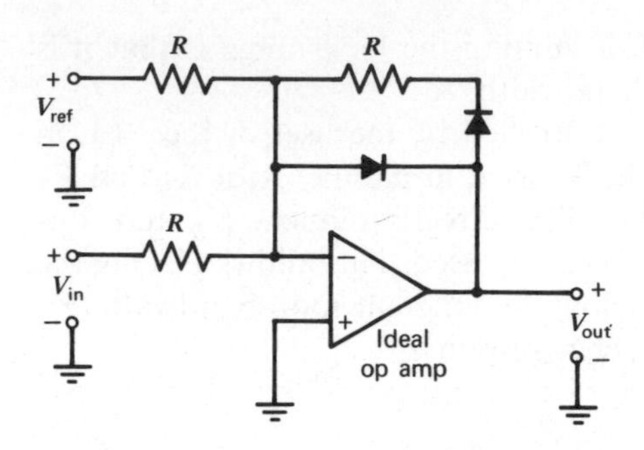

Fig. 10.32 Circuit for Problem 10.3.

10.4 Sketch the dc transfer curve of I_{out} versus V_2 for the Gilbert multiplier of Fig. 10.9 for V_1 equal to $0.1V_T$, $0.5V_T$, and V_T.

10.5 For the emitter-coupled pair of Fig. 10.6, determine the magnitude of the dc differential input voltage required to cause the *slope* of the transfer curve to be different by 1 percent from the slope through the origin.

10.6 Assume a sinusoidal signal is applied to the emitter-coupled pair of Fig. 10.6. Determine the maximum allowable magnitude of the sinusoid such that the magnitude of the third harmonic in the output is less than 1 percent of the fundamental. To work this problem, approximate the transfer characteristic of the pair with the first two terms of the Taylor series for the tanh function. Then assume that all the other harmonics in the output are negligible and that the output is approximately:

$$I_{out}(t) = I_o(\sin \omega_o t + \delta \sin 3\omega_o t)$$

where δ = fractional third-harmonic distortion.

10.7 Determine the worst-case input offset voltage of the voltage-current converter shown in Fig. 10.33. Assume the op amps are ideal, that the resistors mismatch by $\pm.3$ percent, and that transistor I_S mismatch by ± 2 percent. Neglect base currents.

10.8 Determine the dc transfer characteristic of the circuit of Fig. 10.34. Assume that $Z = 0.1XY$ for the multiplier.

10.9 A phase-locked loop has a center frequency of 10^5 rad/sec, a K_O of 10^3 rad/V-sec, and a K_D of 1 V/rad. There is no other gain in the loop. Determine the loop bandwidth in the first-order loop configuration. Determine the single-pole, loop-filter pole location to give the closed-loop poles located on 45° radials from the origin.

10.10 For the same PLL of Problem 10.9, design a loop filter with a zero, which gives a crossover frequency for the loop gain of 100 rad/sec. The loop phase shift at the loop crossover frequency should be −135°.

10.11 Estimate the capture range of the PLL of Problem 10.10, assuming that it is not artificially limited by the VCO frequency range.

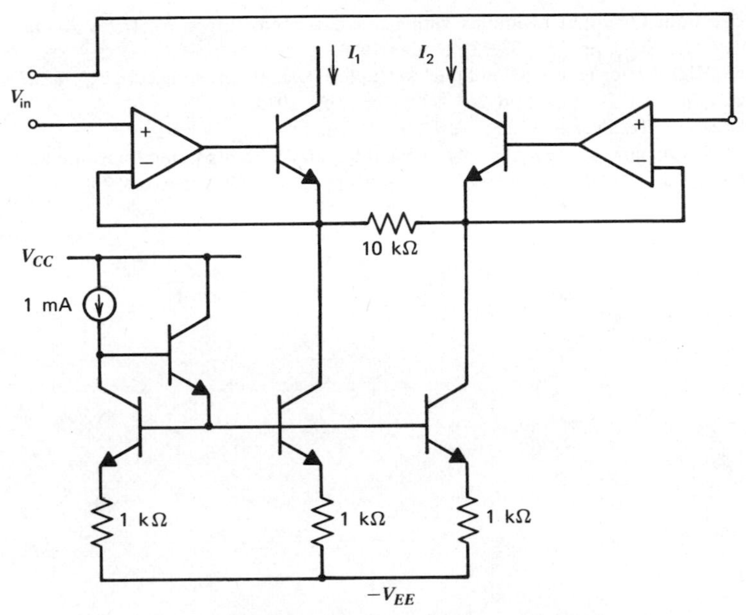

Fig. 10.33 Circuit for Problem 10.7. All op amps are ideal.

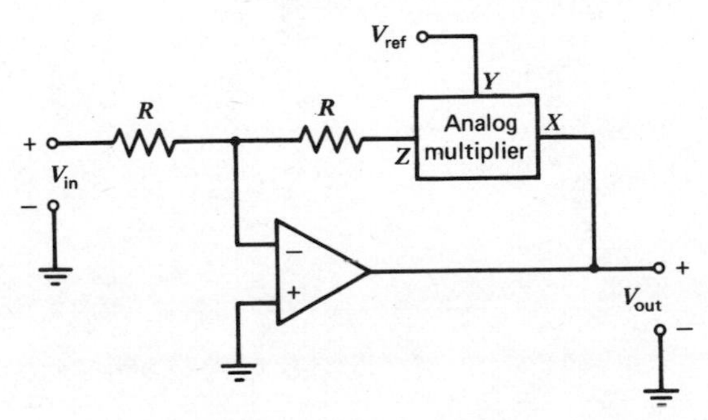

Fig. 10.34 Circuit for Problem 10.8. The op amp is ideal.

10.12 An FM demodulator using the 560B has a center frequency of 2 kHz and is implemented as a first-order loop. The input signal alternates between 1.95 kHz and 2.05 kHz at a rate of 200 Hz with instantaneous transitions between the two frequency values. Sketch the demodulated output voltage waveform.

REFERENCES

1. G. E. Tobey, J. G. Graeme, and L. P. Huelsman. *Operational Amplifiers.* McGraw-Hill, New York, 1971.

2. B. Gilbert. "A Precise Four-Quadrant Multiplier with Subnanosecond Response," *IEEE Journal of Solid-State Circuits*, Vol. SC-3, pp. 365–373, December 1968.
3. B. Gilbert. "A New High-Performance Monolithic Multiplier Using Active Feedback," *IEEE Journal of Solid-State Circuits*, Vol. SC-9, pp. 364–373, December 1974.
4. F. M. Gardner. *Phase-Lock Techniques*. Wiley, New York, 1966.
5. A. B. Grebene and H. R. Camenzind. "Frequency Selective Integrated Circuits Using Phase-Locked Techniques," *IEEE Journal of Solid-State Circuits*, Vol. SC-4, pp. 216–225, August 1969.
6. *Applications of Phase-Locked Loops*, Signetics Corporation, 1974.

CHAPTER 11

NOISE IN INTEGRATED CIRCUITS

11.1 INTRODUCTION

This chapter deals with the effects of *electrical noise* in integrated circuits. The noise phenomena considered here are caused by the small current and voltage fluctuations that are generated within the devices themselves, and we specifically exclude extraneous pickup of man-made signals that can also be a problem in high-gain circuits. The existence of noise is basically due to the fact that electrical charge is not continuous but is carried in discrete amounts equal to the electron charge, and thus noise is associated with fundamental processes in the integrated-circuit devices.

The study of noise is important because it represents a lower limit to the size of electrical signal that can be amplified by a circuit without significant deterioration in signal quality. Noise also results in an upper limit to the useful gain of an amplifier, because if the gain is increased without limit the output stages of the circuit will eventually begin to limit (that is, cut off or saturate) on the amplified noise from the input stages.

In this chapter, the various sources of electronic noise are considered, and the equivalent circuits of common devices including noise generators are described. Methods of circuit analysis with noise generators as inputs are illustrated, and the noise analysis of complex circuits such as op amps is performed. Methods of computer analysis of noise are examined, and, finally, some common methods of specifying circuit noise performance are described.

11.2 SOURCES OF NOISE

11.2.1 Shot Noise[1, 2, 3, 4]

Shot noise is *always* associated with a direct-current flow and is present in diodes and bipolar transistors. The origin of shot noise can be seen by considering the diode of Fig. 11.1*a* and the carrier concentrations in the device in the forward-bias region as shown in Fig. 11.1*b*. As explained in Chapter 1, an electric field $\mathscr{E}$ exists in the depletion region and a voltage $(\psi_0 - V)$ exists between the *p*-type and *n*-type regions, where ψ_0 is the built-in potential and V is the forward bias on the diode. The forward current of the diode, I, is composed of holes from the *p* region and electrons from the *n* region, which have sufficient energy to overcome

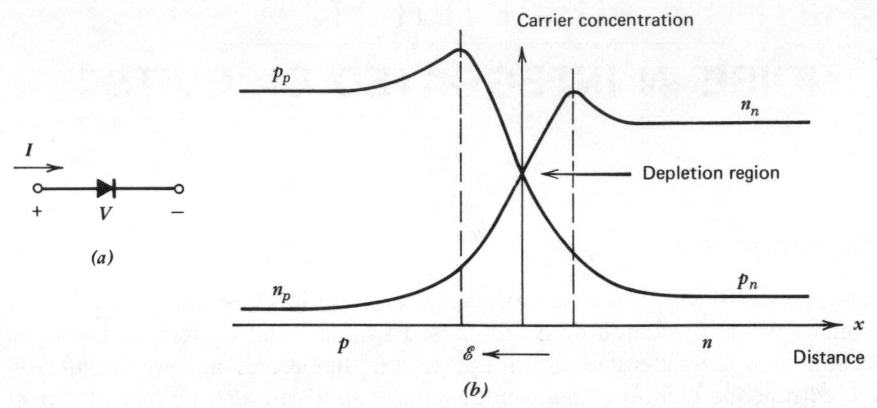

Fig. 11.1 (*a*) Forward-biased *pn* junction diode. (*b*) Carrier concentrations in the diode (not to scale).

the potential barrier at the junction. Once the carriers have crossed the junction, they diffuse away as minority carriers.

The passage of each carrier across the junction is a purely random event and is dependent on the carrier having sufficient energy and a velocity directed towards the junction. Thus external current I, which appears to be a steady current, is, in fact, composed of a large number of random independent current pulses. If the current is examined on a sensitive oscilloscope the trace would appear as in Fig. 11.2, where I_D is the average current.

The fluctuation in I is termed *shot noise* and is generally specified in terms of its mean-square variation about the average value. This is written as $\overline{i^2}$ where

$$\overline{i^2} = \overline{(I - I_D)^2}$$

$$= \lim_{T \to \infty} \frac{1}{T} \int_0^T (I - I_D)^2 \, dt \tag{11.1}$$

It can be shown that if a current I is composed of a series of random independent pulses with average value I_D, then the resulting noise current has a mean-square

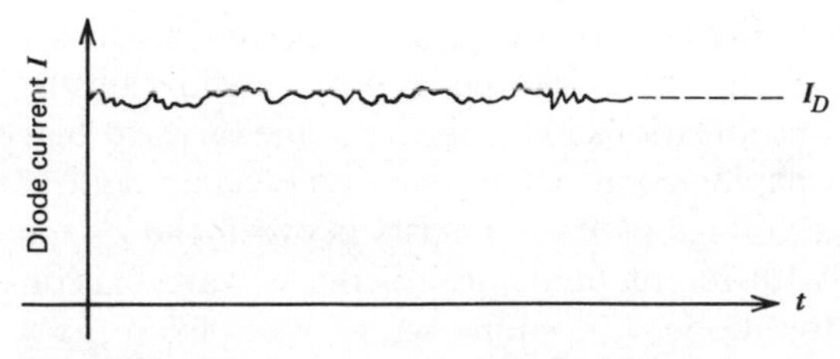

Fig. 11.2 Diode current I as a function of time (not to scale).

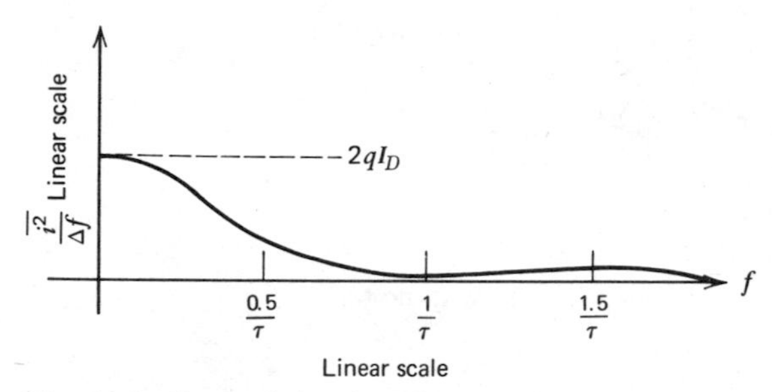

Fig. 11.3 Spectral density of shot noise in a diode with transit time τ. (Not to scale.)

value

$$\overline{i^2} = 2qI_D\,\Delta f(\mathrm{A}^2) \tag{11.2}$$

where q is the electronic charge (1.6×10^{-19} C) and Δf is the bandwidth in hertz. This equation shows that the noise current has a mean-square value that is *directly proportional* to the bandwidth Δf (in hertz) of the measurement. Thus a noise-current *spectral density* $\overline{i^2}/\Delta f$ (with units square amperes per hertz) can be defined that is *constant* as a function of frequency. Noise with such a spectrum is often called *white noise*. Since noise is a purely random signal, the *instantaneous value* of the waveform cannot be predicted at any time. The only information available for use in circuit calculations concerns the *mean-square* value of the signal given by (11.2). Bandwidth Δf in (11.2) is determined by the circuit in which the noise source is acting.

Equation 11.2 is valid until the frequency becomes comparable to $1/\tau$ where τ is the carrier transit time through the depletion region. For most practical electronic devices, τ is extremely small and (11.2) is accurate well into the gigahertz region. A sketch of noise-current spectral density versus frequency for a diode is shown in Fig. 11.3 assuming that the passage of each charge carrier across the depletion region produces a square pulse of current with width τ.

EXAMPLE

Calculate the shot noise in a diode current of 1 mA in a bandwidth of 1 MHz. Using (11.2) we have

$$\overline{i^2} = 2 \times 1.6 \times 10^{-19} \times 10^{-3} \times 10^{6}\ \mathrm{A}^2 = 3.2 \times 10^{-16}\ \mathrm{A}^2$$

and thus

$$i = 1.8 \times 10^{-8}\ \mathrm{A\ rms}$$

where i represents the root-mean-square (rms) value of the noise current.

The effect of shot noise can be represented in the low-frequency, small-signal equivalent circuit of the diode by inclusion of a current generator shunting the

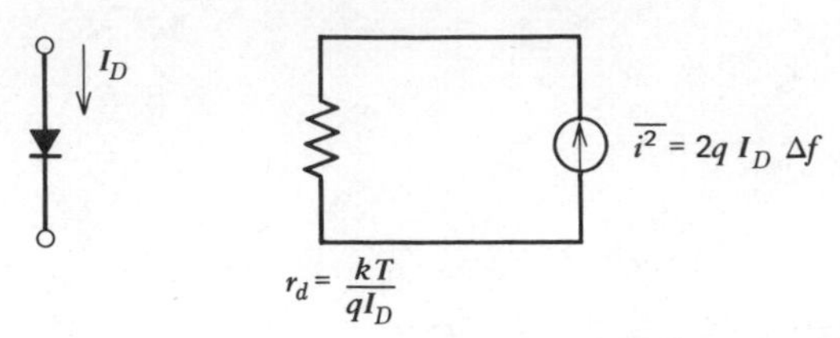

Fig. 11.4 Junction diode small-signal equivalent circuit with noise.

diode as shown in Fig. 11.4. Since this noise signal has random phase and is defined solely in terms of its mean-square value, it also has no polarity. Thus the arrow in the current source in Fig. 11.2 has no significance and is included only to identify the generator as a current source. This practice is followed in this chapter where we deal only with *independent* noise generators having random phase.

The noise-current signal produced by the shot noise mechanism has an amplitude that varies randomly with time and that can only be specified by a *probability-density* function. It can be shown that the amplitude distribution of shot noise is Gaussian and the probability-density function $p(I)$ of the diode current is plotted versus current in Fig. 11.5 (not to scale). The probability that the diode current lies between values I and $(I + dI)$ at any time is given by $p(I)dI$. If σ is the standard deviation of the Gaussian distribution then the diode current amplitude lies between limits $I_D \pm \sigma$ for 68 percent of the time. By definition, variance σ^2 is the mean-square value of $(I - I_D)$ and thus, from (11.1),

$$\sigma^2 = \overline{i^2}$$

and

$$\sigma = \sqrt{2qI_D\, \Delta f} \tag{11.3}$$

using (11.2). Note that, theoretically, the noise amplitude can have positive or negative values approaching infinity. However, the probability falls off very quickly as amplitude increases and an effective limit to the noise amplitude is $\pm 3\,\sigma$. The

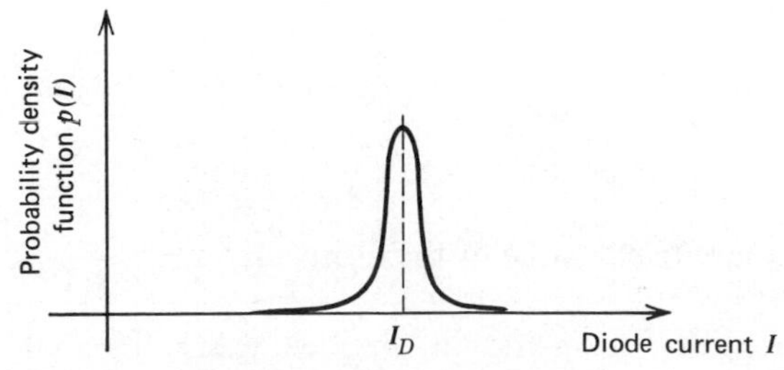

Fig. 11.5 Probability density function for the diode current, I. (Not to scale.)

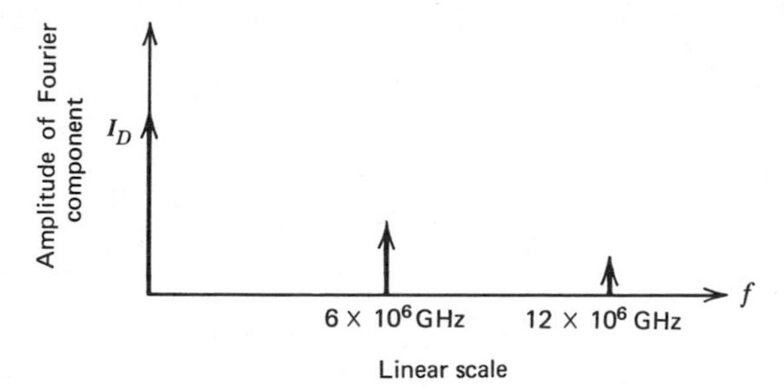

Fig. 11.6 Shot-noise spectrum assuming uniform emission of carriers.

noise signal is within these limits for 99.7 percent of the time. A brief description of the Gaussian distribution is given in Appendix A.3.1 in Chapter 3.

It is important to note that the distribution of noise in frequency as shown in Fig. 11.3 is due to the random nature of the hole and electron transitions across the *pn* junction. Consider the situation if all the carriers made transitions with uniform time separation. Since each carrier has a charge of 1.6×10^{-19} C, a 1-mA current would then consist of current pulses every 1.6×10^{-16} sec. The Fourier analysis of such a waveform would give the spectrum of Fig. 11.6, which shows an average or dc value I_D and harmonics at multiples of $1/\Delta t$, where Δt is the period of the waveform and equals 1.6×10^{-16} sec. Thus the first harmonic is at 6×10^6 GHz, which is far beyond the useful frequency of the device. There would be *no noise produced* in the normal frequency range of operation.

11.2.2 Thermal Noise[1, 3, 5]

Thermal noise is generated by a completely different mechanism from shot noise. In conventional resistors it is due to the random thermal motion of the electrons and is unaffected by the presence or absence of direct current, since typical electron drift velocities in a conductor are much less than electron thermal velocities. Since this source of noise is due to the thermal motion of electrons, we expect that it is related to absolute temperature T. In fact thermal noise is *directly proportional* to T (unlike shot noise, which is *independent of* T) and, as T approaches zero, thermal noise also approaches zero.

In a resistor R, thermal noise can be shown to be represented by a series voltage generator $\overline{v^2}$ as shown in Fig. 11.7*a*, or by a shunt current generator $\overline{i^2}$ as in Fig. 11.7*b*. These representations are equivalent and

$$\overline{v^2} = 4kTR\,\Delta f \tag{11.4}$$

$$\overline{i^2} = 4kT\frac{1}{R}\,\Delta f \tag{11.5}$$

where k is Boltzmann's constant. At room temperature $4kT = 1.66 \times 10^{-20}$ V-C.

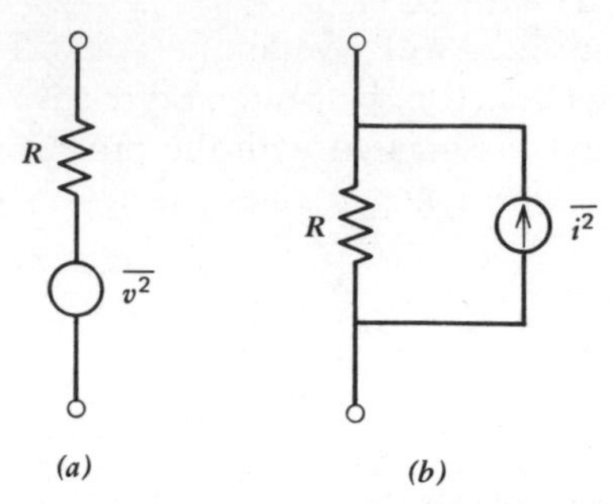

Fig. 11.7 Alternative representations of thermal noise.

Equations 11.4 and 11.5 show that the noise spectral density is again *independent* of frequency and, for thermal noise, this is true up to 10^{13} Hz. Thus thermal noise is another source of white noise. Note that the Norton equivalent of (11.5) can be derived from (11.4) as

$$\overline{i^2} = \frac{\overline{v^2}}{R^2} \tag{11.6}$$

A useful number to remember for thermal noise is that at room temperature (300°K), the thermal noise spectral density in a 1-kΩ resistor is $\overline{v^2}/\Delta f \simeq 16 \times 10^{-18}$ V^2/Hz This can be written in rms form as $v \simeq 4$ nV/$\sqrt{\text{Hz}}$ where the form nV/$\sqrt{\text{Hz}}$ is used to emphasize that the *rms noise voltage* varies as the *square root* of the bandwidth. Another useful equivalence is that the thermal noise-current generator of a 1-kΩ resistor at room temperature is the same as that of 50 μA of direct current exhibiting shot noise.

Thermal noise as described above is a fundamental physical phenomenon and is present in *any* linear passive resistor. This includes conventional resistors and the radiation resistance of antennas, loudspeakers, and microphones. In the case of loudspeakers and microphones, the source of noise is the thermal motion of the air molecules. In the case of antennas, the source of noise is the black-body radiation of the object at which the antenna is directed. In all cases, (11.4) and (11.5) give the mean-square value of the noise.

The amplitude distribution of thermal noise is again Gaussian. Since both shot and thermal noise each have a flat frequency spectrum and a Gaussian amplitude distribution, they are indistinguishable once they are introduced into a circuit. The waveform of shot and thermal noise combined with a sinewave of equal power is shown in Fig. 11.21.

11.2.3 Flicker Noise[6,7,8] (1/*f* Noise)

This is a type of noise found in all active devices, as well as some discrete passive elements such as carbon resistors. The origins of flicker noise are varied, but in bipolar transistors it is caused mainly by traps associated with contamination and crystal defects in the emitter-base depletion layer. These traps capture and release carriers in a random fashion and the time constants associated with the process give rise to a noise signal with energy concentrated at low frequencies.

Flicker noise is always associated with a flow of direct current and displays a spectral density of the form

$$\overline{i^2} = K_1 \frac{I^a}{f^b} \Delta f \tag{11.7}$$

where

Δf	is a small bandwidth at frequency f
I	is a direct current
K_1	is a constant for a particular device
a	is a constant in the range 0.5 to 2
b	is a constant of about unity

If $b = 1$ in (11.7), the noise spectral density has a $1/f$ frequency dependence (hence the alternative name "$1/f$ noise") and is shown in Fig. 11.8. It is apparent that flicker noise is most significant at low frequencies, although in devices exhibiting high flicker noise levels, this noise source may dominate the device noise at frequencies well into the megahertz range.

It was noted above that flicker noise only exists in association with a direct current. Thus, in the case of carbon resistors, no flicker noise is present until a direct current is passed through the resistor (however, *thermal* noise *always* exists in the resistor and is *unaffected* by any direct current as long as the temperature remains constant). Consequently, carbon resistors can be used if required as external elements in low-noise, low frequency integrated circuits as long as they carry no direct current. If the external resistors for such circuits must carry direct current, however, metal film resistors that have no flicker noise should be used.

In earlier sections of this chapter, we saw that shot and thermal noise signals have well-defined mean-square values that can be expressed in terms of current flow, resistance, and a number of well-known physical constants. By contrast, the

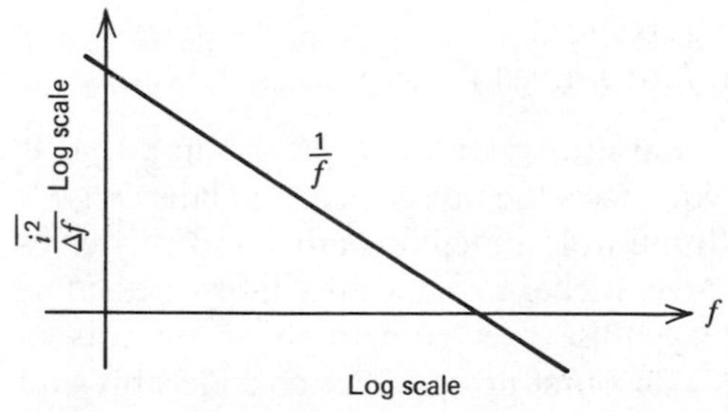

Fig. 11.8 Flicker noise spectral density versus frequency.

mean-square value of a flicker noise signal as given by (11.7) contains an unknown constant K_1. This constant not only varies by orders of magnitude from one device type to the next, but it can also vary widely for different transistors or integrated circuits from the same process wafer that has undergone identical fabrication steps. This is due to the dependence of flicker noise on contamination and crystal imperfections which are factors that can vary randomly even on the same silicon slice. However, experiments have shown that if a typical value of K_1 is determined from measurements on a number of devices from a given process, then this value can be used to predict average or typical flicker noise performance for integrated circuits from that process.[9]

The final characteristic of flicker noise that is of interest is its amplitude distribution, and measurements have shown that this is often non-Gaussian.

11.2.4 Burst Noise[7] ("Popcorn Noise")

This is another type of low-frequency noise found in some integrated circuits and discrete transistors. The source of this noise is not fully understood, although it has been shown to be related to the presence of heavy-metal ion contamination. Gold-doped devices show very high levels of burst noise.

Burst noise is so named because an oscilloscope trace of this type of noise shows bursts of noise on a number (two or more) of discrete levels as illustrated in Fig. 11.9*a*. The repetition rate of the noise pulses is usually in the audio frequency range (a few kilokertz or less) and produces a "popping" sound when played through a loudspeaker. This has led to the name "popcorn noise" for this phenomenon.

The spectral density of burst noise can be shown to be of the form

$$\overline{i^2} = K_2 \frac{I^c}{1 + \left(\dfrac{f}{f_c}\right)^2} \Delta f \tag{11.8}$$

where K_2 is a constant for a particular device

I is a direct current

c is a constant in the range 0.5 to 2

f_c is a particular frequency for a given noise process

This spectrum is plotted in Fig. 11.9*b* and illustrates the typical hump that is characteristic of burst noise. At higher frequencies the noise spectrum falls as $1/f^2$.

Burst noise processes often occur with multiple time constants, and this gives rise to multiple humps in the spectrum. Also, flicker noise is invariably present as well so that the composite low-frequency noise spectrum often appears as in Fig. 11.10. As with flicker noise, factor K_2 for burst noise varies considerably and must be determined experimentally. The amplitude distribution of the noise is also non-Gaussian.

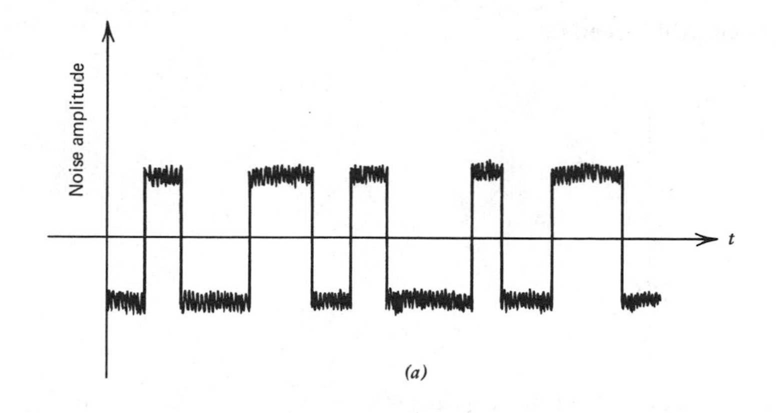

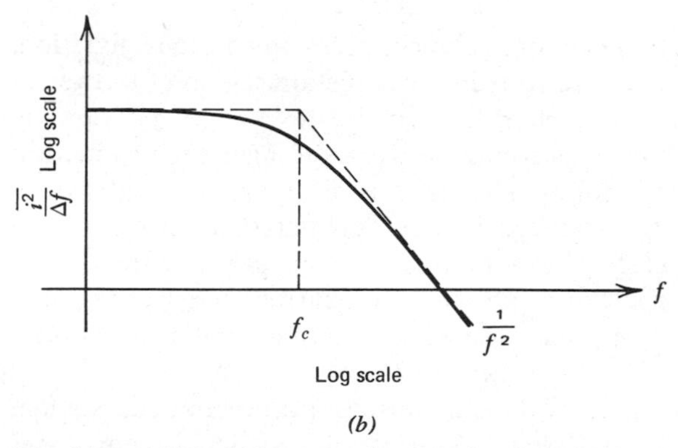

Fig. 11.9 (*a*) Typical burst noise waveform. (*b*) Burst noise spectral density versus frequency.

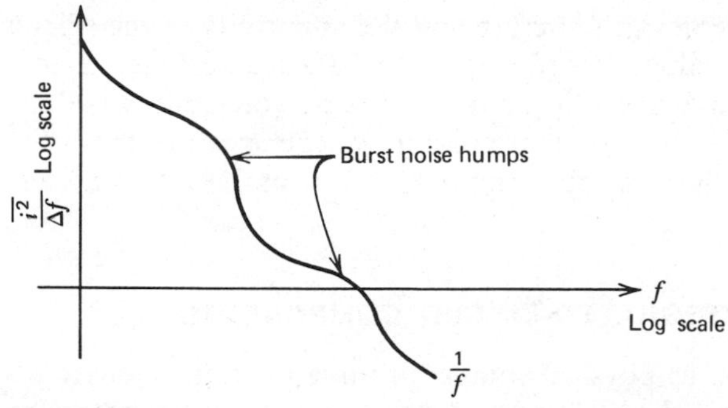

Fig. 11.10 Spectral density of combined multiple burst noise sources and flicker noise.

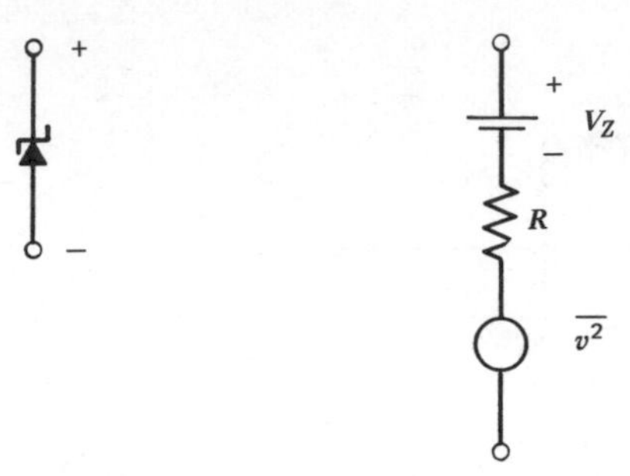

Fig. 11.11 Equivalent circuit of a Zener diode including noise.

11.2.5 Avalanche Noise[10]

This is a form of noise produced by Zener or avalanche breakdown in a *pn* junction. In avalanche breakdown, holes and electrons in the depletion region of a reverse-biased *pn* junction acquire sufficient energy to create hole-electron pairs by colliding with silicon atoms. This process is cumulative, resulting in the production of a random series of large noise spikes. The noise is always associated with a direct-current flow and the noise produced is much greater than shot noise in the same current, as given by (11.2). This is because a single carrier can start an avalanching process that results in the production of a current burst containing many carriers moving together. The total noise is the sum of a number of random bursts of this type.

The most common situation where avalanche noise is a problem occurs when Zener diodes are used in the circuit. These devices display avalanche noise and are generally avoided in low-noise circuits. If Zener diodes are present, the noise representation of Fig. 11.11 can be used where the noise is represented by a series voltage generator $\overline{v^2}$. The dc voltage, V_Z, is the breakdown voltage of the diode and the series resistance, R, is typically 10 to 100 Ω. The magnitude of $\overline{v^2}$ is difficult to predict as it depends on the device structure and the uniformity of the silicon crystal, but a typical measured value is $\overline{v^2}/\Delta f \simeq 10^{-14}\ \text{V}^2/\text{Hz}$ at a dc Zener current of 0.5 mA. Note that this is equivalent to the thermal noise voltage in a 600-kΩ resistor and completely overwhelms thermal noise in R. The spectral density of the noise is approximately flat but the amplitude distribution is generally non-Gaussian.

11.3 NOISE MODELS OF INTEGRATED-CIRCUIT COMPONENTS

In the above sections, the various physical sources of noise in electronic circuits were described. In this section, these sources of noise are brought together to form the small-signal equivalent circuits including noise for diodes and for bipolar and field-effect transistors.

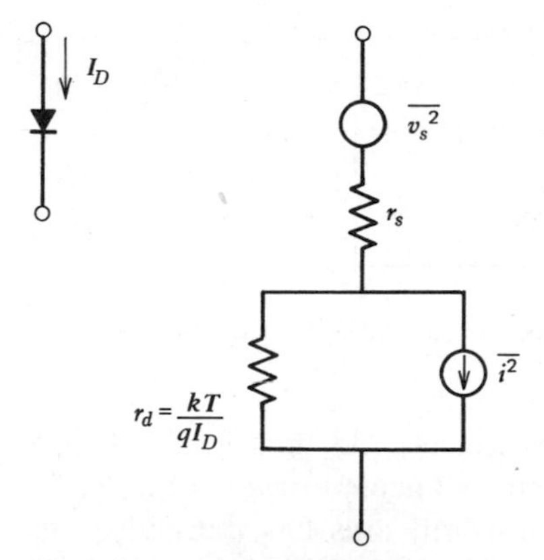

Fig. 11.12 Complete diode small-signal equivalent circuit with noise sources.

11.3.1 Junction Diode

The equivalent circuit for a junction diode was considered briefly in the consideration of shot noise. The basic equivalent circuit of Fig. 11.4 can be made complete by adding series resistance r_s as shown in Fig. 11.12. Since r_s is a physical resistor due to the resistivity of the silicon, it exhibits thermal noise. Experimentally it has been found that any flicker noise present can be represented by a current generator in shunt with r_d, and this is conveniently combined with the shot-noise generator as indicated by (11.10) to give

$$\overline{v_s^2} = 4kTr_s \, \Delta f \tag{11.9}$$

$$\overline{i^2} = 2qI_D \, \Delta f + K \frac{I_D^a}{f} \Delta f \tag{11.10}$$

11.3.2 Bipolar Transistor[11]

In a bipolar transistor in the forward-active region, minority carriers diffuse and drift across the base region to be collected at the collector-base junction. Minority carriers entering the collector-base depletion region are accelerated by the field existing there and swept across this region to the collector. The time of arrival at the collector-base junction of the diffusing (or drifting) carriers is a purely random process, and thus the transistor collector current consists of a series of random current pulses. Consequently, collector current I_C shows *full shot noise* as given by (11.2) and this is represented by a shot noise current generator $\overline{i_c^2}$ from collector to emitter as shown in the equivalent circuit of Fig. 11.13.

Base current I_B in a transistor is due to recombination in the base and base-emitter depletion regions and also to carrier injection from the base into the

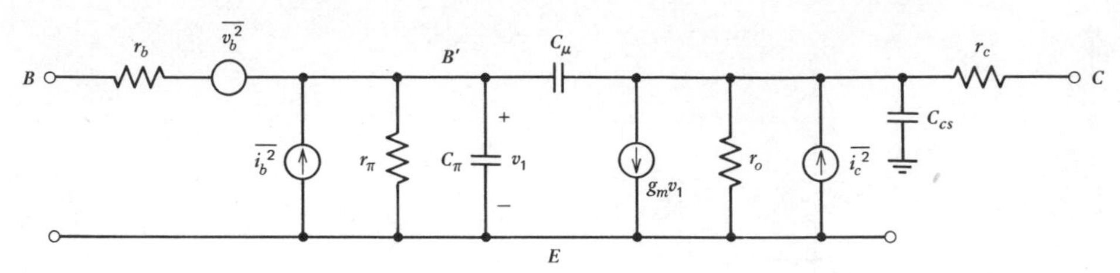

Fig. 11.13 Complete bipolar transistor small-signal equivalent circuit with noise sources.

emitter. All of these are independent random processes, and thus I_B also shows *full shot noise*. This is represented by shot noise current generator $\overline{i_b^2}$ in Fig. 11.13.

Transistor base resistor r_b is a physical resistor and thus has thermal noise. Collector series resistor r_c also shows thermal noise, but since this is in series with the high-impedance collector node, this noise is negligible and is usually not included in the model. Note that resistors r_π and r_o in the model are *fictitious* resistors that are used for modeling purposes only, and they do *not* exhibit thermal noise.

Flicker noise and burst noise in a bipolar transistor have been found experimentally to be represented by current generators across the internal base-emitter junction. These are conveniently combined with the shot noise generator in $\overline{i_b^2}$. Avalanche noise in bipolar transistors is found to be negligible if V_{CE} is kept at least 5V below the breakdown voltage BV_{CEO} and this source of noise will be neglected in subsequent calculations.

The full small-signal equivalent circuit including noise for the bipolar transistor is shown in Fig. 11.13. Since they arise from separate, independent physical mechanisms, all the noise sources are *independent* of each other and have mean-square values:

$$\overline{v_b^2} = 4kTr_b\,\Delta f \tag{11.11}$$

$$\overline{i_c^2} = 2qI_C\,\Delta f \tag{11.12}$$

$$\overline{i_b^2} = \underbrace{2qI_B\,\Delta f}_{\text{Shot noise}} + \underbrace{K_1\frac{I_B^{\,a}}{f}\Delta f}_{\text{Flicker noise}} + \underbrace{K_2\frac{I_B^{\,c}}{1+\left(\dfrac{f}{f_c}\right)^2}\Delta f}_{\text{Burst noise}} \tag{11.13}$$

This equivalent circuit is valid for both *npn* and *pnp* transistors. For *pnp* devices, the magnitudes of I_B and I_C are used in the above equations.

The base-current noise spectrum can be plotted using (11.13) and this has been done in Fig. 11.14 where burst noise has been neglected for simplicity. The

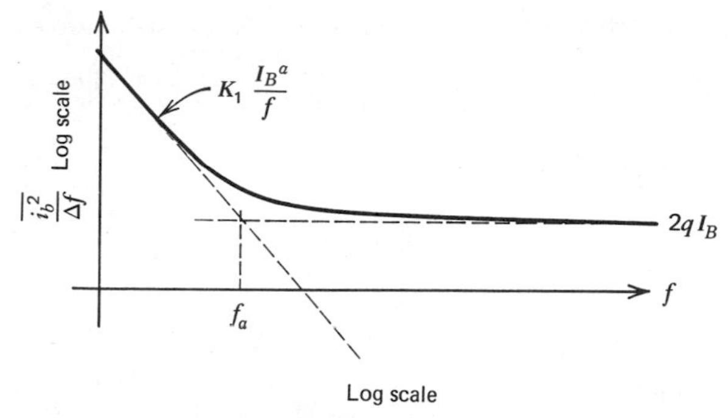

Fig. 11.14 Spectral density of the base-current noise generator in a bipolar transistor.

shot noise and flicker noise asymptotes meet at a frequency f_a, which is called the flicker noise "corner" frequency. In some transistors using careful processing, f_a can be as low as 100 Hz. In other transistors f_a can be as high as 10 MHz.

11.3.3 Field-Effect Transistor[12,13]

The structure of field-effect transistors (FET) was described in Chapter 1. There it was shown that the resistive channel joining source and drain is modulated by the gate-source voltage so that the drain current is controlled by the gate-source voltage. Since the channel material is *resistive* it exhibits *thermal noise* and this is the major source of noise in FETs (both JFET and MOSFET). It can be shown that this noise source can be represented by a noise-current generator $\overline{i_d^2}$ from drain to source in the FET small-signal equivalent circuit of Fig. 11.15. Flicker noise in the FET is also found experimentally to be represented by a drain-source current generator and these two can be lumped into one noise generator $\overline{i_d^2}$. The other source of noise in FETs is shot noise generated by the gate leakage current. This is represented by $\overline{i_g^2}$ in Fig. 11.15 and is usually very small. It becomes significant only when the driving-source impedance connected to the FET gate is very

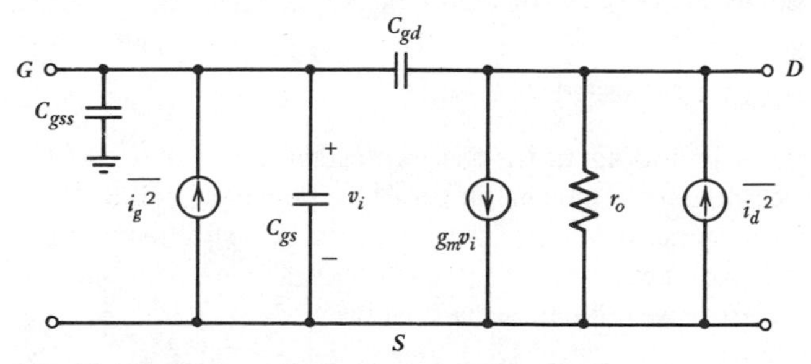

Fig. 11.15 FET small-signal equivalent circuit with noise sources.

large. The generators in Fig. 11.15 are all *independent* of each other and have values

$$\overline{i_g^2} = 2qI_G\,\Delta f \tag{11.14}$$

$$\overline{i_d^2} = \underbrace{4kT\left(\frac{2}{3}g_m\right)\Delta f}_{\text{Thermal noise}} + \underbrace{K\frac{I_D{}^a}{f}\Delta f}_{\text{Flicker noise}} \tag{11.15}$$

where

I_G is the gate leakage current
I_D is the drain bias current
K is a constant for a given device
a is a constant between 0.5 and 2
g_m is the device transconductance at the operating point

11.3.4 Resistors

Monolithic and thin-film resistors display thermal noise as given by (11.4) and (11.5), and the circuit representation of this is shown in Fig. 11.7. As mentioned in Section 11.2.3, discrete carbon resistors also display flicker noise and this should be considered if such resistors are used as external components to the integrated circuit.

11.3.5 Capacitors and Inductors

Capacitors are common elements in integrated circuits, either as unwanted parasitics or as elements introduced for a specific purpose. Inductors in general cannot be realized on the silicon die but are sometimes used as external elements, particularly in integrated communication circuits. There are *no sources of noise* in *ideal* capacitors or inductors. In practice, real components have parasitic resistance that *does* display noise as given by the thermal noise formulas of (11.4) and (11.5). In the case of integrated-circuit capacitors, the parasitic resistance usually consists of a small value in series with the capacitor. Parasitic resistance in inductors can be modeled either by series or shunt elements.

11.4 CIRCUIT NOISE CALCULATIONS[14,15]

The device equivalent circuits including noise that were derived above can be used for the calculation of circuit noise performance. First, however, methods of circuit calculation with noise generators as sources must be established and attention is now given to this problem.

Consider a noise current source with mean-square value

$$\overline{i^2} = S(f)\,\Delta f \tag{11.16}$$

where $S(f)$ is the *noise spectral density*

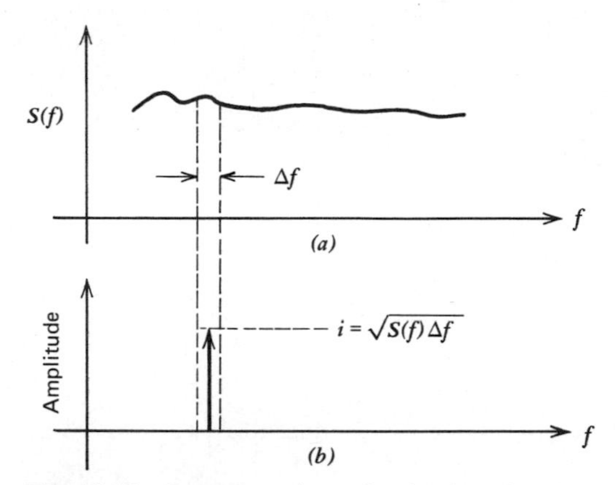

Fig. 11.16 Representation of noise in a bandwidth Δf by an equivalent sinusoid with the same rms value.

The value of $S(f)$ is plotted versus frequency in Fig. 11.16*a* for an arbitrary noise generator. In a small bandwidth Δf, the mean-square value of the noise current is given by (11.16) and the rms value can be written as

$$i = \sqrt{S(f)\,\Delta f} \tag{11.17}$$

The noise current in bandwidth Δf can be represented approximately[14] by a sinusoidal current generator with rms value i as shown in Fig. 11.16*b*. If the noise current in bandwidth Δf is now applied as an input signal to a circuit, its effect can be calculated by substituting the sinusoidal generator and performing circuit analysis in the usual fashion. When the circuit response to the sinusoid is calculated, the mean-square value of the output sinusoid gives the mean-square value of the output noise in bandwidth Δf. *Thus network noise calculations reduce to familiar sinusoidal circuit analysis calculations.* The only difference occurs when multiple noise sources are applied, as is usually the case in practical circuits. Each noise source is then represented by a separate sinusoidal generator and the output contribution of each one is separately calculated. The total output noise in bandwidth Δf is calculated as a *mean-square* value by *adding* the individual *mean-square* contributions from each output sinusoid. This depends, however, on the original noise sources being *independent*, as will be shown below. This requirement is always satisfied if the equivalent noise circuits derived in previous sections are used, as all the noise sources arise from separate mechanisms and are thus independent.

For example, consider two resistors R_1 and R_2 connected in series as shown in Fig. 11.17. Resistors R_1 and R_2 have respective noise generators

$$\overline{v_1^2} = 4kTR_1\,\Delta f \tag{11.18}$$

$$\overline{v_2^2} = 4kTR_2\,\Delta f \tag{11.19}$$

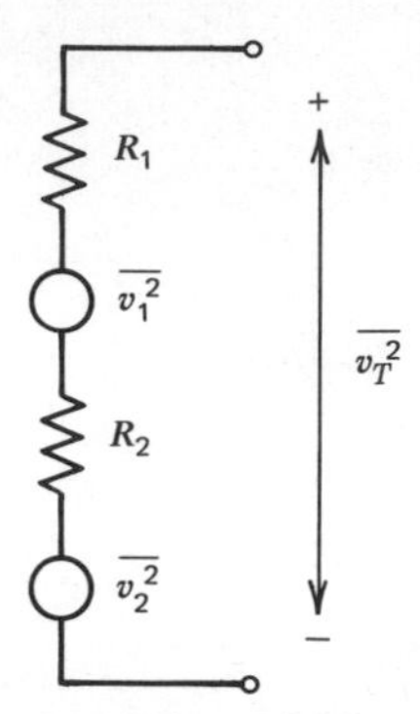

Fig. 11.17 Circuit for the calculation of the total noise $\overline{v_T{}^2}$ produced by two resistors in series.

In order to calculate the mean-square noise voltage $\overline{v_T{}^2}$ produced by the two resistors in series, let $v_T(t)$ be the instantaneous value of the total noise voltage and $v_1(t)$ and $v_2(t)$ the instantaneous values of the individual generators. Then

$$v_T(t) = v_1(t) + v_2(t) \tag{11.20}$$

and thus

$$\begin{aligned}\overline{v_T(t)^2} &= \overline{v_1(t) + v_2(t)^2}\\ &= \overline{v_1(t)^2} + \overline{v_2(t)^2} + \overline{2v_1(t)v_2(t)}\end{aligned} \tag{11.21}$$

Now, since noise generators $v_1(t)$ and $v_2(t)$ arise from separate resistors, they must be *independent*. Thus the *average* value of their product $\overline{v_1(t)v_2(t)}$ will be zero and (11.21) becomes

$$\overline{v_T{}^2} = \overline{v_1{}^2} + \overline{v_2{}^2} \tag{11.22}$$

Thus the mean-square value of the sum of a number of independent noise generators is the sum of the individual mean-square values. Substituting (11.18) and (11.19) in (11.22) gives

$$\overline{v_T{}^2} = 4kT(R_1 + R_2)\,\Delta f \tag{11.23}$$

Equation 11.23 is just the value that would be predicted for thermal noise in a resistor $(R_1 + R_2)$ using (11.4), and thus the results are consistent. These results are also consistent with the representation of the noise generators by *independent* sinusoids as described earlier. It is easily shown that when two or more such generators are connected in series, the mean-square value of the total voltage is equal to the sum of the individual mean-square values.

In the above calculation, two noise voltage sources were considered connected in series. It can be similarly shown that an analogous result is true for independent noise *current* sources connected in *parallel*. The mean-square value of the combination is the sum of the individual mean-square values. This result was assumed in the modeling of Section 11.3 where, for example, three independent noise-current generators (shot, flicker, and burst) were combined into a single base-emitter noise source for a bipolar transistor.

11.4.1 Bipolar Transistor Noise Performance

As an example of the manipulation of noise generators in circuit calculations, consider the noise performance of the simple transistor stage with the ac schematic shown in Fig. 11.18*a*. The small-signal equivalent circuit including noise is shown in Fig. 11.18*b*. (It should be pointed out that, for noise calculations, the equivalent circuit analyzed must be the actual circuit configuration used. That is, Fig. 11.18*a* cannot be used as a half-circuit representation of a differential pair for the purposes of noise calculation because noise sources in each half of a differential pair affect the total output noise.)

In the equivalent circuit of Fig. 11.18*b*, the external input signal v_i has been ignored so that output signal v_o is due to noise generators only. C_μ is assumed small and is neglected. Output resistance r_o is also neglected. The transistor noise generators are as described previously and in addition

$$\overline{v_s^2} = 4kTR_S\,\Delta f \tag{11.24}$$

$$\overline{i_l^2} = 4kT\frac{1}{R_L}\,\Delta f \tag{11.25}$$

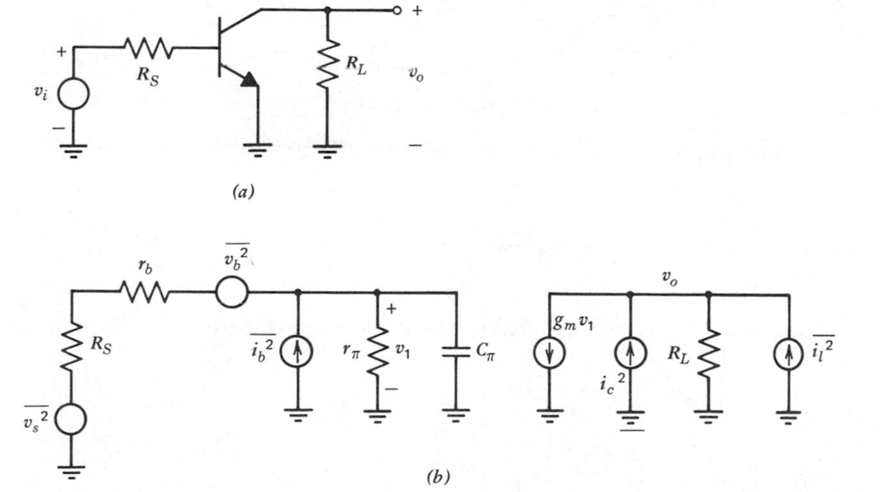

Fig. 11.18 (*a*) Simple transistor amplifier ac schematic. (*b*) Small-signal equivalent circuit with noise sources.

The total output noise can be calculated by considering each noise source in turn and performing the calculation *as if* each noise source were a sinusoid with rms value equal to that of the noise source being considered. Consider first the noise generator v_s due to R_S. Then

$$v_1 = \frac{Z}{Z + r_b + R_S} v_s \tag{11.26}$$

where

$$Z = r_\pi \left|\right| \frac{1}{j\omega C_\pi} \tag{11.27}$$

The output noise voltage due to v_s is

$$v_{o1} = -g_m R_L v_1 \tag{11.28}$$

Use of (11.26) in (11.28) gives

$$v_{o1} = -g_m R_L \frac{Z}{Z + r_b + R_S} v_s \tag{11.29}$$

The phase information contained in (11.29) is irrelevant because the noise signal has random phase and the only quantity of interest is the mean-square value of the output voltage produced by v_s. From (11.29) this is

$$\overline{v_{o1}^2} = g_m^2 R_L^2 \frac{|Z|^2}{|Z + r_b + R_S|^2} \overline{v_s^2} \tag{11.30}$$

By similar calculations it is readily shown that the noise voltage produced at the output by $\overline{v_b^2}$ and $\overline{i_b^2}$ is

$$\overline{v_{o2}^2} = g_m^2 R_L^2 \frac{|Z|^2}{|Z + r_b + R_S|^2} \overline{v_b^2} \tag{11.31}$$

$$\overline{v_{o3}^2} = g_m^2 R_L^2 \frac{(R_S + r_b)^2 |Z|^2}{|Z + r_b + R_S|^2} \overline{i_b^2} \tag{11.32}$$

Noise at the output due to $\overline{i_l^2}$ and $\overline{i_c^2}$ is:

$$\overline{v_{o4}^2} = \overline{i_l^2} R_L^2 \tag{11.33}$$

$$\overline{v_{o5}^2} = \overline{i_c^2} R_L^2 \tag{11.34}$$

Since all five noise generators are *independent*, the total output noise is

$$\overline{v_o^2} = \sum_{n=1}^{5} \overline{v_{on}^2} \tag{11.35}$$

$$= g_m^2 R_L^2 \frac{|Z|^2}{|Z + r_b + R_S|^2} (\overline{v_s^2} + \overline{v_b^2} + (R_S + r_b)^2 \overline{i_b^2}) + R_L^2 (\overline{i_l^2} + \overline{i_c^2}) \tag{11.36}$$

Substituting expressions for the noise generators we obtain

$$\frac{\overline{v_o^2}}{\Delta f} = g_m^2 R_L^2 \frac{|Z|^2}{|Z + r_b + R_S|^2}[4kT(R_S + r_b) + (R_S + r_b)^2 2qI_B] + R_L^2\left(4kT\frac{1}{R_L} + 2qI_C\right) \tag{11.37}$$

where flicker noise has been assumed small and neglected. Substituting for Z from (11.27) in (11.37) we find

$$\frac{\overline{v_o^2}}{\Delta f} = g_m^2 R_L^2 \frac{r_\pi^2}{(r_\pi + R_S + r_b)^2}\frac{1}{1 + \left(\frac{f}{f_1}\right)^2}[4kT(R_S + r_b) + (R_S + r_b)^2 2qI_B] + R_L^2\left(4kT\frac{1}{R_L} + 2qI_C\right) \tag{11.38}$$

where

$$f_1 = \frac{1}{2\pi(r_\pi \| R_S)C_\pi} \tag{11.39}$$

The output noise-voltage spectral density represented by (11.38) has a frequency-dependent part and a constant part. The frequency dependence arises because the gain of the stage begins to fall above frequency f_1, and noise duc to generators $\overline{v_s^2}$, $\overline{v_b^2}$, and $\overline{i_b^2}$, which appears amplified in the output, also begins to fall. The constant term in (11.38) is due to noise generators $\overline{i_l^2}$ and $\overline{i_c^2}$. Note that this noise contribution would also be frequency dependent if the effect of C_μ had not been neglected. The noise-voltage spectral density represented by (11.38) has the form shown in Fig. 11.19.

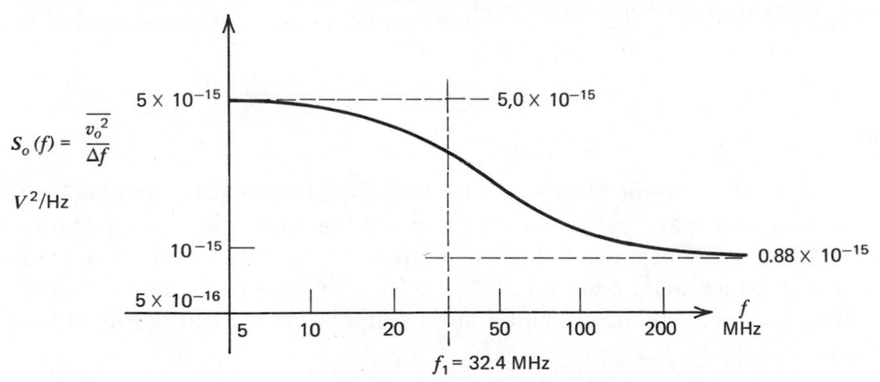

Fig. 11.19 Noise voltage spectrum at the output of the circuit of Fig. 11.18.

EXAMPLE

In order to give an appreciation of the numbers involved, specific values will now be assigned to the parameters of (11.38) and the various terms in the equation will be evaluated. Assume that

$$I_C = 100\ \mu\text{A} \qquad \beta = 100 \qquad r_b = 200\ \Omega$$
$$R_S = 500\ \Omega \qquad C_\pi = 10\text{ pF}$$
$$R_L = 5\text{ k}\Omega$$

Substituting these values in (11.38) and using $4kT = 1.66 \times 10^{-20}$ V-C gives

$$\frac{\overline{v_o^2}}{\Delta f} = \left[5.82 \times 10^{-18} \frac{1}{1 + \left(\dfrac{f}{f_1}\right)^2} (700 + 9.4) + 1.66 \times 10^{-20}(5000 + 48{,}080) \right] \text{V}^2/\text{Hz}$$

$$= \frac{4.13 \times 10^{-15}}{1 + \left(\dfrac{f}{f_1}\right)^2} + 0.88 \times 10^{-15}\ \text{V}^2/\text{Hz} \qquad (11.40)$$

Equation 11.39 gives

$$f_1 = 32.4\text{ MHz} \qquad (11.41)$$

Equation 11.40 shows that the output noise-voltage spectral density is 5.0×10^{-15} V^2/Hz at low frequencies and it approaches 0.88×10^{-15} V^2/Hz at high frequencies. The major contributor to the output noise in this case is the source resistance, R_S, followed by the base resistance of the transistor. The noise spectrum given by (11.40) is plotted in Fig. 11.19.

EXAMPLE

Suppose the amplifier in the above example is followed by later stages that limit the bandwidth to a sharp cutoff at 1 MHz. Since the noise spectrum as shown in Fig. 11.19 does not begin to fall significantly until $f_1 = 32.4$ MHz, the noise spectrum may be assumed constant at 5.0×10^{-15} V^2/Hz over the bandwidth 0 to 1 MHz. Thus the *total* noise voltage at the output of the circuit of Fig. 11.18*a* in a 1-MHz bandwidth is

$$v_{oT}^2 = 5.0 \times 10^{-15} \times 10^6\ \text{V}^2 = 5.0 \times 10^{-9}\ \text{V}^2$$

and thus

$$v_{oT} = 71\ \mu\text{V rms} \tag{11.42}$$

Now suppose that the amplifier of Fig. 11.18*a* is *not* followed by later stages that limit the bandwidth but is fed directly to a wideband detector (this could be an oscilloscope or a voltmeter). In order to find the total output noise voltage in this case, the contribution from each frequency increment Δf must be summed at the output. This reduces to *integration* across the bandwidth of the detector of the noise-voltage spectral-density curve of Fig. 11.19. For example, if the detector had a 0 to 50-MHz bandwidth with a sharp cutoff then the total output noise would be

$$\begin{aligned}\overline{v_{oT}^2} &= \sum_{f=0}^{50\times 10^6} S_o(f)\,\Delta f \\ &= \int_0^{50\times 10^6} S_o(f)\,df \end{aligned} \tag{11.43}$$

where

$$S_o(f) = \frac{\overline{v_o^2}}{\Delta f} \tag{11.44}$$

is the noise spectral density defined by (11.40). In practice, the exact evaluation of such integrals is often difficult and approximate methods are often used. Note that if the integration of (11.43) is done graphically, the noise spectral density versus frequency must be plotted on *linear scales.*

11.4.2 Equivalent Input Noise and the Minimum Detectable Signal

In the previous section, the output noise produced by the circuit of Fig. 11.18 was calculated. The significance of the noise performance of a circuit is, however, the limitation it places on the smallest input signals the circuit can handle before the noise degrades the quality of the output signal. For this reason, the noise performance is usually expressed in terms of an *equivalent input noise signal,* which gives the same output noise as the circuit under consideration. In this way, the equivalent input noise can be compared directly with incoming signals and the effect of the noise on those signals is easily determined. For this purpose, the circuit of Fig. 11.18 can be represented as shown in Fig. 11.20 where $\overline{v_{iN}^2}$ is an input noise-voltage generator that produces the same output noise as all of the original noise generators. All other sources of noise in Fig. 11.20 are considered removed. Using the same equivalent circuit as in Fig. 11.18*b*, we obtain, for the output noise from Fig. 11.20,

$$\overline{v_o^2} = g_m^2 R_L^2 \frac{|Z|^2}{|Z + r_b + R_S|^2}\overline{v_{iN}^2} \tag{11.45}$$

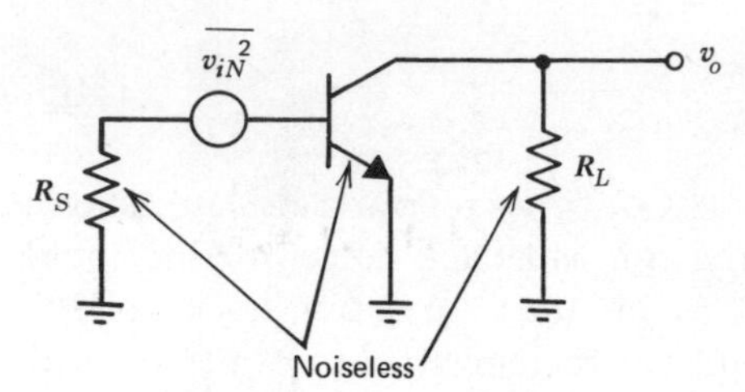

Fig. 11.20 Representation of circuit noise performance by an equivalent input noise voltage.

If this noise expression is equated to $\overline{v_o^2}$ from (11.37), the equivalent input noise voltage for the circuit can be calculated as

$$\frac{\overline{v_{iN}^2}}{\Delta f} = 4kT(R_S + r_b) + (R_S + r_b)^2 2qI_B$$

$$+ \frac{1}{g_m^2 R_L^2} \frac{|Z + r_b + R_S|^2}{|Z|^2} R_L^2 \left(4kT\frac{1}{R_L} + 2qI_C\right) \tag{11.46}$$

Note that the noise-voltage spectral density given by (11.46) *rises* at high frequencies because of the variation of $|Z|$ with frequency. This is due to the fact that as the *gain* of the device falls with frequency, output noise generators $\overline{i_c^2}$ and $\overline{i_l^2}$ have a larger effect when referred back to the input.

EXAMPLE

Calculate the *total* input noise voltage, $\overline{v_{iNT}^2}$, for the circuit of Fig. 11.18 in a bandwidth of 0 to 1 MHz.

This could be calculated using (11.46) derived above. Alternatively, since the total output noise voltage, $\overline{v_{oT}^2}$ has already been calculated, this can be used to calculate $\overline{v_{iNT}^2}$ (in a 1-MHz bandwidth) by dividing by the circuit voltage gain squared. If A_v is the low-frequency, small-signal voltage gain of Fig. 11.18, then

$$A_v = \frac{r_\pi}{r_b + r_\pi + R_S} g_m R_L$$

Use of the previously specified data for this circuit gives

$$A_v = \frac{26{,}000}{200 + 26{,}000 + 500} \frac{5000}{260} = 18.7$$

Since the noise spectrum is flat up to 1 MHz, the low-frequency gain can be used to calculate $\overline{v_{iNT}^2}$ as

$$\overline{v_{iNT}^2} = \frac{\overline{v_{oT}^2}}{A_v^2} = \frac{5 \times 10^{-9}}{18.7^2} \text{V}^2 = 14.3 \times 10^{-12} \text{ V}^2$$

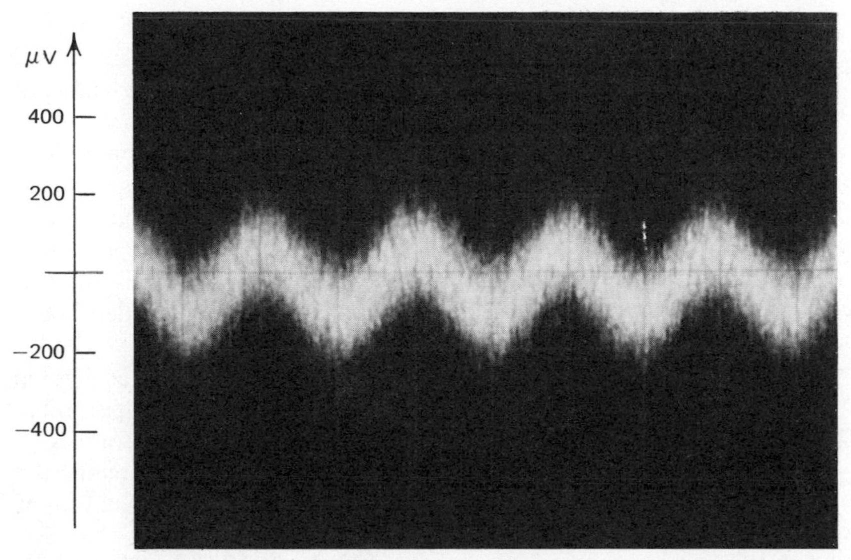

Fig. 11.21 Output voltage waveform of the circuit of Fig. 11.18 with a 3.78-μV rms sinewave applied at the input. The circuit bandwidth is limited to 1 MHz, which gives an equivalent input noise voltage of 3.78 μV rms.

Thus we have

$$v_{iNT} = 3.78 \ \mu\text{V rms}$$

The above example shows that in a bandwidth of 0 to 1 MHz, the noise in the circuit *appears to come* from a 3.78-μV rms noise-voltage source in series with the input. This noise voltage can be used to estimate the smallest signal that the circuit can effectively amplify, sometimes called the *minimum detectable signal* (MDS). This depends strongly on the nature of the signal and the application. If no special filtering or coding techniques are used, the MDS can be taken as equal to the equivalent input noise voltage in the passband of the amplifier. Thus, in this case,

$$\text{MDS} = 3.78 \ \mu\text{V rms}$$

If a sinewave of magnitude 3.78 μV rms were applied to this circuit, and the output in a 1-MHz bandwidth examined on an oscilloscope, the sine wave would be barely detectable in the noise, as shown in Fig. 11.21. The noise waveform in this figure is typical of that produced by shot and thermal noise.

11.5 EQUIVALENT INPUT NOISE GENERATORS[16]

In the previous section, the equivalent input noise voltage for a particular configuration was calculated. This gave rise to an expression for an equivalent input noise-voltage generator that was dependent on the source resistance, R_S, as well as the transistor parameters. This method is now extended to a more general and more useful representation in which the noise performance of *any* two-port network is represented by *two* equivalent input noise generators. The situation is shown in Fig. 11.22 where a two-port network containing noise generators is represented by the *same* network with internal noise sources removed (the *noiseless* network) and with a noise voltage $\overline{v_i^2}$ and current generator $\overline{i_i^2}$ connected at the input. It can be shown that this representation is valid for *any* source impedance, provided that *correlation* between the two noise generators is considered. That is, the two noise generators are not independent in general because they are both dependent on the same set of original noise sources.

The inclusion of correlation in the noise representation results in a considerable increase in the complexity of the calculations, and if correlation is important it is often easier to return to the original network with internal noise sources to perform the calculations. However, in a large number of practical circuits, the correlation is small and may be neglected. In addition, if either equivalent input generator $\overline{v_i^2}$ or $\overline{i_i^2}$ dominates, the correlation may be neglected in any case. The use of this method of representation is then extremely useful, as will become apparent.

The need for both an equivalent input noise voltage generator and an equivalent input noise current generator to represent the noise performance of the circuit for any source resistance can be appreciated as follows. Consider the extreme cases of source resistance R_S equal to zero or infinity. If $R_S = 0$, $\overline{i_i^2}$ in Fig. 11.22 is shorted out, and since the original circuit will still show output noise in general, we need an equivalent input noise voltage $\overline{v_i^2}$ to represent this behavior. Similarly, if $R_S = \infty$, $\overline{v_i^2}$ in Fig. 11.22 cannot produce output noise and $\overline{i_i^2}$ represents the noise performance of the original noisy network. For finite values of R_S, both $\overline{v_i^2}$ and $\overline{i_i^2}$ contribute to the equivalent input noise of the circuit.

The values of the equivalent input generators of Fig. 11.22 are readily determined. This is done by first short circuiting the input of both circuits and equating the output noise in each case to calculate $\overline{v_i^2}$. The value of $\overline{i_i^2}$ is found by open circuiting the input of each circuit and equating the output noise in each case. This will now be done for the bipolar transistor and the field-effect transistor.

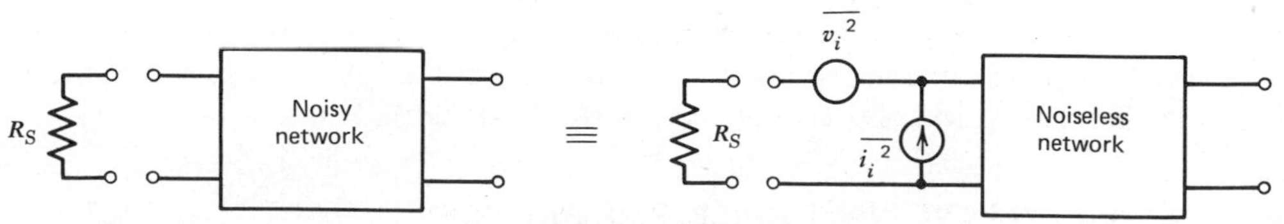

Fig. 11.22 Representation of noise in a two-port network by equivalent input voltage and current generators.

11.5.1 Bipolar Transistor Noise Generators

The equivalent input noise generators for a bipolar transistor can be calculated from the equivalent circuit of Fig. 11.23*a*. The output noise is calculated with a short-circuited load, and $C\mu$ is neglected. This will be justified later. The circuit of Fig. 11.23*a* is to be equivalent to that of Fig. 11.23*b* in that each circuit should give the *same* output noise for *any* source impedance.

The value of v_i^2 can be calculated by short circuiting the input of each circuit and equating the output noise, i_o. We use rms noise quantities in the calculations, but make no attempt to preserve the signs of the noise quantities as the noise generators are all independent and have random phase. The polarity of the noise generators does not affect the answer. Short circuiting the inputs of both circuits in Fig. 11.23, assuming that r_b is small ($\ll r_\pi$) and equating i_o, we obtain

$$g_m v_b + i_c = g_m v_i \tag{11.47}$$

which gives

$$v_i = v_b + \frac{i_c}{g_m} \tag{11.48}$$

Since r_b is small, the effect of $\overline{i_b^2}$ is neglected in this calculation.

Using the fact that v_b and i_c are *independent*, we obtain, from (11.48),

$$\overline{v_i^2} = \overline{v_b^2} + \frac{\overline{i_c^2}}{g_m^2} \tag{11.49}$$

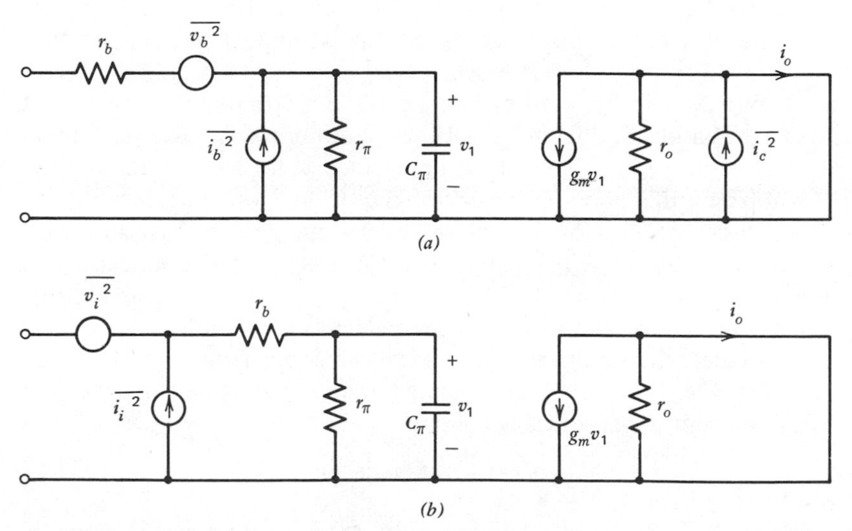

Fig. 11.23 (*a*) Bipolar transistor small-signal equivalent circuit with noise generators. (*b*) Representation of the noise performance of (*a*) by equivalent input generators.

Substituting in (11.49) for v_b^2 and i_c^2 from (11.11) and (11.12) gives

$$\overline{v_i^2} = 4kTr_b\,\Delta f + \frac{2qI_C\,\Delta f}{g_m^2}$$

and thus

$$\frac{\overline{v_i^2}}{\Delta f} = 4kT\left(r_b + \frac{1}{2g_m}\right) \tag{11.50}$$

The equivalent input noise-voltage spectral density of a bipolar transistor thus appears to come from a resistor R_{eq} such that

$$\frac{\overline{v_i^2}}{\Delta f} = 4kTR_{\text{eq}} \tag{11.51}$$

where

$$R_{\text{eq}} = r_b + \frac{1}{2g_m} \tag{11.52}$$

and this is called the "equivalent input noise resistance." Of this fictitious resistance, portion r_b, is, in fact, a physical resistor in series with the input, whereas portion $1/2g_m$ represents the effect of collector current shot noise referred back to the input. Equations 11.50 and (11.52) are extremely useful approximations, although the assumption that r_b is $\ll r_\pi$ may not be valid at high collector bias currents, and the calculation should be repeated without restrictions in those circumstances.

Equation 11.50 allows easy comparison of the relative importance of noise from r_b and I_C in contributing to $\overline{v_i^2}$. For example, if $I_C = 1\mu\text{A}$, then $1/2g_m = 13\ \text{k}\Omega$ and this will dominate typical r_b values of about 100 Ω. Alternately, if $I_C = 10$ mA, then $1/2g_m = 1.3\ \Omega$ and noise from r_b will totally dominate $\overline{v_i^2}$. Since $\overline{v_i^2}$ is the important noise generator for low source impedance (since $\overline{i_i^2}$ then tends to be shorted), it is apparent that good noise performance from a low source impedance requires minimization of R_{eq}. This is achieved by designing the transistor to have a low r_b and running the device at a large collector bias current to reduce $1/2g_m$. Finally, it should be noted from (11.50) that the equivalent input noise-voltage spectral density of a bipolar transistor is independent of frequency.

In order to calculate the equivalent input noise current generator $\overline{i_i^2}$, the inputs of both circuits in Fig. 11.23 are open circuited and output noisc currcnts i_o are equated. Using rms noise quantities, we obtain

$$\beta(j\omega)i_i = i_c + \beta(j\omega)i_b \tag{11.53}$$

which gives

$$i_i = i_b + \frac{i_c}{\beta(j\omega)} \tag{11.54}$$

Since i_b and i_c are independent generators, we obtain, from (11.54),

$$\overline{i_i^2} = \overline{i_b^2} + \frac{\overline{i_c^2}}{|\beta(j\omega)|^2} \tag{11.55}$$

where

$$\beta(j\omega) = \frac{\beta_0}{1 + j\dfrac{\omega}{\omega_\beta}} \tag{11.56}$$

and β_0 is the low-frequency, small-signal current gain. [See (1.122) and (1.126)]
Substituting in (11.55) for $\overline{i_b^2}$ and $\overline{i_c^2}$ from (11.13) and (11.12) gives

$$\frac{\overline{i_i^2}}{\Delta f} = 2q\left[I_B + K_1' \frac{I_B{}^a}{f} + \frac{I_C}{|\beta(j\omega)|^2}\right] \tag{11.57}$$

where

$$K_1' = \frac{K_1}{2q} \tag{11.57a}$$

and the burst noise term has been omitted for simplicity. The last term in parentheses in (11.57) is due to collector current noise referred to the input. At low frequencies this becomes $I_C/\beta_0{}^2$ and is negligible compared with I_B for typical β_0 values. When this is true, $\overline{i_i^2}$ and $\overline{v_i^2}$ do not contain common noise sources and are *totally independent*. At high frequencies, however, the last term in (11.57) increases and can become dominant, and correlation between $\overline{v_i^2}$ and $\overline{i_i^2}$ may then be important since both contain a contribution from $\overline{i_c^2}$.

The equivalent input noise current spectral density given by (11.57) appears to come from a current I_{eq} showing full shot noise, such that

$$\frac{\overline{i_i^2}}{\Delta f} = 2qI_{eq} \tag{11.58}$$

where

$$I_{eq} = I_B + K_1' \frac{I_B{}^a}{f} + \frac{I_C}{|\beta(j\omega)|^2} \tag{11.59}$$

and this is called the "equivalent input shot noise current." This is a fictitious current composed of the base current of the device plus a term representing flicker noise and one representing collector-current noise transformed to the input. It is apparent from (11.59) that I_{eq} is minimized by utilizing low bias currents in the transistor, and also using high-β transistors. Since $\overline{i_i^2}$ is the dominant equivalent input noise generator in circuits where the transistor is fed from a high source impedance, low bias currents and high β are obviously required for good noise performance under these conditions. Note that the requirement for low bias currents to minimize $\overline{i_i^2}$ conflicts with the requirement for *high* bias current to minimize $\overline{v_i^2}$.

Spectral density $\overline{i_i^2}/\Delta f$ of the equivalent input noise current generator can be plotted as a function of frequency using (11.57). This is shown in Fig. 11.24 for typical transistor parameters. In this case, the spectral density is frequency dependent at both low and high frequencies, the low-frequency rise being due to flicker noise and the high-frequency rise being due to collector-current noise referred to the input. This input referred noise rises at high frequencies because the transistor current gain begins to fall, and this is the reason for the degradation in transistor noise performance observed at high frequencies.

Frequency f_b in Fig. 11.24 is the point where the high-frequency noise asymptote intersects the midband asymptote. This can be calculated from (11.57) as follows:

$$\beta(jf) = \frac{\beta_0}{1 + j\dfrac{f}{f_T}\beta_0} \tag{11.60}$$

where β_0 is the low-frequency, small-signal current gain. Thus the collector current noise term in (11.57) is

$$2q\frac{I_C}{|\beta(jf)|^2} = 2q\frac{I_C}{\beta_0^{\,2}}\left(1 + \frac{f^2}{f_T^{\,2}}\beta_0^{\,2}\right) \simeq 2qI_C\frac{f^2}{f_T^{\,2}} \tag{11.61}$$

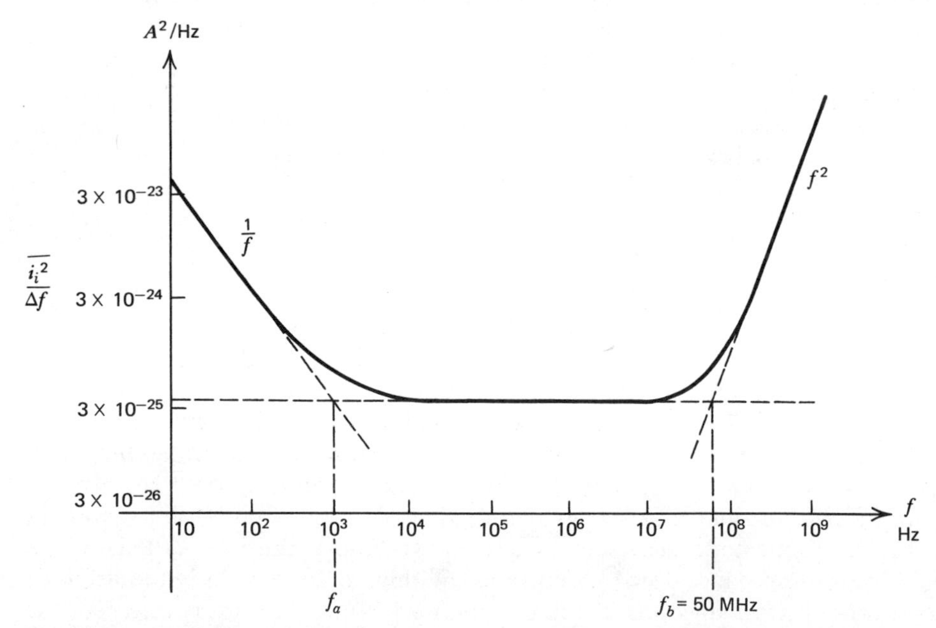

Fig. 11.24 Equivalent input noise current spectral density for a bipolar transistor with $I_C = 100\ \mu\text{A}$, $\beta_0 = \beta_F = 100$, $f_T = 500$ MHz. Typical flicker noise is included.

at high frequencies. Equation 11.61 shows that the equivalent input noise current spectrum rises as f^2 at high frequencies. Frequency f_b can be calculated by equating (11.61) to the midband noise, which is $2q[I_B + (I_C/\beta_0{}^2)]$. For typical values of β_0, this is approximately $2qI_B$, and equating this quantity to (11.61) we obtain

$$2qI_B = 2qI_C \frac{f_b{}^2}{f_T{}^2}$$

and thus

$$f_b = f_T \sqrt{\frac{I_B}{I_C}} \tag{11.62}$$

The large-signal (or dc) current gain is defined as

$$\beta_F = \frac{I_C}{I_B} \tag{11.63}$$

and thus (11.62) becomes

$$f_b = \frac{f_T}{\sqrt{\beta_F}} \tag{11.64}$$

Using the data given in Fig. 11.24, we obtain $f_b = 50$ MHz for that example.

Once the above input noise generators have been calculated, the transistor noise performance with any source impedance is readily calculated. For example, consider the simple circuit of Fig. 11.25*a* with a source resistance R_S. The noise performance of this circuit can be represented by the *total* equivalent noise voltage $\overline{v_{iN}{}^2}$ in series with the input of the circuit as shown in Fig. 11.25*b*. Neglecting noise in R_L (this will be discussed later), and equating the total noise voltage at the base of the transistor in Figs. 11.25*a* and *b*, we obtain

$$v_{iN} = v_s + v_i + i_i R_S$$

If correlation between v_i and i_i is neglected this equation gives

$$\overline{v_{iN}{}^2} = \overline{v_s{}^2} + \overline{v_i{}^2} + \overline{i_i{}^2} R_S{}^2 \tag{11.65}$$

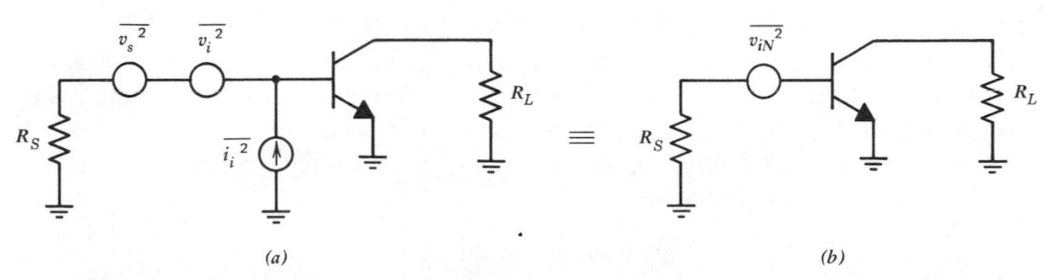

Fig. 11.25 Representation of circuit noise by a single equivalent input noise voltage generator. (*a*) Original circuit. (*b*) Equivalent representation.

Using (11.50) and (11.57) in (11.65) and neglecting flicker noise we find

$$\frac{\overline{v_{iN}^2}}{\Delta f} = 4kTR_S + 4kT\left(r_b + \frac{1}{2g_m}\right) + R_S^2 2q\left[I_B + \frac{I_C}{|\beta(jf)|^2}\right] \tag{11.66}$$

Equation 11.66 is similar to (11.46) if r_b is small, as has been assumed.

EXAMPLE

Using data from the example in Section 11.4.1, calculate the total input noise voltage for the circuit of Fig. 11.25*a* in a bandwidth 0 to 1 MHz neglecting flicker noise and using (11.66). At low frequencies, (11.66) becomes

$$\begin{aligned}\frac{\overline{v_{iN}^2}}{\Delta f} &= 4kT\left(R_S + r_b + \frac{1}{2g_m}\right) + R_S^2 2qI_B \\ &= [1.66 \times 10^{-20}(500 + 200 + 130) + 500^2 \times 3.2 \times 10^{-19} \times 10^{-6}]\ \text{V}^2/\text{Hz} \\ &= (13.8 + 0.08) \times 10^{-18}\ \text{V}^2/\text{Hz} \\ &= 13.9 \times 10^{-18}\ \text{V}^2/\text{Hz}\end{aligned}$$

The total input noise in a 1-MHz bandwidth is

$$\begin{aligned}\overline{v_{iNT}^2} &= 13.9 \times 10^{-18} \times 10^6\ \text{V}^2 \\ &= 13.9 \times 10^{-12}\ \text{V}^2\end{aligned}$$

and thus

$$v_{iNT} = 3.73\ \mu\text{V rms}$$

This is almost identical to the answer obtained in Section 11.4.1. However, the method described above has the advantage that once the equivalent input generators are known for any particular device, the answer can be written down almost by inspection and requires much less labor. Also, the relative contributions of the various noise generators are more easily seen. In this case, for example, the equivalent input noise current is obviously a negligible factor.

11.5.2 Field-Effect Transistor Noise Generators

The equivalent input noise generators for a field-effect transistor (FET) can be calculated from the equivalent circuit of Fig. 11.26*a*. This circuit is to be made equivalent to that of Fig. 11.26*b*. The output noise in each case is calculated with a short-circuit load and C_{gd} is neglected.

If the input of each circuit in Fig. 11.26 is short circuited and resulting output noise currents i_o are equated we obtain

$$i_d = g_m v_i$$

and thus

$$\overline{v_i^2} = \frac{\overline{i_d^2}}{g_m^2} \tag{11.67}$$

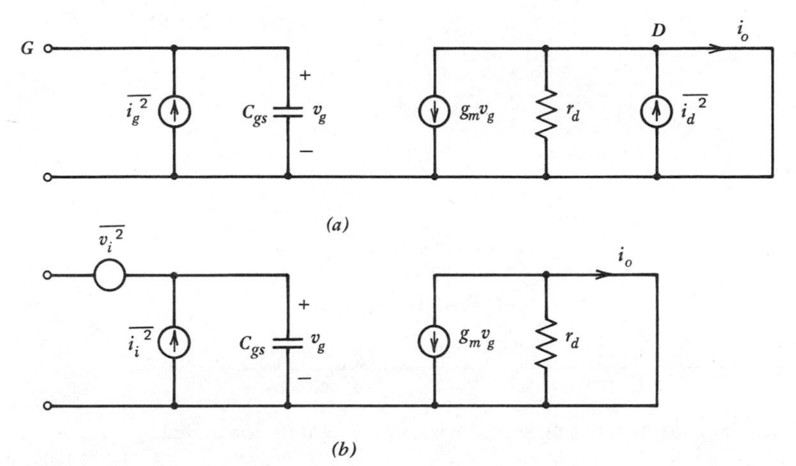

Fig. 11.26 (*a*) FET small-signal equivalent circuit with noise generators. (*b*) Representation of (*a*) by two input noise generators.

Substituting $\overline{i_d^2}$ from (11.15) in (11.67) gives

$$\frac{\overline{v_i^2}}{\Delta f} = 4kT\,\tfrac{2}{3}\frac{1}{g_m} + K\frac{I_D{}^a}{g_m{}^2 f} \tag{11.68}$$

The equivalent input noise resistance, R_{eq}, of the FET is defined as

$$\frac{\overline{v_i^2}}{\Delta f} = 4kTR_{eq}$$

where

$$R_{eq} = \tfrac{2}{3}\frac{1}{g_m} + K'\frac{I_D{}^a}{g_m{}^2 f} \tag{11.69}$$

and

$$K' = \frac{K}{4kT}$$

At frequencies above the flicker noise region, $R_{eq} = (2/3)(1/g_m)$. For $g_m = 1$ mA/V, this gives $R_{eq} = 667\ \Omega$, which is significantly *higher* than for a bipolar transistor at a comparable bias current (about 1 mA). The equivalent input noise-voltage spectral density for a typical FET is plotted versus frequency in Fig. 11.27. Unlike the bipolar transistor, the equivalent input noise-voltage generator for a FET contains flicker noise, and it is not uncommon for the flicker noise to extend well into the megahertz region.

The equivalent input noise-current generator i_i^2 for the FET can be calculated by open circuiting the input of each circuit in Fig. 11.26 and equating the output

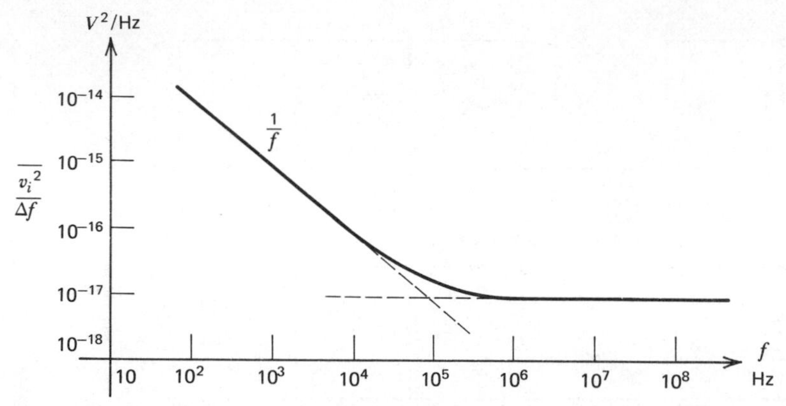

Fig. 11.27 Typical equivalent input noise voltage spectral density for a FET.

noise. This gives

$$i_i \frac{g_m}{j\omega C_{gs}} = i_g \frac{g_m}{j\omega C_{gs}} + i_d$$

and thus

$$i_i = i_g + \frac{j\omega C_{gs}}{g_m} i_d \tag{11.70}$$

Since i_g and i_d represent *independent* generators, (11.70) can be written as

$$\overline{i_i^2} = \overline{i_g^2} + \frac{\omega^2 C_{gs}^2}{g_m^2} \overline{i_d^2} \tag{11.71}$$

Substituting (11.14) and (11.15) in (11.71) gives

$$\frac{\overline{i_i^2}}{\Delta f} = 2qI_G + \frac{\omega^2 C_{gs}^2}{g_m^2}\left(4kT\,\tfrac{2}{3}g_m + K\frac{I_D^a}{f}\right) \tag{11.72}$$

In (11.72) the ac current gain of the FET can be identified as

$$A_I = \frac{g_m}{\omega C_{gs}} \tag{11.73}$$

and thus the noise generators at the output are divided by A_I^2 when referred back to the input. At low frequencies the input noise-current generator is determined by the gate leakage current, I_G, which is very small (as low as 10^{-12} A or less). For this reason, FETs have noise performance that is *much superior* to bipolar transistors when the driving source impedance is large. Under these circumstances the input noise-current generator is dominant and is much smaller for a FET than for a bipolar transistor. It should be emphasized, however, that the input noise-*voltage* generator of a bipolar transistor (Equation 11.50) is typically *smaller* than that of a FET (Equation 11.68) because the bipolar transistor has a larger g_m for a given bias current. Thus for *low* source impedances, a bipolar transistor often has noise performance *superior* to that of a FET.

11.6 EFFECT OF FEEDBACK ON NOISE PERFORMANCE

The representation of circuit noise performance with two equivalent input noise generators is extremely useful in the consideration of the effect of feedback on noise performance. This will be illustrated by considering first the effect of ideal feedback on the noise performance of an amplifier. Practical aspects of feedback and noise performance will then be considered.

11.6.1 Effect of Ideal Feedback on Noise Performance

In Fig. 11.28*a* a series-shunt feedback amplifier is shown where the feedback network is ideal in that the signal feedback to the input is a pure voltage source

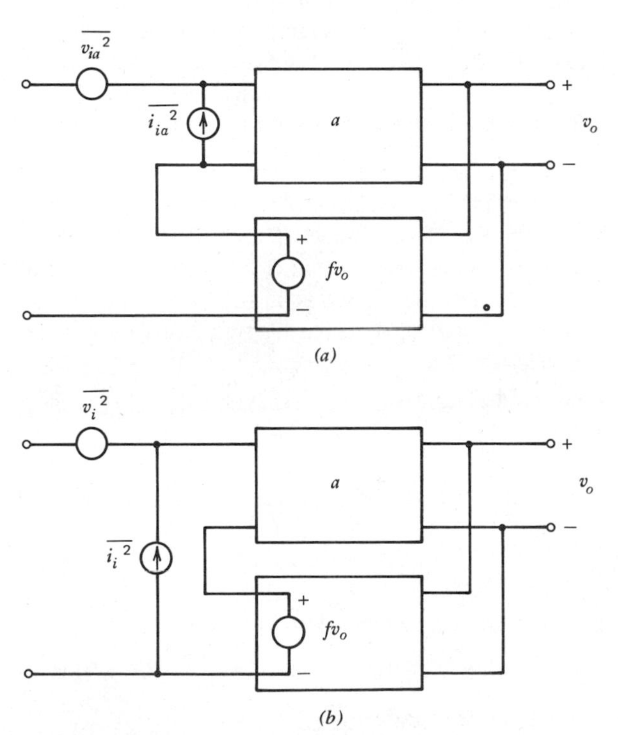

Fig. 11.28 (*a*) Series-shunt feedback amplifier with noise generators. (*b*) Equivalent representation of (*a*) with two input noise generators.

and the feedback network is unilateral. Noise in the basic amplifier is represented by equivalent input generators $\overline{v_{ia}^2}$ and $\overline{i_{ia}^2}$. The noise performance of the overall circuit is represented by equivalent input generators $\overline{v_i^2}$ and $\overline{i_i^2}$ as shown in Fig. 11.28*b*. The value of $\overline{v_i^2}$ can be found by short circuiting the input of each circuit and equating the output signal. However, since the output of the feedback network has a zero impedance, the current generators in each circuit are then short circuited and the two circuits are identical only if

$$\overline{v_i^2} = \overline{v_{ia}^2} \tag{11.74}$$

If the input terminals of each circuit are open circuited, both voltage generators have a floating terminal and thus no effect on the circuit, and for equal outputs it is necessary that

$$\overline{i_i^2} = \overline{i_{ia}^2} \tag{11.75}$$

Thus, for the case of *ideal feedback*, the equivalent input noise generators can be moved *unchanged outside* the feedback loop and the feedback has *no effect* on the circuit noise performance. Since the feedback reduces the circuit gain the *output noise* is reduced by the feedback, but desired signals are reduced by the same amount and the signal-to-noise ratio will be unchanged. The above result is easily shown for all four possible feedback configurations described in Chapter 8.

11.6.2 Effect of Practical Feedback on Noise Performance

The idealized series-shunt feedback circuit considered in the previous section is usually realized in practice as shown in Fig. 11.29*a*. The feedback circuit is a resistive divider consisting of R_E and R_F. If the noise of the basic amplifier is represented by equivalent input noise generators $\overline{i_{ia}^2}$ and $\overline{v_{ia}^2}$ and the thermal noise generators in R_F and R_E are included, the circuit is as shown in Fig. 11.29*b*. The noise performance of the circuit is to be represented by two equivalent input generators $\overline{v_i^2}$ and $\overline{i_i^2}$ as shown in Fig. 11.29*c*.

In order to calculate $\overline{v_i^2}$, consider the inputs of the circuits of Fig. 11.29*b* and *c* short circuited, and equate the output noise. It is readily shown that

$$v_i = v_{ia} + i_{ia}R + \frac{R_F}{R_F + R_E} v_e + \frac{R_E}{R_F + R_E} v_f \tag{11.76}$$

where

$$R = R_F \| R_E \tag{11.77}$$

Assuming that all noise sources in (11.76) are independent we have

$$\overline{v_i^2} = \overline{v_{ia}^2} + \overline{i_{ia}^2}R^2 + 4kTR\,\Delta f \tag{11.78}$$

where the following substitutions have been made:

$$\overline{v_e^2} = 4kTR_E\,\Delta f \tag{11.79}$$

$$\overline{v_f^2} = 4kTR_F\,\Delta f \tag{11.80}$$

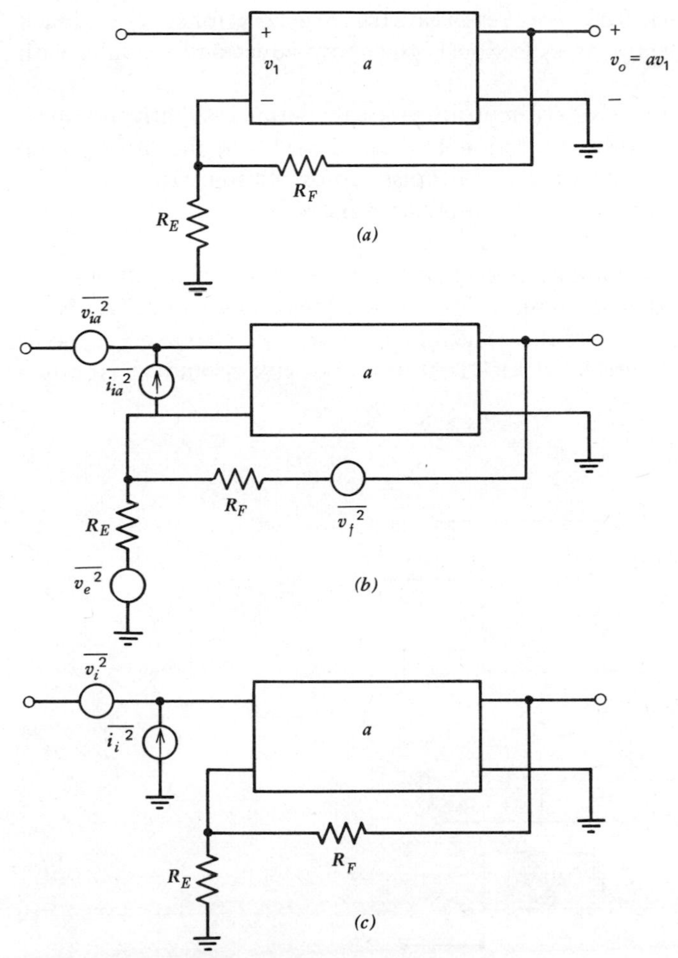

Fig. 11.29 (*a*) Series-shunt feedback circuit. (*b*) Series-shunt feedback circuit including noise generators. (*c*) Equivalent representation of (*b*) with two input noise generators.

Equation 11.78 shows that in this practical case, the equivalent input noise voltage of the overall amplifier contains the input noise voltage of the basic amplifier plus two other terms. The second term in (11.78) is usually negligible, but the third term represents thermal noise in $R = R_E \| R_F \simeq R_E$ and is often significant.

The equivalent input noise current, $\overline{i_i^2}$, is calculated by open circuiting both inputs and equating output noise. It is apparent that

$$\overline{i_i^2} \simeq \overline{i_{ia}^2} \tag{11.81}$$

since noise in the feedback resistors is no longer amplified, but appears only in shunt with the output. Thus the equivalent input noise current is *unaffected* by the application of feedback. The above results are true in general for series feedback at the input. For single-stage series feedback, the above equations are valid with $R_F = \infty$ and $R = R_E$.

If the basic amplifier in Fig. 8.29 is an op amp the calculation is slightly modified. This is due to the fact (shown in Section 11.8 and Fig. 11.39) that an op amp must be considered a three-port device for noise representation. However if the circuit of Fig. 11.39 is used as the basic amplifier in the above calculation, expressions very similar to (11.78) and (11.81) are obtained.

Consider now the case of shunt feedback at the input, and as an example consider the shunt-shunt feedback circuit of Fig. 11.30*a*. This is shown in Fig. 11.30*b* with noise sources $\overline{v_{ia}^2}$ and $\overline{i_{ia}^2}$ of the basic amplifier, and noise source $\overline{i_f^2}$ due to R_F. These noise sources are referred back to the input to give $\overline{v_i^2}$ and $\overline{i_i^2}$ as shown in Fig. 11.30*c*.

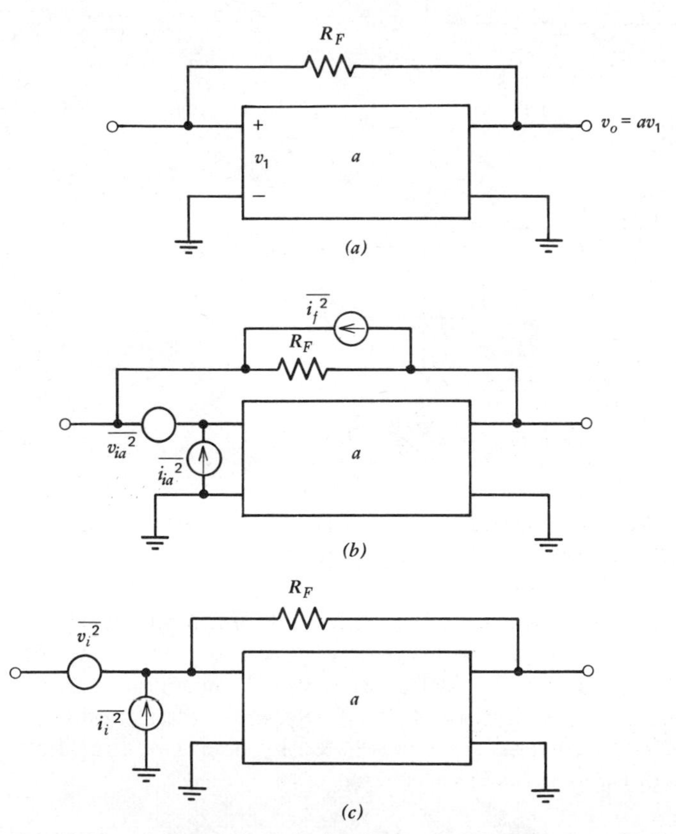

Fig. 11.30 (*a*) Shunt-shunt feedback circuit. (*b*) Shunt-shunt feedback circuit including noise generators. (*c*) Equivalent representation of (*b*) with two input noise generators.

Open circuiting the inputs of Fig. 11.30*b* and *c*, and equating output noise, we calculate

$$i_i = i_{ia} + \frac{v_{ia}}{R_F} + i_f \tag{11.82}$$

Assuming all noise sources in (11.82) are independent we find

$$\overline{i_i^2} = \overline{i_{ia}^2} + \frac{\overline{v_{ia}^2}}{R_F^2} + 4kT\frac{1}{R_F}\Delta f \tag{11.83}$$

Thus the equivalent input noise current with shunt feedback applied consists of the input noise current of the basic amplifier together with a term representing thermal noise in the feedback resistor. The second term in (11.83) is usually negligible. These results are true in general for shunt feedback at the input. *A general rule* for calculating the equivalent input noise contribution due to thermal noise in the feedback resistors is to follow the methods described in Chapter 8 for calculating feedback-circuit loading on the basic amplifier. Once the shunt or series resistors representing feedback loading at the input have been determined, these same resistors may be used to calculate the thermal noise contribution at the input due to the feedback resistors.

If the inputs of the circuits of Fig. 11.30*b* and *c* are short circuited, and the output noise equated, it follows that

$$\overline{v_i^2} \simeq \overline{v_{ia}^2} \tag{11.84}$$

Equations 11.83 and 11.84 are true in general for shunt feedback at the input. They apply directly when the basic amplifier of Fig. 11.30 is an op amp, since one input terminal of the basic amplifier is grounded and the op amp becomes a two-port device.

The above results allow justification of some assumptions made earlier. For example, in the calculation of the equivalent input noise generators for a bipolar transistor in Section 11.5.1, collector-base capacitance C_μ was ignored. This capacitance represents single-stage *shunt feedback* and thus does *not* significantly affect the equivalent input noise generators of a transistor, *even if* Miller effect is dominant. Note that there is no thermal noise contribution from the capacitor as there was from R_F in Fig. 11.30. Also, the second term in (11.83) becomes $\overline{v_{ia}^2}/|Z_F|^2$ where Z_F is the impedance of C_μ. Since $|Z_F|$ is quite large at all frequencies of interest, this term is negligible.

EXAMPLE

As an example of calculations involving noise in feedback amplifiers, consider the wideband current-feedback pair whose ac schematic is shown in Fig. 11.31. The circuit is fed from a current source and the frequency response $|(i_o/i_i)(j\omega)|$ is flat

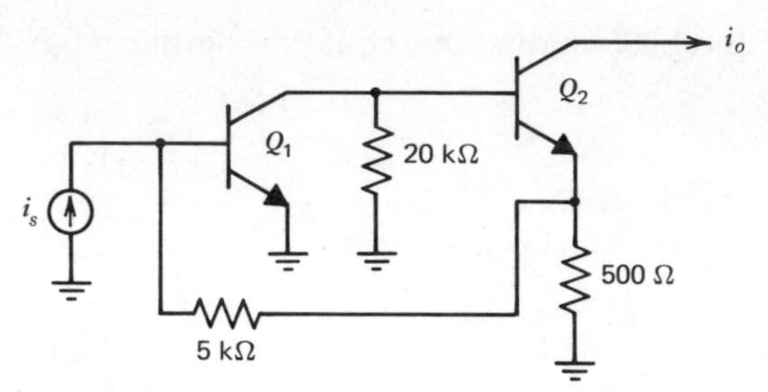

Fig. 11.31 ac schematic of a current feedback pair.

with frequency to 100 MHz where it falls rapidly. We calculate the minimum input signal, i_s, required for an output signal-to-noise ratio greater than 20 dB. Data is as follows: $\beta_1 = \beta_2 = 100$, $f_{T1} = 300$ MHz, $I_{C1} = 0.5$ mA, $I_{C2} = 1$ mA, $f_{T2} = 500$ MHz, $r_{b1} = r_{b2} = 100\ \Omega$. Flicker noise is neglected.

The methods developed above allow the equivalent input noise generators for this circuit to be written down by inspection. A preliminary check shows that the noise due to the 20-kΩ interstage resistor and the base current noise of Q_2 are negligible. Using the rule stated in Section 11.2.2, we find that the 20-kΩ resistor contributes an equivalent noise current of 2.5 μA. The base current of Q_2 is 10 μA. Both of these can be neglected when compared to the 500 μA collector current of Q_1. Thus the input noise generators of the whole circuit are those of Q_1 moved outside the feedback loop, together with the noise contributed by the feedback resistors.

Using the methods of Chapter 8, we can derive the basic amplifier including feedback loading and noise sources for the circuit of Fig. 11.31 as shown in Fig. 11.32. The equivalent input noise-current generator for the overall circuit can be calculated from Fig. 11.32 or by using (11.83) with $R_F = 5.5$ kΩ. Since the circuit is assumed to be driven from a current source, the equivalent input noise voltage is not important. From (11.83)

$$\overline{i_i^2} = \overline{i_{ia}^2} + \frac{\overline{v_{ia}^2}}{(5500)^2} + 4kT\frac{1}{5500}\Delta f \tag{11.85}$$

Fig. 11.32 Basic amplifier for the circuit of Fig. 11.31 including feedback loading and noise sources.

Using (11.57) and neglecting flicker noise, we have, for $\overline{i_{ia}^2}$,

$$\overline{i_{ia}^2} = 2q\left(I_B + \frac{I_C}{|\beta(jf)|^2}\right)\Delta f$$

and thus

$$\frac{\overline{i_{ia}^2}}{\Delta f} = 2q\left(5 + \frac{500}{|\beta|^2}\right) \times 10^{-6}\ \text{A}^2/\text{Hz} \tag{11.86}$$

Substitution of (11.86) in (11.85) gives

$$\frac{\overline{i_i^2}}{\Delta f} = 2q\left(5 + \frac{500}{|\beta|^2}\right) \times 10^{-6} + \frac{\overline{v_{ia}^2}}{(5500)^2\,\Delta f} + 2q(9.1) \times 10^{-6} \tag{11.87}$$

where the noise in the 5.5-kΩ resistor has been expressed in terms of the equivalent noise current of 9.1 μA.

Use of (11.50) gives

$$\frac{\overline{v_{ia}^2}}{\Delta f} = 4kT\left(r_{b1} + \frac{1}{2g_m}\right) = 4kT(126)$$

Division of this equation by $(5500)^2$ gives

$$\frac{\overline{v_{ia}^2}}{(5500)^2\,\Delta f} = 4kT\frac{1}{240{,}000} \tag{11.88}$$

$$= 2q(0.2) \times 10^{-6} \tag{11.89}$$

Thus the term involving $\overline{v_{ia}^2}$ in (11.87) is seen to be equivalent to thermal noise in a 240-kΩ resistor using (11.88), and this can be expressed as noise in 0.2 μA of equivalent noise current as shown in (11.89). This term is negligible in this example, as is usually the case.

Combining all these terms we can express (11.87) as

$$\frac{\overline{i_i^2}}{\Delta f} = 2q\left(5 + \frac{500}{|\beta|^2} + 0.2 + 9.1\right) \times 10^{-6}\ \text{A}^2/\text{Hz}$$

$$= 2q\left(14.3 + \frac{500}{|\beta|^2}\right) \times 10^{-6}\ \text{A}^2/\text{Hz} \tag{11.90}$$

Equation 11.90 shows that the equivalent input noise-current spectral density rises at high frequencies (as $|\beta|$ falls) as expected for a transistor. In a single transistor without feedback, the equivalent input noise current also rises with frequency, but, because the transistor gain falls with frequency, the output noise spectrum of a transistor without feedback always *falls* as frequency rises (see Section 11.4.1). However in this case, the negative feedback holds the gain constant with frequency and thus the *output* noise spectrum of this circuit will *rise* as frequency increases until the amplifier bandedge is reached. This is illustrated in Fig. 11.33 where the input noise-current spectrum, the amplifier frequency

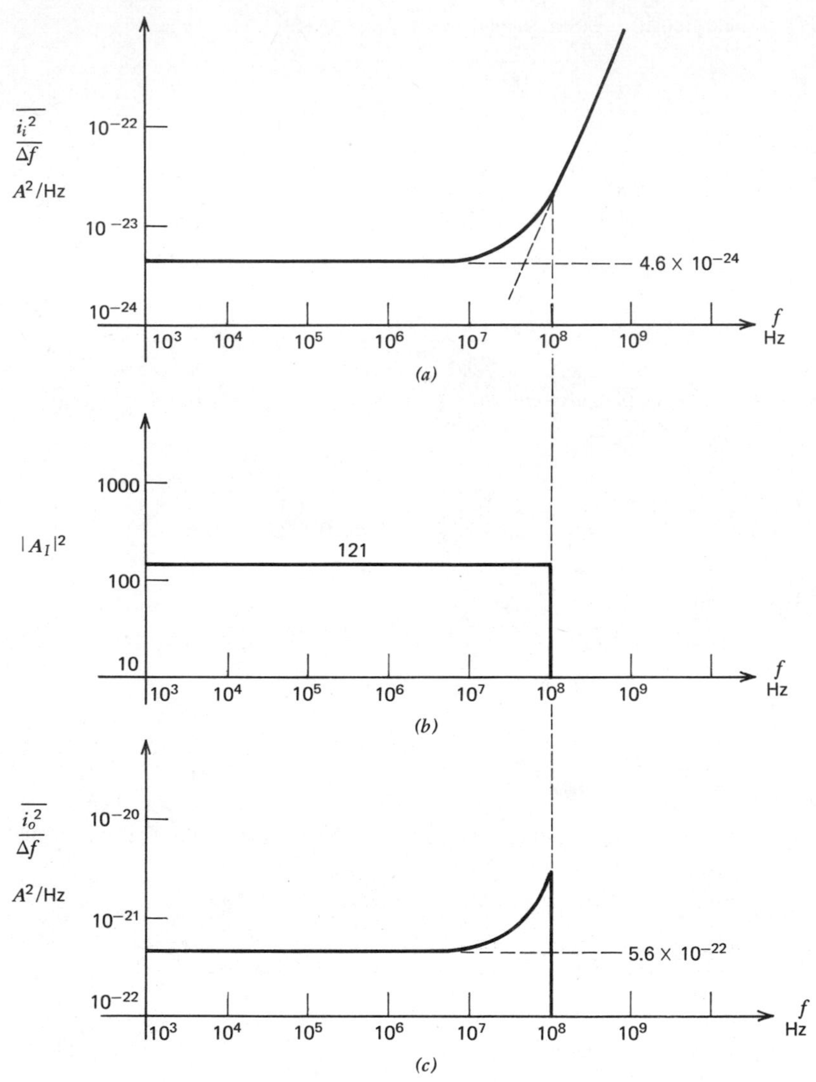

Fig. 11.33 Noise performance of the circuit of Fig. 11.32. (*a*) Equivalent input noise spectrum. (*b*) Frequency response squared. (*c*) Output noise spectrum.

response squared and the output noise-current spectrum (product of the first two) are shown. The current gain of the circuit is $A_I \simeq 11$.

The total output noise from the circuit $\overline{i_{oT}^2}$ is obtained by integrating the output noise spectral density which is

$$\frac{\overline{i_o^2}}{\Delta f} = A_I^2 \frac{\overline{i_i^2}}{\Delta f} \tag{11.91}$$

Thus

$$\overline{i_{oT}^2} = \int_0^B A_I^2 \frac{\overline{i_i^2}}{\Delta f} df$$

$$= \int_0^B 2q\left(14.3 + \frac{500}{|\beta(jf)|^2}\right) \times 10^{-6} df \tag{11.92}$$

where (11.90) has been used and A_I^2 is assumed constant up to $B = 10^8$ Hz as specified earlier. The current gain is

$$\beta(jf) = \frac{\beta_0}{1 + j\dfrac{\beta_0 f}{f_{T1}}} \tag{11.93}$$

and

$$\frac{1}{|\beta(jf)|^2} = \frac{1}{\beta_0^2}\left(1 + \frac{\beta_0^2 f^2}{f_{T1}^2}\right) \tag{11.94}$$

Substitution of (11.94) in (11.92) gives

$$\overline{i_{oT}^2} = A_I^2 2q \times 10^{-6} \int_0^B \left[14.3 + \frac{500}{\beta_0^2}\left(1 + \frac{\beta_0^2 f^2}{f_{T1}^2}\right)\right] df \tag{11.95}$$

$$= A_I^2 2q \times 10^{-6} \left[14.3f + \frac{500}{\beta_0^2} f + \frac{500}{f_{T1}^2}\frac{f^3}{3}\right]_0^B \tag{11.96}$$

Using $\beta_0 = 100$ and $B = 100$ MHz $= f_{T1}/3$ gives

$$\overline{i_{oT}^2} = A_I^2 \times 2q \times 10^{-6}(14.3B + 18.6B) \tag{11.97}$$

$$\overline{i_{oT}^2} = A_I^2 \times 1.05 \times 10^{-15} \text{ A}^2 \tag{11.98}$$

The equivalent input noise current is

$$\overline{i_{iT}^2} = \frac{\overline{i_{oT}^2}}{A_I^2} = 1.05 \times 10^{-15} \text{ A}^2$$

and from this

$$i_{iT} = 32.4 \text{ nA rms} \tag{11.99}$$

Thus the equivalent input noise current is 32.4 nA rms and (11.97) shows that the frequency-dependent part of the equivalent input noise is dominant. For a 20-dB signal-to-noise ratio, input signal current i_s must be greater than 0.32 μA rms.

11.7 NOISE PERFORMANCE OF OTHER TRANSISTOR CONFIGURATIONS

Transistor configurations other than the common-emitter stages considered so far are often used in integrated-circuit design. For example, the 741 operational amplifier has a differential input stage with emitter followers driving common-base stages. The noise performance of these types of configurations will now be considered.

11.7.1 Common-Base Stage Noise Performance

The common-base stage is sometimes used as a low-input-impedance current amplifier and, as mentioned above, is used as a level shift in the 741 op amp input stage. The noise performance of this circuit is thus of some interest.

A common-base stage is shown in Fig. 11.34*a* and the small-signal equivalent circuit is shown in Fig. 11.34*b* together with the equivalent input noise generators derived for a common-emitter stage. Since these noise generators represent the noise performance of the transistor in any connection, Fig. 11.34*b* is a valid representation of common-base noise performance. In Fig. 11.34*c* the noise performance of the common-base stage is represented in the standard fashion with equivalent input noise generators $\overline{v_{iB}^2}$ and $\overline{i_{iB}^2}$. These can be related to the common-emitter input generators by alternately short circuiting and open circuiting the circuits of Fig. 11.34*b* and *c* and equating output noise. It then follows that

$$\overline{i_{iB}^2} = \overline{i_i^2} \tag{11.100}$$

$$\overline{v_{iB}^2} = \overline{v_i^2} \tag{11.101}$$

Thus the equivalent input noise generators of common-emitter and common-base connections are the same and the noise performance of the two configurations is identical, even though their input impedances differ greatly.

Although the noise performance of common-emitter and common-base stages is nominally identical (for the same device parameters) there is one characteristic of the common-base stage that makes it generally *unsuitable* for use as a low-noise input stage. This is due to the fact that its current gain $\alpha \simeq 1$, and thus any noise current at the *output* of the common-base stage, is referred directly back to the input *without* reduction. Thus a 10-kΩ load resistor that has an equivalent noise current of 5 μA produces this amount of equivalent noise current at the input. In many circuits this would be the dominant source of input current noise. The equivalent input noise currents of following stages are also referred back unchanged to the input of the common-base stage. This problem can be overcome in discrete common-base circuits by use of a transformer that gives current gain at the output of the common-base stage. This option is not available in integrated-circuit design unless resort is made to external components.

11.7.2 Emitter-Follower Noise Performance

Consider the emitter follower shown in Fig. 11.35. The noise performance of this circuit can be calculated using the results of previous sections. The circuit can be viewed as a series-feedback stage and the equivalent input noise generators of the

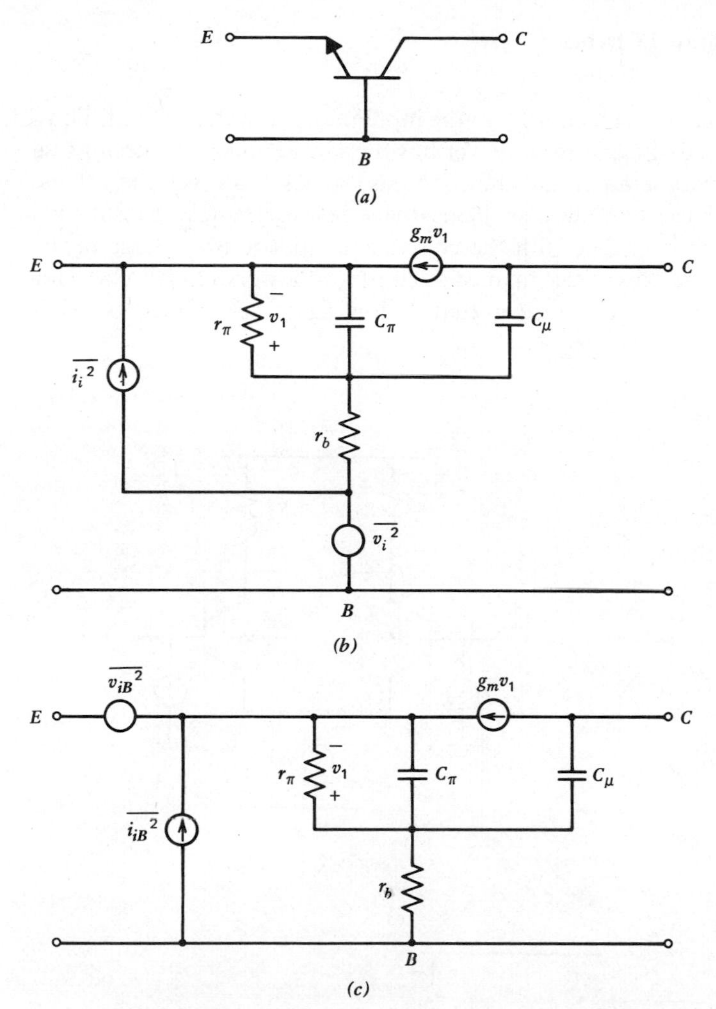

Fig. 11.34 (*a*) Common-base transistor configuration. (*b*) Common-base equivalent circuit with noise generators. (*c*) Common-base equivalent circuit with input noise generators.

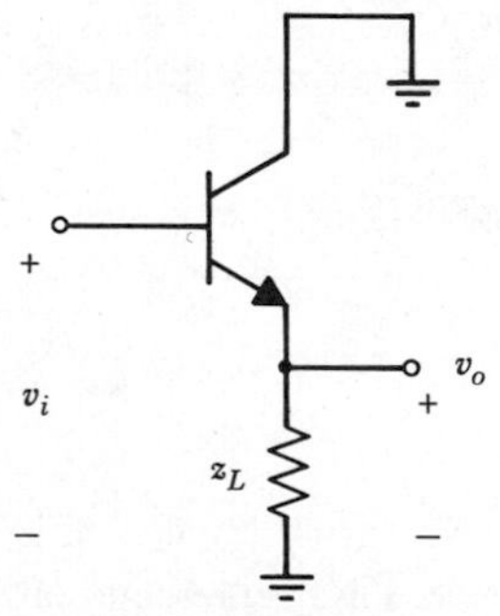

Fig. 11.35 Emitter-follower circuit.

transistor can be moved unchanged back to the input of the complete circuit. Thus, if noise in z_L is neglected, the emitter follower has the same equivalent input noise generators as the common-emitter and common-base stages. However, the voltage noise in load z_L (which can be viewed as the feedback resistor) is now transformed unchanged to the input, together with the equivalent input noise voltage of the following stage. Thus the noise performance of emitter followers tends to be poor because the unity voltage gain means that following stage voltage noise is important.

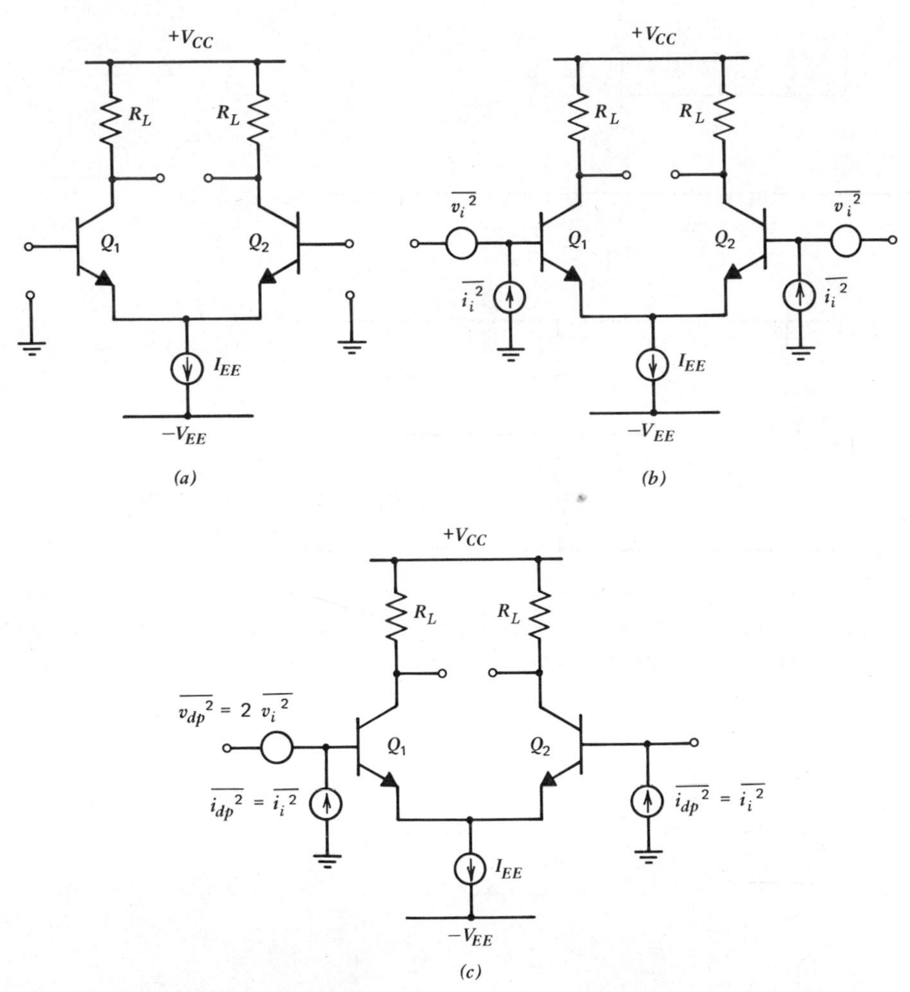

Fig. 11.36 (*a*) Differential-pair circuit. (*b*) Complete differential pair noise representation. (*c*) Simplified noise representation.

11.7.3 Differential-Pair Noise Performance

The differential pair is the basic building block of linear integrated circuits and, as such, its noise performance is of considerable importance. A differential pair is shown in Fig. 11.36*a* and the base of each device is generally independently accessible as shown. Thus this circuit cannot in general be represented as a two-port and its noise performance cannot be represented in the usual fashion by two input noise generators. However, the techniques developed previously can be used to derive an equivalent noise representation of the circuit that employs two noise generators at *each* input. This is illustrated in Fig. 11.36*b* and a simpler version of this circuit, which employs only three noise generators, is shown in Fig. 11.36*c*.

The noise representation of Fig. 11.36*b* can be derived by considering noise due to each device separately. Consider first noise in Q_1, which can be represented by input noise generators $\overline{v_i^2}$ and $\overline{i_i^2}$ as shown in Fig. 11.37*a*. These noise generators are those for a single transistor as given by (11.50) and (11.57). Transistor Q_2 is initially assumed *noiseless* and the impedance seen looking in its emitter is z_{E2}. Note that z_{E2} will be a function of the impedance connected from the base of Q_2 to ground. As described in previous sections, the noise generators of Fig. 11.37*a* can be moved unchanged to the input of the circuit (independent of z_{E2}) as shown in Fig. 11.37*b*. This representation can then be used to calculate the output noise produced by Q_1 in the differential pair for *any* impedances connected from the base of Q_1 and Q_2 to ground.

Now consider noise due to Q_2. In a similar fashion this can be represented by noise generators $\overline{v_i^2}$ and $\overline{i_i^2}$ as shown in Fig. 11.37*c*. In this case Q_1 is assumed

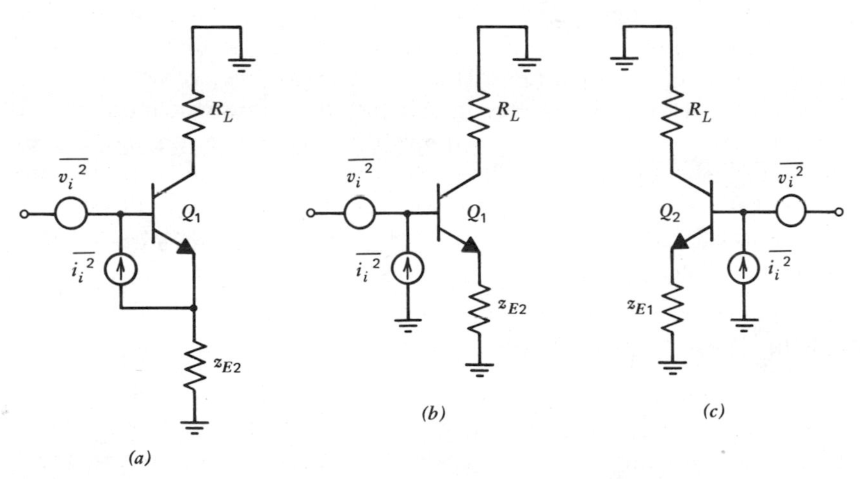

Fig. 11.37 (*a*) ac schematic of a differential pair including noise due to Q_1 only. (*b*) ac schematic of a differential pair with noise due to Q_1 referred to the input. (*c*) ac schematic of a differential pair including noise due to Q_2 only.

noiseless and z_{E1} is the impedance seen looking in at the emitter of Q_1. If Q_1 and Q_2 are identical, the equivalent input noise generators of Fig. 11.37*b* and *c* are identical. However, since they are produced by different transistors, the noise generators of Fig. 11.37*b* and *c* are *independent*. The total noise performance of the differential pair including noise due to both Q_1 and Q_2 can thus be represented as shown in Fig. 11.36*b*, and this representation is valid for *any* source resistance connected to either input terminal. Noise generators $\overline{v_i^2}$ and $\overline{i_i^2}$ are basically those due to each transistor alone. If noise due to R_L or following stages is significant, it should be referred back symmetrically to the appropriate input. In practice, current source I_{EE} will also contain noise and this can be included in the representation. However, if the circuit is perfectly balanced, the current-source noise represents a common-mode signal and will produce no differential output.

The noise representation of Fig. 11.36*b* can be simplified somewhat if the common-mode rejection of the circuit is high. In this case one of the noise-voltage generators can be moved to the other side of the circuit as shown in Fig. 11.36*c*. This can be justified if equal noise-voltage generators are added in series with each input and these generators are chosen such that the noise voltage at the base of Q_2 is canceled. This leaves two independent noise-voltage generators in series with the base of Q_1 and these can be represented as a single noise-voltage generator of value $\overline{2v_i^2}$. Thus for the circuit of Fig. 11.36*c* we can write

$$\overline{v_{dp}^2} = \overline{2v_i^2} \tag{11.102}$$

$$\overline{i_{dp}^2} = \overline{i_i^2} \tag{11.103}$$

where $\overline{v_{dp}^2}$ and $\overline{i_{dp}^2}$ are the equivalent input noise generators of the differential pair.

The differential pair is often operated with the base of Q_2 grounded, and in this case the noise-current generator at the base of Q_2 is short circuited. The noise performance of the circuit is then represented by the two noise generators connected to the base of Q_1 in Fig. 11.36*c*. In this case the equivalent input noise-current generator of the differential pair is simply that due to one transistor alone, whereas the equivalent input noise-voltage generator has a mean-square value *twice* that of either transistor. Thus from a low source impedance, a differential pair has an equivalent input noise voltage 3 dB higher than a common-emitter stage with the same collector current as the devices in the pair.

11.8 NOISE IN OPERATIONAL AMPLIFIERS

Integrated-circuit amplifiers designed for low-noise operation generally use a simple common-emitter or differential-pair input stage with resistive loads. Since the input stage has both current and voltage gain the noise of following stages is generally not significant, and the resistive loads make only a small noise contribution. The noise analysis of such circuits is quite straightforward using the techniques described in this chapter. However, circuits of this type (the 725 op amp is an example) are inefficient in terms of optimizing important op amp parameters such as gain and bandwidth. For example, using active loads as in the 741 allows realization of very high gain in relatively few stages, and this is a significant advantage in circuit design. However, by their very nature, active loads amplify their own internal noise and cause considerable degradation of circuit noise performance. An approximate noise analysis of the 741 will now be made to illustrate these points and show the compromises involved in the design of general-purpose circuits.

A simplified schematic of the input stage of the 741 is shown in Fig. 11.38*a*. Transistor Q_5 may be considered to be diode connected as shown. Components Q_5, Q_6, R_1, and R_3 of the active load generate noise and contribute to the output noise at i_o. Since transistors Q_3 and Q_4 present a high impedance to the active load, the noise due to the active load can be calculated from the isolated portion of the circuit shown in Fig. 11.38*b*. Noise due to Q_6 and R_3 is represented by equivalent input noise generators $\overline{v_{i6}^2}$ and $\overline{i_{i6}^2}$. Since the diode Q_5 and resistor R_1 present a relatively low impedance at the base of Q_6, $\overline{i_{i6}^2}$ may be neglected. Using the results of Section 11.6.2 and neglecting flicker noise we have

$$\frac{\overline{v_{i6}^2}}{\Delta f} = 4kT\left(r_{b6} + \frac{1}{2g_{m6}} + R_3\right) \tag{11.104}$$

The noise current due to diode Q_5 (using Section 11.2.1) is

$$\frac{\overline{i_5^2}}{\Delta f} = 2qI_{C5} \tag{11.105}$$

Since the diode small-signal impedance is $r_d = kT/qI_{C5} = 1/g_{m5}$ the diode noise can be transformed to a Thévenin equivalent voltage in series with the diode

$$\frac{\overline{v_5^2}}{\Delta f} = 2qI_{C5}\frac{1}{g_{m5}^2} = 4kT\frac{1}{2g_{m5}} \tag{11.106}$$

There will also be a noise voltage in series with Q_5 due to its base resistance

$$\frac{\overline{v_{b5}^2}}{\Delta f} = 4kTr_{b5} \tag{11.107}$$

Finally the noise due to R_1 is

$$\frac{\overline{v_1^2}}{\Delta f} = 4kTR_1 \tag{11.108}$$

Combining (11.106), (11.107), (11.108), and (11.104) into one noise-voltage generator $\overline{v_A^2}$ in series with the base of Q_6, we obtain

$$\frac{\overline{v_A^2}}{\Delta f} = 4kT\left(r_{b6} + \frac{1}{2g_{m6}} + R_3 + \frac{1}{2g_{m5}} + r_{b5} + R_1\right) \tag{11.109}$$

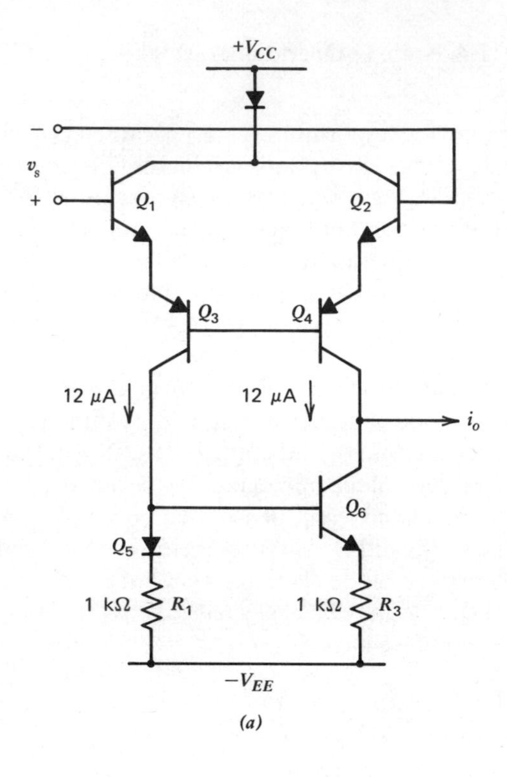

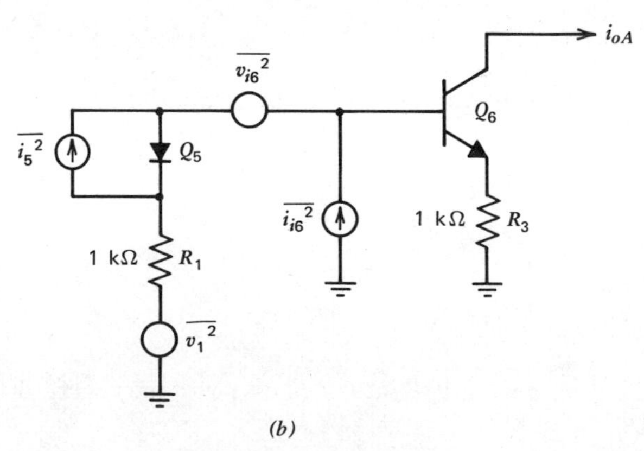

Fig. 11.38 (*a*) Simplified schematic of the 741 op amp input stage. (*b*) ac schematic (including noise) of the active load of the 741 op amp.

This noise generator can be evaluated as follows. The bias current in Q_5 and Q_6 is approximately 12 μA giving $1/2g_{m5} = 1/2g_{m6} = 1.08$ kΩ. Also $R_1 = R_3 = 1$ kΩ and a typical value of r_b for small devices at low currents is 500 Ω. Thus (11.109) becomes

$$\frac{\overline{v_A^2}}{\Delta f} = 4kT(5160) \tag{11.110}$$

This produces a noise output i_{oA} where

$$i_{oA} \simeq \frac{1}{\dfrac{1}{g_{m6}} + R_3} v_A \simeq \frac{v_A}{3160} \tag{11.111}$$

Using (11.110) in (11.111) gives

$$\frac{i_{oA}^2}{\Delta f} = 4kT\frac{5160}{3160^2} \tag{11.112}$$

This can be referred back to the input of the complete circuit of Fig. 11.38*a* in the standard manner. Consider the contribution to the equivalent input noise voltage. A voltage applied at the input of the full circuit of Fig. 11.38*a* gives

$$\frac{i_o}{v_i} = \frac{g_{m1}}{2}$$

and thus

$$\overline{i_o^2} = \frac{g_{m1}^2}{4}\overline{v_i^2} \tag{11.113}$$

Equating output noise current in (11.112) and (11.113) we obtain an equivalent input noise voltage due to $\overline{i_{oA}^2}$ as

$$\frac{\overline{v_{iA}^2}}{\Delta f} = 4kT\frac{5160}{3160^2}\frac{4}{g_{m1}^2} = 4kT\frac{5160}{3160^2}4 \times 2160^2 = 4kT(9640) \tag{11.114}$$

Thus the active load causes a contribution of 9.64 kΩ to the equivalent input noise resistance of the 741 op amp. The remaining important contributions can be written down by inspecting Fig. 11.38*a*. The equivalent input noise voltages of Q_3 and Q_4 at their own emitters were shown in Section 11.7.1 to be that of a common-emitter stage, and the results of Section 11.7.2 show that this noise can then be transformed unchanged to the input of the complete circuit because emitter followers Q_1 and Q_2 have unity voltage gain. The equivalent input noise currents of Q_3 and Q_4 at their own emitters have little influence because the source impedance seen by these devices looking into the emitters of Q_1 and Q_2 is fairly low. Thus these contributions are neglected in this approximate analysis.

Finally, the results of Section 11.7.3 show that the equivalent input noise voltage squared of the differential input stage is just the sum of the values of each half of the circuit. This gives an input noise-voltage contribution due to Q_1–Q_4 of

$$\frac{\overline{v_{iB}^2}}{\Delta f} = 4kT\left(\frac{1}{2g_{m1}} + \frac{1}{2g_{m2}} + \frac{1}{2g_{m3}} + \frac{1}{2g_{m4}} + r_{b1} + r_{b2} + r_{b3} + r_{b4}\right) \quad (11.115)$$

Assuming a collector bias current of 12 μA for each device, and $r_b = 500\ \Omega$ in (11.115), we calculate

$$\frac{\overline{v_{iB}^2}}{\Delta f} = 4kT(6320) \quad (11.116)$$

Combining (11.114) and (11.116) gives, for the total input noise voltage for the 741,

$$\frac{\overline{v_i^2}}{\Delta f} = 4kT(16{,}000) \quad (11.117)$$

Thus the input noise voltage of the 741 is represented by an equivalent input noise resistance of 16 kΩ. This is a large value and is very close to the measured and computer-calculated result. Note that the active load is the main contributor to the noise. The magnitude of the noise voltage can be appreciated by noting that if a 741 is fed from a 1-kΩ source resistance for example, the circuit adds 16 times as much noise power as is contributed by the source resistance itself.

The calculations above concerned the equivalent input noise voltage of the 741. A similar calculation performed to determine the equivalent input noise current shows that this is dominated by the base current of the input device. The current gain of the input emitter followers is sufficient to ensure that following stage noise, including noise in the active loads, gives a negligible contribution to input current noise. Since the circuit is differential, the complete noise representation consists of the equivalent input noise voltage calculated above plus two equivalent input noise-current generators as shown in Fig. 11.39. This follows from the discussion of differential-pair noise performance in Section 11.7.3. The equivalent

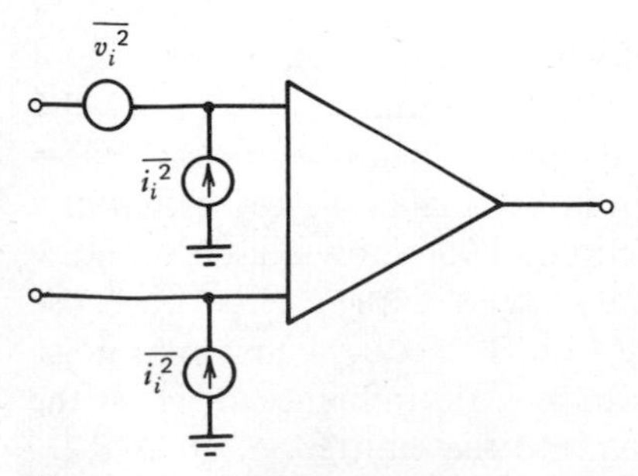

Fig. 11.39 Complete op amp noise representation.

input noise-current generators are (neglecting flicker noise)

$$\frac{\overline{i_i^2}}{\Delta f} \simeq 2qI_B$$

where I_B is the base current of Q_1 or Q_2 in Fig. 11.38*a*. Substitution of typical data gives

$$\frac{\overline{i_i^2}}{\Delta f} \simeq 2q(0.2 \times 10^{-6}) \tag{11.118}$$

In Chapter 6, a number of special op amp configurations were described and the noise performance of these can now be considered. Three techniques were described for the reduction of the input bias current of an op amp, and these result in significantly different noise performance. First, the bias-current cancellation scheme of Fig. 6.35 functions by injecting input bias currents from Q_5 and Q_6 to cancel the base currents of Q_1 and Q_2. However the collector currents of Q_5 and Q_6 contain noise that is *uncorrelated* with the base-current noise of Q_1 and Q_2 and thus the total input noise current is *increased* by this arrangement and the noise performance is actually *degraded.* The equivalent input noise voltage is essentially unaffected and is determined by the bias currents in Q_1 and Q_2 since there is no active load on the input stage of this circuit.

The other two circuits described for the reduction of the input bias current were the use of super-β input devices as in Fig. 6.36 and the use of JFET input devices as in Fig. 6.38. Both these techniques result in significantly *reduced* input noise current and thus *improved* noise performance. The equivalent input noise voltage in these circuits depends on the bias currents of the input devices and the characteristics of the active load.

11.9 NOISE BANDWIDTH

In the noise analysis performed so far, the circuits considered were generally assumed to have simple gain-frequency characteristics with abrupt band edges as shown in Fig. 11.33*b*. The calculation of total circuit noise then reduced to an integration of the noise spectral density across this band. In practice, many circuits do not have such ideal gain-frequency characteristics and the calculation of total circuit noise can be much more complex in those cases. However, if the equivalent input noise spectral density of a circuit is *constant* and independent of frequency (i.e., if the noise is white), we can simplify the calculations using the concept of *noise bandwidth* described below.

Consider an amplifier as shown in Fig. 11.40, and assume it is fed from a low source impedance so that the equivalent input noise voltage $\overline{v_i^2}$ determines the noise performance. Assume initially that the spectral density $\overline{v_i^2}/\Delta f = S_i(f) = S_{i0}$ of the input noise voltage is flat as shown in Fig. 11.41*a*. Further assume that the

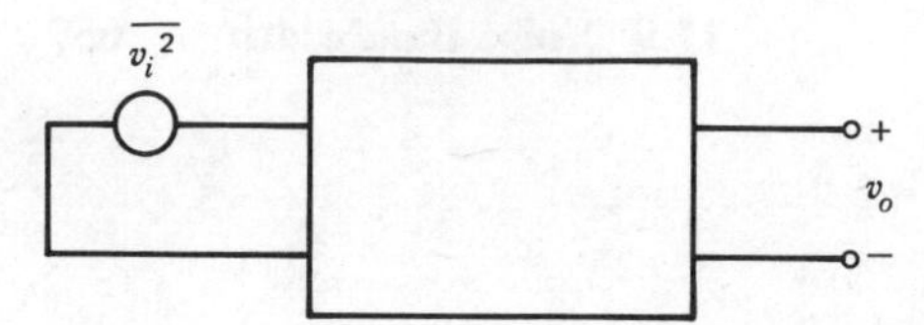

Fig. 11.40 Circuit with equivalent input noise voltage generator.

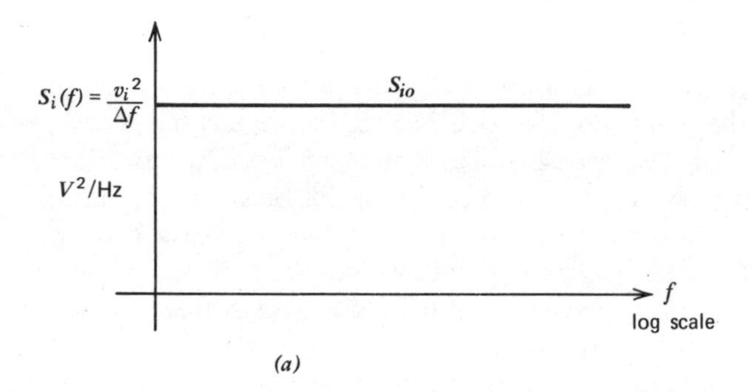

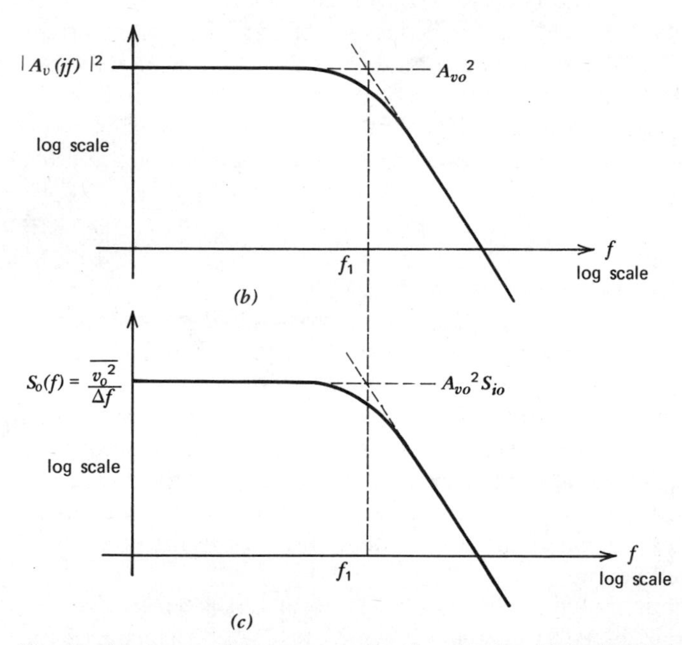

Fig. 11.41 Assumed parameters for the circuit of Fig. 11.40. (*a*) Equivalent input noise-voltage spectral density. (*b*) Circuit transfer function squared. (*c*) Output noise-voltage spectral density.

magnitude squared of the voltage gain $|A_v(jf)|^2$ of the circuit is as shown in Fig. 11.41*b*. The output noise-voltage spectral density $S_o(f) = \overline{v_o^2}/\Delta f$ is the product of the input noise-voltage spectral density and the square of the voltage gain and is shown in Fig. 11.41*c*. The *total* output noise voltage is obtained by summing the contribution from $S_o(f)$ in each frequency increment Δf between zero and infinity to give

$$\overline{v_{oT}^2} = \sum_{f=0}^{\infty} S_o(f)\,\Delta f = \int_0^{\infty} S_o(f)\,df = \int_0^{\infty} |A_v(jf)|^2 S_{i0}\,df$$

$$= S_{i0} \int_0^{\infty} |A_v(jf)|^2\,df \tag{11.119}$$

The evaluation of the integral of (11.119) is often difficult except for very simple transfer functions. However if the problem is transformed into a *normalized* form, the integrals of common circuit functions can be evaluated and tabulated for use in noise calculations. For this purpose, consider a transfer function as shown in Fig. 11.42 with the same low-frequency value A_{v0} as the original circuit but with an abrupt band edge at a frequency f_N. Frequency f_N is chosen to give the *same* total output noise voltage as the original circuit when the same input noise voltage is applied. Thus

$$\overline{v_{oT}^2} = S_{i0} A_{v0}^2 f_N \tag{11.120}$$

If (11.119) and (11.120) are equated we obtain

$$f_N = \frac{1}{A_{v0}^2} \int_0^{\infty} |A_v(jf)|^2\,df \tag{11.121}$$

where f_N is the *equivalent noise bandwidth* of the circuit. Although derived for the case of a voltage transfer function, this result can be used for any type of transfer function. Note that the integration of (11.121) can be performed numerically if measured data for the circuit transfer function is available.

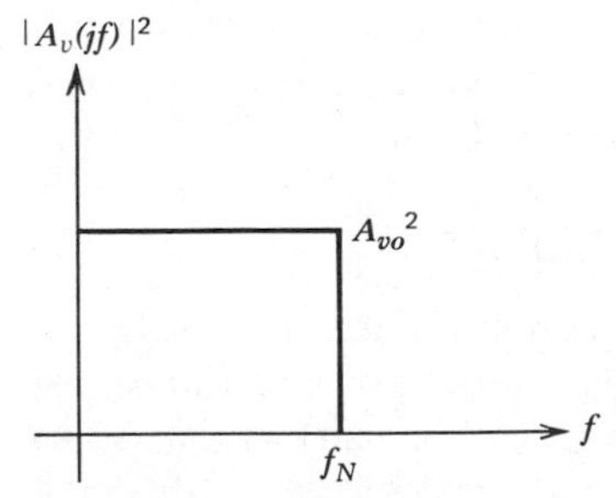

Fig. 11.42 Transfer function of a circuit giving the same output noise as a circuit with a transfer function as specified in Fig. 11.41*b*.

Once the noise bandwidth is evaluated using (11.121) the total output noise of the circuit is readily calculated using (11.120). The advantage of the form of (11.121) is that the circuit gain is *normalized* to its low-frequency value and thus the calculation of f_N concerns only the frequency response of the circuit. This can be done in a general way so that whole classes of circuits are covered by one calculation. For example, consider an amplifier with a single-pole frequency response given by

$$A_v(jf) = \frac{A_{v0}}{1 + j\dfrac{f}{f_1}} \tag{11.122}$$

where f_1 is the -3-dB frequency. The noise bandwidth of this circuit can be calculated from (11.121) as

$$f_N = \int_0^\infty \frac{df}{1 + \left(\dfrac{f}{f_1}\right)^2} = \frac{\pi}{2} f_1 = 1.57 f_1 \tag{11.123}$$

This gives the noise bandwidth of *any* single-pole circuit and shows that it is larger than the -3-dB bandwidth by a factor of 1.57. Thus a circuit with the transfer function of (11.122) produces noise *as if* it had an abrupt band edge at a frequency 1.57 f_1.

As the steepness of the transfer function falloff with frequency becomes greater, the noise bandwidth approaches the -3-dB bandwidth. For example, a two-pole transfer function with complex poles at 45° to the negative real axis has a noise bandwidth only 11 percent greater than the -3-dB bandwidth.

EXAMPLE

As an example of noise bandwidth calculations, suppose a 741 op amp is used in a feedback configuration with a low-frequency gain of $A_{v0} = 100$ and it is desired to calculate the total output noise v_{oT} from the circuit with a zero source impedance and neglecting flicker noise. If the unity-gain bandwidth of the op amp is 1.5 MHz, then the transfer function in a gain of 100 configuration will have a -3-dB frequency of 15 kHz with a single-pole response. From (11.123) the noise bandwidth is

$$f_N = 1.57 \times 15 \text{ kHz} = 23.5 \text{ kHz} \tag{11.124}$$

Assuming that the circuit is fed from a zero source impedance, and using the previously calculated value of 16 kΩ as the equivalent input noise resistance, we can calculate the low-frequency input noise voltage spectral density of the 741 as

$$S_{i0} = \frac{\overline{v_i^2}}{\Delta f} = 4kT(16{,}000) = 2.66 \times 10^{-16} \text{ V}^2/\text{Hz} \tag{11.125}$$

Using $A_{v0} = 100$ together with substitution of (11.124) and (11.125) in (11.120) gives, for the total output noise voltage,

$$\overline{v_{oT}^2} = 2.66 \times 10^{-16} \times 10^4 \times 23.5 \times 10^3 \text{ V}^2 = 6.25 \times 10^{-10} \text{ V}^2$$

and thus $v_{oT} = 25\ \mu\text{V}$

The calculations of noise bandwidth considered above were based on the assumption of a flat input noise spectrum. This is often true in practice and the concept of noise bandwidth is useful in those cases, but there are also many examples where the input noise spectrum varies with frequency. In these cases, the total output noise voltage is given by

$$\overline{v_{oT}^2} = \int_0^\infty S_o(f)\, df \tag{11.126}$$

$$= \int_0^\infty |A_v(jf)|^2 S_i(f)\, df \tag{11.127}$$

where $A_v(jf)$ is the voltage gain of the circuit and $S_i(f)$ is the input noise-voltage spectral density. If the circuit has a voltage gain A_{vS} at the frequency of the applied input signal, then the total equivalent input noise voltage becomes

$$\overline{v_{iT}^2} = \frac{1}{A_{vS}^2} \int_0^\infty |A_v(jf)|^2 S_i(f)\, df \tag{11.128}$$

$$= \int_0^\infty \left| \frac{A_v(jf)}{A_{vS}} \right|^2 S_i(f)\, df \tag{11.129}$$

Equation 11.129 shows that, in general, the total equivalent input noise voltage of a circuit is obtained by integrating the product of the input noise-voltage spectrum and the normalized voltage gain function.

One last topic that should be mentioned in this section is the problem that occurs in calculating the flicker noise in direct-coupled amplifiers. Consider an amplifier with an input noise spectral density as shown in Fig. 11.43*a* and a voltage gain that extends down to dc and up to $f_1 = 10$ kHz with an abrupt cutoff, as shown in Fig. 11.43*b*. Then, using (11.129), we can calculate the total equivalent input noise voltage as

$$\overline{v_{iT}^2} = \int_0^{f_1} S_i(f)\, df$$

$$= \int_0^{f_1} \left(1 + \frac{1000}{f}\right) \times 10^{-16}\, df$$

$$= 10^{-16}[f + 1000 \ln f]_0^{f_1} \tag{11.130}$$

Evaluating (11.130) produces a problem since $\overline{v_{iT}^2}$ is infinite when a lower limit of zero is used on the integration. This suggests infinite power in the $1/f$ noise

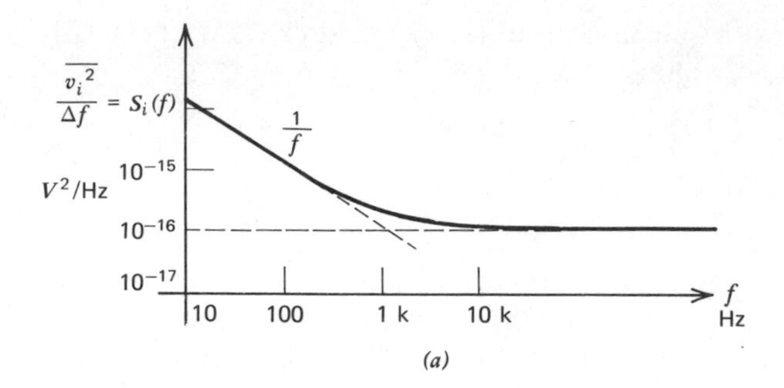

(a)

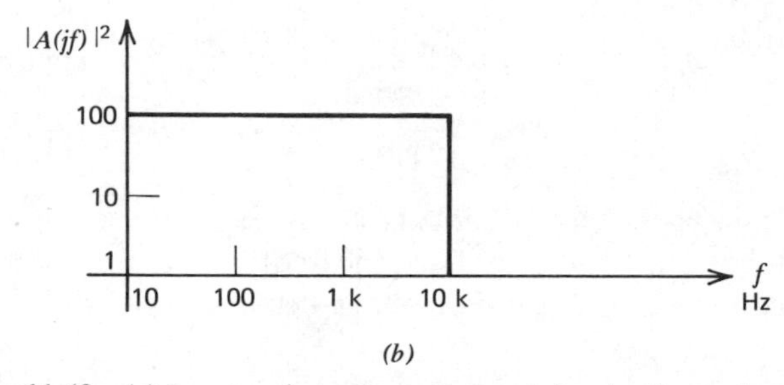

(b)

Fig. 11.43 (*a*) Input noise-voltage spectral density for a circuit. (*b*) Circuit transfer function squared.

signal. In practice, measurements of $1/f$ noise spectra show a continued $1/f$ dependence to as low a frequency as is measured (cycles per day or less). This problem can only be resolved by noting that $1/f$ noise eventually becomes indistinguishable from thermal drift and that the lower limit of the integration must be specified by the period of observation. For example, taking a lower limit to the integration of $f_2 = 1$ cycle/day we have $f_2 = 1.16 \times 10^{-5}$ Hz. Changing the limit in (11.130) we find

$$\overline{v_{iT}^2} = 10^{-16}[f + 1000 \ln f]_{f_2}^{f_1}$$

$$= 10^{-16}\left[(f_1 - f_2) + 1000 \ln \frac{f_1}{f_2}\right] \tag{11.131}$$

Using $f_1 = 10$ kHz and $f_2 = 1.16 \times 10^{-5}$ Hz in (11.131) gives

$$\overline{v_{iT}^2} = 10^{-16}(10{,}000 + 20{,}600)$$
$$= 3.06 \times 10^{-12} \text{ V}^2$$

and thus

$$v_{iT} = 1.75\ \mu\text{V}$$

If the lower limit of integration is changed to 1 cycle/year = 3.2×10^{-8} Hz then (11.131) becomes

$$\overline{v_{iT}^2} = 10^{-16}(10{,}000 + 26{,}500)$$
$$= 3.65 \times 10^{-12}\ \text{V}^2$$

and thus

$$v_{iT} = 1.9\ \mu\text{V}$$

The noise voltage changes very slowly as f_2 is reduced further because of the ln function in (11.131).

11.10 NOISE FIGURE AND NOISE TEMPERATURE

11.10.1 Noise Figure

The most general method of specifying the noise performance of circuits is by specifying input noise generators as described above. However, a number of specialized methods of specifying noise performance have been developed which are convenient in particular situations. Two of these methods are now described.

The *noise figure* (F) is a commonly used method of specifying the noise performance of a circuit or a device. Its disadvantage is that it is limited to situations where the source impedance is resistive, and this precludes its use in many applications where noise performance is important. However, it is widely used as a measure of noise performance in communication systems where the source impedance is often resistive.

The definition of the noise figure of a circuit is

$$F = \frac{\text{input } S/N \text{ ratio}}{\text{output } S/N \text{ ratio}} \tag{11.132}$$

and F is usually expressed in decibels. The utility of the noise-figure concept is apparent from the definition, as it gives a direct measure of the signal-to-noise (S/N) ratio degradation that is caused by the circuit. For example, if the S/N ratio at the input to a circuit is 50 dB, and the circuit noise figure is 5 dB, then the S/N ratio at the output of the circuit is 45 dB.

Consider a circuit as shown in Fig. 11.44, where S represents signal power and N represents noise power. The input noise power N_i is always taken as the

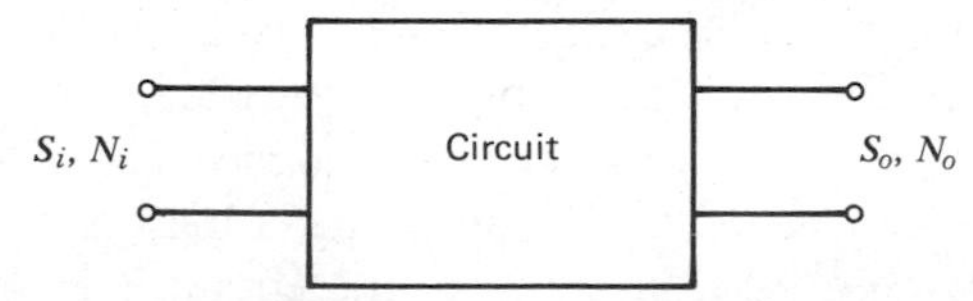

Fig. 11.44 Signal and noise power at the input and output of a circuit.

noise in the *source resistance.* The output noise power N_o is the total output noise including the circuit contribution and noise transmitted from the source resistance. From (11.132) the noise figure is

$$F = \frac{S_i}{N_i}\frac{N_o}{S_o} \tag{11.133}$$

For an *ideal noiseless amplifier*, all output noise comes from the source resistance at the input, and thus if G is the circuit power gain then the output signal S_o and the output noise N_o are given by

$$S_o = GS_i \tag{11.134}$$

$$N_o = GN_i \tag{11.135}$$

Substituting (11.134) and (11.135) in (11.133) gives $F = 1$ or 0 dB in this case.

A useful alternative definition of F may be derived from (11.133) as follows:

$$F = \frac{S_i}{N_i}\frac{N_o}{S_o} = \frac{N_o}{GN_i} \tag{11.136}$$

Equation 11.136 can be written as

$$F = \frac{\text{Total output noise}}{\text{That part of the output noise due to the source resistance}} \tag{11.137}$$

Note that since F is specified by a *power* ratio, the value in decibels is given by $10 \log_{10}$ (numerical ratio).

The calculations of the previous sections have shown that the noise parameters of most circuits vary with frequency, and thus the bandwidth must be specified when the noise figure of a circuit is calculated. The noise figure is often specified for a *small bandwidth* Δf at a frequency f where $\Delta f \ll f$. This is called the *spot noise figure* and applies to tuned amplifiers and also to broadband amplifiers that may be followed by frequency-selective circuits. For broadband amplifiers whose output is utilized over a wide bandwidth, an *average* noise figure is often specified. This requires calculation of the total output noise over the frequency band of interest using the methods described in previous sections.

In many cases, the most convenient way to calculate noise figure is to return to the original equivalent circuit of the device with its basic noise generators to perform the calculation. However, some insight into the effect of circuit parameters on noise figure can be obtained by using the equivalent input noise generator representation of Fig. 11.45. In this figure, a circuit with input impedance z_i and voltage gain $G = v_o/v_x$ is fed from a source resistance R_S and drives a load R_L. The source resistance shows thermal noise $\overline{i_s^2}$, and the noise of the circuit itself is represented by $\overline{i_i^2}$ and $\overline{v_i^2}$, assumed uncorrelated.

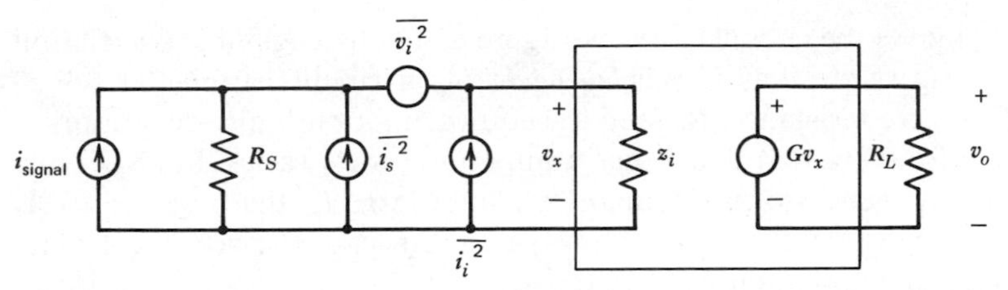

Fig. 11.45 Equivalent input noise representation for the calculation of noise figure.

The noise at the input terminals due to $\overline{v_i^2}$ and $\overline{i_i^2}$ is

$$v_{xA} = v_i \frac{z_i}{z_i + R_S} + i_i \frac{R_S z_i}{R_S + z_i}$$

and thus

$$\overline{v_{xA}^2} = \overline{v_i^2} \frac{|z_i|^2}{|z_i + R_S|^2} + \overline{i_i^2} \frac{|R_S z_i|^2}{|R_S + z_i|^2} \tag{11.138}$$

The noise power in R_L produced by $\overline{v_i^2}$ and $\overline{i_i^2}$ is

$$N_{oA} = \frac{|G|^2}{R_L} \overline{v_{xA}^2} = \frac{|G|^2}{R_L} \left(\overline{v_i^2} \frac{|z_i|^2}{|z_i + R_S|^2} + \overline{i_i^2} \frac{|R_S z_i|^2}{|R_S + z_i|^2} \right) \tag{11.139}$$

The noise power in R_L produced by source resistance noise generator $\overline{i_s^2}$ is

$$N_{oB} = \frac{|G|^2}{R_L} \frac{|R_S z_i|^2}{|R_S + z_i|^2} \overline{i_s^2} \tag{11.140}$$

The noise in the source resistance in a narrow bandwidth Δf is

$$\overline{i_s^2} = 4kT \frac{1}{R_S} \Delta f \tag{11.141}$$

Substituting (11.141) in (11.140) gives

$$N_{oB} = \frac{|G|^2}{R_L} \frac{|R_S z_i|^2}{|R_S + z_i|^2} 4kT \frac{1}{R_S} \Delta f \tag{11.142}$$

Using the definition of noise figure in (11.137) and substituting from (11.142) and (11.139) we find

$$F = \frac{N_{oA} + N_{oB}}{N_{oB}}$$

$$= 1 + \frac{N_{oA}}{N_{oB}} \tag{11.143}$$

$$= 1 + \frac{\overline{v_i^2}}{4kTR_S \Delta f} + \frac{\overline{i_i^2}}{4kT \dfrac{1}{R_S} \Delta f} \tag{11.144}$$

Equation 11.144 gives the circuit *spot* noise figure assuming negligible correlation between $\overline{v_i^2}$ and $\overline{i_i^2}$. Note that F is *independent* of all circuit parameters (G, z_i, R_L) except the source resistance, R_S, and the equivalent input noise generators.

It is apparent from (11.144) that F has a minimum as R_S varies. For very low values of R_S, the $\overline{v_i^2}$ generator is dominant while for large R_S the $\overline{i_i^2}$ generator is the most important. By differentiating (11.144) with respect to R_S, we can calculate the value of R_S giving minimum F:

$$R_{S\,\mathrm{opt}}^2 = \frac{\overline{v_i^2}}{\overline{i_i^2}} \tag{11.145}$$

This result is true in general, even if correlation is significant. A graph of F in decibels versus R_S is shown in Fig. 11.46.

The existence of a minimum in F as R_S is varied is one reason for the widespread use of transformers at the input of low-noise tuned amplifiers. This technique allows the source impedance to be transformed to the value giving lowest noise figure, while at the same time causing minimal loss in the circuit.

For example, consider the noise figure of a bipolar transistor at low-to-moderate frequencies where both flicker noise and high-frequency effects are neglected. From (11.50) and (11.57),

$$\overline{v_i^2} = 4kT\left(r_b + \frac{1}{2g_m}\right)\Delta f$$

$$\overline{i_i^2} = 2qI_B\,\Delta f = 2q\frac{I_C}{\beta_F}\,\Delta f$$

Substitution of these values in (11.145) gives

$$R_{S\,\mathrm{opt}} = \frac{\sqrt{\beta_F}}{g_m}\sqrt{1 + 2g_m r_b} \tag{11.146}$$

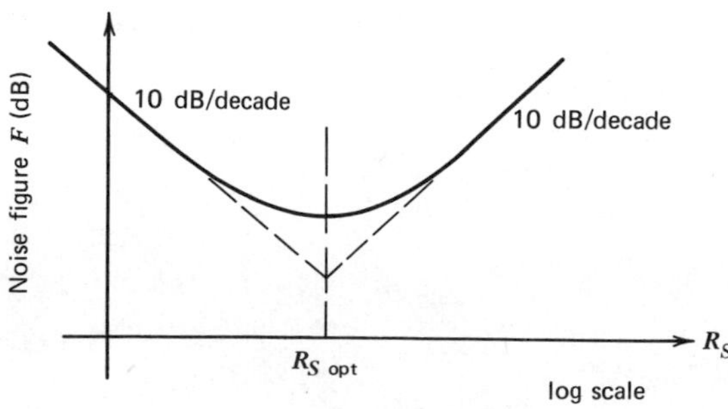

Fig. 11.46 Variation in noise figure F with source resistance R_S.

At this value of R_S, the noise figure is given by (11.144) as

$$F_{opt} \simeq 1 + \frac{1}{\sqrt{\beta_F}}\sqrt{1 + 2g_m r_b} \tag{11.147}$$

At a collector current of $I_C = 1$ mA, and with $\beta_F = 100$ and $r_b = 50$, (11.146) gives $R_{S\,opt} = 572\ \Omega$ and $F_{opt} = 1.22$. In decibels the value is $10 \log_{10} 1.22 = 0.9$ dB. Note that F_{opt} decreases as β_F increases and as r_b and g_m decrease. However, increasing β_F and decreasing g_m result in an *increasing* value of $R_{S\,opt}$ and this may prove difficult to realize in practice.

As another example, consider the FET at low frequencies. Neglecting flicker noise, we can calculate the equivalent input generators from Section 11.5.2 as

$$\overline{v_i^2} \simeq 4kT\,\tfrac{2}{3}\frac{1}{g_m}\,\Delta f \tag{11.148}$$

$$\overline{i_i^2} \simeq 0 \tag{11.149}$$

Using these values in (11.144) and (11.145), we find that $R_{S\,opt} \to \infty$ and $F_{opt} \to 0$ dB. Thus the FET has excellent noise performance from a high source resistance. However, if the source resistance is low (kilohms or less) and transformers cannot be used, the noise figure for the FET may be worse than for a bipolar transistor. For source resistances of the order of megohms or higher, the FET usually has significantly *lower* noise figure than a bipolar transistor.

11.10.2 Noise Temperature

Noise temperature is an alternative noise representation and is closely related to noise figure. The noise temperature, T_n, of a circuit is defined as the temperature at which the source resistance, R_S, must be held so that the noise output from the circuit due to R_S equals the noise output due to the circuit itself. If these conditions are applied to the circuit of Fig. 11.45, the output noise N_{oA} due to the circuit itself is unchanged but the output noise due to the source resistance becomes

$$N'_{oB} = \frac{|G|^2}{R_L}\frac{|R_S z_i|^2}{|R_S + z_i|^2}\,4kT_n\frac{1}{R_S}\,\Delta f \tag{11.150}$$

Substituting for N_{oB} from (11.142) in (11.150) we obtain

$$N'_{oB} = N_{oB}\frac{T_n}{T} \tag{11.151}$$

where T is the circuit temperature at which the noise performance is specified (usually taken as 290°K). Substituting (11.151) in (11.143) gives

$$\frac{T_n}{T} = (F - 1) \tag{11.152}$$

where F is specified as a ratio and is *not* in decibels.

Thus noise temperature and noise figure are directly related. The main application of noise temperature provides a convenient expanded measure of noise performance near $F = 1$ for very low-noise amplifiers. A noise figure of $F = 2$(3 dB) corresponds to $T_n = 290°\text{K}$ and $F = 1.1$ (0.4 dB) corresponds to $T_n = 29°\text{K}$.

PROBLEMS

11.1 Calculate the noise-voltage spectral density in V^2/Hz at v_o for the circuit in Fig. 11.47, and thus calculate the total noise in a 100-kHz bandwidth. Neglect capacitive effects, flicker noise, and series resistance in the diode.

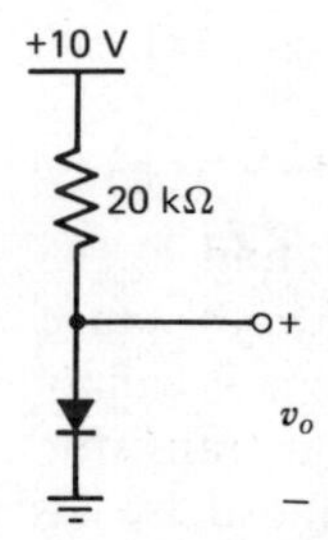

Fig. 11.47 Diode circuit.

11.2 If the diode in Fig. 11.47 shows flicker noise, calculate and plot the output noise voltage spectral density at v_o in V^2/Hz on log scales from $f = 1$ Hz to $f = 10$ MHz. Flicker noise data: In (11.7) use $a = b = 1$, $K_1 = 3 \times 10^{-16}$ A.

11.3 Repeat Problem 11.2 if a 1000-pF capacitor is connected across the diode.

11.4 The ac schematic of an amplifier is shown in Fig. 11.48. The circuit is fed from a current source i_s and data are as follows:

$$R_S = 1\ \text{k}\Omega \qquad R_L = 10\ \text{k}\Omega \qquad I_C = 1\ \text{mA} \qquad \beta = 50 \qquad r_b = 0 \qquad r_o = \infty$$

Neglecting capacitive effects and flicker noise, calculate the total noise voltage spectral density at v_o in V^2/Hz. Thus calculate the MDS at i_s if the circuit bandwidth is limited to a sharp cut-off at 2 MHz.

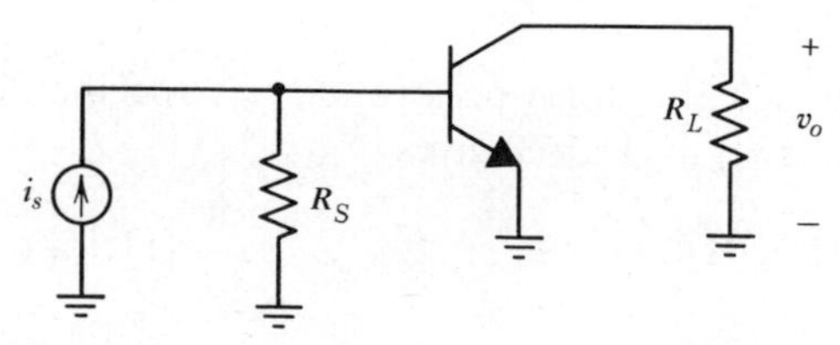

Fig. 11.48 Amplifier ac schematic.

11.5 Repeat Problem 11.4 if the bipolar transistor is replaced by a JFET with $g_m = 0.5$ mA/V and gate leakage current $I_G = 10^{-10}$ A.

11.6 Calculate equivalent input noise voltage and current generators for the circuit of Fig. 11.48 (omitting R_S). Using these results, calculate the total equivalent input noise current in a 2-MHz bandwidth for the circuit of Fig. 11.48 with $R_S = 1$ kΩ and compare with the result of Problem 11.4. Neglect correlation between the noise generators.

11.7 Four methods of achieving an input impedance greater than 100 kΩ are shown in the ac schematics of Fig. 11.49.

(a) Neglecting flicker noise and capacitive effects, derive expressions for the equivalent input noise voltage and current generators of these circuits. For circuit (i) this will be on the *source* side of the 100-kΩ resistor.

(b) Assuming that following stages limit the bandwidth to dc-20 kHz with a sharp cutoff, calculate the magnitude of the total equivalent input noise voltage in each case. Then compare these circuits for use as low-noise amplifiers from low source impedances.

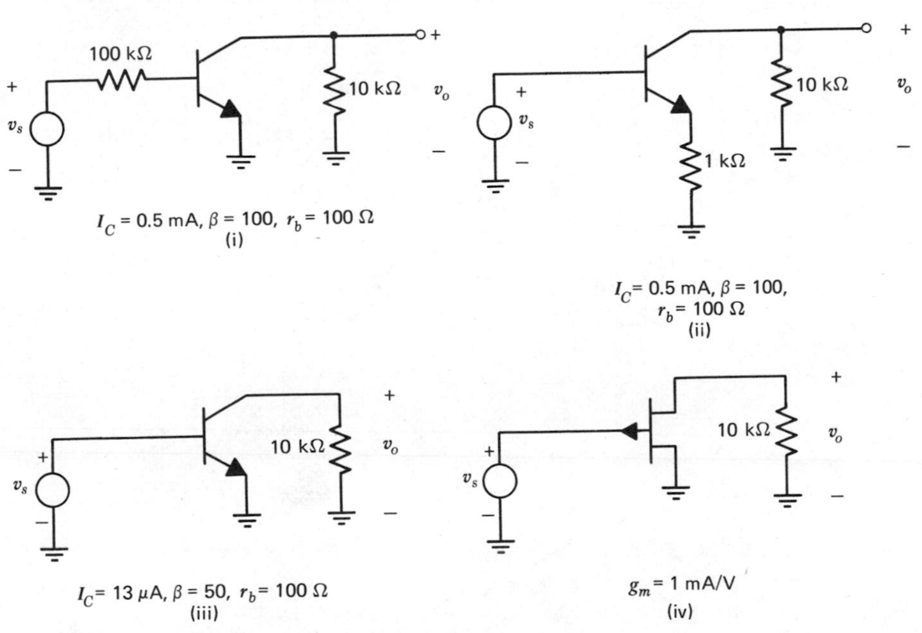

Fig. 11.49 ac schematic of four circuits realizing an input resistance $R_i > 100$ kΩ.

11.8 Neglecting capacitive effects, calculate equivalent input noise voltage and current generators for circuit (iv) of Fig. 11.49, assuming that the spectral density of the flicker noise in the JFET drain current equals that of the thermal noise at 100 kHz. Assuming that following stages limit the bandwidth to 0.001 kHz to 20 kHz with a sharp cut off, calculate the magnitude of the total equivalent input noise voltage.

11.9 The ac schematic of low input impedance common-base amplifier is shown in Fig. 11.50.
 (a) Calculate the equivalent input noise voltage and current generators of this circuit at the emitter of Q_1 using $I_{C1} = I_{C2} = 1$ mA, $r_{b1} = r_{b2} = 0$, $\beta_1 = \beta_2 = 100$, $f_{T1} = f_{T2} = 400$ MHz. Neglect flicker noise but include capacitive effects in the transistors.
 (b) If $R_S = 5$ kΩ, and later stages limit the bandwidth to a sharp cutoff at 150 MHz, calculate the value of i_s giving an output signal-to-noise ratio of 10 dB.

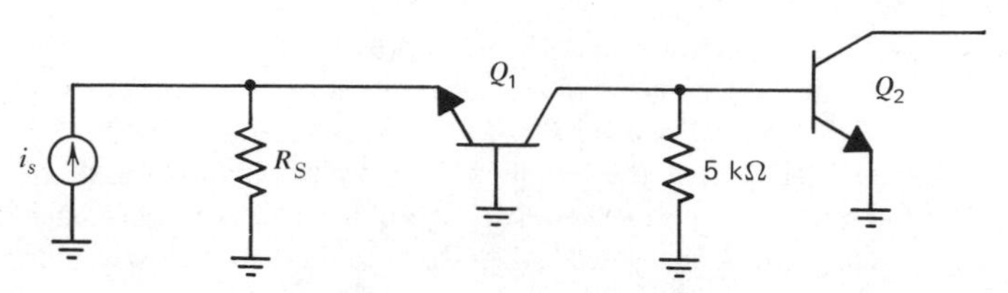

Fig. 11.50 ac schematic of a common-base amplifier.

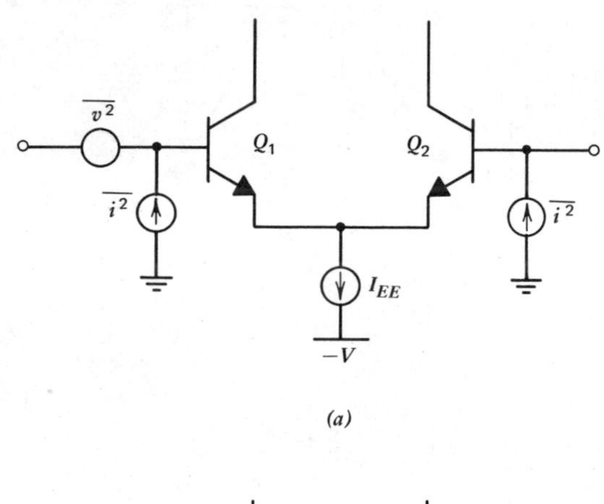

(a)

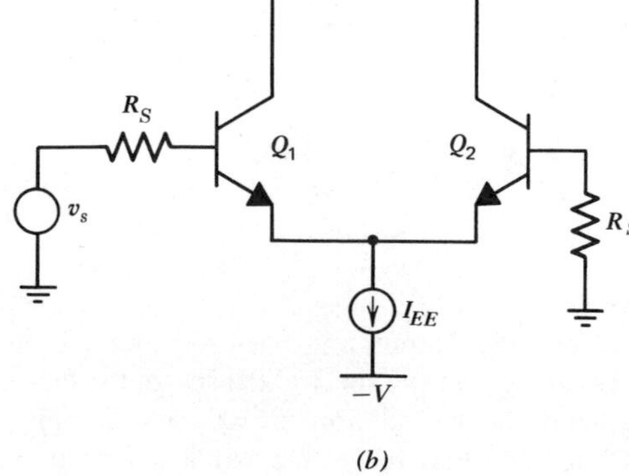

(b)

Fig. 11.51 Super-β input stage.

11.10 A super-β input stage is shown in Fig. 11.51*a*.

(a) Neglecting flicker noise and capacitive effects, calculate the equivalent input noise voltage and current generators $\overline{v^2}$ and $\overline{i^2}$ for this stage.
Data. $I_{EE} = 1\ \mu A$ $\quad \beta_1 = \beta_2 = 5000 \quad r_{b1} = r_{b2} = 500\ \Omega$

(b) If the circuit is fed from a source resistance $R_S = 100\ M\Omega$ as shown in Fig. 11.51*b*, calculate the total equivalent input noise voltage at v_s in a bandwidth of 1 kHz.

11.11 Repeat Problem 11.10 if the active devices are JFETs with $I_G = 10^{-11}$ A and $g_m =$ 0.5 mA/V.

11.12 If a 100-pF capacitor is connected across the diode in Fig. 11.47, calculate the noise bandwidth of the circuit and thus calculate the *total* output noise at v_o. Neglect flicker noise and series resistance in the diode.

11.13 A differential input stage is shown in Fig. 11.52.

(a) Neglecting flicker noise, calculate expressions for the equivalent input noise voltage and current generators at the base of Q_1.

(b) Assuming the circuit has a dominant pole in its frequency response at 30 MHz and $R_S = 50\ \Omega$, calculate the total equivalent input and output noise voltages.
Data. $\beta = 100 \quad r_b = 200$

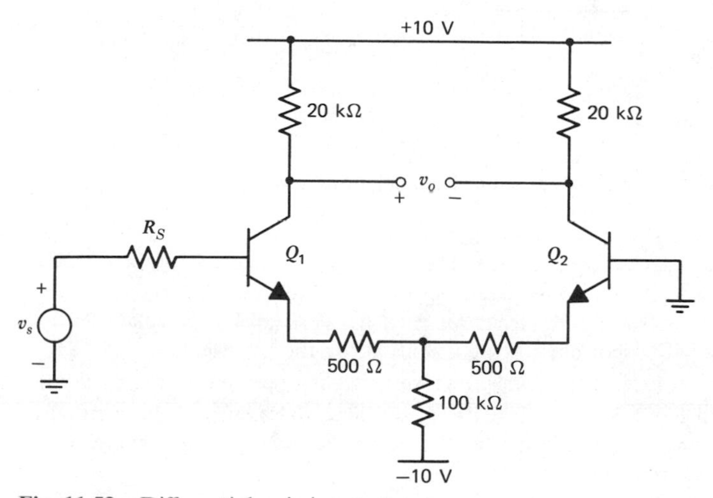

Fig. 11.52 Differential-pair input stage.

11.14 Calculate the source resistance giving minimum noise figure and the corresponding noise figure in decibels for a bipolar transistor with parameters
(a) $I_C = 2$ mA $\quad \beta_F = 50 \quad r_b = 100\ \Omega$
(b) $I_C = 10\ \mu A \quad \beta_F = 100 \quad r_b = 300\ \Omega$
Neglect flicker noise and capacitive effects.

11.15 Repeat Problem 11.14 if the transistor has a flicker noise corner frequency of 1 kHz. Calculate spot noise figure at 500 Hz.

11.16 Repeat Problem 11.14 if the transistor has a 1-kΩ emitter resistor.

11.17 (a) Neglecting flicker noise and capacitive effects, calculate the noise figure in decibels of the circuit of Fig. 11.52 with $R_S = 50\ \Omega$.

(b) If R_S were made equal to (i) 100 Ω or (ii) 200 kΩ would the noise figure increase or decrease? Explain.

(c) If $R_S = 200\ \text{k}\Omega$, and each device has a flicker noise corner frequency of 10 kHz, calculate the low frequency where the circuit spot noise figure is 20 dB.

11.18 (a) A shunt-feedback amplifier is shown in Fig. 11.53. Using equivalent input noise generators for the device, calculate the spot noise figure of this circuit in decibels for $R_S = 10\ \text{k}\Omega$ using the following data.

$$I_C = 0.5\ \text{mA} \qquad \beta = 50 \qquad r_b = 100\ \Omega$$

Neglect flicker noise and capacitive effects.

(b) If the device has $f_T = 500$ MHz, calculate the frequency where the noise figure is 3 dB above its low-frequency value.

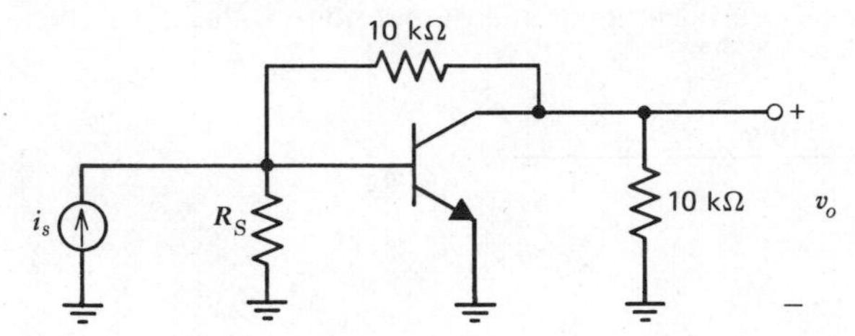

Fig. 11.53 ac schematic of a single-stage shunt-feedback amplifier.

11.19 (a) Neglecting capacitive effects, calculate the noise figure in decibels of the circuit of Fig. 11.50 with $R_S = 5\ \text{k}\Omega$. Use data as in Problem 11.9.

(b) If the flicker noise corner frequency for each device is 1 kHz, calculate the low frequency where the spot noise figure is 3 dB above the value in (a).

11.20 Neglecting flicker noise, calculate the total equivalent input noise voltage for the MC1553 shown in Fig. 8.22*a*. Use $\beta = 100$, $r_b = 100\ \Omega$ and assume a sharp cutoff in the frequency response at 50 MHz. Then calculate the average noise figure of the circuit with a source resistance of 50 Ω.

11.21 Neglecting capacitive effects, calculate the total equivalent input noise current for the shunt-shunt feedback circuit of Fig. 8.40 in a bandwidth 0.01 Hz to 100 kHz. Assume that $\beta = 200$, $r_b = 300\ \Omega$, $I_C = 1$ mA, and that the flicker noise corner frequency for the transistors is $f_a = 5$ kHz.

REFERENCES

1. M. Schwartz. *Information Transmission, Modulation, and Noise.* McGraw-Hill, New York, 1959 Chapter 5.
2. W. B. Davenport, Jr. and W. L. Root. *An Introduction to the Theory of Random Signals and Noise.* McGraw-Hill, New York, 1958, Chapter 7.

3. J. L. Lawson and G. E. Uhlenbeck. *Threshold Signals.* McGraw-Hill, New York, 1950, Chapter 4.
4. A. Van. der Ziel. *Noise.* Prentice-Hall, New York, 1954, Chapter 5.
5. W. B. Davenport, Jr. and W. L. Root, op. cit., Chapter 9.
6. J. L. Plumb and E. R. Chenette. "Flicker Noise in Transistors," *IEEE Trans. Electron Devices,* Vol. ED-10, pp. 304–308, September 1963.
7. R. C. Jaeger and A. J. Broderson. "Low-Frequency Noise Sources in Bipolar Junction Transistors," *IEEE Trans. Electron Devices,* Vol. ED-17, pp. 128–134, February 1970.
8. M. Nishida. "Effects of Diffusion-Induced Dislocations on the Excess Low-Frequency Noise," *IEEE Trans. Electron Devices,* Vol. ED-20, pp. 221–226, March 1973.
9. R. G. Meyer, L. Nagel, and S. K. Lui. "Computer Simulation of 1/f Noise Performance of Electronic Circuits," *IEEE J. Solid-State Circuits,* Vol. SC-8 pp. 237–240, June 1973.
10. R. H. Haitz. "Controlled Noise Generation with Avalanche Diodes," *IEEE Trans. Electron Devices,* Vol. ED-12, pp. 198–207, April 1965.
11. D. G. Peterson. "Noise Performance of Transistors," *IRE Trans. Electron Devices,* Vol. ED-9, pp. 296–303, May 1962.
12. A. Van der Ziel. "Thermal Noise in Field-Effect Transistors," *Proc. IRE,* Vol. 50, pp. 1808–1812, August 1962.
13. C. T. Sah. "Theory of Low-Frequency Generation Noise in Junction-Gate Field-Effect Transistors," *Proc. IEEE,* Vol. 50, pp. 795–814, July 1964.
14. W. R. Bennett. "Methods of Solving Noise Problems," *Proc. IRE,* Vol. 44, pp. 609–638, May 1956.
15. E. M. Cherry and D. E. Hooper. *Amplifying Devices and Low-Pass Amplifier Design.* Wiley, New York, 1968, Chapter 8.
16. H. A. Haus et al. "Representation of Noise in Linear Twoports," *Proc. IRE,* Vol. 48, pp. 69–74, January 1960.

INDEX